A Modern Course in Quantum Field Theory, Volume 2 (Second Edition)

Advanced topics

Online at: https://doi.org/10.1088/978-0-7503-5834-7

A Modern Course in Quantum Field Theory, Volume 2 (Second Edition)

Advanced topics

Badis Ydri

Department of Physics, Faculty of Sciences, Annaba University, Annaba, Algeria

IOP Publishing, Bristol, UK

ISBN   978-0-7503-5834-7 (ebook)
ISBN   978-0-7503-5832-3 (print)
ISBN   978-0-7503-5835-4 (myPrint)
ISBN   978-0-7503-5833-0 (mobi)

DOI   10.1088/978-0-7503-5834-7

Version: 20250601

IOP ebooks

British Library Cataloguing-in-Publication Data: A catalogue record for this book is available from the British Library.

Published by IOP Publishing, wholly owned by The Institute of Physics, London

IOP Publishing, No.2 The Distillery, Glassfields, Avon Street, Bristol, BS2 0GR, UK

US Office: IOP Publishing, Inc., 190 North Independence Mall West, Suite 601, Philadelphia, PA 19106, USA

Contents

Appendix A: Lie algebra representation theory: a primer   **A-1**

Appendix B: On homotopy theory   **B-1**

Prologue

The luminous matter in the Universe is constituted of elementary fermion particles of spin 1/2 (leptons and quarks), which interact via elementary boson particles of spin 1 (gauge vector bosons) mediating the three fundamental interactions of nature: the electromagnetic interaction, the strong nuclear force and the weak nuclear interaction. The fourth fundamental force of nature (the gravitational force) is mediated instead by a tensor particle of spin 2.

These particles are all massless and these forces obey a fundamental symmetry principle called the gauge principle, which can only be broken spontaneously via the Higgs particle (the breaking of the electroweak force into the observed electromagnetic force and weak interactions), which is another (the only) elementary particle of spin 0 in nature. This process of spontaneous symmetry breaking is what gives all elementary particles their measured masses and all the forces their observed strengths.

Quantum field theory (QFT) is a relativistic quantum theory, which describes precisely this luminous matter and its interactions. In fact, it is widely believed that QFT should also describe dark matter and perhaps even dark energy (in terms of vacuum energy). This QFT is perturbatively renormalizable. However, QFT also enjoys non-perturbative formulation either directly (through lattice field theory, the renormalization group equation and conformal field theory) or indirectly by admitting exact solutions (especially in two dimensions but also in four dimensions via the supersymmetric gauge principle).

'Modern quantum field theory' has expanded to encompass gravity via the AdS/CFT correspondence, a cornerstone of the gauge/gravity holographic duality. Additionally, noncommutative field theory provides a vital extension of QFT, where noncommutative geometry not only preserves supersymmetry but is also a unique framework in which we could have spontaneous supersymmetry breaking, thereby broadening the horizons of theoretical physics.

Hence, modern 'QFT', which governs all elementary particles and their interactions as well as gravity, can be summarized in four major sub-theories:

1. **The standard model of elementary particles:** This provides a unified scheme of the electromagnetic force, the weak interaction, and the strong nuclear force, and it is due historically to the work of Weinberg, Abdu Salam and Glashow among many other physicists. The standard model is the most successful (experimentally) QFT to date and perhaps the most successful theory ever (especially its quantum electrodynamics (QED) component). It accounts for a large body of phenomenological effects and observations seen in nature in terms of only a finite (but still relatively large $=19$) number of parameters such as the gauge coupling constants, the Higgs vacuum expectation value, the CKM angles and the theta angle governing CP violation. The standard model is however mostly perturbative and it includes in a fundamental way the phenomena of spontaneous symmetry breaking and is based entirely on the meta-theory of the renormalization group equation.

The standard model consists of two parts. The first part is the electroweak force, which unifies quantum electrodynamics, and quantum flavordynamics, which describes the weak force. The second part of the standard model consists of quantum chromodynamics (QCD), which describes the strong force. QCD admits a non-perturbative definition given typically in terms of a lattice formulation and lattice QCD is arguably the most sophisticated discipline in computational physics.

2. **Supersymmetric gauge theory in four dimensions:** This allows us a non-perturbative formulation (one in which we do not need a small parameter of expansion) of the gauge principle, which can be solved exactly (like the harmonic oscillator in quantum mechanics) in many instances by means of supersymmetry and holomorphy among other things (Witten–Seiberg–Nekrasov theory). This is of paramount importance to strongly coupled systems such as quantum chromodynamics since the strong force is a highly non-perturbative interaction. However, supersymmetric gauge theory also gives a profound understanding of the phenomena of spontaneous gauge symmetry breaking and the associated phenomena of renormalization.

3. **AdS/CFT duality:** As stated above the gravitational force is not mediated via a vector gauge boson but via a tensor particle of spin 2 called the graviton. The AdS/CFT duality is the theory which allows us to bring gravity and black holes into the realm of unitary QFT. Although this theory emerged historically from string theory it is intrinsically a QFT. It relies heavily on conformal field theory, supersymmetry and renormalization. It states simply that supergravity theory (string theory in general) in an Anti-de Sitter (AdS) spacetime, which is five-dimensional, is given precisely by a superconformal gauge field theory (CFT) living on the boundary of AdS, which is an ordinary four-dimensional Minkowski spacetime (a concrete realization of the holographic principle). The AdS/CFT correspondence generalizes to the so-called gauge/gravity duality.

4. **Noncommutative field theory:** Noncommutative field theory (NCFT) is a profound extension of QFT that fundamentally rethinks the structure of spacetime. Unlike conventional QFT, where spacetime coordinates commute and form a smooth continuum, NCFT introduces a noncommutative structure to spacetime, where the coordinates themselves do not commute. This leads to a paradigm shift in how fields, interactions, and symmetries are conceptualized.

   What sets NCFT apart is its dual role in modern theoretical physics.

   Firstly, NCFT is the only framework that seamlessly integrates with supersymmetry, the symmetry connecting bosons and fermions, making it uniquely compatible with theories attempting to unify quantum mechanics and gravity. Furthermore, it is expected to provide a consistent and elegant mechanism for the spontaneous breaking of supersymmetry. This is a crucial feature because it allows for the realization of physical phenomena where supersymmetry is preserved at high energies but broken at observable scales, aligning with the lack of supersymmetric particles detected in experiments.

Secondly, it can be viewed as an 'inverted' approach to QFT. Instead of assuming a fixed spacetime and then quantizing fields, NCFT begins by quantizing spacetime itself. The interplay between this quantized spacetime and the fields defined on it leads to novel physical effects, such as the mixing of ultraviolet (high-energy) and infrared (long-distance) scales, and provides a natural framework for discrete geometries that transition to smooth spacetime at large scales. Hence, NCFT has emerged as a promising approach to quantum gravity, offering insights into the structure of spacetime at the Planck scale, where classical descriptions fail. Noncommutative spacetimes, inspired by principles from string theory and other quantum gravity models, accommodate a fundamental length scale, resolving issues like singularities and infinite densities.

In addition to its theoretical significance, NCFT also offers a rich mathematical structure. It bridges the gap between quantum mechanics and geometry by describing spacetime through operator algebras and noncommutative geometry. This perspective not only reshapes our understanding of spacetime but also serves as a fertile ground for exploring new physical theories that are impossible to realize in the commutative framework.

NCFT thus stands at the crossroads of QFT, quantum gravity, and supersymmetry, providing a unifying platform for addressing some of the most fundamental questions in modern physics.

In the first volume of this book, we focus primarily on the first axis, which addresses gauge interactions and the standard model of elementary particle physics. The second volume delves into the other axes in greater detail. The third axis, which centers on the AdS/CFT correspondence and its implications, is discussed extensively in chapters 16 and 17. In chapter 16, we provide a systematic overview of the AdS/CFT correspondence and demonstrate how Einstein's gravity emerges from quantum entanglement. Chapter 17 explores the emergence of conformal symmetry in the SYK model relevant to the AdS2/CFT1.

The fourth axis, which focuses on noncommutative field theory and their matrix models, is examined in chapters 18 and 19. Chapter 18 discusses the phenomena of emergent geometry in matrix models, while chapter 19 investigates the renormalizability of noncommutative phi-four theory.

The second axis is addressed only partially, with groundwork laid in chapters 13, 14, and 15, which cover exact solutions of quantum field theory, monopoles and instantons, and supersymmetry, respectively. Finally, an extensive study of the renormalization group equation, which is the backbone of quantum field theory, is provided in chapters 11 (volume I) and 12 (volume II). These chapters establish and extend the core principles of renormalization.

This book emphasizes the fundamental physical principle of symmetry, particularly the role of symmetry groups in quantum theory, their representation theory, and the resulting conservation laws. Alongside this, significant focus is placed on the mathematical frameworks of the path integral and the renormalization group equation. The path integral formalism not only simplifies the quantization process but also provides a unified approach to quantum mechanics and quantum field theory.

The renormalization group equation, in addition to addressing traditional quantum field theory problems relevant to particle physics (discussed in chapters 7, 8, 9 and 11), offers powerful tools for exploring three critical physical problems. These include the determination of critical exponents for second-order phase transitions in statistical physics (chapter 11), the renormalizability and continuum limit of the phi-four theory (chapter 12), and the renormalizability of the non-commutative phi-four theory (chapter 19). Chapter 12, in particular, provides a systematic exposition of the functional renormalization group equation, presenting its applications and theoretical underpinnings in detail.

We will start the book in the usual way with canonical quantization of free fields (scalar field of spin 0 and spinor field of spin 1/2) in chapter 3. Chapter 2 contains an initiation to particle physics. Then, we will consider in chapter 4 perturbation theory of phi-four theory where the S-matrix structure of quantum field theory is exhibited explicitly. This is our first fundamental interaction in this book.

Then canonical quantization of the free Abelian vector field of spin 1 is considered in chapter 5 where pure Yang–Mills gauge interactions with $SU(N)$ groups are also introduced. In chapter 6 perturbation theory of quantum electrodynamics (which describes the gauge interaction of a spinor field with a vector field) and its renormalization is considered in great detail. For example, we derive explicitly from the renormalization properties of the theory measurable physical effects such as the electron anomalous magnetic moment. Furthermore, the links to particle physics, i.e., the relations between quantum field theory correlation functions and particle physics cross sections and decay rates, are established explicitly in this chapter, which shows more clearly the S-matrix structure of quantum field theory.

The path integral formalism is introduced in chapters 7 (for scalar fields) and 8 (for spinor and vector fields). In chapter 7 perturbative renormalizability of phi-four theory is considered at the two-loop order using the effective action formalism whereas in chapter 8 the Faddeev–Popov quantization of the Abelian and non-Abelian vector fields is considered. Perturbative renormalizability of $SU(N)$ gauge theory coupled to matter transforming in some representation of the gauge group is then discussed (asymptotic freedom, anomalies, BRST and background field methods, etc).

In chapter 9, we discuss the phenomenology of particle physics, offering a detailed construction of the standard model Lagrangian and explaining the phenomenon of spontaneous symmetry breaking via the Higgs mechanism. Following this, chapter 10 provides an explicit construction of scalar, spinor, and vector fields on the lattice, introducing the Metropolis algorithm in the context of the quenched approximation of quantum electrodynamics.

The final chapter of the book, chapter 20, revisits the Monte Carlo method as applied in quantum field theory, providing further insights into its implementation and significance. Additionally, we introduce neural networks as a promising complementary approach to lattice and matrix quantum field theories, highlighting their potential to address challenges in these areas.

In more detail, this book is then organized as follows:

1. **Relativistic quantum mechanics:** This chapter contains a standard preparatory material. We will present an overview of special relativity [1], relativistic Klein–Gordon and Dirac wave equations and the convention in this book for Dirac spinors [2], and a self-contained discussion of representation theory of the rotation and Lorentz groups [3].

2. **Initiation to particle physics:** This is a quick initiation to elementary particle physics in which we follow the pedagogical and precise book by Griffiths [4]. As everybody knows, particle physics is, historically and conceptually, the very solid bedrock underlying the physical foundation behind quantum field theory. In some sense, we can think of particle physics as the mother of quantum field theory. Thus, a knowledge of the underlying ideas and historical discoveries of particle physics is more than desirable.

3. **Canonical quantization of free fields:** After a brief excursion in classical mechanics [5] we present in this chapter the canonical quantization of free scalar and Dirac fields with a detailed calculation of the corresponding propagators [2, 6]. Then we give a thorough discussion of symmetries starting with discrete symmetries [2], the Poincaré group and its representation theory [3, 6], symmetries in the quantum theory, internal symmetries and the role of Noether's theorem in conservation principles [6, 7].

4. **The phi-four theory:** A detailed discussion of the S-matrix, the Gell-Mann-Low formula, the LSZ reduction formulas, Wick's theorem, Green's functions, Feynman diagrams and the corresponding Feynman rules of quantum Φ^4-theory is presented following [6]. This is our first non-trivial example of an interacting field theory and its canonical quantization.

5. **The electromagnetic field and Yang–Mills gauge interactions:** In this chapter we discuss in great detail the canonical quantization of the electromagnetic gauge field with emphasis on $U(1)$ gauge invariance and the Gupta-Bleuler method. Then a pedagogical introduction to Yang–Mills gauge interactions with $SU(2)$ and $SU(N)$ gauge groups (and even for general gauge groups) is presented. These gauge fields describe spin 1 particles in nature (the so-called vector bosons), which encompass the carriers of the electromagnetic force (the photon γ), the nuclear strong color force (the gluons g) and the nuclear weak radioactive force (the W and Z^0 vector bosons).

   Some good pedagogical references for the canonical quantization of the electromagnetic field are [6, 7].

6. **Quantum electrodynamics:** The goal in this chapter is to develop canonical perturbation theory beyond the free field approximation of quantum electrodynamics (QED), which is an interacting (local gauge) theory of the Dirac field (electrons and positrons) and the gauge vector field (photons). The formalism of canonical quantization of QED is found in [6] whereas radiative corrections and renormalization is found in [2].

7. **Path integral quantization of scalar fields:** In this chapter we will present the path integral method, which is a central tool in quantum field theory, and then give a detailed account of the effective action in the case of a scalar

field theory. A brief discussion of spontaneous symmetry breaking is also given. These are very standrad topics and we have benefited here from the books [2, 8, 9] and the lecture notes [10].

8. **Path integral quantization of Dirac and vector fields:** We develop the powerful and elegant path integral method for spinor fields (Grassmann variables) and gauge fields (gauge fixing, Faddeev–Popov method, ghosts). Then we give two important applications based on the path integral formalism. Firstly, we present a detailed derivation of the one-loop beta function of quantum chromodynamics (QCD) with $SU(N)$ gauge theory and matter fields in the fundamental representation and discuss the resulting phenomena of asymptotic freedom. Secondly, we present the one-loop (and in fact exact) axial or chiral anomaly in quantum electrodynamics (QED) and the Fujikawa path integral method. We also discuss briefly the background field method and symmetries within the path integral method (Schwinger–Dyson equations and Ward identities).

9. **Standard model:**

   The standard model of elementary particle physics describes all known particles and their interactions which are observed in nature. It is based on the following grand theoretical principles:

   (a) Relativistic invariance.

   (b) It is a local gauge theory based on the gauge group $SU(3) \times SU(2)_L \times U(1)_Y$.

   (c) The gauge group is spontaneously broken down to $SU(3) \times U(1)_{em}$. This generates mass in a gauge-invariant way.

   (d) It consists of a lepton sector, a quark sector, a Higgs term and a gauge sector. The matter sector (leptons, quarks and Higgs) are coupled minimally to the gauge sector (which ensures renormalizability). The mechanism by which the symmetry is spontaneously broken is the Higgs mechanism. The Higgs field is coupled to the quarks and leptons via gauge-invariant renormalizable Yukawa couplings.

   (e) It is a chiral gauge theory, i.e., left-handed quarks and leptons couple to the gauge field differently (in the fundamental representation) than right-handed quarks and leptons (singlet representation).

   (f) Renormalizability: The standard model is a renormalizable theory (interaction terms between the gauge fields and the matter fields are given by minimal coupling). The requirement of gauge invariance guarantees renormalizability and unitarity.

   (g) The standard model is not invariant under parity P (nor under CP where C is charge conjugation). But it is invariant under CPT where T is time reversal. This holds in the lepton sector.

   (h) Anomaly cancellation: This is the second quantum consistency check (after renormalizability), which states that any local symmetry like gauge symmetry cannot be allowed to be anomalous. This is satisfied in the standard model since the number of lepton families is equal to the lepton of quark families.

Extensions of the standard model include grand unified theories (GUTs; such $SU(5)$ or $SO(10)$ or any other group that contains the standard model gauge group as a subgroup), supersymmetry (minimal supersymmetric standard model), noncommutative geometry (Connes' standard model) and stringy extensions. Unification of the three forces (color strong, electromagnetic and weak) described by the standard model with gravity is however only achieved in string theory.

In this chapter, and after a brief excursion in the phenomenology of particle physics (isospin symmetry, quark model, neutrino oscillations, etc.), we give a detailed construction of the standard model Lagrangian starting with the Glashow, Weinberg, Salam electroweak theory, then we discuss the Higgs mechanism and spontaneous symmetry breaking, Majorana fermions, neutrino mass and the seesaw mechanism, and then finally we provide an extension to the quark sector and quantum chromodynamics as well as a summary of anomaly cancellation. We will follow the general presentations of [3, 10–12].

10. **Introduction to lattice field theory:** In this chapter a quick excursion into the world of lattice field theory is taken. Scalar, fermion and gauge fields are constructed on the lattice explicitly. Then the two most used Monte Carlo algorithms in numerical simulations on the lattice (the Metropolis and the hybrid Monte Carlo algorithms) are explained within the context of very simple lattice models, namely the scalar phi-four in two dimensions and quenched electrodynamics. The classic textbooks on the subject of lattice field theory are [13–17].

11. **The Callan–Symanzik renormalization group equation:** All second-order phase transitions in nature are described by the Callan–Symanzik renormalization group equations of Euclidean scalar field theory. In this chapter, after a detailed discussion of renormalizability of quantum field theories, in particular the scalar ϕ^4 theory, we present an explicit construction of the Callan–Symanzik renormalization group equations. Then, a detailed calculation of the critical exponents of second-order phase transitions starting from the renormalization properties of scalar ϕ^4 field theory at the tow-loop order is carried out explicitly. We follow closely the book [18].

12. **The Wilson and functional renormalization group equations:**
The renormalization group equation is a central tool of pertrubative and non-perturbative quantum field theory, which is vital for a proper understanding of the renormalizability of the theory and its phase diagram. The Wilson approach [19] to the renormalization group equation is in our opinion the most profound description of the true nature and final goals of quantum field theory. In this chapter, and after a careful review of the original Wilson approach, we describe in great detail the functional renormalization group equation, which is an exact non-perturbative formulation of the Wilson renormalization group equation. The original literature on the functional renormalization group equation includes Polchinski [20] (Polchinski's equation for the effective action) and Wetterich [21] (Wetterich's equation for the average action). See also [22].

13. **Some exact solutions of quantum field theory:** The non-perturbative physics of a quantum field theory (as we have seen) can only be probed by means of Monte Carlo methods on lattices (which can become quite intricate technically and numerically) and/or by means of the exact renormalization group equation (which is always quite intricate analytically and mathematically). But sometimes exact solutions of the quantum field theory model present themselves (lower dimensions and/or high degree of symmetries), which allow us to access the sought-after non-pertrubative physics of the theory directly. In this chapter we present as examples six models in two dimensions, which all enjoy exact solutions, allowing us an unprecedented look at the true heart, i.e., the non-pertrubative reality of a quantum field theory.

14. **The monopoles and instantons:** Monopoles are non-trivial topological gauge field configurations that appear in spontaneously broken gauge theory via the Higgs mechanism. These are particle-like solitonic configurations characterized by stability and finite energy among other properties. Their stability is of a topological origin characterized by the so-called winding numbers or magnetic charges. For this reason monopoles are one of the best examples in which physics and topology become intertwined. The existence of the monopole requires the embedding of electromagnetism, i.e., the group $U(1)$, as a subgroup in a larger non-abelian group G with compact cover which then becomes broken spontaneously via the usual Higgs mechanism.

    The original literature on the subject consists of 't Hooft [23] and Polyakov [24]. Some of the pedagogical (from my perspective) lectures I can mention here: Lenz [25], 't Hooft [26], Coleman [27] and Tong [28]. A comprehensive book is Shnir [29] and a comprehensive review is given by Weinberg and Yi [30].

    Instantons are another fundamental topological gauge configuration, perhaps more fundamental than monopoles, which are given by events localized in spacetime and hence the other name given for them: pseudo-particles (in contrast with particles such as monopoles, which are events localized in space). Instantons are also the gauge field configurations that dominate the path integral in the semi-classical limit with the trivial instanton identified precisely with the perturbative vacuum $A = 0$.

    We will discuss here in great detail the theta term, the role of vacuum degeneracy, the quantization of the topological charge and the role of topology in instanton physics. More precisely, the instanton is defined as a solution of the self-duality equation with zero/finite energy, which happens to saturate the Bogomoln'yi bound. The BPST instanton solution is then derived explicitly. The original literature on the BPST instanton is the paper by Belavin, Polyakov, Schwartz and Tyupkin [31]. We then discuss in some detail the moduli space, the collective coordinates, the zero modes, the ADHM construction, the one-loop quantization in the background of instantons as well as the connection of instantons to quantum tunneling. We have benefited here greatly from the pedagogical presentations found in [28, 32, 33].

15. **Introducing supersymmetry:**
    In this chapter we introduce supersymmetry following mostly [34]. In particular, we will emphasize the formal quantum field theory aspects of the formalism of global $N = 1$ supersymmetry with a detailed calculation of the corresponding F- and d-terms following also [35]. A brief description of $N = 2$ is also given. The classic text on supersymmetry of Wess and Bagger [36] remains in our view one of the best books on quantum field theory. We have also benefited from [37, 38].

16. **The AdS/CFT correspondence:** The goal in this chapter is to provide a pedagogical presentation of the celebrated AdS/CFT correspondence adhering mostly to the language of quantum field theory (QFT). This is certainly possible, and perhaps even natural, if we recall that in this correspondence we are positing that quantum gravity in an anti-de Sitter spacetime AdS_{d+1} is nothing else but a conformal field theory (CFT_d) at the boundary of AdS spacetime. Some of the reviews of the AdS/CFT correspondence that emphasize the QFT aspects and language include Kaplan [39], Zaffaroni [40] and Ramallo [41].

    This chapter contains therefore a thorough introduction to conformal symmetries, anti-de Sitter spacetimes, conformal field theories and the AdS/CFT correspondence. The primary goal however in this chapter is the holographic entanglement entropy. In other words, how spacetime geometry as encoded in Einstein's equations in the bulk of AdS spacetime can emerge from the quantum entanglement entropy of the CFT living on the boundary of AdS.

    A sample of the original literature for the holographic entanglement entropy is [42–45]. However, a very good concise and pedagogical review of the formalism relating spacetime geometry to quantum entanglement due to Van Raamsdonk and collaborators is found in [44] and [46].

17. **Conformal field theory and emergence of conformal symmetry:** In the first section of this chapter we give a brief overview of conformal field theory following [47] then we discuss, in the remainder of the chapter, the celebrated SYK model and the emergence of conformal invariance in its large N IR limit [48, 49].

18. **Novel phenomena in noncommutative field theory: emergent geometry:** Noncommutative field theory (NCFT) [50, 51] is an extension of quantum field theory (QFT) that redefines spacetime, replacing commuting coordinates with a noncommutative structure. This shift fundamentally alters the way fields, interactions, and symmetries are understood. NCFT uniquely integrates with supersymmetry, making it a natural framework for unifying quantum mechanics and gravity. It also provides a consistent mechanism for spontaneous supersymmetry breaking.

    Unlike conventional QFT, which quantizes fields on a fixed spacetime, NCFT begins by quantizing spacetime itself. This perspective reveals novel phenomena, such as ultraviolet-infrared mixing and a natural transition from discrete to continuous geometries. It offers insights into quantum gravity at the Planck scale.

Mathematically, NCFT bridges quantum mechanics and geometry through operator algebras, enabling the exploration of new theories unattainable in traditional frameworks. This dual role cements NCFT as a cornerstone of modern theoretical physics.

19. **Noncommutative scalar field theory and its renormalizability:** In this chapter, and after an efficient introduction to noncommutative scalar field theory, we will apply the Wilson–Polchinski renormalization group equation, discussed in the previous chapter, to the problem of renormalizing noncommutative phi-four theory in two and four dimensions with and without the harmonic oscillator term. Noncommutative field theory is discussed in great detail in our book [52] whereas we follow closely the original programme of Grosse and Wulkenhaar [52–54] in the very difficult problem of renormalization of noncommutative phi-four theory on Moyal–Weyl spaces.

20. **The neural network method for quantum field theory:** In this chapter we propose neural networks as a potentially very powerful alternative to the Monte Carlo method for the non-perturbative study of quantum field theory. In the first section we give a brief overview of the Monte Carlo method following mostly the presentation of [55]. In section two we summarize the main four quantum scalar field theories on commutative and noncommutative spaces, which we have already discussed in great detail in this book. In section three we introduce neural networks and in particular the restricted Boltzmann machine [56–59]. In section four we apply the restricted Boltzmann machine to the noncommutative phi-four theory in the matrix basis and draw an interesting universality conjecture.

This book includes also five appendices (two on classical physics, one on representation theory of Lie groups and Lie algebras, one on homotopy theory and one contains extra exercises given as examination problems throughout the years).

References

[1] Griffiths D 1999 *Introduction to Electromagnetism* 3rd edn (Englewood Cliffs, NJ: Prentice-Hall)

[2] Peskin M E and Schroeder D V 1995 *An Introduction To Quantum Field Theory* (Boca Raton, FL: CRC Press)

[3] Boyarkin O M 2011 *Advanced Particle Physics* **Vol I** (London: Taylor and Francis)

[4] Griffiths D 1987 *Introduction to Elementary Particles* (New York: Wiley)

[5] Goldstein G 1980 *Classical Mechanics* 2nd edn (Reading, MA: Addison-Wesley)

[6] Strathdee J 1995 Course on quantum electrodynamics *ICTP Lecture Notes* IC/95/315 International Centre for Theoretical Physics

[7] Greiner W and Reinhardt J 1996 *Field Quantization* (Berlin: Springer)

[8] Polyakov A M 1987 *Gauge fields and strings Contemporary Concepts in Physics* (New York: Harwood Academic)

[9] Itzykson C and Drouffe J M 1989 *Statistical Field Theory: Volume 1, From Brownian Motion to Renormalization and Lattice Gauge Theory* Cambridge Monographs on Mathematical Physics (Cambridge: Cambridge Univ. Press)

[10] Randjbar-Daemi S Course on quantum field theory *ICTP preprint of 1993–94 HEP-QFT (1)*

[11] Boyarkin O M 2011 *Advanced Particle Physics: Vol II* (London: Taylor and Francis)

[12] Dolan B 2004 *Particle Physics* http://www.thphys.nuim.ie/staff/bdolan/

[13] Gattringer C and Lang C B 2010 Quantum chromodynamics on the lattice *Lect. Notes Phys.* **788** (Berlin: Springer)

[14] Smit J 2002 Introduction to quantum fields on a lattice: a robust mate *Cambridge Lect. Notes Phys.* **15** 1

[15] Rothe H J 1992 Lattice gauge theories: an introduction *World Sci. Lect. Notes Phys.* **43** 1

[16] Creutz M 1985 Quarks, gluons and lattices *Cambridge Monographs on Mathematical Physics* (Cambridge: Cambridge Univ. Press)

[17] Montvay I and Munster G 1994 *Quantum Fields on a Lattice* (Cambridge: Cambridge Univ. Press) 491 (Cambridge Monographs on Mathematical Physics)

[18] Zinn-Justin J 2002 Quantum field theory and critical phenomena *Int. Ser. Monogr. Phys.* **113** 1

[19] Wilson K G and Kogut J B 1974 The renormalization group and the epsilon expansion *Phys. Rep.* **12** 75

[20] Polchinski J 1984 Renormalization and effective Lagrangians *Nucl. Phys.* B **231** 269

[21] Wetterich C 1993 Exact evolution equation for the effective potential *Phys. Lett.* B **301** 90

[22] Kopietz P, Bartosch L and Schutz F 2010 Introduction to the functional renormalization group *Lect. Notes Phys.* **798** 1

[23] 't Hooft G 1974 Magnetic monopoles in unified gauge theories *Nucl. Phys.* B **79** 276

[24] Polyakov A M 1974 Particle spectrum in the quantum field theory *JETP Lett.* **20** 194 [Pisma Zh. Eksp. Teor. Fiz. **20** 430 (1974)]

[25] Lenz F 2005 Topological Concepts in Gauge Theories *Topology and Geometry in Physics* **659** (Berlin: Springer) 7

[26] 't Hooft G 1999 Monopoles, Instantons and Confinement, arXiv:hep-th/0010225

[27] Coleman S R 1975 Classical lumps and their quantum descendents Lectures delivered at Int. School of Subnuclear Physics, Ettore Majorana, Erice, Sicily, July 11–31

[28] Tong D 2005 TASI lectures on solitons: instantons, monopoles, vortices and kinks, arXiv:hep-th/0509216

[29] Shnir Y M 2005 *Magnetic Monopoles* (Berlin: Springer)

[30] Weinberg E J and Yi P 2007 Magnetic monopole dynamics, supersymmetry, and duality *Phys. Rep.* **438** 65 [hep-th/0609055]

[31] Belavin A A, Polyakov A M, Schwartz A S and Tyupkin Y S 1975 Pseudoparticle solutions of the Yang-Mills equations *Phys. Lett.* **59B** 85

[32] Vandoren S and van Nieuwenhuizen P 2008 Lectures on instantons arXiv:0802.1862 [hep-th]

[33] Dorey N, Hollowood T J, Khoze V V and Mattis M P 2002 The calculus of many instantons *Phys. Rep.* **371** 231

[34] Lykken J D 1996 *Introduction to supersymmetry* FERMILAB-PUB-96/445-T FermiLab arXiv:hep-th/9612114

[35] Weinberg S 2005 *The Quantum Theory of Fields, Volume III: Supersymmetry* (Cambridge: Cambridge Univ. Press)

[36] Wess J and Bagger J 1992 *Supersymmetry and Supergravity* (Princeton, NJ: Princeton Univ. Press) 259

[37] West P C 1990 *Introduction to Supersymmetry and Supergravity* (Singapore: World Scientific) 425

[38] Bilal A 2001 *Introduction to Supersymmetry* NEIP-01-001 arXiv:hep-th/0101055

[39] Kaplan J 2013 *Lectures on AdS/CFT from the Bottom Up,* John Hopkins University, https://sites.krieger.jhu.edu/jared-kaplan/

[40] Zaffaroni A 2000 Introduction to the AdS-CFT correspondence *Class. Quant. Grav.* **17** 3571

[41] Ramallo A V 2015 Introduction to the AdS/CFT correspondence *Springer Proc. in Physics Lectures on Particle Physics, Astrophysics and Cosmology* 161 (Berlin: Springer) 411

[42] Lashkari N, McDermott M B and Van Raamsdonk M 2014 Gravitational dynamics from entanglement 'thermodynamics' *JHEP* **1404** 195

[43] Faulkner T, Guica M, Hartman T, Myers R C and Van Raamsdonk M 2014 Gravitation from entanglement in holographic CFTs *JHEP* **1403** 051

[44] Van Raamsdonk M 2016 Lectures on Gravity and Entanglement *New Frontiers in Fields and Strings* (Singapore: World Scientific) 297

[45] Casini H, Huerta M and Myers R C 2011 Towards a derivation of holographic entanglement entropy *JHEP* **1105** 036

[46] Jaksland R 2017 A review of the holographic relation between linearized gravity and the first law of entanglement entropy, arXiv:1711.10854 [hep-th]

[47] Tong D 2009 Introducing conformal field theory https://www.damtp.cam.ac.uk/user/tong/string/string4.pdf

[48] Sachdev S and Ye J 1993 Gapless spin-fluid ground state in a random quantum Heisenberg magnet *Phys. Rev. Lett.* **70** 3339

[49] Kitaev A 2015 A simple model of quantum holography KITP strings seminar and Entanglement 2015 program (February 12, April 7, and May 27, 2015). http://online.kitp.ucsb.edu/online/entangled15/

[50] Ydri B 2018 *Matrix Models of String Theory* (Bristol: IOP Publishing)

[51] Ydri B 2017 Lectures on matrix field theory *Lect. Notes Phys.* **929** (Berlin: Springer) 1–352

[52] Grosse H and Wulkenhaar R 2005 Power counting theorem for nonlocal matrix models and renormalization *Commun. Math. Phys.* **254** 91

[53] Grosse H and Wulkenhaar R 2005 Renormalization of phi**4 theory on noncommutative R**4 in the matrix base *Commun. Math. Phys.* **256** 305

[54] Grosse H and Wulkenhaar R 2003 Renormalization of phi**4 theory on noncommutative R**2 in the matrix base *JHEP* **0312** 019

[55] Ydri B 2017 *Computational Physics: An Introduction to Monte Carlo Simulations of Matrix Field Theory* (Singapore: World Scientific) [arXiv:1506.02567 [hep-lat]]

[56] Hinton E 1668 Boltzmann machine *Scholarpedia* **2** 1668

[57] Hinton G E and Salakhutdinov R R 2006 Reducing the dimensionality of data with neural networks *Science* **313** 504

[58] Hinton G E 2002 Training products of experts by minimizing contrastive divergence *Neural Comput.* **14** 1771–800

[59] Hinton G E 2012 A Practical Guide to Training Restricted Boltzmann Machines in *Neural Networks: Tricks of the Trade* (Berlin: Springer) 599–619 pp

Author biography

Badis Ydri

Badis Ydri is a professor of theoretical physics at the Institute of Physics, Annaba University, Algeria. He earned his PhD from Syracuse University, New York, USA, in 2001. He is also a research associate at the Dublin Institute for Advanced Studies, Ireland. His postdoctoral work includes a Marie Curie fellowship at Humboldt University in Berlin, Germany, and a Hamilton fellowship at the Dublin Institute for Advanced Studies, Ireland. His general areas of expertise encompasses quantum field theory, general relativity, string theory, and philosophy of physics.

His ongoing research explores: 1) matrix quantum mechanics approaches to quantum black holes and quantum gravity, 2) gauge/gravity duality and M-theory, 3) noncommutative geometry and matrix models, 4) renormalization group equation and Monte Carlo methods, 5) artificial intelligence in computational physics, 6) quantum philosophy, and 7) hard physical philosophy of consciousness and being.

He is the author of seven books in theoretical physics.

IOP Publishing

A Modern Course in Quantum Field Theory, Volume 2 (Second Edition)
Advanced topics
Badis Ydri

Chapter 12

The Wilson and functional renormalization group equations

The renormalization group equation is a central tool of perturbative and non-perturbative quantum field theory, which is vital for a proper understanding of the renormalizability of the theory and its phase diagram. The Wilson approach [1] to the renormalization group equation is in our opinion the most profound description of the true nature and final goals of quantum field theory. In this chapter, and after a careful review of the original Wilson approach, we describe in great detail the functional renormalization group equation, which is an exact non-perturbative formulation of the Wilson renormalization group equation. The original literature on the functional renormalization group equation includes Polchinski [2] (Polchinski's equation for the effective action) and Wetterich [3] (Wetterich's equation for the average action). See also [4].

12.1 Wilson renormalization group approach

The primary reference for this section is the classic paper by Wilson and Kogut [1]. The article [5] was very useful and we follow the presentation of [6].

12.1.1 The basic idea

As an example, we consider the Landau theory given by the Hamiltonian

$$E[m] = \int d^d x \mathcal{E}(m(x)), \quad \mathcal{E} = \frac{1}{2}(\vec{\partial}m(x))^2 + \sum_{n=1}^{\infty} K_n m^n(x) + \cdots. \tag{12.1}$$

The field $m(x)$ is the order parameter (magnetization) and the dot stands for higher derivative terms. The coupling constants K_n depend on the cutoff Λ. The most important example for us is the Landau–Wilson model given by

doi:10.1088/978-0-7503-5834-7ch12

$$\mathcal{E} = \frac{1}{2}(\vec{\partial}m(x))^2 + \frac{1}{2}r_0 m^2(x) + u_0 m^4(x). \tag{12.2}$$

The renormalization group transformation consists of three steps:

- **Coarse-Graining**: We integrate in the partition function over the high momentum modes.
- **Rescaling**: We rescale the momenta back to the original domain in order to restore the original coarse-grained 'picture'.
- **Normalization**: We rescale the fields appropriately, which is necessary since the relative size of the fluctuations of the rescaled magnetization is in general different from the original.

These three steps define a renormalization group transformation, which in general is not invertible and thus we are really dealing with a semigroup structure. By repeating these steps we then get highly non-linear equations, which are the renormalization group equations.

The partition function of the Landau theory in momentum space is of the form

$$Z = \int \prod_{|\vec{k}|<\Lambda}^{\prime} [d\tilde{m}(\vec{k})][d\tilde{m}^\dagger(\vec{k})]\exp(-E[\tilde{m}(\vec{k})]). \tag{12.3}$$

The prime indicates that we integrate only over half of the momentum space. The energy functional in momentum space reads (with $K_2 = r_0/2$)

$$E[\tilde{m}] = \frac{1}{2}\int d^d k(\vec{k}^2 + r_0)|\tilde{m}(\vec{k})|^2 + \cdots. \tag{12.4}$$

We now apply the above stated three steps of the renormalization group transformation.

- **Coarse-Graining (Integration)**: We integrate in the partition function over all momentum modes, which satisfy $\Lambda/b \leqslant |\vec{k}| \leqslant \Lambda$ with $b > 1$. Thus the new energy functional should be defined by the expression

$$\exp(-E'[\tilde{m}(\vec{k})]) = \int \prod_{\frac{\Lambda}{b}<|\vec{k}|<\Lambda}^{\prime} [d\tilde{m}(\vec{k})][d\tilde{m}^\dagger(\vec{k})]\exp(-E[\tilde{m}(\vec{k})]). \tag{12.5}$$

Clearly, the new Hamiltonian E' depends only on the fields $\tilde{m}(\vec{k})$ for which $|\vec{k}| < \Lambda/b$. This new Hamiltonian is generally of the same form as the original one but with different coupling constants, viz

$$E'[\tilde{m}] = \frac{1}{2}\int_{|\vec{k}|<\frac{\Lambda}{b}} d^d k(A\vec{k}^2 + \tilde{r}_0)|\tilde{m}(\vec{k})|^2 + \cdots. \tag{12.6}$$

In fact, E' will also contain additional terms that did not exist in the original action and which will be dropped (Wilson truncation). The partition function becomes

$$Z = \int \prod_{|\vec{k}| < \frac{\Lambda}{b}}^{\prime} [d\tilde{m}(\vec{k})][d\tilde{m}^{\dagger}(\vec{k})] \exp(-E'[\tilde{m}(\vec{k})]).$$ (12.7)

- **Rescaling**: We rescale the momenta back to the original domain, i.e., we restore the cutoff back to Λ, as $k_i \longrightarrow k_i' = bk_i$. The action becomes

$$E'[\tilde{m}] = \frac{b^{-d}}{2} \int_{|\vec{k}| < \Lambda} d^d k \left(\frac{A}{b^2} \vec{k}^2 + \tilde{r}_0 \right) \left| \tilde{m}\left(\frac{\vec{k}}{b} \right) \right|^2 + \cdots.$$ (12.8)

- **Normalization**: In order to restore the canonical normalization of the kinetic term we rescale the fields as

$$\tilde{m}'(\vec{k}) = \sqrt{\frac{A}{b^{d+2}}} \, m\left(\frac{\vec{k}}{b} \right).$$ (12.9)

The energy becomes

$$E'[\tilde{m}'(\vec{k})] = \frac{1}{2} \int_{|\vec{k}| < \Lambda} d^d k (\vec{k}^2 + r_0') |\tilde{m}'(k)|^2 + \cdots,$$ (12.10)

where the renormalized mass is given by

$$r_0' = \frac{b^2}{A} \tilde{r}_0.$$ (12.11)

The partition function becomes

$$Z' = \int \prod_{|\vec{k}| < \Lambda}^{\prime} [d\tilde{m}'(\vec{k})][d\tilde{m}'^{\dagger}(\vec{k})] \exp(-E'[\tilde{m}'(\vec{k})]).$$ (12.12)

In summary, the coupling constants $K = (K_2, K_3, \ldots)$ in the original Hamiltonian E will be mapped under the renormalization group transformation to new coupling constant $K' = (K_2', K_3', \ldots)$ in the renormalized Hamiltonian E'. The coupling constants K' are called renormalized coupling constants. We also say that the coupling constants of the theory 'flow' under a change of the energy scale. If we denote the renormalization group transformation by R then this mapping can be written as ($K = K^{(0)}$, $K' = K^{(1)}$)

$$K^{(1)} = R(K^{(0)}).$$ (12.13)

This step can be iterated an arbitrary number of times. The renormalization group transformation is the same for all iterations and as a consequence we can write after $i + 1$ iterations the renormalization group transformation

$$K^{(i+1)} = R(K^{(i)}). \tag{12.14}$$

By inspection, since the partition functions (12.3) and (12.12) are given by the same function except for the replacement $K \longrightarrow K'$, the free energy should satisfy the relation

$$f(K) = f_0(K) + f(K'). \tag{12.15}$$

Next, since the volumes (numbers of degrees of freedom) before and after the renormalization group transformation are related by $(\Lambda'/\Lambda)^d = b^{-d}$, the free energy per degrees of freedom must satisfy the relation

$$F(K) = F_0(K) + b^{-d}F(K'). \tag{12.16}$$

12.1.2 Formalism

We are interested in the most general properties of the following renormalization group transformation

$$K^{(i+1)} = R(K^{(i)}). \tag{12.17}$$

A fixed point K^* in the coupling-constant space is a point which does not move, i.e., it is invariant, under the successive action of the above renormalization group transformation. Thus, we must have

$$K^* = R(K^*). \tag{12.18}$$

Many points K in the coupling-constant space will approach the fixed point under a large number of iterations of the renormalization group transformations. We write this as

$$K^{(i)} \longrightarrow K^*, \quad i \longrightarrow \infty. \tag{12.19}$$

The fixed point K^* describes the system at the critical temperature $T = T_c$. At this point the system becomes scale invariant, i.e., it becomes invariant under a change of length scale. This can also be expressed by the fact that correlation length ξ (inverse mass) diverges at the fixed point.

Now we linearize the above renormalization group equation around the fixed point K^* by writing

$$K_\alpha^{(i+1)} - K_\alpha^* = (K^{(i)} - K^*)_\alpha = W_{\alpha\beta}(K_\beta^{(i)} - K_\beta^*). \tag{12.20}$$

The linearized renormalization group transformation is given by

$$W_{\alpha\beta} = \frac{\partial R_\alpha}{\partial K_\beta}\Big|_{K=K_*}. \tag{12.21}$$

The origin of the coupling-constant space is chosen to be K^*. The left eigenvectors ϕ of the renormalization group matrix W are defined by

$$\phi_\beta W_{\beta\alpha} = \lambda\phi_\alpha. \tag{12.22}$$

For each left eigenvector ϕ we define a scaling field v by the relation

$$v^{(i)} = \sum_\alpha \phi_\alpha (K^{(i)} - K^*)_\alpha. \tag{12.23}$$

It satisfies under the renormalization group transformation the crucial property

$$v^{(i+1)} = \lambda v^{(i)}. \tag{12.24}$$

We say that v scales under the renormalization group transformation with a factor of λ. Thus the scaling field v increases if $\lambda > 1$ and decreases if $\lambda < 1$. In other words, the eigenvalue λ associated with the left eigenvector ϕ is what controls the flow behavior of the corresponding scaling field. Recall that under the renormalization group transformation the length scales as $L \longrightarrow bL$, i.e. it increases since $b > 1$. Thus, if the eigenvalue λ increases under the renormalization group transformation then we may write $\lambda = b^y$ where $y > 0$ whereas for $y < 0$ the eigenvalue λ decreases. The exponent y is called the dimension of the scaling field v. The scaling fields are in some sense the proper coupling constants whereas their exponents are their quantum mass dimensions.

Hence we will choose in the coupling-constant space the origin at K^* whereas the preferred directions are given by the scaling fields v. The fundamental equation (12.16) reads in terms of the scaling fields

$$F(v_1, v_2, \ldots) = F_0(v_1, v_2, \ldots) + b^{-d} F(\lambda_1 v_1, \lambda_2 v_2, \ldots). \tag{12.25}$$

The function F_0 is regular at the fixed point (critical temperature) whereas the second term gives the singular part of the free energy. Some definitions are now in order.

- For $\lambda < 1$ the scaling field is called irrelevant (decreases in the RG transformation) whereas for $\lambda > 1$ it is called relevant (increases in the RG transformation). Thus if we want to remain on a fixed point we must set the relevant scaling fields to zero. The case $\lambda = 1$ corresponds to marginal fields.
- The fixed point lies on a hypersurface called the critical surface in the coupling-constant space. The critical surface is defined by setting all the relevant scaling fields to zero. Thus starting from any point on the critical surface we will approach the fixed point under a large number of iterations of the renormalization group transformation whereas starting from any point off the critical surface we will diverge from the fixed point. All actions (physical systems) located on the critical surface define collectively a so-called universality class.
- The eigenvalues λ determine the critical exponents y. All systems in the same universality class have the same critical exponents.

12.1.3 Critical exponents

The correlation length ξ behaves around the critical temperature as

$$\xi \sim |t|^{-1/D_t}, \quad D_t = \frac{1}{\nu}, \tag{12.26}$$

where ν is the mass critical exponent and D_t is the temperature critical exponent. The gap (order parameter) critical exponent β and the magnetic field critical exponent D_h are given by

$$M\mid_{H=0} \sim \mid t \mid^\beta, \quad \beta = \frac{d - D_h}{D_t}. \tag{12.27}$$

The anomalous exponent η is defined by the behavior of the 2-point function at the critical temperature given by

$$\Gamma(x) \sim \frac{1}{r^p}\exp(-r/\xi), \quad p = d - 2 + \eta = 2d - 2D_h. \tag{12.28}$$

There are three more critical exponents that can be determined solely in terms of D_t and D_h (scaling relations). These are α (specific heat), γ (susceptibility), δ (magnetization at non-zero magnetic field).

12.1.4 The Landau–Wilson model

In the above differential equations we change the variables as

$$x = \frac{r}{\Lambda^2}, \quad y = C_d\frac{u}{\Lambda^{4-d}}. \tag{12.29}$$

We obtain then the RG equations

$$\frac{dx}{d\tau} = 2x + 12y - 12xy, \quad \frac{dy}{d\tau} = \epsilon y - 36y^2. \tag{12.30}$$

The fixed point (x_*, y_*) is defined by

$$0 = 2x_* + 12y_* - 12x_*y_*, \quad 0 = \epsilon y_* - 36y_*^2. \tag{12.31}$$

We have the two solutions

$$\begin{aligned}(x_*, y_*) &= (0, 0), \quad \text{Gaussian} \\ (x_*, y_*) &= (-\epsilon/6, \epsilon/36), \quad \text{Wilson} - \text{Fisher}\end{aligned} \tag{12.32}$$

Linearizing the RG transformation around the fixed point by writing $x = x_* + \delta x$ and $y = y_* + \delta y$ we obtain the differential equations

$$\frac{d\delta x}{d\tau} = (2 - 12y_*)\delta x + (12 - 12x_*)\delta y, \quad \frac{d\delta y}{d\tau} = (\epsilon - 72y_*)\delta y. \tag{12.33}$$

The RG matrix W is given by

$$W = \begin{pmatrix} 2 - 12y_* & 12 - 12x_* \\ 0 & \epsilon - 72y_* \end{pmatrix} \tag{12.34}$$

with eigenvalues given, respectively, by $\lambda_1 = 2 - 12y_*$ and $\lambda_2 = \epsilon - 72y_*$. The corresponding left eigenvectors are given by

$$\phi_1 = \begin{pmatrix} a \\ \dfrac{12(1 - x_*)a}{2 - \epsilon + 60y_*} \end{pmatrix}, \quad \lambda_1 = 2 - 12y_*. \tag{12.35}$$

$$\phi_2 = \begin{pmatrix} 0 \\ 1 \end{pmatrix}, \quad \lambda_1 = \epsilon - 72y_*. \tag{12.36}$$

The scaling fields are given, respectively, by

$$v_1 = a.\, \delta x + \frac{12(1 - x_*)a}{2 - \epsilon + 60y_*} .\, \delta y, \quad \lambda_1 = 2 - 12y_*. \tag{12.37}$$

$$v_2 = 0.\, \delta x + 1.\, \delta y, \quad \lambda_2 = \epsilon - 72y_*. \tag{12.38}$$

These are given explicitly as follows.
- **Gaussian fixed point** $(x_*,\, y_*) = (0,\, 0)$:

$$\begin{aligned} v_1 &= \delta x + 6\delta y, \quad \lambda_1 = 2 \\ v_2 &= \delta y, \quad \lambda = \epsilon. \end{aligned} \tag{12.39}$$

- **Wilson–Fisher fixed point** $(x_*,\, y_*) = (-\epsilon/6,\, \epsilon/36)$:

$$\begin{aligned} v_1 &= \delta x + (6 - \epsilon)\delta y, \quad \lambda_1 = 2 - \frac{\epsilon}{3} \\ v_2 &= \delta y, \quad \lambda_2 = -\epsilon. \end{aligned} \tag{12.40}$$

The scaling fields satisfy

$$\frac{dv_i}{d\tau} = \lambda_i v_i. \tag{12.41}$$

The solution is given by

$$v_i = v_0 b^{\lambda_i}. \tag{12.42}$$

Let us assume now that $d < 4$ and close to 4 ($\epsilon > 0$ and small). In this case we have $\lambda_1 = 2 - \epsilon/3 > 1$ and thus v_1 is a relevant scaling field whereas $\lambda_2 = -\epsilon < 1$ and thus v_2 is an irrelevant scaling field.

The critical surface corresponding to the Wilson–Fisher fixed point is determined by setting the relevant scaling field, i.e., v_1, to zero. This of equivalent to the line in the $x - y$ plane given by

$$y - y_* = -\frac{1}{6 - \epsilon}(x - x_*). \tag{12.43}$$

To the linear order of ϵ this line passes by the other trivial fixed point at the origin. This line or critical surface represents our first preferred direction in the $x - y$ plane.

The critical surface $v_1 = 0$ represents the critical temperature $T = T_C$. This can be seen as follows. The other independent preferred direction should be obtained by setting the irrelevant variable v_2 to 0. This gives $\delta y = 0$ (the direction is parallel to the x axis) and $v_1 = \delta x = \delta r / \Lambda^2$ (this is the direction of increase or decrease of the temperature away from the critical value T_c).

The critical surface $v_1 = 0$ corresponds therefore precisely to the critical temperature $T = T_C$. Hence, starting at a point A with $v_1 < 0$, i.e. with $T < T_C$, the system under RG will approach the low temperature phase where the symmetry is spontaneously broken. Whereas starting at a point B with $v_1 > 0$, i.e. with $T > T_C$, the system will approach under successive RG transformations the high temperature phase.

The mass critical exponent is directly deduced to be (v_1 is the relevant scaling field identified as the temperature and it behaves as $v_1 \sim b^{\lambda_i}$)

$$D_t = \lambda_1 = 2 - \frac{\epsilon}{3} \Rightarrow \nu = \frac{1}{D_t} = \frac{1}{2} + \frac{\epsilon}{12}. \tag{12.44}$$

The magnetic field scaling field was shown in the exercise to be given by the mean field value:

$$D_h = \frac{d + 2}{2}. \tag{12.45}$$

From these two critical exponents we can calculate all other exponents as we have shown in the previous section.

The last remark concerns the dimensions $d > 4$. In this case the two scaling fields are both relevant and therefore the Wilson–Fisher fixed point is unstable. The theory is actually dominated by mean field.

12.2 The Wilson approximate recursion formulas

12.2.1 Kadanoff–Wilson phase space analysis

We start by describing a particular phase space cell decomposition due to Wilson [7, 8], which is largely motivated by Kadanoff block spins [9, 10]. See also [1, 11] and [12, 13].

We assume a hard cutoff 2Λ. Thus if $\phi(x)$ is the field (spin) variable and $\tilde{\phi}(k)$ is its Fourier transform we will assume that $\tilde{\phi}(k)$ is zero for $k > 2\Lambda$.

We expand the field as

$$\phi(x) = \sum_{\vec{m}} \sum_{l=0}^{\infty} \psi_{\vec{m}l}(x) \phi_{\vec{m}l}. \tag{12.46}$$

The wave functions $\psi_{\vec{m}l}(x)$ satisfy the orthonormality condition

$$\int d^d x \, \psi^*_{\vec{m}_1 l_1}(x) \psi_{\vec{m}_2 l_2}(x) = \delta_{\vec{m}_1 \vec{m}_2} \delta_{l_1 l_2}. \tag{12.47}$$

The Fourier transform $\tilde{\psi}_{\vec{m}l}(k)$ is defined by

$$\tilde{\psi}_{\vec{m}l}(k) = \int d^d x \psi_{\vec{m}l}(x)\, e^{ikx}. \tag{12.48}$$

The interpretation of l and $\vec{m}$ is as follows. We decompose momentum space into thin spherical shells, i.e., logarithmically as

$$\frac{1}{2^l} \leqslant \frac{|k|}{\Lambda} \leqslant \frac{1}{2^{l-1}}. \tag{12.49}$$

The functions $\tilde{\psi}_{\vec{m}l}(k)$ for a fixed l are non-zero only inside the shell l, i.e. for $1/2^l \leqslant |k|/\Lambda \leqslant 1/2^{l-1}$. We will assume furthermore that the functions $\tilde{\psi}_{\vec{m}l}(k)$ are constant within its shell and satisfy the normalization condition

$$\int \frac{d^d k}{(2\pi)^d} |\tilde{\psi}_{\vec{m}l}(k)|^2 = 1. \tag{12.50}$$

The functions $\tilde{\psi}_{\vec{m}l}(k)$ and $\psi_{\vec{m}l}(x)$ for a fixed l and a fixed $\vec{m}$ should be thought of as minimal wave packets, i.e., if Δk is the width of $\tilde{\psi}_{\vec{m}l}(k)$ and Δx is the width of $\psi_{\vec{m}l}(x)$ then one must have by the uncertainty principle the requirement $\Delta x \Delta k = (2\pi)^d$. Thus for each shell we divide position space into blocks of equal size each with volume inversely proportional to the volume of the corresponding shell. The volume of the lth momentum shell is proportional to R^d where $R = 1/2^l$, viz $\Delta k = (2\pi)^d 2^{-ld} w$ where w is a constant. Hence the volume of the corresponding position space box is $\Delta x = 2^{ld} w^{-1}$. The functions $\psi_{\vec{m}l}(x)$ are non-zero (constant) only inside this box by construction. This position space box is characterized by the index $\vec{m}$ as is obvious from the normalization condition

$$\int d^d x |\psi_{\vec{m}l}(x)|^2 = \int_{\vec{x}\in\text{box }\vec{m}} d^d x |\psi_{\vec{m}l}(x)|^2 = 1. \tag{12.51}$$

In other words, we have

$$\int d^d x = \sum_{\vec{m}} \int_{\vec{x}\in\text{box }\vec{m}} d^d x. \tag{12.52}$$

The normalization conditions in momentum and position spaces lead to the relations

$$|\tilde{\psi}_{\vec{m}l}(k)| = 2^{ld/2} w^{-1/2}. \tag{12.53}$$

$$|\psi_{\vec{m}l}(x)| = 2^{-ld/2} w^{1/2}. \tag{12.54}$$

Obviously $|\tilde{\psi}_{\vec{m}l}(0)| = 0$. Thus, $\int d^d x \psi_{\vec{m}l}(x) = \tilde{\psi}_{\vec{m}l}(0) = 0$ and as a consequence we will assume that $\psi_{\vec{m}l}(x)$ is equal to $+2^{-ld/2} w^{1/2}$ in one half of the box and $-2^{-ld/2} w^{1/2}$ in the other half.

The meaning of the index $\vec{m}$, which labels the position space boxes, can be clarified further by the following argument. By an appropriate scale transformation in momentum space we can scale the momenta such that the lth shell becomes the largest shell $l = 0$. Clearly the correct scale transformation is $k \longrightarrow 2^l k$ since $1 \leqslant |2^l k|/\Lambda \leqslant 2$. This corresponds to a scale transformation in position space of the form $x \longrightarrow x/2^l$. We obtain therefore the relation $\psi_{\vec{m}l}(x) = \psi_{\vec{m}0}(x/2^l)$. Next we perform an appropriate translation in position space to bring the box $\vec{m}$ to the box $\vec{0}$. This is clearly given by the translation $\vec{x} \longrightarrow \vec{x} - a_0\vec{m}$. We obtain therefore the relation

$$\psi_{\vec{m}l}(x) = \psi_{\vec{0}0}(x/2^l - a_0\vec{m}). \tag{12.55}$$

The functions $\tilde{\psi}_{\vec{m}l}(k)$ and $\psi_{\vec{m}l}(x)$ correspond to a single degree of freedom in phase space occupying a volume $(2\pi)^d$, i.e., a single cell in phase space is characterized by l and $\vec{m}$. Each momentum shell l corresponds to a lattice in the position space with a lattice spacing given by

$$a_l = (\Delta x)^{1/d} = 2^l a_0, \quad a_0 = w^{-1/d}. \tag{12.56}$$

The largest shell $l = 0$ correspond to a lattice spacing a_0 and each time l is increased by 1 the lattice spacing gets doubled, which is the original spin blocking idea of Kadanoff.

We are interested in integrating out only the $l = 0$ modes. We write then

$$\phi(x) = \sum_{\vec{m}}\psi_{\vec{m}0}(x)\phi_{\vec{m}0} + \phi_1(x). \tag{12.57}$$

$$\phi_1(x) = \sum_{\vec{m}}\sum_{l=1}^{\infty}\psi_{\vec{m}l}(x)\phi_{\vec{m}l}. \tag{12.58}$$

From the normalization (12.54) and the scaling law (12.55) we have

$$\psi_{\vec{m}l}(x) = 2^{-d/2}\psi_{\vec{m}l-1}(x/2). \tag{12.59}$$

We define $\phi'(x/2)$ by

$$\phi_1(x) = 2^{-d/2}\alpha_0\phi'(x/2). \tag{12.60}$$

In other words, we have

$$\phi'(x/2) = \sum_{\vec{m}}\sum_{l=1}^{\infty}\psi_{\vec{m}l-1}(x/2)\phi'_{\vec{m}l-1}, \quad \phi'_{\vec{m}l-1} = \alpha_0^{-1}\phi_{\vec{m}l}. \tag{12.61}$$

We have then

$$\phi(x) = \sum_{\vec{m}}\psi_{\vec{m}0}(x)\phi_{\vec{m}0} + 2^{-d/2}\alpha_0\phi'(x/2). \tag{12.62}$$

12.2.2 Recursion formulas

We will be interested in actions of the form

$$S_0(\phi(x)) = \frac{K}{2}\int d^d x R_0(\phi(x))\partial_\mu\phi(x)\partial^\mu\phi(x) + \int d^d x P_0(\phi(x)). \tag{12.63}$$

We will assume that P_0 and R_0 are even polynomials of the field and that $dR_0/d\phi$ is much smaller than P_0 for all relevant configurations. The partition function is

$$
\begin{aligned}
Z_0 &= \int \mathcal{D}\phi(x)\ e^{-S_0(\phi(x))} \\
&= \int \mathcal{D}\phi'(x/2) \int \prod_{\vec{m}} d\phi_{\vec{m}0}\ e^{-S_0(\phi(x))}.
\end{aligned}
\tag{12.64}
$$

The degrees of freedom contained in the fluctuation $\phi'(x/2)$ correspond to momentum shells $l \geqslant 1$ and thus correspond to position space wave packets larger than the box $\vec{m}$ by at least a factor of 2. We can thus assume that $\phi'(x/2)$ is almost constant over the box $\vec{m}$. If x_0 is the center of the box $\vec{m}$ we can expand $\phi'(x/2)$ as

$$
\phi'(x/2) = \phi'(x_0/2) + (x - x_0)^\mu \partial_\mu \phi'(x_0/2) + \frac{1}{2}(x - x_0)^\mu (x - x_0)^\nu \partial_\mu \partial_\nu \phi'(x_0/2) + \cdots
\tag{12.65}
$$

The partition function becomes

$$
Z_0 = \int \mathcal{D}\phi'(x_0/2) \int \prod_{\vec{m}} d\phi_{\vec{m}0}\ e^{-S_0(\phi(x))}.
\tag{12.66}
$$

12.2.2.1 The kinetic term
We compute

$$
\begin{aligned}
\int d^d x R_0 \partial_\mu \phi(x) \partial^\mu \phi(x) = \int d^d x R_0 \Bigg[&\sum_{\vec{m},\vec{m}'} \partial_\mu \psi_{\vec{m}0}(x) \partial^\mu \psi_{\vec{m}'0}(x) \cdot \phi_{\vec{m}0} \phi_{\vec{m}'0} \\
&+ 2^{1-d/2} \alpha_0 \sum_{\vec{m}} \phi_{\vec{m}0} \partial_\mu \psi_{\vec{m}0}(x) \partial_\mu \phi'(x/2) + 2^{-d} \alpha_0^2 \partial_\mu \phi'(x/2) \partial^\mu \phi'(x/2) \Bigg].
\end{aligned}
\tag{12.67}
$$

The integral $\int d^d x$ is $\sum_{\vec{m}} \int_{\vec{x} \in \text{box } \vec{m}}$. In the box $\vec{m}$ the function $R_0(\phi(x))$ can be replaced by the function $R_0(\psi_{\vec{m}0}(x)\phi_{\vec{m}0} + 2^{-d/2}\alpha_0 \phi'(x/2))$. In this box $\psi_{\vec{m}0}(x)$ is approximated by $+w^{1/2}$ in one half of the box and by $-w^{1/2}$ in the other half, i.e., by a step function. Thus the third term in the above equation can be approximated by (dropping also higher derivative corrections and defining $u_{\vec{m}} = 2^{-d/2}\alpha_0 \phi'(x_0/2)$)

$$
2^{-d-1}\alpha_0^2 w^{-1} \sum_{\vec{m}} [R_0(w^{1/2}\phi_{\vec{m}0} + u_{\vec{m}}) + R_0(-w^{1/2}\phi_{\vec{m}0} + u_{\vec{m}})] \partial_\mu \phi'(x_0/2) \partial^\mu \phi'(x_0/2).
\tag{12.68}
$$

The contribution of $\partial_\mu \psi_{\vec{m}0}(x)$ is however only appreciable when $\psi_{\vec{m}0}(x) = 0$, i.e., at the center of the box. In the first and second terms of the above equation we can then replace $R_0(\phi(x))$ by $R_0(u_{\vec{m}})$. The second term in the above equation (12.67) vanishes by conservation of momentum. In the first term we can neglect all the coupling terms $\phi_{\vec{m}0}\phi_{\vec{m}'0}$ with $\vec{m} \neq \vec{m}'$ since $\psi_{\vec{m}'0}(x)$ is zero inside the box $\vec{m}$. We define the integral

$$
\rho = \int d^d x \partial_\mu \psi_{\vec{m}0}(x) \partial^\mu \psi_{\vec{m}0}(x).
\tag{12.69}
$$

This is independent of $\vec{m}$ because of the relation (12.55). The first term becomes therefore $\rho \sum_{\vec{m}} R_0(u_{\vec{m}})\phi_{\vec{m}0}^2$. Equation (12.67) becomes

$$\int d^d x R_0 \partial_\mu \phi(x) \partial^\mu \phi(x) = \rho \sum_{\vec{m}} R_0(u_{\vec{m}}) \phi_{\vec{m}0}^2$$
$$+ 2^{-d-1} \alpha_0^2 w^{-1} \sum_{\vec{m}} [R_0(w^{1/2} \phi_{\vec{m}0} + u_{\vec{m}}) + R_0(-w^{1/2} \phi_{\vec{m}0} + u_{\vec{m}})] \quad (12.70)$$
$$\times \partial_\mu \phi'(x_0/2) \partial^\mu \phi'(x_0/2).$$

12.2.2.2 The interaction term

Next we want to compute

$$\int_{\vec{x} \in \text{box } \vec{m}} d^d x P_0(\phi^{(m)}(x)) = \int_{\vec{x} \in \text{box } \vec{m}} d^d x P_0(\psi_{\vec{m}0}(x) \phi_{\vec{m}0} + 2^{-d/2} \alpha_0 \phi'(x/2)). \quad (12.71)$$

We note again that within the box $\vec{m}$ the field is given by

$$\phi^{(m)}(x) = \psi_{\vec{m}0}(x) \phi_{\vec{m}0} + 2^{-d/2} \alpha_0 \phi'(x/2). \quad (12.72)$$

Introduce $z_0 = \psi_{\vec{m}0}(x) \phi_{\vec{m}0} + u_{\vec{m}}$. We have then

$$\int_{\vec{x} \in \text{box } \vec{m}} d^d x P_0(\phi^{(m)}(x)) = \int_{\vec{x} \in \text{box } \vec{m}} d^d x \left[P_0(z_0) + \left((x - x_0)^\mu \partial_\mu u_{\vec{m}} + \frac{1}{2}(x - x_0)^\mu (x - x_0)^\nu \right) \right.$$
$$\left. \times \partial_\mu \partial_\nu u_{\vec{m}} \frac{dP_0}{dz} \Big|_{z_0} + \frac{1}{2}(x - x_0)^\mu (x - x_0)^\nu \partial_\mu u_{\vec{m}} \partial_\nu u_{\vec{m}} \frac{d^2 P_0}{dz^2} \Big|_{z_0} + \cdots \right]. \quad (12.73)$$

As stated earlier $\psi_{\vec{m}0}(x)$ is approximated by $+w^{1/2}$ in one half of the box and by $-w^{1/2}$ in the other half. Also recall that the volume of the box is w^{-1}. The first term can be approximated by

$$\int_{\vec{x} \in \text{box } \vec{m}} d^d x P_0(z_0) = \frac{w^{-1}}{2} [P_0(w^{1/2} \phi_{\vec{m}0} + u_{\vec{m}}) + P_0(-w^{1/2} \phi_{\vec{m}0} + u_{\vec{m}})]. \quad (12.74)$$

We compute now the third term in (12.73). We start from the obvious identity

$$\int_{\text{box}} d^d x \frac{1}{2}(x - x_0)^\mu (x - x_0)^\nu = \int_{\text{box}_+} d^d x \frac{1}{2}(x - x_0)^\mu (x - x_0)^\nu + \int_{\text{box}_-} d^d x \frac{1}{2}(x - x_0)^\mu (x - x_0)^\nu. \quad (12.75)$$

We will think of the box $\vec{m}$ as a sphere of volume w^{-1}. Thus

$$\int_{\text{box}} d^d x \frac{1}{2}(x - x_0)^\mu (x - x_0)^\nu = \frac{1}{2} V \eta^{\mu\nu}. \quad (12.76)$$

We have the definitions

$$V = \frac{1}{d} \int_{\text{box}} r^2 d^d x, \quad w^{-1} = \int_{\text{box}} d^d x. \quad (12.77)$$

Explicitly we have

$$V = \frac{1}{d+2} \left(\frac{dw^{-1}}{\Omega_{d-1}} \right)^{(d+2)/d} \frac{\Omega_{d-1}}{d}. \quad (12.78)$$

The above identity becomes then

$$\frac{1}{2}V\eta^{\mu\nu} = \int_{\text{box}_+} d^dx \frac{1}{2}(x-x_0)^\mu(x-x_0)^\nu + \int_{\text{box}_-} d^dx \frac{1}{2}(x-x_0)^\mu(x-x_0)^\nu. \qquad (12.79)$$

The sum of these two integrals is rotational invariant. As stated before we think of the box as a sphere divided into two regions of equal volume. The first region (the first half of the box) is a concentric smaller sphere whereas the second region (the second half of the box) is a thin spherical shell. Both regions are spherically symmetric and thus we can assume that

$$\int_{\text{box}_\pm} d^dx \frac{1}{2}(x-x_0)^\mu(x-x_0)^\nu = \frac{1}{2}V_\pm\eta^{\mu\nu}. \qquad (12.80)$$

Clearly $V = V_+ + V_-$. The integral of interest is

$$\mathcal{I}_1 = \int_{\vec{x}\in\text{box}} d^dx \frac{1}{2}(x-x_0)^\mu(x-x_0)^\nu \partial_\mu\partial_\nu u_{\bar{m}} \frac{dP_0}{dz}\Big|_{z_0}. \qquad (12.81)$$

We compute

$$\begin{aligned}
\mathcal{I}_1 &= \frac{dP_0}{dz}\Big|_{w^{1/2}\phi_{\bar{m}0} + u_{\bar{m}}} \partial_\mu\partial_\nu u_{\bar{m}} \int_{\text{box}_+} d^dx \frac{1}{2}(x-x_0)^\mu(x-x_0)^\nu \\
&\quad + \frac{dP_0}{dz}\Big|_{-w^{1/2}\phi_{\bar{m}0} + u_{\bar{m}}} \partial_\mu\partial_\nu u_{\bar{m}} \int_{\text{box}_-} d^dx \frac{1}{2}(x-x_0)^\mu(x-x_0)^\nu \\
&= \frac{V_+}{2}\frac{dP_0}{dz}\Big|_{w^{1/2}\phi_{\bar{m}0} + u_{\bar{m}}} \partial_\mu\partial^\mu u_{\bar{m}} + \frac{V_-}{2}\frac{dP_0}{dz}\Big|_{-w^{1/2}\phi_{\bar{m}0} + u_{\bar{m}}} \partial_\mu\partial^\mu u_{\bar{m}} \\
&= 2^{-1-d/2}\alpha_0 V_+ \frac{dP_0}{dz}\Big|_{w^{1/2}\phi_{\bar{m}0} + u_{\bar{m}}} \partial_\mu\partial^\mu\phi'(x_0/2) + 2^{-1-d/2}\alpha_0 V_- \frac{dP_0}{dz}\Big|_{-w^{1/2}\phi_{\bar{m}0} + u_{\bar{m}}} \partial_\mu\partial^\mu\phi'(x_0/2).
\end{aligned} \qquad (12.82)$$

Similarly the fourth term in (12.73) is computed as follows. We are interested in the integral

$$\mathcal{I}_2 = \int_{\vec{x}\in\text{box }\bar{m}} d^dx \frac{1}{2}(x-x_0)^\mu(x-x_0)^\nu \partial_\mu u_{\bar{m}} \partial_\nu u_{\bar{m}} \frac{d^2P_0}{dz^2}\Big|_{z_0}. \qquad (12.83)$$

We compute

$$\begin{aligned}
\mathcal{I}_2 &= \frac{d^2P_0}{dz^2}\Big|_{w^{1/2}\phi_{\bar{m}0} + u_{\bar{m}}} \partial_\mu u_{\bar{m}}\partial_\nu u_{\bar{m}} \int_{\text{box}_+} d^dx \frac{1}{2}(x-x_0)^\mu(x-x_0)^\nu \\
&\quad + \frac{d^2P_0}{dz^2}\Big|_{-w^{1/2}\phi_{\bar{m}0} + u_{\bar{m}}} \partial_\mu u_{\bar{m}}\partial_\nu u_{\bar{m}} \int_{\text{box}_-} d^dx \frac{1}{2}(x-x_0)^\mu(x-x_0)^\nu \\
&= \frac{V_+}{2}\frac{d^2P_0}{dz^2}\Big|_{w^{1/2}\phi_{\bar{m}0} + u_{\bar{m}}} \partial_\mu u_{\bar{m}}\partial^\mu u_{\bar{m}} + \frac{V_-}{2}\frac{d^2P_0}{dz^2}\Big|_{-w^{1/2}\phi_{\bar{m}0} + u_{\bar{m}}} \partial_\mu u_{\bar{m}}\partial^\mu u_{\bar{m}} \\
&= 2^{-1-d}\alpha_0^2 V_+ \frac{d^2P_0}{dz^2}\Big|_{w^{1/2}\phi_{\bar{m}0} + u_{\bar{m}}} \partial_\mu\phi'(x_0/2)\partial^\mu\phi'(x_0/2) \\
&\quad + 2^{-1-d}\alpha_0^2 V_- \frac{d^2P_0}{dz^2}\Big|_{-w^{1/2}\phi_{\bar{m}0} + u_{\bar{m}}} \partial_\mu\phi'(x_0/2)\partial^\mu\phi'(x_0/2).
\end{aligned} \qquad (12.84)$$

Finally we compute the first term in (12.73). we have

$$\int_{\vec{x}\in\text{box }\vec{m}} d^d x(x-x_0)^\mu \partial_\mu u_{\vec{m}} \frac{dP_0}{dz}\Big|_{z_0} = \frac{dP_0}{dz}\Big|_{w^{1/2}\phi_{\vec{m}0}+u_{\vec{m}}} \partial_\mu u_{\vec{m}} \int_{\text{box}_+} d^d x(x-x_0)^\mu$$
$$+ \frac{dP_0}{dz}\Big|_{-w^{1/2}\phi_{\vec{m}0}+u_{\vec{m}}} \partial_\mu u_{\vec{m}} \int_{\text{box}_-} d^d x(x-x_0)^\mu. \tag{12.85}$$

Clearly we must have

$$\int_{\text{box}_+} d^d x(x-x_0)^\mu + \int_{\text{box}_-} d^d x(x-x_0)^\mu = 0. \tag{12.86}$$

We will assume that both integrals vanish again by the same previous argument. The linear term is therefore 0, viz

$$\int_{\vec{x}\in\text{box }\vec{m}} d^d x(x-x_0)^\mu \partial_\mu u_{\vec{m}} \frac{dP_0}{dz}\Big|_{z_0} = 0. \tag{12.87}$$

The final result is

$$\int_{\vec{x}\in\text{box }\vec{m}} d^d x P_0(\phi^{(m)}(x)) = \frac{w^{-1}}{2}[P_0(w^{1/2}\phi_{\vec{m}0}+u_{\vec{m}}) + P_0(-w^{1/2}\phi_{\vec{m}0}+u_{\vec{m}})]$$
$$+ 2^{-1-d/2}\alpha_0\left[V_+\frac{dP_0}{dz}\Big|_{w^{1/2}\phi_{\vec{m}0}+u_{\vec{m}}} + V_-\frac{dP_0}{dz}\Big|_{-w^{1/2}\phi_{\vec{m}0}+u_{\vec{m}}}\right]\partial_\mu\partial^\mu\phi'(x_0/2) \tag{12.88}$$
$$+ 2^{-1-d}\alpha_0^2\left[V_+\frac{d^2P_0}{dz^2}\Big|_{w^{1/2}\phi_{\vec{m}0}+u_{\vec{m}}} + V_-\frac{d^2P_0}{dz^2}\Big|_{-w^{1/2}\phi_{\vec{m}0}+u_{\vec{m}}}\right]\partial_\mu\phi'\partial^\mu\phi'(x_0/2).$$

12.2.2.3 The action

By putting all the previous results together we obtain the expansion of the action in the following form

$$S_0(\phi(x)) = \frac{K}{2}\int d^d x R_0(\phi(x))\partial_\mu\phi(x)\partial^\mu\phi(x) + \int d^d x P_0(\phi(x))$$
$$= \frac{K\rho}{2}\sum_{\vec{m}} R_0(u_{\vec{m}})\phi_{\vec{m}0}^2$$
$$+ 2^{-d-2}K\alpha_0^2 w^{-1}\sum_{\vec{m}}[R_0(w^{1/2}\phi_{\vec{m}0}+u_{\vec{m}}) + R_0(-w^{1/2}\phi_{\vec{m}0}+u_{\vec{m}})]\partial_\mu\phi'(x_0/2)\partial^\mu\phi'(x_0/2)$$
$$+ \frac{w^{-1}}{2}\sum_{\vec{m}}(P_0(w^{1/2}\phi_{\vec{m}0}+u_{\vec{m}}) + P_0(-w^{1/2}\phi_{\vec{m}0}+u_{\vec{m}})) \tag{12.89}$$
$$+ 2^{-1-d/2}\alpha_0\sum_{\vec{m}}\left(V_+\frac{dP_0}{dz}\Big|_{w^{1/2}\phi_{\vec{m}0}+u_{\vec{m}}} + V_-\frac{dP_0}{dz}\Big|_{-w^{1/2}\phi_{\vec{m}0}+u_{\vec{m}}}\right)\partial_\mu\partial^\mu\phi'(x_0/2)$$
$$+ 2^{-1-d}\alpha_0^2\sum_{\vec{m}}\left(V_+\frac{d^2P_0}{dz^2}\Big|_{w^{1/2}\phi_{\vec{m}0}+u_{\vec{m}}} + V_-\frac{d^2P_0}{dz^2}\Big|_{-w^{1/2}\phi_{\vec{m}0}+u_{\vec{m}}}\right)\partial_\mu\phi'(x_0/2)\partial^\mu\phi'(x_0/2).$$

12.2.2.4 The path integral

We need now to evaluate the path integral

$$Z = \int \prod_{\vec{m}} d\phi_{\vec{m}0}\ e^{-S_0(\phi(x))}. \tag{12.90}$$

We introduce the variables

$$u_{\vec{m}} \longrightarrow z_{\vec{m}} = \left(\frac{K\rho}{2w}\right)^{1/2} u_{\vec{m}}. \tag{12.91}$$

$$\phi_{\vec{m}0} \longrightarrow y_{\vec{m}} = \left(\frac{K\rho}{2}\right)^{1/2} \phi_{\vec{m}0}. \tag{12.92}$$

$$R_0 \longrightarrow W_0(x) = R_0\left(\left(\frac{2w}{K\rho}\right)^{1/2} x\right). \tag{12.93}$$

$$P_0 \longrightarrow Q_0(x) = w^{-1}P_0\left(\left(\frac{2w}{K\rho}\right)^{1/2} x\right). \tag{12.94}$$

We compute then

$$Z = \prod_{\vec{m}} \left(\frac{2}{K\rho}\right)^{1/2} \int dy_{\vec{m}} \, \exp\left(-y_{\vec{m}}^2 W_0(z_{\vec{m}}) - \frac{1}{2}Q_0(y_{\vec{m}} + z_{\vec{m}}) - \frac{1}{2}Q_0(-y_{\vec{m}} + z_{\vec{m}})\right.$$
$$- 2^{-(3+d)/2}\alpha_0(K\rho w)^{1/2}\left[V_+\frac{dQ_0}{dy_{\vec{m}}}(y_{\vec{m}} + z_{\vec{m}}) + V_-\frac{dQ_0}{dy_{\vec{m}}}(-y_{\vec{m}} + z_{\vec{m}})\right]\partial_\mu\partial^\mu\phi'(x_0/2)$$
$$- 2^{-d-2}\alpha_0^2 K\rho\left[V_+\frac{d^2Q_0}{dy_{\vec{m}}^2}(y_{\vec{m}} + z_{\vec{m}}) + V_-\frac{d^2Q_0}{dy_{\vec{m}}^2}(-y_{\vec{m}} + z_{\vec{m}})\right]\partial_\mu\phi'(x_0/2)\partial^\mu\phi'(x_0/2)$$
$$- 2^{-d-2}\alpha_0^2 Kw^{-1}[W_0(y_{\vec{m}} + z_{\vec{m}}) + W_0(-y_{\vec{m}} + z_{\vec{m}})]\partial_\mu\phi'(x_0/2)\partial^\mu\phi'(x_0/2)\bigg). \tag{12.95}$$

Define

$$M_0(z_{\vec{m}}) = \int dy_{\vec{m}} \, \exp\left(-y_{\vec{m}}^2 W_0(z_{\vec{m}}) - \frac{1}{2}Q_0(y_{\vec{m}} + z_{\vec{m}}) - \frac{1}{2}Q_0(-y_{\vec{m}} + z_{\vec{m}})\right.$$
$$- 2^{-(3+d)/2}\alpha_0(K\rho w)^{1/2}\left[V_+\frac{dQ_0}{dy_{\vec{m}}}(y_{\vec{m}} + z_{\vec{m}}) + V_-\frac{dQ_0}{dy_{\vec{m}}}(-y_{\vec{m}} + z_{\vec{m}})\right]\partial_\mu\partial^\mu\phi'(x_0/2)$$
$$- 2^{-d-2}\alpha_0^2 K\rho\left[V_+\frac{d^2Q_0}{dy_{\vec{m}}^2}(y_{\vec{m}} + z_{\vec{m}}) + V_-\frac{d^2Q_0}{dy_{\vec{m}}^2}(-y_{\vec{m}} + z_{\vec{m}})\right]\partial_\mu\phi'(x_0/2)\partial^\mu\phi'(x_0/2)$$
$$- 2^{-d-2}\alpha_0^2 Kw^{-1}[W_0(y_{\vec{m}} + z_{\vec{m}}) + W_0(-y_{\vec{m}} + z_{\vec{m}})]\partial_\mu\phi'(x_0/2)\partial^\mu\phi'(x_0/2)\bigg). \tag{12.96}$$

In this equation $\phi'(x_0/2)$ is given in terms of $z_{\vec{m}}$ by

$$\phi'(x_0/2) = \frac{2^{d/2}}{\alpha_0}\left(\frac{2w}{K\rho}\right)^{1/2} z_{\vec{m}}. \tag{12.97}$$

The remaining dependence on the box $\vec{m}$ is only through the center of the box x_0. Then

$$Z = \prod_{\vec{m}} \left(\frac{2}{K\rho}\right)^{1/2} M_0(z_{\vec{m}})$$

$$= \prod_{\vec{m}} \left(\frac{2}{K\rho}\right)^{1/2} \exp(\ln M_0(z_{\vec{m}})) \tag{12.98}$$

$$= \exp\left(\sum_{\vec{m}} \ln \frac{M_0(z_{\vec{m}})}{I_0(0)}\right) \prod_{\vec{m}} \left(\frac{2}{K\rho}\right)^{1/2} I_0(0).$$

The function $I_0(z)$ is defined by

$$I_0(z) = \int dy \, \exp\left(-y^2 W_0(z) - \frac{1}{2}Q_0(y+z) - \frac{1}{2}Q_0(-y+z)\right). \tag{12.99}$$

In order to compute M_0 we will assume that the derivative terms are small and expand the exponential around the ultra local approximation. We compute

$$M_0(z_{\vec{m}}) = I_0(z_{\vec{m}})\left[1 - 2^{-(3+d)/2}\alpha_0(K\rho w)^{1/2}V\left\langle\frac{dQ_0}{dy_{\vec{m}}}(y_{\vec{m}}+z_{\vec{m}})\right\rangle\partial_\mu\partial^\mu\phi'(x_0/2)\right.$$

$$- 2^{-d-1}\alpha_0^2 K\left[\frac{\rho V}{2}\left\langle\frac{d^2Q_0}{dy_{\vec{m}}^2}(y_{\vec{m}}+z_{\vec{m}})\right\rangle + w^{-1}\langle W_0(y_{\vec{m}}+z_{\vec{m}})\rangle\right]\partial_\mu\phi'(x_0/2)$$

$$\left.\times\, \partial^\mu\phi'(x_0/2) + \cdots\right]. \tag{12.100}$$

The path integral (12.90) becomes

$$Z = \prod_{\vec{m}}\left(\frac{2}{K\rho}\right)^{1/2} I_0(0) \times \exp\left(\sum_{\vec{m}} \ln \frac{I_0(z_{\vec{m}})}{I_0(0)} - 2^{-(3+d)/2}\alpha_0(K\rho w)^{1/2}V\right.$$

$$\times \sum_{\vec{m}}\left\langle\frac{dQ_0}{dy_{\vec{m}}}(y_{\vec{m}}+z_{\vec{m}})\right\rangle\partial_\mu\partial^\mu\phi'(x_0/2) - 2^{-d-1}\alpha_0^2 K\sum_{\vec{m}}\left[\frac{\rho V}{2}\left\langle\frac{d^2Q_0}{dy_{\vec{m}}^2}(y_{\vec{m}}+z_{\vec{m}})\right\rangle\right. \tag{12.101}$$

$$\left.+ w^{-1}\langle W_0(y_{\vec{m}}+z_{\vec{m}})\rangle\right]\partial_\mu\phi'(x_0/2)\partial^\mu\phi'(x_0/2)).$$

We make the change of variable $x_0/2 \longrightarrow x$ (this means that the position space wave packet with $l=1$ corresponding to the highest not integrated momentum will now fit into the box) to obtain

$$Z = \prod_{\vec{m}}\left(\frac{2}{K\rho}\right)^{1/2} I_0(0) \times \exp\left(\sum_{\vec{m}} \ln \frac{I_0(z_{\vec{m}})}{I_0(0)} - 2^{-(3+d)/2}\frac{\alpha_0(K\rho w)^{1/2}V}{4}\right.$$

$$\times \sum_{\vec{m}}\left\langle\frac{dQ_0}{dy_{\vec{m}}}(y_{\vec{m}}+z_{\vec{m}})\right\rangle\partial_\mu\partial^\mu\phi'(x) - 2^{-d-1}\frac{\alpha_0^2 K}{4}\sum_{\vec{m}}\left[\frac{\rho V}{2}\left\langle\frac{d^2Q_0}{dy_{\vec{m}}^2}(y_{\vec{m}}+z_{\vec{m}})\right\rangle\right. \tag{12.102}$$

$$\left.+ w^{-1}\langle W_0(y_{\vec{m}}+z_{\vec{m}})\rangle\right]\partial_\mu\phi'(x)\partial^\mu\phi'(x)).$$

Now $z_{\vec{m}}$ is given by $z_{\vec{m}} = (K\rho/2w)^{1/2}2^{-d/2}\alpha_0\phi'(x)$. It is clear that $dQ_0(y_{\vec{m}}+z_{\vec{m}})/dy_{\vec{m}} = dQ_0(y_{\vec{m}}+z_{\vec{m}})/dz_{\vec{m}}$, etc. Recall that the volume of the box $\vec{m}$ is w^{-1} and thus we can make the identification $w\int d^dx = \sum_{\vec{m}}$. However we have also made the rescaling $x_0 \longrightarrow 2x$ and hence we must make instead the identification $2^d w\int d^dx = \sum_{\vec{m}}$. We obtain

$$Z = \prod_{\bar{m}} \left(\frac{2}{K\rho}\right)^{1/2} I_0(0) \times \exp\left(2^d w \int d^d x \ln \frac{I_0(z)}{I_0(0)} - 2^{-(3+d)/2} \frac{\alpha_0 (K\rho w)^{1/2} V}{4} 2^d w\right.$$

$$\times \int d^d x \left\langle \frac{dQ_0}{dz}(z) \right\rangle \partial_\mu \partial^\mu \phi'(x) - 2^{-d-1} \frac{\alpha_0^2 K}{4} 2^d w \int d^d x \left[\frac{\rho V}{2} \left\langle \frac{d^2 Q_0}{dz^2}(z) \right\rangle \right.$$

$$+ w^{-1} \langle W_0(z) \rangle] \partial_\mu \phi'(x) \partial^\mu \phi'(x)).$$

(12.103)

Now z is given by $z = (K\rho/2w)^{1/2} 2^{-d/2} \alpha_0 \phi'(x)$. The expectation values $\langle O^n(z) \rangle$ are defined by

$$\langle O^n(z) \rangle = \frac{1}{I(z)} \int dy \left(\frac{O(y+z) + O(-y+z)}{2}\right)^n \exp\left(-y^2 W_0(z) - \frac{1}{2} Q_0(y+z)\right.$$

$$\left. - \frac{1}{2} Q_0(-y+z)\right).$$

(12.104)

Next we derive in a straightforward way the formula

$$\frac{d}{dz}\left\langle \frac{dQ_0}{dz}(z) \right\rangle = \left\langle \frac{d^2 Q_0}{dz^2}(z) \right\rangle - \left\langle \left(\frac{dQ_0}{dz}\right)^2 (z) \right\rangle + \left\langle \frac{dQ_0}{dz}(z) \right\rangle^2$$

$$+ \frac{dW_0}{dz}\left(\left\langle \frac{dQ_0}{dz}(z) \right\rangle \frac{1}{I_0(z)} \int dy y^2 e^{\cdots} - \frac{1}{I_0(z)} \int dy \frac{dQ_0}{dz}(y+z) y^2 e^{\cdots}\right).$$

(12.105)

By integrating by part the second term in (12.103) we can see that the first term in (12.105) cancels the third term in (12.103). The last term can be neglected if we assume that dW_0/dz is much smaller than Q_0. We obtain then

$$Z = \prod_{\bar{m}} \left(\frac{2}{K\rho}\right)^{1/2} I_0(0) \times \exp\left(2^d w \int d^d x \ln \frac{I_0(z)}{I_0(0)}\right.$$

$$- \frac{\alpha_0^2 K w}{8} \int d^d x \left[\frac{\rho V}{2}\left(\left\langle \left(\frac{dQ_0}{dz}\right)^2 (z) \right\rangle - \left\langle \frac{dQ_0}{dz}(z) \right\rangle^2\right) + w^{-1}\langle W_0(z) \rangle\right]$$

$$\times \partial_\mu \phi'(x) \partial^\mu \phi'(x)).$$

(12.106)

12.2.2.5 The recursion formulas

The full path integral is therefore given by

$$Z_0 = \int \mathcal{D}\phi'(x_0/2) \int \prod_{\bar{m}} d\phi_{\bar{m}0} \ e^{-S_0(\phi(x))}$$

$$\propto \int \mathcal{D}\phi'(x) \int \exp\left(2^d w \int d^d x \ln \frac{I_0(z)}{I_0(0)} - \frac{\alpha_0^2 K w}{8} \int d^d x \left[\frac{\rho V}{2}\left(\left\langle \left(\frac{dQ_0}{dz}\right)^2 (z) \right\rangle\right.\right.\right.$$

$$\left.\left.\left. - \left\langle \frac{dQ_0}{dz}(z) \right\rangle^2\right) + w^{-1}\langle W_0(z) \rangle\right] \partial_\mu \phi'(x) \partial^\mu \phi'(x)\right).$$

(12.107)

We write this as

$$Z_1 = \int \mathcal{D}\phi'(x)\ e^{-S_1(\phi'(x))}. \tag{12.108}$$

The new action S_1 has the same form as the action S_0, viz

$$S_1(\phi'(x)) = \frac{K}{2} \int d^d x R_1(\phi'(x)) \partial_\mu \phi'(x) \partial^\mu \phi'(x) + \int d^d x P_1(\phi'(x)). \tag{12.109}$$

The new polynomials P_1 and R_1 (or equivalently Q_1 and W_1) are given in terms of the old ones P_0 and R_0 (or equivalently Q_0 and W_0) by the relations

$$W_1(2^{d/2}\alpha_0^{-1}z) = R_1(\phi'(x))$$
$$= \frac{\alpha_0^2 C_d}{8}\left(\left\langle \left(\frac{dQ_0}{dz}\right)^2(z) \right\rangle - \left\langle \frac{dQ_0}{dz}(z) \right\rangle^2 \right) + \frac{\alpha_0^2}{4}\langle W_0(z) \rangle. \tag{12.110}$$

$$Q_1(2^{d/2}\alpha_0^{-1}z) = w^{-1}P_1(\phi'(x))$$
$$= -2^d \ln \frac{I_0(z)}{I_0(0)}. \tag{12.111}$$

The constant C_d is given by

$$C_d = w\rho V. \tag{12.112}$$

Before we write down the recursion formulas we also introduce the notation

$$\phi^{(0)} = \phi, \quad \phi^{(1)} = \phi'. \tag{12.113}$$

The above procedure can be repeated to integrate out the momentum shell $l = 1$ and get from S_1 to S_2. The modes with $l \geqslant 2$ will involve a new constant α_1 and instead of P_0, R_0, S_0 and I_0 we will have P_1, R_1, S_1 and I_1. Aside from this trivial relabeling everything else will be the same including the constants w, V and C_d since $l = 1$ can be mapped to $l = 0$ due to the scaling $x_0 \longrightarrow 2x$ (see equations (12.59) and (12.61)). This whole process can be repeated an arbitrary number of times to get a renormalization group flow of the action given explicitly by the sequences $P_0 \longrightarrow P_1 \longrightarrow P_2 \longrightarrow \cdots \longrightarrow P_i$ and $R_0 \longrightarrow R_1 \longrightarrow R_2 \longrightarrow \cdots \longrightarrow R_i$. By assuming that $dR_i/d\phi^{(i)}$ is sufficiently small compared to P_i for all i the recursion formulas, which relates the different operators at the renormalization group steps i and $i + 1$ are obviously given by

$$W_{i+1}(2^{d/2}\alpha_i^{-1}z) = R_{i+1}(\phi^{(i+1)}(x))$$
$$= \frac{\alpha_i^2 C_d}{8}\left(\left\langle \left(\frac{dQ_i}{dz}\right)^2(z) \right\rangle - \left\langle \frac{dQ_i}{dz}(z) \right\rangle^2 \right) + \frac{\alpha_i^2}{4}\langle W_i(z) \rangle. \tag{12.114}$$

$$Q_{i+1}(2^{d/2}\alpha_i^{-1}z) = w^{-1}P_{i+1}(\phi^{(i+1)}(x))$$
$$= -2^d \ln \frac{I_i(z)}{I_i(0)}. \tag{12.115}$$

The function $I_i(z)$ is given by the same formula (12.99) with the substitutions $I_0 \longrightarrow I_i$, $W_0 \longrightarrow W_i$ and $Q_0 \longrightarrow Q_i$.

The field $\phi^{(i+1)}$ and the variable z are related by $z = (K\rho/2w)^{1/2}2^{-d/2}\alpha_i\phi^{(i+1)}$. The full action at the renormalization group step i is

$$S_i(\phi^{(i)}(x)) = \frac{K}{2}\int d^dx\, R_i(\phi^{(i)}(x))\partial_\mu\phi^{(i)}(x)\partial^\mu\phi^{(i)}(x) + \int d^dx\, P_i(\phi^{(i)}(x)). \quad (12.116)$$

The constants α_i will be determined from the normalization condition

$$W_{i+1}(0) = 1. \quad (12.117)$$

Since Q is even this normalization condition is equivalent to

$$\frac{\alpha_i^2}{4}\langle W_i(z)\rangle|_{z=0} = 1. \quad (12.118)$$

12.2.2.6 The ultra local recursion formula

This corresponds to keeping in the expansion (12.65) only the first term. The resulting recursion formula is obtained from the above recursion formulas by dropping from equation (12.114) the fluctuation term

$$\left\langle\left(\frac{dQ_i}{dz}\right)^2(z)\right\rangle - \left\langle\frac{dQ_i}{dz}(z)\right\rangle^2. \quad (12.119)$$

The recursion formula (12.114) becomes

$$W_{i+1}(2^{d/2}\alpha_i^{-1}z) = \frac{\alpha_i^2}{4}\langle W_i(z)\rangle. \quad (12.120)$$

The solution of this equation together with the normalization condition (12.118) is given by

$$W_i = 1, \quad \alpha_i = 2. \quad (12.121)$$

The remaining recursion formula is given by

$$Q_{i+1}(2^{d/2}2^{-1}z) = -2^d\ln\frac{I_i(z)}{I_i(0)}. \quad (12.122)$$

We state without proof that the use of this recursion formula is completely equivalent to the use in perturbation theory of the Polyakov-Wilson rules given by the approximations:

- We replace every internal propagator $1/(k^2 + r_0^2)$ by $1/(\Lambda^2 + r_0^2)$.
- We replace every momentum integral $\int_{\Lambda/2}^{\Lambda} d^dp/(2\pi)^d$ by the volume $c/4$ where $c = 4\Omega_{d-1}\Lambda^d(1 - 2^{-d})/(d(2\pi)^d)$.

12.2.3 The Wilson–Fisher fixed point within this scheme

Let us start with a ϕ^4 action given by

$$S_0[\phi_0] = \int d^d x \left(\frac{1}{2}(\partial_\mu \phi_0)^2 + \frac{1}{2} r_0 \phi_0^2 + u_0 \phi_0^4 \right). \tag{12.123}$$

The Fourier transform of the field is given by

$$\phi_0(x) = \int_0^\Lambda \frac{d^d p}{(2\pi)^d} \tilde{\phi}_0(p) \; e^{ipx}. \tag{12.124}$$

We will decompose the field as

$$\phi_0(x) = \phi_1'(x) + \Phi(x). \tag{12.125}$$

The background field $\phi_1'(x)$ corresponds to the low frequency modes $\tilde{\phi}_0(p)$ where $0 \leqslant p \leqslant \Lambda/2$ whereas the fluctuation field $\Phi(x)$ corresponds to high frequency modes $\tilde{\phi}_0(p)$ where $\Lambda/2 < p \leqslant \Lambda$, viz

$$\phi_1'(x) = \int_0^{\Lambda/2} \frac{d^d p}{(2\pi)^d} \tilde{\phi}_0(p) \; e^{ipx}, \quad \Phi(x) = \int_{\Lambda/2}^\Lambda \frac{d^d p}{(2\pi)^d} \tilde{\phi}_0(p) \; e^{ipx}. \tag{12.126}$$

The goal is integrate out the high frequency modes from the partition function. The partition function is given by

$$\begin{aligned}
Z_0 &= \int \mathcal{D}\phi_0 \; e^{-S_0[\phi_0]} \\
&= \int \mathcal{D}\phi_1' \int \mathcal{D}\Phi \; e^{-S_0[\phi_1' + \Phi]} \\
&= \int \mathcal{D}\phi_1' \; e^{-S_1[\phi_1']}.
\end{aligned} \tag{12.127}$$

The first goal is to determine the action $S_1[\phi_1']$. We have

$$\begin{aligned}
e^{-S_1[\phi_1']} &= \int d\Phi \; e^{-S_0[\phi_1' + \Phi]} \\
&= e^{-S_0[\phi_1']} \int d\Phi \; e^{-S_0[\Phi]} e^{-u_0 \int d^d x [4\Phi^3 \phi_1' + 4\Phi \phi_1'^3 + 6\Phi^2 \phi_1'^2]}.
\end{aligned} \tag{12.128}$$

We will expand in the field ϕ_1' up to the fourth power. We define expectation values with respect to the partition function

$$Z = \int d\Phi \; e^{-S_0[\Phi]}. \tag{12.129}$$

Let us also introduce

$$\begin{aligned}
V_1 &= -4u_0 \int d^d x \Phi^3 \phi_1' \\
V_2 &= -6u_0 \int d^d x \Phi^2 \phi_1'^2 \\
V_3 &= -4u_0 \int d^d x \Phi \phi_1'^3.
\end{aligned} \tag{12.130}$$

Then we compute

$$e^{-S_1[\phi_1']} = Ze^{-S_0[\phi_1']}\left\langle\left[1 + V_1 + V_2 + V_3 + \frac{1}{2}(V_1^2 + 2V_1V_2 + V_2^2 + 2V_1V_3) + \frac{1}{6}(V_1^3 + 3V_1^2V_2)\right.\right.$$
$$\left.\left. + \frac{1}{24}V_1^4\right]\right\rangle. \tag{12.131}$$

By using the symmetry $\Phi \longrightarrow -\Phi$ we obtain

$$e^{-S_1[\phi_1']} = Ze^{-S_0[\phi_1']}\left\langle\left[1 + V_2 + \frac{1}{2}(V_1^2 + V_2^2 + 2V_1V_3) + \frac{1}{6}(3V_1^2V_2) + \frac{1}{24}V_1^4\right]\right\rangle. \tag{12.132}$$

The term $\langle V_1 V_3 \rangle$ vanishes by momentum conservation. We rewrite the different expectation values in terms of connected functions. We have

$$\langle V_2 \rangle = \langle V_2 \rangle_{\text{co}}$$
$$\langle V_1^2 \rangle = \langle V_1^2 \rangle_{\text{co}}$$
$$\langle V_2^2 \rangle = \langle V_2^2 \rangle_{\text{co}} + \langle V_2 \rangle_{\text{co}}^2 \tag{12.133}$$
$$\langle V_1^2 V_2 \rangle = \langle V_1^2 V_2 \rangle_{\text{co}} + \langle V_1^2 \rangle_{\text{co}}\langle V_2 \rangle_{\text{co}}$$
$$\langle V_1^4 \rangle = \langle V_1^4 \rangle_{\text{co}} + 3\langle V_1^2 \rangle_{\text{co}}^2.$$

By using these results the partition function becomes

$$e^{-S_1[\phi_1']} = Ze^{-S_0[\phi_1']}\, e^{\langle V_2 \rangle_{\text{co}} + \frac{1}{2}(\langle V_1^2 \rangle_{\text{co}} + \langle V_2^2 \rangle_{\text{co}}) + \frac{1}{2}\langle V_1^2 V_2 \rangle_{\text{co}} + \frac{1}{24}\langle V_1^4 \rangle_{\text{co}}}. \tag{12.134}$$

In other words, the partition function is expressible only in terms of irreducible connected functions. This is sometimes known as the cumulant expansion. The action $S_1[\phi_1']$ is given by

$$S_1[\phi_1'] = S_0[\phi_1'] - \langle V_2 \rangle_{\text{co}} - \frac{1}{2}(\langle V_1^2 \rangle_{\text{co}} + \langle V_2^2 \rangle_{\text{co}}) - \frac{1}{2}\langle V_1^2 V_2 \rangle_{\text{co}} - \frac{1}{24}\langle V_1^4 \rangle_{\text{co}}. \tag{12.135}$$

We need the propagator

$$\langle \Phi(x)\Phi(y) \rangle_{\text{co}} = \langle \Phi(x)\Phi(y) \rangle_0 - 12u_0 \int d^d z \langle \Phi(x)\Phi(z) \rangle_0 \langle \Phi(y)\Phi(z) \rangle_0 \langle \Phi(z)\Phi(z) \rangle_0$$
$$+ O(u_0^2). \tag{12.136}$$

The free propagator is obviously given by

$$\langle \Phi(x)\Phi(y) \rangle_0 = \int_{\Lambda/2}^{\Lambda} \frac{d^d p}{(2\pi)^d} \frac{1}{p^2 + r_0} e^{ip(x-y)}. \tag{12.137}$$

Thus

$$\langle \Phi(x)\Phi(y) \rangle_{\text{co}} = \left[\int \frac{d^d p_1}{(2\pi)^d}\frac{1}{p_1^2 + r_0} - 12u_0 \int \frac{d^d p_1}{(2\pi)^d}\frac{d^d p_2}{(2\pi)^d}\frac{1}{(p_1^2 + r_0)^2(p_2^2 + r_0)} + O(u_0^2)\right]e^{ip_1(x-y)}. \tag{12.138}$$

We can now compute

$$-\langle V_2\rangle_{co} = 6u_0 \int d^dx\, \phi_1'^2(x)\langle \Phi^2(x)\rangle_{co}$$

$$= 6u_0 \int \frac{d^dp}{(2\pi)^d}|\tilde{\phi}_1'(p)|^2\left[\int \frac{d^dp_1}{(2\pi)^d}\frac{1}{p_1^2+r_0} - 12u_0 \int \frac{d^dp_1}{(2\pi)^d}\frac{d^dp_2}{(2\pi)^d}\frac{1}{(p_1^2+r_0)^2(p_2^2+r_0)}\right] \tag{12.139}$$

$$+ O(u_0^2)\Big].$$

$$-\frac{1}{2}\langle V_1^2\rangle_{co} = -8u_0^2 \int d^dx_1 d^dx_2\, \phi_1'(x_1)\phi_1'(x_2)\langle \Phi^3(x_1)\Phi^3(x_2)\rangle_{co}$$

$$= -8u_0^2 \int d^dx_1 d^dx_2\, \phi_1'(x_1)\phi_1'(x_2)\Big[6\langle\Phi(x_1)\Phi(x_2)\rangle_0^3 + 3\langle\Phi(x_1)\Phi(x_2)\rangle_0$$

$$\times \langle\Phi(x_1)\Phi(x_1)\rangle_0\langle\Phi(x_2)\Phi(x_2)\rangle_0\Big] \tag{12.140}$$

$$= -8u_0^2 \int \frac{d^dp}{(2\pi)^d}|\tilde{\phi}_1'(p)|^2\left[6\int \frac{d^dp_1}{(2\pi)^d}\frac{d^dp_2}{(2\pi)^d}\frac{1}{(p_1^2+r_0)(p_2^2+r_0)((p+p_1+p_2)^2+r_0)}\right]$$

$$+ O(u_0)\Big].$$

The second term of the second line of the above equation did not contribute because of momentum conservation. Next we compute

$$-\frac{1}{2}\langle V_2^2\rangle_{co} = -18u_0^2 \int d^dx_1 d^dx_2\, \phi_1'^2(x_1)\phi_1'^2(x_2)\langle \Phi^2(x_1)\Phi^2(x_2)\rangle_{co}$$

$$= -36u_0^2 \int d^dx_1 d^dx_2\, \phi_1'^2(x_1)\phi_1'^2(x_2)\langle\Phi(x_1)\Phi(x_2)\rangle_0^2$$

$$= -12u_0^2 \int \frac{d^dp_1}{(2\pi)^d} \cdots \frac{d^dp_3}{(2\pi)^d}\tilde{\phi}_1'(p_1) \cdots \tilde{\phi}_1'(p_3)\tilde{\phi}_1'(-p_1-p_2-p_3)\left[\int \frac{d^dk}{(2\pi)^d}\frac{1}{k^2+r_0}\right. \tag{12.141}$$

$$\left.\times \frac{1}{(k+p_1+p_2)^2+r_0} + 2 \text{ permutations} + O(u_0)\right].$$

The last two terms of the cumulant expansion are of order u_0^3 and u_0^4, respectively, which we are not computing. The action $S_1[\phi_1']$ then reads explicitly

$$S_1[\phi_1'] = \frac{1}{2}\int_0^{\Lambda/2} \frac{d^dp}{(2\pi)^d}|\tilde{\phi}_1'(p)|^2\left\{p^2+r_0+12u_0\int \frac{d^dk_1}{(2\pi)^d}\frac{1}{k_1^2+r_0}\right.$$

$$-144u_0^2 \int \frac{d^dk_1}{(2\pi)^d}\frac{d^dk_2}{(2\pi)^d}\frac{1}{(k_1^2+r_0)^2(k_2^2+r_0)}$$

$$-96u_0^2 \int \frac{d^dk_1}{(2\pi)^d}\frac{d^dk_2}{(2\pi)^d}\frac{1}{(k_1^2+r_0)(k_2^2+r_0)((p+k_1+k_2)^2+r_0)} + O(u_0^3)\Big\} \tag{12.142}$$

$$+\int_0^{\Lambda/2} \frac{d^dp_1}{(2\pi)^d} \cdots \frac{d^dp_3}{(2\pi)^d}\tilde{\phi}_1'(p_1) \cdots \tilde{\phi}_1'(p_3)\tilde{\phi}_1'(-p_1-p_2-p_3)\left[u_0 - 12u_0^2\left[\int \frac{d^dk}{(2\pi)^d}\frac{1}{k^2+r_0}\right.\right.$$

$$\left.\left.\times \frac{1}{(k+p_1+p_2)^2+r_0} + 2 \text{ permutations}\right] + O(u_0^3)\right].$$

The Fourier mode $\tilde{\phi}_1'(p)$ is of course equal $\tilde{\phi}_0(p)$ for $0 \leqslant p \leqslant \Lambda/2$ and 0 otherwise. We scale now the field as

$$\tilde{\phi}_1'(p) = \alpha_0 \tilde{\phi}_1(2p). \tag{12.143}$$

The action becomes

$$
S_1[\phi_1] = \frac{1}{2}\alpha_0^2 2^{-d} \int_0^\Lambda \frac{d^d p}{(2\pi)^d} |\tilde{\phi}_1(p)|^2 \left\{ \frac{p^2}{4} + r_0 + 12u_0 \int \frac{d^d k_1}{(2\pi)^d} \frac{1}{k_1^2 + r_0} - 144u_0^2 \int \frac{d^d k_1}{(2\pi)^d} \frac{d^d k_2}{(2\pi)^d} \right.
$$

$$
\times \frac{1}{(k_1^2 + r_0)^2(k_2^2 + r_0)} - 96u_0^2 \int \frac{d^d k_1}{(2\pi)^d} \frac{d^d k_2}{(2\pi)^d} \frac{1}{(k_1^2 + r_0)(k_2^2 + r_0)\left(\left(\frac{1}{2}p + k_1 + k_2\right)^2 + r_0\right)}
$$

$$
\left. + O(u_0^3) \right\}
$$

$$
+ \alpha_0^4 2^{-3d} \int_0^\Lambda \frac{d^d p_1}{(2\pi)^d} \cdots \frac{d^d p_3}{(2\pi)^d} \tilde{\phi}_1(p_1) \cdots \tilde{\phi}_1(p_3)\tilde{\phi}_1(-p_1 - p_2 - p_3)\left[u_0 - 12u_0^2\left[\int \frac{d^d k}{(2\pi)^d} \frac{1}{k^2 + r_0} \right.\right.
$$

$$
\left.\left. \times \frac{1}{(k + \frac{1}{2}p_1 + \frac{1}{2}p_2)^2 + r_0} + 2 \text{ permutations} \right] + O(u_0^3) \right].
$$

(12.144)

In the above equation the internal momenta k_i are still unscaled in the interval $[\Lambda/2, \Lambda]$. The one-loop truncation of this result is given by

$$
S_1[\phi_1] = \frac{1}{2}\alpha_0^2 2^{-d} \int_0^\Lambda \frac{d^d p}{(2\pi)^d} |\tilde{\phi}_1(p)|^2 \left\{ \frac{p^2}{4} + r_0 + 12u_0 \int \frac{d^d k_1}{(2\pi)^d} \frac{1}{k_1^2 + r_0} + O(u_0^2) \right\}
$$

$$
+ \alpha_0^4 2^{-3d} \int_0^\Lambda \frac{d^d p_1}{(2\pi)^d} \cdots \frac{d^d p_3}{(2\pi)^d} \tilde{\phi}_1(p_1) \cdots \tilde{\phi}_1(p_3)\tilde{\phi}_1(-p_1 - p_2 - p_3)\left[u_0 - 12u_0^2\left[\int \frac{d^d k}{(2\pi)^d} \frac{1}{k^2 + r_0} \right.\right.
$$

(12.145)

$$
\left.\left. \times \frac{1}{(k + \frac{1}{2}p_1 + \frac{1}{2}p_2)^2 + r_0} + 2 \text{ permutations} \right] + O(u_0^3) \right].
$$

We bring the kinetic term to the canonical form by choosing α_0 as

$$
\alpha_0 = 2^{1+d/2}.
$$

(12.146)

Furthermore we truncate the interaction term in the action by setting the external momenta to zero since we are only interested in the renormalization group flow of the operators present in the original action. We get then

$$
S_1[\phi_1] = \frac{1}{2} \int_0^\Lambda \frac{d^d p}{(2\pi)^d} |\tilde{\phi}_1(p)|^2 (p^2 + r_1) + u_1 \int_0^\Lambda \frac{d^d p_1}{(2\pi)^d} \cdots \frac{d^d p_3}{(2\pi)^d} \tilde{\phi}_1(p_1) \cdots \tilde{\phi}_1(p_3)\tilde{\phi}_1(-p_1 - p_2 - p_3).
$$

(12.147)

The new mass parameter r_1 and the new coupling constant u_1 are given by

$$
r_1 = 4r_0 + 48u_0 \int \frac{d^d k_1}{(2\pi)^d} \frac{1}{k_1^2 + r_0} + O(u_0^2).
$$

(12.148)

$$
u_1 = 2^{4-d}\left[u_0 - 36u_0^2 \int \frac{d^d k}{(2\pi)^d} \frac{1}{(k^2 + r_0)^2} + O(u_0^3) \right].
$$

(12.149)

Now we employ the Wilson–Polyakov rules corresponding to the ultra local Wilson recursion formula (12.122) consisting of making the following approximations:

- We replace every internal propagator $1/(k^2 + r_0^2)$ by $1/(\Lambda^2 + r_0^2)$.
- We replace every momentum integral $\int_{\Lambda/2}^{\Lambda} d^d p/(2\pi)^d$ by the volume $c/4$ where $c = 4\Omega_{d-1}\Lambda^d(1 - 2^{-d})/(d(2\pi)^d)$.

The mass parameter r_1 and the coupling constant u_1 become

$$r_1 = 4\left[r_0 + 3c\frac{u_0}{\Lambda^2 + r_0} + O(u_0^2) \right]. \tag{12.150}$$

$$u_1 = 2^{4-d}\left[u_0 - 9c\frac{u_0^2}{(\Lambda^2 + r_0)^2} + O(u_0^3) \right]. \tag{12.151}$$

This is the result of our first renormalization group step. Since the action $S_1[\phi_1]$ is of the same form as the action $S_0[\phi_0]$ the renormalization group calculation can be repeated without any change to go from r_1 and u_1 to a new mass parameter r_2 and a new coupling constant u_2. This whole process can evidently be iterated an arbitrary number of times to define a renormalization group flow $(r_0, u_0) \longrightarrow (r_1, u_1) \longrightarrow \cdots (r_l, u_l) \longrightarrow (r_{l+1}, u_{l+1})\cdots$. The renormalization group recursion equations relating (r_{l+1}, u_{l+1}) to (r_l, u_l) are given precisely by the above equations, viz

$$r_{l+1} = 4\left[r_l + 3c\frac{u_l}{\Lambda^2 + r_l} \right]. \tag{12.152}$$

$$u_{l+1} = 2^{4-d}\left[u_l - 9c\frac{u_l^2}{(\Lambda^2 + r_l)^2} \right]. \tag{12.153}$$

The fixed points of the renormalization group equations is defined obviously by

$$r_* = 4\left[r_* + 3c\frac{u_*}{\Lambda^2 + r_*} \right]. \tag{12.154}$$

$$u_* = 2^{4-d}\left[u_* - 9c\frac{u_*^2}{(\Lambda^2 + r_*)^2} \right]. \tag{12.155}$$

We find the solutions

$$\text{Gaussian fixed point: } r_* = 0, \quad u_* = 0, \tag{12.156}$$

and (by assuming that u_* is sufficiently small)

$$\text{Wilson} - \text{Fisher fixed point: } r_* = -\frac{4cu_*}{\Lambda^2}, \quad u_* = \frac{\Lambda^4}{9c}(1 - 2^{d-4}). \tag{12.157}$$

For $\epsilon = 4 - d$ small the non trivial (interacting) Wilson–Fisher fixed point approaches the trivial (free) Gaussian fixed point as

$$r_* = -\frac{4}{9}\Lambda^2\epsilon \ln 2, \quad u_* = \frac{\Lambda^4}{9c}\epsilon \ln 2. \tag{12.158}$$

The value u_* controls the strength of the interaction of the low energy (infrared) physics of the system.

12.2.4 The critical exponents ν

In the Gaussian model the recursion formula reads simply $r_{l+1} = 4r_l$ and hence we have two possible solutions. At $T = T_c$ the mass parameter r_0 must be zero and hence $r_l = 0$ for all l, i.e. $r_0 = 0$ is a fixed point. For $T \neq T_c$ the mass parameter r_0 is non-zero and hence $r_l = 4^l r_0 \longrightarrow \infty$ for $l \longrightarrow \infty$ ($r_0 = \infty$ is the second fixed point). For T near T_c the mass parameter r_0 is linear in $T - T_c$.

In the ϕ^4 model the situation is naturally more complicated. We can be at the critical temperature $T = T_c$ without having the parameters r_0 and u_0 at their fixed point values. Indeed, as we have already seen, for any value u_0 there will be a critical value $r_{0c} = r_{0c}(u_0)$ of r_0 corresponding to $T = T_c$. At $T = T_c$ we have $r_l \longrightarrow r_*$ and $u_l \longrightarrow u_*$ for $l \longrightarrow \infty$. For $T \neq T_c$ we will have in general a different limit for large l.

The critical exponent ν can be calculated by studying the behavior of the theory only for T near T_c. As stated above $r_l(T_c) \longrightarrow r_*$ and $u_l(T_c) \longrightarrow u_*$ for $l \longrightarrow \infty$. From the analytic property of the recursion formulas we conclude that $r_l(T)$ and $u_l(T)$ are analytic functions of the temperature and hence near T_c we should have
$$r_l(T) = r_l(T_c) + (T - T_c)r_l'(T_c) + \cdots \quad \text{and} \quad u_l(T) = u_l(T_c) + (T - T_c)u_l'(T_c) + \cdots$$
and as a consequence $r_l(T)$ and $u_l(T)$ are close to the fixed point values for sufficiently large l and sufficiently small $T - T_c$. We are thus led in a natural way to studying the recursion formulas only around the fixed point, i.e., to studying the linearized recursion formulas.

12.2.4.1 The linearized recursion formulas
Now we linearize the recursion formulas around the fixed point. We find without any approximation

$$r_{l+1} - r_* = \left[4 - \frac{12cu_*}{(\Lambda^2 + r_l)(\Lambda^2 + r_*)} \right](r_l - r_*) + \frac{12c}{\Lambda^2 + r_*}(u_l - u_*) + 12cu_*\left[\frac{1}{\Lambda^2 + r_l} - \frac{1}{\Lambda^2 + r_*} \right]. \tag{12.159}$$

$$u_{l+1} - u_* = 2^{4-d}\frac{9cu_*^2(2\Lambda^2 + r_* + r_l)}{(\Lambda^2 + r_l)^2(\Lambda^2 + r_*)^2}(r_l - r_*) + 2^{4-d}\left[1 - \frac{9c}{(\Lambda^2 + r_l)^2}(u_l + u_*) \right](u_l - u_*). \tag{12.160}$$

Keeping only linear terms we find

$$r_{l+1} - r_* = \left[4 - \frac{12cu_*}{(\Lambda^2 + r_*)^2} \right](r_l - r_*) + \frac{12c}{\Lambda^2 + r_*}(u_l - u_*). \tag{12.161}$$

$$u_{l+1} - u_* = 2^{4-d}\frac{18cu_*^2}{(\Lambda^2 + r_*)^3}(r_l - r_*) + 2^{4-d}\left[1 - \frac{18cu_*}{(\Lambda^2 + r_*)^2} \right](u_l - u_*). \tag{12.162}$$

This can be put into the matrix form

$$\begin{pmatrix} r_{l+1} - r_* \\ u_{l+1} - u_* \end{pmatrix} = M \begin{pmatrix} r_l - r_* \\ u_l - u_* \end{pmatrix}. \tag{12.163}$$

The matrix M is given by

$$M = \begin{pmatrix} 4 - \dfrac{12cu_*}{(\Lambda^2 + r_*)^2} & \dfrac{12c}{\Lambda^2 + r_*} \\ 2^{4-d}\dfrac{18cu_*^2}{(\Lambda^2 + r_*)^3} & 2^{4-d}\left(1 - \dfrac{18cu_*}{(\Lambda^2 + r_*)^2}\right) \end{pmatrix} = \begin{pmatrix} 4 - \dfrac{4}{3}\epsilon\ln 2 & \dfrac{12c}{\Lambda^2}\left(1 + \dfrac{4}{9}\epsilon\ln 2\right) \\ 0 & 1 - \epsilon\ln 2 \end{pmatrix}. \tag{12.164}$$

After n steps of the renormalization group we will have

$$\begin{pmatrix} r_{l+n} - r_* \\ u_{l+n} - u_* \end{pmatrix} = M^n \begin{pmatrix} r_l - r_* \\ u_l - u_* \end{pmatrix}. \tag{12.165}$$

In other words for large n the matrix M^n is completely dominated by the largest eigenvalue of M.

Let λ_1 and λ_2 be the eigenvalues of M with eigenvectors w_1 and w_2, respectively, such that $\lambda_1 > \lambda_2$. Clearly for $u_* = 0$ we have

$$\begin{aligned} \lambda_1 = 4, \quad w_1 &= \begin{pmatrix} 1 \\ 0 \end{pmatrix} \\ \lambda_2 = 1, \quad w_2 &= \begin{pmatrix} 0 \\ 1 \end{pmatrix}. \end{aligned} \tag{12.166}$$

The matrix M is not symmetric and thus diagonalization is achieved by an invertible (and not an orthogonal) matrix U. We write

$$M = UDU^{-1}. \tag{12.167}$$

The eigenvalues λ_1 and λ_2 can be determined from the trace and determinant, which are given by

$$\lambda_1 + \lambda_2 = M_{11} + M_{22}, \quad \lambda_1\lambda_2 = M_{11}M_{22} - M_{12}M_{21}. \tag{12.168}$$

We obtain immediately

$$\lambda_1 = 4 - \frac{4}{3}\epsilon\ln 2, \quad \lambda_2 = 1 - \epsilon\ln 2. \tag{12.169}$$

The corresponding eigenvectors are

$$w_1 = \begin{pmatrix} 1 \\ 0 \end{pmatrix}, \quad w_2 = \begin{pmatrix} -\dfrac{4c}{\Lambda^2}\left(1 + \dfrac{5}{9}\epsilon\ln 2\right) \\ 1 \end{pmatrix}. \tag{12.170}$$

We write the equation $Mw_k = \lambda_k w_k$ as (with $(w_k)_j = w_{jk}$)

$$M_{ij}w_{jk} = \lambda_k w_{ik}. \tag{12.171}$$

The identity $M = \sum_k \lambda_k |\lambda_k\rangle\langle\lambda_k|$ can be rewritten as

$$M_{ij} = \sum_k \lambda_k w_{ik} v_{kj} = \lambda_1 w_{i1} v_{1j} + \lambda_2 w_{i2} v_{2j}. \tag{12.172}$$

The vectors v_k are the eigenvectors of M^T with eigenvalues λ_k respectively, viz (with $(v_k)_j = v_{kj}$)

$$M^T_{ij} v_{kj} = v_{kj} M_{ji} = \lambda_k v_{ki}. \tag{12.173}$$

We find explicitly

$$v_1 = \begin{pmatrix} 1 \\ \dfrac{4c}{\Lambda^2}\left(1 + \dfrac{5}{9}\epsilon \ln 2\right) \end{pmatrix}, \quad v_2 = \begin{pmatrix} 0 \\ 1 \end{pmatrix}. \tag{12.174}$$

The orthonormality condition is then

$$\sum_j v_{kj} w_{jl} = \delta_{kl}. \tag{12.175}$$

From the result (12.172) we deduce immediately that

$$\begin{aligned} M^n_{ij} &= \lambda^n_1 w_{i1} v_{1j} + \lambda^n_2 w_{i2} v_{2j} \\ &\simeq \lambda^n_1 w_{i1} v_{1j}. \end{aligned} \tag{12.176}$$

The linearized recursion formulas take then the form

$$r'_{l+n} - r_* \simeq \lambda^n_1 w_{11}(v_{11}(r_l - r_*) + v_{12}(u_l - u_*)). \tag{12.177}$$

$$u_{l+n} - u_* \simeq \lambda^n_1 w_{21}(v_{11}(r_l - r_*) + v_{12}(u_l - u_*)). \tag{12.178}$$

Since $r_l = r_l(T)$ and $u_l = u_l(T)$ are close to the fixed point values for sufficiently large l and sufficiently small $T - T_c$ we conclude that $r_l - r_*$ and $u_l - u_*$ are both linear in $T - T_c$ and as a consequence

$$v_{11}(r_l - r_*) + v_{12}(u_l - u_*) = c_l(T - T_c). \tag{12.179}$$

The linearized recursion formulas become

$$r_{l+n} - r_* \simeq c_l \lambda^n_1 w_{11}(T - T_c). \tag{12.180}$$

$$u_{l+n} - u_* \simeq c_l \lambda^n_1 w_{21}(T - T_c). \tag{12.181}$$

12.2.4.2 The critical exponent ν

The correlation length corresponding to the initial action is given by

$$\xi_0(T) = X(r_0(T), u_0(T)). \tag{12.182}$$

After $l + n$ renormalization group steps the correlation length becomes

$$\xi_{l+n}(T) = X(r_{l+n}(T), u_{l+n}(T)). \tag{12.183}$$

At each renormalization group step we scale the momenta as $p \longrightarrow 2p$ which corresponds to scaling the distances as $x \longrightarrow x/2$. The correlation length is a measure of distance and thus one must have

$$X(r_{l+n}, u_{l+n}) = 2^{-l-n}X(r_0, u_0). \tag{12.184}$$

From equations (12.180) and (12.181) we have

$$(r_{l+n+1} - r_*)|_{T-T_c} = \tau/\lambda_1 = (r_{l+n} - r_*)|_{T-T_c} = \tau, \quad (u_{l+n+1} - u_*)|_{T-T_c} = \tau/\lambda_1 = (u_{l+n} - u_*)|_{T-T_c} = \tau. \tag{12.185}$$

Hence

$$X(r_{l+n+1}, u_{l+n+1})|_{T=T_c + \tau/\lambda_1} = X(r_{l+n}, u_{l+n})|_{T=T_c + \tau}. \tag{12.186}$$

By using the two results (12.184) and (12.186) we obtain

$$2^{-l-n-1}\xi_0(T_c + \tau/\lambda_1) = 2^{-l-n}\xi_0(T_c + \tau). \tag{12.187}$$

We expect

$$\xi_0(T_c + \tau) \propto \tau^{-\nu}. \tag{12.188}$$

In other words

$$\frac{1}{2}\left(\frac{\tau}{\lambda_1}\right)^{-\nu} = \tau^{-\nu} \Leftrightarrow \lambda_1^{\nu} = 2 \Leftrightarrow \nu = \frac{\ln 2}{\ln \lambda_1}. \tag{12.189}$$

12.2.5 The critical exponent η

The ultra local recursion formula (12.122) used so far do not lead to a wave function renormalization since all momentum dependence of Feynman diagrams has been dropped and as a consequence the value of the anomalous dimension η within this approximation is 0. This can also be seen from the field scaling (12.143) with the choice (12.146), which are made at every renormalization group step and hence the wave function renormalization is independent of the momentum.

In any case we can see from equation (12.145) that the wave function renormalization at the first renormalization group step is given by

$$Z = \frac{\alpha_0^2}{2^{2+d}}. \tag{12.190}$$

From the other hand we have already established that the scaling behavior of $Z(\lambda)$ for small λ (the limit in which we approach the infrared stable fixed point) is λ^{η}. In our case $\lambda = 1/2$ and hence we must have

$$Z = 2^{-\eta}. \tag{12.191}$$

Let α_* be the fixed value of the sequence α_i. Then from the above two equations we obtain the formula

$$\eta = -\frac{\ln Z}{\ln 2} = d + 2 - \frac{2\ln \alpha_*}{\ln 2}. \tag{12.192}$$

As discussed above since $\alpha_* = 2^{1+d/2}$ for the ultra local recursion formula (12.122) we get immediately $\eta = 0$.

To incorporate a non-zero value of the critical exponent η we must go to the more accurate yet more complicated recursion formulas (12.114) and (12.115). The field scaling at each renormalization group step is a different number α_i. These numbers are determined from the normalization condition (12.118).

Recall that integrating out the momenta $1 \leqslant |k|/\Lambda \leqslant 2$ resulted in the field $\phi_1(x) = \sum_{\vec{m}} \sum_{l=1}^{\infty} \psi_{\vec{m}l}(x) \phi_{\vec{m}l}$, which was expressed in terms of the field $\phi'(x) = \phi^{(1)}$, which appears in the final action as $\phi_1(x) = 2^{-d/2} \alpha_0 \phi'(x/2)$. After n renormalization group steps we integrate out the momenta $2^{1-n} \leqslant |k|/\Lambda \leqslant 2$, which results in the field $\phi_n(x) = \sum_{\vec{m}} \sum_{l=n}^{\infty} \psi_{\vec{m}l}(x) \phi_{\vec{m}l}$. However the action will be expressed in terms of the field $\phi' = \phi^{(n)}$ defined by

$$\phi_n(x) = 2^{-nd/2} \alpha_0 \alpha_1 \cdots \alpha_{n-1} \phi^{(n)}(x/2^n). \tag{12.193}$$

We are interested in the 2-point function

$$\langle \phi_{\vec{m}l} \phi_{\vec{m}'l'} \rangle = \frac{1}{Z} \int \prod_{l_1=0}^{\infty} \prod_{\vec{m}_1} d\phi_{\vec{m}_1 l_1} \, \phi_{\vec{m}l} \, \phi_{\vec{m}'l'} e^{-S_0[\phi]}. \tag{12.194}$$

Let us concentrate on the integral with $l_1 = l$ and $\vec{m}_1 = \vec{m}$ and assume that $l' > l$. We have then the integral

$$\cdots \int d\phi_{\vec{m}l} \phi_{\vec{m}l} \int \prod_{l_1=0}^{l-1} \prod_{\vec{m}_1} d\phi_{\vec{m}_1 l_1} e^{-S_0[\phi]} = \cdots \int d\phi_{\vec{m}l} \phi_{\vec{m}l} \, e^{-S_l[\phi']}. \tag{12.195}$$

We have $\phi' = \phi^{(l)}$ where $\phi^{(l)}$ contains the momenta $|k|/\Lambda \leqslant 1/2^{l-1}$. Since $\phi_{\vec{m}l}$ is not integrated we have $\phi_{\vec{m}l} = \alpha_0 \ldots \alpha_{l-1}(K\rho/2)^{-1/2} y_{\vec{m}}$, which is the generalization of $\phi_{\vec{m}l} = \alpha_0 \phi'_{\vec{m}l-1}$. We want now to further integrate $\phi_{\vec{m}l}$. The final result is similar to (12.95) except that we have an extra factor of $y_{\vec{m}}$ and $z_{\vec{m}}$ contains all the modes with $l_1 > l$. The integral thus clearly vanishes because it is odd under $y_{\vec{m}} \longrightarrow -y_{\vec{m}}$.

We conclude that we must have $l' = l$ and $\vec{m}' = \vec{m}$ otherwise the above 2-point function vanishes. After few more calculations we obtain

$$\langle \phi_{\vec{m}l} \phi_{\vec{m}'l'} \rangle = \delta_{ll'} \delta_{\vec{m}\vec{m}'} \alpha_0^2 \ldots \alpha_{l-1}^2 \left(\frac{K\rho}{2}\right)^{-1} \frac{\int \prod_{l_1=l+1} \prod_{\vec{m}_1} d\phi_{\vec{m}_1 l_1} \prod_{\vec{m}_1} M_l(z_{\vec{m}_1}) . R_l(z_{\vec{m}})}{\int \prod_{l_1=l+1} \prod_{\vec{m}_1} d\phi_{\vec{m}_1 l_1} \prod_{\vec{m}_1} M_l(z_{\vec{m}_1})}. \tag{12.196}$$

The function M_l is given by the same formula (12.96) with the substitutions $M_0 \longrightarrow M_l$, $W_0 \longrightarrow W_l$ and $Q_0 \longrightarrow Q_l$. The variable $z_{\vec{m}}$ is given explicitly by

$$z_{\vec{m}} = \left(\frac{K\rho}{2w}\right)^{1/2} 2^{-d/2} \alpha_l \phi^{(l+1)}(x_0/2). \tag{12.197}$$

The function $R_l(z)$ is defined by

$$R_l(z) = M_l^{-1}(z) \int dy\, y^2 \exp(\ldots). \tag{12.198}$$

The exponent is given by the same exponent of equation (12.96) with the substitutions $M_0 \longrightarrow M_l$, $W_0 \longrightarrow W_l$ and $Q_0 \longrightarrow Q_l$.

An order of magnitude formula for the 2-point function can be obtained by replacing the function $R_l(z)$ by $R_l(0)$. We obtain then

$$\langle \phi_{\vec{m}l} \phi_{\vec{m}'l'} \rangle = \delta_{ll'} \delta_{\vec{m}\vec{m}'} \alpha_0^2 \dots \alpha_{l-1}^2 \left(\frac{K\rho}{2} \right)^{-1} R_l(0). \tag{12.199}$$

At a fixed point of the recursion formulas we must have

$$W_l \longrightarrow W_*, \quad Q_l \longrightarrow Q_* \Leftrightarrow R_l \longrightarrow R_*, \tag{12.200}$$

and

$$\alpha_l \longrightarrow \alpha_*. \tag{12.201}$$

The 2-point function is therefore given by

$$\langle \phi_{\vec{m}l} \phi_{\vec{m}'l'} \rangle \propto \delta_{ll'} \delta_{\vec{m}\vec{m}'} \alpha_*^{2l} \left(\frac{K\rho}{2} \right)^{-1} R_*(0). \tag{12.202}$$

The modes $\langle \phi_{\vec{m}l} \rangle$ correspond to the momentum shell $2^{-l} \leqslant |k|/\Lambda \leqslant 2^{1-l}$, i.e. $k \sim \Lambda 2^{-l}$. From the other hand the 2-point function is expected to behave as

$$\langle \phi_{\vec{m}l} \phi_{\vec{m}'l'} \rangle \propto \delta_{ll'} \delta_{\vec{m}\vec{m}'} \frac{1}{k^{2-\eta}}, \tag{12.203}$$

where η is precisely the anomalous dimension. By substituting $k \sim \Lambda 2^{-l}$ in this last formula we obtain

$$\langle \phi_{\vec{m}l} \phi_{\vec{m}'l'} \rangle \propto \delta_{ll'} \delta_{\vec{m}\vec{m}'} \Lambda^{\eta-2} 2^{l(2-\eta)}. \tag{12.204}$$

By comparing the l-dependent bits in (12.202) and (12.204) we find that the anomalous dimension is given by

$$2^{2-\eta} = \alpha_*^2 \Rightarrow \eta = 2 - \frac{2 \ln \alpha_*}{\ln 2}. \tag{12.205}$$

12.3 Generating functionals

In this section we will re-derive the various generating functional of quantum field theory following the presentation and notation of [4].

12.3.1 The generating functional $\mathcal{G}[J]$

For simplicity, we will only deal with a bosonic real field ϕ. We start from the action and the path integral

$$S[\phi] = S_0[\phi] + S_1[\phi]. \tag{12.206}$$

$$Z = \int d\phi \ \exp(-S[\phi]). \tag{12.207}$$

$S_{\rm I}[\phi]$ is the interaction part while $S_0[\phi]$ is the free Gaussian part, which is of the form

$$S_0[\phi] = -\frac{1}{2}(\phi, G_0^{-1}\phi) = -\frac{1}{2}\int_\alpha \int_\beta \phi_\alpha (G_0^{-1})_{\alpha\beta}\phi_\beta. \tag{12.208}$$

Obviously, $\int_\alpha$ denotes integration over the continuous components and summation over the discrete components of the label α, while the Gaussian propagator G_0 is a real symmetric matrix in label space. We will also use the free path integral

$$Z_0 = \int d\phi \ \exp(-S_0[\phi]). \tag{12.209}$$

The n-point Green's functions are defined by

$$G^{(n)}_{\alpha_1 \cdots \alpha_n} = \langle \phi_{\alpha_n} \cdots \phi_{\alpha_1}\rangle = \frac{\int d\phi \ \exp(-S[\phi]) \ \phi_{\alpha_n} \cdots \phi_{\alpha_1}}{\int d\phi \ \exp(-S[\phi])}. \tag{12.210}$$

The corresponding generating functional is given in terms of a current J by the formula

$$\mathcal{G}[J] = \frac{\int d\phi \ \exp(-S[\phi] + (J, \phi))}{\int d\phi \ \exp(-S[\phi])}. \tag{12.211}$$

Indeed, we compute

$$G^{(n)}_{\alpha_1 \cdots \alpha_n} = \frac{\delta^n \mathcal{G}[J]}{\delta J_{\alpha_n} \cdots \delta J_{\alpha_1}} \Big|_{J=0}. \tag{12.212}$$

Or equivalently

$$\mathcal{G}[J] = \sum_{n=0}^{\infty} \frac{1}{n!} \int_{\alpha_1} \cdots \int_{\alpha_n} G^{(n)}_{\alpha_1 \cdots \alpha_n} J_{\alpha_1} \cdots J_{\alpha_n}. \tag{12.213}$$

12.3.2 The generating functional $\mathcal{G}_c[J]$

The generating functional for connected Green's functions is defined by

$$\mathcal{G}_c[J] = \ln \frac{Z}{Z_0}\mathcal{G}[J], \quad \mathcal{G}[J] = \frac{Z_0}{Z} \exp(\mathcal{G}_c[J]). \tag{12.214}$$

Equivalently

$$\mathcal{G}_c[J] = \ln \left(\frac{\int d\phi \ \exp(-S[\phi] + (J, \phi))}{\int d\phi \ \exp(-S_0[\phi])} \right). \tag{12.215}$$

The connected n-point Green's functions are then given by

$$G^{(n)}_{c\alpha_1 \cdots \alpha_n} = \frac{\delta^n \mathcal{G}_c[J]}{\delta J_{\alpha_n} \cdots \delta J_{\alpha_1}}\Big|_{J=0}. \tag{12.216}$$

Or equivalently

$$\mathcal{G}_c[J] = \sum_{n=0}^{\infty} \frac{1}{n!} \int_{\alpha_1} \cdots \int_{\alpha_n} G^{(n)}_{c\alpha_1 \cdots \alpha_n} J_{\alpha_1} \cdots J_{\alpha_n}. \tag{12.217}$$

Using the language of probability theory the full Green's functions $G^{(n)}_{\alpha_1 \cdots \alpha_n}$ play the role of moments of the probability distribution defined by the Euclidean action whereas the connected Green's functions $G^{(n)}_{c, \alpha_1 \cdots \alpha_n}$ play the role of cumulants.

We can easily check that

$$\mathcal{G}_{0,c}[J] = \ln\left(\frac{\int d\phi \, \exp(-S_0[\phi] + (J, \phi))}{\int d\phi \, \exp(-S_0[\phi])}\right) = -\frac{1}{2}(J, G_0 J). \tag{12.218}$$

Furthermore, we compute (with the substitution $\phi \longrightarrow \delta/\delta J$)

$$\exp(\mathcal{G}_c[J]) = = \frac{Z}{Z_0}\mathcal{G}[J]$$

$$= \frac{\int d\phi \, \exp(-S_0[\phi] - S_1[\phi] + (J, \phi))}{\int d\phi \, \exp(-S_0[\phi])}$$

$$= \exp\left(-S_1\left[\frac{\delta}{\delta J}\right]\right)\frac{\int d\phi \, \exp(-S_0[\phi] + (J, \phi))}{\int d\phi \, \exp(-S_0[\phi])} \tag{12.219}$$

$$= \exp\left(-S_1\left[\frac{\delta}{\delta J}\right]\right)\exp(\mathcal{G}_{0,c}[J])$$

$$= \exp\left(-S_1\left[\frac{\delta}{\delta J}\right]\right)\exp\left(-\frac{1}{2}(J, G_0 J)\right).$$

Thus if $G_0 \longrightarrow 0$, which will be our initial condition for the functional renormalization equation, we have immediately

$$\mathcal{G}_c[J] \longrightarrow 0. \tag{12.220}$$

As it turns out, rewritten in this form, this is not a good initial condition.

12.3.3 The generating functional $\mathcal{G}_{ac}[\bar{\phi}]$

The generating functional for the amputated connected Green's functions is a functional depending on an auxiliary field $\bar{\phi}$ given by the formula

$$\exp(\mathcal{G}_{ac}[\bar{\phi}]) = \frac{\int d\phi \ \exp(-S_0[\phi] - S_1[\phi + \bar{\phi}])}{Z_0}. \tag{12.221}$$

As before, we define the amputated connected n-point Green's functions by

$$G^{(n)}_{ac\alpha_1 \cdots \alpha_n} = \frac{\delta^n \mathcal{G}_{ac}[\bar{\phi}]}{\delta\bar{\phi}_{\alpha_n} \cdots \delta\bar{\phi}_{\alpha_1}}\Big|_{\bar{\phi}=0}. \tag{12.222}$$

Or equivalently

$$\mathcal{G}_{ac}[\bar{\phi}] = \sum_{n=0}^{\infty} \frac{1}{n!} \int_{\alpha_1} \cdots \int_{\alpha_n} G^{(n)}_{ac\alpha_1 \cdots \alpha_n} \bar{\phi}_{\alpha_1} \cdots \bar{\phi}_{\alpha_n}. \tag{12.223}$$

In the above path integral (12.221) we change the variable as $\phi' = \phi + \bar{\phi}$, use the property $G_0^T = G_0$ and we choose

$$\bar{\phi} = -G_0 J. \tag{12.224}$$

We obtain

$$\exp(\mathcal{G}_{ac}[\bar{\phi}]) = \exp\left(\frac{1}{2}(J, G_0 J)\right) \frac{\int d\phi' \ \exp(-S_0[\phi'] - S_1[\phi'] + (J, \phi'))}{Z_0} \tag{12.225}$$
$$= \exp\left(\frac{1}{2}(J, G_0 J)\right) \exp(\mathcal{G}_c[J]).$$

We deduce

$$\mathcal{G}_c[J] = \mathcal{G}_{ac}[\bar{\phi} = -G_0 J] - \frac{1}{2}(J, G_0 J). \tag{12.226}$$

$$\mathcal{G}_{ac}[\bar{\phi}] = \mathcal{G}_c[J = -G_0^{-1}\bar{\phi}] + \frac{1}{2}(\bar{\phi}, G_0^{-1}\bar{\phi}). \tag{12.227}$$

We can easily check that

$$\mathcal{G}_{0,c}[J] = -\frac{1}{2}(J, G_0 J) = -\frac{1}{2}(\bar{\phi}, G_0^{-1}\bar{\phi}). \tag{12.228}$$

Thus

$$\mathcal{G}_{ac}[\bar{\phi}] = (\mathcal{G}_c[J] - \mathcal{G}_{0,c}[J])\big|_{J=-G_0^{-1}\bar{\phi}}. \tag{12.229}$$

Thus for a free field theory the generating functional of the amputated connected Green's functions vanishes identically.

Further, we compute

$$
\begin{aligned}
\exp(\mathcal{G}_{ac}[\bar{\phi}]) &= \frac{\int d\phi \ \exp(-S_0[\phi])\exp(-S_1[\phi + \bar{\phi}])}{Z_0} \\[2ex]
&= \frac{\int d\phi \ \exp(-S_0[\phi])\exp\left(-S_1\left[\frac{\delta}{\delta J}\right]\right)\exp((J, \phi + \bar{\phi}))|_{J=0}}{Z_0} \\[2ex]
&= \exp\left(-S_1\left[\frac{\delta}{\delta J}\right]\right)\frac{\int d\phi \ \exp\left(\frac{1}{2}(\phi, G_0^{-1}\phi)\right)\exp((J, \phi))\exp((J, \bar{\phi}))|_{J=0}}{Z_0} \\[2ex]
&= \exp\left(-S_1\left[\frac{\delta}{\delta J}\right]\right)\exp(G_{0,c}[J])\exp((J, \bar{\phi}))|_{J=0} \\[2ex]
&= \exp\left(-S_1\left[\frac{\delta}{\delta J}\right]\right)\exp\left(-\frac{1}{2}\left(\frac{\delta}{\delta\bar{\phi}}, G_0\frac{\delta}{\delta\bar{\phi}}\right)\right)\exp((J, \bar{\phi}))|_{J=0} \\[2ex]
&= \exp\left(-\frac{1}{2}\left(\frac{\delta}{\delta\bar{\phi}}, G_0\frac{\delta}{\delta\bar{\phi}}\right)\right)\exp(-S_1[\bar{\phi}]).
\end{aligned}
\tag{12.230}
$$

Thus the would-be initial condition $G_0 \longrightarrow 0$ becomes in terms of $\mathcal{G}_{ac}[\bar{\phi}]$ given by the good starting prescription

$$
\mathcal{G}_{ac}[\bar{\phi}] \longrightarrow -\ S_1[\bar{\phi}], \quad G_0 \longrightarrow 0.
\tag{12.231}
$$

12.3.4 The generating functional $\Gamma[\bar{\phi}']$

There are obviously some connected and amputated diagrams which can be separated into two parts by cutting a single line corresponding to a Gaussian propagator. These diagrams are called one-line reducible. Thus, we can decompose the n-point connected Green's functions $G_{c,\,\alpha_1 \ldots \alpha_n}^{(n)}$ into irreducible one-point vertices $\Gamma_{\alpha_1 \ldots \alpha_n}^{(n)}$, which by definition can not be separated into two parts by cutting a single propagator. These vertices are generated from a generating functional $\Gamma[\bar{\phi}']$ known as the effective action as follows

$$
\Gamma_{\alpha_1 \cdots \alpha_n}^{(n)} = \frac{\delta^n \Gamma[\bar{\phi}']}{\delta\bar{\phi}'_{\alpha_n} \cdots \delta\bar{\phi}'_{\alpha_1}} \ |_{\bar{\phi}'=0}.
\tag{12.232}
$$

$$
\Gamma[\bar{\phi}'] = \sum_{n=0}^{\infty} \frac{1}{n!} \int_{\alpha_1} \cdots \int_{\alpha_n} \Gamma_{\alpha_1 \cdots \alpha_n}^{(n)} \bar{\phi}'_{\alpha_1} \cdots \bar{\phi}'_{\alpha_n}.
\tag{12.233}
$$

The generating functional $\Gamma[\bar{\phi}']$ is the Legendre transform of the generating functional $G_c[J]$ given explicitly by

$$
\Gamma[\bar{\phi}'] = \mathcal{L}[\bar{\phi}'] - S_0[\bar{\phi}'].
\tag{12.234}
$$

$$
\mathcal{L}[\bar{\phi}'] = (\bar{\phi}', J) - \mathcal{G}_c[J].
\tag{12.235}
$$

$$\Rightarrow \Gamma[\bar\phi'] = (\bar\phi', J) - \mathcal{G}_c[J] - S_0[\bar\phi'].\tag{12.236}$$

The variable $\bar\phi'$ is the so-called classical field defined by

$$\bar\phi'_\alpha = \langle \phi_\alpha \rangle = \frac{\delta \mathcal{G}_c}{\delta J_\alpha}.\tag{12.237}$$

We note the equation of motion

$$\frac{\delta \mathcal{L}}{\delta \bar\phi'_\alpha} = J_\alpha.\tag{12.238}$$

By using this equation of motion and the chain rule again we obtain

$$\frac{\delta}{\delta \bar\phi'_\alpha} = \int_\beta \frac{\delta^2 \mathcal{L}}{\delta \bar\phi'_\alpha \delta \bar\phi'_\beta} \frac{\delta}{\delta J_\beta} \leftrightarrow \frac{\delta}{\delta \bar\phi} = \left[\frac{\delta}{\delta \bar\phi'} \otimes \frac{\delta}{\delta \bar\phi'} \mathcal{L} \right] \frac{\delta}{\delta J}.\tag{12.239}$$

By applying this operator equation to the definition of the classical field (12.237) we obtain the identity

$$1 = \left[\frac{\delta}{\delta \bar\phi'} \otimes \frac{\delta}{\delta \bar\phi'} \mathcal{L} \right]\left[\frac{\delta}{\delta J} \otimes \frac{\delta}{\delta J} \mathcal{G}_c \right].\tag{12.240}$$

We can express the effective action $\Gamma[\bar\phi']$ in terms of the generating functional $\mathcal{G}_{ac}[\bar\phi']$ as follows. From one hand, we have

$$\Gamma[\bar\phi'] = (\bar\phi', J) - \mathcal{G}_c[J] - S_0[\bar\phi'].\tag{12.241}$$

From the other hand, we have

$$\mathcal{G}_{ac}[\bar\phi] = \mathcal{G}_c[J = -G_0^{-1}\bar\phi] + \frac{1}{2}(\bar\phi, G_0^{-1}\bar\phi).\tag{12.242}$$

We compute immediately the classical field

$$\begin{aligned}
\bar\phi'_\alpha &= \frac{\delta \mathcal{G}_c}{\delta J_\alpha} \\
&= \bar\phi_\alpha - G_{0,\alpha\beta} \frac{\delta \mathcal{G}_{ac}}{\delta \bar\phi_\beta}.
\end{aligned}\tag{12.243}$$

By replacing with equation (12.242) in (12.241), and then using (12.243), we get

$$\begin{aligned}
\Gamma[\bar\phi'] &= (\bar\phi', J) - \mathcal{G}_{ac}[\bar\phi] + \frac{1}{2}(\bar\phi, G_0^{-1}\bar\phi) + \frac{1}{2}(\bar\phi', G_0^{-1}\bar\phi') \\
&= \frac{1}{2}\left(\frac{\delta \mathcal{G}_{ac}}{\delta \bar\phi}, G_0 \frac{\delta \mathcal{G}_{ac}}{\delta \bar\phi} \right) - \mathcal{G}_{ac}[\bar\phi].
\end{aligned}\tag{12.244}$$

Thus the would-be initial condition $G_0 \longrightarrow 0$ becomes in terms of $\Gamma[\bar\phi]$ given by the good starting prescription

$$\Gamma[\bar\phi] \longrightarrow S_1[\bar\phi], \quad G_0 \longrightarrow 0.\tag{12.245}$$

12.3.5 The 2-point function and regularization

We want now to show explicitly that $\mathcal{G}_c[J]$ generates indeed connected Green's functions, and that $\mathcal{G}_{ac}[\bar\phi]$ generates the amputated connected Green's function, and that $\Gamma[\bar\phi']$ generates the irreducible vertices. For simplicity, we will show this for the case of the 2-point function.

The connected 2-point function $G^{(2)}_{c,\,\alpha_1\alpha_2}$ is given in terms of the full 2-point function $G^{(2)}_{\alpha_1\alpha_2}$ by

$$
\begin{aligned}
G^{(2)}_{c,\,\alpha_1\alpha_2} &= \langle \phi_{\alpha_2}\phi_{\alpha_1}\rangle - \langle \phi_{\alpha_2}\rangle\langle\phi_{\alpha_1}\rangle \\
&= G^{(2)}_{\alpha_1\alpha_2} - G^{(1)}_{\alpha_2} G^{(1)}_{\alpha_1}.
\end{aligned}
\tag{12.246}
$$

Thus the connected function is indeed equal to the full function minus the disconnected piece. For the free theory we obtain

$$
\begin{aligned}
G^{(2)}_{0,\,c\alpha_1\alpha_2} &= \langle \phi_{\alpha_2}\phi_{\alpha_1}\rangle_0 \\
&= G^{(2)}_{0,\,\alpha_1\alpha_2} \\
&= - G_{0,\alpha_1\alpha_2}.
\end{aligned}
\tag{12.247}
$$

This last equation can be rewritten as

$$
G_{0,\alpha_1\alpha_2} = -\frac{\delta^2 \mathcal{G}_{0,c}[J]}{\delta J_{\alpha_2}\,\delta J_{\alpha_1}}\Big|_{J=0} = -G^{(2)}_{0,\,c\alpha_1\alpha_2} \leftrightarrow G_0 = -\frac{\delta}{\delta J}\otimes\frac{\delta}{\delta J}\mathcal{G}_{0,c}[J].
\tag{12.248}
$$

By analogy, the exact propagator will be defined by

$$
G_{\alpha_1\alpha_2} = -\frac{\delta^2 \mathcal{G}_c[J]}{\delta J_{\alpha_2}\,\delta J_{\alpha_1}}\Big|_{J=0} = -G^{(2)}_{c\alpha_1\alpha_2} \leftrightarrow G = -\frac{\delta}{\delta J}\otimes\frac{\delta}{\delta J}\mathcal{G}_c[J].
\tag{12.249}
$$

This satisfies the same symmetry property as G_0, i.e. $G^T = G$. We parameterize the difference between the exact propagator G and the free propagator G_0 by a self-energy Σ as

$$
G^{-1} = G_0^{-1} - \Sigma \Rightarrow G = G_0 + G_0\Sigma G \Rightarrow G = G_0 + G_0\Sigma G_0 + G_0\Sigma G_0\Sigma G_0 + \cdots
\tag{12.250}
$$

Next, by taking the second functional derivative of (12.227), we compute

$$
\begin{aligned}
\frac{\delta^2}{\delta\bar\phi_{\alpha_2}\,\delta\bar\phi_{\alpha_1}}\mathcal{G}_{ac} &= (G_0^{-1})_{\beta_1\alpha_1}(G_0^{-1})_{\beta_2\alpha_2}\frac{\delta^2 \mathcal{G}_c}{\delta J_{\beta_2}\,\delta J_{\beta_1}} + (G_0^{-1})_{\alpha_1\alpha_2} \\
&= -(G_0^{-1}GG_0^{-1})_{\alpha_1\alpha_2} + (G_0^{-1})_{\alpha_1\alpha_2}.
\end{aligned}
\tag{12.251}
$$

Or equivalently

$$
\begin{aligned}
G^{(2)}_{ac} &= -G_0^{-1}GG_0^{-1} + G_0^{-1} \\
&= -\Sigma - \Sigma G_0 \Sigma - \Sigma G_0 \Sigma G_0 \Sigma - \cdots
\end{aligned}
\tag{12.252}
$$

Alternatively, we write

$$G = G_0 - G_0 G_{ac}^{(2)} G_0. \tag{12.253}$$

This shows explicitly that $G_{ac}^{(2)}$ is the amputated 2-point function.

Now, by using equation (12.249) in equation (12.240) with $\bar{\phi} = \bar{\phi}'$, we obtain the formula

$$
\begin{aligned}
- G_{\alpha_1\alpha_2}^{-1} &= \frac{\delta^2 \mathcal{L}}{\delta\bar{\phi}_{\alpha_2}\,\delta\bar{\phi}_{\alpha_1}}\,|_{\bar{\phi}=0} \\
&= \Gamma_{\alpha_1\alpha_2}^{(2)} - (G_0^{-1})_{\alpha_1\alpha_2} \Rightarrow \Gamma^{(2)} = G_0^{-1} - G^{-1} = \Sigma.
\end{aligned}
\tag{12.254}
$$

This shows that the 2-point vertex is exactly given by the self-energy, i.e., it is fully irreducible.

We will need to regularize the theory by regularizing the free propagator G_0 in the Gaussian action $S_0[\phi]$. For example, by means of a momentum cutoff Λ. Towards this end, we make the substitution

$$G_0 \longrightarrow G_{0,\Lambda}. \tag{12.255}$$

We require that

$$G_{0,\Lambda} \longrightarrow G_0, \quad \Lambda \longrightarrow 0. \tag{12.256}$$

$$G_{0,\Lambda} \longrightarrow 0, \quad \Lambda \longrightarrow \infty. \tag{12.257}$$

We think of Λ as an infrared cutoff, i.e., we obtain the original theory when we remove the cutoff $\Lambda \longrightarrow 0$. In other words, the fluctuations with momenta below the cutoff $|k| \leqslant \Lambda$ can not propagate and thus they have still to be integrated out, while fluctuations with momenta above the cutoff $|k| \geqslant \Lambda$ have already been integrated out. This is reflected in the fact that we are going to calculate the derivative with respect to the cutoff Λ of various generating functionals, viz $\partial_\Lambda \mathcal{G}_c$, $\partial_\Lambda \mathcal{G}_{ac}$ and $\partial_\Lambda \Gamma$, starting from their initial values at $\Lambda = \infty$. The above condition can be implemented by a sharp momentum cutoff given by

$$G_{0,\Lambda}(k) = \theta(|k| - \Lambda)G_0(k). \tag{12.258}$$

Obviously, for $\Lambda \longrightarrow 0$ we obtain the original theory, viz $G_{0,\Lambda} = G_0$, while for $\Lambda \longrightarrow \infty$ we obtain $G_{0,\Lambda} = 0$, i.e., all modes are frozen. A smooth momentum cutoff is then given by

$$G_{0,\Lambda}(k) = \theta_\Lambda(|k| - \Lambda)G_0(k). \tag{12.259}$$

$$\theta_\Lambda \longrightarrow 1, \quad \Lambda \longrightarrow 0; \quad \theta_\Lambda \longrightarrow 0, \quad \Lambda \longrightarrow \infty. \tag{12.260}$$

In other words, θ_Λ is a smoothed step function with a width ϵ, viz

$$\theta_\Lambda(|k| - \Lambda) = \theta_\epsilon(|k| - \Lambda), \quad \lim \theta_\epsilon(|k| - \Lambda) = \theta(|k| - \Lambda), \quad \epsilon \longrightarrow 0. \tag{12.261}$$

We have already shown that the initial conditions for the generating functionals $\mathcal{G}_c[J]$, $\mathcal{G}_{ac}[\bar{\phi}]$ and $\Gamma[\bar{\phi}]$ are given by equations (12.220), (12.231) and (12.245). These equations can be rewritten as

$$\mathcal{G}_{c,\Lambda}[J] \longrightarrow 0, \quad \Lambda \longrightarrow \infty. \tag{12.262}$$

$$\mathcal{G}_{ac,\Lambda}[\bar{\phi}] \longrightarrow - S_{\mathrm{I}}[\bar{\phi}], \quad \Lambda \longrightarrow \infty. \tag{12.263}$$

$$\Gamma_\Lambda[\bar{\phi}] \longrightarrow S_{\mathrm{I}}[\bar{\phi}], \quad \Lambda \longrightarrow \infty. \tag{12.264}$$

From equation (12.243), it is also obvious that the classical field behaves in this limit as

$$\bar{\phi}' \longrightarrow \bar{\phi}, \quad \Lambda \longrightarrow \infty. \tag{12.265}$$

12.4 The functional renormalization group

The original literature on the functional renormalization group equation, which is an exact non-perturbative formulation of the Wilson renormalization group equation, includes Polchinski [2] and Wetterich [3]. See also [14–16] and [17, 18]. Again we will follow here the presentation of [4].

12.4.1 The FRG equations

The goal now is to derive the functional renormalization group equations (FRG) obeyed by the various generating functionals. Towards this end, we compute the derivative with respect to the cutoff Λ of $\mathcal{G}_\Lambda[J]$, $\mathcal{G}_{c,\Lambda}[J]$, $\mathcal{G}_{ac,\Lambda}[\bar{\phi}]$ and $\Gamma_\Lambda[\phi']$. The computation of the derivative with respect to the cutoff Λ of $\mathcal{G}_\Lambda[J]$ is straightforward. We find (with the replacement $\phi \longrightarrow \delta/\delta J$)

$$\partial_\Lambda \mathcal{G}_\Lambda[J] = \left[\frac{1}{2}\left(\frac{\delta}{\delta J}, \partial_\Lambda G_{0,\Lambda}^{-1} \frac{\delta}{\delta J} \right) - \partial_\Lambda \ln Z_\Lambda \right] \mathcal{G}_\Lambda[J]. \tag{12.266}$$

Next, we recall the defining equation

$$\mathcal{G}_\Lambda[J] = \frac{Z_{0,\Lambda}}{Z_\Lambda} \exp(\mathcal{G}_{c,\Lambda}[J]). \tag{12.267}$$

The derivative with respect to the cutoff Λ of this equation gives

$$\partial_\Lambda \mathcal{G}_\Lambda[J] = \left[\partial_\Lambda \mathcal{G}_{c,\Lambda}[J] + \partial_\Lambda \ln \frac{Z_{0,\Lambda}}{Z_\Lambda} \right] \mathcal{G}_\Lambda[J]. \tag{12.268}$$

By acting with the operator given by the first term in the right-hand side of (12.266) on $\mathcal{G}_\Lambda[J]$ we obtain

$$\frac{1}{2}\left(\frac{\delta}{\delta J}, \partial_\Lambda G_{0,\Lambda}^{-1} \frac{\delta}{\delta J} \right)\mathcal{G}_\Lambda[J] = \left[\frac{1}{2}\left(\frac{\delta \mathcal{G}_{c,\Lambda}}{\delta J}, \partial_\Lambda G_{0,\Lambda}^{-1} \frac{\delta \mathcal{G}_{c,\Lambda}}{\delta J} \right) + \frac{1}{2}Tr\partial_\Lambda G_{0,\Lambda}^{-1}\left(\frac{\delta}{\delta J} \otimes \frac{\delta}{\delta J}\mathcal{G}_{c,\Lambda} \right)^T \right]\mathcal{G}_\Lambda[J]. \tag{12.269}$$

By putting the above two equations together we obtain

$$\partial_\Lambda \mathcal{G}_{c,\Lambda}[J] = \frac{1}{2}\left(\frac{\delta \mathcal{G}_{c,\Lambda}}{\delta J}, \partial_\Lambda G_{0,\Lambda}^{-1}\frac{\delta \mathcal{G}_{c,\Lambda}}{\delta J}\right) + \frac{1}{2}Tr\partial_\Lambda G_{0,\Lambda}^{-1}\left(\frac{\delta}{\delta J}\otimes\frac{\delta}{\delta J}\mathcal{G}_{c,\Lambda}\right)^T - \partial_\Lambda\ln Z_{0,\Lambda}. \qquad (12.270)$$

From equation (12.230) we have

$$\exp(\mathcal{G}_{ac,\Lambda}[\bar\phi]) = \exp\left(-\frac{1}{2}\left(\frac{\delta}{\delta\bar\phi}, G_{0,\Lambda}\frac{\delta}{\delta\bar\phi}\right)\right)\exp(-S_1[\bar\phi]). \qquad (12.271)$$

By differentiating both sides with respect to Λ we obtain

$$\begin{aligned}
\partial_\Lambda \mathcal{G}_{ac,\Lambda}\cdot \exp(\mathcal{G}_{ac,\Lambda}) &= -\frac{1}{2}\left(\frac{\delta}{\delta\bar\phi}, \partial_\Lambda G_{0,\Lambda}\frac{\delta}{\delta\bar\phi}\right)\exp(\mathcal{G}_{ac,\Lambda}) \\
&= \left[-\frac{1}{2}\left(\frac{\delta \mathcal{G}_{ac,\Lambda}}{\delta\bar\phi}, \partial_\Lambda G_{0,\Lambda}\frac{\delta \mathcal{G}_{ac,\Lambda}}{\delta\bar\phi}\right) - \frac{1}{2}Tr\partial_\Lambda G_{0,\Lambda}\left(\frac{\delta}{\delta\bar\phi}\otimes\frac{\delta}{\delta\bar\phi}\mathcal{G}_{ac,\Lambda}\right)^T\right]\exp(\mathcal{G}_{ac,\Lambda}).
\end{aligned} \qquad (12.272)$$

We then get the functional renormalization group equation

$$\partial_\Lambda \mathcal{G}_{ac,\Lambda} = -\frac{1}{2}\left(\frac{\delta \mathcal{G}_{ac,\Lambda}}{\delta\bar\phi}, \partial_\Lambda G_{0,\Lambda}\frac{\delta \mathcal{G}_{ac,\Lambda}}{\delta\bar\phi}\right) - \frac{1}{2}Tr\partial_\Lambda G_{0,\Lambda}\left(\frac{\delta}{\delta\bar\phi}\otimes\frac{\delta}{\delta\bar\phi}\mathcal{G}_{ac,\Lambda}\right)^T. \qquad (12.273)$$

This is the Polchinski equation. Indeed, this FRG equation is identical with the Polchinski equation for the Wilsonian effective action [2].

To obtain the FRG equation for the generating functional $\Gamma_\Lambda[\bar\phi']$ we start from the definitions

$$\Gamma_\Lambda[\bar\phi'] = \mathcal{L}_\Lambda[\bar\phi'] + \frac{1}{2}(\bar\phi', G_{0,\Lambda}^{-1}\bar\phi'). \qquad (12.274)$$

$$\mathcal{L}_\Lambda[\bar\phi'] = (J_\Lambda, \bar\phi') - \mathcal{G}_{c,\Lambda}[J_\Lambda]. \qquad (12.275)$$

Of course, J and $\bar\phi$ are conjugate variables

$$\bar\phi' = \frac{\delta \mathcal{G}_{c,\Lambda}[J]}{\delta J}. \qquad (12.276)$$

We will assume that J depends on the cutoff, i.e., $J = J_\Lambda[\bar\phi']$, in such a way that $\bar\phi'$ is independent of Λ. By taking the derivative with respect to the cutoff Λ we obtain

$$\partial_\Lambda\Gamma_\Lambda[\bar\phi'] = \partial_\Lambda\mathcal{L}_\Lambda[\bar\phi'] + \frac{1}{2}(\bar\phi', \partial_\Lambda G_{0,\Lambda}^{-1}\bar\phi'). \qquad (12.277)$$

$$\partial_\Lambda\mathcal{L}_\Lambda[\bar\phi'] = (\partial_\Lambda J_\Lambda, \bar\phi') - \partial_\Lambda\mathcal{G}_{c,\Lambda} - \int_\beta \partial_\Lambda(J_\Lambda)_\beta\frac{\delta \mathcal{G}_{c,\Lambda}}{\delta(J_\Lambda)_\beta}. \qquad (12.278)$$

The second term arises from the explicit dependence of $\mathcal{G}_{c,\Lambda}$ on Λ whereas the third term arises from the implicit dependence of $\mathcal{G}_{c,\Lambda}$ on Λ through J_Λ. It is obvious that the first and third terms cancel, viz

$$\partial_\Lambda\mathcal{L}_\Lambda[\bar\phi'] = -\partial_\Lambda\mathcal{G}_{c,\Lambda}. \qquad (12.279)$$

By using equation (12.270), (12.276) and (12.240) we obtain immediately

$$\partial_\Lambda \Gamma_\Lambda[\bar\phi'] = -\frac{1}{2} Tr \partial_\Lambda G_{0,\Lambda}^{-1}\left(\frac{\delta}{\delta J} \otimes \frac{\delta}{\delta J} \mathcal{G}_{c,\Lambda}\right) + \partial_\Lambda \ln Z_{0,\Lambda}$$
$$= -\frac{1}{2} Tr \partial_\Lambda G_{0,\Lambda}^{-1}\left(\frac{\delta}{\delta\bar\phi'} \otimes \frac{\delta}{\delta\bar\phi'} \mathcal{L}_\Lambda\right)^{-1} + \partial_\Lambda \ln Z_{0,\Lambda}. \tag{12.280}$$

Or equivalently

$$\partial_\Lambda \Gamma_\Lambda[\bar\phi'] = -\frac{1}{2} Tr \partial_\Lambda G_{0,\Lambda}^{-1}\left(\frac{\delta}{\delta\bar\phi'} \otimes \frac{\delta}{\delta\bar\phi'}\Gamma_\Lambda - G_{0,\Lambda}^{-1}\right)^{-1} + \partial_\Lambda \ln Z_{0,\Lambda}. \tag{12.281}$$

This is the Wetterich equation [3, 14]. The average effective action introduced by Wetterich is given by [3]

$$\Gamma_\Lambda^W[\bar\phi'] = \Gamma_\Lambda[\bar\phi'] - \frac{1}{2}(\bar\phi', G_0^{-1}\bar\phi') - \ln Z_{0,\Lambda}. \tag{12.282}$$

This satisfies the FRG equation [3]

$$\partial_\Lambda \Gamma_\Lambda^W[\bar\phi'] = -\frac{1}{2} Tr \partial_\Lambda G_{0,\Lambda}^{-1}\left(\frac{\delta}{\delta\bar\phi'} \otimes \frac{\delta}{\delta\bar\phi'}\Gamma_\Lambda - G_{0,\Lambda}^{-1}\right)^{-1}. \tag{12.283}$$

12.4.2 The vertex expansion of the Wetterich FRG

One approximation method for solving the FRG equation (12.281) is the vertex expansion [14], which allows us to reduce this equation to an infinite number of coupled integro-differential equations for the irreducible vertices. Before we can explain this method we rewrite (12.281) in a suitable form as follows. First, we note

$$\partial_\Lambda Z_{0,\Lambda} = \int d\phi \frac{1}{2}(\phi, \partial_\Lambda G_{0,\Lambda}^{-1}\phi)\exp(-S_0[\phi])$$
$$= \frac{Z_{0,\Lambda}}{2}\left(\frac{\delta}{\delta J}, \partial_\Lambda G_{0,\Lambda}^{-1}\frac{\delta}{\delta J}\right)\exp\left(-\frac{1}{2}(J, G_{0,\Lambda}J)\right)\Big|_{J=0}. \tag{12.284}$$

Thus

$$\partial_\Lambda \ln Z_{0,\Lambda} = -\frac{1}{2} Tr \partial_\Lambda G_{0,\Lambda}^{-1} . G_{0,\Lambda}. \tag{12.285}$$

Second, we have

$$\frac{\delta}{\delta\bar\phi'} \otimes \frac{\delta}{\delta\bar\phi'}\mathcal{L}_\Lambda = \frac{\delta}{\delta\bar\phi'} \otimes \frac{\delta}{\delta\bar\phi'}\left(\Gamma_\Lambda - \frac{1}{2}(\bar\phi', G_{0,\Lambda}^{-1}\bar\phi')\right)$$
$$= \frac{\delta}{\delta\bar\phi'} \otimes \frac{\delta}{\delta\bar\phi'}\Gamma_\Lambda - \frac{\delta}{\delta\bar\phi'} \otimes \frac{\delta}{\delta\bar\phi'}\Gamma_\Lambda \big|_{\bar\phi'=0} + \Sigma_\Lambda - G_{0,\Lambda}^{-1} \tag{12.286}$$
$$= U_\Lambda[\bar\phi'] - G_\Lambda^{-1}.$$

In the above equation, the second and the third terms are identically equal, viz

$$\Sigma_\Lambda = \frac{\delta}{\delta\bar{\phi}'} \otimes \frac{\delta}{\delta\bar{\phi}'}\Gamma_\Lambda \mid_{\bar{\phi}'=0} = \Sigma_\Lambda = G_{0,\,\Lambda}^{-1} - G_\Lambda^{-1}. \tag{12.287}$$

This equation is equivalent to

$$G_\Lambda^{-1} = G_{0,\,\Lambda}^{-1}(1 - G_{0,\Lambda}\Sigma_\Lambda), \quad G_\Lambda = (1 - G_{0,\Lambda}\Sigma_\Lambda)^{-1}G_{0,\Lambda}. \tag{12.288}$$

Furthermore, we have defined U_Λ as

$$U_\Lambda[\bar{\phi}'] = \frac{\delta}{\delta\bar{\phi}'} \otimes \frac{\delta}{\delta\bar{\phi}'}\Gamma_\Lambda - \frac{\delta}{\delta\bar{\phi}'} \otimes \frac{\delta}{\delta\bar{\phi}'}\Gamma_\Lambda \mid_{\bar{\phi}'=0}. \tag{12.289}$$

By using the above two results (12.285) and (12.286) we rewrite the FRG equation (12.281) in the form

$$\partial_\Lambda\Gamma_\Lambda[\bar{\phi}'] = -\frac{1}{2}Tr\partial_\Lambda G_{0,\,\Lambda}^{-1}\left(\left[U_\Lambda[\bar{\phi}'] - G_\Lambda^{-1}\right]^{-1} + G_{0,\Lambda}\right). \tag{12.290}$$

We compute

$$\begin{aligned}
\left[U_\Lambda[\bar{\phi}'] - G_\Lambda^{-1}\right]^{-1} + G_{0,\Lambda} &= -[1 - G_\Lambda U_\Lambda]^{-1}G_\Lambda + G_{0,\Lambda}\\
&= -[1 + G_\Lambda U_\Lambda + G_\Lambda U_\Lambda. G_\Lambda U_\Lambda + \cdots]G_\Lambda + G_{0,\Lambda}\\
&= -G_\Lambda - G_\Lambda U_\Lambda[1 + G_\Lambda U_\Lambda + G_\Lambda U_\Lambda. G_\Lambda U_\Lambda + \cdots]G_\Lambda + G_{0,\Lambda}\\
&= -G_\Lambda + G_{0,\Lambda} - G_\Lambda U_\Lambda[1 - G_\Lambda U_\Lambda]^{-1}G_\Lambda\\
&= -(1 - G_{0,\Lambda}\Sigma_\Lambda)^{-1}G_{0,\Lambda} + G_{0,\Lambda} - G_\Lambda U_\Lambda[1 - G_\Lambda U_\Lambda]^{-1}G_\Lambda\\
&= -G_{0,\Lambda}\Sigma_\Lambda G_\Lambda - G_\Lambda U_\Lambda[1 - G_\Lambda U_\Lambda]^{-1}G_\Lambda.
\end{aligned} \tag{12.291}$$

In the fifth equation we have used (12.288). By substitution we get

$$\partial_\Lambda\Gamma_\Lambda[\bar{\phi}'] = -\frac{1}{2}Tr\partial_\Lambda G_{0,\,\Lambda}^{-1}(-G_{0,\Lambda}\Sigma_\Lambda G_\Lambda - G_\Lambda U_\Lambda[1 - G_\Lambda U_\Lambda]^{-1}G_\Lambda). \tag{12.292}$$

We define the single-scale propagator $\dot{G}_\Lambda$ by

$$\begin{aligned}
\dot{G}_\Lambda &= -G_\Lambda\partial_\Lambda G_{0,\,\Lambda}^{-1}G_\Lambda\\
&= (1 - G_{0,\Lambda}\Sigma_\Lambda)^{-1}\dot{G}_{0,\Lambda}G_{0,\,\Lambda}^{-1}(1 - G_{0,\Lambda}\Sigma_\Lambda)^{-1}G_{0,\Lambda}\\
&= (1 - G_{0,\Lambda}\Sigma_\Lambda)^{-1}\dot{G}_{0,\Lambda}(1 - \Sigma_\Lambda G_{0,\Lambda})^{-1},
\end{aligned} \tag{12.293}$$

where the single-scale propagator of the free theory is given by

$$\begin{aligned}
\dot{G}_{0,\Lambda} &= -G_{0,\Lambda}\partial_\Lambda G_{0,\,\Lambda}^{-1}G_{0,\Lambda}\\
&= \partial_\Lambda G_{0,\Lambda}.
\end{aligned} \tag{12.294}$$

Using these definitions the above FRG equation becomes

$$\partial_\Lambda\Gamma_\Lambda[\bar{\phi}'] = -\frac{1}{2}Tr(\dot{G}_{0,\Lambda}\Sigma_\Lambda[1 - G_{0,\Lambda}\Sigma_\Lambda]^{-1} + \dot{G}_\Lambda U_\Lambda[\bar{\phi}'][1 - G_\Lambda U_\Lambda[\bar{\phi}']]^{-1}). \tag{12.295}$$

For a free field theory $G_{0,c} = -(J, G_0 J)/2$ and $G_{0,ac} = 0$ and thus $\bar{\phi}' = \bar{\phi}$ and

$$\frac{\delta}{\delta\bar{\phi}'} \otimes \frac{\delta}{\delta\bar{\phi}'}\mathcal{L}_\Lambda = -G_{0,\Lambda}^{-1}$$

$$= \frac{\delta}{\delta\bar{\phi}'} \otimes \frac{\delta}{\delta\bar{\phi}'}\Gamma_\Lambda - G_{0,\Lambda}^{-1} \tag{12.296}$$

$$= U_{0,\Lambda}[\bar{\phi}'] - G_{0,\Lambda}^{-1}.$$

In other words, $U_{0,\Lambda} = \Sigma_{0,\Lambda} = 0$.

We want now to derive the expansion of the functional U_Λ in terms of the irreducible vertices. Towards this end, we recall the expansion of the effective action in terms of irreducible vertices given by

$$\Gamma_\Lambda = \sum_{n=0}^\infty \frac{1}{n!} \int_{\alpha_1} \cdots \int_{\alpha_n} \Gamma_{\Lambda\alpha_1 \dots \alpha_n}^{(n)} \bar{\phi}_{\alpha_1} \cdots \bar{\phi}_{\alpha_n}. \tag{12.297}$$

We take the derivative with respect to $\bar{\phi}_\alpha$ to obtain

$$\frac{\delta\Gamma_\Lambda}{\delta\bar{\phi}_\alpha} = \Gamma_{\Lambda\alpha}^{(1)} + \sum_{n=1}^\infty \frac{1}{n!} \int_{\alpha_1} \cdots \int_{\alpha_n} \Gamma_{\Lambda\alpha\alpha_1 \dots \alpha_n}^{(n+1)} \bar{\phi}_{\alpha_1} \cdots \bar{\phi}_{\alpha_n}. \tag{12.298}$$

We take the derivative a second time with respect to $\bar{\phi}_{\alpha'}$ to obtain

$$\frac{\delta\Gamma_\Lambda}{\delta\bar{\phi}_\alpha \delta\bar{\phi}_{\alpha'}} = \Gamma_{\Lambda\alpha'}^{(1)} + \sum_{n=1}^\infty \frac{1}{n!} \int_{\alpha_1} \cdots \int_{\alpha_n} \Gamma_{\Lambda\alpha\alpha'\alpha_1 \dots \alpha_n}^{(n+2)} \bar{\phi}_{\alpha_1} \cdots \bar{\phi}_{\alpha_n}. \tag{12.299}$$

From this result we obtain the expansion of the functional U_Λ in terms of the irreducible vertices given by

$$(U_\Lambda)_{\alpha\alpha'} = \frac{\delta^2\Gamma_\Lambda}{\delta\bar{\phi}_\alpha \delta\bar{\phi}_{\alpha'}} - \frac{\delta^2\Gamma_\Lambda}{\delta\bar{\phi}_\alpha \delta\bar{\phi}_{\alpha'}} \Big|_{\bar{\phi}=0}$$

$$= \sum_{n=1}^\infty \frac{1}{n!} \int_{\alpha_1} \cdots \int_{\alpha_n} \Gamma_{\Lambda\alpha\alpha'\alpha_1 \dots \alpha_n}^{(n+2)} \bar{\phi}_{\alpha_1} \cdots \bar{\phi}_{\alpha_n}. \tag{12.300}$$

We will define the matrix

$$\left(\mathbf{\Gamma}_{\Lambda\alpha_1 \dots \alpha_n}^{(n+2)}\right)_{\alpha\alpha'} = \Gamma_{\Lambda\alpha\alpha'\alpha_1 \dots \alpha_n}^{(n+2)}. \tag{12.301}$$

We compute from the right-hand side of the **FRG** equation (12.295) the following expansion

$$\partial_\Lambda \Gamma_\Lambda^{(0)} + \sum_{n=1}^\infty \frac{1}{n!} \int_{\alpha_1} \cdots \int_{\alpha_n} \partial_\Lambda \Gamma_{\Lambda\alpha_1 \dots \alpha_n}^{(n)} \bar{\phi}_{\alpha_1} \cdots \bar{\phi}_{\alpha_n}. \tag{12.302}$$

From the left hand side of the **FRG** equation (12.295) we obtain

$$-\frac{1}{2}Tr\dot{G}_{0,\Lambda}\Sigma_\Lambda[1 - G_{0,\Lambda}\Sigma_\Lambda]^{-1} - \frac{1}{2}\sum_{\nu=1}^\infty Tr\dot{G}_\Lambda U_\Lambda(G_\Lambda U_\Lambda)^{\nu-1}. \tag{12.303}$$

We deduce immediately

$$\partial_\Lambda \Gamma_\Lambda^{(0)} = -\frac{1}{2} Tr \dot{G}_{0,\Lambda} \Sigma_\Lambda [1 - G_{0,\Lambda} \Sigma_\Lambda]^{-1}. \tag{12.304}$$

In the expansion of the second term of (12.303) we will adopt the following notation. The fields appearing in the first U_Λ are denoted by the labels $\alpha_1, \ldots, \alpha_{n_1}$, those appearing in the second U_Λ are denoted by $\alpha_{n_1+1}, \ldots, \alpha_{n_1+n_2}, \ldots$, those appearing in the kth U_Λ are denoted by $\alpha_{n_1+\cdots+n_{k-1}+1}, \ldots, \alpha_{n_1+\cdots+n_k}$, etc. We will also employ the notation $n = n_1 + \cdots + n_\nu$. We get

$$
-\frac{1}{2}\sum_{\nu=1}^{\infty} Tr \dot{G}_\Lambda U_\Lambda (G_\Lambda U_\Lambda)^{\nu-1} = -\frac{1}{2}\sum_{n=1}^{\infty}\sum_{\nu=1}^{\infty}\sum_{1n_1=1}^{\infty} \cdots \sum_{n_\nu=1}^{\infty} \frac{\delta_{n,n_1+\cdots+n_\nu}}{n_1! \cdots n_\nu!} \int_{\alpha_1} \cdots \int_{\alpha_{n_1}} \int_{\alpha_{n_1+1}} \cdots \int_{\alpha_{n_1+n_2}} \cdots
$$
$$
\times \int_{\alpha_{n-n_\nu-n_{\nu-1}+1}} \cdots \int_{\alpha_{n-n_\nu}} \int_{\alpha_{n-n_\nu+1}} \cdots \int_{\alpha_n} \bar{\phi}_{\alpha_1} \cdots \bar{\phi}_{\alpha_{n_1}} \bar{\phi}_{\alpha_{n_1+1}} \cdots \bar{\phi}_{\alpha_{n_1+n_2}} \cdots
$$
$$
\times \bar{\phi}_{\alpha_{n-n_\nu-n_{\nu-1}+1}} \cdots \bar{\phi}_{\alpha_{n-n_\nu}} \bar{\phi}_{\alpha_{n-n_\nu+1}} \cdots \bar{\phi}_{\alpha_n} Tr \dot{G}_\Lambda \Gamma^{(n_\nu+2)}_{\Lambda \alpha_{n-n_\nu+1} \cdots \alpha_n}
$$
$$
\times G_\Lambda \Gamma^{(n_{\nu-1}+2)}_{\Lambda \alpha_{n-n_\nu-n_{\nu-1}+1} \cdots \alpha_{n-n_\nu}} \cdots G_\Lambda \Gamma^{(n_2+2)}_{\Lambda \alpha_{n_1+1} \cdots \alpha_{n_1+n_2}} G_\Lambda \Gamma^{(n_1+1)}_{\Lambda \alpha_1 \cdots \alpha_{n_1}}. \tag{12.305}
$$

By symmetrizing the above expression (12.305) in the fields and then comparing (12.302) and (12.305) we obtain

$$
\partial_\Lambda \Gamma^{(n)}_{\Lambda,\, \alpha_1 \cdots \alpha_n} = -\frac{1}{2}\sum_{\nu=1}^{\infty}\sum_{1n_1=1}^{\infty} \cdots \sum_{n_\nu=1}^{\infty} \delta_{n,n_1+\cdots+n_\nu} \mathcal{S}_{\alpha_1 \cdots \alpha_{n_1};\, \ldots;\, \alpha_{n-n_\nu+1} \cdots \alpha_n} Tr \dot{G}_\Lambda \Gamma^{(n_\nu+2)}_{\Lambda,\, \alpha_{n-n_\nu+1} \cdots \alpha_n}
$$
$$
\times G_\Lambda \Gamma^{(n_{\nu-1}+2)}_{\Lambda,\, \alpha_{n-n_\nu-n_{\nu-1}+1} \cdots \alpha_{n-n_\nu}} \cdots G_\Lambda \Gamma^{(n_2+2)}_{\Lambda,\, \alpha_{n_1+1} \cdots \alpha_{n_1+n_2}} G_\Lambda \Gamma^{(n_1+2)}_{\Lambda,\, \alpha_1 \cdots \alpha_{n_1}}. \tag{12.306}
$$

The symmetrizer is obviously defined by

$$
\mathcal{S}_{\alpha_1 \cdots \alpha_{n_1};\, \ldots;\, \alpha_{n-n_\nu+1} \cdots \alpha_n} F_{\alpha_1 \cdots \alpha_n} = \frac{1}{n_1! \cdots n_\nu!}\sum_{P} F_{\alpha_{P(1)} \cdots \alpha_{P(n)}}. \tag{12.307}
$$

As a first example we consider the 2-point vertex $\Gamma^{(2)}_{\Lambda\alpha_1\alpha_2}$. We have

$$
\partial_\Lambda \Gamma^{(2)}_{\Lambda\alpha_1\alpha_2} = -\frac{1}{2}\sum_{\nu=1}^{\infty}\sum_{1n_1=1}^{\infty} \cdots \sum_{n_\nu=1}^{\infty} \delta_{2,n_1+\cdots+n_\nu} \mathcal{S}_{\alpha_1 \cdots \alpha_{n_1};\, \ldots;\, \alpha_{3-n_\nu} \cdots \alpha_2} Tr \dot{G}_\Lambda \Gamma^{(n_\nu+2)}_{\Lambda\alpha_{3-n_\nu} \cdots \alpha_2}
$$
$$
\times G_\Lambda \Gamma^{(n_{\nu-1}+2)}_{\Lambda\alpha_{3-n_\nu-n_{\nu-1}} \cdots \alpha_{2-n_\nu}} \cdots G_\Lambda \Gamma^{(n_2+2)}_{\Lambda\alpha_{n_1+1} \cdots \alpha_{n_1+n_2}} G_\Lambda \Gamma^{(n_1+2)}_{\Lambda\alpha_1 \cdots \alpha_{n_1}}. \tag{12.308}
$$

The first term is directly given by

$$
-\frac{1}{2} Tr \dot{G}_\Lambda \Gamma^{(4)}_{\Lambda\alpha_1\alpha_2}. \tag{12.309}
$$

The second term is also directly given by

$$
-\frac{1}{2}\sum_{n_1=1}^{\infty}\sum_{1n_2=1}^{\infty} \delta_{2,n_1+n_2} \mathcal{S}_{\alpha_1 \cdots \alpha_{n_1};\, \alpha_{n_1+1} \cdots \alpha_2} Tr \dot{G}_\Lambda \Gamma^{(n_2+2)}_{\Lambda\alpha_{n_1+1} \cdots \alpha_2} G_\Lambda \Gamma^{(n_1+2)}_{\Lambda\alpha_1 \cdots \alpha_{n_1}} =
$$
$$
-\frac{1}{2}\Big[\mathcal{S}_{\alpha_1\alpha_2} Tr \dot{G}_\Lambda \Gamma^{(2)}_\Lambda G_\Lambda \Gamma^{(4)}_{\Lambda\alpha_1\alpha_2} + \mathcal{S}_{\alpha_1\alpha_2} Tr \dot{G}_\Lambda \Gamma^{(4)}_{\Lambda\alpha_1\alpha_2} G_\Lambda \Gamma^{(2)}_\Lambda + \tag{12.310}
$$
$$
\mathcal{S}_{\alpha_1;\, \alpha_2} Tr \dot{G}_\Lambda \Gamma^{(3)}_{\Lambda\alpha_1} G_\Lambda \Gamma^{(3)}_{\Lambda\alpha_2} \Big] = -\frac{1}{2} \mathcal{S}_{\alpha_1;\, \alpha_2} Tr \dot{G}_\Lambda \Gamma^{(3)}_{\Lambda\alpha_1} G_\Lambda \Gamma^{(3)}_{\Lambda\alpha_2}.
$$

In the above equation we have used the fact that $\mathbf{\Gamma}_\Lambda^{(2)} = 0$ from (12.300). Higher order corrections vanish for the same reason. We get then

$$\partial_\Lambda \Gamma^{(2)}_{\Lambda \alpha_1 \alpha_2} = -\frac{1}{2} Tr \dot{G}_\Lambda \Gamma^{(4)}_{\Lambda \alpha_1 \alpha_2} - \frac{1}{2} S_{\alpha_1;\ \alpha_2} Tr \dot{G}_\Lambda \Gamma^{(3)}_{\Lambda \alpha_1} G_\Lambda \Gamma^{(3)}_{\Lambda \alpha_2}. \tag{12.311}$$

In the absence of external fields and symmetry breaking all irreducible vertices with an odd number of legs vanish. In this case the above flow equation simplifies to

$$\partial_\Lambda \Gamma^{(2)}_{\Lambda \alpha_1 \alpha_2} = -\frac{1}{2} Tr \dot{G}_\Lambda \Gamma^{(4)}_{\Lambda \alpha_1 \alpha_2}. \tag{12.312}$$

As a second example, we compute the FRG equation of the 4-point vertex. This turns out to involve 8 terms corresponding to $\nu = 1, 2, 3, 4$. Explicitly, we have

$$
\begin{aligned}
\partial_\Lambda \Gamma^{(4)}_{\Lambda \alpha_1 \ldots \alpha_4} = &-\frac{1}{2} Tr \dot{G}_\Lambda \Gamma^{(6)}_{\Lambda \alpha_1 \ldots \alpha_4} - \frac{1}{2}\Big(S_{\alpha_1;\ \alpha_2 \alpha_3 \alpha_4} Tr \dot{G}_\Lambda \Gamma^{(5)}_{\Lambda \alpha_2 \alpha_3 \alpha_4} G_\Lambda \Gamma^{(3)}_{\Lambda \alpha_1} \\
&+ S_{\alpha_1 \alpha_2 \alpha_3;\ \alpha_4} Tr \dot{G}_\Lambda \Gamma^{(3)}_{\Lambda \alpha_4} G_\Lambda \Gamma^{(5)}_{\Lambda \alpha_1 \alpha_2 \alpha_3} + S_{\alpha_1 \alpha_2;\ \alpha_3 \alpha_4} Tr \dot{G}_\Lambda \Gamma^{(4)}_{\Lambda \alpha_3 \alpha_4} G_\Lambda \Gamma^{(4)}_{\Lambda \alpha_1 \alpha_2} \Big) \\
&-\frac{1}{2}\Big(S_{\alpha_1 \alpha_2;\ \alpha_3;\ \alpha_4} Tr \dot{G}_\Lambda \Gamma^{(3)}_{\Lambda \alpha_4} G_\Lambda \Gamma^{(3)}_{\Lambda \alpha_3} G_\Lambda \Gamma^{(4)}_{\Lambda \alpha_1 \alpha_2} + S_{\alpha_1;\ \alpha_2 \alpha_3;\ \alpha_4} Tr \dot{G}_\Lambda \Gamma^{(3)}_{\Lambda \alpha_4} G_\Lambda \Gamma^{(4)}_{\Lambda \alpha_2 \alpha_3} G_\Lambda \Gamma^{(3)}_{\Lambda \alpha_1} \\
&+ S_{\alpha_1;\ \alpha_2;\ \alpha_3 \alpha_4} Tr \dot{G}_\Lambda \Gamma^{(4)}_{\Lambda \alpha_3 \alpha_4} G_\Lambda \Gamma^{(3)}_{\Lambda \alpha_2} G_\Lambda \Gamma^{(3)}_{\Lambda \alpha_1} \Big) \\
&-\frac{1}{2} S_{\alpha_1;\ \alpha_2;\ \alpha_3;\ \alpha_4} Tr \dot{G}_\Lambda \Gamma^{(3)}_{\Lambda \alpha_4} G_\Lambda \Gamma^{(3)}_{\Lambda \alpha_3} G_\Lambda \Gamma^{(3)}_{\Lambda \alpha_2} G_\Lambda \Gamma^{(3)}_{\Lambda \alpha_1}.
\end{aligned}
\tag{12.313}
$$

Again in the case of the absence of external fields and symmetry breaking all irreducible vertices with an odd number of legs vanish and as a consequence the above expression simplifies to

$$
\begin{aligned}
\partial_\Lambda \Gamma^{(4)}_{\Lambda,\ \alpha_1 \ldots \alpha_4} = &-\frac{1}{2} Tr \dot{G}_\Lambda \Gamma^{(6)}_{\Lambda,\ \alpha_1 \ldots \alpha_4} - \frac{1}{2} S_{\alpha_1 \alpha_2;\ \alpha_3 \alpha_4} Tr \dot{G}_\Lambda \Gamma^{(4)}_{\Lambda,\ \alpha_3 \alpha_4} G_\Lambda \Gamma^{(4)}_{\Lambda,\ \alpha_1 \alpha_2} \\
= &-\frac{1}{2} Tr \dot{G}_\Lambda \Gamma^{(6)}_{\Lambda,\ \alpha_1 \ldots \alpha_4} - \frac{1}{2} Tr \dot{G}_\Lambda \Big[\Gamma^{(4)}_{\Lambda,\ \alpha_3 \alpha_4} G_\Lambda \Gamma^{(4)}_{\Lambda,\ \alpha_1 \alpha_2} + \Gamma^{(4)}_{\Lambda,\ \alpha_1 \alpha_2} G_\Lambda \Gamma^{(4)}_{\Lambda,\ \alpha_3 \alpha_4} \\
&+ \Gamma^{(4)}_{\Lambda,\ \alpha_1 \alpha_3} G_\Lambda \Gamma^{(4)}_{\Lambda,\ \alpha_2 \alpha_4} + \Gamma^{(4)}_{\Lambda,\ \alpha_2 \alpha_4} G_\Lambda \Gamma^{(4)}_{\Lambda,\ \alpha_1 \alpha_3} + \Gamma^{(4)}_{\Lambda,\ \alpha_1 \alpha_4} G_\Lambda \Gamma^{(4)}_{\Lambda,\ \alpha_2 \alpha_3} + \Gamma^{(4)}_{\Lambda,\ \alpha_2 \alpha_3} G_\Lambda \Gamma^{(4)}_{\Lambda,\ \alpha_1 \alpha_4} \Big].
\end{aligned}
\tag{12.314}
$$

12.5 The vertex expansion approximation method for ϕ^4

12.5.1 The disordered phase

We consider ϕ^4 in D dimensions. We make the replacements $\alpha \longrightarrow k$, $\phi_\alpha \longrightarrow \phi(k)$ and $\int_\alpha \longrightarrow \int_k = d^D k/(2\pi)^D$. The action is given by

$$S_\Lambda[\phi] = S_{0,\Lambda}[\phi] + S_I[\Phi]. \tag{12.315}$$

We start by writing down the interaction part. This is given by

$$S_I[\Phi] = V f_0 + \frac{r_0}{2} \int_k \phi(-k)\phi(k) + \frac{u_0}{4!} \int_{k_1} \cdots \int_{k_4} (2\pi)^D \delta^D(k_1 + \cdots + k_4)\phi(k_1) \ldots \phi(k_4). \tag{12.316}$$

The free part is

$$S_{0,\Lambda}[\phi] = -\frac{1}{2}(\phi, G_{0,\Lambda}^{-1}\phi) = \frac{1}{2} \int_k G_{0,\Lambda}^{-1}(k)\phi(-k)\phi(k). \tag{12.317}$$

In other words,

$$(G_{0,\Lambda})_{kk'} = -(2\pi)^D \delta^D(k + k')G_{0,\Lambda}(k). \tag{12.318}$$

By translational invariance, we have

$$(G_\Lambda)_{kk'} = -(2\pi)^D \delta^D(k + k')G_\Lambda(k). \tag{12.319}$$

We also have

$$G_\Lambda^{-1}(k) = G_{0,\Lambda}^{-1}(k) + \Sigma_\Lambda(k), \quad G_\Lambda(k) = (1 + G_{0,\Lambda}(k)\Sigma_\Lambda(k))^{-1}G_{0,\Lambda}(k). \tag{12.320}$$

Also

$$(\Sigma_\Lambda)_{kk'} = (2\pi)^D \delta(k + k')\Sigma_\Lambda(k'). \tag{12.321}$$

In the disordered phase, the functional renormalization group equation for the effective action is given by equation (12.295), which in this notation is given by

$$\partial_\Lambda \Gamma_\Lambda = \frac{1}{2} \int_k \dot{G}_\Lambda(k)(U_\Lambda(1 - G_\Lambda U_\Lambda)^{-1})_{-k,k} + \frac{V}{2} \int_k \frac{\dot{G}_{0,\Lambda}(k)\Sigma_\Lambda(k)}{1 + G_{0,\Lambda}(k)\Sigma_\Lambda(k)}. \tag{12.322}$$

We have implicitly used the box convention $(2\pi)^D \delta^D(0) = V$. Obviously, the function U_Λ is given by

$$(U_\Lambda(\bar{\phi}))_{kk'} = \frac{\delta^2 \Gamma_\Lambda(\bar{\phi})}{\delta\bar{\phi}(k)\delta\bar{\phi}(k')} - (2\pi)^D \delta^D(k + k')\Sigma_\Lambda(k). \tag{12.323}$$

The single-scale propagator in this case is

$$\dot{G}_\Lambda(k) = -G_\Lambda(k)\partial_\Lambda G_{0,\Lambda}^{-1} G_\Lambda(k) = \frac{\dot{G}_{0,\Lambda}}{(1 + G_{0,\Lambda}(k)\Sigma_\Lambda(k))^2}, \quad \dot{G}_{0,\Lambda} = \partial_\Lambda G_{0,\Lambda}. \tag{12.324}$$

The FRG equation (12.295) has led to the FRG equation (12.304) for the free energy, the FRG equation (12.312) for the self-energy, and the FRG equation (12.314) for the 4-point vertex. These equations in our notation are given, respectively, by

$$\partial_\Lambda \Gamma_\Lambda^{(0)} = \frac{V}{2} \int_k \frac{\dot{G}_{0,\Lambda}(k)\Sigma_\Lambda(k)}{1 + G_{0,\Lambda}(k)\Sigma_\Lambda(k)}. \tag{12.325}$$

$$\partial_\Lambda \Sigma_\Lambda(p) = \frac{1}{2} \int_k \dot{G}_\Lambda(k)\Gamma_\Lambda^{(4)}(k, -k, p, -p). \tag{12.326}$$

$$\partial_\Lambda \Gamma_\Lambda^{(4)}(p_1, \ldots, p_4) = \frac{1}{2} \int_k \dot{G}_\Lambda(k)\Gamma_\Lambda^{(6)}(k, -k, p_1, \ldots, p_4) - \int_k \left[\dot{G}_\Lambda(k)G_\Lambda(k + p_1 + p_2) \right.$$
$$\left. \times \Gamma_\Lambda^{(4)}(-k - p_1 - p_2, k, p_1, p_2)\Gamma_\Lambda^{(4)}(-k, k + p_1 + p_2, p_3, p_4) + (p_2 \leftrightarrow p_3) \right. \tag{12.327}$$
$$+ (p_2 \leftrightarrow p_4)].$$

In the above equations we have factored out the energy–momentum conserving delta functions contained in the irreducible vertices as follows:

$$\Gamma^{(2)}_{\Lambda,\,p_1 p_2} = (2\pi)^D \delta^D(p_1 + p_2)\Gamma^{(2)}_{\Lambda}(p_1, -p_1), \quad \Gamma^{(2)}_{\Lambda}(p_1, -p_1) = \Sigma_{\Lambda}(p_1). \tag{12.328}$$

$$\Gamma^{(4)}_{\Lambda,\,p_1\,\cdots\,p_4} = (2\pi)^D \delta^D(p_1 + \cdots + p_4)\Gamma^{(4)}_{\Lambda}(p_1, \ldots, p_4). \tag{12.329}$$

$$\Gamma^{(6)}_{\Lambda,\,p_1\,\cdots\,p_6} = (2\pi)^D \delta^D(p_1 + \cdots + p_6)\Gamma^{(6)}_{\Lambda}(p_1, \ldots, p_6). \tag{12.330}$$

Generalization is obvious.

We define the dimensionless n-point vertex $\tilde{\Gamma}^{(n)}_l$ by (use $[\phi] = M^{-(D+1)/2}$, $[\delta/\delta\phi] = M^{(1-D)/2}$, etc.)

$$\tilde{\Gamma}^{(n)}_l(q_1, \ldots, q_n) = \Lambda^{D\left(\frac{n}{2}-1\right)-n}\left(\frac{Z_l}{c_0}\right)^{n/2}\Gamma^{(n)}_{\Lambda}(\Lambda q_1, \ldots, \Lambda q_n). \tag{12.331}$$

The logarithmic scale l is defined by

$$l = -\ln\Lambda/\Lambda_0. \tag{12.332}$$

The dimensionless rescaled momenta are

$$q_i = k_i/\Lambda. \tag{12.333}$$

From $G_{\Lambda}^{-1}(k) = G_{0,\Lambda}^{-1}(k) + \Sigma_{\Lambda}(k)$ we immediately get

$$G_{\Lambda}(k) = \frac{1}{c_0 k^2 + \Sigma_{\Lambda}(k)}. \tag{12.334}$$

We expand as $\Sigma_{\Lambda}(k) = \Sigma_{\Lambda}(0) + k^2 \frac{\partial \Sigma_{\Lambda}}{\partial k^2}\big|_{k^2=0}$ to get

$$G_{\Lambda}(k) = \frac{Z_l}{c_0(k^2 + \zeta^{-2})}. \tag{12.335}$$

The wave function renormalization constant Z_l and the critical length ζ are defined by

$$Z_l = \frac{1}{1 + \dfrac{\partial \Sigma_{\Lambda}}{\partial k^2}\big|_{k^2=0}}, \quad c_0 \zeta^{-2} = Z_l \Sigma_{\Lambda}(0). \tag{12.336}$$

The dimensionless 2-point vertex is

$$\tilde{\Gamma}^{(2)}_l(q, -q) = \frac{Z_l}{c_0 \Lambda^2}\Gamma^{(2)}_{\Lambda}(\Lambda q, -\Lambda q) \Rightarrow \tilde{\Gamma}^{(2)}_l(q) = \frac{Z_l}{c_0 \Lambda^2}\Sigma_{\Lambda}(\Lambda q). \tag{12.337}$$

By using this equation, it is interesting to show that the wave function renormalization constant can be rewritten as

$$Z_l = 1 - \frac{\partial \tilde{\Gamma}^{(2)}_{\Lambda}(q)}{\partial q^2}\big|_{q^2=0} \tag{12.338}$$

We can also see that

$$\tilde{G}_l(q) = \frac{c_0 \Lambda^2}{Z_l} G_\Lambda(\Lambda q).$$

(12.339)

The dimensionless 4-point and 6-point vertices are given by

$$\tilde{\Gamma}_l^{(4)}(q_1, \dots, q_4) = \Lambda^{D-4} \left(\frac{Z_l}{c_0}\right)^2 \Gamma_\Lambda^{(4)}(\Lambda q_1, \dots, \Lambda q_4).$$

(12.340)

$$\tilde{\Gamma}_l^{(6)}(q_1, \dots, q_6) = \Lambda^{2D-6} \left(\frac{Z_l}{c_0}\right)^3 \Gamma_\Lambda^{(6)}(\Lambda q_1, \dots, \Lambda q_6).$$

(12.341)

Thus, we compute (with $k = \Lambda p$)

$$\begin{aligned}
\partial_\Lambda \Sigma_\Lambda(k) &= \partial_\Lambda \Sigma_\Lambda + \partial_\Lambda k \partial_k \Sigma_\Lambda \\
&= \partial_\Lambda \Sigma_\Lambda + \frac{p}{\Lambda} \partial_p \Sigma_\Lambda.
\end{aligned}$$

(12.342)

Thus (with $\partial_l = -\Lambda \partial_\Lambda$)

$$-\frac{Z_l}{c_0 \Lambda} \partial_\Lambda \Sigma_\Lambda(k) = -\frac{Z_l}{c_0 \Lambda} \partial_\Lambda \Sigma_\Lambda - p \partial_p \tilde{\Gamma}_l^{(2)}(p)$$

$$(\eta_l - 2 + \partial_l) \tilde{\Gamma}_l^{(2)}(p) = \dot{\Gamma}_l^{(2)}(p) - p \partial_p \tilde{\Gamma}_l^{(2)}(p).$$

(12.343)

In the above we have defined the anomalous dimension by the usual formula in terms of the wave function renormalization constant given by

$$\eta_l = -\partial_l \ln Z_l.$$

(12.344)

We will write the above equation as

$$\begin{aligned}
\dot{\Gamma}_l^{(2)}(p) &\equiv -\frac{Z_l}{c_0 \Lambda} \partial_\Lambda \Sigma_\Lambda(k) \\
&= (\eta_l - 2 + \partial_l) \tilde{\Gamma}_l^{(2)}(p) + p \partial_p \tilde{\Gamma}_l^{(2)}(p).
\end{aligned}$$

(12.345)

Equivalently,

$$\partial_l \tilde{\Gamma}_l^{(2)}(p) = (2 - \eta_l - p \partial_p) \tilde{\Gamma}_l^{(2)}(p) + \dot{\Gamma}_l^{(2)}(p).$$

(12.346)

The last term is given by the functional renormalization group equation (12.326). We find explicitly

$$\dot{\Gamma}_l^{(2)}(p) = \frac{1}{2} \int_q \dot{\tilde{G}}_l(q) \tilde{\Gamma}_l^{(4)}(q, -q, p, -p).$$

(12.347)

The dimensionless single-scale propagator is given by

$$\dot{\tilde{G}}_l(q) = -\frac{c_0 \Lambda^3}{Z_l} \dot{G}_\Lambda(\Lambda q).$$

(12.348)

Furthermore, we compute

$$\partial_l \tilde{\Gamma}_l^{(4)}(q) = (4 - D - 2\eta_l)\tilde{\Gamma}_l^{(4)}(q) + \Lambda^{D-4}\left(\frac{Z_l}{c_0}\right)^2 \partial_l \tilde{\Gamma}_l^{(4)}(q)$$
$$= (4 - D - 2\eta_l)\tilde{\Gamma}_l^{(4)}(q) - q_i \partial_i \tilde{\Gamma}_l^{(4)}(q) + \dot{\Gamma}_l^{(4)}(q). \tag{12.349}$$

The $\dot{\Gamma}_l^{(4)}(q)$ is given by

$$\dot{\Gamma}_l^{(4)}(q) = -\Lambda^{D-4}\left(\frac{Z_l}{c_0}\right)^2 \Lambda \partial_\Lambda \Gamma_\Lambda^{(4)}(k). \tag{12.350}$$

This is computed from the FRG equation (12.327). We get

$$\dot{\Gamma}_l^{(4)}(q_1, \ldots, q_4) = \frac{1}{2}\int_p \dot{G}_\Lambda(p)\tilde{\Gamma}_l^{(6)}(p, -p, q_1, \ldots, q_4) - \int_p \left[\dot{G}_l(p)\tilde{G}_l(p + q_1 + q_2) \right.$$
$$\left. \times \tilde{\Gamma}_l^{(4)}(-p - q_1 - q_2, p, q_1, q_2)\tilde{\Gamma}_l^{(4)}(-p, p + q_1 + q_2, q_3, q_4) + (q_2 \leftrightarrow q_3) \right.$$
$$\left. + (q_2 \leftrightarrow q_4)\right]. \tag{12.351}$$

12.5.2 The sharp momentum cutoff

In the disordered phase, we can use a multiplicative cutoff given by

$$G_{0,\Lambda}(k) = \theta_\epsilon(|k| - \Lambda)G_0(k), \quad G_0(k) = \frac{1}{c_0 k^2}. \tag{12.352}$$

For a sharp momentum cutoff we have

$$\theta_\epsilon(|k| - \Lambda) = \theta(|k| - \Lambda). \tag{12.353}$$

We compute

$$\dot{G}_\Lambda(k) = \frac{\partial_\Lambda G_{0,\Lambda}(k)}{[1 + G_{0,\Lambda}(k)\Sigma_\Lambda(k)]^2}$$
$$= -\frac{1}{c_0 k^2}\frac{\delta(|k| - \Lambda)}{\left(1 + \dfrac{\theta(|k| - \Lambda)}{c_0 k^2}\Sigma_\Lambda(k)\right)^2}. \tag{12.354}$$

This is an ambiguous expression of the form

$$I(x) = \delta(x)f(\theta(x)). \tag{12.355}$$

In the disordered phase there is no further dependence on the step function in the cutoff dependence of the self-energy $\Sigma_\Lambda(k)$. In other words, the function f is known here. We regularize the integral I as follows

$$I(x) = \lim \, \delta_\epsilon(x)f(\theta_\epsilon(x)); \quad \epsilon \longrightarrow 0. \tag{12.356}$$

We remark that

$$\frac{d}{dx} \int_0^{\theta_\epsilon(x)} dt f(t) = \lim \frac{1}{\delta x} \int_{\theta_\epsilon(x)}^{\theta_\epsilon(x) + \delta x \delta_\epsilon(x)} dt f(t), \quad \delta x \longrightarrow 0$$
$$= \delta_\epsilon(x) f(\theta_\epsilon(x)). \tag{12.357}$$

In other words,

$$\begin{aligned}
I(x) &= \lim \frac{d}{dx} \int_0^{\theta_\epsilon(x)} dt f(t); \quad \epsilon \longrightarrow 0 \\
&= \frac{d}{dx} \lim \int_0^{\theta_\epsilon(x)} dt f(t); \quad \epsilon \longrightarrow 0 \\
&= \frac{d}{dx} \theta(x) \int_0^1 dt f(t) \\
&= \delta(x) \int_0^1 dt f(t).
\end{aligned} \tag{12.358}$$

This is called the Morris lemma. By using this result we obtain

$$\begin{aligned}
\dot{G}_\Lambda(k) &= -\frac{\delta(|k| - \Lambda)}{c_0 k^2} \int_0^1 \frac{dt}{\left(1 + t\dfrac{\Sigma_\Lambda(k)}{c_0 k^2}\right)^2} \\
&= -\frac{\delta(|k| - \Lambda)}{c_0 \Lambda^2 + \Sigma_\Lambda(k)}.
\end{aligned} \tag{12.359}$$

The single-scale propagator becomes

$$\begin{aligned}
\dot{\tilde{G}}_l(q) &= -\frac{c_0 \Lambda^3}{Z_l} \dot{G}_\Lambda(\Lambda q) \\
&= \frac{\delta(|q| - 1)}{Z_l + \tilde{\Gamma}_l^{(2)}(q)}.
\end{aligned} \tag{12.360}$$

The few first terms of the effective action are given by

$$\begin{aligned}
\Gamma_\Lambda = \Gamma_\Lambda^{(0)} &+ \int_{k_1} \int_{k_2} \frac{1}{2!}(2\pi)^D \delta^D(k_1 + k_2)\Gamma_\Lambda^{(2)}(k_1, k_2)\bar{\phi}(k_1)\bar{\phi}(k_2) \\
&+ \int_{k_1} \cdots \int_{k_4} \frac{1}{4!}(2\pi)^D \delta^D(k_1 + \cdots + k_4)\Gamma_\Lambda^{(4)}(k_1, \ldots, k_4)\bar{\phi}(k_1) \ldots \bar{\phi}(k_4) + \cdots
\end{aligned} \tag{12.361}$$

By comparing with the classical action (12.316) we obtain immediately

$$\Gamma_\Lambda^{(2)}(0, 0) = \Sigma_\Lambda(k = 0) = r_\Lambda. \tag{12.362}$$

$$\Gamma_\Lambda^{(4)}(0, 0, 0, 0) = u_\Lambda. \tag{12.363}$$

From the formula (12.331) we can define the dimensionless quantities $\bar{r}_l$ and $\bar{u}_l$ by

$$\bar{r}_l = \frac{Z_l}{c_0\Lambda^2}r_\Lambda \equiv \tilde{\Gamma}_l^{(2)}(0).\tag{12.364}$$

$$\bar{u}_l = K_D\Lambda^{D-4}\left(\frac{Z_l}{c_0}\right)^2 u_\Lambda.\tag{12.365}$$

The constant K_D will be determined shortly. From the FRG equations (12.326) and (12.327) we have

$$\partial_\Lambda r_\Lambda = \frac{1}{2}\int_k \dot{G}_\Lambda(k)\Gamma_\Lambda^{(4)}(k, -k, 0, 0).\tag{12.366}$$

$$\partial_\Lambda u_\Lambda = \frac{1}{2}\int_k \dot{G}_\Lambda(k)\Gamma_\Lambda^{(6)}(k, -k, 0, 0, 0, 0) - 3\int_k \dot{G}_\Lambda(k)G_\Lambda(k)\Gamma_\Lambda^{(4)}(-k, k, 0, 0)\Gamma_\Lambda^{(4)}(0, 0, k, -k).\tag{12.367}$$

By substituting (12.359) into (12.366) we immediately get

$$\partial_\Lambda r_\Lambda = -\frac{\Lambda^{D-1}}{2}\int\frac{d\Omega_D}{(2\pi)^D}\frac{1}{c_0\Lambda^2 + \Sigma_\Lambda(k)}\Big|_{|k|=\Lambda}\Gamma_\Lambda^{(4)}(k, -k, 0, 0)\Big|_{|k|=\Lambda}.\tag{12.368}$$

By making the further approximations

$$\Sigma_\Lambda(k) \simeq r_\Lambda + \cdot s\tag{12.369}$$

$$\Gamma_\Lambda^{(4)}(k, -k, 0, 0) \simeq u_\Lambda + \cdots,\tag{12.370}$$

we obtain (with $K_D = \int d\Omega_D/(2\pi)^D$)

$$\partial_\Lambda r_\Lambda = -\frac{K_D}{2}\frac{\Lambda^{D-1}}{c_0\Lambda^2 + r_\Lambda}u_\Lambda.\tag{12.371}$$

A very important approximation is to truncate the theory by neglecting the contribution coming from the 6-point vertex. We get then

$$\begin{aligned}\partial_\Lambda u_\Lambda &= -3\int_k \dot{G}_\Lambda(k)G_\Lambda(k)\Gamma_\Lambda^{(4)}(-k, k, 0, 0)\Gamma_\Lambda^{(4)}(0, 0, k, -k)\\ &= -3u_\Lambda^2\int_k \dot{G}_\Lambda(k)G_\Lambda(k)\\ &= -3u_\Lambda^2\int\frac{d^D k}{(2\pi)^D}\dot{G}_\Lambda(k)\frac{\theta(|k| - \Lambda)}{c_0 k^2 + \theta(|k| - \Lambda)r_\Lambda}.\end{aligned}\tag{12.372}$$

Because of the delta function in $\dot{G}_\Lambda(k)$, the integrand is of the form $\delta(x)f(\theta(x))$ so we need to employ the Morris lemma. In order to use this lemma just once, we need to go back to the formula (12.354) for the single-scale propagator $\dot{G}_\Lambda(k)$, and after substitution and then using the Morris lemma, we get

$$\partial_\Lambda u_\Lambda = 3u_\Lambda^2 c_0 \Lambda^2 \int \frac{d^D k}{(2\pi)^D} \delta(|k| - \Lambda) \frac{\theta(|k| - \Lambda)}{(c_0 k^2 + \theta(|k| - \Lambda) r_\Lambda)^3}$$
$$= 3u_\Lambda^2 \int \frac{d^D k}{(2\pi)^D} \delta(|k| - \Lambda) \int_0^1 dt \frac{c_0 \Lambda^2 t}{(c_0 \Lambda^2 + t r_\Lambda)^3}$$
$$= 3u_\Lambda^2 \int \frac{d^D k}{(2\pi)^D} \delta(|k| - \Lambda) \frac{1}{2(c_0 \Lambda^2 + r_\Lambda)^2}$$
$$= \frac{3 K_D}{2} \frac{\Lambda^{D-1} u_\Lambda^2}{(c_0 \Lambda^2 + r_\Lambda)^2}.$$
$$(12.373)$$

In terms of the dimensionless quantities we obtain the Wilson momentum shell RG equations

$$\partial_l \bar{r}_l = 2\bar{r}_l + \frac{1}{2} \frac{\bar{u}_l}{1 + \bar{r}_l}. \tag{12.374}$$

$$\partial_l \bar{u}_l = (4 - D)\bar{u}_l - \frac{3}{2} \frac{\bar{u}_l}{(1 + \bar{r}_l)^2}. \tag{12.375}$$

12.5.3 The ordered phase

The vacuum expectation value of the field (classical field) is defined by

$$\bar{\phi}' = \langle \phi_\alpha \rangle = \frac{\delta \mathcal{G}_{c,\Lambda}}{\delta J_\alpha}. \tag{12.376}$$

In the ordered phase, $\lim \langle \phi \rangle \longrightarrow 0$ when $J \longrightarrow 0$. In the ordered phase, we have instead

$$\lim \bar{\phi}' = \lim \langle \phi \rangle = \frac{\delta \mathcal{G}_{c,\Lambda}}{\delta J_\alpha} \Big|_{J=0} = \bar{\phi}^0 \neq 0; \quad J \longrightarrow 0. \tag{12.377}$$

This effect arises in general when spontaneous symmetry braking occurs. We recall the results

$$\mathcal{L}_\Lambda[\bar{\phi}'] = (J_\Lambda[\bar{\phi}'], \bar{\phi}') - \mathcal{G}_{c,\Lambda}[J_\Lambda[\bar{\phi}']], \quad J_\Lambda[\bar{\phi}'] \equiv J. \tag{12.378}$$

$$\frac{\delta \mathcal{L}_\Lambda[\bar{\phi}']}{\delta \bar{\phi}'} = J_\Lambda[\bar{\phi}']. \tag{12.379}$$

Thus, we see that in the limit $J \longrightarrow 0$ the value $\bar{\phi}' = \bar{\phi}^0$ becomes an extremum of the functional $\mathcal{L}_\Lambda$. The fluctuation field will be defined by

$$\delta \bar{\phi} = \bar{\phi}' - \bar{\phi}^0. \tag{12.380}$$

We recall also the definition of the effective action in the disordered phase, which is the generating functional of the irreducible vertices, given by

$$\Gamma_\Lambda[\bar{\phi}'] = \mathcal{L}_\Lambda[\bar{\phi}'] + \frac{1}{2}(\bar{\phi}', G_{0,\Lambda}^{-1}\bar{\phi}'). \tag{12.381}$$

In the ordered phase, we will define this generating functional by [4]

$$
\begin{aligned}
\Gamma_\Lambda[\delta\bar\phi, \bar\phi^0] &= \mathcal{L}_\Lambda[\bar\phi'] + \frac{1}{2}(\delta\bar\phi, G_{0,\Lambda}^{-1}\delta\bar\phi) \\
&= (J_\Lambda[\bar\phi'], \bar\phi') - \mathcal{G}_{c,\Lambda}[J_\Lambda[\bar\phi']] + \frac{1}{2}(\delta\bar\phi, G_{0,\Lambda}^{-1}\delta\bar\phi).
\end{aligned}
\tag{12.382}
$$

The irreducible vertices are therefore given by

$$
\Gamma_\Lambda[\delta\bar\phi, \bar\phi^0] = \sum_{n=0}^{\infty}\frac{1}{n!}\int_{\alpha_1}\cdots\int_{\alpha_n}\Gamma^{(n)}_{\Lambda,\,\alpha_1\ldots\alpha_n}[\bar\phi^0]\delta\bar\phi_{\alpha_1}\cdots\delta\bar\phi_{\alpha_n}.
\tag{12.383}
$$

By following the same steps that led to the FRG equation (12.280) we now instead obtain the FRG equation

$$
\begin{aligned}
\partial_\Lambda\Gamma_\Lambda[\delta\bar\phi, \bar\phi^0] = &-\frac{1}{2}Tr\partial_\Lambda G_{0,\Lambda}^{-1}\left(\frac{\delta}{\delta\bar\phi'}\otimes\frac{\delta}{\delta\bar\phi'}\mathcal{L}_\Lambda\right)^{-1} + \partial_\Lambda\ln Z_{0,\Lambda} \\
&- (\delta\bar\phi, \partial_\Lambda[G_{0,\Lambda}^{-1}\bar\phi^0]) - \frac{1}{2}(\bar\phi^0, \partial_\Lambda[G_{0,\Lambda}^{-1}]\bar\phi^0).
\end{aligned}
\tag{12.384}
$$

In deriving the above equation we have used the fact that the expectation value $\bar\phi^0$ depends on the cutoff and that the RGF equation (12.270) for $\mathcal{G}_{c,\Lambda}[J]$, derived in the disordered phase, is still valid in the ordered phase. Similarly to what we have done in the disordered phase, we define the self-energy Σ_Λ and the U_Λ matrix by

$$
\Sigma_\Lambda[\bar\phi^0] = \frac{\delta}{\delta\bar\phi'}\otimes\frac{\delta}{\delta\bar\phi'}\Gamma_\Lambda[\delta\bar\phi, \bar\phi^0]|_{\delta\bar\phi\,=\,0}.
\tag{12.385}
$$

$$
U_\Lambda[\delta\bar\phi, \bar\phi^0] = \frac{\delta}{\delta\bar\phi'}\otimes\frac{\delta}{\delta\bar\phi'}\Gamma_\Lambda[\delta\bar\phi, \bar\phi^0] - \frac{\delta}{\delta\bar\phi'}\otimes\frac{\delta}{\delta\bar\phi'}\Gamma_\Lambda[\delta\bar\phi, \bar\phi^0]|_{\delta\bar\phi\,=\,0}.
\tag{12.386}
$$

By going through the same steps, which led to equation (12.295), we now obtain the modified FRG equation

$$
\begin{aligned}
\partial_\Lambda\Gamma_\Lambda[\bar\phi'] = &-\frac{1}{2}Tr(\dot{G}_{0,\Lambda}\Sigma_\Lambda[1 - G_{0,\Lambda}\Sigma_\Lambda]^{-1} + \dot{G}_\Lambda U_\Lambda[\bar\phi'][1 - G_\Lambda U_\Lambda[\bar\phi']]^{-1}) \\
&- (\delta\bar\phi, \partial_\Lambda[G_{0,\Lambda}^{-1}\bar\phi^0]) - \frac{1}{2}(\bar\phi^0, \partial_\Lambda[G_{0,\Lambda}^{-1}]\bar\phi^0).
\end{aligned}
\tag{12.387}
$$

The corresponding FRG equations of the irreducible vertices, which generalize equations (12.304) and (12.306), are now given, respectively, by

$$
\begin{aligned}
\partial_\Lambda\Gamma_\Lambda^{(0)} = &-\frac{1}{2}Tr\dot{G}_{0,\Lambda}\Sigma_\Lambda[1 - G_{0,\Lambda}\Sigma_\Lambda]^{-1} \\
&+ \int_\alpha\Gamma^{(1)}_{\Lambda,\,\alpha}[\bar\phi^0]\partial_\Lambda\bar\phi^0_\alpha - \frac{1}{2}(\bar\phi^0, \partial_\Lambda G_{0,\Lambda}^{-1}\bar\phi^0).
\end{aligned}
\tag{12.388}
$$

$$\partial_\Lambda \Gamma^{(n)}_{\Lambda,\,\alpha_1\ldots\alpha_n} = -\frac{1}{2}\sum_{\nu=1}^{\infty}\sum_{n_1=1}^{\infty}\cdots\sum_{n_\nu=1}^{\infty}\delta_{n,n_1+\cdots+n_\nu}\,\mathcal{S}_{\alpha_1\ldots\alpha_{n_1};\,\ldots;\,\alpha_{n-n_\nu+1}\ldots\alpha_n}\,Tr\dot{G}_\Lambda\Gamma^{(n_\nu+2)}_{\Lambda\alpha_{n-n_\nu+1}\ldots\alpha_n}$$

$$\times\,G_\Lambda\Gamma^{(n_{\nu-1}+2)}_{\Lambda\alpha_{n-n_\nu-n_{\nu-1}+1}\ldots\alpha_{n-n_\nu}}\cdots G_\Lambda\Gamma^{(n_2+2)}_{\Lambda\alpha_{n_1+1}\ldots\alpha_{n_1+n_2}}\,G_\Lambda\Gamma^{(n_1+2)}_{\Lambda\alpha_1\ldots\alpha_{n_1}} \tag{12.389}$$

$$+\int_\alpha \Gamma^{(n+1)}_{\Lambda,\,\alpha\alpha_1\ldots\alpha_n}[\bar{\phi}^0]\partial_\Lambda\bar{\phi}^0_\alpha.$$

As opposed to the case of the disordered phase, in the ordered phase the irreducible vertices with odd numbers of external legs do not vanish. In particular, $\Gamma^{(1)}_{\Lambda,\,\alpha_1}[\bar{\phi}^0]$ does not vanish. The corresponding flow equation is found to be given by

$$\partial_\Lambda\Gamma^{(1)}_{\Lambda\alpha_1}[\bar{\phi}^0] = -\frac{1}{2}Tr\dot{G}_\Lambda\Gamma^{(3)}_{\Lambda,\,\alpha_1}[\bar{\phi}^0] + \int_\alpha \Gamma^{(2)}_{\Lambda,\,\alpha\alpha_1}[\bar{\phi}^0]\partial_\Lambda\bar{\phi}^0_\alpha$$

$$-\int_\alpha [G^{-1}_{0,\Lambda}]_{\alpha_1\alpha}\partial_\Lambda\bar{\phi}^0_\alpha - \int_\alpha [\partial_\Lambda G^{-1}_{0,\Lambda}]_{\alpha_1\alpha}\bar{\phi}^0_\alpha. \tag{12.390}$$

As it turns out, it is useful to set the 1-point vertex to zero even in the ordered phase. To this end, we choose the regularized inverse free propagator, by adding appropriate counter terms, such that

$$G^{-1}_{0,\Lambda}\partial_\Lambda\bar{\phi}^0 = 0,\quad \partial_\Lambda G^{-1}_{0,\Lambda}\bar{\phi}^0 = 0. \tag{12.391}$$

The FRG equations for the 0-point and the 1-point vertices become then

$$\partial_\Lambda\Gamma^{(0)}_\Lambda = -\frac{1}{2}Tr\dot{G}_{0,\Lambda}\Sigma_\Lambda[1 - G_{0,\Lambda}\Sigma_\Lambda]^{-1} + \int_\alpha \Gamma^{(1)}_{\Lambda,\,\alpha}[\bar{\phi}^0]\partial_\Lambda\bar{\phi}^0_\alpha. \tag{12.392}$$

$$\partial_\Lambda\Gamma^{(1)}_{\Lambda\alpha_1}[\bar{\phi}^0] = -\frac{1}{2}Tr\dot{G}_\Lambda\Gamma^{(3)}_{\Lambda,\,\alpha_1}[\bar{\phi}^0] + \int_\alpha \Gamma^{(2)}_{\Lambda,\,\alpha\alpha_1}[\bar{\phi}^0]\partial_\Lambda\bar{\phi}^0_\alpha. \tag{12.393}$$

The flow of the vacuum expectation value $\bar{\phi}^0$ is fixed such that the 1-point vertex vanishes identically. In other words, we have

$$\Gamma^{(1)}_{\Lambda\alpha_1}[\bar{\phi}^0] = 0. \tag{12.394}$$

Equivalently

$$\frac{1}{2}Tr\dot{G}_\Lambda\Gamma^{(3)}_{\Lambda,\,\alpha_1}[\bar{\phi}^0] = \int_\alpha \Gamma^{(2)}_{\Lambda,\,\alpha\alpha_1}[\bar{\phi}^0]\partial_\Lambda\bar{\phi}^0_\alpha. \tag{12.395}$$

This equation determines the dependence of the vacuum expectation value on the cutoff, viz $\bar{\phi}^0 = \bar{\phi}^0(\Lambda)$. The flow equation of the free energy is now given by the same formula as in the disordered phase, i.e.,

$$\partial_\Lambda\Gamma^{(0)}_\Lambda = -\frac{1}{2}Tr\dot{G}_{0,\Lambda}\Sigma_\Lambda[1 - G_{0,\Lambda}\Sigma_\Lambda]^{-1}. \tag{12.396}$$

We consider now as before ϕ^4 in D dimension. In other words, we will make the replacements $\alpha \longrightarrow k$, $\phi_\alpha \longrightarrow \phi(k)$ and $\int_\alpha \longrightarrow \int_k = d^Dk/(2\pi)^D$. We will also replace $\bar{\phi}^0_\alpha \longrightarrow \bar{\phi}^0(k)$. We will assume further that the expectation value $\bar{\phi}^0$ is uniform, i.e.,

$$\bar{\phi}^0(k) = (2\pi)^D \delta^D(k)\bar{\phi}_\Lambda^{\,0}. \tag{12.397}$$

By using also $(G_\Lambda)_{k_1 k_2} = -(2\pi)^D \delta^D(k_1 + k_2)G_\Lambda(k_1)$, $\Gamma^{(3)}_{\Lambda,\,k_1 k_2 p} = (2\pi)^D \delta^D(k_1 + k_2 + p)$ $\Gamma^{(3)}_\Lambda(k_1, k_2, p)$, etc., we can show that the condition (12.396) takes the form

$$\Gamma^{(2)}_\Lambda(-p, p)(2\pi)^D \delta^D(p)\partial_\Lambda \bar{\phi}_\Lambda^{\,0} = -\frac{1}{2}\int_k \dot{G}_\Lambda(k)(2\pi)^D \delta^D(p)\Gamma^{(3)}_\Lambda(k, -k, p). \tag{12.398}$$

Equivalently

$$\Sigma_\Lambda(0)\partial_\Lambda \bar{\phi}_\Lambda^{\,0} = -\frac{1}{2}\int_k \dot{G}_\Lambda(k)\Gamma^{(3)}_\Lambda(k, -k, 0). \tag{12.399}$$

The flow equation of the 2-point vertex (self-energy) can be derived from the FRG equation (12.389). This is a generalization of the FRG equation (12.312) or equivalently the FRG equation (12.326) (see also the FRG equation (12.311). We get

$$\partial_\Lambda \Sigma_\Lambda(p) = \frac{1}{2}\int_k \dot{G}_\Lambda(k)\Gamma^{(4)}_\Lambda(k, -k, p, -p) + \partial_\Lambda \bar{\phi}_\Lambda^{\,0}\Gamma^{(3)}_\Lambda(k - k, 0)$$
$$- \int_{k'} \dot{G}_\Lambda(k')G_\Lambda(k' + k)\Gamma^{(3)}_\Lambda(k, -k, -k', k')\Gamma^{(3)}_\Lambda(-k', k + k', -k). \tag{12.400}$$

The re-scaled vacuum expectation value M_l is defined by

$$\Lambda \partial_\Lambda \bar{\phi}_\Lambda^{\,0} = \Lambda^{\frac{D}{2}-1}\left(\frac{Z_l}{c_0}\right)^{1/2}\left[\frac{D - 2 + \eta_l}{2} - \partial_l\right]M_l. \tag{12.401}$$

By using now (12.331) we have the re-scaled objects

$$\tilde{\Gamma}^{(2)}_l(q) = \frac{Z_l}{c_0\Lambda^2}\Sigma_\Lambda(\Lambda q). \tag{12.402}$$

$$\tilde{\Gamma}^{(3)}_l(q_1, \ldots, q_3) = \Lambda^{\frac{D}{2}-3}\left(\frac{Z_l}{c_0}\right)^{3/2}\Gamma^{(3)}_\Lambda(\Lambda q_1, \ldots, \Lambda q_3). \tag{12.403}$$

Also we use the equation (12.348) defining the single-scale propagator. The condition (12.399) becomes

$$\Lambda \partial_\Lambda \bar{\phi}_\Lambda^{\,0} = \Lambda^{\frac{D}{2}-1}\left(\frac{Z_l}{c_0}\right)^{1/2}\frac{1}{2}\int_q \dot{\check{G}}_l(q)\frac{\tilde{\Gamma}^{(3)}_l(q, -q, 0)}{\tilde{\Gamma}^{(2)}_l(0)}. \tag{12.404}$$

By comparing (12.401) and (12.404) we obtain the final result

$$\partial_l M_l = \left[\frac{D - 2 + \eta_l}{2}\right]M_l - \frac{1}{2}\int_q \dot{\check{G}}_l(q)\frac{\tilde{\Gamma}^{(3)}_l(q, -q, 0)}{\tilde{\Gamma}^{(2)}_l(0)}. \tag{12.405}$$

As in the disordered phase, the renormalization group flow equation of the dimensionless 2-point vertex $\tilde{\Gamma}_l^{(2)}(q)$ is given by the equation

$$\partial_l \tilde{\Gamma}_l^{(2)}(q) = (2 - \eta_l - q\partial_q)\tilde{\Gamma}_l^{(2)}(q) + \dot{\Gamma}_l^{(2)}(q), \tag{12.406}$$

where

$$\dot{\Gamma}_l^{(2)}(q) \equiv -\frac{Z_l}{c_0\Lambda}\partial_\Lambda \Sigma_\Lambda(k). \tag{12.407}$$

Now this last quantity is given explicitly from the FRG equation (12.400) by the result

$$\begin{aligned}
\dot{\Gamma}_l^{(2)}(q) = \frac{1}{2}\int_p \dot{\tilde{G}}_l(p)\tilde{\Gamma}_l^{(4)}(p, -p, q, -q) - \tilde{\Gamma}_l^{(3)}(q, -q, 0)\left[\partial_l - \frac{D - 2 + \eta_l}{2}\right]M_l \\
- \int_p \dot{\tilde{G}}_l(p)\tilde{G}_l(p+q)\tilde{\Gamma}_l^{(3)}(q, -q-p, p)\tilde{\Gamma}_l^{(3)}(-p, q+p, -q).
\end{aligned} \tag{12.408}$$

This is a generalization of the FRG equation (12.347). Obviously, we can use equation (12.405) to eliminate M_l from equation (12.408).

12.6 The gradient expansion approximation method

12.6.1 Effective potential

We will consider the $O(N)$ sigma model

$$S[\phi] = \int d^D x \left[\frac{r_0}{2}\phi_a^2 + \frac{c_0}{2}(\partial_\mu \phi_a)^2 + \frac{u_0}{4!}(\phi_a^2)^2\right]. \tag{12.409}$$

The average effective action or the Wetterich's Legendre effective action is defined by

$$\Gamma_\Lambda^W[\bar{\phi}'] = \Gamma_\Lambda[\bar{\phi}'] - \frac{1}{2}(\bar{\phi}', G_0^{-1}\bar{\phi}') - \ln Z_{0,\Lambda}. \tag{12.410}$$

We will employ an additive regularization R_Λ given by

$$G_{0,\Lambda}^{-1} = G_0^{-1} - R_\Lambda. \tag{12.411}$$

In order that, $G_{0,\Lambda} \longrightarrow G_0$ when $\Lambda \longrightarrow 0$, and $G_{0,\Lambda} \longrightarrow 0$ when $\Lambda \longrightarrow \infty$, we must demand that

$$|R_\Lambda| \sim 0, \quad \Lambda \longrightarrow 0; \quad |R_\Lambda| \sim \infty, \quad \Lambda \longrightarrow \infty. \tag{12.412}$$

By comparing with the multiplicative regularization θ_Λ given by $G_{0,\Lambda} = \theta_\Lambda G_0$ we see that $\theta_\Lambda = (1 - R_\Lambda G_0)^{-1}$. By using this additive regularization we can rewrite the Wetterich Legendre effective action as

$$\Gamma_\Lambda^W[\bar{\phi}'] = \mathcal{L}_\Lambda[\bar{\phi}'] - \frac{1}{2}(\bar{\phi}', R_\Lambda\bar{\phi}') - \ln Z_{0,\Lambda}. \tag{12.413}$$

Thus in the limit $\Lambda \longrightarrow 0$ we have the behavior $\Gamma_\Lambda^W \longrightarrow \mathcal{L}$ and thus the name Legendre effective action. Furthermore, from the limiting behavior $\Gamma_\Lambda \longrightarrow S_1$ when $\Lambda \longrightarrow \Lambda_0 = \infty$, we obtain the initial condition

$$\Gamma_\Lambda^W = S, \quad \Lambda \longrightarrow \Lambda_0, \tag{12.414}$$

where S is the above bare action. The Wetterich Legendre effective action satisfies the FRG equation

$$\partial_\Lambda \Gamma_\Lambda^W[\bar{\phi}'] = -\frac{1}{2} Tr \partial_\Lambda G_{0,\Lambda}^{-1}\left(\frac{\delta}{\delta\bar{\phi}'} \otimes \frac{\delta}{\delta\bar{\phi}'}\Gamma_\Lambda - G_{0,\Lambda}^{-1}\right)^{-1}$$

$$= -\frac{1}{2} Tr \partial_\Lambda R_\Lambda\left(\frac{\delta}{\delta\bar{\phi}'} \otimes \frac{\delta}{\delta\bar{\phi}'}\Gamma_\Lambda^W + R_\Lambda\right)^{-1} \tag{12.415}$$

$$= -\frac{1}{2} Tr \frac{\partial_\Lambda R_\Lambda}{\Gamma_\Lambda^{W(2)} + R_\Lambda}.$$

The classical field is defined by $\langle\phi\rangle = \bar{\phi}'$ and it does not depend on Λ. In the limit of vanishing external source $J \equiv J_\Lambda \longrightarrow 0$ it may acquire a non-zero expectation value due to spontaneous symmetry breaking, i.e. $\lim \bar{\phi}' \longrightarrow \bar{\phi}_0$, $J \longrightarrow 0$, and it becomes given by the equation

$$\frac{\delta\mathcal{L}_\Lambda[\bar{\phi}']}{\delta\bar{\phi}'} = 0 = \frac{\delta\Gamma_\Lambda^W[\bar{\phi}']}{\delta\bar{\phi}'} + R_\Lambda\bar{\phi}'. \tag{12.416}$$

$\bar{\phi}' = \bar{\phi}_0$ is a solution of this equation. Since we have the limit $R_\Lambda \longrightarrow 0$ when $\Lambda \longrightarrow 0$ we see that $\bar{\phi}_0$ is a minimum of the Wetterich Legendre effective action Γ_Λ^W. The minimum of the Wetterich action for any Λ will be denoted by $\bar{\phi}_{0,\Lambda}$. Since in the limit $\Lambda \longrightarrow \Lambda_0$ in which $\Gamma_\Lambda^W \longrightarrow S$ we have $\bar{\phi}_{0,\Lambda} \longrightarrow \bar{\phi}_0^{MF}$ where $\bar{\phi}_0^{MF}$ is the mean-field value. Hence, $\bar{\phi}_{0,\Lambda}$ interpolates between the classical value $\bar{\phi}_0^{MF}$ and the quantum value $\bar{\phi}_0$, which contains all fluctuations.

In the derivative or gradient expansion method, we expand the Wetterich Legendre effective action Γ_Λ^W in terms of the derivatives of the field. Clearly, this should consist a good approximation in the low energy limit. In this expansion, the first term will obviously contain no derivatives. We write it as (in the following we will omit for simplicity the prime on the field)

$$\Gamma_\Lambda^W[\bar{\phi}] = \int d^D x\, U_\Lambda(\rho(x)) + \cdots \tag{12.417}$$

In the above equation ρ is the field density defined by

$$\rho = \frac{1}{2}\bar{\phi}_i^2. \tag{12.418}$$

Obviously, for a uniform configuration we obtain exactly

$$\Gamma_\Lambda^W[\bar{\phi}] = V U_\Lambda(\rho). \tag{12.419}$$

In other words, $U_\Lambda[\bar\phi]$ is the effective potential. Thus, $\bar\phi_{0,\Lambda}$ is the minimum of the effective potential $U_\Lambda[\bar\phi]$.

In terms of $U_\Lambda[\bar\phi]$ the above FRG equation (12.415) becomes

$$V\partial_\Lambda U_\Lambda[\bar\phi] = -\frac{1}{2} Tr\frac{\partial_\Lambda R_\Lambda}{\Gamma_\Lambda^{W(2)} + R_\Lambda}. \tag{12.420}$$

We use $\langle k'|R_\Lambda|k\rangle = R_\Lambda(k)(2\pi)^D\delta^D(k+k')$ and $V = (2\pi)^D\delta^D(0)$ and $(G_{0,\Lambda}^{-1})_{kk'} = -(2\pi)^D\delta^D(k+k')G_{0,\Lambda}^{-1}(k)$. Also, we use $G_\Lambda^{-1} = G_{0,\Lambda}^{-1} - \Gamma_\Lambda^{(2)}$ to derive $G_\Lambda^{-1}(k) = \Gamma_\Lambda^{W(2)}(k) + R_\Lambda(k)$. We get

$$\partial_\Lambda U_\Lambda[\bar\phi] = \frac{1}{2}\sum_{i=1}^{N}\int_k \partial_\Lambda R_\Lambda(k)G_{i,\Lambda}(\bar\phi, k), \tag{12.421}$$

where

$$G_{i,\Lambda}(\bar\phi, k) = \left(\frac{1}{V}\frac{\delta^2}{\delta\bar\phi_i(k)\delta\bar\phi_i(-k)}\Gamma_\Lambda^W \big|_{\phi(x)=\bar\phi} + R_\Lambda(k)\right)^{-1}. \tag{12.422}$$

The most general expression of the Wetterich Legendre effective action Γ_Λ^W, which includes up to the quartic term in the derivatives of the field, is of the form

$$\Gamma_\Lambda^W[\bar\phi] = \int d^D x\left[U_\Lambda(\rho(x)) + \frac{c_0}{2}Z_\Lambda^{-1}(\rho(x))(\partial_\mu\phi_i)^2 + \frac{c_0}{4}Y_\Lambda(\rho(x))(\partial_\mu\rho(x))^2 + \cdots\right]. \tag{12.423}$$

We can compute immediately the functional derivative

$$\frac{\delta^2}{\delta\bar\phi_i(k)\delta\bar\phi_i(-k)}\Gamma_\Lambda^W \big|_{\phi(x)=\bar\phi} = V\left(U_\Lambda' + \bar\phi_i^2 U_\Lambda'' + c_0 k^2 Z_\Lambda^{-1} + \frac{c_0}{2}k^2\bar\phi_i^2 Y_\Lambda\right). \tag{12.424}$$

By choosing $O(N)$ invariance we can always choose $\bar\phi$ in such a way that only component is non-zero, say $\bar\phi_1$, and thus we have

$$\bar\phi = \begin{pmatrix}\bar\phi_1\\0\\\vdots\\0\end{pmatrix}. \tag{12.425}$$

In other words, $\rho = \bar\phi_1^2/2$. The longitudinal and transverse propagators become therefore given by

$$G_{l,\Lambda}^{-1}(\rho, k^2) \equiv G_{1,\Lambda}^{-1} = U_\Lambda' + 2\rho U_\Lambda'' + c_0\left(Z_\Lambda^{-1} + \rho^2 Y_\Lambda\right)k^2 + R_\Lambda. \tag{12.426}$$

$$G_{t,\Lambda}^{-1}(\rho, k^2) \equiv G_{i,\Lambda}^{-1} = U_\Lambda' + c_0 Z_\Lambda^{-1}k^2 + R_\Lambda, \quad i = 2, \ldots, N. \tag{12.427}$$

Thus the above FRG equation becomes

$$\partial_\Lambda U_\Lambda[\bar\phi] = \frac{1}{2}\int_k \partial_\Lambda R_\Lambda(k)(G_{l,\Lambda}(\rho, k^2) + (N-1)G_{t,\Lambda}(\rho, k^2))$$
$$= \frac{K_D}{2}\int_0^\infty k^{D-1}dk\, \partial_\Lambda R_\Lambda(k)(G_{l,\Lambda}(\rho, k^2) + (N-1)G_{t,\Lambda}(\rho, k^2)). \tag{12.428}$$

Clearly, in order to solve the above equation, one needs also to determine the flow equations of the wave function renormalization Z_Λ and of Y_Λ. We will employ the local potential approximation to solve this equation in which we set these factors equal to their initial classical values $Z_{\Lambda_0} = 1$ and $Y_{\Lambda_0} = 0$. This allows us then to close the hierarchy of flow equations. Indeed, in this approximation we will only need to deal with the flow equation of the effective potential which depends only on the effective potential itself and nothing else. In this approximation the Wetterich Legendre effective action becomes

$$\Gamma_\Lambda^W[\bar\phi] = \int d^D x \left[U_\Lambda(\rho(x)) + \frac{c_0}{2}(\partial_\mu\phi_i)^2 + \cdots \right]. \tag{12.429}$$

The initial classical value of the effective potential is therefore given by the classical potential, viz

$$U_{\Lambda_0}(\rho) = f_0 + \frac{r_0}{2}\phi_i^2 + \frac{u_0}{4!}(\phi_i^2)$$
$$= \frac{u_0}{6}(\rho - \rho_0)^2. \tag{12.430}$$

In the above equation we have chosen $f_0 = 3r_0^2/2u_0$ and thus $\rho_0 = -3r_0/u_0$. In the following we will use the Litim regulator given by [15]

$$R_\Lambda(k) = c_0(\Lambda^2 - k^2)\theta(\Lambda^2 - k^2). \tag{12.431}$$

Obviously, for $k \leqslant \Lambda$ we have $R_\Lambda(k) = c_0(\Lambda^2 - k^2)$, and hence the longitudinal and transverse propagators become k-independent given by

$$G_{l,\Lambda}^{-1}(\rho, k^2) = U_\Lambda' + 2\rho U_\Lambda'' + c_0\Lambda^2. \tag{12.432}$$

$$G_{t,\Lambda}^{-1}(\rho, k^2) = U_\Lambda' + c_0\Lambda^2. \tag{12.433}$$

We also compute $\partial_\Lambda R_\Lambda(k) = 2c_0\Lambda\theta(\Lambda^2 - k^2)$ and as a consequence the integral over k reduces to $\int_0^\infty k^{D-1}dk(2c_0\Lambda\theta(\Lambda^2 - k^2)(...) = 2c_0\Lambda^{D+1}(...)/D$. The flow equation becomes

$$\partial_\Lambda U_\Lambda(\rho) = \frac{c_0 K_D \Lambda^{D+1}}{D}\left(\frac{1}{U_\Lambda' + 2\rho U_\Lambda'' + c_0\Lambda^2} + \frac{N-1}{U_\Lambda' + c_0\Lambda^2} \right). \tag{12.434}$$

This can be solved numerically for fixed values of the dimensionless parameters $\bar u_0 = u_0/c_0^2\Lambda$, $\bar\rho_0 = c_0\rho_0/\Lambda$ for various values of the cutoff Λ. It is useful to shift the potential by its running minimum, viz $U_\Lambda(\rho) \longrightarrow U_\Lambda(\rho) - U_\Lambda(\rho_0)$, measure it in units of f_0, and plot it as a function of $\bar\phi/\bar\phi^{MF}$. We will be interested in the behavior of the minimum $\bar\phi_{0,\Lambda}$ in the continuum limit $\Lambda \longrightarrow 0$ and whether or not spontaneous symmetry breaking occurs. In the ordered phase we observe that $\bar\phi_{0,\Lambda}$ approaches a non-zero value in the limit $\Lambda \longrightarrow 0$, whereas in the disordered phase it will approach a zero value.

12.6.2 Fixed points and critical exponents

In order to discuss fixed points and critical exponents we introduce the running cutoff $l = \Lambda_0 \exp(-l)$ and the following dimensionless quantities

$$q = k/\Lambda, \quad \tilde{x} = \Lambda x, \quad \tilde\phi(\tilde{x}) = \sqrt{c_0}\,\Lambda^{(2-D)/2}\bar\phi(x), \quad \tilde\rho(x) = c_0\Lambda^{2-D}\rho(x). \qquad (12.435)$$

The dimensionless potential is therefore

$$\tilde{U}_l(\tilde\phi) = \frac{U_\Lambda(\bar\phi)}{\Lambda^D} = \tilde{f}_0 + \frac{\tilde{r}_l}{2}\tilde\phi_i^{\,2} + \frac{\tilde{u}_l}{4!}(\tilde\phi_i^{\,2})^2 + \cdots \qquad (12.436)$$

The definition of the dimensionless parameters $\tilde{f}_l$, $\tilde{r}_l$ and $\tilde{u}_l$ is obvious. By substituting in the above FRG equation we obtain

$$\partial_l \tilde{U}_l(\tilde\rho) = D\tilde{U}_l(\tilde\rho) - (D-2)\tilde\rho\tilde{U}_l'(\tilde\rho) - \frac{K_D}{D}\left(\frac{1}{1 + \tilde{U}_l' + 2\tilde\rho\tilde{U}_l''} + \frac{N-1}{1 + \tilde{U}_l'}\right). \qquad (12.437)$$

We introduce the function

$$\tilde{W}_l(\tilde\rho) = \tilde{U}_l'(\tilde\rho). \qquad (12.438)$$

In terms of this new function, the above FRG equation becomes

$$\partial_l \tilde{W}_l(\tilde\rho) = 2\tilde{W}_l(\tilde\rho) - (D-2)\tilde\rho\tilde{W}_l'(\tilde\rho) + \frac{K_D}{D}\left(\frac{3\tilde{W}_l' + 2\tilde\rho\tilde{W}_l''}{\left[1 + \tilde{W}_l + 2\tilde\rho\tilde{W}_l'\right]^2} + \frac{(N-1)\tilde{W}_l'}{[1 + \tilde{W}_l]^2}\right). \qquad (12.439)$$

The fixed points are solutions of the equation

$$\partial_l \tilde{U}_\Lambda(\tilde\rho_*) = 0 \Rightarrow \partial_l \tilde{W}_l(\tilde\rho_*) = 0. \qquad (12.440)$$

It is well established that there are three fixed points in the $O(N)$ sigma model in D dimensions, which are

- The ferromagnetic low temperature fixed point $\mathcal{F}$, in which the $O(N)$ symmetry is spontaneously broken. In this case, we have a finite vacuum expectation value $\bar\phi_0$, and as a consequence, the rescaled vacuum expectation value $\tilde\phi_0$ diverges with the running cutoff exponentially, viz

$$\tilde\phi_0 = \sqrt{c_0}\,\bar\phi_0\Lambda_0^{-\frac{D-2}{2}}\, e^{\frac{D-2}{2}l} \longrightarrow \infty, \quad l \longrightarrow \infty. \qquad (12.441)$$

$\tilde\phi_0$ is the spontaneous magnetization.

- The paramagnetic high temperature fixed point $\mathcal{P}$, in which the $O(N)$ symmetry is not broken. In this case, both the vacuum expectation value $\bar{\phi}_0$ and the rescaled vacuum expectation value $\tilde{\phi}_0$ vanish. The vanishing of the vacuum expectation value $\bar{\phi}_0$ is driven by fluctuations and it occurs at a value of the cutoff $\Lambda_* > 0$. This gives a measure of the critical length $1/\xi$ of the system.
- The Wilson–Fisher fixed critical point $\mathcal{C}$. This is given by the solution of the equation $\partial_l \tilde{W}_l(\tilde{\rho}_*) = 0$ such that $\tilde{W}_l(\tilde{\rho}_*) \neq 0$, which should be contrasted with the Gaussian trivial fixed point for which the solution is such that $\tilde{W}_l(\tilde{\rho}_*) = 0$. In this case the rescaled vacuum expectation value $\tilde{\phi}_0$ is finite in the limit $\Lambda \longrightarrow 0$, whereas the vacuum expectation value $\bar{\phi}_0$ goes to zero as

$$\bar{\phi}_0 = \frac{1}{\sqrt{c_0}} \tilde{\phi}_0 \Lambda^{\frac{D-2}{2}} \longrightarrow 0, \quad \Lambda \longrightarrow 0. \tag{12.442}$$

There are six critical exponents related by the so-called scaling relations and thus only two of them are linearly independent. We take them to be the anomalous dimension η and the mass critical exponent ν. In the local potential approximation, $\eta = 0$, since the wave function renormalization is set equal to 1. There remains therefore to calculate the mass critical exponent ν. This can be extracted from the behavior of the critical length as a function of the mass parameter (temperature) given by

$$\Lambda_0 \xi \sim |\tilde{r} - \tilde{r}_0^*|^{-\nu}. \tag{12.443}$$

We consider the matrix ϕ^4 action

$$S[\phi] = \int d^D x \, Tr\left[\frac{r_0}{2}\phi^2 + \frac{c_0}{2}(\partial_\mu \phi)^2 + \frac{u_0}{4!}(\phi^2)^2\right]. \tag{12.444}$$

The path integral is

$$Z = \int [d\phi] \, \exp(-S(\phi)). \tag{12.445}$$

By diagonalization $\phi = U\psi U^{-1}$ we obtain the path integral

$$Z = \int [d\psi] \, \exp(-\tilde{S}(\psi)). \tag{12.446}$$

The effective action is given by

$$\tilde{S}(\psi) = S(\psi) - \int d^D x \sum_{i<j} \ln(\psi_i - \psi_j)^2 + F(\psi), \tag{12.447}$$

where the second terms comes from the Vandermonde determinant, and $F(\psi)$ comes from the integration over the $SU(N)$ group, viz

$$\exp(-F(\psi)) = \int [dU] \exp\left(-\frac{c_0}{2}\int d^D x \, Tr([U^{-1}\partial_\mu U, \psi]^2)\right). \tag{12.448}$$

The free action and the interaction part are given by

$$\tilde{S}_0(\psi) = \int d^D x \, Tr \frac{c_0}{2}(\partial_\mu \psi)^2 = -\frac{1}{2}(\psi, G_0^{-1}\psi). \tag{12.449}$$

$$(G_0^{-1})_{kk'} = -c_0 k^2 (2\pi)^D \delta^D(k + k')\delta_{ij}. \tag{12.450}$$

The Wetterich equation in this case is given by

$$\frac{1}{V}\partial_\Lambda \Gamma_\Lambda^W[\bar{\psi}] = \frac{1}{2}\sum_{i=1}^{N}\int_k \partial_\Lambda R_\Lambda(k) G_{i,\Lambda}(\bar{\psi}, k), \tag{12.451}$$

where

$$G_{i,\Lambda}(\bar{\psi}, k) = \left(\frac{1}{V}\frac{\delta^2}{\delta\bar{\psi}_i(k)\delta\bar{\psi}_i(-k)}\Gamma_\Lambda^W\Big|_{\bar{\psi}(x)=\bar{\psi}} + R_\Lambda(k)\right)^{-1}. \tag{12.452}$$

The initial condition is

$$\Gamma_\Lambda^W \longrightarrow \tilde{S}, \quad \Lambda \longrightarrow \Lambda_0. \tag{12.453}$$

We assume the Wetterich effective action to be of the form

$$\Gamma_\Lambda^W[\bar{\psi}] = \int d^D x \, Tr\left[U_\Lambda(\bar{\psi}(x)) + \frac{c_0}{2}Z_\Lambda^{-1}(\bar{\psi})(\partial_\mu\bar{\psi})^2 + \cdots\right]. \tag{12.454}$$

The initial condition evaluated on $\bar{\psi}(x) = \bar{\psi}$ becomes

$$U_\Lambda(\bar{\psi}) \longrightarrow Tr\left(\frac{r_0}{2}\bar{\psi}^2 + \frac{u_0}{4!}(\bar{\psi}^2)^2\right) - \sum_{i<j}\ln(\bar{\psi}_i - \bar{\psi}_j) + \frac{F}{V}, \quad \Lambda \longrightarrow \Lambda_0. \tag{12.455}$$

We compute

$$G_{ij,\,ij,\,\Lambda}^{-1}(\bar{\phi}, k) = U'_{\Lambda,\,ij}\delta_{ij} + \frac{1}{4}\bar{\phi}_{ji}\{\bar{\phi}, U''_\Lambda\}_{ji} + \frac{1}{2}\rho_{ji}U''_{\Lambda,\,ji} + \frac{1}{4}(\phi U''_\Lambda\phi)_{ji}\delta_{ij} + c_0 k^2 Z_{\Lambda,\,ii}^{-1}\delta_{ij} + R_\Lambda(k). \tag{12.456}$$

The expectation value $\bar{\phi} = \langle\phi\rangle$ was assumed to be uniform in space. The analogous assumption in matrix space is to take $\bar{\phi}$ as a diagonal matrix given by

$$\bar{\phi}_i = \lambda_i \delta_{ij}. \tag{12.457}$$

In this case we obtain

$$G_{ij,\,ij,\,\Lambda}^{-1}(\bar{\phi}, k) = \left(U'_{\Lambda,\,i} + 2\rho_i U''_{\Lambda,\,i} + c_0 k^2 Z_{\Lambda,\,i}^{-1}\right)\delta_{ij} + R_\Lambda(k). \tag{12.458}$$

On general ground, we know that $G_{ij,\,kl}^{-1} \sim \delta_{ij,kl}$, and thus

$$G_{ij,\,ij,\,\Lambda}(\bar{\phi}, k) = \frac{1}{\left(U'_{\Lambda,\,i} + 2\rho_i U''_{\Lambda,\,i} + c_0 k^2 Z_{\Lambda,\,i}^{-1}\right)\delta_{ij} + R_\Lambda(k)}. \tag{12.459}$$

12.7 Polchinski's renormalization of ϕ^4 theory

One of the most fundamental applications of the functional renormalization group equation approach is the efficient proof of the renormalizability of phi-four theory in four dimensions given by Polchinski in [2]. Instead of reviewing this result here we will review in the next chapter the proof of the renormalizability of a more general phi-four theory in four dimensions, which is a theory defined on noncommutative Moyal–Weyl space. This will serve the purpose of our main application of the functional renormalization group equation approach in this book as well as allows us the opportunity to introduce noncommutative field theory.

12.8 Exercises

Exercise 1: Consider the Landau–Wilson energy functional in a constant magnetic field

$$E[m] = \int d^d x \left[\frac{1}{2}(\vec{\partial} m(x))^2 + \frac{1}{2} r_0 m^2(x) + u_0 m^4(x) \right] + \int d^d x h m(x). \quad (12.460)$$

We will integrate over a small momentum shell characterized by $b \simeq 1$, i.e., over the spherical shell of radius Λ and thickness $\delta k = \tau \cdot \Lambda$ where $\tau = \ln b$. We expand the field as

$$m(x) = \bar{m}(x) + \delta m(x), \quad \delta m(x) = \sum_{i=1}^{N} c_i \phi_i(x). \quad (12.461)$$

The N wave packets (plane waves!!) $\phi_i(x)$ are of spatial size $\delta x = 1/\delta k$ and they are localized in the momentum shell δk. They are orthonormal and they also satisfy

$$\int d^d x (\phi_i(x))^{2n+1} = 0. \quad (12.462)$$

The background $\bar{m}$ is assumed to be constant over wave packets.

- Show that the RG-transformed energy functional (by integrating out the background at one-loop) is given by

$$E'[\bar{m}] = E[\bar{m}] + \frac{1}{2} \sum_{i=1}^{N} \log\left(1 + \frac{r_0}{\Lambda^2} + \frac{12 u_0}{\Lambda^2} \bar{m}^2(x_i) \right), \quad (12.463)$$

  where x_i is the center of mass of the wave packet i.
- We approximate the sum over x_i by an integral $\int dx$ and then we expand the logarithm. Show that we obtain

$$E'[\bar{m}] = \int d^d x \left[\frac{1}{2}(\vec{\partial} \bar{m}(x))^2 + \frac{1}{2} \tilde{r}_0 \bar{m}^2(x) + \tilde{u}_0 \bar{m}^4(x) \right] + \int d^d x h \bar{m}(x). \quad (12.464)$$

Determine the values of $\tilde{r}_0$ and $\tilde{u}_0$ in terms of r_0 and u_0.

- By performing the second step (rescaling) and the third step (normalization) of the renormalization group transformation show that we can express the RG transformation as

$$r_0' - r_0 = [2r_0 + 12C_d(\Lambda^{d-2}u_0 - \Lambda^{d-4}r_0u_0)]\ln b. \tag{12.465}$$

$$u_0' - u_0 = \left[(4-d)u_0 - 36C_d\Lambda^{d-4}u_0^2)\right]\ln b. \tag{12.466}$$

$$h' = b^{\frac{d+2}{2}}h. \tag{12.467}$$

- Determine the magnetic field critical exponent D_h.
- Show that the linearized renormalized group transformation can be put into the differential form

$$\frac{dr}{d\tau} = 2r + 12C_d(\Lambda^{d-2}u - \Lambda^{d-4}ru), \quad \frac{du}{d\tau} = (4-d)u - 36C_d\Lambda^{d-4}u^2. \tag{12.468}$$

References

[1] Wilson K G and Kogut J B 1974 The renormalization group and the epsilon expansion *Phys. Rep.* **12** 75
[2] Polchinski J 1984 Renormalization and effective Lagrangians *Nucl. Phys.* B **231** 269
[3] Wetterich C 1993 Exact evolution equation for the effective potential *Phys. Lett.* B **301** 90
[4] Kopietz P, Bartosch L and Schutz F 2010 Introduction to the functional renormalization group *Lect. Notes Phys.* **798** 1
[5] Fisher M E 1998 Renormalization group theory: its basis and formulation in statistical physics *Rev. Mod. Phys.* **70** 653
[6] Huang K 1987 *Statistical Mechanics* 2nd edn (New York: Wiley)
[7] Wilson K G 1971 Renormalization group and critical phenomena. 1. Renormalization group and the Kadanoff scaling picture *Phys. Rev.* B 3174
[8] Wilson K G 1971 Renormalization group and critical phenomena. 2. Phase space cell analysis of critical behavior *Phys. Rev.* B 4 3184
[9] Kadanoff L P 1966 Scaling laws for Ising models near T(c) *Physics Physique Fizika* **2** 263
[10] Kadanoff L P *et al* 1967 Static phenomena near critical points: theory and experiment *Rev. Mod. Phys.* **39** 395
[11] Wilson K G 1972 Feynman graph expansion for critical exponents *Phys. Rev. Lett.* **28** 548
[12] Golner G R 1973 Calculation of the critical exponent eta via renormalization-group recursion formulas *Phys. Rev.* B **8** 339
[13] Golner G R 1973 Wave-function renormalization of a scalar field theory in strong coupling *Phys. Rev.* D **8** 3393
[14] Morris T R 1994 The exact renormalization group and approximate solutions *Int. J. Mod. Phys.* A **9** 2411

[15] Litim D F 2001 Optimized renormalization group flows *Phys. Rev.* D **64** 105007

[16] Latorre J I and Morris T R 2000 Exact scheme independence *JHEP* **0011** 004

[17] Berges J, Tetradis N and Wetterich C 2002 Nonperturbative renormalization flow in quantum field theory and statistical physics *Phys. Rep.* **363** 223 [hep-ph/0005122]

[18] Bagnuls C and Bervillier C 2001 Exact renormalization group equations. An Introductory review *Phys. Rep.* **348** 91

A Modern Course in Quantum Field Theory, Volume 2
(Second Edition)
Advanced topics
Badis Ydri

Chapter 13

Some exact solutions of quantum field theory

The non-perturbative physics of a quantum field theory (as we have seen) can only be probed by means of Monte Carlo methods on lattices (which can become quite intricate technically and numerically) and/or by means of the exact renormalization group equation (which is always quite intricate analytically and mathematically). But sometimes exact solutions of the quantum field theory model presents themselves (lower dimensions and/or high degree of symmetries), which allow us to access the sought-after non-perturbative physics of the theory directly. In this chapter we present as examples six models in dimension two which all enjoy exact solutions allowing us an unprecedented look at the true heart, i.e., the non-perturbative reality, of a quantum field theory.

13.1 The linear sigma model and Hartree–Fock approximation

A good reference here is [1].

13.1.1 The mean field theory

We are always interested in the physics near the critical temperature of a second-order phase transition. Let us then consider the action and the partition function

$$S = \int d^d x \left[\frac{1}{2}\phi(-\partial_\mu^2 + r)\phi + u\phi^4 \right]. \tag{13.1}$$

$$Z[J] = \int \mathcal{D}\phi \, e^{-\beta(S - \int d^d x J\phi)}, \quad \beta = \frac{1}{T}. \tag{13.2}$$

The action S is the energy associated with low-energy (long-wavelength) modes with slow spatial variations. It does not describe high-energy (short-wavelength) distortions. Since the physics near the critical temperature of a second-order phase

doi:10.1088/978-0-7503-5834-7ch13   13-1

transition is controlled by long-wavelength fluctuations the action S will be adequate provided we suppress high-energy modes using a cut-off Λ.

The equation of motion is

$$-\partial_\mu^2 \phi_c + r\phi_c + 4u\phi_c^3 = J. \tag{13.3}$$

For spatially uniform configurations we get

$$r\phi_c + 4u\phi_c^3 = J. \tag{13.4}$$

The mass parameter r is related to the temperature by

$$r = a(T - T_*). \tag{13.5}$$

Thus for $J = 0$ we have the solutions

$$\phi_c = 0, \quad T > T_*. \tag{13.6}$$

$$\phi_c = \pm\sqrt{-\frac{r}{4u}}, \quad T < T_*. \tag{13.7}$$

We have a second-order phase transition at $T = T_*$. Both signs have the same free energy. The sign can be fixed by allowing J to go to 0 from above or from below. The up and down phases coexist along the line $J = 0$ and $T < T_*$.

We get the first critical exponent from the behavior of the order parameter at $T = T_*$ given by

$$\langle \phi \rangle = \phi_c \sim (T_* - T)^\beta, \quad \beta = \frac{1}{2}, \quad T < T_*. \tag{13.8}$$

We get the second critical exponent from differentiating the equation of motion with respect to J. We obtain

$$(r + 12u\phi_c^2)\frac{\delta\phi_c}{\delta J} = 1. \tag{13.9}$$

$$\chi = \frac{\delta\phi_c}{\delta J} = \frac{1}{r + 12u\phi_c^2}. \tag{13.10}$$

We compute

$$\chi = \frac{1}{r}, \quad T > T_*$$
$$= \frac{1}{2|r|}, \quad T < T_*. \tag{13.11}$$

In other words,

$$\chi \sim |T - T_*|^\gamma, \quad \gamma = 1. \tag{13.12}$$

The free energy is

$$f = 0, \quad T > T_*$$
$$= -\frac{r^2}{16u}, \quad T < T_*. \tag{13.13}$$

The specific heat

$$c_v = -T\frac{\partial^2 f}{\partial T^2} = 0, \quad T > T_*$$
$$= \frac{a^2}{8u}T, \quad T < T_*. \tag{13.14}$$

By writing the equation of the motion at $T = T_*$ we obtain

$$\phi_c = \left(\frac{J}{4u}\right)^{\frac{1}{\delta}}, \quad \delta = 3. \tag{13.15}$$

In order to find the correlation length we re-introduce the gradient term. In this case the susceptibility is given by

$$\chi(x, x') = \frac{\delta^2 S}{\delta\phi_c(x)\delta\phi_c(x')} = (-\partial_\mu^2 + r + 12u\phi_c^3)\delta^d(x - x'). \tag{13.16}$$

Since ϕ_c is constant we get

$$\chi(p) = \frac{1}{p^2 + r + 12u\phi_c^2}. \tag{13.17}$$

We compute

$$\chi(p) = \frac{\chi(0)}{1 + (p^2\xi^2)}. \tag{13.18}$$

$$\xi = \frac{1}{\sqrt{r + 12u\phi_c^2}} = \frac{1}{\sqrt{r}}, \quad T > T_*$$
$$= \frac{1}{\sqrt{-2r}}, \quad T < T_*. \tag{13.19}$$

Hence

$$\xi \sim |T - T_*|^{-\nu}, \quad \nu = \frac{1}{2}. \tag{13.20}$$

13.1.2 The 1-loop effective action

The fluctuation ψ is defined by

$$\phi = \phi_c + \psi. \tag{13.21}$$

We compute

$$S[\phi] = S[\phi_c] + \int d^dx J\psi + \frac{1}{2\beta} \int d^dx d^dx' \psi(x) G_0^{-1}(x, x')\psi(x') + O(\psi^3). \quad (13.22)$$

$$G_0^{-1}(x, x') = \beta(-\partial_\mu^2 + r + 12u\phi_c^2)\delta^d(x - x') = \frac{\delta J(x)}{\delta\phi_c(x')} = \frac{\delta^2 \beta S}{\delta\phi(x)\delta\phi(x')} \Big|_{\phi=\phi_c}. \quad (13.23)$$

In the disordered phase $\phi_c = 0$. We get

$$G_0^{-1}(x, x') = \beta \int \frac{d^dp}{(2\pi)^d}(p^2 + r)e^{ip(x-x')}$$
$$G_0(x, x') = \frac{1}{\beta} \int \frac{d^dp}{(2\pi)^d}\frac{1}{p^2 + r}e^{ip(x-x')}. \quad (13.24)$$

The classical theory (tree-level) is the mean field theory (saddle point). In this case

$$Z_{MF}[J] = e^{-\beta(S[\phi_c]-\int d^dx J\phi_c)}. \quad (13.25)$$

At one-loop order we have

$$Z[J] = e^{-\beta(S[\phi_c]-\int d^dx J\phi_c)} \int \mathcal{D}\psi e^{-\frac{1}{2}\int d^dx d^dx' \psi(x)G_0^{-1}(x, x')\psi(x')}$$
$$= e^{-\beta(S[\phi_c]-\int d^dx J\phi_c)}\left(\det G_0^{-1}\right)^{-\frac{1}{2}}. \quad (13.26)$$

The vacuum energy (free energy) is defined by

$$Z[J] = e^{-W[J]}. \quad (13.27)$$

We get the free energy

$$W[J] = S[\phi_c] - \int d^dx J\phi_c + \frac{1}{2\beta}Tr \ln G_0^{-1}(\phi_c). \quad (13.28)$$

We define the effective action as usual by the Legendre transform, viz

$$\Gamma[\phi_c] = S[\phi_c] + S_1[\phi_c], \quad S_1[\phi_c] = \frac{1}{2\beta}Tr \ln G_0^{-1}(\phi_c). \quad (13.29)$$

The classical field Φ_c is defined by

$$\Phi_c = -\frac{\delta W[J]}{\delta J}. \quad (13.30)$$

This is actually different from the classical solution ϕ_c.

Indeed, we compute

$$
\begin{aligned}
\Phi_c &= -\frac{\delta W[J]}{\delta J} \\
&= \phi_c - \frac{\delta \Gamma}{\delta J} + \int d^d x' J(x') \frac{\delta \phi_c(x')}{\delta J(x)} \\
&= \phi_c - \frac{\delta S_1}{\delta J} \\
&= \phi_c - \frac{1}{2\beta} \int d^d x_1 d^d x_2 d^d x_3 \frac{\delta \phi_c(x_3)}{\delta J(x)} G_0(x_1, x_2) \frac{\delta G_0^{-1}(x_1, x_2)}{\delta \phi_c(x_3)} \\
&= \phi_c - 12 u G_0(x_2, x_2) \int d^d x_1 G_0(x, x_1)\phi_c(x_1) \\
&= \phi_c + \delta\phi_c.
\end{aligned}
\tag{13.31}
$$

In the disordered phase $\phi_c = \Phi_c = 0$. We can immediately show that up to 2-loop effects we have (where we use the fact that ϕ_c solves the classical equation of motion)

$$
W[J] = S[\Phi_c] - \int d^d x J \Phi_c + \frac{1}{2\beta} Tr \ln G_0^{-1}(\Phi_c).
\tag{13.32}
$$

The effective action as we said is the Legendre transform of $W[J]$:

$$
\Gamma[\Phi_c] = W[J] + \int d^d x J \Phi_c = S[\Phi_c] + \frac{1}{2\beta} Tr \ln G_0^{-1}(\Phi_c).
\tag{13.33}
$$

This satisfies

$$
\frac{\delta \Gamma}{\delta \Phi_c} = J.
\tag{13.34}
$$

The 1-point function is

$$
\langle \phi(x) \rangle = \frac{1}{\beta}\frac{1}{Z}\frac{\delta Z}{\delta J(x)} = -\frac{\delta W}{\delta J} = \Phi_c.
\tag{13.35}
$$

The 2-point function is

$$
\langle \phi(x_1)\phi(x_2) \rangle = \frac{1}{\beta^2}\frac{1}{Z}\frac{\delta^2 Z}{\delta J(x_1)\delta J(x_2)} = G(x_1, x_2) + \langle \phi(x_1) \rangle \langle \phi(x_2) \rangle.
\tag{13.36}
$$

The connected 2-point function is defined by

$$
G(x_1, x_2) = -\frac{1}{\beta}\frac{\delta^2 W}{\delta J(x_1) J(x_2)}.
\tag{13.37}
$$

The 2-point proper vertex is defined by

$$
\Gamma(x_1, x_2) = \beta \frac{\delta^2 \Gamma}{\delta \Phi_c(x_1)\Phi_c(x_2)} = G^{-1}(x_1, x_2).
\tag{13.38}
$$

We write the effective action as

$$\Gamma[\phi] = S[\phi] + \frac{1}{2\beta} Tr \ln G_0^{-1}(\phi).$$

(13.39)

We compute

$$\frac{\delta^2 \Gamma}{\delta\phi(x)\delta\phi(x')} \Big|_{\phi=\Phi_c} = \frac{1}{\beta} G^{-1}(x, x')$$

$$= \frac{1}{\beta} G_0^{-1}(x, x') + \frac{1}{2\beta} Tr G_0 \frac{\delta^2 G_0^{-1}}{\delta\phi(x)\delta\phi(x')} \Big|_{\phi=\Phi_c} + \frac{1}{2\beta} Tr \frac{\delta^2 G_0}{\delta\phi(x')} \Big|_{\phi=\Phi_c} \frac{\delta^2 G_0^{-1}}{\delta\phi(x)} \Big|_{\phi=\Phi_c}.$$

(13.40)

In the disordered phase $\Phi_c = 0$. The 2-point proper vertex is

$$G^{-1}(x, x') = G_0^{-1}(x, x') + \frac{1}{2} Tr G_0 \frac{\delta^2 G_0^{-1}}{\delta\phi(x)\delta\phi(x')} \Big|_{\phi=0}$$

$$= G_0^{-1}(x, x') + \frac{1}{2} \int d^d x_1 d^d x_2 G_0(x_1, x_2) \frac{\delta^2 G_0^{-1}(x_1, x_2)}{\delta\phi(x)\delta\phi(x')} \Big|_{\phi=0}$$

(13.41)

$$= G_0^{-1}(x, x') + 12\beta u \delta^d(x - x') G_0(x, x').$$

In momentum space

$$G^{-1}(x, x') = \int \frac{d^d p}{(2\pi)^d} G^{-1}(p) e^{ip(x-x')}.$$

(13.42)

We then get

$$TG^{-1}(p) = p^2 + r + 12uT \int \frac{d^d k}{(2\pi)^d} \frac{1}{k^2 + r}.$$

(13.43)

Thus the connected 2-point function is defined by

$$G(x, x) = \langle \phi(x)^2 \rangle_{\text{connected}} = \int \frac{d^d p}{(2\pi)^d} G(p)$$

$$= \int \frac{d^d p}{(2\pi)^d} \frac{T}{p^2 + r + 12uG_0(x, x)}.$$

(13.44)

The free connected 2-point function is

$$G_0(x, x) = \int \frac{d^d p}{(2\pi)^d} \frac{T}{p^2 + r}.$$

(13.45)

13.1.3 The self-consistent approximation and critical exponents

The self-consistent Hartree–Fock approximation consists in making the replacement $G_0 \longrightarrow G$. We get the equation

$$G(x, x) = \int \frac{d^d p}{(2\pi)^d} \frac{T}{p^2 + r + 12uG(x, x)}.$$

(13.46)

In this case

$$TG^{-1}(p) = p^2 + r + 12u \int \frac{d^dk}{(2\pi)^d} \frac{T}{k^2 + r + 12uG(x,x)}. \tag{13.47}$$

The susceptibility is defined by

$$\begin{aligned}
\chi^{-1} = TG^{-1}(0) &= r + 12u \int \frac{d^dk}{(2\pi)^d} \frac{T}{k^2 + r + 12uG(x,x)} \\
&= r + 12u \int \frac{d^dk}{(2\pi)^d} \frac{T}{k^2 + \chi^{-1}}.
\end{aligned} \tag{13.48}$$

We introduce the renormalized mass $\chi^{-1} = \tau$. Thus

$$\tau = r + 12u \int \frac{d^dk}{(2\pi)^d} \frac{T}{k^2 + \tau}. \tag{13.49}$$

The integral is restricted to a sphere of radius Λ of the order of the inverse correlation length. We get (with $K_d = \Omega_d/(2\pi)^d$)

$$\tau = r + 12uTK_d \int_0^\Lambda \frac{k^{d-1}dk}{k^2 + \tau}. \tag{13.50}$$

At the critical temperature $T = T_c$ the susceptibility diverges, i.e. $\tau = 0$. We get the critical mass parameter

$$r_c = -12uT_cK_d \int_0^\Lambda \frac{k^{d-1}dk}{k^2} = -\frac{12T_cK_d\Lambda^{d-2}}{d-2}u. \tag{13.51}$$

However $r_c = a(T_c - T_*)$ where the critical temperature T_* is the critical value obtained in mean field theory. Hence

$$T_c = \frac{T_*}{1 + \dfrac{12K_d\Lambda^{d-2}u}{a(d-2)}}. \tag{13.52}$$

The critical value T_c is less than the mean field value T_*. For $d \longrightarrow 2$, $T_c \longrightarrow 0$. In other words, there is no transition in $d < 2$. In this case the critical fluctuations are so violent that no transition is possible.

We compute now

$$\begin{aligned}
\chi^{-1} &= r - r_c + 12uK_d \int k^{d-1}dk \left[\frac{T}{k^2 + \tau} - \frac{T_c}{k^2} \right] \\
&= r - r_c - 12uK_d T\tau I_d(\tau) + 12uK_d(T - T_c) \int k^{d-1}dk \frac{1}{k^2} \\
&= \left(1 + \frac{12uK_d}{a} \frac{\Lambda^{d-2}}{d-2} \right)(r - r_c) - 12uK_d T_c\tau I_d(\tau) - 12uK_d(T - T_c)\tau I_d(\tau) \\
&= (a + 12uK_d \frac{\Lambda^{d-2}}{d-2})(T - T_c) - 12uK_d T_c\tau I_d(\tau) - 12uK_d(T - T_c)\tau I_d(\tau).
\end{aligned} \tag{13.53}$$

We consider first the case $d > 4$. Since $T \longrightarrow T_c$, $\tau \longrightarrow 0$ and therefore

$$
\begin{aligned}
I_d(\tau) &= \int \frac{k^{d-3}dk}{k^2 + \tau} \\
&= \int \frac{k^{d-3}dk}{k^2} \\
&= \frac{\Lambda^{d-4}}{d - 4}.
\end{aligned}
\tag{13.54}
$$

Thus

$$
\chi^{-1} = (a + 12uK_d\frac{\Lambda^{d-2}}{d - 2})(T - T_c) - 12uK_d T_c\frac{1}{\chi}I_d(0) - 12uK_d(T - T_c)\frac{1}{\chi}I_d(0).
\tag{13.55}
$$

The last term can be neglected since it is of order $(T - T_c)^2$. We get then

$$
\chi^{-1} = (a + 12uK_d\frac{\Lambda^{d-2}}{d - 2})(T - T_c) - 12uK_d T_c\frac{1}{\chi}I_d(0).
\tag{13.56}
$$

Equivalently we get

$$
\chi = \frac{1 + 12uK_d T_c I_d(0)}{\left(a + 12uK_d\frac{\Lambda^{d-2}}{d - 2}\right)(T - T_c)}.
\tag{13.57}
$$

The critical exponent is $\gamma = 1$, which is the same as in the mean field theory.

For $d < 4$ the integral $I_d(\tau)$ is divergent at $\tau = 0$. We have (with $\epsilon = 4 - d$)

$$
\begin{aligned}
I_d(\tau) &= \int \frac{k^{d-3}dk}{k^2 + \tau} \\
&= \tau^{\frac{d-4}{2}} \int_0^{\frac{\Lambda}{\sqrt{\tau}}} \frac{x^{d-3}dx}{x^2 + 1} \\
&= \tau^{-\frac{\epsilon}{2}} B_d\left(\frac{\Lambda}{\sqrt{\tau}}\right) \\
&= \tau^{-\frac{\epsilon}{2}}\frac{1}{2}\Gamma\left(\frac{d - 2}{2}\right)\Gamma\left(\frac{\epsilon}{2}\right), \quad \tau \longrightarrow 0.
\end{aligned}
\tag{13.58}
$$

The inverse susceptibility

$$
\chi^{-1} = \tau = \left(a + 12uK_d\frac{\Lambda^{d-2}}{d - 2}\right)(T - T_c) - 12uK_d T_c\tau I_d(\tau).
\tag{13.59}
$$

For temperatures far from the critical temperature τ is large. Thus $\tau^{-\epsilon/2}$ is small and as a consequence we can neglect the second term in the above equation. The susceptibility becomes

$$\chi^{-1} = \left(a + 12uK_d \frac{\Lambda^{d-2}}{d-2} \right)(T - T_c).$$ (13.60)

In other words, $\chi \sim 1/(T - T_c)$, i.e., the critical exponent is the same as in the mean field theory.

For $d < 4$ and $T > T_G$ where T_G is the Ginsberg temperature we get mean field theory prediction. The Ginsberg temperature is the temperature at which perturbation theory breaks down and fluctuations become important. It can be determined from the condition

$$(1 + 12uK_d T_c I_d(\tau))\tau = \left(a + 12uK_d \frac{\Lambda^{d-2}}{d-2} \right)(T - T_c).$$ (13.61)

The mean field behavior corresponds to $12uK_d T_c I_d(\tau) \ll 1$, which is equivalent to

$$\tau \gg (12uK_d T_c B_d)^{-\frac{2}{4-d}}.$$ (13.62)

In other words,

$$r - r_c \gg (12uK_d T_c B_d)^{-\frac{2}{4-d}} \left(1 + \frac{12uK_d}{a} \frac{\Lambda^{d-2}}{d-2} \right)^{-1}.$$ (13.63)

$$T \gg T_G = T_c + \frac{1}{a}(12uK_d T_c B_d)^{-\frac{2}{4-d}} \left(1 + \frac{12uK_d}{a} \frac{\Lambda^{d-2}}{d-2} \right)^{-1}.$$ (13.64)

For $d < 4$ and $T < T_G$ fluctuations dominate and thus $\tau \longrightarrow 0$ and we get

$$0 = \left(a + 12uK_d \frac{\Lambda^{d-2}}{d-2} \right)(T - T_c) - 12uK_d T_c \tau \tau^{-\frac{\epsilon}{2}} B_d.$$ (13.65)

Equivalently

$$\tau^{-1} = \chi \sim (r - r_c)^{-\frac{2}{d-2}}.$$ (13.66)

The critical exponent is

$$\gamma = \frac{2}{d-2}.$$ (13.67)

Let us write the main result one more time as

$$\chi^{-1}(p) = p^2 + r + 12uG(x, x).$$ (13.68)

There is no p-dependent corrections to the mean-field result $p^2 + r$. We then compute

$$\frac{\chi(p)}{\chi(0)} = \frac{1}{1 + \dfrac{p^2}{r + 12uG(x, x)}}.$$ (13.69)

The correlation length ξ is defined by

$$\frac{\chi(p)}{\chi(0)} = \frac{1}{1 + (p\xi)^2}. \tag{13.70}$$

We immediately obtain

$$\xi^2 = \frac{1}{r + 12uG(x, x)} = \frac{1}{\tau} \sim (T - T_c)^{-2\nu}. \tag{13.71}$$

We get

$$\nu = \frac{\gamma}{2} = \frac{1}{d - 2}. \tag{13.72}$$

For $d \longrightarrow 4$ we get $\nu = 1/2$, which is the mean-field result. The mean field theory becomes exact when $d \longrightarrow \infty$. In fact mean field theory describes well second-order phase transitions only for $d > 4$.

13.1.4 The large N saddle point

Let us consider now

$$S = \int d^d x \left[\frac{1}{2}\phi_a(-\partial_\mu^2 + r)\phi_a + u(\phi_a^2)^2 \right]. \tag{13.73}$$

$$Z[J] = \int \mathcal{D}\phi_a e^{-\beta(S - \int d^d x J \phi_a)}. \tag{13.74}$$

We use the Hubbard–Stratonovich transformation given by

$$e^{-\beta u(\phi_a^2)^2} = c \int_{-i\infty}^{i\infty} \mathcal{D}\psi e^{\frac{\beta}{16u}\psi^2 - \frac{\beta}{2}\psi\phi_a^2}. \tag{13.75}$$

The contour of integration is along the imaginary rather than the real axis because the Gaussian integration over ψ has the wrong sign. We compute

$$\begin{aligned}
Z[0] &= \int \mathcal{D}\psi e^{\int d^d x \frac{\beta}{16u}\psi^2} \left[\int \mathcal{D}\phi e^{-\frac{\beta}{2}\int d^d x d^d x' \phi(x)G^{-1}(x, x')\phi(x')} \right]^N \\
&= \int \mathcal{D}\psi e^{\int d^d x \frac{\beta}{16u}\psi^2 - \frac{N}{2}Tr \ln \beta G^{-1}(x, x')}.
\end{aligned} \tag{13.76}$$

$$G^{-1}(x, x') = \left(-\partial_\mu^2 + r + \psi\right)\delta^d(x - x'). \tag{13.77}$$

We take the limit

$$N \longrightarrow \infty, \quad u \longrightarrow 0, \quad g = uN = \text{fixed}. \tag{13.78}$$

We get

$$Z[0] = \int \mathcal{D}\psi e^{N\left[\int d^d x \frac{\beta}{16g}\psi^2 - \frac{1}{2}Tr \ln \beta G^{-1}(x, x')\right]}. \tag{13.79}$$

The integral is dominated by the saddle point solution. The saddle point equation is

$$\frac{\beta}{8g}\psi = \frac{1}{2}TrG\frac{\delta}{\delta\psi}G^{-1}$$
$$= \frac{1}{2}G(x,x). \tag{13.80}$$

Since G is independent of x the configuration ψ is independent of x. Thus

$$\psi = 4gT\int\frac{d^dp}{(2\pi)^d}\frac{1}{p^2+r+\psi}. \tag{13.81}$$

13.2 The non-linear sigma model and the $1/N$ expansion

In this section we follow the presentation of [2] and [3].

13.2.1 The $d=2$ non-linear sigma model

We consider the following action in $d=2$ Minkowskian spacetime

$$S = \int d^2x\mathcal{L}, \quad \mathcal{L} = \frac{1}{2g^2}(\partial_\mu\vec{n})^2, \tag{13.82}$$

where the $O(n)$-vector $\vec{n}$ satisfies also the constraint

$$\vec{n}^2 = \sum_{i=1}^{N}n_i^2 = 1. \tag{13.83}$$

The coupling constant g and the field $\vec{n}$ are both dimensionless. In a perturbative quantization of this model we implement the constraint by writing

$$\vec{n} = (\pi_1, \ldots, \pi_{N-1}, \sigma), \quad \sigma = \sqrt{1-\vec{\pi}^2}, \tag{13.84}$$

and thus

$$(\partial_\mu\vec{n})^2 = (\partial_\mu\vec{\pi})^2 + (\partial_\mu\sigma)^2 = (\partial_\mu\vec{\pi})^2 + \frac{(\vec{\pi}.\,\partial_\mu\vec{\pi})^2}{1-\vec{\pi}^2}. \tag{13.85}$$

The action up to the fourth power in the π-field becomes

$$\mathcal{L} = \frac{1}{2g^2}(\partial_\mu\vec{\pi})^2 + \frac{1}{2g^2}\frac{(\vec{\pi}\cdot\partial_\mu\vec{\pi})^2}{1-\vec{\pi}^2}$$
$$= \frac{1}{2g^2}(\partial_\mu\vec{\pi})^2 + \frac{1}{2g^2}(\vec{\pi}.\,\partial_\mu\vec{\pi})^2 + \cdots \tag{13.86}$$

At this stage the constraint is already being taken into consideration. To derive Feynman rules we write

$$\vec{\pi} = \int\frac{d^2k}{(2\pi)^2}\vec{\pi}(k)e^{ikx} = \vec{\pi}^*, \quad \vec{\pi}(-k) = \vec{\pi}^*(k). \tag{13.87}$$

It is then not difficult to find that the first term in the action is

$$\frac{1}{2g^2}\int d^2x(\partial_\mu\vec{\pi})^2 = \frac{1}{2g^2}\int \frac{d^2k}{(2\pi)^2}k^2|\vec{\pi}(k)|^2 \equiv \frac{1}{2}\int \frac{d^2k}{(2\pi)^2}\frac{d^2p}{(2\pi)^2}\pi_i(k)\left[(2\pi)^2\delta^2(k+p)\left(\frac{k^2}{g^2}\delta_{ij}\right)\right]\pi_j(p). \quad (13.88)$$

The propagator $D_{jl}(p,q) = \langle \pi_j(p)\pi_l(q)\rangle$ should satisfy

$$\int \frac{d^2p}{(2\pi)^2}\sum_j\left[(2\pi)^2\delta^2(k+p)\left(\frac{k^2}{g^2}\delta_{ij}\right)\right]D_{jl}(p,q) = i\delta_{il}(2\pi)^2\delta^2(k-q). \quad (13.89)$$

Hence

$$\langle \pi_i(k)\pi_j(p)\rangle = \frac{ig^2}{k^2}\delta_{ij}(2\pi)^2\delta^2(k+p). \quad (13.90)$$

The interaction is, on the other hand, computed to be given by

$$\frac{1}{2g^2}\int d^2x(\vec{\pi}\partial_\mu\vec{\pi})^2 = \frac{1}{2g^2}\int \frac{d^2k_1}{(2\pi)^2}\cdots\frac{d^2k_4}{(2\pi)^2}(2\pi)^2\delta^2(k_1+\cdots+k_4)\pi_{i_1}(k_1)\ldots\pi_{i_4}(k_4)$$
$$\times(-\delta_{i_1i_2}\delta_{i_3i_4}k_2k_4)$$
$$-\frac{1}{4!g^2}\int \frac{d^2k_1}{(2\pi)^2}\frac{d^2k_4}{(2\pi)^2}(2\pi)^2\delta^2(k_1+\cdots+k_4)\pi_{i_1}(k_1)\ldots\pi_{i_4}(k_4) \quad (13.91)$$
$$\times(\delta_{i_1i_2}\delta_{i_3i_4}(k_1+k_2)(k_3+k_4) + \delta_{i_1i_3}\delta_{i_2i_4}(k_1+k_3)(k_2+k_4)$$
$$+\delta_{i_1i_4}\delta_{i_2i_3}(k_1+k_4)(k_2+k_3)),$$

where we have symmetrized with respect to 1 and 2 then with respect to 3 and 4 and finally with respect to 1, 3 and 1, 4, respectively. The vertex is therefore given by

$$\langle \pi_{i_1}(k_1)\ldots\pi_{i_4}(k_4)\rangle = -\frac{i}{g^2}(\delta_{i_1i_2}\delta_{i_3i_4}(k_1+k_2)(k_3+k_4) + \delta_{i_1i_3}\delta_{i_2i_4}(k_1+k_3)(k_2+k_4)$$
$$+\delta_{i_1i_4}\delta_{i_2i_3}(k_1+k_4)(k_2+k_3))(2\pi)^2\delta^2(k_1+\cdots+k_4). \quad (13.92)$$

The i comes from the exponent e^{iS} in the path integral while the 4! cancels because of the identical contributions from the 4! possible contractions, i.e., when we differentiate the action 4 times with respect to the field.

If one tries however to compute with these Feynman rules we will find that there is an IR divergence coming from the fact that the pion in the action (13.82) is massless. To remove this divergence one can couple the system to a constant external magnetic field h as follows [2]

$$\mathcal{L}_h = \frac{\mu^2}{g^2}\sigma = \frac{\mu^2}{g^2}\left(1 - \frac{\vec{\pi}^2}{2} + \frac{(\vec{\pi}^2)^2}{8} + \cdots\right), \quad (13.93)$$

where as before we have expanded the Lagrangian upto the 4th power in the π-field. We have also defined $\mu^2 \equiv h$ to indicate that the constant magnetic field will act as a mass for the pion and thus regulate the zero momentum singularity. We compute now the action

$$
\begin{aligned}
S_h &= -\frac{\mu^2}{2g^2} \int d^2x\,\vec{\pi}^2 + \frac{\mu^2}{8g^2} \int d^2x (\vec{\pi}^2)^2 + \cdots \\
&= \frac{1}{2} \int \frac{d^2k}{(2\pi)^2} \frac{d^2p}{(2\pi)^2} \pi_i(k) \left[(2\pi)^2 \delta^2(k+p) \left(\frac{-\mu^2}{g^2} \delta_{ij} \right) \right] \pi_j(p) \\
&\quad - \frac{1}{4!g^2} \int \frac{d^2k_1}{(2\pi)^2} \cdots \frac{d^2k_4}{(2\pi)^2} (2\pi)^2 \delta^2(k_1 + \cdots + k_4)\pi_{i_1}(k_1) \ldots \pi_{i_4}(k_4)\left(\delta_{i_1 i_2}\delta_{i_3 i_4} + \delta_{i_1 i_3}\delta_{i_2 i_4} \right. \\
&\quad \left. + \delta_{i_1 i_4}\delta_{i_2 i_3} \right)(-\mu^2).
\end{aligned}
\tag{13.94}
$$

Remark lastly that the Lagrangian $\mathcal{L} + \mathcal{L}_h$ has only $O(N-1)$ symmetry and not the full $O(N)$ symmetry, which is consistent with the fact that this model describes essentially the spontaneous symmetry breaking $O(N) \longrightarrow O(N-1)$.

The action we will study is therefore

$$
S = \frac{1}{2g^2} \int d^2x [(\partial_\mu \vec{\pi})^2 + (\partial_\mu \sigma)^2] + \frac{h}{g^2} \int d^2x\,\sigma.
\tag{13.95}
$$

It is obvious that the first term in this action (i.e., the action (13.82)) is the most general $O(N)$-symmetric Lagrangian, which can be built out of the field $\vec{n}$ with constraint $\vec{n}^2 = 1$ and with dimensionless coupling g. In other words, this term is renormalizable and thus the theory can be completely made finite by renormalizing the coupling constant g and the field strength $|\vec{n}|$ in an $O(N)$-invariant fashion. The origin of the second term is in fact $\vec{h} \cdot \vec{n}$, which is important both to remove the IR divergence of theory as we have just said and also to select a minimum for the model [2]. This term is therefore also $O(N)$-symmetric and the symmetry gets only broken to $O(N-1)$ when the $O(N)$-magnetic field $\vec{h}$ is oriented in the N direction, $\vec{h} = (0, \ldots, 0, h)$. The g, $\vec{\pi}$, σ are all dimensionless and h is of dimension mass squared. If one tries to compute Green's functions using the above action we will find that the theory is UV divergent and thus we need to regularize. We choose in here dimensional regularization (DR) in which the action becomes

$$
S_0 = \frac{\Lambda^{d-2}}{2g_0^2} \int d^dx [(\partial_\mu \vec{\pi}_0)^2 + (\partial_\mu \sigma_0)^2] + \frac{h_0 \Lambda^{d-2}}{g_0^2} \int d^dx\,\sigma_0.
\tag{13.96}
$$

The g_0, $\vec{\pi}_0$, σ_0 are still dimensionless and also h_0 is still of dimension mass squared. The 0 stands for bare quantities and Λ^{d-2} is a mass parameter. Now if one computes anything with this action we will find that it depends on the bare coupling constants g_0 and h_0 as well as on dimensional regularization 'cutoff' which we can define by the pair $(\Lambda, \epsilon = 2 - d)$. Hence the bare Green's functions as opposed to the renormalized correlations are 'cutoff'-dependent, which is quite natural.

As an example we compute the 2-point function of the pion field in the one-loop approximation, which can be easily found to be given by a tadpole diagram. The value of the loop is given by the expression

$$\mathcal{X} = \frac{1}{2} \int \frac{d^d k_3}{(2\pi)^d} \frac{d^d k_4}{(2\pi)^d} (2\pi)^d \delta^d(k_1 + k_2 + k_3 + k_4) \left(-\frac{i}{g^2}\right) \left[\delta_{i_1 i_2} \delta_{i_3 i_4} [(k_1 + k_2)(k_3 + k_4) - \mu^2] \right.$$

$$+ \delta_{i_1 i_3} \delta_{i_2 i_4} [(k_1 + k_3)(k_2 + k_4) - \mu^2] + \delta_{i_1 i_4} \delta_{i_2 i_3} [(k_1 + k_4)(k_2 + k_3) - \mu^2]]$$

$$\times \frac{ig^2}{k_3^2 - \mu^2} \delta_{i_3 i_4} (2\pi)^d \delta^d(k_3 + k_4) \tag{13.97}$$

$$= \delta_{i_1 i_2} (2\pi)^d \delta^d(k_1 + k_2) \int \frac{d^d k_3}{(2\pi)^d} \frac{1}{k_3^2 - \mu^2} \left[-k_1^2 - k_3^2 - \frac{1}{2}\mu^2(N + 1) \right].$$

The factor $\frac{1}{2}$ in front of the first line is a symmetry factor due to the fact that we have one line which starts and ends on the same vertex so the corresponding Feynman diagram is invariant under the interchange of the two ends of this line. Now using the results

$$\int \frac{d^d q}{(2\pi)^d} \frac{1}{q^2 - \mu^2} = -\frac{i}{(4\pi)^{d/2}} \Gamma\left(1 - \frac{d}{2}\right) \frac{1}{(\mu^2)^{1 - \frac{d}{2}}}, \quad \int \frac{d^d q}{(2\pi)^d} \frac{q^2}{q^2 - \mu^2} = \frac{i}{(4\pi)^{d/2}} \frac{d}{2} \Gamma\left(-\frac{d}{2}\right) (\mu^2)^{\frac{d}{2}}, \tag{13.98}$$

and $\Gamma(z + 1) = z\Gamma(z)$ we obtain

$$\mathcal{X} = \delta_{i_1 i_2} (2\pi)^d \delta^d(k_1 + k_2) \frac{ik_1^2}{(4\pi)^{d/2}} \Gamma\left(1 - \frac{d}{2}\right) \frac{1}{(\mu^2)^{1 - \frac{d}{2}}}$$

$$+ \delta_{i_1 i_2} (2\pi)^d \delta^d(k_1 + k_2) \frac{-i(N + 3)}{2(4\pi)^{d/2}} \frac{d}{2} \Gamma\left(-\frac{d}{2}\right) (\mu^2)^{\frac{d}{2}}. \tag{13.99}$$

In here we will assume for simplicity that the coupling constant $h \equiv \mu^2$ does not get renormalized and that its only purpose is to remove the IR divergence. This means that we take $\mu \longrightarrow 0$ wherever is possible and thus the second term in above can be simply neglected since it has no pole at $d = 0$ and for $d > 0$ this term is proportional to a positive power of μ^2. We have then

$$\mathcal{X} = \delta_{i_1 i_2} (2\pi)^d \delta^d(k_1 + k_2) \frac{ik_1^2}{(4\pi)^{d/2}} \Gamma\left(1 - \frac{d}{2}\right) \frac{1}{(\mu^2)^{1 - \frac{d}{2}}}. \tag{13.100}$$

The full 2-point function of the pion field is now given by (by including the amputated external propagators)

$$\langle \pi_i(k) \pi_j(p) \rangle_{1-\text{loop}} = \int \frac{d^d k_1}{(2\pi)^d} \frac{d^d k_2}{(2\pi)^d} \left(\frac{ig^2}{k^2 - \mu^2} \delta_{ii_1} (2\pi)^d \delta(k_1 - k) \right) \left(\delta_{i_1 i_2} (2\pi)^d \delta^d(k_1 + k_2) \right.$$

$$\times \frac{ik_1^2}{(4\pi)^{d/2}} \Gamma\left(1 - \frac{d}{2}\right) \frac{1}{(\mu^2)^{1 - \frac{d}{2}}} \left(\frac{ig^2}{p^2 - \mu^2} \delta_{ji_2} (2\pi)^d \delta(k_2 - p) \right) \tag{13.101}$$

$$= \frac{ig^2}{k^2 - \mu^2} \delta_{ij} (2\pi)^d \delta^d(k + p) \left(-\frac{g^2}{(4\pi)^{d/2}} \frac{k^2}{k^2 - \mu^2} \Gamma\left(1 - \frac{d}{2}\right) \frac{1}{(\mu^2)^{1 - \frac{d}{2}}} \right).$$

The one-loop 2-point function is therefore given by

$$\langle \pi_i(k)\pi_j(p)\rangle = \frac{ig^2}{k^2 - \mu^2}\delta_{ij}(2\pi)^d\delta^d(k+p)\left(1 - \frac{g^2}{(4\pi)^{d/2}}\frac{k^2}{k^2 - \mu^2}\Gamma\left(1 - \frac{d}{2}\right)\frac{1}{(\mu^2)^{1-\frac{d}{2}}}\right)$$
$$= \frac{ig^2}{k^2}\delta_{ij}(2\pi)^d\delta^d(k+p)\left(1 - \frac{g^2}{(4\pi)^{d/2}}\Gamma\left(1 - \frac{d}{2}\right)\frac{1}{(\mu^2)^{1-\frac{d}{2}}}\right),$$

$$(13.102)$$

where again we have set the IR cut-off μ equal to 0 wherever it is possible.

Now in dimensional regularization g^2 stands actually for $\frac{g_0^2}{\Lambda^{d-2}}$, h for h_0 (i.e, μ^2 stands for μ_0^2) and $\pi_i(x)$ for $\pi_{0i}(x)$. In momentum space however $\pi_i(k)$ stands for $\frac{1}{\Lambda^{d-2}}\pi_{i0}(k)$, i.e. $\pi_{i0}(k)$ is of dimension (mass)$^{-2}$, which is the same mass dimension of momentum modes in $d = 2$. Hence the bare 2-point correlation function is given by

$$\langle \pi_{0i}(k)\pi_{0j}(p)\rangle = \frac{ig_0^2}{k^2}\delta_{ij}\Lambda^{d-2}(2\pi)^d\delta^d(k+p)\left(1 - \frac{g_0^2}{(4\pi)^{d/2}}\Gamma\left(1 - \frac{d}{2}\right)\left(\frac{\Lambda^2}{\mu_0^2}\right)^{1-\frac{d}{2}}\right). \qquad (13.103)$$

$\Lambda^{2-d}(2\pi)^d\delta^d(k+p)$ has the same mass dimension as the two-dimensional momentum space delta function $(2\pi)^2\delta^2(k+p)$ and thus we simply identify them in the limit $d \longrightarrow 2$.

Obviously this 2-point function diverges at $d = 2$ (i.e., when we remove dimensional regularization) and thus we need to renormalize the coupling constant g_0 as well as scale the fields $\bar{\pi}_0$ and σ_0 in order to make this 'cut-off' dependent 2-point function finite in the limit $d \longrightarrow 2$. We introduce therefore the renormalized coupling constant g and the renormalized fields π_r and σ_r by

$$g_0^2 = Z_g g_r^2, \quad Z_g = 1 + g_r^2 \bar{Z}_g + \cdots$$
$$\pi_{0i} = \sqrt{Z}\pi_{ri}, \quad Z = 1 + g_r^2 \bar{Z} + \cdots \qquad (13.104)$$
$$\sigma_0 = \sqrt{Z}\sigma_r, \quad \vec{\pi}_r^2 + \sigma_r^2 = z^{-1}.$$

Recall that the expansion in the coupling constant g_0 is equivalent to the $\hbar$-expansion, i.e., to the loop-wise expansion [2]. The renormalization constants Z and Z_g are dimensionless and hence can only depend on the dimensionless coupling constant g_0 (or g_r), on the ratio $\frac{\Lambda}{\mu}$ and on ϵ. It is now not difficult to check that

$$\langle \pi_{ri}(k)\pi_{rj}(p)\rangle = \frac{ig_r^2}{k^2}\delta_{ij}\Lambda^{d-2}(2\pi)^d\delta^d(k+p)\left(1 + g_r^2\left[-\frac{1}{(4\pi)^{d/2}}\Gamma\left(1 - \frac{d}{2}\right)\left(\frac{\Lambda^2}{\mu_0^2}\right)^{1-\frac{d}{2}} - \bar{Z} + \bar{Z}_g\right]\right) + O(g_r^6). \quad (13.105)$$

To determine $\bar{Z}$ and $\bar{Z}_g$ we need to impose at least two renormalization conditions. We impose first the condition that at the scale $\mu_0^2 = M^2$ (the renormalization scale) the 2-point function looks like that of the free theory, namely

$$\langle \pi_{ri}(k)\pi_{rj}(p)\rangle = \frac{ig_r^2}{k^2}\delta_{ij}\Lambda^{d-2}(2\pi)^d\delta^d(k+p), \quad \text{at } \mu_0^2 = M^2. \tag{13.106}$$

This gives us the equation

$$-\frac{1}{(4\pi)^{d/2}}\Gamma\left(1-\frac{d}{2}\right)\left(\frac{\Lambda^2}{M^2}\right)^{1-\frac{d}{2}} - \bar{Z} + \bar{Z}_g = 0, \quad \text{at } \mu_0^2 = M^2. \tag{13.107}$$

Clearly we need one more equation to determine both $\bar{Z}$ and $\bar{Z}_g$ and thus one needs to calculate one more correlation function. We compute the expectation value $\langle\sigma\rangle$ of the field σ, which is essentially like an order parameter if one recalls the spontaneous symmetry breaking, which occurs in the model [3]. We have then

$$\langle\sigma(x)\rangle = \langle\sqrt{1-\vec{\pi}^2(x)}\,\rangle = 1 - \frac{1}{2}\langle\vec{\pi}^2(x)\rangle + \cdots$$

$$= 1 - \frac{1}{2}\int\frac{d^dk}{(2\pi)^d}\frac{d^dp}{(2\pi)^d}e^{i(k+p)x}\delta_{ij}\langle\pi_i(k)\pi_j(p)\rangle$$

$$= 1 - \frac{ig^2(N-1)}{2}\int\frac{d^dk}{(2\pi)^d}\frac{1}{k^2-\mu^2} \tag{13.108}$$

$$= 1 - \frac{g^2(N-1)}{2}\frac{1}{(4\pi)^{d/2}}\Gamma\left(1-\frac{d}{2}\right)\frac{1}{(\mu^2)^{1-\frac{d}{2}}}$$

$$= 1 - \frac{g_0^2(N-1)}{2}\frac{1}{(4\pi)^{d/2}}\Gamma\left(1-\frac{d}{2}\right)\left(\frac{\Lambda^2}{\mu_0^2}\right)^{1-\frac{d}{2}}.$$

Again the bare correlation function is cut-off dependent and therefore it needs renormalization. We get

$$\langle\sigma_r\rangle = 1 - \frac{g_r^2}{2}\left[\bar{Z} + \frac{(N-1)}{(4\pi)^{d/2}}\Gamma\left(1-\frac{d}{2}\right)\left(\frac{\Lambda^2}{\mu_0^2}\right)^{1-\frac{d}{2}}\right] + O(g_r^4). \tag{13.109}$$

We impose now the extra renormalization condition that at the renormalization scale $\mu_0^2 = M^2$ the order parameter $\langle\sigma_r\rangle$ is a unity, namely

$$\langle\sigma_r\rangle = 1, \quad \text{at } \mu_0^2 = M^2. \tag{13.110}$$

We get now the condition

$$\bar{Z} + \frac{(N-1)}{(4\pi)^{d/2}}\Gamma\left(1-\frac{d}{2}\right)\left(\frac{\Lambda^2}{\mu_0^2}\right)^{1-\frac{d}{2}} = 0, \quad \text{at } \mu_0^2 = M^2. \tag{13.111}$$

This is solved by

$$\bar{Z} = -\frac{(N-1)}{(4\pi)^{d/2}}\Gamma\left(1-\frac{d}{2}\right)\left(\frac{\Lambda^2}{M^2}\right)^{1-\frac{d}{2}} = -\frac{(N-1)}{(4\pi)^{d/2}}\left[\frac{2}{\epsilon} - \gamma + ln\frac{\Lambda^2}{M^2}\right], \tag{13.112}$$

and hence the order parameter becomes

$$\langle \sigma_r \rangle = 1 - \frac{g_r^2(N-1)}{8\pi} \ln \frac{M^2}{\mu_0^2} + O(g_r^4).$$

(13.113)

Similarly by putting the solution (13.112) in the renormalization condition (13.107) we obtain the other renormalization constant $\bar{Z}_g$ to be given by

$$\bar{Z}_g = -\frac{N-2}{(4\pi)^{d/2}}\Gamma\left(1 - \frac{d}{2}\right)\left(\frac{\Lambda^2}{M^2}\right)^{1-\frac{d}{2}}.$$

(13.114)

The 2-point function becomes now

$$\langle \pi_{ri}(k)\pi_{rj}(p)\rangle = \frac{ig_r^2}{k^2}\delta_{ij}(2\pi)^2\delta^2(k+p)\left(1 + \frac{g_r^2}{(4\pi)^{d/2}}\Gamma\left(1 - \frac{d}{2}\right)\left[\left(\frac{\Lambda^2}{M^2}\right)^{1-\frac{d}{2}} - \left(\frac{\Lambda^2}{\mu_0^2}\right)^{1-\frac{d}{2}}\right]\right)$$

$$= \frac{ig_r^2}{k^2}\delta_{ij}(2\pi)^2\delta^2(k+p)\left(1 - \frac{g_r^2}{4\pi}\ln\frac{M^2}{\mu_0^2}\right).$$

(13.115)

13.2.2 The Callan–Symanzik equation and the beta function

The bare action we worked with is

$$S_0 = \frac{\Lambda^{d-2}}{2g_0^2}\int d^dx[(\partial_\mu\vec{\pi}_0)^2 + (\partial_\mu\sigma_0)^2] + \frac{h_0\Lambda^{d-2}}{g_0^2}\int d^dx\sigma_0.$$

(13.116)

The renormalized quantities are defined by

$$g_0^2 = Z_g g_r^2, \quad h_0 = Z_h\frac{Z_g}{\sqrt{Z}}h_r$$

$$\pi_{0i} = \sqrt{Z}\pi_{ri}, \quad \sigma_0 = \sqrt{Z}\sigma_r, \quad \vec{\pi}_r^2 + \sigma_r^2 = z^{-1}.$$

(13.117)

Using these definitions in the above action we obtain

$$S_0 = \frac{\Lambda^{d-2}Z}{2g_r^2 Z_g}\int d^dx[(\partial_\mu\vec{\pi}_r)^2 + (\partial_\mu\sigma_r)^2] + \frac{h_r}{g_r^2}\Lambda^{d-2}Z_h\int d^dx\sigma_r.$$

(13.118)

Since the renormalized action does not depend on the cut-off Λ but rather on the renormalization scale M, we can define it simply by

$$S_r = \frac{M^{d-2}Z}{2g_r^2 Z_g}\int d^dx[(\partial_\mu\vec{\pi}_r)^2 + (\partial_\mu\sigma_r)^2] + \frac{h_r}{g_r^2}M^{d-2}\int d^dx\sigma_r.$$

(13.119)

Obviously the overall external constant magnetic field (the coefficient in front of the σ-field) should not get renormalized, i.e., it should be the same in the bare and in the renormalized theory and as a consequence we get the condition $Z_h\Lambda^{d-2} = M^{d-2}$. In fact, one can write this condition in the equivalent form

$$M^{d-2}\frac{h_r}{g_r^2} = \Lambda^{d-2}Z_h\frac{h_r}{g_r^2} \equiv \Lambda^{d-2}\sqrt{Z}\frac{h_0}{g_0^2}. \tag{13.120}$$

Thus the counter term is given by

$$\Delta S_{c.t} = \frac{(\Lambda^{d-2} - M^{d-2})Z}{2g_r^2 Z_g} \int d^d x [(\partial_\mu \vec{\pi}_r)^2 + (\partial_\mu \sigma_r)^2]. \tag{13.121}$$

In equation (13.120), h_0 and g_0 are the bare coupling constants and therefore they do not depend on the renormalization scale M whereas the renormalized coupling constants h_r and g_r as well as the renormalization constant Z do depend on M. This renormalization scale M is largely arbitrary and one would like to know how much does this arbitrariness affects observable physical quantities. To this end we write (13.120) in the form

$$M^{d-2}\frac{h_r}{g_r^2}Z^{-\frac{1}{2}} = \Lambda^{d-2}\frac{h_0}{g_0^2}. \tag{13.122}$$

Taking the logarithm of this equation and then the derivative with respect to the scale M we get

$$d - 2 + M\frac{\partial}{\partial M}\ln h_r - \frac{2}{g_r}M\frac{\partial}{\partial M}g_r + M\frac{\partial}{\partial M}\ln Z^{-\frac{1}{2}} = 0. \tag{13.123}$$

By defining

$$\beta = M\frac{\partial g_r}{\partial M}$$
$$\gamma = M\frac{\partial \ln Z^{-\frac{1}{2}}}{\partial M} \tag{13.124}$$
$$\rho = M\frac{\partial \ln h_r}{\partial M},$$

one can also put the above last result in the form

$$\rho = 2 - d + \frac{2}{g_r}\beta - \gamma. \tag{13.125}$$

The first equation in (13.124) is the beta function and its properties are best understood via the so-called Callan–Symanzik equation. In the renormalization conditions (13.106) and (13.110) the renormalization scale M (as we have just noticed) is completely arbitrary. However and as it turned out this arbitrariness is completely controlled by the beta (as well as the gamma and rho) functions in (13.124), which we now explain in some more detail. Indeed, any bare correlation function can only depend on spacetime points x_i (or equivalently momenta p_i if we were in monetum space), and on the bare coupling constants g_0 and h_0 and on the 'cut-off' (Λ, ϵ) (see for example (13.103) and (13.108)). We write then

$$G_0^{(n)}(x_1, x_2, \ldots, x_n; g_0, h_0, \Lambda, \epsilon) = \langle \phi_{01}(x_1)\phi_{02}(x_2) \ldots \phi_{0n}(x_n) \rangle, \tag{13.126}$$

where ϕ_{0i} stands for the fields π_{0i}'s and σ_0. Bare correlation functions are therefore 'cut-off' dependent, which is essentially their most important characteristic. In particular they do not depend on the renormalization scale M. Renormalized correlations are however 'cut-off' independent and M-dependent and they also depend on the spacetime points x_i and on the renormalized coupling constants g_r and h_r, i.e.,

$$G_r^{(n)}(x_1, x_2, \ldots, x_n; g_r, h_r, M) = \langle \phi_{r1}(x_1)\phi_{r2}(x_2) \ldots \phi_{rn}(x_n) \rangle. \tag{13.127}$$

Obviously by using $\phi_0 = Z^{\frac{1}{2}}\phi_r$ we have the relation

$$G_0^{(n)}(x_1, x_2, \ldots, x_n; g_0, h_0, \Lambda, \epsilon) = Z^{\frac{n}{2}}G_r^{(n)}(x_1, x_2, \ldots, x_n; g_r, h_r, M). \tag{13.128}$$

Since G_0 does not know about the renormalization scale M it must be invariant under any change $M \longrightarrow M'$, i.e., it is unaffected if one chooses to define the renormalized Green's functions at the scale M' instead of M with new renormalized coupling constants g_r', h_r' and a new renormalization constant Z'. This means that the renormalized Green's functions G_r (see for example (13.113) and (13.115)) change as $G_r' = G_r + \delta G_r$ when M changes as $M' = M + \delta M$ and correspondingly g_r and h_r change as $g_r' = g_r + \delta g_r$ and $h_r' = h_r + \delta h_r$ such that

$$\delta G_r^{(n)}(x_1, x_2, \ldots, x_n; g_r, h_r, M) = \delta M \frac{\partial}{\partial M} G_r^{(n)}(x_1, x_2, \ldots, x_n; g_r, h_r, M)$$

$$+ \delta g_r \frac{\partial}{\partial g_r} G_r^{(n)}(x_1, x_2, \ldots, x_n; g_r, h_r, M) \tag{13.129}$$

$$+ \delta h_r \frac{\partial}{\partial h_r} G_r^{(n)}(x_1, x_2, \ldots, x_n; g_r, h_r, M).$$

This change (13.129) in G_r will be balanced by a change in the field strength renormalization Z as follows $Z^{\frac{n}{2}}G_r^{(n)} = Z'^{\frac{n}{2}}G_r'^{(n)}$ so that the bare Green's functions G_0 remain fixed. For infinitesimal change $M \longrightarrow M' = M + \delta M$ we can write $Z' = Z(1 - 2\delta\eta)$ and thus the corresponding field strength shift is defined by

$$\phi_r' = (1 + \delta\eta)\phi_r. \tag{13.130}$$

Hence from equation (13.127) we get

$$G_r^{(n)'}(x_1, x_2, \ldots, x_n; g_r', h_r', M') = \langle \phi_{r1}'(x_1)\phi_{r2}'(x_2) \ldots \phi_{rn}'(x_n) \rangle$$
$$= (1 + n\delta\eta)G_r^{(n)}(x_1, x_2, \ldots, x_n; g_r, h_r, M), \tag{13.131}$$

or equivalently

$$\delta G_r^{(n)}(x_1, x_2, \ldots, x_n; g_r, h_r, M) = n\delta\eta G_r^{(n)}(x_1, x_2, \ldots, x_n; g_r, h_r, M). \tag{13.132}$$

Putting (13.129) and (13.132) together we obtain

$$\left(\delta M\frac{\partial}{\partial M} + \delta g_r\frac{\partial}{\partial g_r} + \delta h_r\frac{\partial}{\partial h_r} - n\delta\eta\right)G_r^{(n)}(x_1, x_2, \ldots, x_n; g_r, h_r, M) = 0. \quad (13.133)$$

It is not difficult to check now that

$$\delta g_r = \frac{\partial g_r}{\partial M}\delta M = \beta\frac{\delta M}{M}$$

$$\delta h_r = \frac{\partial h_r}{\partial M}\delta M = \rho\frac{h_r\delta M}{M} \quad (13.134)$$

$$\delta\eta \equiv \delta lnZ^{-\frac{1}{2}} = \frac{\partial lnZ^{-\frac{1}{2}}}{\partial M}\delta M = \gamma\frac{\delta M}{M}.$$

Hence we get the final result

$$\left(M\frac{\partial}{\partial M} + \beta\frac{\partial}{\partial g_r} + \rho h_r\frac{\partial}{\partial h_r} - n\gamma\right)G_r^{(n)}(x_1, x_2, \ldots, x_n; g_r, h_r, M) = 0. \quad (13.135)$$

The parameters β, ρ and γ are obviously x-independent (which means also that they are p-independent) as well as n-independent. They are furthermore cut-off independent since the renormalized Green's functions G_r do not depend on the cut-off. The renormalization constants $\bar{Z}$ and $\bar{Z}_g$ in (13.112) and (13.114) do not depend on the coupling constant h_r and thus the parameters β, γ and ρ will not depend on h_r [2]. Since the theory is massless, and since β, γ and ρ are p-independent as well as cut-off independent these parameters are also (by dimensional analysis) M-independent. Thus β, γ and ρ can only depend on the dimensionless coupling constant g_r.

The above equation which is called the Callan–Symanzik equation tells us that the three universal functions $\beta(g_r)$, $\rho(h_r)$ and $\gamma(g_r)$ controls, respectively, the shifts of the coupling constants g_r and h_r and the shift in the field strength ϕ_{ri}, which are needed to compensate the shift in the renormalization scale M.

As an example we apply the Callan–Symanzik equation to the results (13.113) and (13.115) to obtain the beta and gamma functions of the non-linear sigma model. We have in this case

$$\left(M\frac{\partial}{\partial M} + \beta(g_r)\frac{\partial}{\partial g_r} - n\gamma(g_r)\right)G_r^{(n)}(x_1, x_2, \ldots, x_n; g_r, M) = 0. \quad (13.136)$$

Where we have dropped for simplicity the h_r-derivative and the h_r-dependence of the renormalized Green's functions, which also means that the bare coupling constant h_0 is now interpreted at the level of correlations as only an IR regulator μ_0^2. Therefore we have first

$$\left(M\frac{\partial}{\partial M} + \beta\frac{\partial}{\partial g_r} - \gamma\right)\langle\sigma_r\rangle = 0, \quad (13.137)$$

which leads to the condition

$$-\frac{g_r^2(N-1)}{4\pi} - \beta g_r \frac{N-1}{4\pi}\ln\frac{M^2}{\mu_0^2} - \gamma + \gamma\frac{g_r^2(N-1)}{8\pi}\ln\frac{M^2}{\mu_0^2} = 0. \qquad (13.138)$$

β as we will show shortly is of order g_r^3 and thus upto order g_r^2 we have for the γ function the result

$$\gamma = -\frac{g_r^2(N-1)}{4\pi} + O(g_r^4). \qquad (13.139)$$

Similarly, we have

$$\left(M\frac{\partial}{\partial M} + \beta\frac{\partial}{\partial g_r} - 2\gamma\right)\langle\pi_{ri}(k)\pi_{rj}(p)\rangle = 0, \qquad (13.140)$$

which leads to

$$-\frac{g_r^4}{2\pi} + 2\beta g_r - \beta\frac{g_r^3}{\pi}\ln\frac{M^2}{\mu_0^2} + \frac{g_r^4(N-1)}{2\pi} - \frac{g_r^6(N-1)}{8\pi^2}\ln\frac{M^2}{\mu_0^2} = 0. \qquad (13.141)$$

Thus the beta function of the non-linear sigma model upto order g_r^3 is given by

$$\beta = -(N-2)\frac{g_r^3}{4\pi} + O(g_r^5). \qquad (13.142)$$

13.2.3 The large N limit analysis

We assume now Euclidean signature in dimension d. The partition function of the non-linear sigma model is given by

$$Z = \int \mathcal{D}\vec{n}\, e^{-\int d^dx \frac{1}{2g_0^2}(\partial_\mu\vec{n})^2}\delta(\vec{n}^2 - 1). \qquad (13.143)$$

The g_0 is the bare coupling constant. By definition the delta function, which enforces the constraint $\vec{n}^2 = 1$, is given by

$$\delta(\vec{n}^2 - 1) = \prod_x \delta(\vec{n}^2(x) - 1). \qquad (13.144)$$

In one dimension we know that the delta function can be represented by

$$\delta(x) = \int_{-\infty}^{\infty}\frac{d\alpha}{2\pi}e^{i\alpha x} = \int_{-\infty}^{\infty}\frac{d\bar{\alpha}}{4g_0^2\pi}e^{-\frac{i}{2g_0^2}\bar{\alpha}x}. \qquad (13.145)$$

Thus we will have

$$\delta(\vec{n}(x) - 1) = \int_{-\infty}^{\infty}\frac{d\alpha(x)}{4g_0^2\pi}e^{-\frac{i}{2g_0^2}\alpha(x)(\vec{n}^2(x)-1)}, \qquad (13.146)$$

and consequently

$$\delta(\vec{n}^2 - 1) = \int \mathcal{D}\alpha e^{-\frac{i}{2g_0^2}\int d^dx \alpha(x)(\vec{n}^2(x)-1)}, \quad \mathcal{D}\alpha = \prod_x \frac{d\alpha(x)}{4g_0^2\pi}. \tag{13.147}$$

The partition function becomes

$$\begin{aligned}
Z &= \int \mathcal{D}\vec{n}\mathcal{D}\alpha e^{-\int d^dx \frac{1}{2g_0^2}(\partial_\mu\vec{n})^2 - \frac{i}{2g_0^2}\int d^dx \alpha(\vec{n}^2 - 1)} \\
&= \int \mathcal{D}\alpha \left[\mathcal{D}\vec{n}e^{\int \frac{d^dx}{2g_0^2}\vec{n}(\partial^2 - i\alpha)\vec{n}}\right] e^{i\int \frac{d^dx}{2g_0^2}\alpha} \\
&= \int \mathcal{D}\alpha \prod_i^N \left[\mathcal{D}n_i e^{\int \frac{d^dx}{2g_0^2}n_i(\partial^2 - i\alpha)n_i}\right] e^{i\int \frac{d^dx}{2g_0^2}\alpha}.
\end{aligned} \tag{13.148}$$

The $\vec{n}$-field measure will be defined by

$$\mathcal{D}n_i = \prod_x \frac{dn_i(x)}{\sqrt{2\pi}}\frac{1}{\sqrt{g_0^2}} = \prod_x \frac{d\tilde{n}_i(x)}{\sqrt{2\pi}}, \tag{13.149}$$

where $\tilde{n}_i = \frac{n_i}{\sqrt{g_0^2}}$. The partition function becomes then

$$Z = \int \mathcal{D}\alpha \prod_i^N \left[\prod_x \frac{d\tilde{n}_i(x)}{\sqrt{2\pi}}e^{-\frac{1}{2}\int d^dx d^dy \tilde{n}_i(x)\delta^d(x-y)(-\partial^2 + i\alpha)_x\tilde{n}_i(y)}\right] e^{i\int \frac{d^dx}{2g_0^2}\alpha}. \tag{13.150}$$

By using now the identity

$$\begin{aligned}
Z_0[J] &= \int \prod_{i=1}^N \frac{dq_i}{\sqrt{2\pi}}e^{-\frac{1}{2}\sum_{n,m}q_n M_{nm}q_m + \sum_n J_n q_n} \\
&= [\det M]^{-\frac{1}{2}} e^{\frac{1}{2}\sum_{n,m}J_n M_{nm}^{-1}J_m},
\end{aligned} \tag{13.151}$$

we obtain

$$\begin{aligned}
Z &= \int \mathcal{D}\alpha[\det (\delta^d(x - y)(-\partial^2 + i\alpha)_x)]^{-\frac{N}{2}} e^{i\int \frac{d^dx}{2g_0^2}\alpha} \\
&= \int \mathcal{D}\alpha[\det(-\partial^2 + i\alpha(x))]^{-\frac{N}{2}} e^{i\int \frac{d^dx}{2g_0^2}\alpha} \\
&= \int \mathcal{D}\alpha e^{-\frac{N}{2}Tr \ln(-\partial^2 + i\alpha)+\frac{i}{2g_0^2}\int d^dx\alpha}.
\end{aligned} \tag{13.152}$$

In the second line we have used the fact that $\det(AB) = \det(A)\det(B)$ and then absorbed for simplicity the infinite factor $[\det(\delta^d(x - y))]^{-\frac{N}{2}}$ into the measure $\mathcal{D}\alpha$. In the last line we have used however the identity $\det A = e^{Tr \ln A}$.

In the large N limit and if we keep $g_0^2 N = h_0^2$ fixed it is not difficult to see that both terms in the exponent in the above partition function (13.152) are of order N.

One can then reasonably use the method of steepest descents to evaluate this partition function Z by effectively dominating the path integral by the configuration $\alpha(x)$, which minimizes the exponent. We write then

$$Z = \int \mathcal{D}\alpha e^{\int d^dx F(\alpha(x))}, \quad F(\alpha(x)) = -\frac{N}{2}\langle x|\ln(-\partial^2 + i\alpha(x))|x\rangle + \frac{i}{2g_0^2}\alpha(x). \tag{13.153}$$

The condition

$$\frac{\partial F(\alpha(x))}{\partial \alpha(x)} = 0. \tag{13.154}$$

takes also the form

$$\frac{N}{2}\langle x|\frac{1}{-\partial^2 + i\alpha(x)}|x\rangle = \frac{1}{2g_0^2}. \tag{13.155}$$

If we introduce now the basis $\{|p\rangle\}$ of the operator $P_\mu = -i\partial_\mu$, which satisfies

$$[x_\mu, P_\nu] = i\delta_{\mu\nu}$$

$$P_\mu|p\rangle = p_\mu|p\rangle, \quad \int d^d p\,|p\rangle\langle p| = 1, \quad \langle p|p'\rangle = \delta^d(p - p'), \quad \langle x|p\rangle = \frac{1}{(2\pi)^{\frac{d}{2}}}e^{ixp}. \tag{13.156}$$

Then we can write the above equation as

$$\begin{aligned}
\frac{1}{2g_0^2} &= \frac{N}{2}\int d^d p\, d^d p'\,\langle x|p\rangle\langle p|\frac{1}{P^2 + i\alpha(x)}|p'\rangle\langle p'|x\rangle \\
&= \frac{N}{2}\int \frac{d^d p}{(2\pi)^d}\frac{1}{p^2 + i\alpha(x)}.
\end{aligned} \tag{13.157}$$

This means that the configuration $\alpha(x)$ is such that at each point x in spacetime this configuration is a pure imaginary constant, i.e.,

$$\alpha(x) = -im^2, \tag{13.158}$$

and thus m^2 is the solution of the equation

$$\frac{1}{2g_0^2} = \frac{N}{2}\int \frac{d^d p}{(2\pi)^d}\frac{1}{p^2 + m^2}. \tag{13.159}$$

From the first line of (13.148) it is obvious that the parameter m plays exactly the role of a mass for the field $\vec{n}$. In $d = 2$ and with a cut-off Λ one can compute

$$\begin{aligned}
\frac{1}{g_0^2} &= N\int_\Lambda \frac{d^d p}{(2\pi)^d}\frac{1}{p^2 + m^2} \\
&= \frac{N}{2\pi}\ln\sqrt{1 + \frac{\Lambda^2}{m^2}} \\
&= \frac{N}{2\pi}\ln\frac{\Lambda}{m},
\end{aligned} \tag{13.160}$$

where we have assumed that the cut-off is much larger than the mass m. Motivated by the perturbative result $g_0^2 = Z_g g_r^2$, which can also be put in the form

$$\frac{1}{g_0^2} = \frac{1}{g_r^2} - \bar{Z}_g = \frac{1}{g_r^2} + \frac{N-2}{2\pi} \ln \frac{\Lambda}{M} + \frac{N-2}{4\pi}\left(\frac{2}{\epsilon} - \gamma\right), \qquad (13.161)$$

we will renormalize in this large N expansion the bare coupling constant g_0^2 as follows

$$\frac{1}{g_0^2} = \frac{1}{g_r^2} + \frac{N}{2\pi} \ln \frac{\Lambda}{M}. \qquad (13.162)$$

The infinite third term in (13.161) can be thought of as being absorbed in the infinite bare coupling constant $1/g_0^2$, i.e., we assume for simplicity that the bare coupling constants in perturbation theory and in the large N expansion are actually different. More precisely we assume that the infinite bare coupling constant, which appears in the path integral Z is in fact

$$\frac{1}{g_0^2} - \frac{N-2}{4\pi}\left(\frac{2}{\epsilon} - \gamma\right)$$

with $1/g_0^2$ in this last equation stands for the bare coupling constant we worked with in perturbation theory. Putting (13.160) and (13.162) together we can solve for the mass m, namely

$$m = M e^{-\frac{2\pi}{N g_r^2}}. \qquad (13.163)$$

This is an $O(N)$-symmetric mass for the (essentially) unconstrained N components of the field $\vec{n}$. This mass cannot vanish for all the non-zero values of the coupling constant $h_r^2 = N g_r^2$ and thus in $d = 2$ there is no critical value for this coupling constant and as a consequence the symmetry is not broken, i.e., $\langle \vec{n} \rangle = 0$ always. This mass depends on the arbitrary renormalization scale M, which is also a reflection of the arbitrariness of the renormalization condition (13.162). If one changes the renormalization scale M as $M \longrightarrow M'$ then from equation (13.162) the renormalized coupling constant g_r^2 must also change appropriately as $g_r^2 \longrightarrow g_r'^2$ such that the bare coupling constant g_0^2 is kept fixed. Correspondingly and from (13.160) the mass m remains fixed under this change of M and hence it does depend only on the cut-off and on the bare coupling constant. We write this fact in the form

$$\left[M \frac{\partial}{\partial M} + \beta(g_r) \frac{\partial}{\partial g_r} \right] m(g_r, M) = 0. \qquad (13.164)$$

Using the solution (13.163) in this last equation we get the beta function

$$\beta(g_r) = -\frac{g_r^3 N}{4\pi}, \qquad (13.165)$$

which is consistent with the large N limit of (13.142). However this last result is exact to all orders in $h_r^2 = Ng_r^2$ [3]. This is also obvious by rewriting the above beta function in the form

$$\beta(h_r) = -\frac{h_r^3}{4\pi}. \tag{13.166}$$

13.3 The Ising model and the Onsager solution

This is perhaps the most important exact solution in all quantum field theory. The original solution is found in [4] but here we follow the presentation of [5].

13.3.1 The Ising model in d dimensions

We consider a d-dimensional periodic lattice with n points in every direction so that there are $N = n^d$ points in total in this lattice. In every point (lattice site) we put a spin variable s_i ($i = 1, \ldots, N$), which can take either the value $+1$ or -1. A configuration of this system of N spins is therefore specified by a set of numbers $\{s_i\}$. In the Ising model the energy of this system of N spins in the configuration $\{s_i\}$ is given by

$$E_I\{s_i\} = -\sum_{\langle ij \rangle} \epsilon_{ij} s_i s_j - H \sum_{i=1}^{N} s_i. \tag{13.167}$$

The parameter H is the external magnetic field. The symbol $\langle ij \rangle$ stands for nearest neighbor spins. The sum over $\langle ij \rangle$ extends over $\frac{\gamma N}{2}$ terms where γ is the number of nearest neighbors. In 2, 3, 4 dimensions $\gamma = 4, 6, 8$. The parameter ϵ_{ij} is the interaction energy between the spins i and j. For isotropic interactions $\epsilon_{ij} = \epsilon$. For $\epsilon > 0$ we obtain ferromagnetism while for $\epsilon < 0$ we obtain antiferromagnetism. We consider only $\epsilon > 0$. The energy is

$$E_I\{s_i\} = -\epsilon \sum_{\langle ij \rangle} s_i s_j - H \sum_{i=1}^{N} s_i. \tag{13.168}$$

The partition function is given by

$$Z = \sum_{s_1} \sum_{s_2} \cdots \sum_{s_N} e^{-\beta E_I\{s_i\}}. \tag{13.169}$$

There are 2^N terms in the sum and $\beta = \frac{1}{k_B T}$.

13.3.2 The Ising model in 2 dimensions

We consider $d = 2$ in the remainder. There are $N = n^2$ spins in the square lattice. The configuration $\{s_i\}$ can be viewed as an $n \times n$ matrix. We impose periodic boundary condition as follows. We consider $(n + 1) \times (n + 1)$ matrix where the $(n + 1)$ th row is identified with the first row and the $(n + 1)$ th column is identified with the first column. The square lattice is therefore a torus. We introduce the notation

$$\mu_\alpha = \{s_1, s_2, \ldots, s_n\}_{\alpha\text{th row}}, \quad \alpha = 1, \ldots, n. \tag{13.170}$$

The toroidal boundary condition means that in each row $s_{n+1} = s_n$. It also means that the $(n + 1)$ th row is identical with the first row, viz $\mu_1 = \mu_{n+1}$.

The interaction energy between the spins of the αth and $(\alpha + 1)$ th rows is

$$E(\mu_\alpha, \mu_{\alpha+1}) = -\epsilon \sum_{k=1}^{n} s_k^{(\alpha)} s_k^{(\alpha+1)}. \tag{13.171}$$

The interaction energy between the spins of the αth row plus the energy due to the external magnetic field is

$$E(\mu_\alpha) = -\epsilon \sum_{k=1}^{n} s_k^{(\alpha)} s_{k+1}^{(\alpha)} - H \sum_{k=1}^{n} s_k^{(\alpha)}. \tag{13.172}$$

The total energy of the lattice is

$$E_I\{s_i\} \equiv E_I\{\mu_1, \ldots, \mu_n\} = \sum_{\alpha=1}^{n} [E(\mu_\alpha, \mu_{\alpha+1}) + E(\mu_\alpha)]. \tag{13.173}$$

The partition function is

$$Z = \sum_{\mu_1} \sum_{\mu_2} \cdots \sum_{\mu_n} e^{-\beta \sum_{\alpha=1}^{n} [E(\mu_\alpha, \mu_{\alpha+1}) + E(\mu_\alpha)]}. \tag{13.174}$$

We define the $2^n \times 2^n$ matrix P by

$$\langle \mu | P | \nu \rangle = e^{-\beta[E(\mu, \nu) + E(\mu)]}$$
$$= \prod_{k=1}^{n} e^{\beta H s_k^{(\mu)}} e^{\beta \epsilon s_k^{(\mu)} s_{k+1}^{(\mu)}} e^{\beta \epsilon s_k^{(\mu)} s_k^{(\nu)}} \tag{13.175}$$
$$= \langle \mu | V_3 V_2 V_1' | \nu \rangle.$$

$$\langle \mu | V_1' | \nu \rangle = \prod_{k=1}^{n} e^{\beta \epsilon s_k^{(\mu)} s_k^{(\nu)}}$$
$$\langle \mu | V_2 | \nu \rangle = \delta_{\mu\nu} \prod_{k=1}^{n} e^{\beta \epsilon s_k^{(\mu)} s_{k+1}^{(\mu)}} \tag{13.176}$$
$$\langle \mu | V_3 | \nu \rangle = \delta_{\mu\nu} \prod_{k=1}^{n} e^{\beta H s_k^{(\mu)}}.$$

Thus

$$Z = \sum_{\mu_1} \sum_{\mu_2} \cdots \sum_{\mu_n} \langle \mu_1 | P | \mu_2 \rangle \langle \mu_2 | P | \mu_3 \rangle \cdots \langle \mu_{n-1} | P | \mu_n \rangle \langle \mu_n | P | \mu_1 \rangle \tag{13.177}$$
$$= Tr P^n.$$

Let us introduce now the direct product of matrices defined by

$$\langle i_1 i_2 \ldots i_n | A_1 \times A_2 \cdots \times A_n | j_1 j_2 \ldots j_n \rangle = \langle i_1 | A_1 | j_1 \rangle \langle i_2 | A_2 | j_2 \rangle \cdots \langle i_n | A_n | j_n \rangle. \tag{13.178}$$

Define the 2×2 matrix a by its components

$$\langle s^{(\mu)}|a|s^{(\nu)}\rangle = e^{\beta \epsilon s^{(\mu)} s^{(\nu)}}. \tag{13.179}$$

In other words

$$
\begin{aligned}
a &= e^{\beta \epsilon} 1 + e^{-\beta \epsilon} X \\
&= \frac{e^{\beta \epsilon}}{\cosh \theta}\, e^{\theta X} \\
&= \sqrt{2 \sinh 2\beta \epsilon}\ e^{\theta X}.
\end{aligned}
\tag{13.180}
$$

In above X is the Pauli matrix σ_1 and θ is defined by

$$\tanh \theta = e^{-2\beta \epsilon}. \tag{13.181}$$

Then

$$
\begin{aligned}
\langle \mu |V_1'|\nu\rangle &= \prod_{k=1}^{n}\ e^{\beta \epsilon s_k^{(\mu)} s_k^{(\nu)}} \\
&= \langle s_1^{(\mu)}|a|s_1^{(\nu)}\rangle \langle s_2^{(\mu)}|a|s_2^{(\nu)}\rangle \cdots \langle s_n^{(\mu)}|a|s_n^{(\nu)}\rangle \\
&= \langle s_1^{(\mu)} s_2^{(\mu)} \ldots s_n^{(\mu)}|a \times a \ldots \times a|s_1^{(\nu)} s_2^{(\nu)} \ldots s_n^{(\nu)}\rangle.
\end{aligned}
\tag{13.182}
$$

In other words,

$$
\begin{aligned}
V_1' &= a \times a \ldots \times a \\
&= (2 \sinh 2\beta \epsilon)^{\frac{n}{2}}\, e^{\theta X} \times e^{\theta X} \ldots \times e^{\theta X}.
\end{aligned}
\tag{13.183}
$$

We introduce the $2^n \times 2^n$ matrices

$$X_\alpha = 1 \times 1 \times \cdots \times X \times \cdots \times 1, \quad \alpha = 1, \ldots, n. \tag{13.184}$$

The X appears in the αth factor. Then we have

$$
\begin{aligned}
V_1' &= (2 \sinh 2\beta \epsilon)^{\frac{n}{2}}\, e^{\theta X} \times e^{\theta X} \ldots \times e^{\theta X} \\
&= (2 \sinh 2\beta \epsilon)^{\frac{n}{2}}\, e^{\theta X_1} e^{\theta X_2} \ldots e^{\theta X_n} \\
&= (2 \sinh 2\beta \epsilon)^{\frac{n}{2}}\, V_1,
\end{aligned}
\tag{13.185}
$$

where

$$V_1 = \prod_{\alpha=1}^{n}\ e^{\theta X_\alpha}. \tag{13.186}$$

Let Y and Z be the other Pauli matrices σ_2 and σ_3 respectively. Let us introduce also the $2^n \times 2^n$ matrices

$$Y_\alpha = 1 \times 1 \times \cdots \times Y \times \cdot \times 1, \quad \alpha = 1, \ldots, n. \tag{13.187}$$

$$Z_\alpha = 1 \times 1 \times \cdots \times Z \times \cdots \times 1, \quad \alpha = 1, \ldots, n. \tag{13.188}$$

Now we find V_2 defined by the matrix elements

$$\langle \mu |V_2|\nu\rangle = \delta_{\mu\nu} \prod_{k=1}^{n}\ e^{\beta \epsilon s_k^{(\mu)} s_{k+1}^{(\mu)}}. \tag{13.189}$$

Let us remark that

$$Z_\alpha Z_{\alpha+1} = 1 \times \cdots \times Z \times Z \times \cdots \times 1. \tag{13.190}$$

The Z appears in the factors α and $\alpha + 1$. Since $(Z_\alpha Z_{\alpha+1})^2 = 1$ we must have

$$e^{\beta \epsilon Z_\alpha Z_{\alpha+1}} = \cosh \beta\epsilon + \sinh \beta\epsilon \; Z_\alpha Z_{\alpha+1}. \tag{13.191}$$

In other words,

$$\begin{aligned}
\langle \mu | e^{\beta \epsilon Z_\alpha Z_{\alpha+1}} | \nu \rangle &= \cosh \beta\epsilon \; \delta_{\mu\nu} + \sinh \beta\epsilon \langle s_1^{(\mu)} | s_1^{(\nu)} \rangle \cdots \langle s_\alpha^{(\mu)} | Z | s_\alpha^{(\nu)} \rangle \langle s_{\alpha+1}^{(\mu)} | Z | s_{\alpha+1}^{(\nu)} \rangle \cdots \\
&\quad \times \langle s_n^{(\mu)} | s_n^{(\nu)} \rangle \\
&= \delta_{\mu\nu} \left(\cosh \beta\epsilon + \sinh \beta\epsilon \; s_\alpha^{(\mu)} s_{\alpha+1}^{(\mu)} \right) \\
&= \delta_{\mu\nu} \; e^{\beta \epsilon s_\alpha^{(\mu)} s_{\alpha+1}^{(\mu)}}.
\end{aligned} \tag{13.192}$$

Thus we can immediately conclude that

$$V_2 = \prod_{\alpha=1}^{n} \; e^{\beta \epsilon Z_\alpha Z_{\alpha+1}}. \tag{13.193}$$

In the above equation we have used the boundary condition $Z_{n+1} = Z_n$.
Finally from the matrix elements of V_3 defined by

$$\langle \mu | V_3 | \nu \rangle = \delta_{\mu\nu} \prod_{k=1}^{n} \; e^{\beta H s_k^{(\mu)}}, \tag{13.194}$$

we can show that

$$V_3 = \prod_{\alpha=1}^{n} \; e^{\beta H Z_\alpha}. \tag{13.195}$$

We obtain thus the result

$$P = (2 \sinh 2\beta\epsilon)^{\frac{n}{2}} \, V_3 V_2 V_1. \tag{13.196}$$

For $H = 0$ we have $V_3 = 1$ and hence

$$P = (2 \sinh 2\beta\epsilon)^{\frac{n}{2}} \, V_2 V_1. \tag{13.197}$$

13.3.3 The Clifford algebra in $2n$ dimensions

A Clifford algebra in $2n$ dimensions is given in terms of $2n$ (gamma) matrices γ_μ, $\mu = 1, 2, \ldots, 2n$, which satisfy the conditions

$$\{\gamma_\mu, \gamma_\nu\} = 2\delta_{\mu\nu}. \tag{13.198}$$

The irreducible representation is 2^n-dimensional. We consider the explicit representation

$$\gamma_{2\alpha-1} = X_1 X_2 \ldots X_{\alpha-1} Z_\alpha, \quad \gamma_{2\alpha} = X_1 X_2 \ldots X_{\alpha-1} Y_\alpha, \quad \alpha = 1, 2, \ldots, n. \tag{13.199}$$

Let ω be a rotation in the $2n$-dimensional space. It must satisfy $\omega^T \omega = 1$. The matrices γ_μ transform under ω to γ'_μ, which are given by

$$\gamma'_\mu = \sum_{\nu=1}^{2n} \omega_{\mu\nu} \gamma_\nu. \tag{13.200}$$

It is not difficult to show that γ'_μ satisfy also the Clifford algebra (13.198), i.e $\{\gamma'_\mu, \gamma'_\nu\} = 2\delta_{\mu\nu}$. Thus there must exist a non-singular matrix $S(\omega)$ such that

$$\gamma'_\mu = \sum_{\nu=1}^{2n} \omega_{\mu\nu} \gamma_\nu = S(\omega)\gamma_\mu S(\omega)^{-1}. \tag{13.201}$$

The $2^n \times 2^n$ matrix $S(\omega)$ is the spin representation of the rotation ω.

Let $\omega(\mu\nu|\theta)$ be the rotation in the plane $\mu\nu$ ($\mu \neq \nu$) through an angle θ, i.e.,

$$\begin{aligned}
\gamma'_\lambda &= \gamma_\lambda, \quad \lambda \neq \mu, \quad \lambda \neq \nu \\
\gamma'_\mu &= \gamma_\mu \cos\theta - \gamma_\nu \sin\theta \\
\gamma'_\mu &= \gamma_\mu \sin\theta + \gamma_\nu \cos\theta.
\end{aligned} \tag{13.202}$$

The corresponding spin representation $S_{\mu\nu}(\theta)$ is given by

$$S_{\mu\nu}(\theta) = \exp\left(-\frac{\theta}{2}\gamma_\mu\gamma_\nu\right). \tag{13.203}$$

Indeed, we check that

$$e^{-\frac{\theta}{2}\gamma_\mu\gamma_\nu}\gamma_\lambda e^{\frac{\theta}{2}\gamma_\mu\gamma_\nu} = \gamma_\lambda, \quad \lambda \neq \mu, \quad \lambda \neq \nu. \tag{13.204}$$

Also since $(\gamma_\mu\gamma_\nu)^2 = -1$ we have

$$e^{-\frac{\theta}{2}\gamma_\mu\gamma_\nu} = \cos\frac{\theta}{2} - \gamma_\mu\gamma_\nu \sin\frac{\theta}{2}. \tag{13.205}$$

Thus

$$\begin{aligned}
e^{-\frac{\theta}{2}\gamma_\mu\gamma_\nu}\gamma_\mu e^{\frac{\theta}{2}\gamma_\mu\gamma_\nu} &= \left(\cos\frac{\theta}{2} - \gamma_\mu\gamma_\nu \sin\frac{\theta}{2}\right)\gamma_\mu e^{\frac{\theta}{2}\gamma_\mu\gamma_\nu} \\
&= \gamma_\mu e^{\theta\gamma_\mu\gamma_\nu} \\
&= \gamma_\mu(\cos\theta - \gamma_\mu\gamma_\nu \sin\theta) \\
&= \gamma_\mu \cos\theta - \gamma_\nu \sin\theta.
\end{aligned} \tag{13.206}$$

Next we compute the eigenvalues of $S_{\mu\nu}(\theta)$. We consider the representation

$$\gamma_{2\alpha-1} = Z_1 Z_2 \ldots Z_{\alpha-1} X_\alpha, \quad \gamma_{2\alpha} = Z_1 Z_2 \ldots Z_{\alpha-1} Y_\alpha, \quad \alpha = 1, 2, \ldots, n. \tag{13.207}$$

We take γ_μ and γ_ν to be the matrices

$$\gamma_\mu = Z_1 X_2, \quad \gamma_\nu = Z_1 Y_2. \tag{13.208}$$

Thus

$$\gamma_\mu \gamma_\nu = X_2 Y_2 = iZ_2 = i1 \times Z \times \cdots \times 1. \tag{13.209}$$

Hence

$$S_{\mu\nu}(\theta) = \cos\frac{\theta}{2} - \gamma_\mu \gamma_\nu \sin\frac{\theta}{2} = 1 \times \begin{pmatrix} e^{-i\frac{\theta}{2}} & 0 \\ 0 & e^{i\frac{\theta}{2}} \end{pmatrix} \times \cdots \times 1. \tag{13.210}$$

Thus $S_{\mu\nu}(\theta)$ has eigenvalue $e^{-i\frac{\theta}{2}}$ with multiplicity 2^{n-1} and eigenvalue $e^{i\frac{\theta}{2}}$ with multiplicity 2^{n-1}.

We consider the product of n commuting plane rotations, vis.

$$\omega = \omega(\mu_1 \nu_1 | \theta_1) \cdots \omega(\mu_n \nu_n | \theta_n). \tag{13.211}$$

The $2n$ eigenvalues of ω are

$$e^{\pm i\theta_1}, \ldots, e^{\pm i\theta_n}. \tag{13.212}$$

The corresponding spin representation is given by

$$S(\omega) = \exp\left(-\frac{\theta_1}{2}\gamma_{\mu_1}\gamma_{\nu_1}\right) \cdots \exp\left(-\frac{\theta_n}{2}\gamma_{\mu_n}\gamma_{\nu_n}\right). \tag{13.213}$$

The set $\{\mu_1, \nu_1, \ldots, \mu_n, \nu_n\}$ is a permutation of the set $\{1, 2, \ldots, 2n-1, 2n\}$. The 2^n eigenvalues of $S(\omega)$ are

$$e^{\frac{i}{2}(\pm\theta_1 \pm \ldots \pm\theta_n)}. \tag{13.214}$$

Now we go back to our original problem. We need the matrices

$$V_1 = \prod_{\alpha=1}^{n} e^{\theta X_\alpha}, \quad \tanh\theta = e^{-2\beta\epsilon}. \tag{13.215}$$

$$V_2 = \prod_{\alpha=1}^{n} e^{\beta\epsilon Z_\alpha Z_{\alpha+1}}. \tag{13.216}$$

We use the representation

$$\gamma_{2\alpha-1} = X_1 X_2 \ldots X_{\alpha-1} Z_\alpha, \quad \gamma_{2\alpha} = X_1 X_2 \ldots X_{\alpha-1} Y_\alpha, \quad \alpha = 1, 2, \ldots, n. \tag{13.217}$$

First we remark that $X_\alpha = -i\gamma_{2\alpha}\gamma_{2\alpha-1}$, i.e.

$$V_1 = \prod_{\alpha=1}^{n} e^{-i\theta\gamma_{2\alpha}\gamma_{2\alpha-1}}. \tag{13.218}$$

In other words, V_1 is a spin representation of a product of commuting plane rotations. Next we remark that for $\alpha = 1, \ldots, n-1$ we have

$$\begin{aligned}
\gamma_{2\alpha+1}\gamma_{2\alpha} &= (X_1 X_2 \ldots X_\alpha Z_{\alpha+1})(X_1 X_2 \ldots X_{\alpha-1} Y_\alpha) \\
&= X_\alpha Z_{\alpha+1} Y_\alpha \\
&= i Z_\alpha Z_{\alpha+1}.
\end{aligned} \tag{13.219}$$

For $\alpha = n$ we have

$$\begin{aligned}
\gamma_{2n+1}\gamma_{2n} = \gamma_1 \gamma_{2n} &= (Z_1)(X_1 X_2 \ldots X_{n-1} Y_n) \\
&= -i Z_1 Z_n (X_1 X_2 \ldots X_n).
\end{aligned} \tag{13.220}$$

In the above equation, we have used $\gamma_{2n+1} = \gamma_1$ and $Y_n = i X_n Z_n$. Next by using $Z_{n+1} = Z_1$ we obtain

$$\begin{aligned}
V_2 &= \prod_{\alpha=1}^{n-1} e^{\beta \epsilon Z_\alpha Z_{\alpha+1}} \; e^{\beta \epsilon Z_n Z_1} \\
&= e^{\beta \epsilon Z_n Z_1} \prod_{\alpha=1}^{n-1} e^{\beta \epsilon Z_\alpha Z_{\alpha+1}} \\
&= e^{i\beta \epsilon U \gamma_1 \gamma_{2n}} \prod_{\alpha=1}^{n-1} e^{-i\beta \epsilon \gamma_{2\alpha+1} \gamma_{2\alpha}}.
\end{aligned} \tag{13.221}$$

We have also defined the chirality operator $U = X_1 X_2 \ldots X_n$. For $U = 0$ the matrix V_2 is also a spin representation of a product of commuting plane rotations. The fact that $U \neq 0$ is due to the toroidal boundary condition. We can check that $U = i^n \gamma_1 \gamma_2 \ldots \gamma_{2n}$, $U^2 = 1$, $[U, \gamma_1 \gamma_{2n}] = 0$ and $(\gamma_1 \gamma_{2n})^2 = -1$. Furthermore we compute

$$\begin{aligned}
e^{i\beta \epsilon U \gamma_1 \gamma_{2n}} &= \cosh \beta \epsilon + i \sinh \beta \epsilon \; U \gamma_1 \gamma_{2n} \\
&= \frac{1+U}{2} e^{i\beta \epsilon \gamma_1 \gamma_{2n}} + \frac{1-U}{2} e^{-i\beta \epsilon \gamma_1 \gamma_{2n}}.
\end{aligned} \tag{13.222}$$

Thus

$$V_2 = \frac{1+U}{2} e^{i\beta \epsilon \gamma_1 \gamma_{2n}} \prod_{\alpha=1}^{n-1} e^{-i\beta \epsilon \gamma_{2\alpha+1} \gamma_{2\alpha}} + \frac{1-U}{2} e^{-i\beta \epsilon \gamma_1 \gamma_{2n}} \prod_{\alpha=1}^{n-1} e^{-i\beta \epsilon \gamma_{2\alpha+1} \gamma_{2\alpha}}. \tag{13.223}$$

In summary we obtain (with $\phi = \beta \epsilon$, $\theta = \tanh^{-1} e^{-2\phi}$)

$$Z = \operatorname{Tr} P^n. \tag{13.224}$$

$$P = (2 \sinh 2\phi)^{\frac{n}{2}} V. \tag{13.225}$$

$$V = V_2 V_1 = \frac{1+U}{2} V_+ + \frac{1-U}{2} V_-. \tag{13.226}$$

$$V_\pm = e^{\pm i\phi \gamma_1 \gamma_{2n}} \prod_{\alpha=1}^{n-1} e^{-i\phi \gamma_{2\alpha+1} \gamma_{2\alpha}} \prod_{\beta=1}^{n} e^{-i\theta \gamma_{2\beta} \gamma_{2\beta-1}}. \tag{13.227}$$

Let λ_α, $\alpha = 1, \ldots, 2^n$ be the eigenvalues of V. Then

$$Z = (2 \sinh 2\phi)^{\frac{n^2}{2}} \sum_{\alpha=1}^{2^n} \lambda_\alpha^n. \tag{13.228}$$

Equivalently

$$\frac{1}{n^2} \log Z = \frac{1}{2} \log(2 \sinh 2\phi) + \frac{1}{n^2} \log\left(\sum_{\alpha=1}^{2^n} \lambda_\alpha^n\right). \tag{13.229}$$

If all the eigenvalues λ_α are positive and if $\lambda_{\max}$ is the largest eigenvalue of V then we must have

$$\lambda_{\max}^n \leqslant \sum_{\alpha=1}^{2^n} \lambda_\alpha^n \leqslant 2^n \lambda_{\max}^n. \tag{13.230}$$

As a consequence we must have

$$\frac{1}{2}\log(2\sinh 2\phi) + \frac{1}{n}\log \lambda_{\max} \leqslant \frac{1}{n^2}\log Z \leqslant \frac{1}{2}\log(2\sinh 2\phi) + \frac{1}{n}\log \lambda_{\max} + \frac{1}{n}\log 2. \tag{13.231}$$

In other words,

$$\frac{1}{n^2} \log Z = \frac{1}{2} \log(2 \sinh 2\phi) + \frac{1}{n} \log \lambda_{\max}, \quad n \longrightarrow \infty. \tag{13.232}$$

It is expected that $\lambda_{\max}$ is of order n in the limit.

13.3.4 Diagonalization

The matrices U, V_+ and V_- are commuting matrices and thus they can be diagonalized simultaneously. Since $U^2 = 1$ the eigenvalues of U are either $+1$ or -1. It is not difficult to verify that they occur with the same multiplicity. We work in the basis in which U takes the form

$$U = \begin{pmatrix} 1 & 0 \\ 0 & -1 \end{pmatrix}. \tag{13.233}$$

The identity 1 is 2^{n-1}-dimensional. In this basis we have also

$$V_\pm = \begin{pmatrix} A_\pm & 0 \\ 0 & B_\pm \end{pmatrix}. \tag{13.234}$$

The $A_\pm$ and $B_\pm$ are $2^{n-1} \times 2^{n-1}$ matrices, which are not necessarily diagonal. We then compute in this basis

$$V = \begin{pmatrix} A_+ & 0 \\ 0 & B_- \end{pmatrix}. \tag{13.235}$$

Clearly we need to diagonalize V_+ and V_- separately and independently. Instead of $V_\pm$ given by equation (13.227) we will consider

$$V_\pm = \Gamma\left(e^{\pm i\phi \gamma_1 \gamma_{2n}} \prod_{\alpha=1}^{n-1} e^{-i\phi \gamma_{2\alpha+1}\gamma_{2\alpha}} \prod_{\beta=1}^{n} e^{-i\theta \gamma_{2\beta}\gamma_{2\beta-1}}\right)\Gamma^{-1}. \tag{13.236}$$

$$\Gamma = \prod_{\beta=1}^{n} e^{-i\frac{\theta}{2}\gamma_{2\beta}\gamma_{2\beta-1}}. \tag{13.237}$$

Under the trace the extra factors Γ and Γ^{-1} will cancel. Next we write the rotations corresponding to the spin representations (13.236). These read

$$\omega_{\pm} = \Delta \chi_{\pm} \Delta. \tag{13.238}$$

$$\begin{aligned}
\chi_{\pm} &= \omega(1\ 2n|\mp 2i\phi)\prod_{\alpha=1}^{n-1} \omega(2\alpha+1,\ 2\alpha|2i\phi) \\
&= \omega(1\ 2n|\mp 2i\phi)\omega(23|-2i\phi)\omega(45|-2i\phi)\ \dots\ \omega(2n-2\ 2n-1|-2i\phi).
\end{aligned} \tag{13.239}$$

$$\begin{aligned}
\Delta &= \prod_{\beta=1}^{n} \omega(2\beta,\ 2\beta-1|i\theta) \\
&= \omega(12|-i\theta)\omega(34|-i\theta)\ \dots\ \omega(2n-1\ 2n|-i\theta).
\end{aligned} \tag{13.240}$$

We do the example $n = 3$ explicitly. We have

$$\begin{aligned}
\Delta &= \omega(12|-i\theta)\omega(34|-i\theta)\omega(56|-i\theta) \\
&= \begin{pmatrix} J & 0 & 0 \\ 0 & J & 0 \\ 0 & 0 & J \end{pmatrix}.
\end{aligned} \tag{13.241}$$

In the above equation, J is the $n = 1$ rotation through an angle $-i\theta$, viz

$$J = \begin{pmatrix} \cos(-i\theta) & \sin(-i\theta) \\ -\sin(-i\theta) & \cos(-i\theta) \end{pmatrix} = \begin{pmatrix} \cosh(\theta) & -i\sinh(\theta) \\ i\sinh(\theta) & \cosh(\theta) \end{pmatrix}. \tag{13.242}$$

Similarly

$$\begin{aligned}
\chi_{\pm} &= \omega(16|\mp 2i\phi)\omega(23|-2i\phi)\omega(45|-2i\phi) \\
&= \begin{pmatrix} a & 0 & 0 & \mp b \\ 0 & K & 0 & 0 \\ 0 & 0 & K & 0 \\ \pm b & 0 & 0 & a \end{pmatrix}.
\end{aligned} \tag{13.243}$$

In the above equation, K is the $n = 1$ rotation through an angle $-2i\phi$, viz

$$K = \begin{pmatrix} \cos(-2i\phi) & \sin(-2i\phi) \\ -\sin(-2i\phi) & \cos(-2i\phi) \end{pmatrix} = \begin{pmatrix} \cosh(2\phi) & -i\sinh(2\phi) \\ i\sinh(2\phi) & \cosh(2\phi) \end{pmatrix} \equiv \begin{pmatrix} a & -b \\ b & a \end{pmatrix}. \tag{13.244}$$

The matrices $\chi_{\pm}$ can also be put into the form

$$\chi_{\pm} = \begin{pmatrix} a 1_2 & V & T_{\pm} \\ V^{+} & a 1_2 & V \\ T_{\pm}^{+} & V^{+} & a 1_2 \end{pmatrix}. \tag{13.245}$$

$$T_\pm = \begin{pmatrix} 0 & \mp b \\ 0 & 0 \end{pmatrix}, \quad V = \begin{pmatrix} 0 & 0 \\ -b & 0 \end{pmatrix}. \tag{13.246}$$

We can immediately calculate

$$\omega_\pm = \begin{pmatrix} aJ^2 & JVJ & JT_\pm J \\ (JVJ)^+ & aJ^2 & JVJ \\ (JT_\pm J)^+ & (JVJ)^+ & aJ^2 \end{pmatrix} = \begin{pmatrix} A & B & \mp B^+ \\ B^+ & A & B \\ \mp B & B^+ & A \end{pmatrix}. \tag{13.247}$$

$$A = aJ^2 = \cosh 2\phi \begin{pmatrix} \cosh 2\theta & -i \sinh 2\theta \\ i \sinh 2\theta & \cosh 2\theta \end{pmatrix}$$

$$B = JVJ = \mp(JTJ)^+ = \sinh 2\phi \begin{pmatrix} -\dfrac{1}{2}\sinh 2\theta & i \sinh^2 \theta \\ -i \cosh^2 \theta & -\dfrac{1}{2}\sinh 2\theta \end{pmatrix}. \tag{13.248}$$

Generalization of this result to $n > 3$ is straightforward. We get the rotation

$$\omega_\pm = \begin{pmatrix} A & B & 0 & \cdots & \mp B^+ \\ B^+ & A & B & \cdots & 0 \\ 0 & B^+ & A & B & \cdots \\ & \cdot & & & \\ \mp B & 0 & \cdots & B^+ & A \end{pmatrix}. \tag{13.249}$$

Now we find the eigenvalues of $\omega_\pm$. We take the eigenvectors to be of the form

$$\psi = \begin{pmatrix} zu \\ z^2 u \\ \cdot \\ \cdot \\ z^n u \end{pmatrix}, \quad u = \begin{pmatrix} u_1 \\ u_2 \end{pmatrix}. \tag{13.250}$$

We get the equations

$$\begin{aligned}
(Az + Bz^2 \mp B^+ z^n)u &= \lambda z u \\
(B^+ z + Az^2 + Bz^3)u &= \lambda z^2 u \\
(B^+ z^2 + Az^3 + Bz^4)u &= \lambda z^3 u \\
&\vdots \\
(B^+ z^{n-2} + Az^{n-1} + Bz^n)u &= \lambda z^{n-1} u \\
(\mp Bz + B^+ z^{n-1} + Az^n)u &= \lambda z^n u.
\end{aligned} \tag{13.251}$$

It is not difficult to observe that only three equations are independent. These are given by the first and the last equations and anyone of the others, viz.

$$\begin{aligned}
(A + Bz \mp B^+ z^{n-1})u &= \lambda u \\
(B^+ z^{-1} + A + Bz)u &= \lambda u \\
(\mp Bz^{1-n} + B^+ z^{-1} + A)u &= \lambda u.
\end{aligned} \tag{13.252}$$

A solution is given by

$$z^n = \mp 1. \tag{13.253}$$

Equivalently

$$z = e^{\frac{i\pi k}{n}}, \quad k = 0, 1, \ldots, 2n - 1 \tag{13.254}$$

where

$$\begin{aligned}
k &= 1, 3, 5, \ldots, 2n - 1, \quad \text{for } \omega_+ \\
k &= 0, 2, 4, \ldots, 2n - 2, \quad \text{for } \omega_-.
\end{aligned} \tag{13.255}$$

The eigenvalues λ_k are solutions of the remaining equation

$$(A + Bz_k + B^+ z_k^{-1})u = \lambda_k u. \tag{13.256}$$

Thus we get for each value of k two eigenvalues λ_k. We can check that

$$\det(A + Bz_k + B^+ z_k^{-1}) = 1. \tag{13.257}$$

We can immediately conclude that

$$\lambda_k = e^{\pm r_k}. \tag{13.258}$$

Next we compute

$$\begin{aligned}
\frac{1}{2}\,\mathrm{Tr}(A + Bz_k + B^+ z_k^{-1}) &= \cosh 2\phi\ \cosh 2\theta - \frac{1}{2}\sinh 2\phi\ \sinh 2\theta\ \left(z_k + \frac{1}{z_k}\right) \\
&= \cosh 2\phi\ \cosh 2\theta - \sinh 2\phi\ \sinh 2\theta\ \cos\frac{\pi k}{n}.
\end{aligned} \tag{13.259}$$

On the other hand, we have

$$\frac{1}{2}\,\mathrm{Tr}(A + Bz_k + B^+ z_k^{-1}) = \frac{1}{2}(e^{r_k} + e^{-r_k}) = \cosh r_k. \tag{13.260}$$

Thus r_k must solve the equation

$$\cosh r_k = \cosh 2\phi\ \cosh 2\theta - \sinh 2\phi\ \sinh 2\theta\ \cos\frac{\pi k}{n}. \tag{13.261}$$

Since $\lambda_k = e^{\pm r_k}$ we can restrict r_k to be the positive solution of the above equation, viz.

$$r_k \geqslant 0. \tag{13.262}$$

It is also obvious that

$$r_{2n-k} = r_k. \tag{13.263}$$

We compute

$$\frac{\partial r_k}{\partial k}\sinh r_k = \frac{\pi}{n}\sinh 2\phi\ \sinh 2\theta\ \sin\frac{\pi k}{n}. \tag{13.264}$$

Since $r_k \geqslant 0$, $\theta \geqslant 0$ and $\phi \geqslant 0$ we have $\sinh r_k \geqslant 0$, $\sinh 2\theta \geqslant 0$ and $\sinh 2\phi \geqslant 0$. Thus $\frac{\partial r_k}{\partial k} \geqslant 0$ for all $k \leqslant n$. In other words,

$$r_0 < r_1 < \cdots < r_n. \tag{13.265}$$

In summary, we have found that the $2n$ eigenvalues of the rotation ω_+ are $e^{\pm r_k}$, $k = 1, 3, 5, \ldots, 2n - 1$ whereas the $2n$ eigenvalues of the rotation ω_- are $e^{\pm r_k}$, $k = 0, 2, 4, \ldots, 2n - 2$. The eigenvalues of the spin representations V_+ and V_- are therefore given by

$$\begin{aligned} V_+: & \quad e^{\frac{1}{2}(\pm r_1 \pm r_3 \pm \ldots \pm r_{2n-1})} \\ V_-: & \quad e^{\frac{1}{2}(\pm r_0 \pm r_2 \pm \ldots \pm r_{2n-2})}. \end{aligned} \tag{13.266}$$

The next step is to find the set of eigenvalues of

$$V = \frac{1 + U}{2} V_+ + \frac{1 - U}{2} V_-$$

where U is the chirality operator. Since the chirality operator is an idempotent the two operators $\frac{1+U}{2}$ and $\frac{1-U}{2}$ are projection operators. Hence half of the eigenvalues of V (i.e. 2^{n-1} of them) will come from the eigenvalues of V_+ while the other half (the other 2^{n-1}) will come from the eigenvalues of V_-. The eigenvalues of V are clearly positive since the eigenvalues of $V_\pm$ are all positive. In other words, the arguments that lead to equation (13.232) are correct and we only need the largest eigenvalue $\lambda_{\max}$ of V. The largest eigenvalues of V_+ and V_- are

$$\begin{aligned} V_+: & \quad \lambda_{\max +} = e^{\frac{1}{2}(+r_1 + r_3 + \cdots + r_{2n-1})} \\ V_-: & \quad \lambda_{\max -} = e^{\frac{1}{2}(+r_0 + r_2 + \cdots + r_{2n-2})}. \end{aligned} \tag{13.267}$$

Since $r_k \geqslant 0$, $r_0 < r_1 < \cdots < r_n$ and $r_{2n-k} = r_k$ the largest eigenvalue of V is $\lambda_{\max +}$, viz.

$$\lambda_{\max} \equiv \lambda_{\max +} = e^{\frac{1}{2}(+r_1 + r_3 + \cdots + r_{2n-1})}. \tag{13.268}$$

Hence

$$\frac{1}{n} \log \lambda_{\max} = \frac{1}{2n}(+r_1 + r_3 + \cdots + r_{2n-1}) = \frac{1}{2n} \sum_{k=1}^{n} r_{2k-1}. \tag{13.269}$$

13.3.5 The largest eigenvalue $\lambda_{\max}$

We are interested in the continuum limit $n \longrightarrow \infty$. We can then use the following continuous variables

$$\begin{aligned} \nu &= \frac{\pi}{n}(2k - 1) \\ r(\nu) &= r_{2k-1}. \end{aligned} \tag{13.270}$$

We make the replacement

$$\frac{1}{n}\sum_{k=1}^{n} r_{2k-1} = \frac{1}{2\pi}\int_0^{2\pi} d\nu\ r(\nu). \tag{13.271}$$

We get

$$\mathcal{L} = \lim_{n\longrightarrow\infty}\frac{1}{n}\log\lambda_{\max} = \frac{1}{4\pi}\int_0^{2\pi} d\nu\ r(\nu). \tag{13.272}$$

The condition $r_k = r_{2n-k}$ becomes $r(\nu) = r(2\pi - \nu)$. Thus

$$\mathcal{L} = \lim_{n\longrightarrow\infty}\frac{1}{n}\log\lambda_{\max} = \frac{1}{2\pi}\int_0^{\pi} d\nu\ r(\nu). \tag{13.273}$$

The $r(\nu)$ must solve the equation

$$\cosh r(\nu) = \cosh 2\phi\ \cosh 2\theta - \sinh 2\phi\ \sinh 2\theta\ \cos\nu. \tag{13.274}$$

Recall that $\phi = \beta\epsilon$ and $\tanh\theta = e^{-2\phi}$ from which we compute $\sinh 2\theta = 1/\sinh 2\phi$ and $\cosh 2\theta = \coth 2\phi$. Thus

$$\cosh r(\nu) = \cosh 2\phi\ \coth 2\phi - \cos\nu. \tag{13.275}$$

Before we proceed any further we introduce the integral

$$I(z) = \frac{1}{\pi}\int_0^{\pi} dt\ \log(2\cosh z - 2\cos t). \tag{13.276}$$

We compute

$$\frac{dI(z)}{dz} = \frac{1}{\pi}\int_0^{\pi} dt\frac{\sinh z}{\cosh z - \cos t}. \tag{13.277}$$

We use the result

$$\frac{1}{\pi}\int_0^{\pi} dt\frac{1}{1 + a\cos t} = \frac{1}{\sqrt{1 - a^2}}. \tag{13.278}$$

Thus

$$\frac{dI(z)}{dz} = \frac{\sinh z}{|\sinh z|} = \frac{z}{|z|}. \tag{13.279}$$

In other words,

$$I(z) = |z| \tag{13.280}$$

An immediate application of this result is the following

$$r(\nu) = \frac{1}{\pi}\int_0^{\pi} d\nu'\ \log(2\cosh r(\nu) - 2\cos\nu')$$
$$= \frac{1}{\pi}\int_0^{\pi} d\nu'\ \log(2\cosh 2\phi\ \coth 2\phi - 2\cos\nu - 2\cos\nu'). \tag{13.281}$$

$$\mathcal{L} = \frac{1}{2\pi^2} \int_0^\pi d\nu \int_0^\pi d\nu' \log(2\cosh 2\phi \ \coth 2\phi - 2\cos\nu - 2\cos\nu'). \quad (13.282)$$

We introduce the variables $\delta_1 = \frac{\nu + \nu'}{2}$ and $\delta_2 = \nu - \nu'$. We have $4\cos\delta_1 \cos\frac{\delta_2}{2} = 2\cos\nu + 2\cos\nu'$ and $d\nu d\nu' = d\delta_1 d\delta_2$. We obtain (with $D = \cosh 2\phi \ \coth 2\phi$)

$$
\begin{aligned}
\mathcal{L} &= \frac{1}{2\pi^2} \int_0^\pi d\delta_1 \int_0^\pi d\delta_2 \log(2D - 4\cos\delta_1 \cos\frac{\delta_2}{2}) \\
&= \frac{1}{\pi^2} \int_0^\pi d\delta_1 \int_0^{\frac{\pi}{2}} d\delta_2 \log(2D - 4\cos\delta_1 \cos\delta_2) \\
&= \frac{1}{\pi^2} \int_0^\pi d\delta_1 \int_0^{\frac{\pi}{2}} d\delta_2 \log(2\cos\delta_2) + \frac{1}{\pi^2} \int_0^\pi d\delta_1 \int_0^{\frac{\pi}{2}} d\delta_2 \log\left(\frac{D}{\cos\delta_2} - 2\cos\delta_1\right) \\
&= \frac{1}{\pi} \int_0^{\frac{\pi}{2}} d\delta_2 \log(2\cos\delta_2) + \frac{1}{\pi} \int_0^{\frac{\pi}{2}} d\delta_2 \cosh^{-1}\frac{D}{2\cos\delta_2}.
\end{aligned}
\quad (13.283)
$$

Using $\cosh^{-1}x = \log(x + \sqrt{x^2 - 1})$ we obtain (with $\kappa = 2/D$)

$$\mathcal{L} = \frac{1}{\pi} \int_0^{\frac{\pi}{2}} d\delta_2 \log[D(1 + \sqrt{1 - \kappa^2 \cos^2\delta})]. \quad (13.284)$$

We make the change of variable $\delta_2 = \frac{\pi}{2} - \rho$. We get

$$
\begin{aligned}
\mathcal{L} &= \frac{1}{\pi} \int_0^{\frac{\pi}{2}} d\rho \log[D(1 + \sqrt{1 - \kappa^2 \sin^2\rho})] \\
&= \frac{1}{2} \log\frac{2\cosh^2 2\beta\epsilon}{\sinh 2\beta\epsilon} + \frac{1}{2\pi} \int_0^\pi d\rho \log\left[\frac{1}{2}(1 + \sqrt{1 - \kappa^2 \sin^2\rho})\right].
\end{aligned}
\quad (13.285)
$$

13.3.6 Thermodynamical quantities

We have already found that

$$
\begin{aligned}
\frac{1}{n^2} \log Z &= \frac{1}{2} \log(2\sinh 2\beta\epsilon) + \frac{1}{n} \log\lambda_{\max} \\
&= \frac{1}{2} \log(2\sinh 2\beta\epsilon) + \mathcal{L} \\
&= \frac{1}{2} \log 4\cosh^2 2\beta\epsilon + \frac{1}{2\pi} \int_0^\pi d\rho \log\left[\frac{1}{2}(1 + \sqrt{1 - \kappa^2 \sin^2\rho})\right].
\end{aligned}
\quad (13.286)
$$

Thus the Helmholtz free energy is given by

$$-\frac{\beta F}{n^2} = \frac{1}{n^2} \log Z = \log 2\cosh 2\beta\epsilon + \frac{1}{2\pi} \int_0^\pi d\rho \log\left[\frac{1}{2}(1 + \sqrt{1 - \kappa^2 \sin^2\rho})\right]. \quad (13.287)$$

The internal energy is given by (with $\Delta = \sqrt{1 - \kappa^2 \sin^2 \rho}$)

$$
\begin{aligned}
-\frac{U}{n^2} = -\frac{\partial}{\partial \beta}\left(\frac{\beta F}{n^2}\right) &= 2\epsilon \tanh 2\beta\epsilon - \frac{\kappa}{2\pi}\frac{\partial \kappa}{\partial \beta}\int_0^\pi d\rho \frac{\sin^2 \rho}{\Delta(1 + \Delta)} \\
&= 2\epsilon \tanh 2\beta\epsilon + \frac{1}{2\kappa}\frac{\partial \kappa}{\partial \beta} - \frac{1}{2\pi\kappa}\frac{\partial \kappa}{\partial \beta}\int_0^\pi \frac{d\rho}{\sqrt{1 - \kappa^2 \sin^2 \rho}} \\
&= 2\epsilon \tanh 2\beta\epsilon + \frac{1}{2\kappa}\frac{\partial \kappa}{\partial \beta} - \frac{1}{\pi\kappa}\frac{\partial \kappa}{\partial \beta}\int_0^{\frac{\pi}{2}} \frac{d\rho}{\sqrt{1 - \kappa^2 \sin^2 \rho}}.
\end{aligned}
\tag{13.288}
$$

We introduce the complete elliptic integral of the first kind

$$
K_1(\kappa) = \int_0^{\frac{\pi}{2}} \frac{d\rho}{\sqrt{1 - \kappa^2 \sin^2 \rho}}
\tag{13.289}
$$

We remark that

$$
\kappa = \frac{2 \sinh 2\beta\epsilon}{\cosh^2 2\beta\epsilon}.
\tag{13.290}
$$

$$
\frac{1}{\kappa}\frac{\partial \kappa}{\partial \beta} = -2\epsilon\kappa' \coth 2\beta\epsilon, \quad \kappa' = 2\tanh^2 2\beta\epsilon - 1.
\tag{13.291}
$$

$$
\kappa^2 + \kappa'^2 = 1.
\tag{13.292}
$$

Hence

$$
-\frac{U}{n^2} = -\frac{\partial}{\partial \beta}\left(\frac{\beta F}{n^2}\right) = \epsilon \coth 2\beta\epsilon \left(1 + \frac{2\kappa'}{\pi}K_1(\kappa)\right).
\tag{13.293}
$$

The specific heat is given by

$$
\frac{C_v}{n^2} = \frac{\partial}{\partial T}\left(\frac{U}{n^2}\right) = -k_B\beta^2\frac{\partial}{\partial \beta}\left(\frac{U}{n^2}\right).
\tag{13.294}
$$

A straightforward calculation yields

$$
\frac{C_v}{n^2} = k_B\frac{2}{\pi}(\beta\epsilon \coth 2\beta\epsilon)^2\left(2K_1(\kappa) - 2E_1(\kappa) - (1 - \kappa')\left(\frac{\pi}{2} + \kappa'K_1(\kappa)\right)\right).
\tag{13.295}
$$

The $E_1(\kappa)$ is the complete elliptic integral of the second kind given by

$$
E_1(\kappa) = \int_0^{\frac{\pi}{2}} d\rho\sqrt{1 - \kappa^2 \sin^2 \rho}.
\tag{13.296}
$$

The complete elliptic integral of the first kind is divergent at $\kappa = 1$. Indeed we compute

$$
K_1(1) = \int_0^\alpha \frac{d\rho}{\sqrt{1 - \sin^2 \rho}} = \log\tan\left(\frac{\pi}{4} + \frac{\alpha}{2}\right).
\tag{13.297}
$$

This divergence is logarithmic. This can be checked as follows. We compute

$$K_1(\kappa) = \int_0^1 \frac{dx}{\sqrt{(1-x^2)(1-\kappa^2 x^2)}}$$
$$= \frac{1}{k} \int_0^k \frac{dy}{\sqrt{(1-y^2)\left(1 - \dfrac{y^2}{\kappa^2}\right)}}. \tag{13.298}$$

Near $\kappa = 1$ we can approximate the integral with

$$K_1(\kappa) = \int_0^k \frac{dy}{1-y^2}$$
$$= -\frac{1}{2}\log(1-k) + \frac{1}{2}\log 2. \tag{13.299}$$

Near $\kappa = 1$ the complete elliptic integral of the second kind is finite given by

$$E_1(1) = 1. \tag{13.300}$$

The value $\kappa = 1$ or equivalently $\kappa' = 0$ corresponds to a temperature T_c, which is given by the solution of any of the following equivalent equations

$$\frac{2\sinh 2\beta_c\epsilon}{\cosh^2 2\beta_c\epsilon} = 1, \quad 2\tanh^2 2\beta_c\epsilon = 1. \tag{13.301}$$

$$\sinh 2\beta_c\epsilon = 1, \quad \cosh 2\beta_c\epsilon = \sqrt{2}. \tag{13.302}$$

$$e^{2\beta_c\epsilon} = 1 + \sqrt{2}. \tag{13.303}$$

The behavior near $\kappa = 1$ is given by

$$e^{2\beta\epsilon} = (1 + \sqrt{2})(1 + 2\epsilon\beta\frac{T_c - T}{T_c} + \cdots). \tag{13.304}$$

$$\sinh 2\beta\epsilon = 1 + \sqrt{2}(2\epsilon\beta)\frac{T_c - T}{T_c} + \cdot s \tag{13.305}$$

Thus

$$K_1(\kappa) = -\frac{1}{2}\log(1-k) + \frac{1}{2}\log 2$$
$$= -\log\frac{|\sinh 2\beta\epsilon - 1|}{\cosh 2\beta\epsilon} + \frac{1}{2}\log 2 \tag{13.306}$$
$$= -\log|1 - \frac{T}{T_c}| + \log\frac{k_B T}{2\epsilon} + \frac{1}{2}\log 2.$$

The specific heat near $\kappa = 1$ behaves as

$$\frac{C_v}{n^2} = k_B \frac{2}{\pi} \left(\frac{2\epsilon}{k_B T_c} \right)^2 \left(-\log \left| 1 - \frac{T}{T_c} \right| + \log \frac{k_B T_c}{2\epsilon} + \frac{1}{2} \log 2 - 1 - \frac{\pi}{4} \right). \tag{13.307}$$

This diverges logarithmically at the critical point $T = T_c$. It is not difficult to observe that the internal energy is continuous at $T = T_c$. Thus there is no latent heat and the transition is second order.

13.3.7 Spontaneous magnetization

This is defined by

$$\langle M \rangle = \left\langle \sum_i s_i \right\rangle = \frac{1}{\beta} \frac{\partial \log Z}{\partial H} = -\frac{\partial F}{\partial H}. \tag{13.308}$$

$$Z = \mathrm{Tr}\, P^n. \tag{13.309}$$

$$P = (2 \sinh 2\beta\epsilon)^{\frac{n}{2}}\, V_3 V_2 V_1. \tag{13.310}$$

$$V_1 = \prod_{\alpha=1}^n e^{\theta X_\alpha}, \quad V_2 = \prod_{\alpha=1}^n e^{\beta\epsilon Z_\alpha Z_{\alpha+1}}, \quad V_3 = \prod_{\alpha=1}^n e^{\beta H Z_\alpha}. \tag{13.311}$$

A very long calculation yields at $H = 0$ the spontaneous magnetization given by [6]

$$\frac{\langle M \rangle}{n^2} = 0, \quad T > T_c. \tag{13.312}$$

$$\frac{\langle M \rangle}{n^2} = \left(1 - \frac{1}{\sinh^4 2\beta\epsilon} \right)^{\frac{1}{8}}, \quad T < T_c. \tag{13.313}$$

13.4 QED$_2$ and the Thi-ring and Sine-Gordon models

The original solution of quantum electrodynamics in two dimensions is found in [7, 8] but here we will follow the presentation of [9, 10].

13.4.1 QED in two dimensions

Quantum electrodynamics in two dimensions, also knows as the Schwinger model, is given by the path integral

$$Z[J] = \int [dA][d\bar\psi][d\psi] e^{i\int d^2x \mathcal{L} + i \int d^2 x J_\mu(x) A_\mu(x)}$$

$$\mathcal{L} = -\frac{1}{4} F_{\mu\nu} F^{\mu\nu} + \bar\psi i \gamma^\mu D_\mu \psi \tag{13.314}$$

$$D_\mu = \partial_\mu + i e A_\mu, \quad F_{\mu\nu} = \partial_\mu A_\nu - \partial_\nu A_\mu.$$

The metric is $g = (1, -1)$ and the coordinates are x^0 and x^1 ($x = (x^0, x^1)$). The Gamma matrices satisfy $\{\gamma^\mu, \gamma^\nu\} = 2g^{\mu\nu}$. Explicitly we have

$$\gamma^0 = \begin{pmatrix} 0 & -i \\ i & 0 \end{pmatrix}, \quad \gamma^1 = \begin{pmatrix} 0 & i \\ i & 0 \end{pmatrix}. \tag{13.315}$$

The chirality operator is $\gamma^5 = \gamma^0\gamma^1 = \mathrm{diag}(1, -1)$. We will also need the epsilon symbol defined by $\epsilon^{01} = -\epsilon^{10} = 1$.

The classical equations of motion are

$$\begin{aligned}
\frac{\partial\mathcal{L}}{\partial\bar\psi} - \partial_\nu\frac{\partial\mathcal{L}}{\partial_\nu\bar\psi} &= 0 \Leftrightarrow i\gamma^\mu D_\mu\psi = 0 \\
\frac{\partial\mathcal{L}}{\partial A_\mu} - \partial_\nu\frac{\partial\mathcal{L}}{\partial_\nu A_\mu} &= 0 \Leftrightarrow \partial_\nu F^{\nu\mu} = j^\mu - J^\mu, \quad j^\mu = e\bar\psi\gamma^\mu\psi.
\end{aligned} \tag{13.316}$$

The quantum equations of motion (the Dyson–Scwhinger equations) corresponding to the second equation of (13.316) reads

$$\left[(\partial^2 g_{\mu\nu} - \partial_\mu\partial_\nu)\frac{1}{i}\frac{\delta}{\delta J_\nu} + \left[J_\mu - j_\mu\left(\frac{1}{i}\frac{\delta}{\delta J}\right)\right]\right]Z[J] = 0. \tag{13.317}$$

In the above equation, we have made the substitution $A_\mu \longrightarrow \frac{1}{i}\frac{\delta}{\delta J^\mu}$ and also taken into account the fact that the current j_μ becomes in the quantum theory a function of the gauge field A_μ. In the Lorentz gauge $\partial_\mu A^\mu = 0$ the quantum equations of motion become

$$\left[(\partial^2)\frac{1}{i}\frac{\delta}{\delta J^\mu} + \left[J_\mu - j_\mu\left(\frac{1}{i}\frac{\delta}{\delta J}\right)\right]\right]Z[J] = 0. \tag{13.318}$$

Let $|\Omega\rangle$ be the vacuum of the theory. The fermion 2-point function is defined by

$$\begin{aligned}
G(x, y)_{\mu\alpha} &= \langle\Omega|T\psi_\mu(x)\bar\psi_\alpha(y)|\Omega\rangle \\
&= \frac{\int [dA][d\bar\psi][d\psi]\psi_\mu(x)\bar\psi_\alpha(y)e^{i\int d^2x\mathcal{L}+i\int d^2x J_\mu(x)A_\mu(x)}}{\int [dA][d\bar\psi][d\psi]e^{i\int d^2x\mathcal{L}+i\int d^2x J_\mu(x)A_\mu(x)}}.
\end{aligned} \tag{13.319}$$

Physically this is the amplitude of probability for propagation between x and y. T is the usual time-ordering operator defined by $T\psi_\mu(x)\bar\psi_\alpha(y) = \psi_\mu(x)\bar\psi_\alpha(y)$ if $x^0 > y^0$ and $T\psi_\mu(x)\bar\psi_\alpha(y) = -\bar\psi_\alpha(y)\psi_\mu(x)$ if $y^0 > x^0$.

The fermion 2-point function in a background gauge field is given by

$$G(x, y)^A_{\mu\alpha} = \langle\Omega|T\psi_\mu(x)\bar\psi_\alpha(y)|\Omega\rangle^A = \frac{\int [d\bar\psi][d\psi]\psi_\mu(x)\bar\psi_\alpha(y)e^{i\int d^2x\bar\psi i\gamma^\mu D_\mu\psi}}{\int [d\bar\psi][d\psi]e^{i\int d^2x\bar\psi i\gamma^\mu D_\mu\psi}}. \tag{13.320}$$

This satisfies the equations of motion

$$i\gamma^{\rho}_{\lambda\mu}D_{\rho}(x)G(x, y)^{A}_{\mu\alpha} = i\delta^{2}(x - y)\delta_{\lambda\alpha}. \tag{13.321}$$

These are the quantum equations corresponding to the first equation of (13.316). They can be derived from the invariance under $\bar{\psi} \longrightarrow \bar{\psi}' = \bar{\psi} + \bar{\epsilon}$ (keeping ψ fixed) of the 1-point function

$$\langle \Omega | \bar{\psi}_{\alpha}(y) | \Omega \rangle^{A} = \frac{\int [d\bar{\psi}][d\psi]\bar{\psi}_{\alpha}(y)e^{i\int d^{2}x\bar{\psi}i\gamma^{\mu}D_{\mu}\psi}}{\int [d\bar{\psi}][d\psi]e^{i\int d^{2}x\bar{\psi}i\gamma^{\mu}D_{\mu}\psi}}. \tag{13.322}$$

The free propagator is the solution of (13.321) with $A_{\mu} = 0$. This reads

$$\gamma^{\mu}\partial_{\mu}G_{0}(x, y) = \delta^{2}(x - y). \tag{13.323}$$

We get immediately the solution

$$G_{0}(x, y) = \int \frac{d^{2}p}{(2\pi)^{2}} \frac{\gamma^{\mu}p_{\mu}}{p^{2}} e^{-ip(x-y)}$$
$$= i\gamma^{\mu}\partial_{\mu}\left(\int \frac{d^{2}p}{(2\pi)^{2}} \frac{1}{p^{2}} e^{-ip(x-y)}\right). \tag{13.324}$$

In Euclidean signature ($x^{0} = -ix^{0}_{E}$) we get when $y \longrightarrow x$ the result

$$G_{0}(x, y)^{E} = -\gamma^{\mu}\partial_{\mu}\left(\int \frac{d^{2}\vec{p}}{4\pi^{2}} \frac{1}{\vec{p}^{2}} e^{i\vec{p}\,(\vec{x}-\vec{y})}\right)$$
$$= -\gamma^{\mu}\partial_{\mu}\left(-\frac{1}{4\pi}\ln(\vec{x} - \vec{y})^{2} + \cdots\right). \tag{13.325}$$

The terms $\cdots$ will not contribute after we perform the differentiation (see below for detailed calculation). Rotating back to Minkowskian signature we get

$$G_{0}(x, y) = -\gamma^{\mu}\partial_{\mu}\left(\frac{i}{4\pi}\ln(x - y)^{2} + \cdots\right)$$
$$= -\frac{i}{2\pi}\frac{\gamma^{\mu}(x_{\mu} - y_{\mu})}{(x - y)^{2}}. \tag{13.326}$$

13.4.2 Schwinger's solution

This consists in finding the solutions of (13.318) and (13.321) for any configuration A_{μ}. A general gauge field in two dimensions can be written in the form (this does not hold in four dimensions or noncommutative two dimensions)

$$A^{\mu} = \epsilon^{\mu\nu}\partial_{\mu}\phi + \partial^{\mu}\Lambda. \tag{13.327}$$

The scalar field ϕ is the physical degree of freedom whereas Λ is a pure gauge. Since we have already fixed the gauge we can set $\Lambda = 0$. The general solution of (13.321) is given by the 2-point function (for any A_μ)

$$G(x, y)^A = G_0(x, y)e^{ie\gamma^5(\phi(x)-\phi(y))}. \tag{13.328}$$

Indeed by using the identities $\{G_0, \gamma^5\} = 0$ and $\gamma^\mu\gamma^5 = -\epsilon^{\mu\nu}\gamma_\nu$ we calculate

$$\gamma^\mu\partial_\mu G(x, y)^A = \delta^2(x - y) - ie\gamma^\mu A_\mu(x)G(x, y)^A. \tag{13.329}$$

Next we need to evaluate the current $j^\mu = e\bar{\psi}\gamma^\mu\psi$ in the background gauge field A_μ. This quantity is needed in (13.318). The gauge-invariant definition of this current employs a Wilson line, viz.

$$\langle\Omega|j^\mu(x)|\Omega\rangle^A = e\langle\Omega|\bar{\psi}(y)\gamma^\mu e^{-ie\int_x^y dz_\mu A^\mu}\psi(x)|\Omega\rangle^A, \quad y \longrightarrow x$$
$$= -e\ tr_2\left(\gamma^\mu e^{-ie\int_x^y dz_\mu A^\mu}G(x, y)^A\right) \tag{13.330}$$

The limit $y \longrightarrow x$ should be taken in a symmetrical way, i.e. $\epsilon^\mu/\epsilon^2 \longrightarrow 0$ and $\epsilon^\mu\epsilon^\nu/\epsilon^2 \longrightarrow g^{\mu\nu}/d$ when $\epsilon = y - x \longrightarrow 0$. Recall that gauge transformations are given by $A_\mu \longrightarrow A_\mu + \partial_\mu\Lambda$, $\psi \longrightarrow \exp(-i.\,e.\,\Lambda)\psi$ and $\bar{\psi} \longrightarrow \bar{\psi}\exp(ie\Lambda)$. This dictates the form of the above expression. Now we compute

$$\langle\Omega|j^\mu(x)|\Omega\rangle^A = -e\ tr_2\left(\gamma^\mu(1 - ie\epsilon^\mu A_\mu)\left(\frac{i}{2\pi}\frac{\gamma^\nu\epsilon_\nu}{\epsilon^2}\right)(1 - ie\gamma^5\epsilon^\mu\partial_\mu\phi)\right)$$
$$= -\frac{ie}{2\pi}\frac{\epsilon_\nu}{\epsilon^2}[tr_2\gamma^\mu\gamma^\nu - ie\epsilon^\rho\partial_\rho\phi tr_2\gamma^\mu\gamma^\nu\gamma^5 - ie\epsilon^\rho A_\rho tr_2\gamma^\mu\gamma^\nu + O(\epsilon^2)] \tag{13.331}$$
$$= -\frac{e^2}{2\pi}(A^\mu + \epsilon^{\mu\nu}\partial_\nu\phi).$$

We have used $tr_2\gamma^\mu\gamma^\nu = 2g^{\mu\nu}$, $tr_2\gamma^5\gamma^\mu\gamma^\nu = 2\epsilon^{\mu\nu}$. By using (13.327) we obtain

$$\langle\Omega|j^\mu(x)|\Omega\rangle^A = -\frac{e^2}{\pi}A^\mu. \tag{13.332}$$

By using this result we can rewrite equation (13.318) as

$$\left(\partial^2 + \frac{e^2}{\pi}\right)\frac{1}{i}\frac{\delta Z[J]}{\delta J^\mu(x)} + J_\mu(x)Z[J] = 0. \tag{13.333}$$

Introduce the 2-point function

$$D(x, y) = \int\frac{d^2p}{(2\pi)^2}\frac{1}{p^2 - \dfrac{e^2}{\pi}}e^{-ip(x-y)}. \tag{13.334}$$

This describes the propagation of a scalar particle with mass e^2/π. It satisfies the equation of motion

$$(-\partial^2 - \frac{e^2}{\pi})D(x, y) = \delta^2(x - y). \tag{13.335}$$

The solution of the schwinger model is given exactly by

$$Z[J] = e^{\frac{i}{2}\int d^2x d^2y J_\mu(x) D(x,y) J^\mu(y)}.$$ (13.336)

13.4.3 The chiral anomaly

The axial current is defined by

$$j^{\mu 5} = e\bar{\psi}\gamma^\mu\gamma^5\psi.$$ (13.337)

Since $\gamma^\mu\gamma^5 = -\epsilon^{\mu\nu}\gamma_\nu$ we must have

$$j^{\mu 5} = -\epsilon^{\mu\nu}j_\nu.$$ (13.338)

Hence

$$\langle\Omega|j^{\mu 5}(x)|\Omega\rangle^A = \frac{e^2}{\pi}\epsilon^{\mu\nu}A_\nu.$$ (13.339)

Immediately we get

$$\langle\Omega|\partial_\mu j^{\mu 5}(x)|\Omega\rangle^A = \frac{e^2}{2\pi}\epsilon^{\mu\nu}F_{\mu\nu}.$$ (13.340)

13.4.4 The Sine-Gordon model

This is given by the Lagrangian density

$$\begin{aligned}
\mathcal{L} &= \frac{1}{2}\partial_\mu\phi\partial^\mu\phi + \frac{\alpha_0}{\beta^2}\cos\beta\phi + \gamma_0 \\
&= \frac{1}{2}\partial_\mu\phi\partial^\mu\phi + \left(\frac{\alpha_0}{\beta^2} - \frac{1}{2}\alpha_0\phi^2 + \frac{\alpha_0\beta^2}{4!}\phi^4 + \cdots\right) + \gamma_0.
\end{aligned}$$ (13.341)

The parameters α_0 and β are positive real numbers. The parameter α_0 is a 'squared mass' associated with small oscillations about the minimum $\phi = 0$ and β measures the strength of the interaction. We will give the scalar field a small mass μ in order to remove the IR divergences, which plague any perturbation calculations about a massless theory (especially about a massless scalar field theory in two dimensions). The Lagrangian density is then given by

$$\mathcal{L} = \mathcal{L}_0 + \frac{\alpha_0}{\beta^2}\cos\beta\phi = \frac{1}{2}\partial_\mu\phi\partial^\mu\phi - \frac{1}{2}\mu^2\phi^2 + \frac{\alpha_0}{\beta^2}\cos\beta\phi + \gamma_0.$$ (13.342)

The corresponding Hamiltonian density

$$\mathcal{H} = \frac{1}{2}\pi^2 + \frac{1}{2}(\partial_1\phi)^2 + \frac{1}{2}\mu^2\phi^2 - \frac{\alpha_0}{\beta^2}\cos\beta\phi - \gamma_0.$$ (13.343)

Clearly we want the limit $\mu \longrightarrow 0$ where (obviously) we will get the usual IR divergences. We handle this problem by multiplying the interaction by a spacetime function $f(x)$ of compact support. To find the desired Green's functions we must (1) compute all orders in perturbation theory, (2) take the limit $\mu \longrightarrow 0$, (3) sum up the series then (4) set $f = 1$.

Next we need to renormalize the above theory. For any scalar field theory in two dimensions with non-derivative interactions the only UV divergences that occur in any order of perturbation theory come from tadpole graphs (those which contain a closed loop consisting of a single internal line). Hence all UV divergences can be removed by normal ordering the Hamiltonian. We will denote by N_m the normal ordering operator defined by a mass m which is generally different from the mass μ appearing in the original model.

We will use the Wick's theorem in the form

$$T e^{i \int d^2 x J(x)\phi(x)} = N_m e^{i \int d^2 x J(x)\phi(x)} e^{-\frac{1}{2}\int J(x)\Delta(x-y;m)J(y)d^2 x d^2 y}. \tag{13.344}$$

The proof will be by computing some examples. Towards this end define $J\phi = i \int d^2 x J(x)\phi(x)$ and $\Delta = -(\int d^2 x d^2 y J(x)\Delta(x - y; m)J(y))/2$. We can then compute

$$T(J\phi) = N_m(J\phi)$$

$$T\frac{(J\phi)^2}{2!} = N_m \frac{(J\phi)^2}{2!} + \Delta$$

$$T\frac{(J\phi)^3}{3!} = N_m \frac{(J\phi)^3}{3!} + N_m \frac{(J\phi)^1}{1!}\Delta$$

$$T\frac{(J\phi)^4}{4!} = N_m \frac{(J\phi)^4}{4!} + N_m \frac{(J\phi)^2}{2!}\Delta + \frac{\Delta^2}{2!} \tag{13.345}$$

$$T\frac{(J\phi)^5}{5!} = N_m \frac{(J\phi)^5}{5!} + N_m \frac{(J\phi)^3}{3!}\Delta + N_m \frac{(J\phi)^1}{1!}\frac{\Delta^2}{2!}.$$

This is sufficient to show the desired structure. In the above equation, $\Delta(x - y; m)$ is the contraction of two fields $\phi(x)$ and $\phi(y)$, i.e., it is the Feynman propagator (if we remove the time ordering operator T then we will get the Wightman function instead). It is given by the equation

$$\Delta(x - y; m) = \int \frac{d^2 p}{(2\pi)^2} \frac{1}{p^2 - m^2} e^{-ip(x-y)}. \tag{13.346}$$

To stress one more time we are using in the normal ordering a mass m different from the mass μ appearing in the original Lagrangian density. The Euclidean version of this integral is (with $y \neq x$)

$$\Delta(x - y; m)^E = i \int \frac{d^2 \vec{p}}{(2\pi)^2} \frac{1}{\vec{p}^2 + m^2} e^{i\vec{p}\,(\vec{x}-\vec{y})}$$

$$= i \int_0^\infty \int_{-\frac{\pi}{2}}^{\frac{\pi}{2}} \frac{dp^2 d\alpha}{4\pi^2} \frac{\cos(p|\vec{x} - \vec{y}|\cos\alpha)}{p^2 + m^2} \tag{13.347}$$

$$= i \int_0^\infty \int_{-\frac{\pi}{2}}^{\frac{\pi}{2}} \frac{dp^2 d\alpha}{4\pi^2} \frac{\cos(p\cos\alpha)}{p^2 + m^2 |\vec{x} - \vec{y}|^2}.$$

In the limit $y \longrightarrow x$ we can set the cosine equal to 1 and we find the following dominant contribution

$$\Delta(x - y; m)^E = i \int \frac{d^2\vec{p}}{(2\pi)^2} \frac{1}{\vec{p}^2 + m^2 |\vec{x} - \vec{y}|^2}. \tag{13.348}$$

In other words, we have (with $d = 2 - \epsilon$) [3]

$$\begin{aligned}
\Delta(x - y; m)^E &= i \int \frac{d^d\vec{p}}{(2\pi)^d} \frac{1}{\vec{p}^2 + m^2 |\vec{x} - \vec{y}|^2} \\
&= \frac{i}{4\pi} \Gamma\left(\frac{\epsilon}{2}\right) \left(\frac{4\pi}{m^2 |\vec{x} - \vec{y}|^2}\right)^{\frac{\epsilon}{2}} \\
&= \frac{i}{4\pi} \left(\frac{2}{\epsilon} - \gamma\right) \left(1 + \frac{\epsilon}{2} \log\left(\frac{4\pi}{m^2 |\vec{x} - \vec{y}|^2}\right)\right) \\
&= \frac{i}{4\pi} \left(\frac{2}{\epsilon} - \log cm^2 |\vec{x} - \vec{y}|^2 + O(\epsilon^2)\right).
\end{aligned} \tag{13.349}$$

Hence

$$\Delta(x - y; m) = \frac{1}{4\pi} \left(\frac{2}{\epsilon} - \log(-cm^2(x - y)^2) + O(\epsilon^2)\right). \tag{13.350}$$

By dropping the singular term we get when $y \longrightarrow x$ the propagator

$$\Delta(x - y; m) = -\frac{1}{4\pi} \log\left(-cm^2(x - y)^2\right) + O((x - y)^2), \quad c = \frac{e^\gamma}{4\pi}. \tag{13.351}$$

Let us do this calculation for $x = y$. From (13.347) we have

$$\Delta(0; m)^E = i \int \frac{d^2\vec{p}}{(2\pi)^2} \frac{1}{\vec{p}^2 + m^2}. \tag{13.352}$$

This gives using an ordinary cutoff Λ the result

$$\Delta(0; m) = -\frac{1}{4\pi} \log \frac{m^2}{\Lambda^2}, \quad y = x. \tag{13.353}$$

This agrees with Coleman's result although he got it using a different method. Now we use this result in Wick's theorem (13.344) with $J(x) = \beta\delta(x - x_1)$. We get immediately

$$e^{i\beta\phi(x_1)} = \left(\frac{m^2}{\Lambda^2}\right)^{\frac{\beta^2}{8\pi}} N_m e^{i\beta\phi(x_1)}. \tag{13.354}$$

Thus

$$\alpha_0 \cos \beta\phi = \alpha N_m \cos \beta\phi, \quad \alpha = \alpha_0 \left(\frac{m^2}{\Lambda^2}\right)^{\frac{\beta^2}{8\pi}}. \tag{13.355}$$

We remark the identity (where N_μ is the normal ordering operator defined by the mass μ)

$$N_m \cos \beta\phi = \left(\frac{\mu^2}{m^2}\right)^{\frac{\beta^2}{8\pi}} N_\mu \cos \beta\phi. \tag{13.356}$$

Let us now introduce

$$\phi(x^1) = \phi_+(x^1) + \phi_-(x^1)$$

$$\phi_+(x^1) = \int \frac{dk^1}{2\pi} \frac{1}{\sqrt{2\omega(k^1, m)}} a(k^1, m) e^{-ik^1 x_1}$$

$$\phi_-(x^1) = \int \frac{dk^1}{2\pi} \frac{1}{\sqrt{2\omega(k^1, m)}} a^+(k^1, m) e^{ik^1 x_1}. \tag{13.357}$$

As usual $\omega(k^1, m)^2 = (k^0)^2 = (k^1)^2 + m^2$ and $[a(k^1, m), a^+(p^1, m)] = 2\pi\delta(k^1 - p^1)$. We can compute

$$T(\partial_1\phi)^2 = N_m(\partial_1\phi)^2 + [\partial_1\phi_+, \partial_1\phi_-] = N_m(\partial_1\phi)^2 + \int \frac{dk^1}{2\pi} \frac{(k^1)^2}{2\omega(k^1, m)}$$

$$T(\phi)^2 = N_m(\phi)^2 + [\phi_+, \phi_-] = N_m(\phi)^2 + \int \frac{dk^1}{2\pi} \frac{1}{2\omega(k^1, m)}. \tag{13.358}$$

Similarly

$$\pi(x^1) = \pi_+(x^1) + \pi_-(x^1)$$

$$\pi_+(x^1) = -i \int \frac{dk^1}{2\pi} \sqrt{\frac{\omega(k^1, m)}{2}} a(k^1, m) e^{-ik^1 x_1}, \quad \pi_-(x^1) = i \int \frac{dk^1}{2\pi} \sqrt{\frac{\omega(k^1, m)}{2}} a^+(k^1, m) e^{ik^1 x_1}. \tag{13.359}$$

Then

$$T(\pi)^2 = N_m(\pi)^2 + [\pi_+, \pi_-] = N_m(\pi)^2 + \int \frac{dk^1}{2\pi} \frac{(k^1)^2 + m^2}{2\omega(k^1, m)}. \tag{13.360}$$

Hence we get

$$T\left(\frac{1}{2}\pi^2 + \frac{1}{2}(\partial_1\phi)^2 + \frac{1}{2}\mu^2\phi^2 - \gamma_0\right) = N_m\left(\frac{1}{2}\pi^2 + \frac{1}{2}(\partial_1\phi)^2 + \frac{1}{2}\mu^2\phi^2 - \gamma\right). \tag{13.361}$$

The renormalized coupling γ is given by

$$\gamma = \gamma_0 - \int \frac{dk^1}{8\pi} \frac{2(k^1)^2 + m^2 + \mu^2}{\omega(k^1, m)}. \tag{13.362}$$

The renormalized Hamiltonian is

$$\mathcal{H} = N_m\left(\frac{1}{2}\pi^2 + \frac{1}{2}(\partial_1\phi)^2 + \frac{1}{2}\mu^2\phi^2 - \frac{\alpha}{\beta^2}\cos \beta\phi - \gamma\right). \tag{13.363}$$

The mass μ does not get renormalized (anyway we want to take the limit $\mu \longrightarrow 0$ at the end of calculation). Similarly the parameter β does not get renormalized. The parameters α_0 and γ_0 were renormalized (multiplicatively and additively, respectively) to obtain the finite values α and γ. All that was required is normal ordering. In all this the coupling α is the only crucial parameter.

We now compute the following quantity (where ϕ is a field of mass μ)

$$\left\langle 0, \mu \Big| T \prod_i N_m e^{i\beta_i \phi(x_i)} \Big| 0, \mu \right\rangle = \left(\frac{\mu^2}{m^2} \right)^{\sum_i \frac{\beta_i^2}{8\pi}} \left\langle 0, \mu \Big| T \prod_i N_\mu e^{i\beta_i \phi(x_i)} \Big| 0, \mu \right\rangle$$

$$= \left(\frac{\Lambda^2}{m^2} \right)^{\sum_i \frac{\beta_i^2}{8\pi}} \left\langle 0, \mu \Big| T \prod_i e^{i\beta_i \phi(x_i)} \Big| 0, \mu \right\rangle . \tag{13.364}$$

We have used the identities (13.356), (13.354). Thus we have using Wick's theorem (13.344) the result

$$\left\langle 0, \mu \Big| T \prod_i N_m e^{i\beta_i \phi(x_i)} \Big| 0, \mu \right\rangle = \left(\frac{\Lambda^2}{m^2} \right)^{\sum_i \frac{\beta_i^2}{8\pi}} \left\langle 0, \mu | T e^{i \sum_i \beta_i \phi(x_i)} | 0, \mu \right\rangle$$

$$= \left(\frac{\Lambda^2}{m^2} \right)^{\sum_i \frac{\beta_i^2}{8\pi}} e^{-\sum_{i,j} \frac{1}{2} \beta_i \beta_j \Delta(x_i - x_j; \mu)} . \tag{13.365}$$

Assuming that the spacetime points x_i's are restricted to a finite region (determined by the regulator $f(x)$) we can use the short-distance formula (13.351). We get by neglecting contractions of fields at the same point the expression

$$\left\langle 0, \mu \Big| T \prod_i N_m e^{i\beta_i \phi(x_i)} \Big| 0, \mu \right\rangle = \left(\frac{\Lambda^2}{m^2} \right)^{\sum_i \frac{\beta_i^2}{8\pi}} \prod_{i>j} (-c\mu^2(x_i - x_j)^2)^{\frac{\beta_i \beta_j}{4\pi}} . \tag{13.366}$$

This expression is proportional to $\mu^{(\sum_i \beta_i)^2/4\pi}$ (including the contribution from the neglected contractions). Hence if $\sum_i \beta_i \neq 0$ we get the result that this expression must vanish when $\mu \longrightarrow 0$. In other words, when $\mu \longrightarrow 0$ we must have $\sum_i \beta_i = 0$ otherwise expectation value is 0.

We are interested in the perturbation series of the theory in powers of α_0 given by (with $A_\pm = exp(\pm i\beta\phi)$)

$$\int [d\phi] e^{i \int d^2x \mathcal{L}_0} e^{i \frac{\alpha_0}{2\beta^2} \int d^2x (A_+ + A_-)} =$$

$$\sum_n \frac{(i\frac{\alpha_0}{2\beta^2})^{2n}}{(n!)^2} \prod_{i=1}^n \int d^2x_i d^2y_i \left\langle 0, \mu \Big| T \prod_{i=1}^n A_+(x_i) A_-(y_i) \Big| 0, \mu \right\rangle , \tag{13.367}$$

where

$$\left\langle 0, \mu \Big| T \prod_{i=1}^n A_+(x_i) A_-(y_i) \Big| 0, \mu \right\rangle = \int [d\phi] e^{i \int d^2x \mathcal{L}_0} A_+(x_i) A_-(y_i) . \tag{13.368}$$

We have used the fact that the only non-zero terms in the expansion are those with equal numbers of A_+ and A_-. We have by using Wick's theorem (13.344) and by dropping contractions at the same point the result

$$\left\langle 0, \mu \left| T \prod_i A_+(x_i) A_-(y_i) \right| 0, \mu \right\rangle = \left\langle 0, \mu \left| T e^{i\beta \sum_i (\phi(x_i) - \phi(y_i))} \right| 0, \mu \right\rangle$$

$$= e^{-\frac{\beta^2}{2} \sum_{i,j} [\Delta(x_i - x_j; m) - \Delta(x_i - y_j; m) - \Delta(y_i - x_j; m) + \Delta(y_i - y_j; m)]}$$

$$= \frac{\prod_{i>j} [c^2 m^4 (x_i - x_j)^2 (y_i - y_j)^2]^{\frac{\beta^2}{4\pi}}}{\prod_{i,j} [cm^2 (x_i - y_j)^2]^{\frac{\beta^2}{4\pi}}}. \tag{13.369}$$

The last calculation we want to do is to compute the commutators

$$[\partial_\nu \phi(x), A_\pm(y)] = e^{\pm i\beta \phi_+(y)}[\partial_\nu \phi_+(x), e^{\pm i\beta \phi_-(y)}] + [\partial_\nu \phi_-(x), e^{\pm i\beta \phi_+(y)}] e^{\pm i\beta \phi_-(y)}. \tag{13.370}$$

In the above equation, we have used the fact that $\phi = \phi_+ + \phi_-$, $\partial_\nu \phi = \partial_\nu \phi_+ + \partial_\nu \phi_-$. Using the expressions (with $k^0 = \omega(k^1, m)$)

$$\phi_+(x) = \int \frac{dk^1}{2\pi} \frac{1}{\sqrt{2\omega(k^1, m)}} a(k^1, m) e^{-ikx}, \quad \phi_-(x) = \int \frac{dk^1}{2\pi} \frac{1}{\sqrt{2\omega(k^1, m)}} a^+(k^1, m) e^{ikx}. \tag{13.371}$$

The expressions (13.371) are the fields in the Heisenberg picture whereas (13.357) are the fields in the Schrodinger picture. We can immediately compute (with $k^0 = \omega(k^1, m)$)

$$[\phi_+(x), \phi_-(y)] = \int \frac{dk^1}{2\pi} \frac{1}{2\omega(k^1, m)} e^{-ik(x-y)}$$

$$= i\Delta_+(x - y). \tag{13.372}$$

Thus

$$[\phi_+(x), \phi_-(y)^n] = in\phi_-^{n-1}(y) \Delta_+(x - y). \tag{13.373}$$

In other words, we have

$$[\phi_+(x), e^{\pm i\beta \phi_-(y)}] = \mp \beta \Delta_+(x - y) e^{\pm i\beta \phi_-(y)}. \tag{13.374}$$

Similarly, we compute

$$[\phi_-(x), e^{\pm i\beta \phi_+(y)}] = \mp \beta \Delta_-(x - y) e^{\pm i\beta \phi_+(y)}. \tag{13.375}$$

Now (with $k^0 = -\omega(k^1, m)$)

$$-[\phi_-(x), \phi_+(y)] = \int \frac{dk^1}{2\pi} \frac{1}{2\omega(k^1, m)} e^{-ik(x-y)}$$

$$= -i\Delta_-(x - y). \tag{13.376}$$

Hence

$$[\phi(x), A_\pm(y)] = \mp \beta \Delta(x - y) A_\pm(y). \tag{13.377}$$

We have used the identity

$$
\begin{aligned}
i\Delta_+(x - y) + i\Delta_-(x - y) &= \int \frac{dk^1}{2\pi} \frac{1}{2\omega(k^1, m)} e^{-ik(x-y)}\big|_{k^0 = \omega(k^1, m)} \\
&\quad + \int \frac{dk^1}{2\pi} \frac{1}{-2\omega(k^1, m)} e^{-ik(x-y)}\big|_{k^0 = -\omega(k^1, m)} \\
&= i\Delta(x - y).
\end{aligned} \tag{13.378}
$$

The propagator $\Delta(x - y)$ is given by (13.346). The sign of (13.377) agrees with Coleman's sign in his paper on the Schwinger model but not with his sign in his paper on the Sine-Gordon model.

From (13.377) we deduce

$$[\epsilon^{\mu\nu}\partial_\nu\phi(x), A_\pm(y)] = \mp\beta\epsilon^{\mu\nu}\partial_\nu\Delta(x - y)A_\pm(y). \tag{13.379}$$

The final point we mention about the Sine-Gordon model is that the quantum theory does not make sense for $\beta > 8\pi$ since the energy becomes unbounded from below and the theory has no ground state.

13.4.5 The Thirring model

The massless Thirring model is given by the action

$$\mathcal{L} = \bar{\psi}i\gamma_\mu\partial^\mu\psi - \frac{1}{2}g\hat{j}_\mu\hat{j}^\mu, \quad \hat{j}^\mu = \bar{\psi}\gamma^\mu\psi. \tag{13.380}$$

By requiring that the current $\hat{j}^\mu$ must obey the appropriate Ward identities we find that the Hamiltonian of the above massless Thirring model requires no further renormalization. The model is exactly solvable and makes sense for all $g > -\pi$. We are interested in the scalar density $\bar{\psi}\psi$. The renormalized scalar density involve a cutoff-dependent constant Z, viz.

$$\sigma = Z\bar{\psi}\psi, \quad \sigma_+ = Z\bar{\psi}_0\psi_0, \quad \sigma_- = Z\bar{\psi}_1\psi_1. \tag{13.381}$$

The massive Thirring model is defined by

$$\mathcal{L}_m = \mathcal{L} - m'\sigma. \tag{13.382}$$

The perturbation series of the theory in powers of m' is well defined (aside from trivial infrared divergences associated with the massless Thirring model). We are thus interested in perturbation theory in m', viz

$$
\begin{aligned}
&\int [d\psi][d\bar{\psi}]e^{i\int d^2x\mathcal{L}}e^{-im'\int d^2x(\sigma_+ + \sigma_-)} = \\
&\sum_n \frac{(-im')^{2n}}{(n!)^2} \prod_{i=1}^n \int d^2x_i d^2y_i \langle 0 \,|\, T\prod_{i=1}^n \sigma_+(x_i)\sigma_-(y_i) \,|\, 0 \rangle,
\end{aligned} \tag{13.383}
$$

where

$$\left\langle 0 \left| T\prod_{i=1}^{n} \sigma_+(x_i)\sigma_-(y_i) \right| 0 \right\rangle = \int [d\psi][d\bar{\psi}]e^{i\int d^2x \mathcal{L}}\sigma_+(x_i)\sigma_-(y_i). \tag{13.384}$$

In the above equation, all terms which contain unequal numbers of σ_+ and σ_- vanish by chiral invariance of the massless Thirring model. This is obvious at least for $g = 0$ where we can use the free propagators given by (13.387). Coleman computes these expectation values using Klaiber's results and he finds

$$\left\langle 0 \left| T\prod_{i=1}^{n} \sigma_+(x_i)\sigma_-(y_i) \right| 0 \right\rangle = \left(\frac{1}{2}\right)^{2n} \frac{\prod_{i>j} [M^4(x_i - x_j)^2(y_i - y_j)^2]^{1+\frac{b}{\pi}}}{\prod_{i,j} [M^2(x_i - y_j)^2]^{1+\frac{b}{\pi}}}. \tag{13.385}$$

The parameter M is an arbitrary mass, which is the reflection of the arbitrary finite renormalization in the definition of σ. In other words, σ in this formula should be given by equation (13.381). The Klaiber's parameter is given by

$$b = -\frac{g}{1 + \frac{g}{\pi}}. \tag{13.386}$$

Let us say that this result should be easily checked for $g = 0$, $b = 0$. For $g = 0$ we have computed the following 2-point functions (from (13.326))

$$\langle 0|T\psi_0(x)\bar{\psi}_1(y)|0\rangle = -\frac{1}{2\pi}\frac{(x_0 - y_0) - (x_1 - y_1)}{(x - y)^2}$$

$$\langle 0|T\psi_1(x)\bar{\psi}_0(y)|0\rangle = \frac{1}{2\pi}\frac{(x_0 - y_0) + (x_1 - y_1)}{(x - y)^2} \tag{13.387}$$

$$\langle 0|T\psi_0(x)\bar{\psi}_0(y)|0\rangle = 0$$

$$\langle 0|T\psi_1(x)\bar{\psi}_1(y)|0\rangle = 0.$$

By comparing (13.368), (13.369) with (13.384), (13.385) we see that the perturbative series of the Thirring model in m' is identical to the perturbative series of the Sine-Gordon model in α_0 provided we match the couplings and fields in the two theories as follows

$$1 + \frac{b}{\pi} = \frac{1}{1 + \frac{g}{\pi}} = \frac{\beta^2}{4\pi}. \tag{13.388}$$

$$-m'\sigma = \frac{\alpha_0}{\beta^2}\cos\beta\phi \Leftrightarrow -m'Z\bar{\psi}\psi = \frac{\alpha}{\beta^2}N_m\cos\beta\phi. \tag{13.389}$$

These two matching conditions do not depend on renormalization conventions. In other words, they do not depend on the equation

$$M^2 = cm^2. \tag{13.390}$$

We can choose

$$m' = \frac{\alpha_0}{\beta^2}$$

$$-\sigma_\pm = \frac{1}{2}A_\pm \Leftrightarrow -\sigma = -Z\bar{\psi}\psi = \cos\beta\phi.$$

(13.391)

or

$$m' = \frac{\alpha}{\beta^2}$$

$$-\sigma_\pm = \frac{1}{2}N_m A_\pm \Leftrightarrow -\sigma = -Z\bar{\psi}\psi = N_m \cos\beta\phi.$$

(13.392)

Coleman computes also the commutators

$$[\hat{j}^\mu(x),\, \sigma_\pm(y)] = \mp \frac{2}{1 + \frac{g}{\pi}}\epsilon^{\mu\nu}\partial_\nu\Delta(x - y)\sigma_\pm(y).$$

(13.393)

Comparing with (13.379) we get the extra identity

$$\hat{j}^\mu = \bar{\psi}\gamma^\mu\psi = \frac{\beta}{2\pi}\epsilon^{\mu\nu}\partial_\nu\phi.$$

(13.394)

The sign does not agree with Coleman's sign in the paper on the Sine-Gordon model. This is related to the sign of (13.379). However, this sign agrees with his sign in the paper on the Schwinger model. Compare this equation with (13.332).

13.4.6 Coleman's solution of QED in 2D

We take the model

$$\mathcal{L} = -\frac{1}{4}F_{\mu\nu}F^{\mu\nu} + \bar{\psi}(i\gamma^\mu\partial_\mu - e\gamma^\mu A_\mu - m)\psi.$$

(13.395)

We use the (axial or radiation) gauge

$$A_1 = 0.$$

(13.396)

The equation of motion of the remaining field A_0 becomes a constraint equation given by

$$\partial_1^2 A_0 = -j_0 = -e\psi^+\psi.$$

(13.397)

In one spatial dimension there is no photon since there are no transverse dimensions. We solve the above equation by introducing the following Green's function

$$-\partial_1^2 D(x^1, y^1) = \delta(x^1 - y^1).$$

(13.398)

It has the solution

$$D(x^1, y^1) = \int \frac{dp_1}{2\pi}\frac{1}{p_1^2}e^{-ip_1(x^1 - y^1)}.$$

(13.399)

Then we have the solutions for A_0 and F_{01} given by

$$A_0 = \int D(x^1, y^1) j_0(y^1) dy^1 - Fx^1 - G, \quad F_{01} = -\int \partial_1 D(x^1, y^1) j_0(y^1) dy^1 + F. \qquad (13.400)$$

The constant G is irrelevant but F (a background electric field) is physically significant. It is essentially the famous theta angle. Indeed, as we will see we can make the identification

$$F = e\frac{\theta}{2\pi}. \qquad (13.401)$$

The parameter θ, which is an angle in $[-\pi, \pi]$, is an extra parameter in the model, which is completely independent of the charge e and the mass m. In four dimensions a background electric field will be cancelled by pair production from the vacuum.

We need the quantity

$$-\partial_1 D(x^1, y^1) = \int \frac{dp_1}{2\pi} \frac{i}{p_1} e^{-ip_1(x^1 - y^1)} = \frac{1}{2}\theta(x^1 - y^1). \qquad (13.402)$$

This can be checked by using the identity

$$\int dy^1 \theta(x^1 - y^1) = 2x^1. \qquad (13.403)$$

Hence we must have

$$D(x^1, y^1) = -\frac{1}{2}|x^1 - y^1|. \qquad (13.404)$$

The canonical momentum associated with the field ψ is $\bar{\psi}i\gamma^0$. All other canonical momenta are zero. Thus we have the Hamiltonian density (by using also the equation of motion and integration by parts)

$$\mathcal{H} = \bar{\psi}(i\gamma_1\partial_1 + m)\psi + \frac{1}{2}(\partial_1 A_0)^2. \qquad (13.405)$$

Restriction on the states of charge zero (which seems to be necessary for translational invariance) means that we must have

$$\int j_0(x^1) dx^1 = 0. \qquad (13.406)$$

The Hamiltonian takes on these states the form

$$H = \int \bar{\psi}(i\gamma_1\partial_1 + m)\psi - \frac{1}{4}\int dx^1 dy^1 j_0(x^1) j_0(y^1)|x^1 - y^1| - \frac{F}{2}\int dx^1 j_0(x^1)x^1. \qquad (13.407)$$

Let us specialize our results of the Sine-Gordon and Thirring models to the values $g = 0$ (the Thirring model becomes a theory of a free massive Dirac field) and $\beta^2 = 4\pi$ (in the Sine-Gordon model). The identity (13.388) is trivially satisfied. The identity (13.389) reads with $m'Z = m$ (which is the mass used in normal ordering) as follows

$$-m\bar{\psi}\psi = \frac{\alpha}{4\pi}N_m\cos 2\sqrt{\pi}\,\phi. \tag{13.408}$$

If we choose

$$\frac{\alpha}{4\pi} \equiv cm^2 = M^2. \tag{13.409}$$

Then we will obtain instead the relation

$$\bar{\psi}\psi = -cmN_m\cos 2\sqrt{\pi}\,\phi. \tag{13.410}$$

We get therefore the two equivalent theories (we put also $\mu = 0$ and $\gamma = 0$ in our expressions)

$$\mathcal{L} = \bar{\psi}(i\gamma_\mu\partial^\mu - m)\psi \Leftrightarrow \mathcal{H} = \bar{\psi}(i\gamma_1\partial_1 + m)\psi. \tag{13.411}$$

$$\begin{aligned}
\mathcal{H} &= N_m\left(\frac{1}{2}\pi^2 + \frac{1}{2}(\partial_1\phi)^2 - \frac{\alpha}{4\pi}\cos 2\sqrt{\pi}\,\phi\right) \\
&= N_m\left(\frac{1}{2}\pi^2 + \frac{1}{2}(\partial_1\phi)^2 - cm^2\cos 2\sqrt{\pi}\,\phi\right).
\end{aligned} \tag{13.412}$$

We have also the equivalence between the local operators in the two theories

$$\bar{\psi}\gamma^\mu\psi = \frac{1}{\sqrt{\pi}}\epsilon^{\mu\nu}\partial_\nu\phi. \tag{13.413}$$

States of charge 0 (analogue of states of color 0 in QCD) satisfy

$$\int dx^1 j^0(x^1) = e\int dx^1\bar{\psi}\gamma^0\psi = \frac{e}{\sqrt{\pi}}\int dx^1\partial_1\phi = 0 \tag{13.414}$$

On these states we can immediately compute

$$\begin{aligned}
F_{01} &= -e\int \partial_1 D(x^1, y^1)\bar{\psi}(y^1)\gamma^0\psi(y^1)dy^1 + \frac{e\theta}{2\pi} \\
&= -\frac{e}{\sqrt{\pi}}\int \partial_1 D(x^1, y^1)\partial_1\phi(y^1)dy^1 + \frac{e\theta}{2\pi} \\
&= -\frac{e}{\sqrt{\pi}}\left(\phi - \frac{\theta}{2\sqrt{\pi}}\right).
\end{aligned} \tag{13.415}$$

The Hamiltonian (13.405) becomes

$$\mathcal{H} = N_m\left(\frac{1}{2}\pi^2 + \frac{1}{2}(\partial_1\phi)^2 - cm^2\cos 2\sqrt{\pi}\,\phi + \frac{e^2}{2\pi}\left(\phi - \frac{\theta}{2\sqrt{\pi}}\right)^2\right). \tag{13.416}$$

Finally we make the shift $\phi \longrightarrow -\phi + \frac{\theta}{2\sqrt{\pi}}$ and set

$$\mathcal{M}^2 = \frac{e^2}{\pi}. \tag{13.417}$$

We get

$$
\begin{aligned}
\mathcal{H} &= N_m\left(\frac{1}{2}\pi^2 + \frac{1}{2}(\partial_1\phi)^2 - cm^2\cos(2\sqrt{\pi}\,\phi - \theta) + \frac{\mathcal{M}^2}{2}\phi^2\right) \\
&= N_{\mathcal{M}}\left(\frac{1}{2}\pi^2 + \frac{1}{2}(\partial_1\phi)^2 - cm\mathcal{M}\cos(2\sqrt{\pi}\,\phi - \theta) + \frac{\mathcal{M}^2}{2}\phi^2\right).
\end{aligned}
\tag{13.418}
$$

We get exactly the correct Schwinger mass $\mathcal{M} = e/\sqrt{\pi}$. From this we can see immediately that θ is indeed an angle. The physics is periodic in θ. We can restrict this angle to the interval $[-\pi, \pi]$. This angle is independent from the mass m and the electromagnetic coupling e^2 (or $\mathcal{M}^2$) and thus it is another fundamental parameter of QED$_2$.

13.4.7 Flavor and condensate

The two-flavor model is given by

$$
\mathcal{L} = -\frac{1}{4}F_{\mu\nu}F^{\mu\nu} + \sum_{i=1}^{2}\bar{\psi}_i(i\gamma^\mu\partial_\mu - e\gamma^\mu A_\mu - m)\psi_i.
\tag{13.419}
$$

We use the (axial or radiation) gauge $A_1 = 0$. The equation of motion of the remaining field A_0 becomes a constraint equation given by

$$
\partial_1^2 A_0 = -j_0 = -e\sum_{i=1}^{2}\psi_i^+\psi_i.
\tag{13.420}
$$

Only the definition of j_0 has changed. Thus as before we have the solutions for A_0 and F_{01} given by

$$
A_0 = \int D(x^1, y^1)j_0(y^1)dy^1 - Fx^1 - G, \quad F_{01} = -\int \partial_1 D(x^1, y^1)j_0(y^1)dy^1 + F, \quad F = e\frac{\theta}{2\pi}.
\tag{13.421}
$$

The canonical momentum associated with the fields ψ_i is $\bar{\psi}_i i\gamma^0$. All other canonical momenta are zero. Thus we have the Hamiltonian density

$$
\mathcal{H} = \sum_{i=1}^{2}\bar{\psi}_i(i\gamma_1\partial_1 + m)\psi_i + \frac{1}{2}(\partial_1 A_0)^2.
\tag{13.422}
$$

We introduce two scalar fields ϕ_i by the equations

$$
-m\bar{\psi}_1\psi_1 = \frac{\alpha}{4\pi}N_m\cos 2\sqrt{\pi}\,\phi_1, \quad -m\bar{\psi}_2\psi_2 = \frac{\alpha}{4\pi}N_m\cos 2\sqrt{\pi}\,\phi_2.
\tag{13.423}
$$

As before we choose

$$
\frac{\alpha}{4\pi} = cm^2 = \mathcal{M}^2.
\tag{13.424}
$$

We get therefore the two equivalent theories

$$\mathcal{L} = \sum_{i=1}^{2} \bar{\psi}_i(i\gamma_\mu \partial^\mu - m)\psi_i \Leftrightarrow \mathcal{H} = \sum_{i=1}^{2} \bar{\psi}_i(i\gamma_1 \partial_1 + m)\psi_i. \tag{13.425}$$

$$\mathcal{H} = N_m\left(\frac{1}{2}\pi_1^2 + \frac{1}{2}\pi_2^2 + \frac{1}{2}(\partial_1\phi_1)^2 + \frac{1}{2}(\partial_1\phi_2)^2 - \frac{\alpha}{4\pi}\cos 2\sqrt{\pi}\,\phi_1 - \frac{\alpha}{4\pi}\cos 2\sqrt{\pi}\,\phi_1\right)$$

$$= N_m\left(\frac{1}{2}\pi_1^2 + \frac{1}{2}\pi_2^2 + \frac{1}{2}(\partial_1\phi_1)^2 + \frac{1}{2}(\partial_1\phi_2)^2 - cm^2\cos 2\sqrt{\pi}\,\phi_1 - cm^2\cos 2\sqrt{\pi}\,\phi_1\right). \tag{13.426}$$

We have also the equivalence between the local operators in the two theories

$$\bar{\psi}_1\gamma^\mu\psi_1 = \frac{1}{\sqrt{\pi}}\epsilon^{\mu\nu}\partial_\nu\phi_1, \quad \bar{\psi}_2\gamma^\mu\psi_2 = \frac{1}{\sqrt{\pi}}\epsilon^{\mu\nu}\partial_\nu\phi_2. \tag{13.427}$$

States of charge 0 satisfy

$$\int dx^1 j^0(x^1) = e\sum_{i=1}^{2}\int dx^1 \bar{\psi}_i\gamma^0\psi_i = \frac{e}{\sqrt{\pi}}\sum_{i=1}^{2}\int dx^1 \partial_1\phi_i = 0 \tag{13.428}$$

On these states we can immediately compute

$$F_{01} = -e\sum_{i=1}^{2}\int \partial_1 D(x^1, y^1)\bar{\psi}_i(y^1)\gamma^0\psi_i(y^1)dy^1 + \frac{e\theta}{2\pi}$$

$$= -\frac{e}{\sqrt{\pi}}\sum_{i=1}^{2}\int \partial_1 D(x^1, y^1)\partial_1\phi_i(y^1)dy^1 + \frac{e\theta}{2\pi} \tag{13.429}$$

$$= -\frac{e}{\sqrt{\pi}}(\phi_1 + \phi_2 - \frac{\theta}{2\sqrt{\pi}}).$$

The Hamiltonian (13.422) becomes

$$\mathcal{H} = N_m\left(\frac{1}{2}\pi_1^2 + \frac{1}{2}\pi_2^2 + \frac{1}{2}(\partial_1\phi_1)^2 + \frac{1}{2}(\partial_1\phi_2)^2 - cm^2\cos 2\sqrt{\pi}\,\phi_1 - cm^2\cos 2\sqrt{\pi}\,\phi_1\right.$$

$$\left. + \frac{e^2}{2\pi}(\phi_1 + \phi_2 - \frac{\theta}{2\sqrt{\pi}})^2\right). \tag{13.430}$$

Let us define

$$\phi_+ = \frac{1}{\sqrt{2}}(\phi_1 + \phi_2 - \frac{\theta}{2\sqrt{\pi}}), \quad \phi_- = \frac{1}{\sqrt{2}}(\phi_1 - \phi_2) \tag{13.431}$$

and

$$\mathcal{M}^2 = \frac{2e^2}{\pi}. \tag{13.432}$$

We then get the Hamiltonian

$$\mathcal{H} = N_m\left(\frac{1}{2}\pi_+^2 + \frac{1}{2}\pi_-^2 + \frac{1}{2}(\partial_1\phi_+)^2 + \frac{1}{2}(\partial_1\phi_-)^2 - 2cm^2\cos(\sqrt{2\pi}\,\phi_+ + \frac{\theta}{2})\cos(\sqrt{2\pi}\,\phi_-)\right.$$

$$\left. + \frac{\mathcal{M}^2}{2}\phi_+^2\right). \tag{13.433}$$

In the weak coupling region $m \gg e$ there is no difference between the one-flavor and the two-flavor models (see Coleman). Indeed, when e/m goes to zero (i.e. $m \gg e$) the one-flavor and two-flavor models become the exactly soluble free Dirac theories with one Dirac fermion and two Dirac fermions, respectively.

For $m \ll e$ (or equivalently e/m goes to infinity) the one-flavor model becomes the soluble Schwinger model. For the two-flavor model the physics of the strong coupling region $m \ll e$ is significantly different. Indeed, for $m \ll e$ we see from the above Hamiltonian (13.433) that we have 2 scalar fields (the ϕ_- is light whereas the ϕ_+ is heavy) with a weak interaction. Perturbation theory will not work here (in contrast to the case of one-flavor) since the scale of the ϕ_- mass and the magnitude of the ϕ_- coupling are given by the same small parameter m. Although we are sure of the existence of a ϕ_+ meson with mass close to $\mathcal{M}$ it is very difficult to compute corrections to this mass (naive calculations will show that the one-loop correction is of the same order as the tree-level). However, if we look for particles of mass much less than $\mathcal{M}$ and restrict our attention to ϕ_- Green's functions then graphs with internal ϕ_+ are down by powers of $m/\mathcal{M}$. Thus, they can be neglected and perturbation theory will be reliable. The only possible source of problems are tadpole graphs involving ϕ_+, which again can be removed by normal ordering, viz (with $\beta = \sqrt{2\pi}$)

$$N_m \cos(\beta\phi_+ + \frac{\theta}{2}) = \left(\frac{\mathcal{M}^2}{m^2}\right)^{\frac{\beta^2}{8\pi}} N_{\mathcal{M}} \cos(\beta\phi_+ + \frac{\theta}{2}) = \left(\frac{\mathcal{M}}{m}\right)^{\frac{1}{2}} N_{\mathcal{M}} \cos(\beta\phi_+ + \frac{\theta}{2}). \tag{13.434}$$

The Hamiltonian becomes (the remaining normal ordering N_m acts on the field ϕ_- only)

$$\mathcal{H} = N_m \left(\frac{1}{2} N_{\mathcal{M}} \pi_+^2 + \frac{1}{2}\pi_-^2 + \frac{1}{2}N_{\mathcal{M}}(\partial_1\phi_+)^2 + \frac{1}{2}(\partial_1\phi_-)^2 - 2cm^{\frac{3}{2}}\mathcal{M}^{\frac{1}{2}} N_{\mathcal{M}} \cos(\sqrt{2\pi}\,\phi_+ \right.$$
$$\left. + \frac{\theta}{2}) \cdot \cos(\sqrt{2\pi}\,\phi_-) + \frac{\mathcal{M}^2}{2}\phi_+^2 \right). \tag{13.435}$$

Clearly we can set $\phi_+ = 0$ in this regime. We get the Hamiltonian

$$\mathcal{H} = N_m \left(\frac{1}{2}\pi_-^2 + \frac{1}{2}(\partial_1\phi_-)^2 - 2cm^{\frac{3}{2}}\mathcal{M}^{\frac{1}{2}} \cos\left(\frac{\theta}{2}\right) \cos(\sqrt{2\pi}\,\phi_-) \right)$$
$$= N_{\mathcal{M}'} \left(\frac{1}{2}\pi_-^2 + \frac{1}{2}(\partial_1\phi_-)^2 - \mathcal{M}'^2 \cos\frac{\theta}{2} \cos(\sqrt{2\pi}\,\phi_-) \right) \tag{13.436}$$

where

$$\mathcal{M}' = (2\,cm\mathcal{M}^{\frac{1}{2}} \cos\frac{\theta}{2})^{\frac{2}{3}}. \tag{13.437}$$

The only effect of θ is through an overall mass scale (in contrast to the one-flavor model). This model is a special case of the Sine-Gordon theory (again in contrast to the one-flavor model).

Another (more important) difference between the one-flavor model and the two-flavor model is with regard to the chiral condensate and spontaneous symmetry breaking of chiral symmetry. In the two-flavor model there is an extra $SU(2)_A$ (axial, flavor, isospin) chiral symmetry. Indeed, although it is not obvious from the above Hamiltonians, these Hamiltonians are in fact all isospin invariant. Since we are in two dimensions this $SU(2)_A$ symmetry cannot suffer from spontaneous symmetry breaking due to Coleman's theorem (this applies only in two dimensions). Thus, the chiral condensate $\langle \bar{\psi}\psi \rangle$ is protected from acquiring a non-zero vacuum value. In the one-flavor model, the condensate acquires a non-zero value because the axial $U(1)_A$ symmetry is anomalous –meaning it is not an actual symmetry –and there is no additional symmetry to prevent the formation of the condensate. Thus, we get in the one-flavor model $\langle \bar{\psi}\psi \rangle \neq 0$. In the two-flavor model there is also an axial $U(1)_A$ symmetry, which is also anomalous but since $\langle \bar{\psi}\psi \rangle \neq 0$ would also mean that the $SU(2)_A$ is spontaneously broken (which it cannot be in 2 dimensions) we conclude that we will have a zero condensate despite the presence of this anomaly. Seeing it this way, the condensate is really tied to the spontaneous symmetry breakdown of chiral $SU(2)_A$ symmetry more than it is associated with the chiral $U(1)_A$ anomaly. Explicitly we have the following mass terms

$$m\bar{\psi}\psi = -cm^2 N_m \cos(2\sqrt{\pi}\,\phi - \theta), \quad N_f = 1$$

$$m\sum_{i=1}^{2} \bar{\psi}_i\psi_i = -2cm^{\frac{3}{2}}\mathcal{M}^{\frac{1}{2}} \cos\frac{\theta}{2} N_m \cos(\sqrt{2\pi}\,\phi_-), \quad N_f = 2. \tag{13.438}$$

By differentiating with respect to m we get the condensates

$$\bar{\psi}\psi = -2c\mathcal{M}N_{\mathcal{M}} \cos(2\sqrt{\pi}\,\phi - \theta), \quad N_f = 1$$

$$\sum_{i=1}^{2} \bar{\psi}_i\psi_i = -\frac{3}{2}m^{\frac{1}{3}}(2c\mathcal{M}^{\frac{1}{2}} \cos\frac{\theta}{2})^{\frac{4}{3}} N_{\mathcal{M}'} \cos(\sqrt{2\pi}\,\phi_-), \quad N_f = 2. \tag{13.439}$$

This is indeed the correct behavior. For the one-flavor model the condensate approaches a constant value $-2c\mathcal{M}$ as m is sent to zero (due to the $U(1)_A$ anomaly) whereas in the two-flavor model the condensate vanishes as $m^{\frac{1}{3}}$ when $m \longrightarrow 0$ (due to the exact $SU(2)_A$ chiral symmetry).

The axial (chiral) symmetry $U(2)_A = U(1)_A \times SU(2)_A$ in four dimensions is fundamental to QCD with two flavors u and d (the gauge group is now non-Abelian $SU(3)$). There is also a vector $U(2)_V = U(1)_V \times SU(2)_V$, which is always exact. The chiral $SU(2)_A$ symmetry is not afflicted with anomalies and it is spontaneously broken in the quantum theory (the vacuum condensate acquires a non-zero value and the massless quarks become effectively massive). The corresponding Goldtstone's bosons are the three pions (they are massless in the limit of vanishing mass of the quarks). The symmetry $U(1)_A$ (related to the famous 'U(1) problem') is actually anomalous in QCD (due to instanton effects) and thus it is not really a symmetry of the action and there is neither spontaneous symmetry breaking nor Goldstone's bosons. If there was no instanton (i.e., if this symmetry was not anomalous) then it would spontaneously break as the $SU(2)_A$ and the corresponding boson would be the η. The difference between the spontaneously broken $SU(2)_A$ and the anomalous $U(1)_A$ is the reason why the η particle is much heavier than the pions.

The quarks of QCD couple also to electromagnetic interactions and as a consequence the $SU(2)_A$ is electromagnetically anomalous. This anomaly provides the dominant contribution to the electromagnetic decay $\pi^0 \longrightarrow 2\gamma$.

13.5 Other exactly solvable models

Other very interesting exactly solvable models include for example:

- (1) The Schwinger model on the sphere [11], which can be thought of as a regularization/compactification of the Schwinger model studied in this chapter.
- (2) 't Hooft model [12], which deals with quantum chromodynamics in two dimensions in the large N limit.
- (3) Yang–Mills theory in two dimensions [13–15] (see also [16] for a comparison between the exact solution and perturbation theory), which is intimately related to string theory [17–19]. This is a very important case, which is as important as the Ising model and perhaps even more.

All these solutions are in fact possible because they all live in two dimensions (one time and one space direction) and perhaps because of the large N expansion in many instances. By going to four dimensions (one time and three space directions) the possibility of an exact solution becomes so rare that we can single out only (aside from large N solutions) supersymmetric gauge theory with $\mathcal{N} = 2$, which is due originally to Seiberg and Witten.

The fundamental question is always: can we solve exactly quantum chromodynamics, which is the gauge theory of the strong nuclear force?

The answer is no. This is a highly non-perturbative theory with highly non-trivial properties such as confinement, the mass gap, chiral symmetry breaking and asymptotic freedom (in the perturbative regime). At this time only the powerful tools of Monte Carlo methods on lattices and the renormalization group equation are available to us to tackle this fundamental theory.

But another powerful complete trick was discovered which allows us to solve the theory. Namely, by adding a lot of supersymmetry (more precisely we add $\mathcal{N} = 2$ supersymmetry) and using holomorphy the theory becomes exactly soluble as shown explicitly by Seiberg and Witten in 1994 [20, 21].

Nekrasov in 2002 rederived the Seiberg-Witten prepotential, which encodes the full low-energy physics of the system, from microscopic considerations and first principles [22]. He managed to compute the path integral/partition integral of the system exactly and explicitly. The prepotential was found to be equal to a perturbative part, computed from renormalization group equation considerations, and a non-perturbative part computed by summing up all instanton sectors.

It was the first time a sum over all instanton contributions of some gauge theory can be preformed so explicitly. Some of the technical machineries used by Nekrasov to achieve this feat are 1) supergravity, the so-called omega background, to regulate the infrared behavior of the theory and 2) noncommutative geometry to smooth out the moduli space of instantons.

Thus, by means of a lot of supersymmetry, holomorphy, instantons (and monopoles) we can also solve models in four dimensions exactly.

13.6 Exercises

Exercise 1: Set up the explicit and detailed solution of the following models:
- (1) The Schwinger model on the sphere [11].
- (2) 't Hooft model [12].
- (3) Yang–Mills theory in two dimensions [13–15].

References

[1] Chaikin P M and Lubensky T C 1995 *Principles of Condensed Matter Physics* (Cambridge: Cambridge University Press)

[2] Zinn-Justin J 2002 Quantum field theory and critical phenomena *Int. Ser. Monogr. Phys.* **113** 1

[3] Peskin M E and Schroeder D V 1995 *An Introduction to Quantum Field Theory* (Reading, MA: Addison-Wesley)

[4] Onsager L 1944 Crystal statistics. 1. A two-dimensional model with an order disorder transition *Phys. Rev.* **65** 117

[5] Huang K 1987 *Statistical Mechanics* 2nd edn (New York: Wiley)

[6] Yang C N 1952 The spontaneous magnetization of a two-dimensional Ising model *Phys. Rev.* **85** 808

[7] Schwinger J S 1962 Gauge invariance and mass *Phys. Rev.* **125** 397

[8] Schwinger J S 1962 Gauge invariance and mass. 2 *Phys. Rev.* **128** 2425

[9] Coleman S R 1975 The quantum Sine-Gordon equation as the massive thirring model *Phys. Rev.* D **11** 2088

[10] Coleman S R, Jackiw R and Susskind L 1975 Charge shielding and quark confinement in the massive Schwinger model *Ann. Phys.* **93** 267

[11] Jayewardena C 1988 Schwinger model on S(2) *Helv. Phys. Acta* **61** 636

[12] 't Hooft G 1974 *Nucl. Phys.* B **75** 461

[13] Migdal A A 1975 Recursion equations in gauge theories *Sov. Phys. JETP* **42** 413 [Zh. Eksp. Teor. Fiz. **69** 810 (1975)]

[14] Witten E 1991 On quantum gauge theories in two-dimensions *Commun. Math. Phys.* **141** 153

[15] Levy T 2001 Yang-Mills measure on compact surfaces math/0101239

[16] Nguyen T 2015 arXiv:1508.06305 [math-ph]

[17] 't Hooft G 1974 A planar diagram theory for strong interactions *Nucl. Phys.* B **72** 461

[18] Kazakov V A and Kostov I K 1980 Nonlinear strings in two-dimensional U(infinity) gauge theory *Nucl. Phys.* B **176** 199

[19] Gross D J and Taylor W 1993 *Nucl. Phys.* B **400** 181 [hep-th/9301068]

[20] Seiberg N and Witten E 1994 Electric–magnetic duality, monopole condensation, and confinement in N = 2 supersymmetric Yang–Mills theory *Nucl. Phys.* B **426** 19 Erratum: [Nucl. Phys. B **430** 485 (1994)] [hep-th/9407087]

[21] Seiberg N and Witten E 1994 Monopoles, duality and chiral symmetry breaking in N = 2 supersymmetric QCD *Nucl. Phys.* B **431** 484 [hep-th/9408099]

[22] Nekrasov N A 2003 *Adv. Theor. Math. Phys.* **7** 831 [hep-th/0206161]

IOP Publishing

A Modern Course in Quantum Field Theory, Volume 2 (Second Edition)

Advanced topics

Badis Ydri

Chapter 14

The monopoles and instantons

Monopoles are non-trivial topological gauge field configurations, which appear in spontaneously broken gauge theory via the Higgs mechanism. These are particle-like solitonic configurations characterized by stability and finite energy among other properties. Their stability is of a topological origin characterized by the so-called winding numbers or magnetic charges. For this reason monopoles are one of the best examples in which physics and topology become intertwined. The existence of the monopole requires the embedding of electromagnetism, i.e., the group $U(1)$, as a subgroup in a larger non-Abelian group G with compact cover, which then becomes broken spontaneously via the usual Higgs mechanism.

The original literature on the subject consists of 't Hooft [1] and Polyakov [2]. Some of the pedagogical (from my perspective) lectures I can mention here: Lenz [3], 't Hooft [4], Coleman [5] and Tong [6]. A comprehensive book is Shnir [7] and a comprehensive review is given by Weinberg and Yi [8].

Instantons are another type of fundamental topological gauge configurations, perhaps more fundamental than monopoles, which are given by events localized in spacetime and hence the other name given for them: pseudo-particles (in contrast with particles such as monopoles, which are events localized in space). Instantons are also the gauge field configurations that dominate the path integral in the semi-classical limit with the trivial instanton identified precisely with the perturbative vacuum $A = 0$.

We will discuss here in great detail the theta term, the role of vacuum degeneracy, the quantization of the topological charge and the role of topology in instanton physics. More precisely, the instanton is defined as a solution of the self-duality equation with zero/finite energy, which happens to saturate the Bogomoln'yi bound. The BPST instanton solution is then derived explicitly. The original literature on the BPST instanton is the paper by Belavin, Polyakov, Schwartz and Tyupkin [9]. We then discuss in some detail the moduli space, the collective coordinates, the zero

doi:10.1088/978-0-7503-5834-7ch14

modes, the ADHM construction, the one-loop quantization in the background of instantons as well as the connection of instantons to quantum tunneling. We have benefited here greatly from the pedagogical presentations found in [6, 10, 11].

14.1 Monopoles

14.1.1 The Georgi–Glashow model

Since the Higgs mechanism plays an essential role here, we will consider a non-abelian gauge field A_μ coupled to a scalar field ϕ in the adjoint representation, with a Lagrangian density given by the Georgi–Glashow model [12]. We will assume an $SU(N)$ gauge group with gauge coupling constant g. The covariant derivative and the field strength are given, respectively, by $D_\mu = \partial_\mu + igA_\mu$ and $F_{\mu\nu} = \partial_\mu A_\nu - \partial_\nu A_\mu + ig[A_\mu, A_\nu]$. Explicitly, the Georgi–Glashow model is given by

$$\mathcal{L}_{GG} = -\frac{1}{2}trF_{\mu\nu}F^{\mu\nu} + tr(D_\mu\phi)(D^\mu\phi) - V(\phi). \tag{14.1}$$

The first term is the Yang–Mills Lagrangian density for the pure gauge field A_μ, whereas the second term is the covariant kinetic term for the scalar field ϕ, and since the scalar field is in the adjoint representation, the covariant derivative is defined via commutators, viz $D_\mu\phi = \partial_\mu\phi + ig[A_\mu, \phi]$. The Yang–Mills term can also be rewritten in terms of the electric and magnetic fields as

$$-\frac{1}{2}trF_{\mu\nu}F^{\mu\nu} = tr(\vec{E}^2 - \vec{B}^2), \quad E^i = -F^{0i}, \quad B^i = -\frac{1}{2}\varepsilon^{ijk}F^{jk}. \tag{14.2}$$

The third and final term in the above Lagrangian density is the potential term, which specifies the self-interaction of the scalar Higgs field ϕ and is taken to be a ϕ^4-interaction, viz

$$V(\phi) = \frac{1}{4}\lambda(\phi^a\phi^a - F^2)^2, \quad \lambda > 0. \tag{14.3}$$

Since this is a non-Abelian gauge theory, the gauge field A_μ, the field strength $F_{\mu\nu}$, the scalar Higgs field ϕ and the covariant derivative $D_\mu\phi$ are actually matrix-valued functions of spacetime. If we let λ_a be the generators of the gauge group $SU(N)$, viz $tr\lambda_a\lambda_b = 2\delta_{ab}$, and $[\lambda^a, \lambda^b] = 2if_{abc}\lambda^c$, where f_{abc} are the structure constants of the group, then we can expand all the fields as

$$A_\mu = A_\mu^a\frac{\lambda^a}{2}, \quad \phi = \phi^a\frac{\lambda^a}{2}, \quad F_{\mu\nu} = F_{\mu\nu}^a\frac{\lambda^a}{2}, \quad D_\mu\phi = (D_\mu\phi)^a\frac{\lambda^a}{2}. \tag{14.4}$$

The index a is called the color index, which runs from 1 to $N^2 - 1$, transforms as a vector under the gauge group. The components A_μ^a, ϕ^a, $(D_\mu\phi)^a$ and $F_{\mu\nu}^a$ are called the color components of the fields A_μ, ϕ, $(D_\mu\phi)$ and $F_{\mu\nu}$, respectively. For $SU(2)$ gauge group we have three generators, which are precisely the Pauli matrices, and $f_{abc} = \varepsilon_{abc}$.

The equations of motion are given by the inhomogeneous field equations

$$[D_\mu, F^{\mu\nu}] = j^\nu, \quad j^\nu = g f^{abc} \phi^b (D^\nu \phi)^c, \tag{14.5}$$

and the homogeneous field equations (Jacobi identities)

$$[D_\mu, \tilde{F}^{\mu\nu}] = 0, \quad \tilde{F}^{\mu\nu} = \frac{1}{2} \varepsilon^{\mu\nu\alpha\beta} F_{\alpha\beta}. \tag{14.6}$$

The field A_μ transforms under the gauge transformation $U \in SU(N)$ in such a way as to make the transformation law for the matter field $D_\mu \phi$ covariant, i.e., as to make $D_\mu \phi$ transforms in the same way as ϕ. This transformation law is given explicitly by

$$A_\mu \longrightarrow U(A_\mu + \frac{1}{ig}\partial_\mu)U^\dagger. \tag{14.7}$$

The fact that the fields ϕ, $D_\mu \phi$ and $F_{\mu\nu}$ transform in the adjoint representation under the gauge transformation $U \in SU(N)$ means that

$$\phi \longrightarrow U\phi U^\dagger, \quad D_\mu \phi \longrightarrow U D_\mu \phi U^\dagger, \quad F_{\mu\nu} \longrightarrow U F_{\mu\nu} U^\dagger. \tag{14.8}$$

If we want to apply the Hamiltonian or canonical formalism to this problem, then we must fix the gauge, i.e., fix the symmetry under the above gauge transformations. The Hamiltonian is the Legendre transformation of the Lagrangian. And, then we need to compute the conjugate momenta given by the derivatives of the Lagrangian with respect to the time derivatives of the fields. The first immediate observation is that the time derivative $\partial_0 A_0$ does not appear in the Lagrangian density and hence the conjugate momenta associated with A_0 vanishes identically. This means that A_0 is actually a Lagrange multiplier, i.e., it is not an independent dynamical variable, and hence it should be eliminated. The corresponding equation of motion (the 0-component of (14.5)) can be rewritten as $[D_i, E^i] = j^0$, and since it does not involve a time derivative, i.e., it is strictly speaking a constraint equation (Gauss's law) not an equation of motion, eliminating A_0 can be done explicitly by choosing the Weyl or temporal gauge given by

$$A_0 = 0. \tag{14.9}$$

This fixes the gauge, i.e., it eliminates A_0, as we can check. This gauge is unchanged under time-independent gauge transformations, which constitute the residual gauge symmetry in this case. This residual time-independent gauge symmetries are generated infinitesimally by the Gauss operator $Q = [D_i, E^i]$, which also commutes with the Hamiltonian. The fact that states which differ by time-independent gauge transformations are gauge equivalent translates to the requirement that the physical states (the gauge invariant states) of the theory must satisfy Gauss's law, viz

$$[D_i, E^i]|\psi\rangle = j^0 |\psi\rangle. \tag{14.10}$$

After gauge fixing and the implementation of the Gauss's law as a constraint, only a global gauge symmetry group remains, which we will also denote by $G = SU(N)$.

The conjugate momenta associated with the fields A_i and ϕ are given, respectively, by

$$\frac{\delta \mathcal{L}_{GG}}{\delta(\partial_0 A_i^a)} = -F^{0ia} = E^{ia}, \quad \frac{\delta \mathcal{L}_{GG}}{\delta(\partial_0 \phi^a)} = (D_0 \phi)^a = \pi^a. \tag{14.11}$$

The Hamiltonian density is then found to be given by

$$\mathcal{H} = \frac{1}{2}(\vec{E}^{a2} + \vec{B}^{a2}) + \frac{1}{2}\pi^a \pi^a + \frac{1}{2}((D_i \phi)^a)^2 + V(\phi). \tag{14.12}$$

14.1.2 Ground state configurations

We will consider the gauge group $G = SU(2)$ for simplicity.

We remark that the Hamiltonian density is a sum of positive definite terms This means that the energy density of the field configurations with the lowest energy vanishes identically. The vanishing of $\vec{E}$ and π signifies that the lowest energy field configurations are in fact static. They must also have vanishing magnetic field $\vec{B}$ and minimize the potential. The vanishing of the magnetic field can be solved by $\vec{A} = 0$ while the vanishing of the potential can be solved by a constant ϕ which satisfies $V(\phi) = 0$. We have then the lowest energy field configurations

$$\vec{A} = 0, \quad \phi = \phi_0, \quad V(\phi_0) = 0. \tag{14.13}$$

These are static solutions of the Euler–Lagrange equations. There is in this case a vacuum degeneracy, i.e., there exists an infinite number of field configurations, which solve these equations and thus minimize the energy density. Indeed, we have

$$V(\phi) = 0 \Rightarrow \phi^a \phi^a = F^2. \tag{14.14}$$

Thus, ϕ^a lives on a sphere $\mathbf{S}^2$ of radius F. This is the manifold of zeroes of the potential energy. In other words, the vanishing of the potential V determines completely the modulus F of the Higgs field ϕ^a and only leaves its orientation n^a in color space arbitrary, i.e., $\phi^a = \phi_0^a = Fn^a$.

The Higgs field exhibits therefore a spontaneous orientation in color space which breaks the remaining global symmetry group $G = SU(2)$ down to the stability (little) isotropy group of transformations, which leave ϕ_0 invariant. Obviously, these residual symmetries form a subgroup H of the group of global rotations in color space $G = SU(2)$, which rotate the Higgs field ϕ_0 around its axis n^a. Since the sphere is a homogeneous space any point on it can be reached starting from ϕ_0 by the application of a rotation $g \in G$. In other words, the action of $G = SU(2)$ leaves the gauge field and the modulus of the Higgs field invariant whereas the action of H leaves the orientation of the Higgs field invariant as well. The stability group is given explicitly by

$$H = \{h \in G \colon h\phi_0 = \phi_0\} = U(1) \otimes \mathbf{Z}_2. \tag{14.15}$$

Z_2 is the center element of $G = SU(2)$ consisting of the two elements 1 and -1. This division by the center is due to the fact that rotations on the sphere actually form the group $SO(3)$ (and not $SU(2)$), which is not simply connected, i.e., $SO(3) = SU(2)/\mathbf{Z}_2$.

The space of zeroes of V given by the sphere $\mathbf{S}^2$ and the coset space G/H are homeomorphic to each other. We write $G/H = SU(2)/(U(1) \otimes \mathbf{Z}_2) \sim \mathbf{S}^2$. This homeomorphism is given explicitly by

$$f \; : \; G/H \longrightarrow \mathbf{S}^2$$
$$\tilde{g} \longrightarrow f(\tilde{g}) = g\phi_0 = \phi. \tag{14.16}$$

The group element g is a representative of the coset $\tilde{g}$. The above mapping is bijective (injective or one-to-one and subjective or onto), continuous, and invertible, and hence it is a homeomorphism.

The above spontaneous symmetry breaking of $SU(2)$ down to $U(1) \otimes \mathbf{Z}_2$ is therefore a consequence of vacuum degeneracy. It can be shown (in the unitary gauge) that the gauge field acquires a mass that is determined by the value of the Higgs field and as a consequence we obtain short range gauge interactions. This is the Higgs phase to be contrasted with the more familiar Coulomb phase.

14.1.3 't Hooft–Polyakov monopole

14.1.3.1 Non-trivial configurations

The above trivial field configurations are characterized by lowest energy and thus they define the ground state. The topologically non-trivial field configurations are characterized by finite energy (among other things). These finite energy configurations can be obtained from (14.12) if, as before, $\vec{E} = \pi = 0$ (static configurations), and $\vec{B}$ and $(D_i \phi)^a$ vanish asymptotically, and furthermore the potential is also minimized asymptotically. We write the requirement that the potential energy is asymptotically minimized as

$$\phi(\vec{x}) \longrightarrow Fn^a(\vec{x}), \;\; |\vec{x}| \longrightarrow \infty. \tag{14.17}$$

Now, the unit vector n^a (which characterizes the orientation of the Higgs field) is a function on the sphere $\mathbf{S}^2$ since $|\vec{x}| \longrightarrow \infty$. Therefore $\phi(\vec{x})$ is a mapping from the sphere, which bounds space at infinity to the sphere of vacuum field configurations (which are elements of the algebra $G = SU(2)$ with fixed length F). In other words,

$$\phi \colon \; \mathbf{S}^2 \longrightarrow \mathbf{S}^2 \sim G/H. \tag{14.18}$$

This mapping is a topological mapping, which will be characterized by a winding number (degree of the mapping), which is an integer. Indeed, the winding number is the number of times the mapping ϕ winds around its sphere (the sphere of vacuum configurations) as we wind around the boundary sphere once. We will also show that this degree or winding number is equal to the magnetic charge of the monopole.

Recall that ϕ is actually an element of the coset space $G/H = SU(2)/(U(1) \otimes \mathbf{Z}_2)$. Thus, mathematically, the winding number (the degree or the magnetic charge) is related to the second homotopy group of $G/H = SU(2)/(U(1) \otimes \mathbf{Z}_2)$. This second homotopy group is equal to the first homotopy group of $U(1)$ where this $U(1)$ is the connected component of $H = U(1) \otimes \mathbf{Z}_2$ to the identity element. The second homotopy group is the group of mappings from the sphere into the manifold whereas the first homotopy group is the group of mappings from the circle into the manifold. The above mapping is then characterized by the second homotopy group given explicitly by

$$\pi_2(SU(2)/(U(1) \otimes \mathbf{Z}_2)) = \pi_1(U(1)) = \mathbf{Z}. \tag{14.19}$$

Thus, the above mapping is characterized by a single topological object given by a monopole carrying a quantized (since it is proportional to the winding number) magnetic charge g_m under the unbroken abelian subgroup $U(1)$. The non-triviality of this homotopy group is what guarantees the stability of the corresponding monopole configurations.

The other requirements for finite energy configurations (in addition to (14.17)) are

$$\vec{B} \longrightarrow 0, \quad |\vec{x}| \longrightarrow \infty. \tag{14.20}$$

$$(D_i\phi)^a(x) \longrightarrow 0, \quad |\vec{x}| \longrightarrow \infty. \tag{14.21}$$

In the last equation we see that the gauge and the Higgs fields on the sphere which bounds space at infinity are correlated such that

$$(D_i\phi)^a(x) = \partial_i\phi^a(x) - g\varepsilon^{abc}A_i^b(x)\phi^c(x) = 0, \quad |\vec{x}| \longrightarrow \infty. \tag{14.22}$$

We decompose the gauge field $A_i^a(x)$ in color space (index a) into parallel and perpendicular components with respect to the direction of the Higgs field ϕ. A solution of the above equation is immediately given by

$$A_i^a(x) = \frac{1}{F}\phi^a(x)\mathcal{A}_i(x) + \frac{1}{gF^2}\varepsilon^{abc}\phi^b(x)\partial_i\phi^c(x), \quad |\vec{x}| \longrightarrow \infty. \tag{14.23}$$

The first term is the parallel component while the second term is the perpendicular component. The field $\mathcal{A}_i$ is the abelian gauge field corresponding to the remaining unbroken $U(1)$ group living in the monopole background. Indeed, we calculate the field strength

$$F_{ij}^a(x) = \frac{1}{F}\phi^a(x)F_{ij}(x), \quad F_{ij} = \partial_i\mathcal{A}_j - \partial_j\mathcal{A}_i - \frac{1}{gF^3}\varepsilon^{abc}\phi^a\partial_i\phi^b\partial_j\phi^c, \quad \partial^i F_{ij} = 0. \tag{14.24}$$

We compute the magnetic charge q_m as the flux of the magnetic field through the boundary sphere, viz (with $n^a = \phi^a/F$)

$$\begin{aligned}
q_m &= \int_{\mathbf{S}^2} \vec{B} \cdot d\vec{\sigma} \\
&= -\int_{\mathbf{S}^2} (\vec{\nabla} \times \vec{\mathcal{A}}) \cdot d\vec{\sigma} + \frac{1}{2g}\int_{\mathbf{S}^2} \varepsilon^{ijk}\varepsilon^{abc}n^a\partial_j n^b\partial_k n^c d\sigma^i \\
&= \frac{4\pi n}{g}.
\end{aligned} \tag{14.25}$$

The first term in the second line in the above equation vanishes by using Stockes' theorem and noting that the sphere has no boundary, whereas in the second term we have used the fact that the winding number of the mapping $n^a = \phi^a/F$, which by construction is equal to an integer n, is given explicitly by the relation

$$n = \frac{1}{8\pi} \int_{S^2} \varepsilon^{ijk} \varepsilon^{abc} n^a \partial^j n^b \partial^k n^c d\sigma^i. \tag{14.26}$$

Hence, we conclude that the magnetic charge is quantized.

14.1.3.2 't Hooft–Polyakov monopole

An extremely important monopole solution is the 't Hooft–Polyakov monopole, which is a hedgehog-like spherically symmetric regular solution of the form

$$\phi^a(\vec{x}) = F\frac{x^a}{r} = Fn^a, \quad \sum_{a=1}^{3} n^a n^a = 1. \tag{14.27}$$

The coordinates x^a on the asymptotic sphere are given by the coordinates σ^i. This is a mapping with winding number equal 1, i.e., this solution winds around the sphere of vacuum solutions once as we wind around the asymptotic sphere once. This can be calculated as follows (where we use $\varepsilon^{ijk} d\sigma^i = d\sigma^j \wedge d\sigma^k$)

$$\begin{aligned}
n &= \frac{1}{8\pi} \int_{S^2} \varepsilon^{ijk} \varepsilon^{abc} n^a \partial^j n^b \partial^k n^c d\sigma^i \\
&= \frac{1}{8\pi} \int_{S^2} \varepsilon^{abc} n^a dn^b \wedge dn^c \\
&= \frac{1}{8\pi} \int_{S^2} (-2d\cos\theta \wedge d\phi) \\
&= 1.
\end{aligned} \tag{14.28}$$

The magnetic charge of this monopole is therefore given by

$$q_m = \int_{S^2} \vec{B}.\,d\vec{\sigma} = \frac{4\pi}{g}. \tag{14.29}$$

The corresponding field strength tensor is given by (we use $\partial^i \phi^a = F(\delta^{ia} - n^i n^a)/r$ and setting $\mathcal{A}_i = 0$)

$$F^{ij} = -\frac{1}{gr^2} n^a \varepsilon^{aij}. \tag{14.30}$$

We obtain immediately the magnetic field

$$B^i = -\frac{1}{2}\varepsilon^{ijk} F^{jk} = +\frac{1}{gr^2} n^i = \frac{q_m}{4\pi r^2} n^i. \tag{14.31}$$

Thus, for an observer on the asymptotic sphere it looks like we have a magnetic field generated by a magnetic monopole with charge q_m sitting at the origin. The corresponding gauge potential can be found from the condition (14.22). We have

$$\partial_i \phi^a = g\varepsilon^{abc} A_i^b \phi^c \Rightarrow \delta^{ai} - n^a n^i = g\varepsilon^{abc} A_i^b x^c. \tag{14.32}$$

We get immediately

$$A_i^a = -\frac{1}{gr^2}\varepsilon^{aij} x^j. \tag{14.33}$$

14.1.4 Dirac quantization condition and Wu–Yang potential

We could try also to find the gauge potential associated with the magnetic field

$$\vec{B} = \frac{g_m}{r^2}\hat{r} \tag{14.34}$$

by the usual expression $\vec{B} = \vec{\nabla} \times \vec{A}$ (with $g_m = q_m/4\pi$). However, this expression leads immediately to $\vec{\nabla}\vec{B} = 0$ which is contradictory to the fact that we have a magnetic charge at the origin. We discuss this crucial point in some more detail next.

Since the magnetic field is spherically symmetric, i.e., $\vec{B}$ is in the direction of $\hat{r}$, and since we are assuming the relation $\vec{B} = \vec{\nabla} \times \vec{A}$ anyway, the corresponding gauge field $\vec{A}$ must be circular, i.e.,

$$\vec{A} = -g_m \frac{1}{r}A(\theta)\hat{\phi}. \tag{14.35}$$

We must then have

$$\frac{g_m}{r^2}\hat{r} = \frac{1}{r\sin\theta}\frac{\partial}{\partial\theta}(A_\phi \sin\theta)\hat{r} - \frac{1}{r}\frac{\partial}{\partial r}(rA_\phi)\hat{\theta} \Rightarrow A(\theta) = \frac{1 + \cos\theta}{\sin\theta}. \tag{14.36}$$

We immediately get the solution

$$\vec{A} = g_m \frac{1}{r}\left(\frac{1 + \cos\theta}{\sin\theta}\sin\phi, -\frac{1 + \cos\theta}{\sin\theta}\cos\phi, 0\right) = g_m \frac{1}{r}\frac{\vec{r} \times \hat{z}}{r\,r - \vec{r}\hat{z}}. \tag{14.37}$$

This is the Dirac monopole gauge potential [13]. This is obviously singular along the line $\theta = 0$, i.e., along the positive z-axis. This is the famous Dirac string. Its appearance is only due to the fact that we are forcing the use of the law $\vec{B} = \vec{\nabla} \times \vec{A}$ in spite of the presence of a monopole at the origin. Indeed, the magnetic field associated with the above gauge potential is not (14.31) but it is given by [7]

$$B^i = g_m\left(\frac{n^i}{r^2} - 4\pi k^i\theta(z)\delta(x)\delta(y)\right). \tag{14.38}$$

We can verify that the flux of the Coulomb field (first term) cancels exactly the flux of the Dirac string (step function term) and hence the Dirac equation $\vec{\nabla}.\,\vec{B} = 0$ is maintained.

Although, the existence of the Dirac string is gauge invariant, its position is gauge dependent. To see this we consider a $U(1)$ gauge transformation (with $U = \exp(ie\Lambda)$)

$$\vec{A} \longrightarrow \vec{A}' = \vec{A} - \frac{i}{e}U^{-1}\vec{\nabla}U = \vec{A} + \vec{\nabla}\Lambda. \tag{14.39}$$

The gauge function Λ is typically taken to be single-valued. However, it can also be taken to be a multivalued function thus giving rise to extra singular terms in the transformed gauge potential $\vec{A}'$ and its associated magnetic field $\vec{B}'$. Indeed, we can check that the magnetic flux $\Phi = \int_\Sigma \vec{B} d\vec{S}$ is constant only if the gauge function satisfies $\Lambda(\phi + 2\pi) = \Lambda(\phi)$, i.e., it is single-valued, otherwise $\Delta\Phi = \int_\Sigma (\vec{B}' - \vec{B})d\vec{S} \neq 0$ [7].

Let us then consider a multivalued gauge function $\Lambda = 2g_m\phi$. We will denote $\vec{A} = \vec{A}^S$ and $\vec{A}' = \vec{A}^N$. Then we calculate

$$\vec{A}^N = \vec{A}^S + 2g_m\vec{\nabla}\phi = -\frac{g_m}{r}\frac{1+\cos\theta}{\sin\theta}\hat{\phi} + 2g_m\frac{\hat{\phi}}{r\sin\theta} = \frac{g_m}{r}\frac{1-\cos\theta}{\sin\theta}\hat{\phi}. \quad (14.40)$$

This potential is now singular along the line $\theta = \pi$. The Dirac string is now sitting along the negative z-axis. In other words, the position of the Dirac string is gauge dependent.

In conclusion, the monopole gauge potential can not be smoothly defined everywhere in space although the corresponding field strength is globally defined. However, a non-singular monopole gauge potential can still be constructed following Wu and Yang [14], which provides one of the most interesting examples of the profound connection existing between physics and topology. This potential is obtained by first re-parametrizing the space $\mathbf{R}^3/\{0\}$ surrounding the monopole by two overlapping coordinate patches (not a single coordinate patch as usual) covering $z > -\varepsilon$ (north hemisphere $\mathbf{R}^N$) and $z < +\varepsilon$ (south hemisphere $\mathbf{R}^S$), then taking the gauge potentials in these two regions to be $\vec{A}^N$ and $\vec{A}^S$ respectively, viz

$$\begin{aligned}
\vec{A}^N &= +\frac{g_m}{r}\frac{1-\cos\theta}{\sin\theta}\hat{\phi}, \; 0 \leqslant \theta \leqslant \frac{\pi}{2} + \frac{\varepsilon}{2}: \; \mathbf{R}^N \\
\vec{A}^S &= -\frac{g_m}{r}\frac{1+\cos\theta}{\sin\theta}\hat{\phi}, \; \frac{\pi}{2} - \frac{\varepsilon}{2} \leqslant \theta \leqslant 0: \; \mathbf{R}^S.
\end{aligned} \quad (14.41)$$

These potentials are now free from singularities since their domain of definition is restricted to the regions where there are no Dirac strings. In the overlap region $\mathbf{R}^N \cap \mathbf{R}^S$ these two potentials are well defined and they are related by the above constructed singular gauge transformation (14.40).

We consider now the dynamics of an electric charge e, of mass m and described by a position vector $\vec{r}$, in the above monopole gauge field $\vec{A}$. The Lagrangian of the system electron-monopole is of the form

$$L = \frac{1}{2}m\dot{\vec{r}}^2 + e\dot{\vec{r}}\vec{A}. \quad (14.42)$$

Obviously, the second term is the interaction term, i.e., the interaction action is given by (with $\vec{r}(T) = \vec{r}(0)$)

$$S_{\text{int}} = e\int_0^T \dot{\vec{r}}\vec{A}\,dt = e\oint d\vec{r}.\,\vec{A}. \quad (14.43)$$

The wave function ψ of the electric charge can be represented as

$$\psi(\vec{r}) = \psi_0(\vec{r})\exp(ie \int_0^T \vec{r}\vec{A}\,dt),\tag{14.44}$$

where ψ_0 is the free wave function, i.e., the interaction action is just a phase.

Under singular gauge transformations the interaction action S_{int} is not necessarily invariant, i.e., it may pick up additional phase factors, but the quantum mechanical transition amplitude given by the wave function must always remain invariant. In other words, the wave function is always a continuous function of $\vec{r}$ although its phase can be discontinuous. This leads immediately to Dirac's charge quantization condition. This is the condition that the change of the phase factor along a closed path must be a multiple of 2π, viz

$$e \oint d\vec{r}.\,\vec{A} = 2\pi n.\tag{14.45}$$

We do this extremely important calculation explicitly. Let l be a closed path in the overlap region $\mathbf{R}^N \cap \mathbf{R}^S$. When the electric charge e is transported along this closed path once its wave function picks up precisely the phase factor (14.43) with $\vec{A}$ being either $\vec{A}^N$ and $\vec{A}^S$, which are both well defined in the overlap region. By employing now Stokes's theorem in the northern hemisphere we have

$$e \oint_l d\vec{r}.\,\vec{A}^N = e \int_{\mathbf{R}^N} d\vec{S}.\,\vec{B} = e \int_{\mathbf{R}^N} r^2 d\Omega \frac{g_m}{r^2} = 2\pi e g_m.\tag{14.46}$$

But by employing Stokes's theorem in the southern hemisphere we obtain an extra minus sign due to the opposite orientation of the element of surface $d\vec{S}$. We have

$$e \oint_l d\vec{r}.\,\vec{A}^S = -e \int_{\mathbf{R}^S} d\vec{S}.\,\vec{B} = -e \int_{\mathbf{R}^S} r^2 d\Omega \frac{g_m}{r^2} = -2\pi e g_m.\tag{14.47}$$

Hence the variation of the interaction action under the singular gauge transformation, which takes us from $\vec{A}^N$ to $\vec{A}^S$ is

$$\Delta S_{\text{int}} = e \int_{\mathbf{R}^N \cup \mathbf{R}^S} d\vec{S}.\,\vec{B} = 4\pi e g_m.\tag{14.48}$$

This is not physically observable and hence we must have the quantization condition

$$eq_m = 2\pi n.\tag{14.49}$$

This is identical to the topological quantization condition (14.25) with the identification $g \leftrightarrow 2e$. This can also be put into the form

$$g_m = \frac{n}{g}.\tag{14.50}$$

The above Dirac–Wu–Yang Abelian magnetic monopole is embedded into $SU(2)$ as $A_\mu \equiv A_\mu^{\text{Dirac}} \sigma_3/3$, i.e. $A_\mu^{\text{Dirac}} = A_\mu^3$. This will allow us to relate the singular Dirac–Wu–Yang Abelian magnetic monopole with the regular 't Hooft–Polyakov non-Abelian monopole via a singular gauge transformation. In other words, the string singularity of the Dirac monopole cancels exactly the singularity of the gauge transformation (provided the charge quantization condition is satisfied). We have explicitly the Dirac–Wu–Yang Abelian magnetic monopole

$$A_\mu^1 = A_\mu^2 = 0$$
$$A_0^3 = A_r^3 = A_\theta^3 = 0 \tag{14.51}$$
$$A_\phi^3 = -\frac{g_m}{r}\frac{1+\cos\theta}{\sin\theta}.$$

This configuration can be rewritten in a compact form as

$$A_\mu^a = -g_m\delta^{a3}(1+\cos\theta)\partial_\mu\phi. \tag{14.52}$$

The four-vector $\partial_\mu\phi$ is singular along the entire z-axis and it is given explicitly by (think about the gradient in spherical coordinates)

$$\partial_\mu\phi = \frac{1}{r\sin\theta}(0, -\sin\phi, \cos\phi, 0). \tag{14.53}$$

Under an $SU(2)$ gauge rotation U (characterized by three angles θ, ϕ and α on $\mathbf{S}^3$) this transforms as

$$A_\mu \longrightarrow A_\mu' = UA_\mu U^{-1} - \frac{i}{g}U\partial_\mu U^{-1}. \tag{14.54}$$

We consider the gauge transformation corresponding to the rotation of the unit vector on the sphere $\mathbf{S}^2$ to the third axis in color space, i.e., θ is the polar angle and ϕ is the azimuthal angle on the ordinary sphere $\mathbf{S}^2$ and $\alpha = 0$. This rotation is given explicitly by

$$U = \exp(i\sigma_3\frac{\phi}{2})\exp\left(i\sigma_2\frac{\theta}{2}\right)\exp\left(-i\sigma_3\frac{\phi}{2}\right)$$
$$= \begin{pmatrix} \cos\frac{\theta}{2} & -\sin\frac{\theta}{2}\exp(-i\phi) \\ \sin\frac{\theta}{2}\exp(i\phi) & \cos\frac{\theta}{2} \end{pmatrix}. \tag{14.55}$$

After some calculation we arrive at the result [1, 15, 16]

$$A_i^{a'} = \varepsilon_{iaj}\frac{x_j}{gr^2}. \tag{14.56}$$

This is the hedgehog 't Hooft–Polyakov non-Abelian monopole (14.33).

14.1.5 BPS states, Bogomol'nyi bound and duality

14.1.5.1 BPS states and Bogomol'nyi bound

We have seen that the behavior of the Higgs and gauge fields on the asymptotic sphere which bounds space is given by

$$\phi^a(\vec{x}) = F\frac{x^a}{r}, \quad A_i^a = -\frac{1}{gr^2}\varepsilon^{aij}x^j, \quad |\vec{x}| \longrightarrow \infty. \tag{14.57}$$

Inside the monopole it is natural to write down the spherically symmetric ansatz

$$\phi^a(\vec{x}) = F\frac{x^a}{r}H(Fgr), \quad A_i^{\,a} = -\frac{1}{gr^2}\varepsilon^{aij}x^j[1 - K(agr)].$$

(14.58)

Clearly the boundary conditions are

$$H(\infty) = 1, \quad K(\infty) = 1.$$

(14.59)

But also, in order for the energy to be finite in the core of the monopole, we must require

$$H(0) = 0, \quad K(0) = 1.$$

(14.60)

This monopole solution is a local minimum of the energy and also it solves the field equations of motion. Thus, the functions H and K can be determined from the field equations or from the requirement that the energy must be minimal. The resulting system of non-linear coupled differential equations has no analytic solution except in the celebrated limiting case known as the Bogomol'nyi–Prasad–Sommerfield (BPS) limit.

Let us start by writing down the energy functional for static configurations as (using the equation of motion $\vec{D}\vec{B} = 0$, Stokes' theorem, the magnetic charge (14.25), and recalling that the covariant derivative acts in the adjoint representation, i.e., via commutators)

$$
\begin{aligned}
E &= \int d^3\vec{x}\left[\frac{1}{2}\vec{B}^{a2} + \frac{1}{2}\vec{D}\phi^{a2} + V(\phi)\right] \\
&= \int d^3\vec{x}\left[\frac{1}{2}(\vec{B}^a \pm (\vec{D}\phi)^a)^2 + V(\phi) \mp \vec{B}^a(\vec{D}\phi)^a\right] \\
&= \int d^3\vec{x}\left[\frac{1}{2}(\vec{B}^a \pm (\vec{D}\phi)^a)^2 + V(\phi) \mp 2tr\vec{B}(\vec{D}\phi)\right] \\
&= \int d^3\vec{x}\left[\frac{1}{2}(\vec{B}^a \pm (\vec{D}\phi)^a)^2 + V(\phi) \mp 2tr\vec{D}(\vec{B}\phi)\right] \\
&= \int d^3\vec{x}\left[\frac{1}{2}(\vec{B}^a \pm (\vec{D}\phi)^a)^2 + V(\phi) \mp 2tr\vec{\partial}(\vec{B}\phi)\right] \\
&= \int d^3\vec{x}\left[\frac{1}{2}(\vec{B}^a \pm (\vec{D}\phi)^a)^2 + V(\phi)\right] \mp \int_{S^2} d\vec{\sigma}\vec{B}^a\phi^a \\
&= \int d^3\vec{x}\left[\frac{1}{2}(\vec{B}^a \pm (\vec{D}\phi)^a)^2 + V(\phi)\right] \mp \int_{S^2} d\vec{\sigma}\left(\frac{\phi^a}{F}B^i\right)\phi^a \\
&= \int d^3\vec{x}\left[\frac{1}{2}(\vec{B}^a \pm (\vec{D}\phi)^a)^2 + V(\phi)\right] \mp Fq_m.
\end{aligned}
$$

(14.61)

We choose the sign according to the sign of the magnetic charge or winding number to obtain (of course $n = 1$ in this particular case)

$$E = \int d^3\vec{x} \left[\frac{1}{2}(\vec{B}^a \pm (\vec{D}\phi)^a)^2 + V(\phi) \right] + \frac{4\pi F}{g}|n|. \tag{14.62}$$

The energy (a sum of positive terms) is thus bound by the third term, which is proportional to the winding number, viz

$$E \geqslant \frac{4\pi F}{g}|n| = \frac{4\pi}{g^2} M_W |n|. \tag{14.63}$$

In other words, the energy of the monopole is bounded from below by the mass of the charged vector particle (or W-boson) $M_W = gF$ corresponding to the broken generators of the gauge group. This is the famous Bogomol'nyi bound. This system in this limit is therefore characterized by one single length scale, which is given by the wavelength of the vector boson $1/M_W$ as opposed to the generic case where two different length scales characterize the changes in the Higgs and gauge fields.

This bound is saturated for vanishing Higgs potential

$$V = 0, \quad \leftrightarrow \lambda = 0. \tag{14.64}$$

However, as we take the limit $\lambda \longrightarrow 0$, the asymptotic of the Higgs scalar field must remain the same. The vanishing of the Higgs potential occurs naturally in super-symmetric gauge theories with two and four supercharges.

The saturation of the Bogomol'nyi bound requires also a field satisfying the so-called BPS monopole equations given by the equation

$$\vec{B}^a \pm (\vec{D}\phi)^a = 0. \tag{14.65}$$

This is a system of first-order differential equations for H and K with a solution given by the Prasad–Sommerfield monopole

$$H(x) = x \coth x - 1, \quad K(x) = \frac{x}{\sinh x}. \tag{14.66}$$

These functions reach their asymptotic values very fast. This is in fact a general result.

The region where the asymptotic behavior is obtained defines the Higgs vacuum where we have spontaneous symmetry breaking of $SU(2)$, there is no difference between the 't Hooft–Polyakov and the Dirac monopoles, and there is one single length scale given by the mass of the vector boson M_W (the asymptotic scalar field does not depend on λ). Inside the core of the monopole, which is of size $R_c = 1/M_W$, the original $SU(2)$ symmetry is restored and we obtain the perturbative phase of the Georgi–Glashow model with two independent length scales (related to g and λ).

14.1.5.2 Duality

The spectrum of the non-Abelian Georgi–Glashow model with $SU(2)$ gauge group consists therefore of the following particles:

- A Higgs scalar particle of mass $m = 0$ and electric/magnetic charges $(q_e, q_m) = (0, 0)$.

- A vector particle (photon) of mass $m = 0$ and electric/magnetic charges $(q_e, q_m) = (0, 0)$.
- Two charged vector particles $A_\pm$ of mass $M_W = gF$ and electric/magnetic charges $(q_e, q_m) = (g, 0)$.
- A magnetic monopole, which is a spin 0 particle, of mass $M = 4\pi F/g$ and electric/magnetic charges $(q_e, q_m) = (0, 4\pi/g)$.

All these states saturate the Bogomol'nyi bound with mass

$$M = F\sqrt{q_e^2 + q_m^2}. \tag{14.67}$$

It is then obvious that in the weak coupling regime $g \longrightarrow 0$ (Coulomb phase) the magnetic monopole is very heavy whereas the gauge and Higgs fields are very light, whereas in the strong coupling regime $g \longrightarrow 0$ (Higgs phase) the magnetic monopole is light whereas the gauge and Higgs fields are very heavy. The monopole is thus a non-perturbative topological configuration of the field whereas the gauge and Higgs fields are perturbative excitations over the vacuum only in the weak coupling regime.

The electric/magnetic Montonen–Olive duality [17] is defined by the exchange of the electric and magnetic fields and charges as $\vec{E} \longrightarrow \vec{B}, \vec{B} \longrightarrow -\vec{E}$ and

$$q_e = g \leftrightarrow q_m = \frac{4\pi}{g}. \tag{14.68}$$

This clearly exchanges the non-perturbative topological sector with the perturbative Noether sector, i.e. the massive vector bosons are mapped to the monopoles and vice versa. This is an example of a strong-weak coupling duality, i.e., an S-duality. Hence, under this duality, monopoles become perturbative fluctuations of some dual gauge theory whereas the Higgs and gauge particles will constitute the non-perturbative sector of this dual theory. The Montonen–Olive conjecture states that in the strong coupling regime the theory is described by a spontaneously broken 'magnetic' gauge theory with some dual gauge group $G^\star$ (as opposed to the 'electric' gauge theory which describes the weak coupling regime). See [7, 18] for a brief and pedagogical discussion.

Also, it is important to note that the above Montonen–Olive duality transformation is actually only a particular $SO(2)$ rotation in the complex plane $q_e + iq_m$ by an angle $\theta = \pi/2$. The Montonen–Olive duality can also be extended to a full $SL(2, Z)$ duality by also including the θ parameter. The Georgi–Glashow model is self-dual under the above duality transformation since the Lagrangian and the gauge group remain invariant, i.e., $G^\star = SU(2)$. However, the invariance under duality is different from the invariance under symmetry since it actually relates two different theories describing weak and strong coupling regimes.

14.1.6 Other results

14.1.6.1 Moduli space and zero modes

The moduli space is the space of solutions $\mathcal{M} = \{\vec{A}, \phi\}$ modulo gauge transformations. It consists of disconnected components labeled by the winding number

n, viz $\mathcal{M}_n$. These spaces are described by collective coordinates. Because of translational invariance, i.e., the fact that the monopole can be located anywhere in space, lead immediately to zero modes. The dimensionality of the moduli space in each topological sector, i.e., of $\mathcal{M}_n$, is exactly given by the number of square-integrable zero modes. This can be computed by calculating the index of the Dirac operator in the monopole background. The answer is very simply given by [19, 20].

$$\mathcal{M}_n = 4|n|. \tag{14.69}$$

The factor of 4 corresponds to three translations plus one phase associated with the unbroken $U(1)$ gauge rotations, which is the number of collective coordinates of the fundamental monopole with $n = 1$. The $n = 1$ moduli space is $\mathbf{R}^3 \times \mathbf{S}^1$ [21].

14.1.6.2 Coupling to fermions

We can introduce massless fermions in the background monopole gauge field by coupling them through the usual minimal and Yukawa interactions. The 't Hooft–Polyakov monopole will then induce n fermionic zero modes where n is the winding number associated with the magnetic charge q_m as given by equation (14.25).

14.1.6.3 The spin/isospin coupling

The group of invariance of the energy functional with static fields is given by $SO(3) \times SO(3)$ where the first $SO(3)$ is spatial rotations (spin) whereas the second $SO(3)$ is color rotations (isospin). However, the static 't Hooft–Polyakov hedgehog monopole solution enjoys only a diagonal $SO(3)$ in which the space and color indices are mixed. For example, remark that the hedgehog solution (14.33) does not transform covariantly under spatial rotations generated by $\vec{L}$. Thus, the monopole is actually invariant only under the combined rotations generated by the

$$\vec{K} = \vec{J} + \vec{I}, \tag{14.70}$$

where $\vec{J} = \vec{L} + \vec{S}$ is the total angular momentum and $\vec{I}$ corresponds to isospin or color rotations. The correct angular momentum operator should then be identified with $\vec{K}$, i.e.,

$$K = |J - I|, ..., J + I - 1, J + I. \tag{14.71}$$

On the other hand, electric charges in the Georgi–Glashow model arise from $SU(2)$ representations. We have already seen that the minimal $U(1)$ electric charge q_e (written e in equation (14.42)) must be equal to $g/2$ where g is the $SU(2)$ gauge coupling constant. The Dirac quantization condition $gq_m = 4\pi$ becomes then $q_e q_m = 2\pi$. This minimal charge must transform in the fundamental representation under $SU(2)$ isospin/color rotations, viz

$$I = \frac{1}{2}, \quad q_e = \frac{1}{2}g. \tag{14.72}$$

This generalizes to particles in other $SU(2)$ representations as

$$I = n + \frac{1}{2}, \quad q_e = \left(n + \frac{1}{2}\right)g \tag{14.73}$$
$$I = n, \quad q_e = ng.$$

This is consistent with Dirac quantization condition $q_e q_m = 2\pi n$.

We consider now a charged particle with isospin $I = 1/2$ in the monopole background. This particle binds to the magnetic monopole in such a way that the total angular momentum of the bound state is $S \pm 1/2$ where S is the spin of the particle (the spin of the monopole is 0 and we are assuming zero orbital angular momenta). If the particle is a fermion then the bound state is a boson whereas if the particle is a boson the bound state is a fermion, i.e. we can get fermions from bosons and bosons from fermions in the background of a 't Hooft–Polyakov magnetic field. This is an example of the anomalous spin addition theorem for the charge-monopole bound state. This conversion of spin into isospin can be traced to the existence of Dirac strings. The original literature on this topic is [22–24].

14.1.6.4 Condition for the existence of monopoles
This is given by the requirement that the gauge group G must be simply connected group such that the unbroken subgroup H contains $U(1)$ as a factor. In this case the second homotopy group is

$$\pi_2(G/(\tilde{H}\otimes U(1))) = \pi_1(\tilde{H}) \otimes \pi_1(U(1)) = \pi_1(\tilde{H}) \otimes \mathbf{Z}^{N-1}. \tag{14.74}$$

For example, magnetic monopoles fail to appear in the standard model because $SU(2) \times U(1)$ is not a simply connected group, i.e., since $\pi_2(SU(2)) = 0$.

14.1.6.5 Dyons
A magnetic monopole with an added electric charge is called a dyon [25].

14.1.6.6 Generalization to SU(N)
We have found that for $SU(2)$ the gauge symmetry is broken spontaneously to $U(1)$. As it turns out, $U(1)$ is the maximally abelian subgroup of $SU(2)$. Indeed, we can always choose the asymptotic solution $\phi \longrightarrow Fn^a$, which lies on a sphere $\mathbf{S}^2$, along the third axis and hence $n^1 = n^2 = 0$ and $n^3 = 1$, i.e. $\phi = (\phi_1, \phi_2) = F\sigma^3/2$ where $\phi_1 + \phi_2 = 0$ and $\phi_1 = F/2$.

For $SU(N)$ gauge group the symmetry will also be spontaneously broken down to the maximally Abelian subgroup, which is in this case given by the torus $U(1)^{N-1}$. The asymptotic solution lies on a sphere $\mathbf{S}^{N^2-2}$. In particular, in the BPS limit $\lambda \longrightarrow 0$, there is no Higgs potential and as such we can use gauge symmetry to rotate the Higgs field to its diagonal form. The asymptotic solution will take then the form

$$\phi = \text{diag}(\phi_1,..., \phi_N) = \phi_a H^a, \quad \sum_a \phi_a = 0. \tag{14.75}$$

The N matrices H^a are the generators of the Cartan subalgebra of $U(N)$ while ϕ^a are the components of a (root) vector in the space of the Cartan subalgebra. With the requirement $\phi_a \neq \phi_b$ for $a \neq b$ the gauge group $SU(N)$ breaks down to $U(1)^{N-1}$, which is the group that leaves the above configuration invariant. The spectrum consists therefore of $N-1$ massless photons, $N-1$ massless scalars, and $N(N-1)/2$ massive gauge bosons with mass $M_W = |\phi_a - \phi_b|$.

The topological classification of monopole solutions is then connected with the maps from the boundary sphere $\mathbf{S}^2$ onto the coset space $SU(N)/U(1)^{N-1}$, i.e., they are field configurations winding non-trivially around the asymptotic $\mathbf{S}^2$. Hence, we must consider the maps

$$\mathbf{S}^2 \longrightarrow SU(N)/U(1)^{N-1}. \tag{14.76}$$

As a consequence, the topological charge (winding number or magnetic charge) of the monopole is given by the elements of the second homotopy group

$$\pi_2(SU(N)/U(1)^{N-1}) = \pi_1(U(1)^{N-1}) = \mathbf{Z}^{N-1}. \tag{14.77}$$

Thus, this $SU(N)$ magnetic monopole carries a magnetic charge under each of the $N-1$ unbroken abelian gauge fields. This is also Dirac's charge quantization condition for $SU(N)$ case in mathematical terms.

14.1.6.7 Vortices

These are the analogous topological excitations in $1+2$ dimensions relevant for superconductivity in the same way that monopoles are the topological excitations in $1+3$ dimensions relevant for confinement. These configurations appear in the Abelian Georgi–Glashow-Higgs model with a $U(1)$ gauge group. This Abelian group becomes spontaneously broken and the photon acquires a mass in the Higgs phase, which determines the London penetration depth of the magnetic field. Indeed, the magnetic field is excluded from the superconducting region (Meissner effect).

The manifold of vacuum solutions in this case is also non-trivial given by a circle $\mathbf{S}^1$. Topological solutions are static configurations determined by the requirement of finite energy and their flux is quantized proportional to the winding number n of the mapping $\theta: \mathbf{S}^1 \longrightarrow \mathbf{S}^1_\infty$ where θ is the phase of the Higgs field and $\mathbf{S}^1_\infty$ is the asymptotic circle bounding the plane. Also the energy is determined to satisfy the Bogomol'nyi bound $E \geqslant 2\pi |n|$. This energy is saturated in some BPS-like limit in which the energy becomes given by the magnetic charge and the fields satisfy the Euler–Lagrange equations of motion.

14.2 Instantons

14.2.1 The theta term

The Yang–Mills action with the topological theta term is given by

$$S = -\frac{1}{2g^2} \int d^4x \, \mathrm{Tr} \, F_{\mu\nu} F^{\mu\nu} - \frac{\theta}{16\pi^2} \int d^4x \, \mathrm{Tr} \, F_{\mu\nu} \tilde{F}^{\mu\nu}. \tag{14.78}$$

Recall that $F_{\mu\nu} = \partial_\mu A_\nu - \partial_\nu A_\mu - i[A_\mu, A_\nu]$, $\tilde{F}^{\mu\nu} = \varepsilon^{\mu\nu\alpha\beta}F_{\alpha\beta}/2$ and $D_\alpha = \partial_\alpha - i[A_\alpha,...]$. The path integral of interest is

$$Z = \int DA_\mu \, \exp(iS). \tag{14.79}$$

This is invariant under the finite gauge transformations $A_\mu \longrightarrow U^{-1}A_\mu U + iU^{-1}\partial_\mu U$ with $U = \exp(i\Lambda) \in G$ (we will consider mostly $SU(N)$).

We Wick rotate to Euclidean signature as $x^0 \longrightarrow x^4 = ix^0$ and as a consequence $d^4x \longrightarrow d_E^4 x = id^4x$, $\partial_0 \longrightarrow \partial_4 = -i\partial_0$ and $A_0 \longrightarrow A_4 = -iA_0$. We compute $F_{\mu\nu}F^{\mu\nu} \longrightarrow (F_{\mu\nu}^2)_E$ and $F_{\mu\nu}\tilde{F}^{\mu\nu} \longrightarrow i(F_{\mu\nu}\tilde{F}_{\mu\nu})_E$. We get then

$$Z_E = \int DA_\mu \, \exp(-S_E). \tag{14.80}$$

$$S_E = \frac{1}{2g^2} \int (d^4x)_E \, \mathrm{Tr}(F_{\mu\nu}^2)_E + \frac{i\theta}{16\pi^2} \int (d^4x)_E \, \mathrm{Tr}(F_{\mu\nu}\tilde{F}_{\mu\nu})_E. \tag{14.81}$$

In the following we will drop the subscript E for simplicity. Remark that the theta term is imaginary and, as we will show shortly, this term is in fact a surface term and thus it is actually proportional to an integer number k.

The path integral/partition function is of the form

$$\int DA_\mu \, \exp\left(-\frac{1}{2g^2} \int d^4x \, \mathrm{Tr}\, F_{\mu\nu}^2 - ik\theta\right). \tag{14.82}$$

In this form, it is clear that the semi-classical approximation $\hbar \longrightarrow 0$ is given by $g^2 \longrightarrow 0$. In this limit the path integral is clearly dominated by the trivial gauge configuration $A = 0$ with zero Euclidean action, i.e. the global minimum.

However, it is also clear that the above path integral will receive very important contributions from other minima, i.e. gauge configurations characterized by finite Euclidean action which are known as instantons. The trivial instanton is precisely the perturbative vacuum and the expansion around it is the perturbative loop expansion defined by the usual Feynman diagrams. The goal of instanton calculus is, among other things, to develop a systematic expansion around each instanton configuration of finite action $S_{cl} = \hat{S}_{cl}/g^2$. Thus, the leading contribution of an instanton configuration is proportional to $\exp(-\hat{S}_{cl}/g^2)$, which leads to corrections that are further suppressed by powers of g^2 (to be contrasted with the perturbative expansion around the trivial sector where the exponential factor is 1 and cannot be observed).

For finite theories, such as the $\mathcal{N} = 4$ supersymmetric Yang–Mills theory, instantons seem to be the dominant configurations and the semi-classical approximation is valid since the coupling constant g^2 does not run[1]. For asymptotically free theories[2], the coupling constant g^2 runs with energy and in particular it becomes

---

[1] Another crucial example in which instantons give the complete solution in the semi-classical approximation is compact QED and Abelian gauge theories in general.

[2] For example four-dimensional pure Yang–Mills gauge theories and (the very similar) two-dimensional principal chiral theories given by the actions

$$S = \frac{1}{2g^2} \int d^2x \, \mathrm{Tr}\, \partial_\mu U^+ \partial_\mu U, \quad UU^\dagger = 1, \quad U \in G. \tag{14.83}$$

large in the infrared and as a consequence instantons do not dominate the path integral. In the presence of scalar fields the gauge group can become spontaneously broken and thus the running of the coupling constant g^2 may be appropriately cutoff and the semi-classical approximation and thus instantons can still play a crucial role. In three dimensions it seems that instantons play a decisive role whereas in four dimensions they play a major role.

Now we find the equations of motion in the trivial sector $\theta = 0$. We have

$$
\begin{aligned}
\delta S_E &= \frac{1}{g^2} \int d^4x \ \mathrm{Tr} \, F_{\mu\nu} \delta F_{\mu\nu} \\
&= \frac{2}{g^2} \int d^4x \ \mathrm{Tr} \, F_{\mu\nu} D_\mu \delta A_\nu \\
&= -\frac{2}{g^2} \int d^4x \ \mathrm{Tr} \, D_\mu F_{\mu\nu}. \ \delta A_\nu + \frac{2}{g^2} \int d^4x \ \mathrm{Tr} \, D_\mu(F_{\mu\nu} \delta A_\nu) \\
&= -\frac{2}{g^2} \int d^4x \ \mathrm{Tr} \, D_\mu F_{\mu\nu}. \ \delta A_\nu + \frac{2}{g^2} \int d^4x \ \mathrm{Tr} \, \partial_\mu(F_{\mu\nu} \delta A_\nu).
\end{aligned}
\tag{14.84}
$$

The equations of motion for variations of the gauge field, which vanish at infinity, are therefore given by

$$
D_\mu F_{\mu\nu} = 0.
\tag{14.85}
$$

Next we consider the general sector $\theta \neq 0$. First we show that the second term in the action S_E does not affect the equations of motion. In other words, the theta term is only a surface term. We define

$$
\mathcal{L}_\theta = \frac{1}{16\pi^2} \, \mathrm{Tr} \, F_{\mu\nu} \tilde{F}_{\mu\nu}.
\tag{14.86}
$$

We compute the variation

$$
\begin{aligned}
\delta \mathcal{L}_\theta &= \frac{1}{16\pi^2} \varepsilon_{\mu\nu\alpha\beta} \, \mathrm{Tr} \, F_{\mu\nu} \delta F_{\alpha\beta} \\
&= \frac{1}{8\pi^2} \varepsilon_{\mu\nu\alpha\beta} \, \mathrm{Tr} \, F_{\mu\nu} D_\alpha \delta A_\beta.
\end{aligned}
\tag{14.87}
$$

We use the Jacobi identity

$$
\begin{aligned}
\varepsilon_{\mu\nu\alpha\beta} D_\alpha F_{\mu\nu} &= \varepsilon_{\mu\nu\alpha\beta}(\partial_\alpha F_{\mu\nu} - i[A_\alpha, F_{\mu\nu}]) \\
&= -\varepsilon_{\mu\nu\alpha\beta}[A_\alpha, [A_\mu, A_\nu]] \\
&= 0.
\end{aligned}
\tag{14.88}
$$

Thus

$$
\begin{aligned}
\delta \mathcal{L}_\theta &= \frac{1}{8\pi^2} \varepsilon_{\mu\nu\alpha\beta} \, \mathrm{Tr} \, D_\alpha(F_{\mu\nu} \delta A_\beta) \\
&= \frac{1}{8\pi^2} \varepsilon_{\mu\nu\alpha\beta} \, \mathrm{Tr} \, (\partial_\alpha(F_{\mu\nu} \delta A_\beta) - i[A_\alpha, F_{\mu\nu} \delta A_\beta]) \\
&= \partial_\alpha \delta \mathcal{K}_\alpha,
\end{aligned}
\tag{14.89}
$$

where

$$\delta\mathcal{K}_\alpha = \frac{1}{8\pi^2}\varepsilon_{\alpha\mu\nu\beta}\,TrF_{\mu\nu}\,\delta A_\beta.\tag{14.90}$$

This shows explicitly that the theta term will not contribute to the equations of motion for variations of the gauge field, which vanish at infinity.

In order to find the current $\mathcal{K}_\alpha$ itself we consider a one-parameter family of gauge fields $A_\mu(x, \tau) = \tau A_\mu(x)$ with $0 \leqslant \tau \leqslant 1$ [26]. By using the above result we have immediately

$$\begin{aligned}
\frac{\partial}{\partial\tau}\mathcal{K}_\alpha &= \frac{1}{8\pi^2}\varepsilon_{\alpha\mu\nu\beta}\,TrF_{\mu\nu}(x,\,\tau)\frac{\partial}{\partial\tau}A_\beta\\
&= \frac{1}{8\pi^2}\varepsilon_{\alpha\mu\nu\beta}\,Tr(\tau\partial_\mu A_\nu - \tau\partial_\nu A_\mu - i\tau^2[A_\mu,\,A_\nu]).\,A_\beta(x).
\end{aligned}\tag{14.91}$$

By integrating both sides with respect to τ between $\tau = 0$ and $\tau = 1$ and setting $\mathcal{K}_\alpha(x, 1) = \mathcal{K}_\alpha(x)$ and $\mathcal{K}_\alpha(x, 0) = 0$ we get the so-called Chern–Simons current

$$\begin{aligned}
\mathcal{K}_\alpha &= \frac{1}{8\pi^2}\varepsilon_{\alpha\mu\nu\beta}\,Tr\left(\frac{1}{2}\partial_\mu A_\nu - \frac{1}{2}\partial_\nu A_\mu - \frac{i}{3}[A_\mu,\,A_\nu]\right)\cdot A_\beta(x)\\
&= \frac{1}{8\pi^2}Tr\left(\tilde{F}_{\alpha\beta}A_\beta + \frac{i}{3}\varepsilon_{\alpha\mu\nu\beta}A_\mu A_\nu A_\beta\right).
\end{aligned}\tag{14.92}$$

14.2.2 Vacuum degeneracy and Bogomoln'yi bound

14.2.2.1 Vacuum degeneracy

The existence of instantons, like the existence of monopoles, can be traced down to the property of vacuum degeneracy. Instantons are static solutions of the classical equations of motion of Yang–Mills gauge theory which are also configurations of finite energy [9, 27, 28]. We will employ again the Weyl or temporal gauge $A_0 = 0$. Static solution will correspond as before to $\vec{E}^a = 0$ while vanishing energy corresponds to $\vec{B}^a = 0$. The vanishing of the magnetic field can be solved by $\vec{A}^a = 0$. The Weyl gauge is unchanged under time-independent gauge transformations which changes $\vec{A}^a = 0$ to the pure gauges

$$\vec{A} = iU^{-1}(\vec{x})\vec{\nabla}U(\vec{x}).\tag{14.93}$$

Thus, the manifold of zero-energy vaccuum solutions is $U \in SU(2) \sim \mathbf{S}^3$. Hence, a given solution $U(\vec{x}) \in SU(2)$ defines a map from $\mathbf{R}^3$ to $\mathbf{S}^3$. As usual we will impose the condition that at infinity the group element $U(\vec{x})$ approaches the identity in the group, viz

$$U(\vec{x}) \longrightarrow 1, \quad |\vec{x}| \longrightarrow \infty.\tag{14.94}$$

This effectively compactifies the space $\mathbf{R}^3$, i.e., identify the points at infinity, which produces the sphere $\mathbf{S}^3_\infty$, which is actually the boundary of $\mathbf{R}^4$ (we are implicitly assuming Euclidean signature). In other words, $U(\vec{x})$ is a map from a 3-sphere to a 3-sphere, namely

$$U(\vec{x}): \quad \mathbf{S}^3_\infty \longrightarrow SU(N)|_{N=2} \sim \mathbf{S}^3. \tag{14.95}$$

These maps can be classified by the third homotopy group. They can then be attributed a winding number known as the Pontryagin number or second Chern class given by

$$k \in \pi_3(SU(N)). \tag{14.96}$$

This number counts how many times we cover the group 3-sphere if $\vec{x}$ covers the space 3-sphere once, i.e., how many times the group wraps itself around spatial infinity $\mathbf{S}^3_\infty$.

The above winding number can be expressed in terms of the gauge field as follows [29, 30]

$$\begin{aligned}
k &= \frac{1}{16\pi^2} \int d^4x \, \mathrm{Tr} \, F_{\mu\nu} \tilde{F}_{\mu\nu} \\
&= \int d^4x \partial_\alpha \mathcal{K}_\alpha \\
&= \int_{\mathbf{S}^3_\infty} d^3\sigma_\alpha \mathcal{K}_\alpha \\
&= \frac{1}{16\pi^2} \varepsilon_{\alpha\mu\nu\beta} \int_{\mathbf{S}^3_\infty} d^3\sigma_\alpha \left(\partial_\mu A^a_\nu A^a_\beta + \frac{1}{3} f_{abc} A^a_\mu A^b_\nu A^c_\beta \right).
\end{aligned} \tag{14.97}$$

The first line is precisely the theta term. The last line, as we will show explicitly, gives an integer only of A_μ is a pure gauge. The winding number k can also be rewritten in terms of the gauge element U as

$$k = \frac{1}{24\pi^2} \varepsilon_{\alpha\mu\nu\beta} \int_{\mathbf{S}^3_\infty} d^3\sigma_\alpha \, \mathrm{Tr}(U^{-1}\partial_\mu U)(U^{-1}\partial_\nu U)(U^{-1}\partial_\beta U). \tag{14.98}$$

We have therefore an infinite number of vacuum solutions with vanishing energy characterized by an integer number k. These topological sectors are connected by large gauge transformations which can change the winding number k. We note that 'large gauge transformations', in contrast with the usual transformations which are also termed 'small gauge transformations', are not continuously connected to the identity of the gauge group.

In order to get finite not vanishing energy we will require the group elements $U(\vec{x})$ to only approach the pure gauges (14.93) as we approach the spatial infinity $\mathbf{S}^3_\infty = \partial \mathbf{R}^4$. We will have then an infinite number of vacuum solutions with finite energy characterized by the winding number k.

14.2.2.2 Bogomoln'yi bound

We go back to our original action

$$\begin{aligned}
S &= \frac{1}{2g^2} \int d^4x \, \mathrm{Tr} \, F^2_{\mu\nu} + \frac{i\theta}{16\pi^2} \int d^4x \, \mathrm{Tr} \, F_{\mu\nu} \tilde{F}_{\mu\nu} \\
&= \frac{1}{2g^2} \int d^4x \, \mathrm{Tr} \, F^2_{\mu\nu} + i\theta k.
\end{aligned} \tag{14.99}$$

Using $\tilde{F}_{\mu\nu}^2 = \varepsilon_{\mu\nu\alpha\beta}\varepsilon_{\mu\nu\rho\sigma}F_{\alpha\beta}F_{\rho\sigma}/4 = F_{\mu\nu}^2/2$ we have

$$\frac{1}{2g^2}\int d^4x\,\mathrm{Tr}\,F_{\mu\nu}^2 = \frac{1}{4g^2}\int d^4x\,\mathrm{Tr}(F_{\mu\nu}\mp\tilde{F}_{\mu\nu})^2 \pm \frac{1}{2g^2}\int d^4x\,\mathrm{Tr}\,F_{\mu\nu}\tilde{F}_{\mu\nu}$$

$$= \frac{1}{4g^2}\int d^4x\,\mathrm{Tr}(F_{\mu\nu}\mp\tilde{F}_{\mu\nu})^2 \pm \frac{8\pi^2 k}{g^2}. \tag{14.100}$$

We remark that for $F_{\mu\nu} = \pm\tilde{F}_{\mu\nu}$ the action takes the value $\pm 8\pi^2 k/g^2$. Thus for $F_{\mu\nu} = +\tilde{F}_{\mu\nu}$ and $F_{\mu\nu} = -\tilde{F}_{\mu\nu}$ we must have $k > 0$ and $k < 0$, respectively, since the action is positive definite. We conclude immediately therefore that the Yang–Mills action in an instanton configuration with topological charge k is bounded by

$$\frac{1}{2g^2}\int d^4x\,\mathrm{Tr}\,F_{\mu\nu}^2 \geqslant \frac{8\pi^2|k|}{g^2}. \tag{14.101}$$

This is the Bogomoln'yi bound. The equality holds if and only if the so-called duality equations (instanton equations) are satisfied, viz

$$F_{\mu\nu} = \tilde{F}_{\mu\nu}, \quad k > 0, \quad \text{instantons}. \tag{14.102}$$

$$F_{\mu\nu} = -\tilde{F}_{\mu\nu}, \quad k < 0, \quad \text{anti} - \text{instantons}. \tag{14.103}$$

We will focus on the self-dual solution corresponding to $F_{\mu\nu} = \tilde{F}_{\mu\nu}$ since parity maps positive k to negative k and vice versa. As we already noted the self-dual and anti-self-dual solutions minimize the action in a given instanton solution (i.e., for a fixed k) but also they actually solve the equations of motion.

Indeed, it is obvious that since $F_{\mu\nu} = \pm\tilde{F}_{\mu\nu}$ we have $D_\mu F_{\mu\nu} = D_\mu \tilde{F}_{\mu\nu} = \varepsilon_{\mu\nu\alpha\beta}D_\mu F_{\alpha\beta}/2 = 0$, which is the Bianchi identity. In some sense the duality equations, which are first order, are the square root of the full equations of motion. A solution of the equations of motion does not necessarily solve the duality equations. The values of the Yang–Mills action in the self-dual and anti-self-dual solutions are

$$S = \frac{8\pi^2|k|}{g^2} + i\theta k. \tag{14.104}$$

Alternatively

$$S = 2\pi ik\tau^*, \quad k > 0, \quad S = 2\pi ik\tau, \quad k < 0. \tag{14.105}$$

The complex coupling is

$$\tau = \frac{4i\pi}{g^2} + \frac{\theta}{2\pi}. \tag{14.106}$$

14.2.3 The topological charge or instanton number

Before we find the explicit one-instanton solution due to Belavin, Polyakov, Schwartz and Tyupkin [9] we discuss a little further the topological charge, i.e., the integral over spacetime of the theta term given by

$$k = \frac{1}{16\pi^2} \int d^4x \, \mathrm{Tr}\, F_{\mu\nu}\tilde{F}_{\mu\nu}$$

$$= \frac{1}{16\pi^2}\varepsilon_{\alpha\mu\nu\beta}\int_{\mathbf{S}^3_\infty} d^3\sigma_\alpha\left(\partial_\mu A_\nu^a A_\beta^a + \frac{1}{3}f_{abc}A_\mu^a A_\nu^b A_\beta^c\right). \tag{14.107}$$

A Yang–Mills instanton is a solution of the equations of motion, which has finite action. In order to have a finite action the field strength $F_{\mu\nu}$ must approach 0 at infinity faster than $1/x^2$, viz[3]

$$F_{\mu\nu}^I(x) = O\left(\frac{1}{|\vec{x}\,|^2}\right), \quad |\vec{x}\,|^2 \longrightarrow \infty. \tag{14.108}$$

We can immediately deduce that the gauge field must approach a pure gauge at infinity, viz

$$A_\mu^I(x) = iU^{-1}\partial_\mu U + O\left(\frac{1}{|\vec{x}|}\right), \quad |\vec{x}| \longrightarrow \infty. \tag{14.109}$$

This can be checked by simple substitution in $F_{\mu\nu} = \partial_\mu A_\nu - \partial_\nu A_\mu - i[A_\mu, A_\nu]$. As we have discussed, we can imagine that the four-dimensional Euclidean spacetime $\mathbf{R}^4$ is bounded by a large three-sphere $\mathbf{S}^3_\infty$ in the same way that we can imagine that the plane is bounded by a large $\mathbf{S}^1_\infty$, viz

$$\partial\mathbf{R}^4 = \mathbf{S}^3_\infty. \tag{14.110}$$

Now a gauge configuration $A_\mu^I(x)$ at infinity (on the sphere $\mathbf{S}^3_\infty$) defines a group element U which satisfies (from the above asymptotic behavior) the equation $\partial_\mu U^{-1} = iA_\mu^I U^{-1}$ or equivalently

$$\frac{d}{ds}U^{-1}(x(s), x_0) = i\frac{dx^\mu}{ds}A_\mu^I(x(s))U^{-1}(x(s), x_0). \tag{14.111}$$

The solution is given by the path-ordered Wilson line

$$U^{-1}(x, x_0) = \mathcal{P}\exp\left(i\int_0^1 ds\frac{dy^\mu}{ds}A_\mu^I(y(s))\right). \tag{14.112}$$

The path is labeled by the parameter s which runs from $s = 0$ ($y = x_0$) to $s = 1$ ($y = x$) and the path-ordering operator $\mathcal{P}$ is defined such that terms with higher values of s are always put on the left in every order in the Taylor expansion of the exponential.

In the above formula for U^{-1} the points x and x_0 are both at infinity, i.e., on the sphere $\mathbf{S}^3_\infty$. In other words, gauge configurations $A_\mu^I(x)$ with finite action (instanton configurations) define a map from $\mathbf{S}^3_\infty$ into G, viz

$$U^{-1}\colon \mathbf{S}^3_\infty \longrightarrow G. \tag{14.113}$$

---

[3] The requirement of finite action can be neatly satisfied if we compactify $\mathbf{R}^4$ by adding one point (at) ∞ to obtain the four-sphere $\mathbf{S}^4$.

These maps are classified by homotopy theory. As an example we concentrate on the group $G = SU(2)$. The group $SU(2)$ is topologically a three-sphere since any element $U \in SU(2)$ can be expanded (in the fundamental representation) as

$$U = n_4 + i\vec{n}\vec{\tau} \tag{14.114}$$

and as a consequence the unitarity condition $U^+ U = 1$ becomes $n_4^2 + \vec{n}^2 = 1$. In this case we have therefore maps from the three-sphere to the three-sphere, viz

$$U^{-1} \colon \mathbf{S}_\infty^3 \longrightarrow SU(2) = \mathbf{S}^3. \tag{14.115}$$

These maps are characterized by an integer k (known variously as the Pontryagin class, the winding number, the instanton number and the topological charge) defined by

$$k = \frac{1}{16\pi^2} \int d^4x \; \mathrm{Tr} \, F_{\mu\nu} \tilde{F}_{\mu\nu}. \tag{14.116}$$

This number measures how many times the second $\mathbf{S}^3$ (group) wraps (covers) the first sphere $\mathbf{S}_\infty^3$ (space). In fact this is the underlying reason why k must be quantized. In other words, k is an element of the third homotopy group $\pi_3(\mathbf{S}^3)$, viz[4]

$$k \in \pi_3(SU(2)) = \pi_3(\mathbf{S}^3) = \mathbf{Z}. \tag{14.117}$$

To see this more explicitly we write (14.116) as

$$\begin{aligned} k &= \int_{\partial \mathbf{R}^4 \, = \, \mathbf{S}_\infty^3} d^3\sigma_\alpha \mathcal{K}_\alpha \\ &= \frac{i}{24\pi^2} \varepsilon_{\alpha\mu\nu\beta} \int_{\partial \mathbf{R}^4 \, = \, \mathbf{S}_\infty^3} d^3\sigma_\alpha \, \mathrm{Tr} \, A_\mu A_\nu A_\beta. \end{aligned} \tag{14.118}$$

In the above equation we have used the current (14.92) and the fact that on $\mathbf{S}_\infty^3$ we have $F_{\mu\nu} = 0$. We can also use the fact that $A_\mu = A_\mu^I = iU^{-1}\partial_\mu U \equiv iL_\mu$ on $\mathbf{S}_\infty^3$ and also (where τ_a are Pauli matrices)

$$U^{-1} = n_4 - in_a\tau_a = -in_\mu\tau_\mu, \quad \tau_\mu = (\tau_a, i). \tag{14.119}$$

$$U = n_4 + in_a\tau_a = in_\mu\bar{\tau}_\mu, \quad \bar{\tau}_\mu = (\tau_a, -i). \tag{14.120}$$

Thus by using the fact that $U^{-1}. \, \partial_\mu U = -\partial_\mu U^{-1}. \, U$ we have $L_\mu = n_\alpha \partial_\mu n_\beta \tau_{\alpha\beta}$ where

$$\tau_{\alpha\beta} = \frac{1}{2}(\tau_\alpha\bar{\tau}_\beta - \tau_\beta\bar{\tau}_\alpha) = -\tau_{\beta\alpha}. \tag{14.121}$$

The spinor representation $\tau_{\alpha\beta}$ of $SU(2)$ is related to the vector representation τ_a of $SU(2)$ by $\tau_{\mu\nu} = i\bar{\eta}_{\mu\nu}^a \tau_a$ where $\bar{\eta}_{\mu\nu}^a$ are the t'Hooft symbols (more on this below). We conclude immediately that $\mathrm{Tr} \, \tau_{\mu\nu} = 0$. We will also use

$$[\tau_{AB}, \tau_{CD}] = -2(\delta_{AC}\tau_{BD} + \delta_{BD}\tau_{AC} - \delta_{AD}\tau_{BC} - \delta_{BC}\tau_{AD}). \tag{14.122}$$

$$\{\tau_{AB}, \tau_{CD}\} = -2(\delta_{AC}\delta_{BD} - \delta_{AD}\delta_{BC} - \varepsilon_{ABCD}). \tag{14.123}$$

---

[4] In general $\pi_n(\mathbf{S}^n) = \mathbf{Z}$. It is obvious that $\pi_1(\mathbf{S}^1) = \pi_2(\mathbf{S}^2) = \mathbf{Z}$.

We compute immediately

$$
\begin{aligned}
\varepsilon_{\alpha\mu\nu\beta}\,\mathrm{Tr}\,A_\mu A_\nu A_\beta &= -\,i\varepsilon_{\alpha\mu\nu\beta}\,\mathrm{Tr}\,L_\mu L_\nu L_\beta \\
&= -\frac{i}{2}\varepsilon_{\alpha\mu\nu\beta}(n_A\partial_\mu n_B)(n_C\partial_\nu n_D)(n_E\partial_\beta n_F)\mathrm{Tr}[\tau_{AB},\tau_{CD}]\tau_{EF} \\
&= -\frac{i}{2}\varepsilon_{\alpha\mu\nu\beta}(n_A\partial_\mu n_B)(n_C\partial_\nu n_D)(n_E\partial_\beta n_F)\mathrm{Tr}\,(-\delta_{AC}\{\tau_{BD},\tau_{EF}\} - \delta_{BD}\{\tau_{AC},\tau_{EF}\} \\
&\quad + \delta_{AD}\{\tau_{BC},\tau_{EF}\} + \delta_{BC}\{\tau_{AD},\tau_{EF}\}).
\end{aligned}
\tag{14.124}
$$

By inspection all the terms in the above equation are zero because of either $n_A\partial_\mu n_A = 0$ (since $n_A^2 = 1$) or $\varepsilon_{\alpha\mu\nu\beta}\partial_\mu n_A\partial_\nu n_A = 0$. The only non-zero contribution comes from the term proportional to the epsilon tensor in the first anticommutator, which gives

$$
\begin{aligned}
\varepsilon_{\alpha\mu\nu\beta}\,\mathrm{Tr}\,A_\mu A_\nu A_\beta &= -\frac{i}{2}\varepsilon_{\alpha\mu\nu\beta}(n_A\partial_\mu n_B)(n_C\partial_\nu n_D)(n_E\partial_\beta n_F)\mathrm{Tr}\,(-2\delta_{AC}\varepsilon_{BDEF}) \\
&= 2i\varepsilon_{\alpha\mu\nu\beta}\varepsilon_{ABCD}n_A\partial_\mu n_B\partial_\nu n_C\partial_\beta n_D.
\end{aligned}
\tag{14.125}
$$

We get then

$$
k = -\frac{1}{12\pi^2}\varepsilon_{\alpha\mu\nu\beta}\int_{\partial\mathbf{R}^4\,=\,S_\infty^3} d^3\sigma_\alpha\,\varepsilon_{ABCD}n_A\partial_\mu n_B\partial_\nu n_C\partial_\beta n_D.
\tag{14.126}
$$

Let us consider the so-called fundamental map given by

$$
n_A = x_A/r,\ r = \sqrt{x_A^2}.
\tag{14.127}
$$

This corresponds to $k = -1$, i.e., the $\mathbf{S}^3$ (group) wraps the $\mathbf{S}_\infty^3$ (space) only once in the negative direction. We compute $\partial_\mu n_B = (\delta_{\mu B} - n_\mu n_B)/r$ and as a consequence

$$
\varepsilon_{\alpha\mu\nu\beta}\varepsilon_{ABCD}n_A\partial_\mu n_B\partial_\nu n_C\partial_\beta n_D = \varepsilon_{\alpha\mu\nu\beta}\varepsilon_{A\mu\nu\beta}n_A/r^3 = 6n_\alpha/r^3.
\tag{14.128}
$$

We get then

$$
\begin{aligned}
-k &= \frac{1}{2\pi^2}\int_{\partial\mathbf{R}^4\,=\,S_\infty^3} d^3\sigma_\alpha\frac{x_\alpha}{r^4} \\
&= \frac{1}{2\pi^2 r_\infty^4}\int_{\partial\mathbf{R}^4\,=\,S_\infty^3} d^3\sigma_\alpha x_\alpha \\
&= \frac{1}{2\pi^2 r_\infty^4}\int d^4x\,\partial_\alpha x_\alpha \\
&= \frac{2}{\pi^2 r_\infty^4}\int d^4x \\
&= \frac{2}{\pi^2 r_\infty^4}\frac{1}{2}\pi^2 r_\infty^4 \\
&= +1.
\end{aligned}
\tag{14.129}
$$

The topological charge $k = +1$ can be obtained by making the replacements $U \longrightarrow U^{-1}$, $U^{-1} \longrightarrow U$. Equivalently we should consider the anti-fundamental map $U = n_4 - in_a\tau_a = -in_\mu\tau_\mu$, $U^{-1} = n_4 + in_a\tau_a = in_\mu\bar{\tau}_\mu$ and as a consequence $L_\mu = n_\alpha\partial_\mu n_\beta\bar{\tau}_{\alpha\beta}$ where $\bar{\tau}_{\mu\nu} = (\bar{\tau}_\mu\tau_\nu - \bar{\tau}_\nu\tau_\mu)/2 = i\eta^a_{\mu\nu}\tau_a$ is the other spinor representation of $SU(2)$ and $\eta^a_{\mu\nu}$ are also t'Hooft symbols (see below). We have thus $\mathrm{Tr}\,\bar{\tau}_{\mu\nu} = 0$ and the commutators and anticommutators of $\bar{\tau}_{\mu\nu}$ are again given by (14.122) and (14.123) with the replacements $\tau_{\mu\nu} \longrightarrow \bar{\tau}_{\mu\nu}$ and $\varepsilon_{ABCD} \longrightarrow -\varepsilon_{ABCD}$. It is obvious that this last point will give rise to a the value $k = +1$.

The equation (14.126) can be rewritten as

$$k = -\frac{1}{12\pi^2}\varepsilon_{abc}\int_{\partial\mathbf{R}^4 = S^3_\infty} d^3\xi\,\varepsilon_{ABCD}n_A\partial_a n_B\partial_b n_C\partial_c n_D. \tag{14.130}$$

This can be shown as follows. By introducing a local parametrization $\xi_a = \xi_a(x)$ of the $SU(2)$ group elements we can rewrite k as

$$k = \frac{1}{24\pi^2}\varepsilon_{\alpha\mu\nu\beta}\int_{\partial\mathbf{R}^4 = \mathbf{S}^3_\infty} d^3\sigma_\alpha\frac{\partial\xi_a}{\partial x_\mu}\frac{\partial\xi_b}{\partial x_\nu}\frac{\partial\xi_c}{\partial x_\beta}\,\mathrm{Tr}\,L_a L_b L_c, \quad L_a = U^{-1}\partial_a U. \tag{14.131}$$

By using $d^3\sigma_\alpha = (\varepsilon_{\alpha\mu\nu\beta}dx_\mu \wedge dx_\nu \wedge dx_\beta)/6$ and $\varepsilon_{\alpha\mu\nu\beta}\varepsilon_{\alpha\mu'\nu'\beta'} = \delta^{\mu'\nu'\beta'}_{[\mu\nu\beta]} = \delta^{\mu'}_\mu(\delta^{\nu'}_\nu\delta^{\beta'}_\beta - \delta^{\nu'}_\beta\delta^{\beta'}_\nu) + \delta^{\mu'}_\nu(\delta^{\nu'}_\beta\delta^{\beta'}_\mu - \delta^{\nu'}_\mu\delta^{\beta'}_\beta) + \delta^{\mu'}_\beta(\delta^{\nu'}_\mu\delta^{\beta'}_\nu - \delta^{\nu'}_\nu\delta^{\beta'}_\mu)$ we get

$$k = \frac{1}{24\pi^2}\frac{1}{6}\delta^{\mu'\nu'\beta'}_{[\mu\nu\beta]}\int_{\partial\mathbf{R}^4 = \mathbf{S}^3_\infty} dx_{\mu'} \wedge dx_{\nu'} \wedge dx_{\beta'}\frac{\partial\xi_a}{\partial x_\mu}\frac{\partial\xi_b}{\partial x_\nu}\frac{\partial\xi_c}{\partial x_\beta}\,\mathrm{Tr}\,L_a L_b L_c$$

$$= \frac{1}{24\pi^2}\int_{\partial\mathbf{R}^4 = \mathbf{S}^3_\infty} d\xi_a \wedge d\xi_b \wedge d\xi_c\,\mathrm{Tr}\,L_a L_b L_c \tag{14.132}$$

$$= \frac{1}{24\pi^2}\int_{\partial\mathbf{R}^4 = \mathbf{S}^3_\infty} d^3\xi\,\varepsilon_{abc}\,\mathrm{Tr}\,L_a L_b L_c.$$

This then leads to equation (14.130) by using the result

$$-\varepsilon_{abc}\,\mathrm{Tr}\,L_a L_b L_c = 2\varepsilon_{abc}\varepsilon_{ABCD}n_A\partial_a n_B\partial_b n_C\partial_c n_D. \tag{14.133}$$

The result (14.132) is precisely equation (14.98) if we also employ the replacement $\varepsilon_{\alpha\mu\nu\beta}d^3\sigma_\alpha \longrightarrow \varepsilon_{abc}d^3\xi$ one more time.

14.2.4 BPST instanton

14.2.4.1 't Hooft symbols
The Euclidean Lorentz group is $SO(4) = SU(2)_L \times SU(2)_R$. The vector representation of $SU(2)$ is generated by Pauli matrices τ_a. We will work with $\tau_\mu = (\tau_a, i)$, $\bar{\tau}_\mu = (\tau_a, -i)$. These satisfy

$$\bar{\tau}_\alpha\tau_\beta + \bar{\tau}_\beta\tau_\alpha = 2\delta_{\alpha\beta}, \quad \tau_\alpha\bar{\tau}_\beta + \tau_\beta\bar{\tau}_\alpha = 2\delta_{\alpha\beta}. \tag{14.134}$$

There are two inequivalent spinor representations of $SU(2)$ given by

$$\tau_{\alpha\beta} = \frac{1}{2}(\tau_\alpha\bar{\tau}_\beta - \tau_\beta\bar{\tau}_\alpha) = -\tau_{\beta\alpha}, \quad \bar{\tau}_{\alpha\beta} = \frac{1}{2}(\bar{\tau}_\alpha\tau_\beta - \bar{\tau}_\beta\tau_\alpha) = -\bar{\tau}_{\beta\alpha}. \tag{14.135}$$

These are anti-Hermitian matrices. The vector and spinor representations are related by t'Hooft symbols, viz

$$\tau_{\mu\nu} = i\bar{\eta}^a_{\mu\nu}\tau_a, \quad \bar{\tau}_{\mu\nu} = i\eta^a_{\mu\nu}\tau_a. \tag{14.136}$$

Explicitly we compute

$$\bar{\eta}^c_{ab} = \varepsilon_{abc}, \quad \bar{\eta}^b_{4a} = \delta^b_a = -\bar{\eta}^b_{a4}. \tag{14.137}$$

$$\eta^c_{ab} = \varepsilon_{abc}, \quad \eta^b_{4a} = -\delta^b_a = -\eta^b_{a4}. \tag{14.138}$$

We have $\tau_{ab} = i\varepsilon_{abc}\tau_c$, $\tau_{4a} = -\tau_{a4} = i\tau_a$. We compute $[\tau_{ab}, \tau_{c4}] = -2(\delta_{ac}\tau_{b4} - \delta_{bc}\tau_{a4})$ and $[\tau_{ab}, \tau_{cd}] = -2i(\varepsilon_{cda}\tau_b - \varepsilon_{cdb}\tau_a)$, which collectively can be rewritten as

$$[\tau_{AB}, \tau_{CD}] = -2(\delta_{AC}\tau_{BD} + \delta_{BD}\tau_{AC} - \delta_{AD}\tau_{BC} - \delta_{BC}\tau_{AD}). \tag{14.139}$$

We also compute $\{\tau_{ab}, \tau_{c4}\} = 2\varepsilon_{abc}$ and $\{\tau_{ab}, \tau_{cd}\} = -2\varepsilon_{abi}\varepsilon_{cdi}$, which can be rewritten collectively as

$$\{\tau_{AB}, \tau_{CD}\} = -2(\delta_{AC}\delta_{BD} - \delta_{AD}\delta_{BC} - \varepsilon_{ABCD}). \tag{14.140}$$

Similarly, we compute

$$[\bar{\tau}_{AB}, \bar{\tau}_{CD}] = -2(\delta_{AC}\bar{\tau}_{BD} + \delta_{BD}\bar{\tau}_{AC} - \delta_{AD}\bar{\tau}_{BC} - \delta_{BC}\bar{\tau}_{AD}). \tag{14.141}$$

$$\{\bar{\tau}_{AB}, \bar{\tau}_{CD}\} = -2(\delta_{AC}\delta_{BD} - \delta_{AD}\delta_{BC} + \varepsilon_{ABCD}). \tag{14.142}$$

By using $\tau_{ab} = i\varepsilon_{abc}\tau_c = \bar{\tau}_{ab}$, $\tau_{4a} = -\tau_{a4} = i\tau_a$, $\bar{\tau}_{4a} = -\bar{\tau}_{a4} = -i\tau_a$ and $\varepsilon_{abc4} = \varepsilon_{abc} = -\varepsilon_{4abc}$ we can check that $\tau_{\mu\nu}$ is anti-self-dual and $\bar{\tau}_{\mu\nu}$ is self-dual, viz

$$\tau_{\mu\nu} = -\tilde{\tau}_{\mu\nu} = -\frac{1}{2}\varepsilon_{\mu\nu\alpha\beta}\tau_{\alpha\beta}. \tag{14.143}$$

$$\bar{\tau}_{\mu\nu} = \tilde{\bar{\tau}}_{\mu\nu} = \frac{1}{2}\varepsilon_{\mu\nu\alpha\beta}\bar{\tau}_{\alpha\beta}. \tag{14.144}$$

We can also compute

$$\begin{aligned}
\varepsilon_{\mu\nu\alpha\beta}\tau_{\beta\rho} &= \frac{1}{2}\varepsilon_{\mu\nu\alpha\beta}\varepsilon_{\rho\sigma\gamma\beta}\tau_{\sigma\gamma} \\
&= \frac{1}{2}\delta^{\rho\sigma\gamma}_{[\mu\nu\alpha]}\tau_{\sigma\gamma} \\
&= \delta_{\mu\rho}\tau_{\nu\alpha} + \delta_{\nu\rho}\tau_{\alpha\mu} + \delta_{\alpha\rho}\tau_{\mu\nu}.
\end{aligned} \tag{14.145}$$

Similarly, we compute

$$\varepsilon_{\mu\nu\alpha\beta}\bar{\tau}_{\beta\rho} = -\delta_{\mu\rho}\bar{\tau}_{\nu\alpha} - \delta_{\nu\rho}\bar{\tau}_{\alpha\mu} - \delta_{\alpha\rho}\bar{\tau}_{\mu\nu}. \tag{14.146}$$

14.2.4.2 BPST instanton

Next we find explicit solutions for the duality equations for $SU(2)$ gauge theory with $k = \pm 1$. First we recall that the fundamental map $U = n_4 + in_a\tau_a = in_\mu\bar{\tau}_\mu$ corresponds to $k = -1$. The corresponding anti-instanton gauge field is $A^I_\mu = iU^{-1}\partial_\mu U$.

Explicitly, we compute (using $\bar{\eta}^{a}_{\alpha\beta} = -\bar{\eta}^{a}_{\beta\alpha}$, i.e., the 't Hooft symbol is antisymmetric and $\tau_{\mu\nu} = i\bar{\eta}^{a}_{\mu\nu}\tau_a$)

$$
\begin{aligned}
A^{I}_{\mu} &= in_\alpha \partial_\mu n_\beta \tau_{\alpha\beta} \\
&= \frac{i}{r} n_\alpha (\delta_{\mu\beta} - n_\mu n_\beta)\tau_{\alpha\beta} \\
&= -\frac{1}{r} n_\alpha (\delta_{\mu\beta} - n_\mu n_\beta)\bar{\eta}^{a}_{\alpha\beta}\tau_a \\
&= \frac{1}{r}\bar{\eta}^{a}_{\mu\alpha} n_\alpha \tau_a \\
&= -\frac{i}{r}\tau_{\mu\alpha} n_\alpha, \quad (k = -1, \text{anti} - \text{instanton}).
\end{aligned}
\tag{14.147}
$$

For $k = +1$ we should consider the anti-fundamental map $U = n_4 - in_a\tau_a = -in_\mu\tau_\mu$ and we compute the gauge field

$$
\begin{aligned}
A^{I}_{\mu} &= \frac{1}{r}\eta^{a}_{\mu\alpha} n_\alpha \tau_a \\
&= -\frac{i}{r}\bar{\tau}_{\mu\alpha} n_\alpha, \quad (k = +1, \text{instanton}).
\end{aligned}
\tag{14.148}
$$

The first observation is the fact that this configuration is not invariant under the rotational symmetry group $SU(2)_L$ of $SO(4)$ unless we also rotate the configuration under the gauge (isospin) symmetry group $SU(2)$. The instanton configuration breaks the product of the two groups $SU(2)_L$ and $SU(2)$ down to a diagonal subgroup. Indeed, we have the symmetry breaking pattern

$$
SU(2)_L \times SU(2) \longrightarrow SU(2)_{\text{diagonal}}.
\tag{14.149}
$$

The second observation concerns the behavior of the above instanton configuration when approaching the origin, which is obviously very singular. The $k = +1$ (instanton) and $k = -1$ (anti-instanton) configurations can also be rewritten as $A^{I}_{\mu} = -i\bar{\tau}_{\mu\alpha}\partial_\alpha \ln r$ and $A^{I}_{\mu} = -i\tau_{\mu\alpha}\partial_\alpha \ln r$, respectively. The anti-self-duality of the anti-instanton configuration ($k = -1$) and the self-duality of the instanton configuration ($k = +1$), respectively, cannot be expected to hold since $A^{I}_{\mu} = iU^{-1}\partial_\mu U$ is only an asymptotic behavior.

A systematic method of finding instanton and anti-instanton configurations is as follows. We concentrate on the instanton and leave the anti-instanton as an exercise. We try the more general ansatz

$$
-A^{In}_{\mu} = -i\alpha\tau_{\mu\alpha}\partial_\alpha \ln \phi(r).
\tag{14.150}
$$

We compute immediately

$$
\begin{aligned}
-F^{In}_{\alpha\beta} = &-i\alpha\tau_{\beta\mu}\partial_\alpha\partial_\mu \ln \phi \\
&+ i\alpha\tau_{\alpha\mu}\partial_\beta\partial_\mu \ln \phi - 2i\alpha^2\tau_{\mu\beta}\partial_\mu \ln \phi\partial_\alpha \ln \phi + 2i\alpha^2\tau_{\mu\alpha}\partial_\mu \ln \phi\partial_\beta \ln \phi + 2i\alpha^2\tau_{\alpha\beta}(\partial_\mu \ln \phi)^2.
\end{aligned}
\tag{14.151}
$$

By using the properties (14.143) and (14.145) of 't Hooft symbols we compute the dual field strength

$$-\tilde{F}^{In}_{\alpha\beta} = -\,i\alpha\tau_{\beta\mu}\partial_\mu\partial_\alpha \ln\phi + i\alpha\tau_{\alpha\mu}\partial_\mu\partial_\beta \ln\phi - i\alpha\tau_{\alpha\beta}\partial_\mu^2 \ln\phi$$
$$+ 2i\alpha^2\tau_{\beta\mu}\partial_\alpha \ln\phi\partial_\mu \ln\phi - 2i\alpha^2\tau_{\alpha\mu}\partial_\beta \ln\phi\partial_\mu \ln\phi. \tag{14.152}$$

The solution of the self-duality equation $-F^{In}_{\alpha\beta} = -\tilde{F}^{In}_{\alpha\beta}$ satisfies the equation

$$-F^{In}_{\alpha\beta} = -\tilde{F}^{In}_{\alpha\beta} \Rightarrow 2\alpha(\partial_\mu \ln\phi)^2 = -\partial_\mu^2 \ln\phi. \tag{14.153}$$

Without any loss of generality we can choose $\alpha = 1/2$ because we can always perform the rescaling $\phi \longrightarrow \phi \longrightarrow \phi^{1/2\alpha}$. As a consequence we get the equation

$$\frac{\partial_\mu^2\phi}{\phi} = 0. \tag{14.154}$$

A solution is given by $\phi = 1/r^2$. It is straightforward to check that for $x \neq 0$ we have $\partial_\mu^2(1/r^2) = 0$ whereas for $x = 0$ we must use $\partial_\mu^2(1/r^2) = -4\pi^2\delta^4(x)$ which also leads to $\partial_\mu^2\phi/\phi = 0$. The generalization of the solution $\phi = 1/r^2$ is

$$\phi = 1 + \frac{\rho^2}{(x-a)^2}. \tag{14.155}$$

The additive constant has been set equal to one so that for large r the instanton gauge field approaches 0. The instanton gauge field then takes the form

$$\begin{aligned}
A^{In}_\mu &= \frac{i}{2}\tau_{\mu\alpha}\partial_\alpha \ln\left(1 + \frac{\rho^2}{(x-a)^2}\right) \\
&= \frac{i}{2}\tau_{\mu\alpha}\frac{-2\rho^2(x-a)_\alpha}{(x-a)^2[(x-a)^2 + \rho^2]} \\
&= -\,i\tau_{\mu\alpha}\frac{\rho^2(x-a)_\alpha}{(x-a)^2[(x-a)^2 + \rho^2]}, \quad k = 1 \text{ (singular)}.
\end{aligned} \tag{14.156}$$

This is singular at $x = a$. Near $x = a$ we have

$$A^{In}_\mu = -\,i\tau_{\mu\alpha}\frac{(x-a)_\alpha}{(x-a)^2}, \quad x \longrightarrow a. \tag{14.157}$$

This is a pure gauge field configuration, viz

$$A^{In}_\mu = iU^{-1}\partial_\mu U, \quad U = \frac{i(x-a)_\mu}{|x-a|}\bar{\tau}_\mu, \quad x \longrightarrow a. \tag{14.158}$$

The generic configuration A^{In}_μ can then be put in the form

$$A^{In}_\mu = (iU^{-1}\partial_\mu U)\frac{\rho^2}{(x-a)^2 + \rho^2}, \quad k = 1 \text{ (singular)}. \tag{14.159}$$

The singular behavior at $x = a$ can be removed by a singular gauge transformation (given precisely by U^{-1}), which maps the singularity to $x \longrightarrow \infty$. Indeed, we compute

$$A_\mu^{In'} = U(A_\mu^{In} + i\partial_\mu)U^{-1}$$
$$= (iU\partial_\mu U^{-1})\frac{(x - a)^2}{(x - a)^2 + \rho^2}, \quad k = 1 \text{ (regular).} \tag{14.160}$$

This have the correct asymptotic behavior, viz

$$A_\mu^{In'} = iU\partial_\mu U^{-1}$$
$$= iU'^{-1}\partial_\mu U', \quad x \longrightarrow \infty. \tag{14.161}$$

The group element U' is given by

$$U' = -\frac{i(x - a)_\mu}{|x - a|}\tau_\mu \longrightarrow -in_\mu\tau_\mu, \quad x \longrightarrow \infty. \tag{14.162}$$

This is precisely the behavior which leads to the $k = +1$ configuration (14.148). We check now explicitly that the field strengths are self-dual. We will assume that $a = 0$ for simplicity. We have then the regular gauge field

$$A_\mu^{In'} = (iU\partial_\mu U^{-1})\phi, \quad \phi = 1 - \frac{\rho^2}{r^2 + \rho^2}. \tag{14.163}$$

We compute

$$F_{\mu\nu}^{In'} = \partial_\mu A_\nu^{In'} - \partial_\nu A_\mu^{In'} - i[A_\mu^{In'}, A_\nu^{In'}]$$
$$= i\phi(1 - \phi)[\partial_\mu U\partial_\nu U^{-1} - \partial_\nu U\partial_\mu U^{-1}] + i\partial_r\phi[n_\mu U\partial_\nu U^{-1} - n_\nu U\partial_\mu U^{-1}]$$
$$= \frac{2ir^2\rho^2}{(r^2 + \rho^2)^2}\partial_\mu n_\alpha \partial_\nu n_\beta \bar\tau_{\alpha\beta} + \frac{2ir\rho^2}{(r^2 + \rho^2)^2}n_\alpha(n_\mu\partial_\nu n_\beta - n_\nu\partial_\mu n_\beta)\bar\tau_{\alpha\beta} \tag{14.164}$$
$$= \frac{2i\rho^2}{(r^2 + \rho^2)^2}\bar\tau_{\mu\nu}, \quad k = 1 \text{ (regular).}$$

This is clearly self-dual. From this result we can deduce immediately the field strength of the singular gauge field A_μ^{In} by simply applying the singular gauge transformation g as follows

$$F_{\mu\nu}^{In} = U^{-1}F_{\mu\nu}^{In'}U$$
$$= \frac{(x - a)_\alpha}{|x - a|}\tau_\alpha \cdot \frac{2i\rho^2\bar\tau_{\mu\nu}}{(r^2 + \rho^2)^2} \cdot \frac{(x - a)_\beta}{|x - a|}\bar\tau_\beta, \quad k = 1 \text{ (singular).} \tag{14.165}$$

For the anti-instanton configuration with topological charge $k = -1$ we proceed in the same way. We start from the ansatz

$$A_\mu^{a-In} = -i\beta\bar\tau_{\mu\alpha}\partial_\alpha \ln \phi(r). \tag{14.166}$$

We impose the anti-self-duality condition $F_{\alpha\beta}^{a-In} = -\tilde{F}_{\alpha\beta}^{a-In}$, which allows us to determine both the constant β and the function ϕ. This leads immediately to the singular anti-instanton gauge field. Next we determine the singular gauge transformation U^{-1}, which moves the singularity at $x = a$ to $x \longrightarrow \infty$. This singular gauge transformation can be used to construct the regular anti-instanton gauge configuration as $A_\mu^{a-In'} = U(A_\mu^{a-In} + i\partial_\mu)U^{-1}$. This will have the correct asymptotic behavior, which we have discussed in the beginning of this section and as a consequence it will have a topological charge $k = -1$. The field strengths for the regular and singular anti-instanton gauge fields can then be found following the method used for the instanton fields.

14.2.5 Tunneling and large gauge transformations

Instantons, which strictly live in Euclidean space, can be also viewed as mediating quantum tunneling between different vacua in Minkowski spacetime. We go back then to Minkowski signature.

Recall also that in the temporal gauge $A_0 = 0$ the surviving gauge invariance of a Yang–Mills action is given by the time-independent unitary matrices

$$\Omega = \Omega(\vec{x}). \tag{14.167}$$

Clearly these transformations commute with the Hamiltonian, i.e., $[H, \Omega] = 0$, and hence they can be diagonalized simultaneously, viz

$$H|E\rangle = E|E\rangle, \quad \Omega(\vec{x})|E\rangle = \omega(\vec{x})|E\rangle. \tag{14.168}$$

By Lorentz invariance the eigenvalues $\omega(\vec{x})$ must also be constant in space, i.e., $\omega(\vec{x}) = \omega$. They originate from two different classes of gauge transformations given by:

- **Small gauge transformations:** These are the transformations which are homotopic, i.e., connected, to the identity and which form a closed subgroup of the full gauge group. They are in fact the gauge transformations which are targeted by the gauge fixing procedure and they form the physical equivalency class. They must have no winding or else they would be singular. Since they are connected to the identity and they have no winding we must therefore have $\omega = 1$.
- **Large gauge transformations:** These transformations are not connected to the identity, they are not gauge-fixed, and they also change the winding number. They are in fact the gauge transformations corresponding to instanton configurations. Indeed, if we assume that $\Omega(\vec{x})$ takes the same limit when going to spatial infinity in any direction, then it can be viewed as a function on $\mathbf{S}^3 = \mathbf{R}^3 \cup \{\infty\}$. Hence for the gauge group $SU(2)$ the large gauge transformations $\Omega(\vec{x})$ are maps from $\mathbf{S}^3$ (space) to $\mathbf{S}^3$ (group), which are classified by the third homotopy group $\pi_3(\mathbf{S}^3) = \mathbf{Z}$.
  Remark that this $\mathbf{S}^3$ (space) is not the boundary of $\mathbf{R}^4$ as before but it is a spacelike surface corresponding to some fixed time t. Indeed, we will consider a foliation of Minkowski spacetime in which all spatial surfaces at fixed t are compactified to spheres $\mathbf{S}^3$ by assuming that $\Omega(x)$ takes the same limit at

spatial infinity in any direction on each surface. We can thus make the identification $\Omega(x)|_{t=\text{fixed}} = \Omega(\vec{x})$.

The one-to-one map $\Omega_1(x)$ is characterized by winding number 1 while the one-to-n map $\Omega_n(x) = (\Omega_1(x))^n$ is characterized by winding number n. We have explicitly (back in Euclidean signature)

$$\Omega_1(x) = \frac{x_4 1_2 + i x_i \tau_i}{|x|}. \tag{14.169}$$

Their eigenvalues are characterized by an angle $\theta \in [0, 2\pi[$ as

$$\Omega_1(x)|E\rangle = e^{i\theta}|E\rangle, \quad \Omega_n(x)|E\rangle = e^{in\theta}|E\rangle. \tag{14.170}$$

The angle θ is a Lorentz invariant number, which provides another fundamental parameter of the theory called the instanton angle (think of it as spin).

We now show that large gauge transformations change the winding number.

We consider initial and final configurations at the infinite past $t = -T \longrightarrow -\infty$ and the infinite future $t = +T \longrightarrow +\infty$, which describe vacua of the theory, i.e., they have vanishing energies and definite winding numbers. They are therefore pure gauges. The winding number in Minkowski spacetime is defined in the same as in the Euclidean space because it is an affine quantity, which does not depend on the metric.

We consider a continuous line of gauge fields $A_i(\lambda, \vec{x})$, parameterized by a real number $\lambda \in [0, 1]$, which connects the two vacua at $t = -T$ and $t = +T$. In other words, $A_i(0, \vec{x})$ is the gauge field configuration at $t = -T$ and $A_i(1, \vec{x})$ is the gauge field configuration at $t = +T$. We can think of λ as a function of t which for $\lambda(t) = t$ gives the configurations $A_i(t, \vec{x})$ but different $\lambda(t)$ yield paths which run at different speeds through the same sequence of configurations $A_i(t, \vec{x})$ [10].

We will further assume, for simplicity, that these two configurations are related by the one-to-one map $\Omega_1(T, \vec{x})$, viz

$$A_i(1, \vec{x}) = \Omega_1^{-1} A_i(0, \vec{x})\Omega_1 + i\Omega_1^{-1}\partial_i\Omega_1. \tag{14.171}$$

They describe therefore the same physical situation.

The winding numbers at the surfaces $t = -T$ and $t = +T$ are integers (0 and $+1$ respectively) since the gauge fields there are pure gauges everywhere on these spacelike surfaces whereas for any $0 < \lambda < 1$ the gauge field is not pure gauge everywhere on the corresponding spacelike surface and hence the winding number is not an integer. More precisely, the winding number can be re-defined by integrating from $t = -T$ to only up to a time t, and thus the continuous winding number will be seen to interpolate between the values of the winding number at the surfaces $t = -T$ and $t = +T$. Moreover, the gauge field $A_i(\lambda, \vec{x})$ will also be found to be closely related to the instanton with a topological charge given by the difference between the winding numbers at the surfaces $t = -T$ and $t = +T$.

Indeed, if we assume that the configuration for $\lambda = 0$ has vanishing energy then, by the residual gauge symmetry (14.167), we can choose $A_i(0, \vec{x}) = 0$ and as a consequence $A_i(1, \vec{x})$ is effectively the limit of the instanton configuration with topological charge $n = +1$ when $t \longrightarrow +T$. Furthermore, by requiring that the paths from $t = -T$ to $t = +T$ have finite energies the gauge fields $A_i(\lambda, \vec{x})$ for $0 < \lambda < 1$ must vanish only for $|\vec{x}| \longrightarrow \infty$ and as a consequence they are not pure gauges everywhere, i.e. they must have continuous winding numbers. We write

$$A_i \longrightarrow i\Omega^{-1}(x)\partial_i\Omega(x), \quad \text{on boundary.} \tag{14.172}$$

Where

$$\Omega(x) = \Omega_1(x), \quad t \longrightarrow +T$$
$$\Omega(x) = 1, \quad \text{elsewhere on boundary.} \tag{14.173}$$

The winding number can be written as an integral over the boundary of the box (figure 14.1). The contribution to the winding number coming from the sides vanish because of the gauge condition $A_0 = 0$ whereas the contribution from the bottom of the box vanishes since $A_i = 0$ there. Hence all contributions to the winding number come from the top of the box which is obviously equal to $n = 1$. Thus, the winding number can be changed continuously from 0 at $t - -T$ to $+1$ at $t - +T$ by means of the large gauge transformation $\Omega_1(x)$.

The winding number is a gauge-invariant quantity and in general it will be equal to the difference between the winding numbers of the spacelike surfaces at the top and the bottom of the box, viz

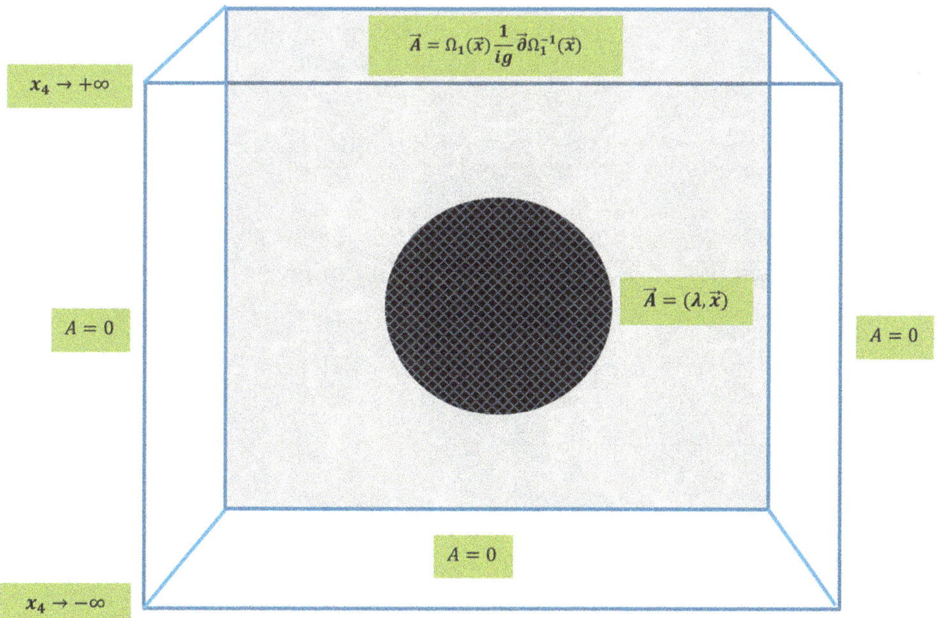

Figure 14.1. Tunneling through instantons.

$$k = \Delta k = n\,|_{t=+T} - n\,|_{t=-T}. \tag{14.174}$$

On the other hand, since the gauge configurations for $\lambda = 0$ and $\lambda = 1$ are vacua of the Yang–Mills system, i.e., with vanishing energies, the energy for $0 < \lambda < 1$ must be necessarily higher. Alternatively, for $0 < \lambda < 1$ the fields $A_i(\lambda, \vec{x})$ cannot be solutions of the equation of motion otherwise we can not change the winding, i.e., these paths have positive energy at some intermediate times. In other words, there is a potential barrier in-between the vacua $\lambda = 0$ and $\lambda = 1$ and as a consequence the system can tunnel through this barrier via the large gauge transformation (instanton) $\Omega_1(x)$.

By substituting the paths $A_i(\lambda, \vec{x})$ in the Yang–Mills Lagrangian we obtain the Lagrangian for a point particle $\lambda = \lambda(t)$ given explicitly by [10]

$$L = \frac{1}{2}m(\lambda)\dot{\lambda}^2 - V(\lambda). \tag{14.175}$$

The position-dependent mass $m(\lambda)$ and the potential barrier $V(\lambda)$ are both positive definite. In particular $V(\lambda)$ depends only on the magnetic field. The corresponding Hamiltonian is obviously (with $p(\lambda) = m(\lambda)\dot{\lambda}$)

$$H = \frac{1}{2m(\lambda)}p(\lambda)^2 + V(\lambda). \tag{14.176}$$

This is vanishing at $t = -T$ and $t = +T$ and thus as we go from the vacuum at $t = -T$ with winding number zero to the vacuum at $t = +T$ with a non-trivial winding number the energy must necessarily become positive at some intermediate times and tunneling is then required.

In the quantum theory we should solve the Schrodinger equation $H|\psi\rangle = E|\psi\rangle$. However, semi-classical treatment of tunneling is generically sufficient. We have $p = \sqrt{2m(E - V)}$ and $p = -i\partial/\partial\lambda$ and hence

$$p|\psi\rangle = -i\frac{\partial}{\partial\lambda}|\psi\rangle \Rightarrow \psi \propto \exp\left(i\int \sqrt{2m(E - V)}\,d\lambda\right). \tag{14.177}$$

This gives the transition amplitude in the allowed region $E > V$. In the forbidden region $E < V$ there is still a possibility for transition given by the tunneling wave function

$$\psi \propto \exp\left(-\int \sqrt{2m(V - E)}\,d\lambda\right). \tag{14.178}$$

Since we are considering tunneling from one vacuum with $E = 0$ to another vacuum with $E = 0$ we can see that

$$\int \sqrt{2m(E - V)}\,d\lambda = \int p\dot{\lambda}\,dt = \int L\,dt = S. \tag{14.179}$$

Note that the sign switch $V - E \longrightarrow E - V$ is equivalent to $p \longrightarrow ip$ which, since $p = m\dot{\lambda}$, is equivalent to $t \longrightarrow x_4 = it$. In other words, we are working in Euclidean space and $S = iS_E$ where S_E is the Euclidean action. Hence in the semi-classical approximation the dominant contribution to the tunneling rate is given by the minima of the Euclidean action, i.e., by instantons. The tunneling rate is given explicitly by

$$R = \exp(-2S_E), \quad S_E = \int_{t=-T}^{t+T} \sqrt{2m(V-E)}\, d\lambda. \tag{14.180}$$

The tunneling rate is bounded from above by

$$R = \exp(-2S_E) \leqslant \exp\left(-\frac{16\pi^2}{g^2}|n|\right). \tag{14.181}$$

This is due to the Bogomoln'yi bound, which also holds in Minkowski spacetime for the paths $A_i(\lambda, \vec{x})$ [10]. As it turns out, the paths with the same winding number k and with the smallest action S_E are those with parallel electric and magnetic field. These are the most probable paths which connect the two vacua at $t = -T$ and $t = +T$ with different winding numbers. Among these paths there is one particular path when viewed in Euclidean space it is precisely an instanton, i.e., a gauge field configuration $A_i(\lambda, \vec{x})$ satisfying the duality equation[5]. See [10] and references therein for a very careful treatment and more detail.

Thus the vacuum with vanishing energy $E = 0$ consists of several topological sectors with winding numbers n and instantons with topological charges k mediate vacuum tunneling between these vacua provided $k = \Delta n$. The true θ-vacuum is then a linear combination of the form

$$|\theta\rangle = \sum_n \exp(-in\theta)|n\rangle. \tag{14.182}$$

It is almost obvious that the vacuum-to-vacuum amplitudes should be given by in terms of the theta term as follows

$$\langle\theta'|\exp(-itH)|\theta\rangle = \delta(\theta' - \theta)\sum_n \int [dA]_n \exp\left(-i\int d^4x \mathcal{L}_{\text{eff}}\right). \tag{14.183}$$

$$\mathcal{L}_{\text{eff}} = \mathcal{L}_{\text{YM}} + \frac{\theta}{16\pi^2} tr(F_{\mu\nu}\tilde{F}^{\mu\nu}). \tag{14.184}$$

14.2.6 The moduli space and ADHM construction

In this section we will follow [6, 10, 11].

14.2.6.1 Collective coordinates, zero modes and moduli space

We have found that the Yang–Mills action in an instanton configuration is bound from below by the Bogomoln'yi bound given by

---

[5] Remark that this path in Minkowski spacetime does not solve the field equation, in order to be able to change the winding number, but when viewed in Euclidean space it is an instanton which solves the duality equation.

$$\frac{1}{2g^2} \int d^4x \, \mathrm{Tr} \, F_{\mu\nu}^2 \geq \frac{8\pi^2 |k|}{g^2}.$$ (14.185)

The topological charge k is given by the theta term as

$$k = \frac{1}{16\pi^2} \int d^4x \, \mathrm{Tr} \, F_{\mu\nu} \tilde{F}_{\mu\nu}.$$ (14.186)

A Yang–Mills instanton is a solution of the equations of motion, i.e., it minimizes the action, which has either a self-dual or anti-self-dual field strength. Indeed, the Bogomoln'yi bound is saturated (in fact by minimum action solution) if and only if the so-called duality equations (instanton equations) are satisfied, viz

$$F_{\mu\nu} = \tilde{F}_{\mu\nu}, \quad k > 0, \quad \text{instantons.}$$ (14.187)

$$F_{\mu\nu} = -\tilde{F}_{\mu\nu}, \quad k < 0, \quad \text{anti} - \text{instantons.}$$ (14.188)

These equations are properly understood in Euclidean space.

A finite action solution of the Euclidean classical equations of motion or equivalently of the above duality equations for $SU(2)$ gauge group with topological charge $k = +1$, i.e., a regular one-instanton solution, was already constructed in the previous section. It is given explicitly by equation (14.160) or equivalently

$$A_\mu^{In'} = - i \frac{(x - a)_\alpha \bar{\tau}_{\mu\alpha}}{(x - a)^2 + \rho^2}.$$ (14.189)

The $SU(2)$ spinor representation $\bar{\tau}_{\mu\alpha}$ is an anti-Hermitian matrix given in terms of 't Hooft symbols $\eta_{\mu\alpha}^a$ by $\bar{\tau}_{\mu\alpha} = i\eta_{\mu\alpha}^a \tau_a$.

The parameters a_μ and ρ correspond to the position and the size of the instanton and they are called collective coordinates. Thus, there are five collective coordinates (4 translations and 1 scale). However, there are three extra collective coordinates corresponding to the gauge orientation given by a global $SU(2)$ rotation U, viz

$$A_\mu^{In'}(x; a, \rho, \theta) = U^{-1}(\theta) A_\mu^{In'}(x; a, \rho) U(\theta).$$ (14.190)

In total we have therefore 8 collective coordinates for the $SU(2)$ instanton with $k = +1$ charge. For a general gauge group $SU(N)$ we will have instead $4N$ collective coordinates for the one-instanton solution whereas for higher topological charge k in $SU(2)$ we will have $8k$ collective coordinates. In general, the number of collective coordinates for a generic instanton with topological charge k in an $SU(N)$ gauge group is given by the formula

$$\dim \, \mathcal{I}_{k,N} = 4kN.$$ (14.191)

The charge k instanton can be viewed in many circumstances as k charge 1 (well separated) instantons where each one of them is characterized by a position and a size and a gauge orientation. This is the origin of the factor k. On the other hand, the origin of the factor N comes down to the fact that the one-instanton $SU(2)$ solution

(14.189) suitably embedded in $SU(N)$ is the most general one-instanton solution in an $SU(N)$ gauge theory.

The space of all solutions of the self-duality equation $F_{\mu\nu} = \tilde{F}_{\mu\nu}$ modulo gauge transformations in a given topological sector k in an $SU(N)$ gauge theory is called the moduli space and it is denoted by $\mathcal{I}_{k,N}$. The dimension of this space is precisely $4kN$. The coordinates on the moduli space $\mathcal{I}_{k,N}$ are given precisely by the collective coordinates. Equivalently the coordinate on the moduli space $\mathcal{I}_{k,N}$ are given by the so-called zero modes defined as follows. Let $A_\mu = A_\mu(x_\mu, X^\alpha)$, where X^α are the collective coordinates, be a solution of $F_{\mu\nu} = \tilde{F}_{\mu\nu}$. If we require that the perturbed configuration $A'_\mu = A_\mu + \delta A_\mu$ is still a solution of $F_{\mu\nu} = \tilde{F}_{\mu\nu}$ then the perturbation δA_μ must solve the linearized self-duality equation

$$\mathcal{D}_\mu \delta A_\nu - \mathcal{D}_\nu \delta A_\mu = \varepsilon_{\mu\nu\alpha\beta} \mathcal{D}^\alpha \delta A^\beta. \tag{14.192}$$

The covariant derivative $\mathcal{D}_\mu$ is evaluated on the background solution A_μ. The solutions of this equation are called the zero modes. Obviously each collective coordinate X^α is associated with a zero mode which satisfies equation (14.192) given by

$$\delta_\alpha A_\mu = \frac{\partial A_\mu}{\partial X^\alpha}. \tag{14.193}$$

It is also obvious that any local gauge transformation is a solution to the linearized self-duality equation and thus the general solution is

$$\delta_\alpha A_\mu = \frac{\partial A_\mu}{\partial X^\alpha} + \mathcal{D}_\mu \Omega_\alpha. \tag{14.194}$$

Remark that each zero mode is associated with a different infinitesimal gauge transformation labeled by α. We gauge fix by choosing Ω_α such that the zero mode $\delta_\alpha A_\mu$ is orthogonal to any other gauge transformation, viz

$$\int d^4x \, Tr \delta_\alpha A_\mu \mathcal{D}_\mu \eta = 0, \quad \forall \, \eta. \tag{14.195}$$

This leads to the gauge fixing condition

$$\mathcal{D}_\mu \delta_\alpha A_\mu = 0. \tag{14.196}$$

The Atiyah–Singer index theorem states that the number of zero modes, i.e., solutions of (14.192), subjected to the gauge fixing condition (14.196), is exactly given by $4kN$, i.e., equal to the dimension of the moduli space $\mathcal{I}_{k,N}$.

The moduli space $\mathcal{I}_{k,N}$ is a smooth manifold except at possible localized points corresponding to small instantons $\rho \longrightarrow 0$. This manifold is actually a hyper-Kahler manifold with a reduced holonomy group $Sp(kN) \in SO(4kN)$ and three complex structures J^i satisfying $J^i J^j = -\delta^{ij} + \varepsilon^{ijk} J^k$. The metric is given by the overlap of the zero modes, viz

$$g_{\alpha\beta} = \frac{1}{2g^2} \int d^4x \, Tr \delta_\alpha A_\mu \delta_\beta A_\mu. \tag{14.197}$$

As a simple example it can be shown that the moduli space $\mathcal{I}_{1,\,2}$ of a single instanton in an $SU(2)$ gauge theory is given by the hyper-Kahler manifold

$$\mathcal{I}_{1,\,2} = \mathbf{R}^4 \times \mathbf{R}^4/\mathbf{Z}_2. \tag{14.198}$$

If fermions, such as for example adjoint Weyl spinors λ and $\bar{\lambda}$ with kinetic term $iTr\bar{\lambda}\bar{\sigma}_\mu\mathcal{D}^\mu\lambda$, are coupled to the gauge field then in the background of an instanton $F = \tilde{F}$ the spinor λ will contribute fermion zero modes while the spinor $\bar{\lambda}$ will not lead to any further zero modes. In the background of an anti-instanton $F = -\tilde{F}$ it is $\bar{\lambda}$ that picks up fermion zero modes while λ does not pick up any zero modes. Thus, fermions lead to extra collective coordinates.

14.2.6.2 ADHM construction

The so-called ADHM (Atiyah, Drinfeld, Hitchin and Manin) construction allows us to produce all solutions of the self-dual Yang–Mills equation $F = \tilde{F}$ using twistor space [31]. A more physical approach based on type II string theory and D-brane physics can be found in [32]. In here we will adopt the pedagogical streamlined version of the ADHM construction found in [11], which does not rely on string theory, which is beyond the scope of this chapter.

The spacetime coordinate will be denoted by x_n. A quaternionic structure of spacetime is introduced by viewing x_n as a $(2, 2)$ representation of the group $SU(2)_L \times SU(2)_R$ (which is the covering group of $SO(4)$) with components $x_{\alpha\dot\alpha}$ or $\bar{x}^{\dot\alpha\alpha}$ given, respectively, by

$$x_{\alpha\dot\alpha} = x_n\sigma_{n\alpha\dot\alpha}, \quad \bar{x}^{\dot\alpha\alpha} = x_n\bar{\sigma}_n^{\dot\alpha\alpha}. \tag{14.199}$$

The indices α and $\dot\alpha$ take the two values 1 and 2 and they are the spinor indices of $SU(2)_L$ and $SU(2)_R$, respectively. The matrices $\sigma_{n\alpha\dot\alpha}$ and $\bar{\sigma}_n^{\dot\alpha\alpha}$ are given by $\sigma_n = (i\vec{\tau}, 1)$ and $\bar{\sigma}_n = (-i\vec{\tau}, 1)$ where τ^c are Pauli matrices. Indices are raised and lowered with the ε-tensor, viz $\bar{x}^{\dot\alpha\alpha} = \varepsilon^{\alpha\beta}\varepsilon^{\dot\alpha\dot\beta}x_{\beta\dot\beta}$.

We introduce now an $(N + 2k) \times 2k$ complex-valued matrix $\Delta_{\lambda i\dot\alpha}$, which is linear in x by the relation

$$\Delta_{\lambda i\dot\alpha}(x) = a_{\lambda i\dot\alpha} + b_{\lambda i}^\alpha x_{\alpha\dot\alpha}. \tag{14.200}$$

The conjugate is the matrix

$$\bar{\Delta}_i^{\dot\alpha\lambda}(x) = (\Delta_{\lambda i\dot\alpha}(x))^* = \bar{a}_i^{\dot\alpha\lambda} + \bar{x}^{\dot\alpha\alpha}\bar{b}_{i\alpha}^\lambda. \tag{14.201}$$

The index λ is called the ADHM index and runs from 1 to $N + 2k$ whereas the index i is called the instanton index and runs from 1 to k. The complex-valued matrices a and b provides an over-complete set of collective coordinates on the moduli space satisfying the so-called ADHM constraints.

The matrix $\Delta_{\dot\alpha}(x)$ can also be viewed as a map $\Delta_{\dot\alpha}(x)$: $\mathbf{C}^k \longrightarrow \mathbf{C}^{N+2k}$, which is assumed to be injective whereas the matrix $\bar{\Delta}^{\dot\alpha}(x)$ should be viewed as a map $\bar{\Delta}^{\dot\alpha}(x)$: $\mathbf{C}^{N+2k} \longrightarrow \mathbf{C}^k$, which is assumed to be subjective. These are the so-called non-degeneracy conditions.

The matrix $\bar{\Delta}(x)$ is $2k \times (N + 2k)$ so it has $2k$ rows and $N + 2k$ columns, i.e., it has N fewer rows than columns. As a consequence the matrix $\bar{\Delta}(x)$ will have N null eigenvectors denoted by the $(N + 2k) \times N$-dimensional complex-valued matrix $U_{\lambda u}$ where $u = 1, 2,\ldots, N$. Explicitly we have

$$\bar{\Delta}_i^{\dot{\alpha}\lambda} U_{\lambda u} = 0. \tag{14.202}$$

The Hermitian conjugate of this equation is given by

$$\bar{U}_u^{\lambda} \Delta_{\lambda i \dot{\alpha}} = 0. \tag{14.203}$$

The eigenvectors U are orthonormalized according to

$$\bar{U}_u^{\lambda} U_{\lambda v} = \delta_{uv}. \tag{14.204}$$

The ADHM ansatz construct the $SU(N)$ gauge field A_n in the instanton sector k from the eigenvectors U as formally given by the pure gauge

$$(A_n)_{uv} = \frac{1}{g} \bar{U}_u^{\lambda} \partial_n U_{\lambda v}. \tag{14.205}$$

This configuration solves the self-duality equation provided the matrices Δ and $\bar{\Delta}$ satisfy the requirement

$$\bar{\Delta}_i^{\dot{\alpha}\lambda} \Delta_{\lambda j\beta} = \delta_{\beta}^{\dot{\alpha}}(f^{-1})_{ij}. \tag{14.206}$$

The function f is a Hermitian $k \times k$ matrix, which is x-dependent. Obviously, $f(x) \longrightarrow 1_k/x^2$ when $|x| \longrightarrow \infty$. Its inverse f^{-1} exists by virtue of the non-degeneracy conditions. Equations (14.202), (14.203) and (14.206) allow us to derive the projector or completeness relation

$$\mathcal{P}_{\lambda}^{\mu} = U_{\lambda u} \bar{U}_u^{\mu} = \delta_{\lambda}^{\mu} - \Delta_{\lambda i \dot{\alpha}} f_{ij} \bar{\Delta}_j^{\dot{\alpha}\mu}. \tag{14.207}$$

This identity will allow us to show that the field strength F_{mn} associated with the above ADHM ansatz is indeed self-dual, viz

$$\begin{aligned}
F_{mn} &= \partial_m A_n - \partial_n A_m + g[A_m, A_n] \\
&= \frac{4}{g} \bar{U} b \sigma_{mn} f \bar{b} U.
\end{aligned} \tag{14.208}$$

The σ_{mn} are the Lorentz generators defined by $\sigma_{mn} = (\sigma_m \bar{\sigma}_n - \sigma_n \bar{\sigma}_m)/4$, which are by construction self-dual and as a consequence F_{mn} is self-dual.

The instanton number associated with the ADHM can also be checked to be equal to k by using Osborn's formula [33]. We have

$$-\frac{g^2}{16\pi^2} \int d^4x \, tr_N F_{mn}^2 = \frac{1}{16\pi^2} \int d^4x \, \Box^2 \, tr_k \log f. \tag{14.209}$$

14.2.7 The one-loop calculation

The one-loop calculation around the instanton solution involves the reduction of the path integral measure over the instanton to an integral over the moduli space of collective coordinates. This in turns requires the determination of the exact normalization of the zero modes and their measure. We consider for simplicity only bosonic collective coordinates and follow closely [10]. See also references therein.

First, we start by decomposing the field into a background and a fluctuation in the usual way, viz

$$A_\mu = A_\mu^{\text{cl}}(\gamma) + A_\mu^{\text{qu}}, \tag{14.210}$$

where γ_i, $i = 1,\dots, 4kN$, are the collective coordinates. Then we add gauge fixing and ghost terms where we choose the gauge condition

$$D_\mu^{\text{cl}} A_\mu^{\text{qu}} = 0. \tag{14.211}$$

The expanded quadratic action is given by

$$S = S_{\text{cl}} + \frac{1}{g^2} \int d^4x tr\left(A_\mu^{\text{qu}} M_{\mu\nu} A_\nu^{\text{qu}} + b M_{\text{gh}} c\right), \quad S_{\text{cl}} = \frac{8\pi^2}{g^2}|k|. \tag{14.212}$$

The operators $M_{\mu\nu}$ and M_{gh} are given by (by dropping also the subscript cl)

$$M_{\mu\nu} = D^2\delta_{\mu\nu} + 2F_{\mu\nu}, \quad M_{\text{gh}} = D^2. \tag{14.213}$$

The zero modes are precisely the normalizable eigenfunctions of the operator $M_{\mu\nu}$ with zero eigenvalues. They are given explicitly by the definition

$$Z_\mu^{(i)} = \frac{\partial A_\mu^{\text{cl}}}{\partial \gamma_i} + D_\mu^{\text{cl}} \Lambda^i. \tag{14.214}$$

They must satisfy the gauge condition

$$D_\mu^{\text{cl}} Z_\mu^{(i)} = 0. \tag{14.215}$$

Compare (14.214) and (14.215) with (14.194) and (14.196). The inner product of zero modes is defined by the relation

$$U^{ij} = \langle Z^{(i)}|Z^{(j)}\rangle = -\frac{2}{g^2} \int d^4x tr Z_\mu^{(i)} Z^{\mu(j)} = \frac{1}{g^2} \int d^4x Z_\mu^{(i)a} Z^{\mu(j)a}. \tag{14.216}$$

Let us consider for simplicity the case of the anti-instanton with topological charge $k = -1$ and $SU(2)$ gauge group. In this case we have $4|k|N = 4.1.2 = 8$ zero modes. There are four translational zero modes, one dilatational zero mode, three gauge modes. If we generalize to $SU(N)$ then the number of zero modes becomes $4N$: 4 translational, 1 dilatational and $4N - 5$ gauge orientations. Recall that the $SU(N)$ solution is obtained by embedding the $SU(2)$ solution in an $N \times N$ matrix and then rotating by a unitary matrix U as

$$A_\mu^{SU(N)} = U\begin{pmatrix} 0 & 0 \\ 0 & A_\mu^{SU(2)} \end{pmatrix} U^\dagger. \tag{14.217}$$

The stability group is obviously $SU(N-2) \times U(1)$ and thus the $4N-5$ gauge orientations find their origin in the set of all independent unitary matrices U, which is given by the coset or Grassmannian

$$SU(N)/(SU(N-2) \times U(1)). \tag{14.218}$$

The four translational zero modes and the one dilatational zero mode and their normalization are given respectively by (with collective coordinates denoted by x_0^μ (below we should always shift $x \longrightarrow x - x_0$) and ρ, respectively)

$$Z_\mu^{(\nu)} = F_{\mu\nu}^{\mathrm{cl}} \to U^{\mu\nu} = S_{\mathrm{cl}}\delta^{\mu\nu}. \tag{14.219}$$

$$Z_\mu^{(\rho)} = -2\frac{\rho\bar{\sigma}_{\mu\nu}x_\nu}{(x^2 + \rho^2)^2} \to U^{\mu\nu} = 2S_{\mathrm{cl}}. \tag{14.220}$$

The gauge zero modes are obtained from (14.190). The gauge orientations are encoded in the matrix $U(\theta)$, which is an element of the Grassmannian (14.218). We take $U(\theta) = \exp(\theta^a T_a)$. The group vielbein $e_\alpha^a(\theta)$, where the index α is curved in contrast to the index a, which is flat, is defined by $U^{-1}\partial U/\partial\theta^\alpha = e_\alpha^a T_a$. The vielbein is essentially the square root of the group metric, viz $g_{\alpha\beta} = e_\alpha^a e_\beta^a$. The Haar measure and the group volume are then given by

$$\mu(\theta) = \det e_\alpha^a d^3\theta = \sqrt{\det g_{\alpha\beta}}\, d^3\theta, \quad V = \int \mu(\theta). \tag{14.221}$$

After some calculation we find that the gauge zero modes and their normalization are given by [10]

$$Z_\mu^{(\alpha)} = U^{-1}\left[D_\mu(A(\theta = 0))\left(\frac{x^2}{x^2 + \rho^2}\partial^\alpha U.\, U^{-1} \right) \right] U \to U^{\alpha\beta} = g_{\alpha\beta}\left(\frac{1}{2}\rho^2 S_{\mathrm{cl}} \right). \tag{14.222}$$

The metric U^{ij} in the space of zero modes is therefore an 8×8 diagonal matrix given by

$$U^{ij} = U\begin{pmatrix} S_{\mathrm{cl}}\delta^{\mu\nu} & 0 & 0 \\ 0 & 2S_{\mathrm{cl}} & 0 \\ 0 & 0 & g_{\alpha\beta}\left(\frac{1}{2}\rho^2 S_{\mathrm{cl}}\right) \end{pmatrix} U^\dagger. \tag{14.223}$$

We compute immediately the measure

$$\sqrt{\det U}\, d^4x_0 d\rho d^3\theta = \frac{2^{11}\pi^8\rho^3}{g^8}\sqrt{\det g_{\alpha\beta}(\theta)}\, d^4x_0 d\rho d^3\theta. \tag{14.224}$$

Generalization to $SU(N)$ reads

$$\sqrt{\det U}\, d^4x_0 d\rho d^{4N-5}\theta = \frac{2^{2N+7}\pi^{4N}\rho^{4N-5}}{g^{4N}}\sqrt{\det g_{\alpha\beta}(\theta)}\, d^4x_0 d\rho d^{4N-5}\theta. \tag{14.225}$$

We see from the action (14.212) that the integration over the fluctuation field A_μ^{qu} is Gaussian yielding, therefore an amputated determinant $(\det{}'M)^{-1/2}$, i.e., the determinant of M leaving out its zero eigenvalues corresponding to the zero modes. The remaining path integral over the zero modes can then be converted to an integral over the collective coordinates as

$$\int \mathcal{D}A_\mu \,\ldots\, \exp(-S\,\ldots\,) = \sum_{k=-\infty}^{+\infty} \int \prod_{i=1}^{4N} \frac{d\gamma_i}{\sqrt{2\pi}} \sqrt{\det U}\, \exp(-S_{\text{cl}})(\det{}'M)^{-1/2}\ldots \quad (14.226)$$

Obviously, the dots stand for terms depending on the ghost fields.

For the case of the anti-instanton with topological charge $k = -1$ we have $\prod_{i=1}^{4N} d\gamma_i = d^4x_0\, d\rho\, d^{4N-5}\theta$ and hence we obtain the contribution

$$\int \mathcal{D}A_\mu \,\ldots\, \exp(-S\,\ldots\,) = \frac{1}{(2\pi)^{2N}}$$
$$\frac{2^{2N+7}\pi^{4N}}{g^{4N}} \int d^4x_0 \int \rho^{4N-5}d\rho \int d^{4N-5}\theta \sqrt{\det g}\; \exp(-S_{\text{cl}})(\det{}'M)^{-1/2}\ldots \quad (14.227)$$

If we are only interested in calculating gauge invariant correlators then we can integrate out gauge orientations yielding the volume of the Grassmannian (14.218), viz

$$\int d^{4N-5}\theta \sqrt{\det g} = \text{Volume}\left(\frac{SU(N)}{SU(N-2)\times U(1)}\right) = \frac{2^{4N-5}\pi^{2N-2}}{(N-1)!(N-2)!}. \quad (14.228)$$

Hence the contribution of the anti-instanton with $k = -1$ to the path integral reduces to

$$\int \mathcal{D}A_\mu \,\ldots\, \exp(-S\,\ldots\,) = \frac{1}{(N-1)!(N-2)!}$$
$$\frac{2^{4N+2}\pi^{4N-2}}{g^{4N}} \int d^4x_0 \int \rho^{4N-5}d\rho\; \exp(-S_{\text{cl}})(\det{}'M)^{-1/2}\ldots \quad (14.229)$$

If fermions are added to the theory then there will be extra zero modes and collective coordinates which should be treated in the same way.

The one-loop correction due to the non-zero modes is captured by the product of the determinants corresponding to the gauge and the ghost fields which are given respectively by

$$(\det{}'M)^{-1/2} = (\det{}'\Delta_-)^{-1}, \quad \Delta_- = -D^2 - \frac{1}{2}\sigma_{\mu\nu}F_{\mu\nu}. \quad (14.230)$$

$$\det{}'M_{\text{gf}} = (\det \Delta_+)^{+1/2}, \quad \Delta_+ = -D^2. \quad (14.231)$$

If n real adjoint scalars and $\mathcal{N}$ Majorana spinors are included in the action an extra product of three determinants is reproduced, which is given by

$$(\det \Delta_+)^{-n/4}(\det{}'\Delta_-)^{\mathcal{N}/4}(\det \Delta_+)^{+\mathcal{N}/4}. \quad (14.232)$$

The complete one-loop correction due to the non-zero modes is then

$$(\det {}'\Delta_-)^{-1+\mathcal{N}/4}(\det \Delta_+)^{(2+\mathcal{N}-n)/4}. \tag{14.233}$$

This result reduces to the ratio of the two determinants Δ_+ and Δ_- (and hence equals 1 at least formally) for the values of n and $\mathcal{N}$ such that $\mathcal{N} = 1 + n/2$, which correspond to supersymmetric Yang–Mills gauge theories where $\mathcal{N}$ is exactly the number of supersymmetries. See the next chapter for the construction of these remarkable theories.

14.3 Exercises

Exercise 1:
- Rewrite the Yang–Mills action in terms of the electric and magnetic fields (equation (14.2)).
- Derive the field equations of motion (14.5) and (14.6).
- Show explicitly that A_0 is not an independent dynamical variable.
- Derive the Hamiltonian density (14.12).

Exercise 2: Show that the mapping (14.16) is a homeomorphism.

Exercise 3:
- Show that the solution of the monopole equation

$$\partial_i \phi^a(x) = g\varepsilon^{abc} A_i^b(x)\phi^c(x) \tag{14.234}$$

  is given by the monopole field (16.366). Compute the corresponding field strength. Extract the Abelian gauge field corresponding to the unbroken $U(1)$ group.
- Verify that the flux of the Coulomb field in (14.38) cancels exactly the flux of the Dirac string (step function term) and hence the Dirac equation $\vec{\nabla}\cdot\vec{B} = 0$ is maintained.
- Show that for a multivalued gauge function Λ the magnetic flux $\Phi = \int_\Sigma \vec{B}d\vec{S}$ is not constant.

Exercise 4:
- Show that the wave functions in the northern and southern hemispheres are connected via the singular gauge transformation $U = \exp(2ieg_m\phi)$ (see equation (14.40)) by the relation

$$\psi^N = U\psi^S. \tag{14.235}$$

- Construct explicitly the singular gauge rotation in $SU(2)$ that transforms the singular Dirac–Wu–Yang Abelian magnetic monopole into the regular 't Hooft–Polyakov non-Abelian monopole. Show in particular that the Dirac string singularity cancels under this gauge transformation.

Exercise 5: Verify the results

$$\partial_\mu^2(1/r^2) = -4\pi^2\delta^4(x).$$ (14.236)

$$-i\tau_{\mu\alpha}\frac{(x-a)_\alpha}{(x-a)^2} = iU^{-1}\partial_\mu U, \quad U = \frac{i(x-a)_\mu}{|x-a|}\bar\tau_\mu.$$ (14.237)

Exercise 6: Solve the anti-self-duality condition $F_{\alpha\beta}^{a-In} = -\tilde{F}_{\alpha\beta}^{a-In}$ for the anti-instanton ansatz

$$A_\mu^{a-In} = -\,i\beta\bar\tau_{\mu\alpha}\partial_\alpha\ln\phi(r).$$ (14.238)

Determine the constant β and the function ϕ. Determine the singular gauge transformation U^{-1}, which moves the singularity at $x = a$ to $x \longrightarrow \infty$. Construct the regular anti-instanton gauge configuration as $A_\mu^{a-In'} = U(A_\mu^{a-In} + i\partial_\mu)U^{-1}$. Determine the asymptotic behavior and verify explicitly that the topological charge is indeed $k = -1$. Verify the anti-self-duality condition for the regular configuration.

    Exercise 6: Show that instantons do not curve Euclidean spacetime because of their topological nature. Hint: calculate the energy–momentum tensor for a self-dual field configuration.

Exercise 7: Show that the moduli space $\mathcal{I}_{1,2}$ of a single instanton in an $SU(2)$ gauge theory is given by the hyper-Kahler manifold

$$\mathcal{I}_{1,2} = \mathbf{R}^4 \times \mathbf{R}^4/\mathbf{Z}_2.$$ (14.239)

Exercise 8: Show explicitly that the field strength F_{mn} associated with the ADHM ansatz is self-dual by showing that

$$F_{mn} = \partial_m A_n - \partial_n A_m + g[A_m, A_n]$$
$$= \frac{4}{g}\bar U b\sigma_{mn}f\bar b U.$$ (14.240)

Exercise 9:
- Compute the expanded quadratic action (14.212).
- Show explicitly that $Z_\mu^{(i)}$ given by (14.214) is an eigenfunctions of the operator $M_{\mu\nu}$ with zero eigenvalue.
- Show that the four translational zero modes and the one dilatational zero mode and their normalization are given, respectively, by

$$Z_\mu^{(\nu)} = F_{\mu\nu}^{\mathrm{cl}} \longrightarrow U^{\mu\nu} = S_{\mathrm{cl}}\delta^{\mu\nu}.$$ (14.241)

$$Z_\mu^{(\rho)} = -2\frac{\rho\bar{\sigma}_{\mu\nu}x_\nu}{(x^2 + \rho^2)^2} \rightarrow U^{\mu\nu} = 2S_{\text{cl}}. \tag{14.242}$$

- Show that the gauge zero modes and their normalization are given by

$$Z_\mu^{(\alpha)} = U^{-1}\left[D_\mu(A(\theta = 0))\left(\frac{x^2}{x^2 + \rho^2}\partial^\alpha U.\, U^{-1}\right)\right]U \rightarrow U^{\alpha\beta} = g_{\alpha\beta}\left(\frac{1}{2}\rho^2 S_{\text{cl}}\right). \tag{14.243}$$

Exercise 10: There is a connection between instantons and tunneling trajectories in quantum mechanics. Consider for example the quantum mechanical ϕ^4 model

$$S = \int dt\left(\frac{1}{2}\dot{\phi}^2 - \frac{\mu^2}{2}\phi^2 + \frac{\lambda}{4}\phi^4\right). \tag{14.244}$$

The trivial minimum is given by $\phi_\pm = \pm\mu/\sqrt{\lambda}$. The one-instanton solutions parameterized by the initial time t_0 (here we are really dealing with kinks) is

$$\phi = \frac{\mu}{\sqrt{\lambda}}\tanh\frac{\mu(t - t_0)}{\sqrt{2}}. \tag{14.245}$$

These interpolate between ϕ_+ at $t = +\infty$ and ϕ_- at $t = -\infty$. In other words these solutions allow tunneling between the two wells of the potential.

Show that their contribution to the path integral is

$$Z = e^{-\frac{2\sqrt{2}\mu^3}{3\lambda}}\int dt_0. \tag{14.246}$$

This is very large in the large time limit but negligible in the limit of small time in accord with the fact that these solutions are tunneling trajectories.

References

[1] 't Hooft G 1974 Magnetic monopoles in unified gauge theories *Nucl. Phys.* B **79** 276

[2] Polyakov A M 1974 Particle spectrum in the quantum field theory *JETP Lett.* **20** 194 [Pisma Zh. Eksp. Teor. Fiz. **20** 430 (1974)]

[3] Lenz F 2005 Topological concepts in gauge theories *Topology and Geometry in Physics* **659** ed E. Bick and F. D. Steffen (Berlin: Springer) 7

[4] 't Hooft G 1999 Monopoles, instantons and confinement arXiv:hep-th/0010225

[5] Coleman S R 1975 Classical lumps and their quantum descendents Lectures delivered at Int. School of Subnuclear Physics, Ettore Majorana, Erice, Sicily, Jul 11-31

[6] Tong D 2005 TASI lectures on solitons: instantons, monopoles, vortices and kinks arXiv: hep-th/0509216

[7] Shnir Y M 2005 *Magnetic Monopoles* (Berlin: Springer)

[8] Weinberg E J and Yi P 2007 Magnetic monopole dynamics, supersymmetry, and duality *Phys. Rep.* **438** 65

[9] Belavin A A, Polyakov A M, Schwartz A S and Tyupkin Y S 1975 Pseudoparticle solutions of the Yang-Mills equations *Phys. Lett.* **59B** 85

[10] Vandoren S and van Nieuwenhuizen P 2008 Lectures on instantons arXiv:0802.1862 [hep-th]

[11] Dorey N, Hollowood T J, Khoze V V and Mattis M P 2002 The calculus of many instantons *Phys. Rep.* **371** 231

[12] Georgi H and Glashow S L 1972 Unified weak and electromagnetic interactions without neutral currents *Phys. Rev. Lett.* **28** 1494

[13] Dirac P A M 1931 Quantized singularities in the electromagnetic field *Proc. R. Soc. Lond.* A **133** 60

[14] Wu T T and Yang C N 1975 Concept of nonintegrable phase factors and global formulation of gauge fields *Phys. Rev.* D **12** 3845

[15] Bais F A 1976 SO(3) Monopoles and dyons with multiple magnetic charge *Phys. Lett.* **64B** 465

[16] Arafune J, Freund P G O and Goebel C J 1975 Topology of higgs fields *J. Math. Phys.* **16** 433

[17] Montonen C and Olive D I 1977 Magnetic monopoles as gauge particles? *Phys. Lett.* 117

[18] Bais F.A. 2005 *50 Years of Yang-Mills Theory* (Singapore: World Scientific)

[19] Callias C *Index theorem on Open Spaces* (Cambridge, MA: MIT Press) 31

[20] Weinberg E J 1979 Parameter counting for multi-monopole solutions *Phys. Rev.* D **20** 936

[21] Mottola E 1978 Zero modes of the 't Hooft–Polyakov monopole *Phys. Lett.* **79B** 242 Erratum: 1979 *Phys. Lett.* **80B** 433

[22] Jackiw R and Rebbi C 1976 Solitons with Fermion number 1/2 *Phys. Rev.* D **13** 3398

[23] Jackiw R and Rebbi C 1976 Spin from isospin in a gauge theory *Phys. Rev. Lett.* **36** 1116

[24] Hasenfratz P and 't Hooft G 1976 A Fermion-Boson puzzle in a gauge theory *Phys. Rev. Lett.* **36** 1119

[25] Julia B and Zee A 1975 Poles with both magnetic and electric charges in nonabelian gauge theory *Phys. Rev.* D **11** 2227

[26] Polyakov A M 1987 Gauge Fields and Strings *Contemp. Concepts Phys.* **3** 1

[27] Jackiw R 1980 Introduction to the Yang-Mills quantum theory *Rev. Mod. Phys.* **52** 661

[28] Schwartz A S 1993 *Quantum Field Theory and Topology* (Berlin: Springer)

[29] Jackiw R 1983 Topological investigations of quantized gauge theories *Conf. Proc.* C **8306271** 221

[30] Rajaraman R 1982 *Solitons and Instantons. An Introduction to Solitons and Instantons in Quantum Field Theory* (North-Holland: Amsterdam) p 409

[31] Atiyah M F, Hitchin N J, Drinfeld V G and Manin Y I 1978 Construction of instantons *Phys. Lett.* A **65** 185

[32] Douglas M R 1995 Branes within Branes *Strings, Branes and Dualities* ed L. Beaulieu (Dordrecht: Springer)

[33] Osborn H 1981 Semiclassical functional integrals for selfdual gauge fields *Ann. Phys.* **135** 373

IOP Publishing

A Modern Course in Quantum Field Theory, Volume 2 (Second Edition)

Advanced topics

Badis Ydri

Chapter 15

Introducing supersymmetry

In this chapter we introduce supersymmetry following mostly [1]. In particular, we will emphasize the formal quantum field theory aspects of the formalism of global $N = 1$ supersymmetry with a detailed calculation of the corresponding F- and d-terms, also following [2]. A brief description of $N = 2$ is also given. The classic text on supersymmetry of Wess and Bagger [3] remains in our view one of the best books on quantum field theory. We have also benefited from [4, 5].

15.1 Lorentz symmetry revisited

15.1.1 Lorentz/Poincaré invariance

The spacetime metric is taken to be

$$g_{\mu\nu} = (+1, -1, -1, -1). \tag{15.1}$$

We will use the Weyl or chiral representation of Dirac matrices given by

$$\gamma^\mu = \begin{pmatrix} 0 & \sigma^\mu \\ \bar{\sigma}^\mu & 0 \end{pmatrix}, \quad \gamma_5 = i\gamma^0\gamma^1\gamma^2\gamma^3 = \begin{pmatrix} 1 & 0 \\ 0 & -1 \end{pmatrix} \tag{15.2}$$

where we have introduced the spin matrices

$$\sigma^\mu = (1, \sigma^i) = (1, -\sigma_i) = \bar{\sigma}_\mu, \quad \bar{\sigma}^\mu = (1, -\sigma^i) = (1, \sigma_i) = \sigma_\mu. \tag{15.3}$$

The Pauli matrices are

$$\sigma^1 = \begin{pmatrix} 0 & 1 \\ 1 & 0 \end{pmatrix}, \quad \sigma^2 = \begin{pmatrix} 0 & -i \\ i & 0 \end{pmatrix}, \quad \sigma^3 = \begin{pmatrix} 1 & 0 \\ 0 & -1 \end{pmatrix}. \tag{15.4}$$

doi:10.1088/978-0-7503-5834-7ch15   15-1

There are six generators of the Lorentz group $SO(1, 3)$. The rotation generators are J_i and the boost generators are K_i, $i = 1, 2, 3$. The commutation relations are

$$[J_i, J_j] = i\varepsilon_{ijk}J_k, \quad [J_i, K_j] = i\varepsilon_{ijk}K_k, \quad [K_i, K_j] = i\varepsilon_{ijk}K_k. \tag{15.5}$$

This Lie algebra is a complexified $SU(2) \times SU(2)$. Indeed by defining the complex generators $J_i^{\pm} = (J_i \pm iK_i)/2$ we can replace the above commutation relations by

$$[J_i^{\pm}, J_j^{\pm}] = i\varepsilon_{ijk}J_k^{\pm}, \quad [J_i^{\pm}, J_j^{\mp}] = 0. \tag{15.6}$$

This is the group $SL(2, C)$. The group $SL(2, C)$ is the universal cover of the Lorentz group $SO(1, 3)$ in the same way that $SU(2)$ is the universal cover of the rotation group $SO(3)$.

The Poincaré group is a semi-direct product of the Lorentz group and the group of translations. The generators of translations are the components of the energy–momentum 4-vector operator denoted $P_\mu = (P_0, P_i)$. The energy operator P_0 is a scalar while the momentum operator P_i is an ordinary vector and hence we must have the commutation relations

$$[J_i, P_j] = i\varepsilon_{ijk}P_k, \quad [J_i, P_0] = 0, \quad [K_i, P_j] = -iP_0, \quad [K_i, P_0] = -iP_j. \tag{15.7}$$

We must also have

$$[P_\mu, P_\nu] = 0. \tag{15.8}$$

The generators of the Lorentz group can also be defined by $M_{0i} = -M_{i0} = K_i$ and $M_{ij} = \varepsilon_{ijk}J_k$ such that the Poincaré algebra becomes

$$[P_\mu, P_\nu] = 0. \tag{15.9}$$

$$[M_{\mu\nu}, M_{\rho\sigma}] = ig_{\nu\rho}M_{\mu\sigma} - ig_{\mu\rho}M_{\nu\sigma} - ig_{\nu\sigma}M_{\mu\rho} + ig_{\mu\sigma}M_{\nu\rho}. \tag{15.10}$$

$$[M_{\mu\nu}, P_\rho] = -ig_{\rho\mu}P_\nu + ig_{\rho\nu}P_\mu. \tag{15.11}$$

The Dirac fermion is a 4-(complex)component spinor, which transforms reducibly under Lorentz transformations. It transforms as the reducible $(1/2, 0) \oplus (0, 1/2)$ representation of the Lorentz group.

The chiral Dirac (Weyl) fermion is a 2-component complex spinor, which transforms irreducibly under Lorentz transformations.

A left-handed Weyl spinor ψ transforms as the irreducible $(1/2, 0)$ representation of the Lorentz group whereas a right-handed Weyl spinor $\bar{\lambda}$ transforms as the irreducible $(0, 1/2)$ representation of the Lorentz group. The Dirac spinor is a direct sum of a left-handed and a right-handed Weyl spinors, viz

$$\psi_D = \begin{pmatrix} \psi \\ \bar{\lambda} \end{pmatrix}. \tag{15.12}$$

A Majorana (real) fermion is a Dirac fermion ψ_D with $\lambda = \psi$. Spacetime dimension four is special in the sense that Majorana (real) and Weyl (chiral) fermions are equivalent.

The left-handed 2-component Weyl spinor ψ transforms under the action of an element $\mathcal{M}$ of $SL(2, C) \simeq SO(1, 3)$ as

$$\psi_\alpha \longrightarrow \psi'_\alpha = \mathcal{M}_\alpha{}^\beta \psi_\beta. \tag{15.13}$$

A general element of $SL(2, C)$ is of the form

$$\mathcal{M} = \exp(iz_i\sigma_i), \quad z_i \in C. \tag{15.14}$$

The right-handed 2-component Weyl spinor $\bar{\lambda}$ transforms under the action of $\mathcal{M}$ as

$$\bar{\lambda}_{\dot\alpha} \longrightarrow \bar{\lambda}'_{\dot\alpha} = (\mathcal{M}^*)_{\dot\alpha}{}^{\dot\beta}\bar{\lambda}_{\dot\beta}. \tag{15.15}$$

Thus the representation $\mathcal{M}$ and its complex conjugate $\mathcal{M}^*$ are not equivalent (in contrast with $SU(2)$).

We introduce the Levi-Civita antisymmetric tensors

$$\varepsilon^{\alpha\beta} = \varepsilon^{\dot\alpha\dot\beta} = \begin{pmatrix} 0 & 1 \\ -1 & 0 \end{pmatrix}, \quad \varepsilon_{\alpha\beta} - \varepsilon_{\dot\alpha\dot\beta} - \begin{pmatrix} 0 & -1 \\ 1 & 0 \end{pmatrix}. \tag{15.16}$$

Then we will raise and lower indices as follows

$$\psi^\alpha = \varepsilon^{\alpha\beta}\psi_\beta, \quad \psi_\alpha = \varepsilon_{\alpha\beta}\psi^\beta. \tag{15.17}$$

$$\bar{\lambda}^{\dot\alpha} = \varepsilon^{\dot\alpha\dot\beta}\bar{\lambda}_{\dot\beta}, \quad \bar{\lambda}_{\dot\alpha} = \varepsilon_{\dot\alpha\dot\beta}\bar{\lambda}^{\dot\beta}. \tag{15.18}$$

By using the facts $\varepsilon^{\alpha\beta} = (i\sigma^2)^{\alpha\beta}, \varepsilon_{\alpha\beta} = (-i\sigma^2)_{\alpha\beta}, \sigma_2\sigma_i^T\sigma_2 = -\sigma_i$ and also the fact that $\mathcal{M} = a1 + b_i\sigma_i$ we can show that

$$\psi'^\alpha = \psi^\beta(\sigma_2\mathcal{M}^T\sigma_2)_\beta{}^\alpha = \psi^\beta(\mathcal{M}^{-1})_\beta{}^\alpha. \tag{15.19}$$

Similarly, we can show that

$$\bar{\lambda}'^{\dot\alpha} = \bar{\lambda}^{\dot\beta}(\mathcal{M}^{*-1})_{\dot\beta}{}^{\dot\alpha}. \tag{15.20}$$

Explicitly the Dirac spinor is given by

$$(\psi_D)_{\hat\alpha} = \begin{pmatrix} \psi_\alpha \\ \bar{\lambda}^{\dot\alpha} \end{pmatrix}. \tag{15.21}$$

The undotted index $\alpha = 1, 2$ is a chiral index, the dotted index $\dot\alpha = \dot1, \dot2$ is an anti-chiral index whereas the hatted index $\hat\alpha = 1, 2, \dot1, \dot2$ is a Dirac index.

The Dirac conjugate of the Dirac spinor ψ_D is

$$(\bar{\psi}_D)^{\hat\alpha} = (\lambda^\alpha \; \bar{\psi}_{\dot\alpha}). \tag{15.22}$$

The dotted and undotted indices are related by Hermitian conjugation. Indeed we can see from the transformation laws that

$$(\psi_\alpha)^+ = \bar{\psi}_{\dot{\alpha}}, \quad (\lambda^\alpha)^+ = \bar{\lambda}^{\dot{\alpha}}. \tag{15.23}$$

The spin matrices σ^μ and $\bar{\sigma}^\mu$ have undotted-dotted and dotted-undotted indices, respectively. In other words we have $(\sigma^\mu)_{\alpha\dot{\beta}}$ and $(\bar{\sigma}^\mu)^{\dot{\alpha}\beta}$.

The Lorentz generators in the spinor representation are given by

$$\Sigma^{\mu\nu} = \frac{i}{4}[\gamma^\mu, \gamma^\nu] = \begin{pmatrix} \sigma^{\mu\nu} & 0 \\ 0 & \bar{\sigma}^{\mu\nu} \end{pmatrix}. \tag{15.24}$$

The $SL(2, C)$ generators for ψ_α and $\bar{\lambda}^{\dot{\alpha}}$ are $\sigma^{\mu\nu}$ and $\bar{\sigma}^{\mu\nu}$, respectively. They are given explicitly by

$$\begin{aligned}
(\sigma^{\mu\nu})_\alpha{}^\beta &= \frac{i}{4}(\sigma^\mu_{\alpha\dot{\gamma}}\bar{\sigma}^{\nu\dot{\gamma}\beta} - \sigma^\nu_{\alpha\dot{\gamma}}\bar{\sigma}^{\mu\dot{\gamma}\beta}) \\
(\bar{\sigma}^{\mu\nu})^{\dot{\alpha}}{}_{\dot{\beta}} &= \frac{i}{4}(\bar{\sigma}^{\mu\dot{\alpha}\gamma}\sigma^\nu_{\gamma\dot{\beta}} - \bar{\sigma}^{\nu\dot{\alpha}\gamma}\sigma^\mu_{\gamma\dot{\beta}}).
\end{aligned} \tag{15.25}$$

There are two Casimir operators of the Poincaré group. These are $P^2 = P_\mu P^\mu$ and $W^2 = W_\mu W^\mu$ where W^μ is the Pauli–Lubanski vector defined by

$$W^\mu = \frac{1}{2}\varepsilon^{\mu\nu\rho\sigma}P_\nu M_{\rho\sigma}. \tag{15.26}$$

Therefore, irreducible representations of the Poincaré group are characterized by two numbers.

For massive particles we can go to the rest frame where $P_\mu = (m, 0, 0, 0)$ and thus $P^2 = m^2$ and $W^2 = -m^2 s(s + 1)$ where s is the spin of the particle. The irreducible representations of the Poincaré group are then characterized by the mass m and the spin s.

For massless particles there is no a rest frame and usually we work in the reference frame where $P_\mu = (E, 0, 0, E)$ and thus $P^\mu P_\mu = 0$ and $P^\mu W_\mu = 0$. We can also verify that $W^\mu W_\mu = 0$ and thus we must have $W^\mu = hP^\mu$ where h is called the helicity, which is the projection of angular momentum along the direction of motion. Thus irreducible representations of the Poincaré group are characterized by a single number, which is the helicity.

15.1.2 Spinor calculus

We can suppress spinor indices consistently by adopting the summation convention

$$^\alpha{}_\alpha \quad \text{and} \quad {}_{\dot{\alpha}}{}^{\dot{\alpha}}. \tag{15.27}$$

For example

$$\psi\chi = \psi^\alpha \chi_\alpha = \varepsilon^{\alpha\beta}\psi_\beta\chi_\alpha = -\varepsilon^{\alpha\beta}\psi_\alpha\chi_\beta = -\psi_\alpha\chi^\alpha = \chi^\alpha\psi_\alpha = \chi\psi. \tag{15.28}$$

$$\bar{\psi}\bar{\chi} = \bar{\psi}_{\dot{\alpha}}\bar{\chi}^{\dot{\alpha}} = \varepsilon_{\dot{\alpha}\dot{\beta}}\bar{\psi}^{\dot{\beta}}\bar{\chi}^{\dot{\alpha}} = -\varepsilon_{\dot{\alpha}\dot{\beta}}\bar{\psi}^{\dot{\alpha}}\bar{\chi}^{\dot{\beta}} = -\bar{\psi}^{\dot{\alpha}}\bar{\chi}_{\dot{\alpha}} = \bar{\chi}_{\dot{\alpha}}\bar{\psi}^{\dot{\alpha}} = \bar{\chi}\bar{\psi}. \tag{15.29}$$

$$(\psi\chi)^+ = (\psi^\alpha\chi_\alpha)^+ = (\chi_\alpha)^+(\psi^\alpha)^+ = \bar\chi_{\dot\alpha}\bar\psi^{\dot\alpha} = \bar\chi\bar\psi. \tag{15.30}$$

Also (by using $(\sigma^\mu)^*_{\alpha\dot\beta} = (\sigma^\mu)_{\beta\dot\alpha}$)

$$(\chi\sigma^\mu\bar\psi)^+ = (\chi^\alpha\sigma^\mu_{\alpha\dot\beta}\bar\psi^{\dot\beta})^+ = (\bar\psi^{\dot\beta})^+(\sigma^\mu)^*_{\alpha\dot\beta}(\chi^\alpha)^+ = \psi^\beta(\sigma^\mu)_{\beta\dot\alpha}\bar\chi^{\dot\alpha} = \psi\sigma^\mu\bar\chi. \tag{15.31}$$

By using $(\sigma^\mu)^*_{\alpha\dot\beta} = (\sigma^\mu)_{\beta\dot\alpha}$ and $(\bar\sigma^\mu)^{*\dot\alpha\beta} = (\bar\sigma^\mu)^{\dot\beta\alpha}$ we can show that $(\sigma^{\mu\nu})_\alpha{}^\beta = (\bar\sigma^{\mu\nu})^{\dot\beta}{}_{\dot\alpha}$ and thus

$$(\xi\sigma^{\mu\nu}\theta)^+ = \bar\theta\bar\sigma^{\mu\nu}\bar\xi. \tag{15.32}$$

Some of the Fierz identities we will use extensively are

$$\theta^\alpha\theta^\beta = -\frac{1}{2}\varepsilon^{\alpha\beta}\theta\theta, \quad \bar\theta^{\dot\alpha}\bar\theta^{\dot\beta} = \frac{1}{2}\varepsilon^{\dot\alpha\dot\beta}\bar\theta\bar\theta. \tag{15.33}$$

$$\phi\sigma^\mu\bar\chi = -\bar\chi\bar\sigma^\mu\phi. \tag{15.34}$$

$$(\theta\phi)(\theta\psi) = -\frac{1}{2}(\theta\theta)(\phi\psi). \tag{15.35}$$

$$(\bar\theta\bar\phi)(\bar\theta\bar\psi) = -\frac{1}{2}(\bar\phi\bar\psi)(\bar\theta\bar\theta). \tag{15.36}$$

$$(\theta\phi)(\chi\sigma^\mu\bar\eta) = -\frac{1}{2}[(\theta\sigma^\mu\bar\eta)(\chi\phi) - 2i(\theta\sigma_\nu\bar\eta)(\chi\sigma^{\mu\nu}\phi)]. \tag{15.37}$$

$$(\theta\phi)(\bar\chi\bar\eta) = -\frac{1}{2}(\theta\sigma^\mu\bar\eta)(\bar\chi\bar\sigma_\mu\phi). \tag{15.38}$$

$$(\theta\sigma^\mu\bar\theta)(\sigma^\nu\bar\theta)_\alpha = \frac{1}{2}\eta^{\mu\nu}\theta_\alpha(\bar\theta\bar\theta) + i(\sigma^{\mu\nu}\theta)_\alpha(\bar\theta\bar\theta). \tag{15.39}$$

$$(\theta\sigma^\mu\bar\theta)(\theta\sigma^\nu)_{\dot\alpha} = \frac{1}{2}\eta^{\mu\nu}\bar\theta_{\dot\alpha}(\theta\theta) - i(\bar\theta\bar\sigma^{\mu\nu})_{\dot\alpha}(\theta\theta). \tag{15.40}$$

$$(\theta\sigma^\mu\bar\theta)(\theta\sigma^\nu\bar\theta) = \frac{1}{2}\eta^{\mu\nu}(\bar\theta\bar\theta)(\theta\theta). \tag{15.41}$$

15.2 Supersymmetry algebra and representations

It is well established (the Coleman–Mandula theorem [6]) that the only symmetries of the S-matrix are:

- Poincaré invariance.
- The discrete symmetries C, P, T.
- Internal symmetries.

A central assumption underlying the Coleman–Mandula theorem is the requirement that the symmetry Lie algebra involves only commutators. It can be shown (the Haag–Lopuszanski–Sohnius theorem [7]) that supersymmetry, which involves anticommuting generators as well as commuting generators is the only possible extension of Poincaré symmetry.

The new supersymmetry generators will be denoted by Q_α^A, $A = 1,..., N$. They must be anticommuting and thus it is natural to assume that they transform under the Lorentz group as the undotted spinors ψ_α and ψ^α, i.e. as the $(1/2, 0)$ representation. The adjoint of these generators are $\bar{Q}_{\dot\alpha A}$, which transform under the Lorentz group as the dotted spinors $\bar\lambda^{\dot\alpha}$ and $\bar\lambda_{\dot\alpha}$, i.e., as the $(0, 1/2)$ representation. Explicitly we must have

$$[M_{\mu\nu}, Q_\alpha^A] = -(\sigma_{\mu\nu})_\alpha{}^\beta Q_\beta^A, \quad [M_{\mu\nu}, \bar{Q}_A^{\dot\alpha}] = -(\bar\sigma_{\mu\nu})^{\dot\alpha}{}_{\dot\beta} Q_A^{\dot\beta}. \tag{15.42}$$

The anticommutator $\{Q_\alpha^A, \bar{Q}_{\dot\beta B}\}$ must transform as the tensor product $(1/2, 0) \otimes (0, 1/2)$, i.e., as the vector representation $(1/2, 1/2)$. It must therefore be proportional to P^μ. We write $\{Q_\alpha^A, \bar{Q}_{\dot\beta B}\} = 2(\sigma^\mu)_{\alpha\dot\beta} P_\mu C_B^A$. It is almost obvious that C is a positive definite Hermitian matrix. Hence we can choose the supersymmetry generators such as $C_B^A = \delta_B^A$. We get then

$$\{Q_\alpha^A, \bar{Q}_{\dot\beta B}\} = 2(\sigma^\mu)_{\alpha\dot\beta} P_\mu \delta_B^A. \tag{15.43}$$

The supersymmetry generators commute with the generators of translations, viz [1]

$$[P_\mu, Q_\alpha^A] = [P_\mu, \bar{Q}_A^{\dot\alpha}] = 0. \tag{15.44}$$

Next we would like to compute the anticommutators $\{Q_\alpha^A, Q_\beta^B\}$. This must transform as the tensor product $(1/2, 0) \otimes (1/2, 0) = (0, 0) \oplus (1, 0)$. Recall that $M_{\mu\nu}$ transforms in the $(1, 0)$ representation. Thus we have in general $\{Q_\alpha^A, Q_\beta^B\} = \varepsilon_{\alpha\beta} X^{AB} + \varepsilon_{\beta\gamma}(\sigma^{\rho\sigma})_\alpha{}^\gamma M_{\rho\sigma} Y^{AB}$. The two Lorentz scalars matrices X^{AB} and Y^{AB} are antisymmetric and symmetric, respectively, under the exchange $A \leftrightarrow B$. Since the supersymmetry generators commute with the generators of translations we must have $\varepsilon_{\beta\gamma}(\sigma^{\rho\sigma})_\alpha{}^\gamma [M_{\rho\sigma}, P_\theta] Y^{AB} = 0$ and thus $Y^{AB} = 0$. We get then

$$\{Q_\alpha^A, Q_\beta^B\} = \varepsilon_{\alpha\beta} X^{AB}. \tag{15.45}$$

Similarly

$$\{\bar{Q}_{\dot\alpha A}, \bar{Q}_{\dot\beta B}\} = \varepsilon_{\dot\alpha\dot\beta}(X^{AB})^+. \tag{15.46}$$

The complex Lorentz scalars X^{AB} (which can also be checked to commute with $\bar{Q}_{\dot\alpha}^A$ and $\bar{Q}_{\dot\beta}^B$) are called central charges.

Let B_a be the generators of the internal global symmetry group, viz

$$[B_a, B_b] = i f_{ab}^c B_c. \tag{15.47}$$

Clearly these generators must be Lorentz scalars under Poincaré transformations, viz $[P^\mu, B_a] = [M_{\mu\nu}, B_a] = 0$. Under supersymmetry generators we must have

$$[Q_\alpha^A, B_a] = S_a^A{}_B Q_\alpha^B, \quad [\bar{Q}_{\dot\alpha A}, B_a] = -(S_a^*)_A{}^B \bar{Q}_{\dot\alpha B}. \tag{15.48}$$

The central charges X^{AB} generate an Abelian invariant subalgebra of the compact Lie algebra $\mathcal{A}$ generated by B_a. To see this we first use the Jacobi identity $[B_a, \{Q_\alpha^A, Q_\beta^B\}] - \{Q_\beta^B, [B_a, Q_\alpha^A]\} + \{Q_\alpha^A, [Q_\beta^B, B_a]\} = 0$ to deduce

$$[B_a, X^{AB}] + S_a^B{}_C X^{AC} - S_a^A{}_C X^{BC} = 0. \tag{15.49}$$

This means that X^{AB} (which can always be rewritten as linear combinations of the B_a) generates an invariant subalgebra of $\mathcal{A}$.

Next it is not difficult to show that $[X^{AB}, X^{CD}] = 0$ which means that the X^{AB} generate an Abelian invariant subalgebra of $\mathcal{A}$. In general, the algebra $\mathcal{A}$ is a direct sum of a semisimple algebra $\mathcal{A}_1$ and an Abelian algebra $\mathcal{A}_2$. Hence X^{AB} must be in $\mathcal{A}_2$ and as a consequence $[B_a, X^{AB}] = 0$.

As pointed above we can write

$$X^{AB} = (a^a)^{AB} B_a, \quad (a^a)^{AB} = -(a^a)^{BA}, \quad (a^a)^{AB} = -(a^a)_{AB}. \tag{15.50}$$

Substituting in (15.49) we get

$$S_a^B{}_C (a^b)^{AC} = S_a^A{}_C (a^b)^{BC}. \tag{15.51}$$

By using the fact that S_a is Hermitian we rewrite this equation as

$$S_a^A{}_C (a^b)^{CB} = -(a^b)^{AC} (S_a^*)_C{}^B. \tag{15.52}$$

The coefficients $S_a^A{}_B$ furnish a representation of the algebra $\mathcal{A}$ while the coefficients $(a^a)^{AB}$ intertwines the representation S_a with its complex conjugate $-S_a^*$.

Since $X^{AB} = (a^a)^{AB} B_a$ we have

$$(X^{AB})^+ = -(a^a)^*_{AB} B_a. \tag{15.53}$$

For $N=1$ the central charges X^{AB} (and as a consequence the coefficients $(a^a)^{AB} = 0$) vanish whereas the coefficients S_a are real. We have then $[Q_\alpha, B_a] = S_a Q_\alpha$ and $[\bar{Q}_{\dot\alpha}, B_a] = -S_a \bar{Q}_{\dot\alpha}$. Furthermore from the Jacobi identity $[[Q_\alpha, B_a], B_b] + [[B_b, Q_\alpha], B_a] + [[B_a, B_b], Q_\alpha] = 0$ we conclude that we must have $f_{ab}^c = 0$ and thus the algebra $\mathcal{A}$ is Abelian. We can always rescale the Abelian generators B_a such that

$$[Q_\alpha, B_a] = Q_\alpha, \quad [\bar{Q}_{\dot\alpha}, B_a] = -\bar{Q}_{\dot\alpha}. \tag{15.54}$$

Clearly any linear combination of the Abelian generators B_a will satisfy the above equation. We can check that there is only one independent combination of the B_a (which we will call R), which acts non-trivially on the supersymmetry generators whereas all other combinations commute with Q_α and $\bar{Q}_{\dot\alpha}$. We have then

$$[Q_\alpha, R] = Q_\alpha, \quad [\bar{Q}_{\dot\alpha}, R] = -\bar{Q}_{\dot\alpha}. \tag{15.55}$$

This global $U(1)$ symmetry is known as R-symmetry and obviously the supersymmetry generators Q_α and $\bar{Q}_{\dot\alpha}$ carry R-charges $+1$ and -1 respectively.

The R-symmetry group for $N > 1$ supersymmetry is $SU(N)_R$.

The supersymmetric Casimirs required to characterize the irreducible representations of supersymmetry are the mass operator $P^2 = P^\mu P_\mu$ (as before) and $C^2 = C_{\mu\nu} C^{\mu\nu}$, $C_{\mu\nu} = B_\mu P_\nu - B_\nu P_\mu$ where B_μ is the supersymmetric Pauli–Lubanski vector define by

$$B_\mu = W_\mu - \frac{1}{4}\bar{Q}_{\dot\alpha}\bar{\sigma}_\mu^{\dot\alpha\beta}Q_\beta. \tag{15.56}$$

It is quite clear that finding representations of the supersymmetry algebra on single particle states depends on (1) the number of supersymmetries, (2) whether the states are massive or massless and (3) whether or not there are central charges. We will not do this lengthy exercise at this stage but we will simply quote the most relevant results. First we note that for massive states we will work in the rest frame $P_\mu = (m, 0, 0, 0)$ whereas for massless states we will work in the light-cone frame $P_\mu = (E, 0, 0, E)$.

We have then the following possibilities:

- An $N = 1$ massless supermultiplet (no central charges) contains four states: The two states with helicities $(\lambda, \lambda + 1/2)$ and their CPT conjugates with helicities $(-\lambda - 1/2, -\lambda)$. The most important $N = 1$ massless supermultiplets are:
  - The chiral multiplet (ψ_a, ϕ) corresponding to $\lambda = 0$. The spinor ψ_a is a Weyl fermion and ϕ a complex scalar. This is a matter multiplet transforming in some representation of the gauge group.
  - The $N = 1$ gauge multiplet (A_μ, λ_a) corresponding to $\lambda = 1/2$ (or $\lambda = -1/2$). The spinor λ_α is known as a gaugino field and it is a Weyl fermion while A_μ is a massless gauge field. They must be in the adjoint representation of the gauge group.
- An $N = 2$ massless supermultiplet (no central charges) contains eight states: The four states with helicities $(\lambda, \lambda + 1/2, \lambda + 1/2, \lambda + 1)$ and their CPT conjugates with helicities $(-\lambda - 1, -\lambda - 1/2, -\lambda - 1/2, -\lambda)$. The most important $N = 2$ massless supermultiplets are:
  - The $N = 2$ gauge multiplet $(A_\mu, \lambda_a^+, \lambda_a^-, \phi)$ corresponding to $\lambda = 0$. The spinor $\lambda_\alpha^\pm$ are Weyl fermions, A_μ is a massless gauge field while ϕ is a complex scalar (gauge scalar). They must be in the adjoint representation of the gauge group. The spinors $\lambda_\alpha^\pm$ transform as a doublet under $SU(2)_R$ while A_μ and ϕ transform as singlets. The $N = 2$ gauge multiplet is equivalent to the $N = 1$ gauge multiplet plus the $N = 1$ chiral multiplet all in the adjoint representation of the gauge group.
  - The hypermultiplet $(\psi_\alpha^+, H_+, H_-, \psi_\alpha^+)$ corresponding to $\lambda = -1/2$. The spinor $\psi_\alpha^\pm$ are Weyl fermions while $H^\pm$ are complex scalars[1]. This is a

---

[1] The state $(-1/2, 0, 0, -1/2)$ on its own may be CPT self-conjugate in which case we call this supermultiplet a half-hypermultiplet.

matter multiplet transforming in some representation of the gauge group. The scalar fields $H^{\pm}$ transform as a doublet under $SU(2)_R$ while $\psi_{\alpha}^{\pm}$ transform as singlets.

- An $N = 4$ massless supermultiplet (no central charges) contains 32 states: The 16 states with helicities $(\lambda, \lambda + 1/2(4 \text{ times}), \lambda + 1(6 \text{ times}), \lambda + 3/2(4 \text{ times}), \lambda + 2)$ and their CPT conjugates. The most important case corresponds to $\lambda = -1$, which is actually CPT self-conjugate and thus contains 16 states only. This is precisely the $N = 4$ gauge multiplet containing a massless gauge field, 4 Weyl fermions and 6 real scalars, viz $(A_{\mu}, \lambda_a^A, \phi^I)$, $A = 1,\ldots, 4$ and $I = 1,\ldots, 6$. The $N = 4$ gauge multiplet is equivalent to the $N = 2$ gauge multiplet plus the $N = 2$ hypermultiplet all in the adjoint representation of the gauge group. The fermions λ^A transform in the fundamental representation of $SU(4)_R$, the scalar fields ϕ^I transform in the 6 of $SO(6) \simeq \text{spin}(6) = SU(4)$ while A_{μ} is a singlet.
- It is well known that $N = 3$ supersymmetry coincides with $N = 4$ super-symmetry because of the CPT theorem.

15.3 $N = 1$ supersymmetry

15.3.1 $N = 1$ superspace

We introduce four anticommuting Grassmannian coordinates θ^{α}, $\bar{\theta}_{\dot{\alpha}}$ such that $\{\theta^{\alpha}, \theta^{\beta}\} = \{\bar{\theta}_{\dot{\alpha}}, \bar{\theta}_{\dot{\beta}}\} = \{\theta^{\alpha}, \bar{\theta}_{\dot{\beta}}\} = 0$. Now we will enlarge the 4-dimensional space-time labeled by x^{μ} to a $(4 + 4)$-dimensional superspace labeled by $(x^{\mu}, \theta^{\alpha}, \bar{\theta}_{\dot{\alpha}})$ known as $N = 1$ rigid superspace. This is a coset space defined as the $N = 1$ super-Poincaré group modded out by the Lorentz group in analogy with spacetime, which can be defined as the Poincaré group modded out by the Lorentz group.

Next we rewrite the $N = 1$ supersymmetry algebra as a Lie algebra. We compute immediately

$$[\theta Q, \bar{\theta}\bar{Q}] = 2\theta^{\alpha}(\sigma^{\mu})_{\alpha}{}^{\dot{\alpha}}\bar{\theta}_{\dot{\alpha}}P_{\mu} = 2\theta\sigma^{\mu}\bar{\theta}P_{\mu}. \tag{15.57}$$

$$[\theta Q, \theta Q] = [\bar{\theta}\bar{Q}, \bar{\theta}\bar{Q}] = 0. \tag{15.58}$$

Remark that

$$(\theta Q)^+ = (\theta^{\alpha}Q_{\alpha})^+ = (Q_{\alpha})^+(\theta^{\alpha})^+ = \bar{Q}_{\dot{\alpha}}\bar{\theta}^{\dot{\alpha}} = \bar{Q}\bar{\theta} = \bar{\theta}\bar{Q}. \tag{15.59}$$

Superspace derivatives are defined by (recall that we raise and lower indices on spinors with the symbol ε, viz $\theta_{\alpha} = \varepsilon_{\alpha\beta}\theta^{\beta}$, $\bar{\theta}^{\dot{\alpha}} = \varepsilon^{\dot{\alpha}\dot{\beta}}\bar{\theta}_{\dot{\beta}}$)

$$\partial_{\alpha} = \frac{\partial}{\partial\theta^{\alpha}}, \quad \partial^{\alpha} = \frac{\partial}{\partial\theta_{\alpha}} = -\varepsilon^{\alpha\beta}\partial_{\beta}. \tag{15.60}$$

$$\bar{\partial}^{\dot{\alpha}} = \frac{\partial}{\partial\bar{\theta}_{\dot{\alpha}}}, \quad \bar{\partial}_{\dot{\alpha}} = \frac{\partial}{\partial\bar{\theta}^{\dot{\alpha}}} = -\varepsilon_{\dot{\alpha}\dot{\beta}}\bar{\partial}^{\dot{\beta}}. \tag{15.61}$$

The action of these derivatives on the Grassmannian coordinates is quite obvious. We note however the results

$$\partial_\alpha(\theta^\beta\theta^\gamma) = \partial_\alpha\theta^\beta.\,\theta^\gamma - \theta^\beta.\,\partial_\alpha\theta^\gamma = \delta_\alpha^\beta\theta^\gamma - \delta_\alpha^\gamma\theta^\beta. \tag{15.62}$$

$$\partial_\rho(\theta\theta) = \delta_\rho^\alpha.\,\theta_\alpha - \theta^\alpha.\,\partial_\rho\theta_\alpha = \theta_\rho - \theta^\alpha\varepsilon_{\alpha\beta}\delta_\rho^\beta = 2\theta_\rho. \tag{15.63}$$

$$\partial^2(\theta\theta) = \partial^\alpha\partial_\alpha(\theta\theta) = 2\partial^\alpha\theta_\alpha = 4. \tag{15.64}$$

$$\bar{\partial}^2(\bar\theta\bar\theta) = -4. \tag{15.65}$$

$$\partial_{\dot\alpha}(\bar\theta\bar\theta) = -2\bar\theta_{\dot\alpha}. \tag{15.66}$$

Superspace integration will be given by Berezin integration. A general function for a single Grassmann variable θ is $f(\theta) = f_0 + \theta f_1$. The Berezin integral in this case is defined by

$$\int d\theta\theta = 1, \quad \int d\theta = 0. \tag{15.67}$$

The integral of a general function is

$$\int d\theta f(\theta) = f_1. \tag{15.68}$$

Thus integration is equivalent to differentiation since

$$\frac{df(\theta)}{d\theta} = f_1. \tag{15.69}$$

Also we conclude that

$$\int d\theta\frac{df(\theta)}{d\theta} = 0. \tag{15.70}$$

Furthermore the Berezin integration is translationally invariant since

$$\int d\theta f(\theta) = \int d(\theta + \xi)f(\theta + \xi). \tag{15.71}$$

A delta function can be defined by

$$\delta(\theta) = \theta. \tag{15.72}$$

We define the Berezin measure for $N = 1$ superspace by

$$d^2\theta = -\frac{1}{4}\varepsilon_{\alpha\beta}d\theta^\alpha d\theta^\beta = \frac{1}{2}d\theta^1 d\theta^2, \quad d^2\bar\theta = -\frac{1}{4}\varepsilon^{\dot\alpha\dot\beta}d\bar\theta_{\dot\alpha}d\bar\theta_{\dot\beta} = \frac{1}{2}d\bar\theta^{\dot1}d\bar\theta^{\dot2}. \tag{15.73}$$

This is because

$$\int d^2\theta(\theta\theta) = \int \frac{1}{2}d\theta^1 d\theta^2(2\theta^2\theta^1) = 1, \quad \int d^2\bar\theta(\bar\theta\bar\theta) = \int \frac{1}{2}d\bar\theta^{\dot1}d\bar\theta^{\dot2}(2\bar\theta^{\dot2}\bar\theta^{\dot1}) = 1. \tag{15.74}$$

The full measure on the $N = 1$ superspace is

$$d^4\theta = d^2\theta d^2\bar\theta. \tag{15.75}$$

15.3.2 Scalar superfield

A scalar superfield is the most general scalar function $\Phi(x, \theta, \bar\theta)$ on the $N = 1$ superspace. There are finite terms in the Taylor expansion of this field in powers of θ^α and $\bar\theta_{\dot\alpha}$ given by

$$\begin{aligned}
\Phi(x, \theta, \bar\theta) &= f(x) + (\theta\phi(x)) + (\bar\theta\bar\chi(x)) + (\theta\theta)m(x) + (\bar\theta\bar\theta)n(x) + (\theta\sigma^\mu\bar\theta)v_\mu(x) \\
&\quad + (\theta\theta)(\bar\theta\bar\lambda(x)) + (\bar\theta\bar\theta)(\theta\psi(x)) + (\theta\theta)(\bar\theta\bar\theta)d(x).
\end{aligned} \tag{15.76}$$

It is clear that $(\bar\theta\bar\sigma^\mu\theta)v'_\mu$ and $(\theta\phi)(\bar\theta\bar\psi)$ are redundant while $(\theta\theta)(\theta\lambda'(x))$ and $(\bar\theta\bar\theta)(\bar\theta\bar\psi'(x))$ are zero. The fields f, m, n, v_μ, d are bosonic while ϕ, $\bar\chi$, $\bar\lambda$ and ψ are fermionic.

The action of the ordinary translation generator P_μ on Φ is as usual given by the differential operator $i\partial_\mu$, viz

$$(1 + i\varepsilon^\mu P_\mu)\Phi(x, \theta, \bar\theta) = \Phi(x + \varepsilon, \theta, \bar\theta) \rightarrow P_\mu = -i\partial_\mu. \tag{15.77}$$

Next we need to find the representation of the supersymmetry generators Q^α and $\bar Q_{\dot\alpha}$ as superspace differential operators. Let ξ^α be a constant Weyl spinor and consider the generator $\delta_\xi = i\xi Q + i\bar\xi\bar Q$. We expect that the action of $i\xi Q$ on the superfield $\Phi(x, \theta, \bar\theta)$ is to translate the spinor θ by ξ while the action of $i\bar\xi\bar Q$ is to translate the spinor $\bar\theta$ by $\bar\xi$. Since $\{Q_\alpha, \bar Q_{\dot\beta}\} = 2(\sigma^\mu)_{\alpha\dot\beta}P_\mu$ we expect that x^μ will also be translated. As a consequence Q_α and $\bar Q_{\dot\alpha}$ must be of the form

$$Q_\alpha = -i(\partial_\alpha - ic(\sigma^\mu)_{\alpha\dot\beta}\bar\theta^{\dot\beta}\partial_\mu). \tag{15.78}$$

By using $Q_\alpha^+ = \bar Q_{\dot\alpha}$, $\partial_\alpha^+ = \bar\partial_{\dot\alpha}$, $(i\partial_\mu)^+ = i\partial_\mu$, $(\sigma^\mu)^*_{\alpha\dot\beta} = (\sigma^\mu)_{\beta\dot\alpha}$ and $(\bar\theta^{\dot\beta})^+ = \theta^\beta$ we get

$$\bar Q_{\dot\alpha} = i(\bar\partial_{\dot\alpha} - ic^*\theta^\beta(\sigma^\mu)_{\beta\dot\alpha}\partial_\mu). \tag{15.79}$$

A direct calculation shows that

$$\{Q_\alpha, \bar Q_{\dot\beta}\}\Phi = -i(c + c^*)(\sigma^\mu)_{\alpha\dot\beta}\partial_\mu\Phi \tag{15.80}$$

$$(1 + i\xi Q + i\bar\xi\bar Q)\Phi(x^\mu, \theta^\alpha, \bar\theta^{\dot\alpha}) = \Phi(x^\mu - ic\xi\sigma^\mu\bar\theta + ic^*\theta\sigma^\mu\bar\xi, \theta^\alpha + \xi^\alpha, \bar\theta^{\dot\alpha} + \bar\xi^{\dot\alpha}). \tag{15.81}$$

Thus we must choose $c = 1$.

Next we can compute the action of linear supersymmetry transformations on the components of the superfield Φ. The variation of the first term in the expansion of Φ is

$$\delta_\xi f = -i(\xi\sigma^\mu\bar\theta)\partial_\mu f + i(\theta\sigma^\mu\bar\xi)\partial_\mu f. \tag{15.82}$$

The variation of the linear terms in θ and $\bar{\theta}$ are given by (by using the Fierz identities (15.35), (15.36) and (15.37))

$$\delta_\xi(\theta\phi) = -i(\theta\partial_\mu\phi)(\xi\sigma^\mu\bar{\theta}) + i(\theta\partial_\mu\phi)(\theta\sigma^\mu\bar{\xi}) + \xi\phi$$
$$= \frac{i}{2}[(\theta\sigma^\mu\bar{\theta})(\xi\partial_\mu\phi) - 2i(\theta\sigma_\nu\bar{\theta})(\xi\sigma^{\mu\nu}\partial_\mu\phi)] - \frac{i}{2}(\theta\theta)(\partial_\mu\phi\sigma^\mu\bar{\xi}) + \xi\phi. \tag{15.83}$$

$$\delta_\xi(\bar{\theta}\bar{\chi}) = -i(\bar{\theta}\partial_\mu\bar{\chi})(\xi\sigma^\mu\bar{\theta}) + i(\bar{\theta}\partial_\mu\bar{\chi})(\theta\sigma^\mu\bar{\xi}) + \bar{\xi}\bar{\chi}$$
$$= \frac{i}{2}(\bar{\theta}\bar{\theta})(\xi\sigma^\mu\partial_\mu\bar{\chi}) - \frac{i}{2}[(\theta\sigma^\mu\bar{\theta})(\xi\partial_\mu\chi) - 2i(\theta\sigma_\nu\bar{\theta})(\xi\sigma^{\mu\nu}\partial_\mu\chi)]^+ + \bar{\xi}\bar{\chi}. \tag{15.84}$$

The variation of the cubic terms in θ and $\bar{\theta}$ are given by (here we use also the Fierz identity (15.38))

$$\delta_\xi((\theta\theta)(\bar{\theta}\bar{\lambda})) = -i(\theta\theta)(\bar{\theta}\partial_\mu\bar{\lambda})(\xi\sigma^\mu\bar{\theta}) + (\theta\theta)(\bar{\xi}\bar{\lambda}) + 2(\theta\xi)(\bar{\theta}\bar{\lambda})$$
$$= \frac{i}{2}(\theta\theta)(\bar{\theta}\bar{\theta})(\xi\sigma^\mu\partial_\mu\bar{\lambda}) + (\theta\theta)(\bar{\xi}\bar{\lambda}) + (\theta\sigma^\mu\bar{\theta})(\xi\sigma_\mu\bar{\lambda}). \tag{15.85}$$

$$\delta_\xi((\bar{\theta}\bar{\theta})(\theta\psi)) = i(\bar{\theta}\bar{\theta})(\theta\partial_\mu\psi)(\theta\sigma^\mu\bar{\xi}) + (\bar{\theta}\bar{\theta})(\xi\psi) + 2(\bar{\theta}\bar{\xi})(\theta\psi)$$
$$= -\frac{i}{2}(\theta\theta)(\bar{\theta}\bar{\theta})(\partial_\mu\psi\sigma^\mu\bar{\xi}) + (\bar{\theta}\bar{\theta})(\xi\psi) + (\theta\sigma^\mu\bar{\theta})(\psi\sigma_\mu\bar{\xi}). \tag{15.86}$$

The variation of the quadratic terms in θ and $\bar{\theta}$ are given by

$$\delta_\xi((\theta\theta)m(x)) = -i(\theta\theta)(\xi\sigma^\mu\bar{\theta})\partial_\mu m + 2(\theta\xi)m. \tag{15.87}$$

$$\delta_\xi((\bar{\theta}\bar{\theta})n(x)) = i(\bar{\theta}\bar{\theta})(\theta\sigma^\mu\bar{\xi})\partial_\mu n + 2(\bar{\theta}\bar{\xi})n. \tag{15.88}$$

Also by using the Fierz identity (15.39) we compute

$$\delta_\xi((\theta\sigma^\mu\bar{\theta})v_\mu(x)) = i\partial_\nu v_\mu(\theta\sigma^\mu\bar{\theta})(\theta\sigma^\nu\bar{\xi}) - i\partial_\nu v_\mu(\theta\sigma^\mu\bar{\theta})(\xi\sigma^\nu\bar{\theta}) + (\xi\sigma^\mu\bar{\theta})v_\mu + (\theta\sigma^\mu\bar{\xi})v_\mu$$
$$= \left[-\frac{i}{2}\partial^\mu v_\mu(\xi\theta)(\bar{\theta}\bar{\theta}) - \partial_\mu v_\nu(\xi\sigma^{\mu\nu}\theta)(\bar{\theta}\bar{\theta})\right]^+ - \frac{i}{2}\partial^\mu v_\mu(\xi\theta)(\bar{\theta}\bar{\theta}) - \partial_\mu v_\nu(\xi\sigma^{\mu\nu}\theta)(\bar{\theta}\bar{\theta}) \tag{15.89}$$
$$+ (\xi\sigma^\mu\bar{\theta})v_\mu + (\theta\sigma^\mu\bar{\xi})v_\mu.$$

The variation of the last term in the expansion of Φ is (by using the Fierz identities (15.34), (15.35) and (15.36))

$$\delta_\xi((\theta\theta)(\bar{\theta}\bar{\theta})d(x)) = 2(\theta\theta)(\bar{\theta}\bar{\xi})d - 2i(\theta\theta)(\bar{\theta}\bar{\xi})(\xi\sigma^\mu\bar{\theta})\partial_\mu d + 2(\bar{\theta}\bar{\theta})(\theta\xi)d + 2i(\bar{\theta}\bar{\theta})(\theta\xi)(\theta\sigma^\mu\bar{\xi})\partial_\mu d.$$
$$= 2(\theta\theta)(\bar{\theta}\bar{\xi})d + 2(\bar{\theta}\bar{\theta})(\theta\xi)d. \tag{15.90}$$

By using the above results we can derive immediately the following supersymmetry transformation laws

$$\delta_\xi f = \xi\phi + \bar{\xi}\bar{\chi}. \tag{15.91}$$

$$\delta_\xi\phi_\alpha = 2\xi_\alpha m + (\sigma^\mu)_{\alpha\beta}\bar{\xi}^{\dot{\beta}}(i\partial_\mu f + v_\mu). \tag{15.92}$$

$$\delta_\xi \bar{\chi}^{\dot\alpha} = 2\bar{\xi}^{\dot\alpha} n + \xi^\beta (\sigma^\mu)_{\beta\dot\gamma}\varepsilon^{\dot\gamma\dot\alpha}(i\partial_\mu f - v_\mu). \tag{15.93}$$

$$\delta_\xi m = \bar{\xi}\bar{\lambda} - \frac{i}{2}\partial_\mu \phi \sigma^\mu \bar{\xi}. \tag{15.94}$$

$$\delta_\xi n = \xi\psi + \frac{i}{2}\xi\sigma^\mu\partial_\mu\bar{\chi}. \tag{15.95}$$

$$\delta_\xi v_\mu = \xi\sigma_\mu\bar{\lambda} + \psi\sigma_\mu\bar{\xi} + \frac{i}{2}\xi\partial_\mu\phi - \frac{i}{2}\partial_\mu\bar{\chi}\bar{\xi} - \xi\sigma^{\mu\nu}\partial_\nu\phi - \partial_\nu\bar{\chi}\bar{\sigma}^{\mu\nu}\bar{\xi}. \tag{15.96}$$

$$\delta_\xi d = \frac{i}{2}\partial_\mu(\xi\sigma^\mu\bar{\lambda} - \psi\sigma^\mu\bar{\xi}). \tag{15.97}$$

$$\delta_\xi \psi_\alpha = 2\xi_\alpha d + i(\sigma^\mu\bar{\xi})_\alpha\partial_\mu n - \frac{i}{2}\partial^\mu v_\mu \xi_\alpha - \frac{1}{2}(\partial_\mu v_\nu - \partial_\nu v_\mu)(\xi\sigma^{\mu\nu})^\beta\varepsilon_{\alpha\beta}. \tag{15.98}$$

$$\delta_\xi \bar{\lambda}^{\dot\alpha} = 2\bar{\xi}^{\dot\alpha}d + i\xi^\beta(\sigma^\mu)_{\beta\dot\gamma}\varepsilon^{\dot\gamma\dot\alpha}\partial_\mu m + \frac{i}{2}\partial^\mu v_\mu\bar{\xi}^{\dot\alpha} - \frac{1}{2}(\partial_\mu v_\nu - \partial_\nu v_\mu)(\bar{\sigma}^{\mu\nu}\bar{\xi})^{\dot\alpha}. \tag{15.99}$$

The four last transformation laws are slightly different from those obtained by Lykken.

15.3.3 Chiral superfield

Clearly the general scalar field Φ cannot correspond to an irreducible representation of the $N = 1$ supersymmetry algebra since it obviously contains too many components. The scalar superfield Φ provides a reducible representation although not a fully reducible one, i.e., it is not given by a direct sum of irreducible representations. We would like to reduce the number of components of Φ in a supersymmetric covariant fashion and thus we must impose covariant constraints on Φ.

Towards this end we introduce the so-called covariant derivatives D_α and $\bar{D}_{\dot\alpha}$, which anticommute with the supersymmetry generators Q_α and $\bar{Q}_{\dot\alpha}$, viz

$$\{D_\alpha, Q_\beta\} = \{D_\alpha, \bar{Q}_{\dot\beta}\} = 0. \tag{15.100}$$

$$\{\bar{D}_{\dot\alpha}, Q_\beta\} = \{\bar{D}_{\dot\alpha}, \bar{Q}_{\dot\beta}\} = 0. \tag{15.101}$$

Then if Φ is a scalar superfield then $D_\alpha\Phi$ and $\bar{D}_{\dot\alpha}\Phi$ are also superfields in the sense that $(1 + \delta_\xi)D_\alpha\Phi(x, \theta, \bar{\theta}) = D_\alpha\phi(x - i\xi\sigma^\mu\bar{\theta} + i\theta\sigma^\mu\bar{\xi}, \theta + \xi, \bar{\theta} + \bar{\xi})^2$. Furthermore the condition $D_\alpha\Phi = 0$ is clearly covariant under supersymmetry, i.e.

---

[2] Also $\partial_\mu\Phi$ is a superfield (in an obvious way) while $\partial_\alpha\Phi$, $\partial_{\dot\alpha}\Phi$ are not superfields since they do not contain a d-term.

$D_\alpha \delta_\xi \Phi = \delta_\xi(D_\alpha \Phi) = 0$. It is obvious that D_α, $\bar{D}_{\dot\alpha}$ are generalizations of the derivatives ∂_α, $\bar{\partial}_{\dot\alpha}$. Indeed we find

$$D_\alpha = \partial_\alpha + i(\sigma^\mu)_{\alpha\dot\beta}\bar{\theta}^{\dot\beta}\partial_\mu. \tag{15.102}$$

$$\bar{D}_{\dot\alpha} = \bar{\partial}_{\dot\alpha} + i\theta^\beta(\sigma^\mu)_{\beta\dot\alpha}\partial_\mu. \tag{15.103}$$

It is clear that

$$\{D_\alpha, \bar{D}_{\dot\beta}\} = 2i(\sigma^\mu)_{\alpha\dot\beta}\partial_\mu. \tag{15.104}$$

An $N = 1$ chiral superfield is a general scalar superfield fulfilling the supersymmetric covariant constraint

$$\bar{D}_{\dot\alpha}\Phi = 0. \tag{15.105}$$

This has less components than the general scalar superfield and in fact provides an irreducible representation of the $N = 1$ supersymmetry algebra. Writing this condition in components we get

$$\bar{\chi}_{\dot\alpha} = 0, \quad n = 0, \quad v_\mu = i\partial_\mu f, \quad \psi = 0, \quad \bar{\lambda}_{\dot\alpha} = -\frac{i}{2}\partial_\mu\phi^\beta(\sigma^\mu)_{\beta\dot\alpha}, \quad d = -\frac{1}{4}\partial_\mu\partial^\mu f. \tag{15.106}$$

We can check directly that these conditions are supersymmetric invariant by using the supersymmetry transformations derived in the previous section. We get

$$\delta_\xi f = \xi\phi, \quad \delta_\xi \phi_\alpha = 2\xi_\alpha m + 2i(\sigma^\mu)_{\alpha\dot\beta}\bar{\xi}^{\dot\beta}\partial_\mu f, \quad \delta_\xi \bar{\chi}^{\dot\alpha} = 0, \quad \delta_\xi n = 0, \quad \delta_\xi \psi = 0. \tag{15.107}$$

Equivalently

$$\Phi(x, \theta, \bar{\theta}) = f + \theta\phi + (\theta\theta)m + i(\theta\sigma^\mu\bar{\theta})\partial_\mu f - \frac{i}{2}(\theta\theta)(\partial_\mu\phi\sigma^\mu\bar{\theta}) - \frac{1}{4}(\theta\theta)(\bar{\theta}\bar{\theta})\partial_\mu\partial^\mu f. \tag{15.108}$$

By using

$$\sigma^\mu\bar{\sigma}^\nu + \sigma^\nu\bar{\sigma}^\mu = 2\eta^{\mu\nu}, \quad \bar{\sigma}^\mu\sigma^\nu + \bar{\sigma}^\nu\sigma^\mu = 2\eta^{\mu\nu}, \tag{15.109}$$

and

$$\hat{\varepsilon}(\sigma^\mu)^T = -\bar{\sigma}^\mu\tilde{\varepsilon}, \quad (\sigma^\mu)^T\hat{\varepsilon} = -\tilde{\varepsilon}\bar{\sigma}^\mu, \tag{15.110}$$

where $\hat{\varepsilon}^{\alpha\beta} = \varepsilon^{\alpha\beta}$ and $\tilde{\varepsilon}_{\alpha\beta} = \varepsilon_{\alpha\beta}$, we get

$$\delta_\xi d = -\frac{1}{4}\partial_\mu\partial^\mu\phi\xi, \quad \delta\bar{\lambda}^{\dot\alpha} = \partial_\mu(i\xi^\beta(\sigma_\mu)_{\beta\dot\gamma}\varepsilon^{\dot\gamma\dot\alpha}m - \partial_\mu f\bar{\xi}^{\dot\alpha}). \tag{15.111}$$

We also get[3]

$$\delta_\xi v^\mu = i\partial_\mu\phi\xi - \partial_\nu\phi\sigma^{\mu\nu}\xi - \xi\sigma^{\mu\nu}\partial_\nu\phi. \tag{15.112}$$

---

[3] First show that $(\sigma^{\mu\nu})^T = \tilde{\varepsilon}\sigma^{\mu\nu}\tilde{\varepsilon}$. Then show that $\xi\sigma^{\mu\nu}\phi = \xi^\alpha(\sigma^{\mu\nu})_\alpha{}^\beta\phi_\beta = \phi_\beta(\hat{\varepsilon}\sigma^{\mu\nu}\tilde{\varepsilon})_\alpha{}^\beta\xi^\alpha = -\theta^\beta(\sigma^{\mu\nu})_\beta{}^\alpha\xi_\alpha$.

A more elegant approach is to determine the most general form of a chiral superfield by solving directly the constraint $\bar{D}_{\dot\alpha}\Phi = 0$. Let us introduce the new bosonic coordinates

$$y^\mu = x^\mu + i\theta\sigma^\mu\bar\theta, \quad \bar{y}^\mu = x^\mu - i\theta\sigma^\mu\bar\theta. \tag{15.113}$$

Then

$$\bar{D}_{\dot\alpha}y^\mu = D_\alpha\bar{y}^\mu = 0. \tag{15.114}$$

Remark also that

$$\bar{D}_{\dot\alpha}\theta^\beta = D_\alpha\bar\theta^{\dot\beta} = 0. \tag{15.115}$$

Then clearly the most general solution of $\bar{D}_{\dot\alpha}\Phi = 0$ is $\Phi = \Phi(y, \theta)$. The expansion of this superfield in powers of θ is then very simple given by

$$\Phi(y, \theta) = A(y) + \sqrt{2}\,\theta\psi(y) + \theta\theta F(y). \tag{15.116}$$

This superfield consists of two complex scalar fields A and F and a single complex left-handed Weyl spinor and therefore contains $4 + 4 = 8$ real off-shell field components. This is twice the number of degrees of freedom in the on-shell fundamental $N = 1$ massive irreducible representation of $N = 1$ supersymmetry.

Similarly we can define antichiral superfields by the condition $D_\alpha\Phi = 0$ and thus $\Phi = \Phi(\bar{y}^\mu, \bar\theta)$. Clearly if Φ is a chiral (left-handed) superfield then Φ^+ is an anti-chiral (right-handed) superfield.

In terms of y the supersymmetry generators and the covariant derivatives are given by (we simply make the replacements $\partial_\mu \longrightarrow \partial^y_\mu$, $\partial_\alpha \longrightarrow \partial_\alpha + i(\sigma^\mu\bar\theta)_\alpha\partial^y_\mu$, $\bar\partial_{\dot\alpha} \longrightarrow \bar\partial_{\dot\alpha} - i(\theta\sigma^\mu)_{\dot\alpha}\partial^y_\mu$)

$$\begin{aligned}
D_\alpha &= \partial_\alpha + 2i(\sigma^\mu\bar\theta)_\alpha\partial^y_\mu, \quad \bar{D}_{\dot\alpha} = \bar\partial_{\dot\alpha} \\
Q_\alpha &= -i\partial_\alpha, \quad \bar{Q}_{\dot\alpha} = i(\bar\partial_{\dot\alpha} - 2i(\theta\sigma^\mu)_{\dot\alpha}\partial_\mu).
\end{aligned} \tag{15.117}$$

The supersymmetry transformation of a chiral superfield is

$$\delta\Phi(y^\mu, \theta) = i(\xi Q + \bar\xi\bar{Q})\Phi(y^\mu, \theta). \tag{15.118}$$

$$(\Phi + \delta\Phi)(y^\mu, \theta) = \Phi(y^\mu + 2i\theta\sigma^\mu\bar\xi + i\xi\sigma^\mu\bar\xi, \theta + \xi). \tag{15.119}$$

Explicitly we have the linear supersymmetry transformations

$$\delta\Phi(y^\mu, \theta) = 2i\partial_\mu A(\theta\sigma^\mu\bar\xi) - i\sqrt{2}\,(\theta\theta)(\partial_\mu\psi\sigma^\mu\bar\xi) + \sqrt{2}\,\xi\psi + 2(\theta\xi)F. \tag{15.120}$$

Equivalently

$$\begin{aligned}
\delta_\xi A &= \sqrt{2}\,\xi\psi \\
\delta_\xi F &= -i\sqrt{2}\,\partial_\mu\psi\sigma^\mu\bar\xi \\
\delta_\xi\psi^\alpha &= \sqrt{2}\,\xi^\alpha F + \sqrt{2}\,i\partial_\mu A(\sigma^\mu\bar\xi)^\alpha.
\end{aligned} \tag{15.121}$$

15.3.4 Vector superfield

A vector superfield is a general scalar field, which satisfies the following covariant reality condition

$$V(x, \theta, \bar{\theta}) = V^+(x, \theta, \bar{\theta}). \tag{15.122}$$

In terms of components this condition reads

$$f = f^+, \quad \chi = \phi, \quad m = n^+, \quad v_\mu = v_\mu^+, \quad \lambda = \psi, \quad d = d^+. \tag{15.123}$$

Thus

$$V = f + \theta\chi + \bar{\theta}\bar{\chi} + (\theta\theta)n^+ + (\bar{\theta}\bar{\theta})n + (\theta\sigma^\mu\bar{\theta})v_\mu + (\theta\theta)(\bar{\theta}\bar{\lambda}) + (\bar{\theta}\bar{\theta})(\theta\lambda) + (\theta\theta)(\bar{\theta}\bar{\theta})d. \tag{15.124}$$

An example of a vector superfield is $\Phi + \Phi^+$ where Φ is a chiral superfield. This is given by (where we will set $f = A$, $\phi = \sqrt{2}\psi$ and $m = F$ in (15.108))

$$\Phi + \Phi^+ = A + A^+ + \sqrt{2}\theta\psi + \sqrt{2}\bar{\theta}\bar{\psi} + \theta\theta F + \bar{\theta}\bar{\theta}F^+ + i(\theta\sigma^\mu\bar{\theta})\partial_\mu(A - A^+)$$
$$- \frac{i}{\sqrt{2}}(\theta\theta)(\partial_\mu\psi\sigma^\mu\bar{\theta}) + \frac{i}{\sqrt{2}}(\bar{\theta}\bar{\theta})(\theta\sigma^\mu\partial_\mu\bar{\psi}) - \frac{1}{4}(\theta\theta)(\bar{\theta}\bar{\theta})\partial_\mu\partial^\mu(A + A^+). \tag{15.125}$$

An infinitesimal Abelian gauge transformation can be defined by

$$V \longrightarrow V' = V + \Phi + \Phi^+. \tag{15.126}$$

In terms of components we have

$$\begin{aligned}
f &\longrightarrow f + A + A^+ \\
\chi &\longrightarrow \chi + \sqrt{2}\psi \\
n &\longrightarrow n + F^+ \\
v_\mu &\longrightarrow v_\mu + i\partial_\mu(A - A^+) \\
\lambda &\longrightarrow \lambda + \frac{i}{\sqrt{2}}\sigma^\mu\partial_\mu\bar{\psi} \\
d &\longrightarrow d - \frac{1}{4}\partial_\mu\partial^\mu(A + A^+).
\end{aligned} \tag{15.127}$$

It is clear that any action that is invariant under (15.126) will not depend on many components of V. The reason is very simple. The five first components of $\Phi + \Phi^+$ are completely unconstrained and thus they can be used to gauge away the first five components of V. Indeed we can choose

$$f + A + A^+ = 0, \quad \chi + \sqrt{2}\psi = 0, \quad n + F^+ = 0. \tag{15.128}$$

Then the transformed superfield V' becomes

$$V' = (\theta\sigma^\mu\bar{\theta})v'_\mu + (\theta\theta)(\bar{\theta}\bar{\lambda}') + (\bar{\theta}\bar{\theta})(\theta\lambda') + (\theta\theta)(\bar{\theta}\bar{\theta})d', \tag{15.129}$$

where

$$v'_\mu = v_\mu + i\partial_\mu(A - A^+), \quad \lambda' = \lambda + \frac{i}{\sqrt{2}}\sigma^\mu\partial_\mu\bar{\psi}, \quad d' = d - \frac{1}{4}\partial_\mu\partial^\mu(A + A^+). \tag{15.130}$$

The gauge-fixed superfield V' is called the Wess–Zumino gauge-fixed superfield. We write

$$V' = V_{\text{WZ}}. \tag{15.131}$$

Thus any vector superfield V can be written as

$$V = V_{\text{WZ}} - (\Phi + \Phi^+). \tag{15.132}$$

As it turns out fixing the Wess–Zumino gauge does not fix the Abelian gauge symmetry. Indeed we can shift v'_μ in $V' = V_{\text{WZ}}$ and $A - A^+$ in $\Phi + \Phi^+$ as $v'_\mu \longrightarrow v'_\mu - i\partial_\mu(A' - A'^+)$ and $A - A^+ \longrightarrow A - A^+ - i\partial_\mu(A' - A'^+)$ without changing V.

15.3.5 Spinor superfield

The Wess–Zumino gauge given by equation (15.128) is not preserved by supersymmetry transformations. Thus any action written in terms of the vector superfield V will necessarily contain unphysical degrees of freedom. The spinor superfield will, on the other hand, only contain the Wess–Zumino gauge-fixed fields v_μ, λ and d. It is a chiral superfield defined by

$$W_\alpha = -\frac{1}{4}(\bar{D}\bar{D})D_\alpha V. \tag{15.133}$$

This is a chiral superfield, viz $\bar{D}_{\dot\alpha} W_\beta = 0$. The antichiral spinor superfield $\bar{W}_{\dot\alpha}$ is then obviously defined by

$$\bar{W}_{\dot\alpha} = -\frac{1}{4}(DD)\bar{D}_{\dot\alpha} V. \tag{15.134}$$

Furthermore we can check that

$$\bar{D}_{\dot\alpha} \bar{W}^{\dot\alpha} = D^\alpha W_\alpha. \tag{15.135}$$

Thus W_α is not a general chiral superfield. More importantly is the fact that W_α is invariant under the infinitesimal Abelian gauge transformation (15.126). Indeed we compute (by using $D_\alpha \Phi^+ = 0$, $\bar{D}_{\dot\alpha}\Phi = 0$, $[\bar{D}_{\dot\alpha}, \partial_\mu] = 0$ and $\{D_\alpha, \bar{D}_{\dot\beta}\} = 2i(\sigma^\mu)_{\alpha\dot\beta}\partial_\mu$)

$$\begin{aligned}
W'_\alpha &= -\frac{1}{4}(\bar{D}\bar{D})D_\alpha V' \\
&= W_\alpha - \frac{1}{4}(\bar{D}\bar{D})D_\alpha(\Phi + \Phi^+) \\
&= W_\alpha - \frac{1}{4}(\bar{D}\bar{D})D_\alpha \Phi \\
&= W_\alpha - \frac{1}{4}\bar{D}_{\dot\beta}\{\bar{D}^{\dot\beta}, D_\alpha\}\Phi \\
&= W_\alpha - \frac{i}{2}(\sigma^\mu)_{\alpha\dot\beta}\partial_\mu\bar{D}^{\dot\beta}\Phi \\
&= W_\alpha.
\end{aligned} \tag{15.136}$$

Thus without any loss of generality we can compute the components of W_α in the Wess–Zumino gauge, viz

$$W_\alpha = -\frac{1}{4}(\bar{D}\bar{D})D_\alpha V_{WZ}. \tag{15.137}$$

$$\bar{W}_{\dot\alpha} = -\frac{1}{4}(DD)\bar{D}_{\dot\alpha} V_{WZ}. \tag{15.138}$$

The chiral superfield $W_{\dot\alpha}$ can only be a function of y (not x) and θ, $\bar\theta$. Thus the covariant derivatives can be given simply by

$$D_\alpha = \partial_\alpha + 2i(\sigma^\mu\bar\theta)_\alpha\partial^y_\mu, \quad \bar{D}_{\dot\alpha} = \bar\partial_{\dot\alpha}. \tag{15.139}$$

Furthermore

$$V_{WZ}(x^\mu, \theta, \bar\theta) = V_{WZ}(y^\mu - i\theta\sigma^\mu\bar\theta, \theta, \bar\theta). \tag{15.140}$$

Recall that

$$V_{WZ} = (\theta\sigma^\mu\bar\theta)v_\mu(x) + (\theta\theta)(\bar\theta\bar\lambda(x)) + (\bar\theta\bar\theta)(\theta\lambda(x)) + (\theta\theta)(\bar\theta\bar\theta)d(x). \tag{15.141}$$

Thus in terms of y we obtain

$$V_{WZ} = (\theta\sigma^\mu\bar\theta)v_\mu(y) + (\theta\theta)(\bar\theta\bar\lambda(y)) + (\bar\theta\bar\theta)(\theta\lambda(y)) + (\theta\theta)(\bar\theta\bar\theta)(d(y) - \frac{i}{2}\partial_\mu v^\mu(y)). \tag{15.142}$$

We compute

$$D_\alpha((\theta\sigma^\mu\bar\theta)v_\mu(y)) = (\sigma^\mu\bar\theta)_\alpha v_\mu + i\theta_\alpha(\bar\theta\bar\theta)\partial^\mu v_\mu + (\sigma^{\mu\nu}\theta)_\alpha(\bar\theta\bar\theta)(\partial_\mu v_\nu - \partial_\nu v_\mu). \tag{15.143}$$

$$D_\alpha((\theta\theta)(\bar\theta\bar\theta)(d(y) - \frac{i}{2}\partial_\mu v^\mu(y))) = 2\theta_\alpha(\bar\theta\bar\theta)(d - \frac{i}{2}\partial_\mu v^\mu). \tag{15.144}$$

$$D_\alpha((\bar\theta\bar\theta)(\theta\lambda(y))) = \lambda_\alpha(\bar\theta\bar\theta). \tag{15.145}$$

$$D_\alpha((\theta\theta)(\bar\theta\bar\lambda(y))) = 2\theta_\alpha(\bar\theta\bar\lambda) - i(\bar\theta\bar\theta)(\theta\theta)(\sigma^\mu\partial_\mu\bar\lambda)_\alpha. \tag{15.146}$$

Thus

$$-\frac{1}{4}(DD)D_\alpha((\theta\sigma^\mu\bar\theta)v_\mu(y)) = i\theta_\alpha\partial^\mu v_\mu + (\sigma^{\mu\nu}\theta)_\alpha(\partial_\mu v_\nu - \partial_\nu v_\mu). \tag{15.147}$$

$$-\frac{1}{4}(DD)D_\alpha((\theta\theta)(\bar\theta\bar\theta)(d(y) - \frac{i}{2}\partial_\mu v^\mu(y))) = 2\theta_\alpha(d - \frac{i}{2}\partial_\mu v^\mu). \tag{15.148}$$

$$-\frac{1}{4}(DD)D_\alpha((\bar\theta\bar\theta)(\theta\lambda(y))) = \lambda_\alpha. \tag{15.149}$$

$$-\frac{1}{4}(DD)D_\alpha((\theta\theta)(\bar\theta\bar\lambda(y))) = -i(\theta\theta)(\sigma^\mu\partial_\mu\bar\lambda)_\alpha. \tag{15.150}$$

The final formula is

$$W_\alpha = \lambda_\alpha + 2\theta_\alpha d + (\sigma^{\mu\nu}\theta)_\alpha(\partial_\mu v_\nu - \partial_\nu v_\mu) - i(\theta\theta)(\sigma^\mu\partial_\mu\bar\lambda)_\alpha. \tag{15.151}$$

15.3.6 Non-Abelian chiral, vector and spinor superfields

The infinitesimal Abelian gauge transformation (15.126) corresponds to the finite transformation

$$e^V \longrightarrow e^{V\prime} = e^{\Phi^+}e^V e^\Phi. \tag{15.152}$$

The non-Abelian finite transformation will be given by this form with the understanding that $V = V_a T^a$, $\Phi = \Phi_a T^a$ where $[T^a, T^b] = i f^{abc} T^c$ and $Tr T^a T^b = \delta^{ab}$.

Next we compute the infinitesimal non-Abelian gauge transformation. We employ the Baker–Campbell–Hausdorff formula for $e^A e^B$ which includes all linear terms in B given by

$$e^A e^B = e^{A + \mathcal{L}_{A/2}.\,[B + \coth(\mathcal{L}_{A/2}).B] + \cdots} \tag{15.153}$$

The Lie derivative is given by

$$\mathcal{L}_{A/2}.\,B = [A/2, B]. \tag{15.154}$$

The hyperbolic cotangent is defined by its Taylor series expansion. For example the order n is given by

$$(\mathcal{L}_{A/2})^n.\,B = [A/2, [A/2, [A/2,..., [A/2, B] \,...\,]]], \quad n \text{ commutators.} \tag{15.155}$$

We can immediately compute

$$e^V e^\Phi = e^{V + \mathcal{L}_{V/2}.\,[\Phi + \coth(\mathcal{L}_{V/2}).\Phi] + \cdots} \tag{15.156}$$

$$e^{\Phi^+} e^V = e^{V + \mathcal{L}_{V/2}.\,[-\Phi^+ + \coth(\mathcal{L}_{V/2}).\Phi^+] + \cdots} \tag{15.157}$$

Thus

$$e^{V\prime} = e^{V + \mathcal{L}_{V/2}.\,[-\Phi^+ + \coth(\mathcal{L}_{V/2}).\Phi^+] + \mathcal{L}_{V/2}.\,[\Phi + \coth(\mathcal{L}_{V/2}).\Phi] + \cdots} \tag{15.158}$$

In other words, we have

$$\delta V = \mathcal{L}_{V/2}.\,[\Phi - \Phi^+] + \mathcal{L}_{V/2} \cdot \coth(\mathcal{L}_{V/2}) \cdot [\Phi + \Phi^+]. \tag{15.159}$$

Recall that

$$x \coth x = 1 + \frac{x^2}{3} - \frac{x^4}{45} + \cdots \tag{15.160}$$

As it turns out we can use the infinitesimal non-Abelian gauge transformation (15.159) to bring the vector superfield V into the Wess–Zumino form, viz

$$V_{\mathrm{WZ}} = (\theta\sigma^\mu\bar\theta)v_\mu(x) + (\theta\theta)(\bar\theta\bar\lambda(x)) + (\bar\theta\bar\theta)(\theta\lambda(x)) + (\theta\theta)(\bar\theta\bar\theta)d(x). \tag{15.161}$$

The only difference between the Abelian and non-Abelian cases is the fact that the relations (15.128) between the components of V and $\Phi + \Phi^+$ which define the

Wess–Zumino gauge become very complicated due to the non-linearity of the infinitesimal non-Abelian gauge transformation (15.159). Clearly, the Wess–Zumino gauge in the non-Abelian case, as in the Abelian case, is not preserved by supersymmetry. Also, the Wess–Zumino gauge does not fix the non-Abelian gauge freedom generated, as in the Abelian case, by the scalar component of $\Phi - \Phi^+$.

We will therefore only consider vector superfields in the Wess–Zumino gauge, viz $V = V_{\mathrm{WZ}}$. Furthermore we will only consider chiral superfields Φ such that $\Phi - \Phi^+$ generate the remaining non-Abelian gauge freedom, i.e., those $\Phi - \Phi^+$ with only the scalar component non-zero. In other words

$$\Phi - \Phi^+ = 2A - \frac{1}{2}(\theta\theta)(\bar{\theta}\bar{\theta})\partial_\mu\partial^\mu A. \tag{15.162}$$

The corresponding vector superfield $\Phi + \Phi^+$ is given by

$$\Phi + \Phi^+ = 2i(\theta\sigma^\mu\bar{\theta})\partial_\mu A. \tag{15.163}$$

It is then clear that the infinitesimal non-Abelian gauge transformation (15.159) reduces to

$$\delta V_{\mathrm{WZ}} = \frac{1}{2}[V_{\mathrm{WZ}}, \Phi - \Phi^+] + \Phi + \Phi^+. \tag{15.164}$$

In terms of components this reads

$$\delta(v_\mu/2) = [v_\mu/2, A] + i\partial_\mu A, \quad \delta\lambda_\alpha = [\lambda_\alpha, A], \quad \delta\bar{\lambda}^{\dot{\alpha}} = [\bar{\lambda}^{\dot{\alpha}}, A], \quad \delta d = [d, A]. \tag{15.165}$$

The spinor superfields W_α and $\bar{W}_{\dot{\alpha}}$ in the non-Abelian case are given by

$$W_\alpha = -\frac{1}{4}(\bar{D}\bar{D})e^{-V}D_\alpha e^V. \tag{15.166}$$

$$\bar{W}_{\dot{\alpha}} = -\frac{1}{4}(DD)e^{-V}\bar{D}_{\dot{\alpha}}e^V. \tag{15.167}$$

These obviously reduce to the previous formulas when the vector superfield V is Abelian. Next we check (using $D_\alpha\Phi^+ = 0$) that

$$\begin{aligned}
e^{-V'}D_\alpha e^{V'} &= e^{-\Phi}e^{-V}e^{-\Phi^+}D_\alpha e^{\Phi^+}e^V e^\Phi \\
&= e^{-\Phi}e^{-V}(D_\alpha e^V)e^\Phi + e^{-\Phi}(D_\alpha e^\Phi).
\end{aligned} \tag{15.168}$$

Then (using $\bar{D}_{\dot{\alpha}}\Phi = 0$)

$$\begin{aligned}
\bar{D}\bar{D}(e^{-V'}D_\alpha e^{V'}) &= e^{-\Phi}\bar{D}\bar{D}(e^{-V}(D_\alpha e^V)e^\Phi) + e^{-\Phi}\bar{D}\bar{D}(D_\alpha e^\Phi) \\
&= e^{-\Phi}\bar{D}\bar{D}(e^{-V}(D_\alpha e^V))e^\Phi + e^{-\Phi}\bar{D}(\{\bar{D}, D_\alpha\}e^\Phi)
\end{aligned} \tag{15.169}$$

The second term is obviously zero. The first term yields the result

$$W'_\alpha = e^{-\Phi}W_\alpha e^\Phi. \tag{15.170}$$

Similarly we should find

$$\bar{W}'_{\dot\alpha} = e^{\Phi^+}\bar{W}_{\dot\alpha}e^{-\Phi^+}. \tag{15.171}$$

The next task is to compute W_α and $\bar{W}_{\dot\alpha}$ in terms of component fields in the Wess–Zumino gauge (for simplicity and without any loss of generality). First we note

$$e^{V_{\text{WZ}}} = 1 + V_{\text{WZ}} + \frac{1}{2}V_{\text{WZ}}^2, \quad e^{-V_{\text{WZ}}} = 1 - V_{\text{WZ}} + \frac{1}{2}V_{\text{WZ}}^2. \tag{15.172}$$

Then

$$\begin{aligned}
e^{-V_{\text{WZ}}}D_\alpha e^{V_{\text{WZ}}} &= D_\alpha V_{\text{WZ}} + \frac{1}{2}D_\alpha V_{\text{WZ}}^2 - V_{\text{WZ}}D_\alpha V_{\text{WZ}} \\
&= D_\alpha V_{\text{WZ}} + \frac{1}{2}[D_\alpha V_{\text{WZ}}, V_{\text{WZ}}].
\end{aligned} \tag{15.173}$$

The second term in this last equation will be absent in the Abelian case. We have already computed $D_\alpha V_{\text{WZ}}$ in equations (15.143), (15.144), (15.145) and (15.146), viz

$$\begin{aligned}
D_\alpha V_{\text{WZ}} &= (\sigma^\mu\bar\theta)_\alpha v_\mu + 2\theta_\alpha(\bar\theta\bar\theta)d + (\sigma^{\mu\nu}\theta)_\alpha(\bar\theta\bar\theta)(\partial_\mu v_\nu - \partial_\nu v_\mu) + \lambda_\alpha(\bar\theta\bar\theta) + 2\theta_\alpha(\bar\theta\bar\lambda) \\
&\quad - i(\bar\theta\bar\theta)(\theta\theta)(\sigma^\mu\partial_\mu\bar\lambda)_\alpha.
\end{aligned} \tag{15.174}$$

Next we compute

$$\begin{aligned}
[(\sigma^\mu\bar\theta)_\alpha v_\mu, V_{\text{WZ}}] &= (\sigma^\mu\bar\theta)_\alpha(\theta\sigma^\nu\bar\theta)[v_\mu, v_\nu] + \theta\theta[v_\mu, (\sigma^\mu\bar\theta)_\alpha(\bar\theta\bar\lambda)] \\
&= -i(\sigma^{\mu\nu}\theta)_\alpha(\bar\theta\bar\theta)[v_\mu, v_\nu] - \frac{1}{2}(\theta\theta)(\bar\theta\bar\theta)(\sigma^\mu)_{\alpha\dot\beta}[v_\mu, \bar\lambda^{\dot\beta}].
\end{aligned} \tag{15.175}$$

$$\begin{aligned}
[2\theta_\alpha(\bar\theta\bar\lambda), V_{\text{WZ}}] &= [2\theta_\alpha(\bar\theta\bar\lambda), \theta\sigma^\mu\bar\theta v_\mu] \\
&= -\bar\theta\bar\theta[\theta_\alpha(\theta\sigma^\mu\bar\lambda), v_\mu] \\
&= -\frac{1}{2}(\theta\theta)(\bar\theta\bar\theta)(\sigma^\mu)_{\alpha\dot\beta}[v_\mu, \bar\lambda^{\dot\beta}].
\end{aligned} \tag{15.176}$$

$$[\lambda_\alpha(\bar\theta\bar\theta), V_{\text{WZ}}] = [2\theta_\alpha(\bar\theta\bar\theta)d, V_{\text{WZ}}] = [(\sigma^{\mu\nu}\theta)_\alpha(\bar\theta\bar\theta)(\partial_\mu v_\nu - \partial_\nu v_\mu), V_{\text{WZ}}] = 0. \tag{15.177}$$

Thus

$$[D_\alpha V_{\text{WZ}}, V_{\text{WZ}}] = -i(\sigma^{\mu\nu}\theta)_\alpha(\bar\theta\bar\theta)[v_\mu, v_\nu] - (\theta\theta)(\bar\theta\bar\theta)(\sigma^\mu)_{\alpha\dot\beta}[v_\mu, \bar\lambda^{\dot\beta}]. \tag{15.178}$$

Hence

$$\begin{aligned}
e^{-V_{\text{WZ}}}D_\alpha e^{V_{\text{WZ}}} &= (\sigma^\mu\bar\theta)_\alpha v_\mu + 2\theta_\alpha(\bar\theta\bar\theta)d + (\sigma^{\mu\nu}\theta)_\alpha(\bar\theta\bar\theta)(\partial_\mu v_\nu - \partial_\nu v_\mu - \frac{i}{2}[v_\mu, v_\nu]) \\
&\quad + \lambda_\alpha(\bar\theta\bar\theta) + 2\theta_\alpha(\bar\theta\bar\lambda) - i(\bar\theta\bar\theta)(\theta\theta)(\sigma^\mu\partial_\mu\bar\lambda - \frac{i}{2}\sigma^\mu[v_\mu, \bar\lambda])_\alpha.
\end{aligned} \tag{15.179}$$

The remaining calculation is identical to the Abelian case. We get

$$W_\alpha = \lambda_\alpha + 2\theta_\alpha d + (\sigma^{\mu\nu})_\alpha{}^\beta\theta_\beta(\partial_\mu v_\nu - \partial_\nu v_\mu - \frac{i}{2}[v_\mu, v_\nu]) - i(\theta\theta)(\sigma^\mu)_{\alpha\dot\beta}(\partial_\mu\bar\lambda^{\dot\beta} - \frac{i}{2}[v_\mu, \bar\lambda^{\dot\beta}]). \tag{15.180}$$

15.3.7 The Wess–Zumino model

Recall that the d-term of a vector superfield transforms by a total derivative. Thus given a vector superfield V a globally $N = 1$ supersymmetric action can be given by

$$\int d^4x \int d^4\theta \; V(x, \theta, \bar{\theta}). \tag{15.181}$$

Very simple vector superfields V can be constructed from a chiral superfield Φ as $V = \Phi^+\Phi$. We will be interested in actions of the form

$$\int d^4x \int d^4\theta \; \Phi^+\Phi. \tag{15.182}$$

This action will generate kinetic terms for the component fields of Φ. This term is called the Kahler potential.

Also the F-term of a chiral superfield transforms by a total derivative. Thus given a chiral superfield Φ a globally $N = 1$ supersymmetric action can be given by

$$\int d^4x \int d^2\theta \; \Phi(y, \theta) + \int d^4x \int d^2\bar{\theta} \; \Phi^+(y^+, \bar{\theta}). \tag{15.183}$$

Any holomorphic function $W(\Phi)$ of a chiral superfield Φ is also a chiral superfield. Thus we will be interested in actions of the form

$$\int d^4x \left[\int d^2\theta \; W(\Phi) + \text{h. c} \right]. \tag{15.184}$$

This will generate mass terms and interactions for the component fields of Φ. The holomorphic function $W(\Phi)$ is called the superpotential.

A general action of interested can be constructed from an elementary chiral superfield Φ as

$$S = \int d^4x \int d^4\theta \; \Phi^+\Phi - \int d^4x \left[\int d^2\theta \; W(\Phi) + \text{h. c} \right]. \tag{15.185}$$

The Wess–Zumino model is the simplest globally $N = 1$ supersymmetric theory in four dimensions corresponding to $W(\Phi) = m\Phi^2/2 + g\Phi^3/3$. The total action is given by

$$S_{\text{WZ}} = \int d^4x \int d^4\theta \; \Phi^+\Phi - \int d^4x \left[\int d^2\theta \left[\frac{m}{2}\Phi^2 + \frac{g}{3}\Phi^3 \right] + \text{h. c} \right]. \tag{15.186}$$

As it turns out the component field F of Φ is auxiliary and therefore it can be set equal to its classical value. As a consequence the bosonic part of the Wess–Zumino model will only depend on the component field A and it reads

$$(S_{\text{WZ}})_{\text{bosonic}} = \int d^4x [\partial^\mu A^+ \partial_\mu A - V_F(A^+, A)]. \tag{15.187}$$

$$V_F(A^+, A) = F^+F = (mA + gA^2)^+(mA + gA^2). \tag{15.188}$$

This involves a quartic scalar interaction. Thus the Wess–Zumino model is the most general renormalizable globally $N = 1$ supersymmetric theory in four dimensions. Indeed the bosonic part of the more general model S reads

$$(S)_{\text{bosonic}} = \int d^4x \left[\partial^\mu A^+ \partial_\mu A - \left(\frac{\delta W}{\delta \Phi}\right)^+ \left(\frac{\delta W}{\delta \Phi}\right) \right]. \qquad (15.189)$$

The potential in this case will involve higher powers in A and thus S is not renormalizable in 4 dimensions.

A generalization of the action S can be constructed from a set of elementary chiral superfields Φ^I as

$$S = \int d^4x \int d^4\theta \; K(\Phi^{I+}, \Phi^J) - \int d^4x \left[\int d^2\theta \; W(\Phi^I) + \text{h. c} \right]. \qquad (15.190)$$

The Kahler potential $K(\Phi^{I+}, \Phi^J)$ is a vector superfield. This will always generate kinetic terms for the components fields, i.e., terms with no more than two derivatives. This is the most general globally $= 1$ supersymmetric theory in four dimensions (excluding higher derivative theories).

15.3.8 The $N = 1$ super Yang–Mills theory

Recall that the spinor superfield (also known as the gaugino superfield) W_α is given by (with $F_{\mu\nu} = \partial_\mu v_\nu - \partial_\nu v_\mu - i[v_\mu, v_\nu]/2$ and $D_\mu = \partial_\mu(\ldots) - i[v_\mu,\ldots]/2$)

$$W_\alpha = \lambda_\alpha + 2\theta_\alpha d + (\sigma^{\mu\nu})_\alpha{}^\beta \theta_\beta F_{\mu\nu} - i(\theta\theta)(\sigma^\mu)_{\alpha\dot\beta} D_\mu \bar\lambda^{\dot\beta}. \qquad (15.191)$$

The component fields $\bar\lambda$, λ, d and $F_{\mu\nu}$ are functions of y^μ. Since W_α is a chiral superfield, the square $W^\alpha W_\alpha$ is also a chiral superfield and as a consequence their F-terms will transform by total derivatives under supersymmetry transformations. The F-term of the square $W^\alpha W_\alpha$ is given by

$$[\text{Tr}\, W^\alpha W_\alpha]_F \theta\theta = \varepsilon^{\alpha\beta} \text{Tr}\left[-2i\lambda_\beta \cdot (\theta\theta)(\sigma^\mu)_{\alpha\dot\gamma} D_\mu \bar\lambda^{\dot\gamma} + 2\theta_\beta d \cdot 2\theta_\alpha d + 4\theta_\beta d \cdot (\sigma^{\mu\nu})_\alpha{}^\gamma \theta_\gamma F_{\mu\nu} \right.$$
$$\left. + (\sigma^{\mu\nu})_\beta{}^\rho \theta_\rho F_{\mu\nu} \cdot (\sigma^{\mu'\nu'})_\alpha{}^\gamma \theta_\gamma F_{\mu'\nu'} \right]. \qquad (15.192)$$

The third term is zero because $\theta\sigma^{\mu\nu}\theta = 0$. By using $\theta_\alpha \theta_\beta = \varepsilon_{\alpha\beta}\theta\theta/2$, $\hat\varepsilon = -\tilde\varepsilon$, $\tilde\varepsilon(\sigma^{\mu\nu})^T\tilde\varepsilon = \sigma^{\mu\nu}$ and

$$\text{tr}_2 \sigma^{\mu\nu}\sigma^{\mu'\nu'} = \frac{1}{2}(\eta^{\mu\mu'}\eta^{\nu\nu'} - \eta^{\mu\nu'}\eta^{\nu\mu'}) + \frac{i}{2}\varepsilon^{\mu\nu\mu'\nu'}, \qquad (15.193)$$

we compute

$$\varepsilon^{\alpha\beta}[(\sigma^{\mu\nu})_\beta{}^\rho \theta_\rho F_{\mu\nu} \cdot (\sigma^{\mu'\nu'})_\alpha{}^\gamma \theta_\gamma F_{\mu'\nu'}] = \frac{1}{2}\theta\theta \cdot \varepsilon^{\alpha\beta}(\sigma^{\mu\nu})_\beta{}^\rho \varepsilon_{\rho\gamma}(\sigma^{\mu'\nu'})_\alpha{}^\gamma \cdot F_{\mu\nu} F_{\mu'\nu'}$$
$$= -\frac{1}{2}\theta\theta \cdot \text{tr}_2 \sigma^{\mu\nu}\tilde\varepsilon(\sigma^{\mu'\nu'})^T \tilde\varepsilon \cdot F_{\mu\nu} F_{\mu'\nu'}$$
$$= -\frac{1}{2}\theta\theta \cdot \text{tr}_2 \sigma^{\mu\nu}\sigma^{\mu'\nu'} \cdot F_{\mu\nu} F_{\mu'\nu'} \qquad (15.194)$$
$$= -\frac{1}{4}\theta\theta[2F_{\mu\nu} F^{\mu\nu} + i\varepsilon^{\mu\nu\mu'\nu'}F_{\mu\nu} F_{\mu'\nu'}].$$

We get then

$$[\text{Tr } W^\alpha W_\alpha]_F = \text{Tr}\left[-2i\lambda\sigma^\mu D_\mu\bar\lambda + 4d^2 - \frac{1}{2}F_{\mu\nu}F^{\mu\nu} - \frac{i}{4}\varepsilon^{\mu\nu\alpha\beta}F_{\mu\nu}F_{\alpha\beta}\right]. \quad (15.195)$$

Let g and θ be the Yang–Mills gauge coupling and the Yang–Mills theta angle, respectively. We define the complex coupling τ by

$$\tau = \frac{4\pi i}{g^2} + \frac{\theta}{2\pi}. \quad (15.196)$$

The $N = 1$ super Yang–Mills action is then given by

$$S_{YM}^{N=1} = \frac{1}{8\pi}\int d^4x \ \text{Im}\left(\tau\int d^2\theta \ \text{Tr } W^\alpha W_\alpha\right)$$
$$= \frac{1}{8\pi}\int d^4x \ \text{Im}\left[\tau \ \text{Tr } W^\alpha W_\alpha\right]_F. \quad (15.197)$$

This is obviously invariant under the gauge transformation (the chiral superfield, which generates a finite gauge transformation is now denoted by Λ)

$$W_\alpha \longrightarrow e^{-\Lambda}W_\alpha e^{\Lambda}. \quad (15.198)$$

Indeed explicitly we have (using the fact that the fermion term is real including the factor of i and $\tilde F^{\mu\nu} = \varepsilon^{\mu\nu\alpha\beta}F_{\alpha\beta}/2$)

$$S_{YM}^{N=1} = \frac{1}{g^2}\int d^4x \ \text{Tr}\left[-\frac{1}{4}F_{\mu\nu}F^{\mu\nu} - i\lambda\sigma^\mu D_\mu\bar\lambda + 2d^2\right] - \frac{\theta}{32\pi^2}\int d^4x \ \text{Tr } F_{\mu\nu}\tilde F^{\mu\nu}. \quad (15.199)$$

Next, we add matter in the form of a chiral superfield Φ in some representation R of the gauge group. In other words, under a gauge transformation we must have

$$\Phi \longrightarrow e^{-\Lambda}\Phi, \quad \Phi^+ \longrightarrow \Phi^+ e^{-\Lambda^+}, \quad (15.200)$$

with $\Lambda = \Lambda_a T_R^a$ where T_R^a are the generators of the gauge group in the representation R. The precise meaning of these gauge transformations is

$$\Phi_a \longrightarrow (e^{-\Lambda})_a{}^b\Phi_b, \quad \Phi_a^+ \longrightarrow \Phi_b^+(e^{-\Lambda^+})_{ba}, \quad (15.201)$$

Recall that the vector superfield transforms under gauge transformations as

$$e^V \longrightarrow e^{\Lambda^+}e^V e^{\Lambda}. \quad (15.202)$$

Thus V must be a matrix in the representation R, viz $V = V_a T_R^a$. Properly speaking the vector superfield is actually V_a, which carries an adjoint representation index and is independent of the representation R.

It is then obvious that the combination $\Phi^+ e^V \Phi$ is invariant under gauge transformations, viz

$$\Phi^+ e^V \Phi \longrightarrow \Phi^+ e^V \Phi. \quad (15.203)$$

The $N = 1$ supersymmetric matter action can then be given by the gauge-covariant Wess–Zumino model, viz

$$S_{\text{matter}} = \int d^4x \int d^4\theta \; \Phi^+ e^V \Phi - \int d^4x \left[\int d^2\theta \left[\frac{m_{ab}}{2} \Phi_a \Phi_b + \frac{g_{abc}}{3} \Phi_a \Phi_b \Phi_c \right] + \text{h. c} \right]$$
$$= \int d^4x [\Phi^+ e^V \Phi]_d - \int d^4x \left[\frac{m_{ab}}{2} \Phi_a \Phi_b + \frac{g_{abc}}{3} \Phi_a \Phi_b \Phi_c + \text{h. c} \right]_F .$$

$$(15.204)$$

The mass and cubic terms are allowed by gauge invariance only if m_{ab} and g_{abc} are totally symmetric invariant tensors with respect to the gauge group. Equivalently m_{ab} and g_{abc} can only be (by gauge invariance) non-zero if the tensor products $R \times R$ and $R \times R \times R$, respectively, contain the singlet representation of the gauge group. Another way of stating this result is that one must contract the indices a, b and c correctly and properly in order to have gauge invariance.

The full $N = 1$ super Yang–Mills action with a chiral superfield in a representation R of the gauge group is therefore given by

$$S^{N=1} = \frac{1}{8\pi} \int d^4x \; \text{Im} \, [\tau \, \text{Tr} \, W^\alpha W_\alpha]_F + \int d^4x [\Phi^+ e^V \Phi]_d - \int d^4x [W(\Phi) + \text{h. c}]_F . \quad (15.205)$$

As before the component field F of Φ is auxiliary and therefore it can be set equal to its classical value. In fact the F-term contribution to the scalar potential of the field components A and A^+ remains unchanged, viz

$$V_F(A^+, A) = F^+ F = (mA + gA^2)^+(mA + gA^2). \quad (15.206)$$

As it turns out the component field d of the gaugino spinor superfield W_α is also an auxiliary field. Thus d can be set to its classical value and one ends up with the extra contribution to the scalar potential

$$V_D(A^+, A) = \frac{2}{g^2} d^2 = \frac{g^2}{8}[A, A^+]^2 . \quad (15.207)$$

The total potential is therefore

$$V(A^+, A) = V_F(A^+, A) + V_D(A^+, A) = F^+ F + \frac{2}{g^2} d^2 . \quad (15.208)$$

For renormalizability in $d = 4$ the superpotential $W(\Phi)$ must be given by at most a cubic potential. If we forget about renormalizability we can take $W(\Phi)$ to be a generic superpotential. A further generalization can be obtained by considering several flavors of chiral superfields Φ^I transforming in different representations R^I of the gauge group and also by replacing the kinetic term of the chiral superfields by a general Kahler potential. The most general $N = 1$ super Yang–Mills action with chiral superfields Φ^I in representations R^I is of the form

$$S^{N=1} = \frac{1}{8\pi} \int d^4x \; \text{Im} \, [\tau \, \text{Tr} f(\Phi^I) W^\alpha W_\alpha]_F + \int d^4x [K(\Phi^{I+} e^V, \Phi^J)]_d - \int d^4x [W(\Phi^I) + \text{h. c}]_F . \quad (15.209)$$

The holomorphic function $f(\Phi)$, which is called the gauge kinetic function, is assumed to transform under the symmetrized square of the adjoint representation. This simply means that $f(\phi)$ must transform like W_α, viz

$$f(\Phi^I) \longrightarrow e^{-\Lambda} f(\Phi^I) e^\Lambda. \tag{15.210}$$

The last remark concerns the so-called Fayet–Iliopoulos term. For an Abelian gauge theory, we observe that the d-term of the vector superfield V, which by construction transforms under supersymmetry transformations by a total derivative, transforms also by a total derivative under gauge transformations. Thus another gauge invariant $N = 1$ supersymmetric action can be given by

$$
\begin{aligned}
S_{FI} &= \int d^4x \ \xi \int d^4\theta V \\
&= \int d^4x \ \xi [V]_d.
\end{aligned}
\tag{15.211}
$$

Generalization to non-Abelian semi-simple gauge groups (groups which are products of simple groups and $U(1)$ factors) is straightforward.

15.4 $N = 2$ Supersymmetry

We introduce eight anticommuting Grassmannian coordinates θ^α, $\bar{\theta}_{\dot{\alpha}}$ and $\tilde{\theta}^\alpha$, $\bar{\tilde{\theta}}_{\dot{\alpha}}$. By the global $SU(2)$ R-symmetry the two Weyl spinors θ^α and $\tilde{\theta}^\alpha$ form a doublet in the fundamental representation of $SU(2)$ while $\bar{\theta}_{\dot{\alpha}}$ and $\bar{\tilde{\theta}}_{\dot{\alpha}}$ form a doublet in the anti-fundamental representation of $SU(2)$. There will be therefore four supercharges Q_α, $\bar{Q}_{\dot{\alpha}}$ and $\tilde{Q}_\alpha$, $\bar{\tilde{Q}}_{\dot{\alpha}}$ and four covariant derivatives D_α, $\bar{D}_{\dot{\alpha}}$ and $\tilde{D}_\alpha$, $\bar{\tilde{D}}_{\dot{\alpha}}$. The Berezin measure for $N = 2$ superspace is defined in obvious way by

$$\int d^2\theta d^2\bar{\theta} d^2\tilde{\theta} d^2\bar{\tilde{\theta}}. \tag{15.212}$$

We define a new bosonic coordinate $\tilde{y}^\mu$ by

$$\tilde{y}^\mu = x^\mu + i\theta\sigma^\mu\bar{\theta} + i\tilde{\theta}\sigma^\mu\bar{\tilde{\theta}}. \tag{15.213}$$

Clearly

$$\bar{D}_{\dot{\alpha}}\tilde{y}^\mu = \bar{\tilde{D}}_{\dot{\alpha}}\tilde{y}^\mu = 0. \tag{15.214}$$

A scalar superfield is the most general scalar function $\Psi(x, \theta, \bar{\theta}, \tilde{\theta}, \bar{\tilde{\theta}})$ on the $N = 2$ superspace. An $N = 2$ chiral superfield is an $N = 2$ scalar superfield $\Psi(x, \theta, \bar{\theta}, \tilde{\theta}, \bar{\tilde{\theta}})$ which is a singlet under the global $SU(2)$ R-symmetry and which satisfies the constraints

$$\bar{D}_{\dot{\alpha}}\Psi(x, \theta, \bar{\theta}, \tilde{\theta}, \bar{\tilde{\theta}}) = \bar{\tilde{D}}_{\dot{\alpha}}\Psi(x, \theta, \bar{\theta}, \tilde{\theta}, \bar{\tilde{\theta}}) = 0. \tag{15.215}$$

The most general solution to these constraints is $\Psi = \Psi(\tilde{y}, \theta, \tilde{\theta})$. Expanding in powers of $\tilde{\theta}$ we get

$$\Psi = \Phi(\tilde{y}, \theta) + \sqrt{2}\,\tilde{\theta}^{\alpha} W_{\alpha}(\tilde{y}, \theta) + \tilde{\theta}\tilde{\theta} G(\tilde{y}, \theta)$$
$$= \Phi(y + i\tilde{\theta}\sigma^{\mu}\bar{\tilde{\theta}}, \theta) + \sqrt{2}\,\tilde{\theta}^{\alpha} W_{\alpha}(y + i\tilde{\theta}\sigma^{\mu}\bar{\tilde{\theta}}, \theta) + \tilde{\theta}\tilde{\theta} G(y + i\tilde{\theta}\sigma^{\mu}\bar{\tilde{\theta}}, \theta). \tag{15.216}$$

It is obvious that Φ and G are $N = 1$ chiral superfields while W_{α} is a chiral spinor superfield. We compute the F term (with respect to $\tilde{\theta}$)

$$[\mathrm{Tr}\ \Psi^2]_{\tilde{\theta}\tilde{\theta}}\,\tilde{\theta}\tilde{\theta} = \mathrm{Tr}\ (W^{\alpha}W_{\alpha} + 2G\Phi)\tilde{\theta}\tilde{\theta}. \tag{15.217}$$

The $N = 2$ Yang–Mills action is therefore

$$S^{N=2} = \frac{1}{8\pi} \int d^4x\ \mathrm{Im}\,[\tau\ \mathrm{Tr}\ \Psi^2]_F$$
$$= \frac{1}{8\pi} \int d^4x\ \mathrm{Im}\left[\tau \int d^2\theta d^2\tilde{\theta}\,\mathrm{Tr}\ \Psi^2\right] \tag{15.218}$$
$$= \frac{1}{8\pi} \int d^4x\ \mathrm{Im}\left[\tau \int d^2\theta\ \mathrm{Tr}\ (W^{\alpha}W_{\alpha} + 2G\Phi)\right].$$

The $N = 2$ gauge multiplet is equivalent to the $N = 1$ gauge multiplet plus the $N = 1$ chiral multiplet all in the adjoint representation of the gauge group. The spinor component λ of W and the spinor component ψ of Φ form a doublet under $SU(2)_R$ whereas the gauge field v_{μ} and the auxiliary field d of W_{α} and the auxiliary field F of Φ transform as singlets under $SU(2)_R$.

The first term in equation (15.218) is precisely the $N = 1$ Yang–Mills action. The central point is that the above action must be invariant under $N = 1$ supersymmetry transformations considered in the previous subsections, and as a consequence we should think of the chiral superfield G as an auxiliary superfield. In other words, the superfield G must satisfy a particular constraint such that the second term in equation (15.218) becomes the action of a chiral superfield Φ in the adjoint representation of the gauge group (without a superpotential). By comparing with the first term of (15.204) we see that the chiral superfield G must be given by

$$G(y, \theta) = \frac{1}{2} \int d^2\bar{\theta}[\Phi(y - i\theta\sigma^{\mu}\bar{\theta}, \theta, \bar{\theta})]^{+}\ e^{V(y - i\theta\sigma^{\mu}\bar{\theta},\, \theta,\, \bar{\theta})}. \tag{15.219}$$

This can be deduced from the constraints on the $N = 2$ superfield Ψ given by (with $D^a = (D, \tilde{D})$ and $\bar{D}^a = (\bar{D}, \bar{\tilde{D}})$)

$$D^{a\alpha}D_{\alpha}^{b}\Psi = \bar{D}^{a\dot{\alpha}}\bar{D}_{\dot{\alpha}}^{b}\Psi^{+}. \tag{15.220}$$

The $N = 2$ Yang–Mills action becomes

$$S^{N=2} = \frac{1}{8\pi} \int d^4x\ \mathrm{Im}\left[\tau \int d^2\theta\ \mathrm{Tr}\ W^{\alpha}W_{\alpha} + \tau \int d^2\theta \int d^2\bar{\theta}\ \mathrm{Tr}\ \Phi^{+}e^{V}\Phi\right]. \tag{15.221}$$

15.5 Simple supersymmetry in more detail

In this section we will give a detailed and explicit calculation of the F- and d-terms of simple $N = 1$ supersymmetry. We will employ the notation of [2].

15.5.1 The F- and d-terms

The most general real scalar superfield is given by

$$S(x,\, \theta) = C(x) - i(\bar{\theta}\gamma_5\omega(x)) - \frac{1}{2}(\bar{\theta}\theta)N(x) - \frac{i}{2}(\bar{\theta}\gamma_5\theta)M(x) + \frac{i}{2}(\bar{\theta}\gamma_5\gamma_\mu\theta)V^\mu(x)$$
$$- i(\bar{\theta}\gamma_5\theta)\left(\bar{\theta}\left(\lambda(x) + \frac{1}{2}\gamma_\mu\partial^\mu\omega(x)\right)\right) - \frac{1}{4}(\bar{\theta}\gamma_5\theta)^2\left(D(x) + \frac{1}{2}\Delta C(x)\right).$$

$\tag{15.222}$

C, M, N and D are real scalar fields, V_μ is a real vector field, ω and λ are Majorana spinors satisfying $s^* = -\beta\varepsilon\gamma_5 s$.

A four-component Majorana spinor s is formed from a $(1/2, 0)$ two-component spinor u as $s = (u, -eu^*)$. It can be formed from a $(0, 1/2)$ two-component spinor v as $s = (ev^*, v)$. The matrix e is defined by $e = i\sigma_2 = -e^T$. We note that under supersymmetry transformations only the d-component of S transforms in a trivial way, i.e., it changes by a total derivative. A supersymmetric action can thus be constructed as the integral over spacetime of the d-term of a general real scalar superfield.

A chiral superfield is a reduced superfield obtained by setting

$$\lambda = D = 0, \quad V_\mu = \partial_\mu Z \tag{15.223}$$

The corresponding supersymmetry transformations (with the infinitesimal Majorana four-component constant spinor α as a parameter) are given by

$$\omega \longrightarrow \omega' = \omega + (-i\gamma_5\gamma_\mu\partial^\mu C + i\gamma_5 N - M + \gamma_\mu\partial^\mu Z)\alpha$$
$$C \longrightarrow C' = C + i\bar{\alpha}\gamma_5\omega$$
$$N \longrightarrow N' = N + i\bar{\alpha}\gamma_5(\gamma_\mu\partial^\mu\omega) \tag{15.224}$$
$$M \longrightarrow M' = M - \bar{\alpha}(\gamma_\mu\partial^\mu\omega)$$
$$V_\mu \longrightarrow Z' = Z + \bar{\alpha}\omega.$$

A chiral superfield X can also be given by $X = (\Phi + \tilde{\Phi})/\sqrt{2}$ where Φ and $\tilde{\Phi}$ are left-chiral and right-chiral superfields, respectively. If the chiral superfield X is real then $\tilde{\Phi} = \Phi^+$. In general the left-chiral superfield Φ and the right-chiral superfield $\tilde{\Phi}$ of a chiral superfield X are completely independent superfields. However, any right-chiral superfield is the complex conjugate of a left-chiral superfield and thus there is no any loss of generality in considering only left-chiral superfields.

A left-chiral superfield can be written in two equivalent ways. The first is

$$\Phi(x,\, \theta) = \phi(x) - \sqrt{2}(\bar{\theta}\psi_L(x)) + \mathcal{F}(x)\left(\bar{\theta}\frac{1 + \gamma_5}{2}\theta\right) + \frac{1}{2}(\bar{\theta}\gamma_5\gamma_\mu\theta)\partial^\mu\phi(x)$$
$$- \frac{1}{\sqrt{2}}(\bar{\theta}\gamma_5\theta)(\bar{\theta}\gamma_\mu\partial^\mu\psi_L(x)) - \frac{1}{8}(\bar{\theta}\gamma_5\theta)^2\Delta\phi(x).$$

$\tag{15.225}$

The second is

$$\Phi(x,\, \theta) = \phi(x_+) - \sqrt{2}\left(\theta_L^T\varepsilon\ \psi_L(x_+)\right) + \mathcal{F}(x_+)\left(\theta_L^T\varepsilon\theta_L\right). \tag{15.226}$$

We have introduced the coordinates $x_{\pm}^{\mu} = x^{\mu} \pm \frac{1}{2}\bar{\theta}\gamma_5\gamma^{\mu}\theta$ and for the Majorana spinor θ we have used the identity $\bar{\theta} = \theta^T \varepsilon \gamma_5$ where ε is the matrix $\varepsilon = -\gamma_2\beta\gamma_5$. Let us also recall the following simple supersymmetry transformations

$$\psi_L \longrightarrow \psi'_L = \psi_L + \sqrt{2}(\gamma_\mu \partial^\mu \phi \alpha_R + \mathcal{F}\alpha_L)$$
$$\mathcal{F} \longrightarrow \mathcal{F}' = \mathcal{F} + \sqrt{2}\,\bar{\alpha}_L \gamma_\mu \partial^\mu \psi_L \qquad (15.227)$$
$$\phi \longrightarrow \phi' = \phi + \sqrt{2}\,\bar{\alpha}_R \psi_L.$$

The Kahler potential K is an arbitrary real scalar superfield. The kinetic part of a supersymmetric action is given by the d-term of K

$$\frac{1}{2}\int d^4x [K]_d. \qquad (15.228)$$

K is a function of elementary left-chiral superfields Φ_i, their complex conjugates Φ_i^+ and their superderivatives and spacetime derivatives. For a renormalizable theory K can at most be quadratic in the superfields Φ_i and Φ_i^+ of the form $K = g_{ij}\Phi_i^+\Phi_j$. In this case K does not depend on the superderivatives and spacetime derivatives. For a single left-chiral superfield we have

$$K = \Phi^+\Phi \qquad (15.229)$$

We note also that under supersymmetry transformations the F-component of the left-chiral superfield Φ changes by a total derivative. Another supersymmetric action can thus be constructed as the integral over spacetime of the real part of the F-term of a general left-chiral superfield f. In general f is a function of elementary left-chiral superfields Φ_i but not their superderivatives or spacetime derivatives. It is known as the superpotential. Remark that f does not depend on Φ_i^+ nor on its superderivatives and spacetime derivative. The corresponding supersymmetric action for a single left-chiral superfield Φ is

$$\frac{1}{2}\int d^4x [f(\Phi)]_F + \frac{1}{2}\int d^4x [f(\Phi)]_F^*. \qquad (15.230)$$

15.5.2 Extended gauge symmetry

Gauge transformations leave the simple supersymmetry generator Q invariant. Indeed, since Q is a Majorana spinor it can only transform in the trivial representation of any semi-simple gauge group. Gauge transformations act on the left-chiral superfield Φ given by equation (15.226) in the usual way

$$\Phi(x, \theta) \longrightarrow \Phi'(x, \theta) = e^{it_A\Lambda^A(x_+)}\Phi(x, \theta). \qquad (15.231)$$

In the above equation t_A are the Hermitian generators of the gauge group and Λ^A are left-chiral superfields with no $\psi_L(x_+)$- and $\mathcal{F}(x_+)$-components. Remark that the functions Λ^A depend on x_+ and not directly on the coordinates x and hence we can conclude that the components $\phi(x_+)$, $\psi_L(x_+)$, $\mathcal{F}(x_+)$ of the left-chiral superfield Φ must transform precisely in the same way as in (15.231).

Let us remark that the superpotential term is gauge invariant if and only if it is invariant under the (global) gauge transformation (15.231) with Λ independent of x_+. Gauge superfields are needed explicitly only in the construction of gauge-invariant Kähler potential. In these terms the fields Φ and Φ^+ transform as $\Phi'(x, \theta) = e^{it_A\Lambda^A(x_+)}\Phi(x, \theta)$ and $\Phi'^+(x, \theta) = \Phi^+(x, \theta)e^{-it_A\Lambda^A(x_-)}$. We have used the fact that $x_+^* = x_-$ (since θ is a Majorana spinor) and hence $(\Lambda^A(x_+))^+ = \Lambda^A(x_-)$. Because x_+^μ and x_-^μ are different the combination $\Phi^+\Phi$ is not gauge invariant. We introduce therefore a gauge connection $\Gamma(x, \theta)$ with the transformation rule

$$\Gamma(x, \theta) \longrightarrow \Gamma'(x, \theta) = e^{it_A\Lambda^A(x_-)}\Gamma(x, \theta)e^{-it_A\Lambda^A(x_+)}. \tag{15.232}$$

We immediately notice that the field $\Phi^+\Gamma$ transforms in exactly the desired way, viz

$$\Phi^+(x, \theta)\Gamma(x, \theta) \longrightarrow \Phi'^+(x, \theta)\Gamma'(x, \theta) = \Phi^+(x, \theta)\Gamma(x, \theta)e^{-it_A\Lambda^A(x_+)}. \tag{15.233}$$

Hence the combination $\Phi^+\Gamma\Phi$ is precisely gauge invariant. This is the gauge invariant version of $\Phi^+\Phi$. In fact $\Phi^+\Gamma\Phi$ will also be invariant under the following extended gauge transformations

$$\Phi(x, \theta) \longrightarrow \Phi'(x, \theta) = e^{it_A\Omega^A(x, \theta)}\Phi(x, \theta)$$
$$\Gamma(x, \theta) \longrightarrow \Gamma'(x, \theta) = e^{it_A\Omega^A(x, \theta)^+}\Gamma(x, \theta)e^{-it_A\Omega^A(x, \theta)}. \tag{15.234}$$

Ω^A are general left-chiral superfields. The choice of the connection Γ is not unique. Given a left-chiral superfield Σ with the transformation property $\Sigma'(x, \theta) = e^{it_A\Lambda^A(x_+)}\Sigma(x, \theta)e^{-it_A\Lambda^A(x_+)}$, we can easily show that $\Gamma\Sigma$ is also a connection. We use this freedom to choose Γ to be Hermitian of the form

$$\Gamma(x, \theta) = e^{-2t_A V^A(x, \theta)}. \tag{15.235}$$

The gauge superfields $V^A(x, \theta)$ are real superfields, which by construction do not depend on the representation t_A of the gauge group.

By using equation (15.235) in the second equation of (15.234) we obtain

$$e^{-2t_A V'^A(x, \theta)} = e^{it_A\Omega^A(x, \theta)^+}e^{-2t_A V^A(x, \theta)}e^{-it_A\Omega^A(x, \theta)}. \tag{15.236}$$

The extended gauge transformations of the gauge superfields $V^A(x, \theta)$ are given by

$$V'^A(x, \theta) = V^A(x, \theta) + \frac{i}{2}(\Omega^A(x, \theta) - \Omega^A(x, \theta)^+) + \cdots \tag{15.237}$$

The gauge superfields $V^A(x, \theta)$ are real superfields of the form (15.222), namely

$$V^A(x, \theta) = C^A(x) - i(\bar{\theta}\gamma_5\omega^A(x)) - \frac{1}{2}(\bar{\theta}\theta)N^A(x) - \frac{i}{2}(\bar{\theta}\gamma_5\theta)M^A(x)$$
$$+ \frac{i}{2}(\bar{\theta}\gamma_5\gamma^\mu\theta)V^A_\mu(x) - i(\bar{\theta}\gamma_5\theta)\left(\bar{\theta}\left(\lambda^A(x) + \frac{1}{2}\gamma_\mu\partial^\mu\omega^A(x)\right)\right) \tag{15.238}$$
$$- \frac{1}{4}(\bar{\theta}\gamma_5\theta)^2\left(D^A(x) + \frac{1}{2}\Delta C^A(x)\right).$$

We can use the extended gauge transformations (15.237) to put the gauge superfields $V^A(x, \theta)$ into a simple form known as the Wess–Zumino gauge obtained by setting $C^A = \omega^A = N^A = M^A = 0$, i.e.,

$$V^A(x, \theta) = \frac{i}{2}(\bar\theta\gamma_5\gamma^\mu\theta)V_\mu^A(x) - i(\bar\theta\gamma_5\theta)(\bar\theta\lambda^A(x)) - \frac{1}{4}(\bar\theta\gamma_5\theta)^2 D^A(x). \tag{15.239}$$

The supersymmetry transformations for the components of the gauge superfields $V^A(x, \theta)$ become

$$V_\mu^A \longrightarrow V_\mu'^A = V_\mu^A + \bar\alpha\gamma_\mu\lambda^A$$

$$\lambda^A \longrightarrow \lambda'^A = \lambda^A + \left(i\gamma_5 D^A + \frac{1}{2}[\partial_\mu\gamma^\nu V_\nu^A,\, \gamma^\mu]\right)\alpha \tag{15.240}$$

$$D^A \longrightarrow D'^A = D^A + i\bar\alpha\gamma_5\gamma_\mu\partial^\mu\lambda^A.$$

And

$$
\begin{aligned}
C^A = 0 &\longrightarrow C'^A = 0 \\
\omega^A = 0 &\longrightarrow \omega'^A = \gamma^\mu V_\mu^A \alpha \\
N^A = 0 &\longrightarrow N'^A = i\bar\alpha\gamma_5\lambda^A \\
M^A = 0 &\longrightarrow M'^A = -\bar\alpha\lambda^A.
\end{aligned}
\tag{15.241}
$$

We can immediately notice that under supersymmetry the three conditions $\omega^A = N^A = M^A = 0$ of the Wess–Zumino gauge are not stable. We conclude that the Wess–Zumino gauge is not supersymmetric. Thus in the Wess–Zumino gauge the action is not invariant under both extended gauge transformations and supersymmetry transformations. However, the action is still invariant under supersymmetry transformations, which take us out of the Wess–Zumino gauge, followed by particular extended gauge transformations, which take us back to the Wess–Zumino gauge.

We get ordinary gauge transformations from extended gauge transformations when Ω^A are taken to be left-chiral superfields without the $\psi_L(x)$- and $\mathcal{F}(x)$-components. In particular infinitesimal ordinary gauge transformations are left-chiral superfields Ω^A with $\phi(x_+)$-components given by infinitesimal real functions $\Lambda^A(x_+)$. Thus

$$\Omega^A(x_+) = \Lambda^A(x) + \frac{1}{2}(\bar\theta\gamma_5\gamma_\mu\theta)\partial^\mu\Lambda^A(x) - \frac{1}{8}(\bar\theta\gamma_5\theta)^2\Delta\Lambda^A(x) \equiv \Lambda^A(x_+) \tag{15.242}$$

The transformation law (15.236) becomes

$$e^{-2t_A V'^A(x,\,\theta)} = e^{it_A\Lambda^A(x_+)*}e^{-2t_A V^A(x,\,\theta)}e^{-it_A\Lambda^A(x_+)}. \tag{15.243}$$

In the Wess–Zumino gauge the gauge superfield $V^A(x, \theta)$ is of the form (15.239). We will use the identity

$$e^a e^X e^b = e^{(X + L_X(b-a) + L_X \coth L_X(b+a) + \cdots)} \tag{15.244}$$

where quadratic and higher order terms in a and/or b are neglected and $L_X f = \frac{1}{2}[X, f]$. We have

$$X = -2t_A V^A(x, \theta) = -2t_A\left(\frac{i}{2}(\bar\theta\gamma_5\gamma^\mu\theta)V_\mu^A(x) - i(\bar\theta\gamma_5\theta)(\bar\theta\lambda^A(x)) - \frac{1}{4}(\bar\theta\gamma_5\theta)^2 D^A(x)\right). \quad (15.245)$$

$$a = it_A\Lambda^A(x_+)^* = it_A\left(\Lambda^A(x) - \frac{1}{2}(\bar\theta\gamma_5\gamma^\mu\theta)\partial_\mu\Lambda^A - \frac{1}{8}(\bar\theta\gamma_5\theta)^2\Delta\Lambda^A(x)\right) \quad (15.246)$$

$$b = -it_A\Lambda^A(x_+) = -it_A\left(\Lambda^A(x) + \frac{1}{2}(\bar\theta\gamma_5\gamma^\mu\theta)\partial_\mu\Lambda^A - \frac{1}{8}(\bar\theta\gamma_5\theta)^2\Delta\Lambda^A(x)\right). \quad (15.247)$$

We can easily compute

$$a + b = -it_A(\bar\theta\gamma_5\gamma^\mu\theta)\partial_\mu\Lambda^A$$
$$a - b = 2it_A\left(\Lambda^A(x) - \frac{1}{8}(\bar\theta\gamma_5\theta)^2\Delta\Lambda^A(x)\right). \quad (15.248)$$

Let us recall that $\theta = \theta_L + \theta_R$ where θ_L and θ_R are two-dimensional Majorana spinors. Hence we compute $\bar\theta\gamma_5\gamma^\mu\theta = \bar\theta_L\gamma_5\gamma^\mu\theta_R + \bar\theta_R\gamma_5\gamma^\mu\theta_L$, $(\bar\theta\gamma_5\theta)\bar\theta = (\bar\theta_L\gamma_5\theta_L)\bar\theta_R + (\bar\theta_R\gamma_5\theta_R)\bar\theta_L$ and $(\bar\theta\gamma_5\theta)^2 = 2(\bar\theta_L\gamma_5\theta_L)(\bar\theta_R\gamma_5\theta_R)$. We remark that (1) every term in X has at least one factor of θ_L and at least one factor of θ_R, (2) the sum $a + b$ is proportional to the product of one θ_L and one θ_R. It is therefore evident that $L_X^2(a + b) = 0$ and hence $L_X \coth L_X(a + b) = a + b = -it_A(\bar\theta\gamma_5\gamma^\mu\theta)\partial_\mu\Lambda^A$. Next we compute $L_X(b - a) = \frac{1}{2}[X, b - a]$. Again every term in X has at least one factor of θ_L and at least one factor of θ_R whereas the second term in $b - a$ is proportional to the product of two θ_Ls and two θ_Rs. Hence the second term in $b - a$ will not contribute and only the first term will contribute. Explicitly we compute $L_X(b - a) = -2C_{AB}^C t_C V^A\Lambda^B$. The structure constants are defined by $[t_A, t_B] = iC_{AB}^C t_C$. From the transformation law (15.243) we get the final result

$$V'^C = V^C + C_{AB}^C V^A\Lambda^B + \frac{i}{2}(\bar\theta\gamma_5\gamma^\mu\theta)\partial_\mu\Lambda^C. \quad (15.249)$$

Remark that under ordinary gauge transformations the Wess–Zumino gauge is stable. Indeed, the gauge transformed superfield V'^C given by (15.249) is of the same form as the superfield V^C, in other words it is of the form (15.239). In terms of the components we have

$$V'^C_\mu = V^C_\mu + C_{AB}^C V^A_\mu\Lambda^B + \partial_\mu\Lambda^C$$
$$\lambda'^C = \lambda^C + C_{AB}^C \lambda^A\Lambda^B \quad (15.250)$$
$$D'^C = D^C + C_{AB}^C D^A\Lambda^B.$$

The first equation is precisely the usual gauge transformation rule $V'_\mu = gV_\mu g^+ + ig\partial_\mu g^+$ where the gauge group element g is related to the gauge parameter Λ by $g = 1 + i\Lambda$. The second and third equation indicate that the fields χ^C and D^C are matter fields, which transform in the adjoint representation of the gauge group. The Majorana field χ^C is known as the gaugino field whereas D^C will turn out to be an auxiliary field.

15.5.3 Calculation of the F-term in the Abelian theory

To construct $N = 1$ supersymmetric $U(1)$ pure gauge action the starting point is the spinor superfield

$$W_\alpha(x, \theta) = \frac{i}{4}(\mathcal{D}^T \varepsilon \mathcal{D})\mathcal{D}_\alpha V(x, \theta). \tag{15.251}$$

The real superfield $V(x, \theta)$ represents the vector gauge superfield. The superderivatives are defined by

$$\mathcal{D}_\alpha = (\gamma_5\varepsilon)_{\alpha\beta}\frac{\partial}{\partial\theta_\beta} - (\gamma^\mu\theta)_\alpha\partial_\mu = -\frac{\partial}{\partial\bar\theta_\alpha} - (\gamma^\mu\theta)_\alpha\partial_\mu$$

$$\bar{\mathcal{D}}_\beta = \mathcal{D}_\gamma(\gamma_5\varepsilon)_{\gamma\beta} = \frac{\partial}{\partial\theta_\beta} + (\gamma_5\varepsilon\gamma^\mu\theta)_\beta\partial_\mu. \tag{15.252}$$

They satisfy

$$\{\mathcal{D}_\alpha, \bar{\mathcal{D}}_\beta\} = -2\gamma^\mu_{\alpha\beta}\partial_\mu. \tag{15.253}$$

The chirality matrix γ_5 is defined by $\gamma_5 = i\gamma_0\gamma_1\gamma_2\gamma_3 = \mathrm{diag}(1, 1, -1, -1)$. The matrix ε is defined by $\varepsilon = \mathrm{diag}(e, e)$, $e = i\sigma_2$. We have $\varepsilon^2 = -1$, $\gamma_5^2 = 1$, $\gamma_5\varepsilon = \varepsilon\gamma_5$, $\varepsilon^T = -\varepsilon$, $\gamma_5^T = \gamma_5$. In the above equation we have also used the fact that θ is a Majorana spinor, viz $\bar\theta_\alpha = (\gamma_5\varepsilon)_{\beta\alpha}\theta_\beta$. We can compute

$$\mathcal{D}_\alpha = \mathcal{D}_{L\alpha} + \mathcal{D}_{R\alpha}$$
$$\mathcal{D}^T\varepsilon\mathcal{D} = \mathcal{D}_L^T\varepsilon\mathcal{D}_L + \mathcal{D}_R^T\varepsilon\mathcal{D}_R. \tag{15.254}$$

$$\mathcal{D}_{L\alpha} = \left(\frac{1 + \gamma_5}{2}\mathcal{D}\right)_\alpha = \varepsilon_{\alpha\beta}\frac{\partial}{\partial\theta_{L\beta}} - (\gamma^\mu\theta_R)_\alpha\partial_\mu$$

$$\mathcal{D}_{R\alpha} = \left(\frac{1 - \gamma_5}{2}\mathcal{D}\right)_\alpha = -\varepsilon_{\alpha\beta}\frac{\partial}{\partial\theta_{R\beta}} - (\gamma^\mu\theta_L)_\alpha\partial_\mu. \tag{15.255}$$

A left-chiral superfield Φ must satisfy $\mathcal{D}_{R\alpha}\Phi = 0$ whereas a right-chiral superfield $\tilde\Phi$ must satisfy $\mathcal{D}_{L\alpha}\tilde\Phi = 0$. We remark that $\mathcal{D}_{R\alpha}x_+^\mu = 0$ and $\mathcal{D}_{L\alpha}x_-^\mu = 0$. By using $\gamma_\mu^T = -C\gamma_\mu C^{-1}$ where $C = -\gamma_5\varepsilon$ and $C^{-1} = \varepsilon\gamma_5$ we can show after a long calculation that

$$\{\mathcal{D}_\alpha, \mathcal{D}_\beta\} = 2(\gamma^\mu\varepsilon\gamma_5)_{\alpha\beta}\partial_\mu. \tag{15.256}$$

We can also verify after some algebra that

$$\{\mathcal{D}_{L\alpha}, \mathcal{D}_{L\beta}\} = \{\mathcal{D}_{R\alpha}, \mathcal{D}_{R\beta}\} = 0$$
$$\{\mathcal{D}_{L\alpha}, \mathcal{D}_{R\beta}\} = -2(\gamma^\mu \varepsilon)_{\alpha\beta} \partial_\mu \tag{15.257}$$
$$\{\mathcal{D}_{R\alpha}, \mathcal{D}_{L\beta}\} = 2(\gamma^\mu \varepsilon)_{\alpha\beta} \partial_\mu.$$

Therefore

$$(\mathcal{D}^T \varepsilon \mathcal{D})\mathcal{D} = (\mathcal{D}_L^T \varepsilon \mathcal{D}_L)\mathcal{D}_R + (\mathcal{D}_R^T \varepsilon \mathcal{D}_R)\mathcal{D}_L. \tag{15.258}$$

This follows partially from the fact that there are only two independent left superderivatives, which anticommute and only two independent right superderivatives, which also anticommute. Hence the spinor superfield W can be decomposed as follows

$$W_\alpha = W_{L\alpha} + W_{R\alpha}. \tag{15.259}$$

$$W_{L\alpha} = \frac{i}{4}(\mathcal{D}_R^T \varepsilon \mathcal{D}_R)\mathcal{D}_{L\alpha} V, \quad W_{R\alpha} = \frac{i}{4}(\mathcal{D}_L^T \varepsilon \mathcal{D}_L)\mathcal{D}_{R\alpha} V. \tag{15.260}$$

In other words, the spinor superfield W is chiral. Extended Abelian gauge transformations are given by

$$V(x, \theta) \longrightarrow V(x, \theta) + \delta V(x, \theta), \quad \delta V(x, \theta) = \frac{i}{2}(\Omega(x, \theta) - \Omega^+(x, \theta)). \tag{15.261}$$

$\Omega(x, \theta)$ is an arbitrary left-chiral superfield so it is a function of x_+ and θ_L. Because Ω is a left-chiral superfield (viz $\mathcal{D}_R \Omega = \mathcal{D}_L \Omega^+ = 0$) we have $W_{L\alpha} \longrightarrow W_{L\alpha} + \delta W_{L\alpha}$ where

$$\delta W_{L\alpha} = \frac{i}{4}(\mathcal{D}_R^T \varepsilon \mathcal{D}_R)\mathcal{D}_{L\alpha}\left(\frac{i}{2}\Omega\right)$$
$$= -\frac{1}{8}[\varepsilon_{\mu\nu}\mathcal{D}_{R\mu}\mathcal{D}_{R\nu}, \mathcal{D}_{L\alpha}]\Omega. \tag{15.262}$$

But we compute

$$[\varepsilon_{\mu\nu}\mathcal{D}_{R\mu}\mathcal{D}_{R\nu}, \mathcal{D}_{L\alpha}] = -4(\gamma^\mu \partial_\mu \mathcal{D}_R)_\alpha. \tag{15.263}$$

Hence $\delta W_{L\alpha} = 0$. Similarly, we can show that $\delta W_{R\alpha} = 0$. Therefore the spinor superfield W_α is gauge invariant. In the Wess–Zumino gauge the gauge superfield V takes the form

$$V = \frac{i}{2}(\bar{\theta}\gamma_5 \gamma^\mu \theta) V_\mu(x) - i(\bar{\theta}\gamma_5 \theta)(\bar{\theta}\lambda(x)) - \frac{1}{4}(\bar{\theta}\gamma_5 \theta)^2 D(x). \tag{15.264}$$

We need to compute

$$W_\alpha = \frac{i}{4}\varepsilon_{\mu\nu}\mathcal{D}_\mu \mathcal{D}_\nu \mathcal{D}_\alpha V. \tag{15.265}$$

This is a very long calculation. We first compute

$$\mathcal{D}_\alpha V = (\gamma_5 \varepsilon)_{\alpha\beta} \frac{\partial V}{\partial \theta_\beta} - (\gamma^\mu \theta)_\alpha \partial_\mu V. \tag{15.266}$$

$$\mathcal{D}_\nu \mathcal{D}_\alpha V = (\gamma_5\varepsilon)_{\nu\nu'}(\gamma_5\varepsilon)_{\alpha\alpha'}\frac{\partial^2 V}{\partial\theta_{\nu'}\partial\theta_{\alpha'}} + (\gamma^\sigma\varepsilon\gamma_5)_{\alpha\nu}\partial_\sigma V + (\gamma_5\varepsilon)_{\nu\rho}(\gamma^\sigma\theta)_\alpha\frac{\partial}{\partial\theta_\rho}\partial_\sigma V$$
$$- (\gamma^{\nu'}\theta)_\nu(\gamma_5\varepsilon)_{\alpha\alpha'}\partial_{\nu'}\frac{\partial V}{\partial\theta_{\alpha'}} + (\gamma^{\nu'}\theta)_\nu(\gamma^{\alpha'}\theta)_\alpha\partial_{\nu'}\partial_{\alpha'}V. \tag{15.267}$$

We act with one more superderivative $\mathcal{D}_\mu$ and contract with $\varepsilon_{\mu\nu}$ to get

$$\varepsilon_{\mu\nu}\mathcal{D}_\mu\mathcal{D}_\nu\mathcal{D}_\alpha V = 2\varepsilon_{\mu\nu}(\gamma^{\alpha'}\varepsilon\gamma_5)_{\alpha\mu}(\gamma^{\nu'}\theta)_\nu\partial_{\nu'}\partial_{\alpha'}V$$
$$+ 2\varepsilon_{\mu\nu}(\gamma_5\varepsilon)_{\mu\mu'}(\gamma^{\nu'}\theta)_\nu(\gamma^{\alpha'}\theta)_\alpha\partial_{\nu'}\partial_{\alpha'}\frac{\partial V}{\partial\theta_{\mu'}}$$
$$+ 2\varepsilon_{\mu\nu}(\gamma_5\varepsilon)_{\mu\mu'}(\gamma_5\varepsilon)_{\alpha\alpha'}(\gamma^{\nu'}\theta)_\nu\partial_{\nu'}\frac{\partial^2 V}{\partial\theta_{\mu'}\partial\theta_{\alpha'}}$$
$$+ 2\varepsilon_{\mu\nu}(\gamma^{\nu'}\varepsilon\gamma_5)_{\alpha\nu}(\gamma_5\varepsilon)_{\mu\mu'}\partial_{\nu'}\frac{\partial V}{\partial\theta_{\mu'}} \tag{15.268}$$
$$+ \varepsilon_{\mu\nu}(\gamma_5\varepsilon)_{\mu\mu'}(\gamma_5\varepsilon)_{\nu\nu'}(\gamma_5\varepsilon)_{\alpha\alpha'}\frac{\partial^3 V}{\partial\theta_{\mu'}\partial\theta_{\nu'}\partial\theta_{\alpha'}}$$
$$+ \varepsilon_{\alpha'\nu}(\gamma^{\mu'}\theta)_\alpha\partial_{\mu'}\frac{\partial^2 V}{\partial\theta_{\nu'}\partial\theta_{\alpha'}}$$
$$+ \varepsilon_{\mu\nu}(\gamma_5\varepsilon)_{\alpha\alpha'}(\gamma^{\mu'}\theta)_\mu(\gamma^{\nu'}\theta)_\nu\partial_{\mu'}\partial_{\nu'}\frac{\partial V}{\partial\theta_{\alpha'}}.$$

The different terms are computed separately as follows

$$2\varepsilon_{\mu\nu}(\gamma^{\alpha'}\varepsilon\gamma_5)_{\alpha\mu}(\gamma^{\nu'}\theta)_\nu\partial_{\nu'}\partial_{\alpha'}V = 2(\gamma_5\theta)_\alpha\partial^2 V$$
$$= \frac{i}{2}(\theta^T\varepsilon\theta)^2\partial^2\lambda_\alpha - i(\bar\theta\gamma_5\theta)(\gamma_5\gamma^\mu\theta)_\alpha\partial^2 V_\mu. \tag{15.269}$$

$$2\varepsilon_{\mu\nu}(\gamma_5\varepsilon)_{\mu\mu'}(\gamma^{\nu'}\theta)_\nu(\gamma^{\alpha'}\theta)_\alpha\partial_{\nu'}\partial_{\alpha'}\frac{\partial V}{\partial\theta_{\mu'}} = 2(\gamma_5\gamma^{\nu'}\theta)_{\mu'}(\gamma^{\alpha'}\theta)_\alpha\partial_{\nu'}\partial_{\alpha'}\frac{\partial V}{\partial\theta_{\mu'}}$$
$$= -2i(\bar\theta\gamma_5\theta)(\gamma^\nu\gamma_5\theta)_\alpha\partial^\mu\partial_\nu V_\mu - 2i(\bar\theta\gamma_5\gamma^\nu\theta)\theta_\alpha(\bar\theta\gamma_5\gamma^\mu\partial_\mu\partial_\nu\lambda)$$
$$= i(\theta^T\varepsilon\theta)(\gamma_5\gamma^\mu\gamma^\nu\gamma^\sigma\theta)_\alpha\partial_\sigma F_{\mu\nu} + 2i(\theta^T\varepsilon\theta)(\gamma_5\gamma^\mu\theta)_\alpha\partial^2 V_\mu$$
$$- 2i(\bar\theta\gamma_5\gamma^\nu\theta)\theta_\alpha(\bar\theta\gamma_5\gamma^\mu\partial_\mu\partial_\nu\lambda) \tag{15.270}$$
$$= i(\theta^T\varepsilon\theta)(\gamma_5\gamma^\mu\gamma^\nu\gamma^\sigma\theta)_\alpha\partial_\sigma F_{\mu\nu} + 2i(\theta^T\varepsilon\theta)(\gamma_5\gamma^\mu\theta)_\alpha\partial^2 V_\mu$$
$$- \frac{i}{2}(\bar\theta\gamma_5\theta)^2(\partial^2\lambda)_\alpha.$$

In the third line we have used the Bianchi identity whereas in the last line we have used, among other things, the identity $(\gamma^\nu\theta)_\alpha(\bar\theta\gamma_5\theta) = -\theta_\alpha(\bar\theta\gamma_5\gamma^\nu\theta)$.

We also compute

$$2\varepsilon_{\mu\nu}(\gamma^{\nu'}\varepsilon\gamma_5)_{\alpha\nu}(\gamma_5\varepsilon)_{\mu\mu'}\partial_{\nu'}\frac{\partial V}{\partial\theta_{\mu'}} = 2(\gamma^{\nu'}\varepsilon)_{\alpha\mu'}\partial_{\nu'}\frac{\partial V}{\partial\theta_{\mu'}}$$

$$= -i(\gamma^\mu\gamma^\nu\theta)_\alpha F_{\mu\nu} - 2i\theta_\alpha\partial^\mu V_\mu + 4i(\gamma^\mu\theta)_\alpha(\theta^T\varepsilon\gamma_5\partial_\mu\lambda)$$

$$+ 2i(\theta^T\varepsilon\theta)(\gamma^\mu\gamma_5\partial_\mu\lambda)_\alpha + 2(\gamma^\mu\theta)_\alpha(\theta^T\varepsilon\theta)\partial_\mu D. \tag{15.271}$$

$$2\varepsilon_{\mu\nu}(\gamma_5\varepsilon)_{\mu\mu'}(\gamma_5\varepsilon)_{\alpha\alpha'}(\gamma^{\nu'}\theta)_\nu\partial_{\nu'}\frac{\partial^2 V}{\partial\theta_{\mu'}\partial\theta_{\alpha'}} = 2(\gamma_5\varepsilon)_{\alpha\alpha'}(\gamma_5\gamma^{\mu'}\theta)_\nu\partial_{\mu'}\frac{\partial^2 V}{\partial\theta_{\nu'}\partial\theta_{\alpha'}}$$

$$= -i(\gamma^\mu\gamma^\nu\theta)_\alpha F_{\mu\nu} + 2i\theta_\alpha\partial^\mu V_\mu + 4i(\gamma^\mu\theta)_\alpha(\theta^T\varepsilon\gamma_5\partial_\mu\lambda)$$

$$+ 4i(\gamma_5\theta)_\alpha(\theta^T\varepsilon\gamma^\mu\partial_\mu\lambda) + 2(\gamma^\mu\theta)_\alpha(\theta^T\varepsilon\theta)\partial_\mu D$$

$$= -i(\gamma^\mu\gamma^\nu\theta)_\alpha F_{\mu\nu} + 2i\theta_\alpha\partial^\mu V_\mu - 2i(\gamma^\mu\partial_\mu\lambda)_\alpha(\theta^T\varepsilon\gamma_5\theta) \tag{15.272}$$

$$- 2i(\gamma_5\partial_\mu\lambda)_\alpha(\bar\theta\gamma_5\gamma^\mu\theta) + 2(\gamma^\mu\theta)_\alpha(\theta^T\varepsilon\theta)\partial_\mu D.$$

$$\varepsilon_{\mu\nu}(\gamma_5\varepsilon)_{\mu\mu'}(\gamma_5\varepsilon)_{\nu\nu'}(\gamma_5\varepsilon)_{\alpha\alpha'}\frac{\partial^3 V}{\partial\theta_{\mu'}\partial\theta_{\nu'}\partial\theta_{\alpha'}} = \varepsilon_{\mu'\nu'}(\gamma_5\varepsilon)_{\alpha\alpha'}\frac{\partial^3 V}{\partial\theta_{\mu'}\partial\theta_{\nu'}\partial\theta_{\alpha'}} \tag{15.273}$$

$$= -4i\lambda_\alpha - 4(\gamma_5\theta)_\alpha D.$$

$$\varepsilon_{\alpha'\nu'}(\gamma^{\mu'}\theta)_\alpha\partial_{\mu'}\frac{\partial^2 V}{\partial\theta_{\nu'}\partial\theta_{\alpha'}} = -4i(\gamma^{\mu'}\theta)_\alpha(\theta^T\varepsilon\gamma_5\partial_{\mu'}\lambda) - 2(\gamma^{\mu'}\theta)_\alpha(\theta^T\varepsilon\theta)\partial_{\mu'}D. \tag{15.274}$$

$$\varepsilon_{\mu\nu}(\gamma_5\varepsilon)_{\alpha\alpha'}(\gamma^{\mu'}\theta)_\mu(\gamma^{\nu'}\theta)_\nu\partial_{\mu'}\partial_{\nu'}\frac{\partial V}{\partial\theta_{\alpha'}} = (\gamma_5\varepsilon)_{\alpha\alpha'}(\theta^T\varepsilon\theta)\frac{\partial}{\partial\theta_{\alpha'}}\partial^2 V$$

$$= \frac{i}{2}(\bar\theta\gamma_5\theta)^2(\partial^2\lambda)_\alpha - i(\gamma_5\gamma^\mu\theta)_\alpha(\theta^T\varepsilon\theta)\partial^2 V_\mu. \tag{15.275}$$

The spinor superfield W_α takes the form

$$W_\alpha = \lambda_\alpha + \frac{1}{2}(\gamma^\mu\gamma^\nu\theta)_\alpha F_{\mu\nu} - i(\gamma_5\theta)_\alpha D - \frac{1}{2}(\theta^T\varepsilon\theta)(\gamma^\mu\gamma_5\partial_\mu\lambda)_\alpha + \frac{1}{2}(\theta^T\varepsilon\gamma_5\theta)(\gamma^\mu\partial_\mu\lambda)_\alpha$$

$$+ \frac{1}{2}(\theta^T\varepsilon\gamma^\mu\theta)(\gamma_5\partial_\mu\lambda)_\alpha + \frac{i}{2}(\gamma^\mu\theta)_\alpha(\theta^T\varepsilon\theta)\partial_\mu D - \frac{1}{4}(\theta^T\varepsilon\theta)(\gamma_5\gamma^\mu\gamma^\nu\gamma^\sigma\theta)_\alpha\partial_\sigma F_{\mu\nu} - \frac{1}{8}(\theta^T\varepsilon\theta)^2\partial^2\lambda_\alpha. \tag{15.276}$$

We also need to compute the left-chiral and right-chiral spinor superfields $W_{L\alpha}$ and $W_{R\alpha}$. We have

$$\frac{1}{2}(\gamma^\mu\gamma^\nu\theta_L)_\alpha F_{\mu\nu}(x_+) = \left(\frac{1+\gamma_5}{2}\right)_{\alpha\beta}\left[\frac{1}{2}(\gamma^\mu\gamma^\nu\theta)_\beta F_{\mu\nu} - \frac{1}{4}(\theta^T\varepsilon\theta)(\gamma_5\gamma^\mu\gamma^\nu\gamma^\sigma\theta)_\beta\partial_\sigma F_{\mu\nu}\right]. \tag{15.277}$$

$$\lambda_{L\alpha}(x_+) = \left(\frac{1+\gamma_5}{2}\right)_{\alpha\beta}\left[\lambda_\beta + \frac{1}{2}(\theta^T\varepsilon\gamma^\mu\theta)(\gamma_5\partial_\mu\lambda)_\beta - \frac{1}{8}(\theta^T\varepsilon\theta)^2\partial^2\lambda_\beta\right]. \tag{15.278}$$

$$(\theta_L^T\varepsilon\theta_L)(\gamma^\mu\partial_\mu\lambda_R(x_+))_\alpha = \left(\frac{1+\gamma_5}{2}\right)_{\alpha\beta}\left[-\frac{1}{2}(\theta^T\varepsilon\theta)(\gamma^\mu\gamma_5\partial_\mu\lambda)_\beta + \frac{1}{2}(\theta^T\varepsilon\gamma_5\theta)(\gamma^\mu\partial_\mu\lambda)_\beta\right]. \tag{15.279}$$

$$-i\theta_{L\alpha}D(x_+) = \left(\frac{1+\gamma_5}{2}\right)_{\alpha\beta}\left[-(i\gamma_5\theta)_\beta D + \frac{i}{2}(\theta^T\varepsilon\theta)(\gamma^\mu\theta)_\beta\partial_\mu D\right]. \tag{15.280}$$

Thus

$$W_{L\alpha} \equiv \left(\frac{1+\gamma_5}{2}\right)_{\alpha\beta}W_\beta = \lambda_{L\alpha}(x_+) + \frac{1}{2}(\gamma^\mu\gamma^\nu\theta_L)_\alpha F_{\mu\nu}(x_+) + (\theta_L^T\varepsilon\theta_L)(\gamma^\mu\partial_\mu\lambda_R)_\alpha(x_+) - i\theta_{L\alpha}D(x_+). \tag{15.281}$$

Similarly we obtain

$$W_{R\alpha} = \lambda_{R\alpha}(x_-) + \frac{1}{2}(\gamma^\mu\gamma^\nu\theta_R)_\alpha F_{\mu\nu}(x_-) - (\theta_R^T\varepsilon\theta_R)(\gamma^\mu\partial_\mu\lambda_L)_\alpha(x_-) - i\theta_{R\alpha}D(x_-). \tag{15.282}$$

Let us now compute the following F-term:

$$[\varepsilon_{\alpha\beta}W_{L\alpha}W_{L\beta}]_F. \tag{15.283}$$

Another long calculation yields

$$-\frac{1}{2}(\theta_L^T\varepsilon\theta_L)[\varepsilon_{\alpha\beta}W_{L\alpha}W_{L\beta}]_F = (\theta_L^T\varepsilon\theta_L)\left[-\frac{1}{2}\lambda^T\varepsilon\gamma^\mu\partial_\mu\lambda - \frac{1}{2}\lambda^T\gamma_5\varepsilon\gamma^\mu\partial_\mu\lambda + \frac{1}{2}D^2 - \frac{1}{4}F_{\mu\nu}F^{\mu\nu}\right. $$
$$\left. + \frac{i}{8}\varepsilon^{\mu\nu\alpha\beta}F_{\mu\nu}F_{\alpha\beta}\right]. \tag{15.284}$$

The supersymmetric $U(1)$ action is taken to be the real part of this F-term, viz

$$\mathcal{L}_{\text{gauge}} = -\frac{1}{2}\,\text{Re}[\varepsilon_{\alpha\beta}W_{L\alpha}W_{L\beta}]_F = \left[-\frac{1}{2}\bar{\lambda}\gamma^\mu\partial_\mu\lambda + \frac{1}{2}D^2 - \frac{1}{4}F_{\mu\nu}F^{\mu\nu}\right]. \tag{15.285}$$

The supersymmetry transformations for the components of the gauge superfields V are

$$V_\mu \longrightarrow V_\mu' = V_\mu + \bar{\alpha}\gamma_\mu\lambda$$
$$\lambda \longrightarrow \lambda' = \lambda + \left(i\gamma_5 D + \frac{1}{4}F_{\mu\nu}[\gamma^\nu, \gamma^\mu]\right)\alpha \tag{15.286}$$
$$D \longrightarrow D' = D + i\bar{\alpha}\gamma_5\gamma_\mu\partial^\mu\lambda.$$

We can immediately verify the invariance of the action by using the identities

$$\delta F_{\mu\nu} = \bar{\alpha}((\partial_\mu\gamma_\nu - \partial_\nu\gamma_\mu)\lambda)$$
$$\delta(\bar{\lambda}\gamma^\mu\partial_\mu\lambda) = 2iD(\bar{\alpha}\gamma_5\gamma^\mu\partial_\mu\lambda) + F_{\mu\nu}\bar{\alpha}((\gamma^\mu\partial^\nu - \gamma^\nu\partial^\mu)\lambda). \tag{15.287}$$

15.5.4 Calculation of the F-term in the non-Abelian theory

Under generalized gauge transformations the non-Abelian gauge superfield $V(x, \theta) = t_A V_A(x, \theta)$ transforms as

$$e^{-2t_A V_A} \longrightarrow e^{-it_A\Omega_A}e^{-2t_A V_A}e^{it_A\Omega_A^*}$$
$$e^{+2t_A V_A} \longrightarrow e^{-it_A\Omega_A^*}e^{+2t_A V_A}e^{it_A\Omega_A}. \tag{15.288}$$

Ω_A is a left-chiral superfield ($\mathcal{D}_{R\alpha}\Omega_A = 0$) whereas Ω_A^* is a right-chiral superfield ($\mathcal{D}_{L\alpha}\Omega_A^* = 0$). Thus

$$e^{-2t_A V_A}\mathcal{D}_{L\alpha}e^{2t_A V_A} \longrightarrow e^{-it_A\Omega_A}e^{-2t_A V_A}\mathcal{D}_{L\alpha}e^{2t_A V_A}e^{it_A\Omega_A}. \tag{15.289}$$

In above $\mathcal{D}_{L\alpha}$ still acts on $e^{it_A\Omega_A}$. By analogy with the Abelian case we define the left-chiral spinor superfields $W_{AL\alpha}$ by

$$2t_A W_{AL\alpha} = \frac{i}{4}\varepsilon_{\beta\gamma}\mathcal{D}_{R\beta}\mathcal{D}_{R\gamma}e^{-2t_A V_A}\mathcal{D}_{L\alpha}e^{2t_A V_A}. \tag{15.290}$$

Obviously these are left-chiral superfields ($\mathcal{D}_{R\beta}W_{AL\alpha} = 0$) since the product of 3 $\mathcal{D}_R$ s vanishes. The above definition of the spinor superfields is motivated by the requirement of covariance under gauge transformations. Indeed since Ω_A is left-chiral we have

$$
\begin{aligned}
2t_A W_{AL\alpha} = \frac{i}{4}\varepsilon_{\beta\gamma}\mathcal{D}_{R\beta}\mathcal{D}_{R\gamma}e^{-2t_A V_A}\mathcal{D}_{L\alpha}e^{2t_A V_A}\longrightarrow \quad & \frac{i}{4}e^{-it_A\Omega_A}\varepsilon_{\beta\gamma}\mathcal{D}_{R\beta}\mathcal{D}_{R\gamma}e^{-2t_A V_A}\mathcal{D}_{L\alpha}e^{2t_A V_A}e^{it_A\Omega_A} \\
\longrightarrow \quad & e^{-it_A\Omega_A}2t_A W_{AL\alpha}e^{it_A\Omega_A} \\
+ \quad & \frac{i}{4}e^{-it_A\Omega_A}[\varepsilon_{\beta\gamma}\mathcal{D}_{R\beta}\mathcal{D}_{R\gamma}, \mathcal{D}_{L\alpha}]e^{it_A\Omega_A} \\
\longrightarrow \quad & e^{-it_A\Omega_A}2t_A W_{AL\alpha}e^{it_A\Omega_A}.
\end{aligned}
\tag{15.291}
$$

In the last line we have also used the identity (15.263). Thus the full spinor superfield $W_{L\alpha} = t_A W_{AL\alpha}$ transforms as

$$W_{L\alpha} \longrightarrow e^{-it_A\Omega_A}W_{L\alpha}e^{it_A\Omega_A}. \tag{15.292}$$

In the Wess–Zumino gauge the gauge superfields are given by

$$V_A(x,\theta) = \frac{i}{2}(\bar{\theta}\gamma_5\gamma_\mu\theta)V_A^\mu(x) - i(\bar{\theta}\gamma_5\theta)(\bar{\theta}\lambda_A(x)) - \frac{1}{4}(\bar{\theta}\gamma_5\theta)^2 D_A(x). \tag{15.293}$$

As in the Abelian case it is crucial to compute the spinor superfields $W_{AL\alpha}$ in terms of the components of the gauge superfields V_A. We will only calculate the spinor field at some point $x^\mu = X^\mu$. Towards this end we will adopt a version of the Wess–Zumino gauge in which

$$V_A^\mu(X) = 0. \tag{15.294}$$

We have (with $V = t_A V_A$)

$$(\mathcal{D}_R^T\varepsilon\mathcal{D}_R)e^{-2t_A V_A}\mathcal{D}_\alpha e^{2t_A V_A} = 2(\mathcal{D}_R^T\varepsilon\mathcal{D}_R)(1 - 2V + 2V^2 + \cdots)\mathcal{D}_\alpha(V + V^2 + \cdots) \tag{15.295}$$

The terms $2(\mathcal{D}_R^T\varepsilon\mathcal{D}_R)\mathcal{D}_\alpha(V^2)$ and $2(\mathcal{D}_R^T\varepsilon\mathcal{D}_R)(2V^2)\mathcal{D}_\alpha(V)$ will lead to zero in the gauge (15.293) together with (15.294) and hence higher order terms must also be zero in this gauge. The calculation reduces to

$$(\mathcal{D}_R^T\varepsilon\mathcal{D}_R)e^{-2t_A V_A}\mathcal{D}_\alpha e^{2t_A V_A} = 2(\mathcal{D}_R^T\varepsilon\mathcal{D}_R)(1 - 2V)\mathcal{D}_\alpha(V) \tag{15.296}$$

The term $2(\mathcal{D}_R^T\varepsilon\mathcal{D}_R)(-2V)\mathcal{D}_\alpha(V)$ is similar to $2(\mathcal{D}_R^T\varepsilon\mathcal{D}_R)\mathcal{D}_\alpha(V^2)$ and hence it will also lead to zero. We get the result

$$(\mathcal{D}_R^T \varepsilon \mathcal{D}_R) e^{-2t_A V_A} \mathcal{D}_\alpha e^{2t_A V_A} = 2(\mathcal{D}_R^T \varepsilon \mathcal{D}_R) \mathcal{D}_\alpha(V) \tag{15.297}$$

Hence in the gauge (15.293) together with (15.294) the spinor superfields are given by

$$W_{AL\alpha} = \frac{i}{4}(\mathcal{D}_R^T \varepsilon \mathcal{D}_R) \mathcal{D}_{L\alpha} V_A. \tag{15.298}$$

In other words, in this gauge we have a number of copies of the Abelian theory labeled A. The calculation of this superfield proceeds therefore as in the Abelian case with the result

$$W_{AL\alpha} = \lambda_{AL\alpha}(X_+) + \frac{1}{2}(\gamma_\mu \gamma_\nu \theta_L)_\alpha (\partial^\mu V_A^\nu(X_+) - \partial^\nu V_A^\mu(X_+)) + (\theta_L^T \varepsilon \theta_L)(\gamma^\mu \partial_\mu \lambda_{AR}(X_+))_\alpha \tag{15.299}$$
$$- i\theta_{L\alpha} D_A(X_+).$$

Since $W_{L\alpha}$ is gauge-covariant, at a general point we must have

$$W_{AL\alpha} = \lambda_{AL\alpha}(x_+) + \frac{1}{2}(\gamma_\mu \gamma_\nu \theta_L)_\alpha F_A^{\mu\nu}(x_+) + (\theta_L^T \varepsilon \theta_L)(\gamma^\mu (D_\mu \lambda_R)_A)_\alpha(x_+) - i\theta_{L\alpha} D_A(x_+). \tag{15.300}$$

The curvature $F_A^{\mu\nu}$ and the covariant derivatives of the Majorana spinors $(D^\mu \lambda_R)_A$ are defined by

$$F_A^{\mu\nu} = \partial^\mu V_A^\nu - \partial^\nu V_A^\mu + C_{ABC} V_B^\mu V_C^\nu. \tag{15.301}$$

$$(D^\mu \lambda_R)_A = \partial^\mu \lambda_{RA} + C_{ABC} V_B^\mu \lambda_{RC}. \tag{15.302}$$

Let us now compute the following F-term:

$$(\theta_L^T \varepsilon \theta_L)[\varepsilon_{\alpha\beta} W_{AL\alpha} W_{AL\beta}]_\mathcal{F}. \tag{15.303}$$

We have

$$[\varepsilon_{\alpha\beta} W_{AL\alpha} \lambda_{AL\beta}]_{\theta_L^2} = (\theta_L^T \varepsilon \theta_L)(\lambda_{AL}^T \varepsilon \gamma^\mu (D_\mu \lambda_R)_A). \tag{15.304}$$

$$\left[\varepsilon_{\alpha\beta} W_{AL\alpha} \frac{1}{2}(\gamma^\mu \gamma^\nu \theta_L)_\beta F_{A\mu\nu}\right]_{\theta_L^2} = \frac{1}{4}[(\gamma^\mu \gamma^\nu \theta_L)^T \varepsilon (\gamma^\alpha \gamma^\beta \theta_L)] F_{A\mu\nu} F_{A\alpha\beta} - i D_A \left[\theta_L^T \varepsilon \gamma^\mu \gamma^\nu \theta_L\right] F_{A\mu\nu}. \tag{15.305}$$

We use among other things the identities $\bar{\theta}\gamma_5 = \theta^T \varepsilon$, $\gamma_\mu^T = -C\gamma^\mu C^{-1}$ with $C = -C^{-1} = \varepsilon\gamma^5$ and $\varepsilon^2 = -1$ to show that $\bar{\theta}[\gamma^\mu, \gamma^\nu]\theta = \bar{\theta}\gamma_5[\gamma^\mu, \gamma^\nu]\theta = 0$. Hence the second contribution in (15.305) vanishes. The first contribution can be put in the form

$$\left[\varepsilon_{\alpha\beta} W_{AL\alpha} \frac{1}{2}(\gamma^\mu \gamma^\nu \theta_L)_\beta F_{A\mu\nu}\right]_{\theta_L^2} = -\frac{1}{16}[(\bar{\theta}_L[\gamma^\mu, \gamma^\nu][\gamma^\alpha, \gamma^\beta]\theta_L)] F_{A\mu\nu} F_{A\alpha\beta}$$

$$= -\frac{1}{4}(\theta_L^T \varepsilon \theta_L)(-\eta^{\mu\alpha}\eta^{\nu\beta} + \eta^{\mu\beta}\eta^{\nu\alpha} + i\varepsilon^{\mu\nu\alpha\beta}) F_{A\mu\nu} F_{A\alpha\beta} \tag{15.306}$$

$$= -\frac{1}{4}(\theta_L^T \varepsilon \theta_L)(-2 F_{A\mu\nu} F^{A\mu\nu} + i\varepsilon^{\mu\nu\alpha\beta} F_{A\mu\nu} F_{A\alpha\beta}).$$

Finally we compute

$$[\varepsilon_{\alpha\beta} W_{AL\alpha}(\theta_L^T \varepsilon \theta_L)(\gamma^\mu (D_\mu \lambda_R)_A)_\beta]_{\theta_L^2} = (\theta_L^T \varepsilon \theta_L)(\lambda_{AL}^T \varepsilon \gamma^\mu (D_\mu \lambda_R)_A). \tag{15.307}$$

$$[\varepsilon_{\alpha\beta} W_{AL\alpha}(-i\theta_{L\beta} D_A)]_{\theta_L^2} = (\theta_L^T \varepsilon \theta_L) D_A^2 \tag{15.308}$$

Hence we obtain the F-term

$$-\frac{1}{2}(\theta_L^T \varepsilon \theta)[\varepsilon_{\alpha\beta} W_{AL\alpha} W_{AL\beta}]_F = -\frac{1}{2}(\theta_L^T \varepsilon \theta)\Bigg[\bar{\lambda}_A \gamma^\mu (D_\mu(1-\gamma_5)\lambda)_A + \frac{1}{2}F_{A\mu\nu}F^{A\mu\nu}$$
$$-\frac{i}{4}\varepsilon^{\mu\nu\alpha\beta}F_{A\mu\nu}F_{A\alpha\beta} - D_A^2 \Bigg]. \tag{15.309}$$

As in the Abelian case the gauge action is given by the real part of this term, viz

$$\mathcal{L}_{\text{gauge}} = -\frac{1}{2}\,\text{Re}[\varepsilon_{\alpha\beta} W_{AL\alpha} W_{AL\beta}]_F = -\frac{1}{2}\bar{\lambda}_A(\gamma^\mu D_\mu \lambda)_A - \frac{1}{4}F_{A\mu\nu}F^{A\mu\nu} + \frac{1}{2}D_A^2. \tag{15.310}$$

Let us say that $\bar{\lambda}_A \gamma^\mu (D_\mu \gamma_5 \lambda)_A = -(\varepsilon\gamma^\mu)_{\alpha\beta}\lambda_{A\alpha}(D_\mu\lambda)_{A\beta}$. But since $(\varepsilon\gamma^\mu)^T = -\varepsilon\gamma^\mu$ we get $\bar{\lambda}_A\gamma^\mu(D_\mu\gamma_5\lambda)_A = -\partial_\mu(\bar{\lambda}_A\gamma_5\gamma^\mu\lambda_A)$. In other words, this term is a total derivative and it is pure imaginary since $(\bar{\lambda}_A\gamma_5\gamma^\mu\lambda_A)^+ = -\bar{\lambda}_A\gamma_5\gamma^\mu\lambda_A$. The imaginary part of the above F-term is therefore given by

$$-\frac{1}{2}\,\text{Im}[\varepsilon_{\alpha\beta} W_{AL\alpha} W_{AL\beta}]_F = -\frac{i}{2}\bar{\lambda}_A\gamma^\mu(D_\mu\gamma_5\lambda)_A + \frac{1}{8}\varepsilon^{\mu\nu\alpha\beta}F_{A\mu\nu}F_{A\alpha\beta} \tag{15.311}$$

This is essentially the famous theta term. Although it is a total derivative, it does not vanish due to instantons. The theta term is defined by

$$\mathcal{L}_\theta = -\frac{g^2\theta}{16\pi^2}\,\text{Im}[\varepsilon_{\alpha\beta} W_{AL\alpha} W_{AL\beta}]_F. \tag{15.312}$$

The structure constants are defined by $C_{ABC} = gf_{ABC}$. By absorbing a factor g into the gauge field the Lagrangian density for the gauge field becomes multiplied by an over-all factor $1/g^2$. The total action becomes

$$\mathcal{L}_{\text{gauge}} + \mathcal{L}_\theta = -\text{Re}\left[\frac{\tau}{8\pi i}\varepsilon_{\alpha\beta} W_{AL\alpha} W_{AL\beta}\right]_F. \tag{15.313}$$

The complex coupling parameter τ is defined by

$$\tau = \frac{4\pi i}{g^2} + \frac{\theta}{2\pi}. \tag{15.314}$$

The supersymmetry transformations of the action (15.310) are

$$\delta V_\mu^A = \bar{\alpha}\gamma_\mu\lambda^A$$
$$\delta\lambda^A = \left(i\gamma_5 D^A - \frac{1}{4}[\gamma^\mu, \gamma^\nu]F_{\mu\nu}^A\right)\alpha \tag{15.315}$$
$$\delta D^A = i\bar{\alpha}\gamma_5\gamma_\mu\partial^\mu\lambda^A.$$

15.5.5 Calculation of the d-term

The starting point is the left-chiral superfields (where n is a color index)

$$\Phi_n(x,\theta) = \phi_n(x) - \sqrt{2}\,(\bar\theta\psi_{nL}(x)) + \mathcal{F}_n(x)\left(\bar\theta\frac{1+\gamma_5}{2}\theta\right) + \frac{1}{2}(\bar\theta\gamma_5\gamma_\mu\theta)\partial^\mu\phi_n(x)$$
$$- \frac{1}{\sqrt{2}}(\bar\theta\gamma_5\theta)(\bar\theta\gamma_\mu\partial^\mu\psi_{nL}(x)) - \frac{1}{8}(\bar\theta\gamma_5\theta)^2\partial^2\phi_n(x). \tag{15.316}$$

Their complex conjugates are given by

$$\Phi_n^*(x,\theta) = \phi_n^*(x) - \sqrt{2}\,(\bar\psi_{nL}(x)\theta) + \mathcal{F}_n^*(x)\left(\bar\theta\frac{1-\gamma_5}{2}\theta\right) - \frac{1}{2}(\bar\theta\gamma_5\gamma_\mu\theta)\partial^\mu\phi_n^*(x)$$
$$- \frac{1}{\sqrt{2}}(\bar\theta\gamma_5\theta)(\partial^\mu\bar\psi_{nL}(x)\gamma^\mu\theta) - \frac{1}{8}(\bar\theta\gamma_5\theta)^2\partial^2\phi_n^*(x). \tag{15.317}$$

We want to compute the following d-term:

$$[\Phi^+\Gamma\Phi]_d = [\Phi_n^*\Gamma_{nm}\Phi_m]_d. \tag{15.318}$$

The Kahler potential $\Phi^+\Gamma\Phi$ is by construction a real superfield. The gauge connection is given by

$$\Gamma(x,\theta) = e^{-2t_A V^A(x,\theta)} = e^{-2t_A\left(\frac{i}{2}(\bar\theta\gamma_5\gamma^\mu\theta)V_\mu^A - i(\bar\theta\gamma_5\theta)(\bar\theta\lambda^A) - \frac{1}{4}(\bar\theta\gamma_5\theta)^2 D^A\right)}$$
$$= 1 - 2t_A V^A + \frac{1}{2}t_A t_B(\bar\theta\gamma_5\theta)^2 V_\mu^A V^{\mu B}. \tag{15.319}$$

We have immediately

$$\Phi^+\Gamma\Phi = \Phi^+\Phi - 2\Phi^+ t_A\Phi V^A + \frac{1}{2}\Phi^+ t_A t_B\Phi(\bar\theta\gamma_5\theta)^2 V_\mu^A V^{\mu B}. \tag{15.320}$$

Thus

$$[\Phi^+\Gamma\Phi]_{\theta^0} = [\Phi^+\Phi]_{\theta^0} = \phi_n^*\phi_n. \tag{15.321}$$

$$[\Phi^+\Gamma\Phi]_{\theta^4} = [\Phi^+\Phi]_{\theta^4} - 2(t_A)_{nm}[\Phi_n^*\Phi_m V^A]_{\theta^4} + \frac{1}{2}\phi_n^*(t_A t_B)_{nm}\phi_m(\bar\theta\gamma_5\theta)^2 V_\mu^A V^{\mu B}. \tag{15.322}$$

The first contribution is computed as follows

$$[\Phi^+\Phi]_{\theta^4} = -\frac{1}{8}(\bar\theta\gamma_5\theta)^2\left(\partial^2\phi_n^*\cdot\phi_n + \phi_n^*\cdot\partial^2\phi_n\right) + (\bar\theta\gamma_5\theta)\partial_\mu(\bar\psi_{nL}\gamma^\mu\theta)(\bar\theta\psi_{nL}) + (\bar\psi_{nL}\theta)(\bar\theta\gamma_5\theta)(\bar\theta\gamma^\mu\partial_\mu\psi_{nL})$$
$$+ (\bar\theta\frac{1-\gamma_5}{2}\theta)(\bar\theta\frac{1+\gamma_5}{2}\theta)\mathcal{F}_n^*\mathcal{F}_n - \frac{1}{4}(\bar\theta\gamma_5\gamma_\mu\theta)(\bar\theta\gamma_5\gamma_\nu\theta)\partial^\mu\phi_n^*\partial^\nu\phi_n. \tag{15.323}$$

In the above equation we have used the fact

$$(\bar\theta\frac{1-\gamma_5}{2}\theta)(\bar\theta\gamma_5\gamma_\mu\theta)\mathcal{F}_n^*\partial^\mu\phi_n = (\bar\theta\frac{1+\gamma_5}{2}\theta)(\bar\theta\gamma_5\gamma_\mu\theta)\mathcal{F}_n\partial^\mu\phi_n^* = 0. \tag{15.324}$$

We need also to use the results

$$(\bar{\theta}\gamma_5\theta)\partial_\mu(\bar{\psi}_{nL}\gamma^\mu\theta)(\bar{\theta}\psi_{nL}) = \frac{1}{4}(\bar{\theta}\gamma_5\theta)^2\partial_\mu\bar{\psi}_{nL}\gamma_5\gamma^\mu\psi_{nL}. \tag{15.325}$$

$$(\bar{\psi}_{nL}\theta)(\bar{\theta}\gamma_5\theta)(\bar{\theta}\gamma^\mu\partial_\mu\psi_{nL}) = \frac{1}{4}(\bar{\theta}\gamma_5\theta)^2\bar{\psi}_{nL}\gamma^\mu\partial_\mu\psi_{nL}. \tag{15.326}$$

$$(\bar{\theta}\frac{1-\gamma_5}{2}\theta)(\bar{\theta}\frac{1+\gamma_5}{2})\mathcal{F}_n^*\mathcal{F}_n = -\frac{1}{2}(\bar{\theta}\gamma_5\theta)^2\mathcal{F}_n^*\mathcal{F}_n. \tag{15.327}$$

$$-\frac{1}{4}(\bar{\theta}\gamma_5\gamma_\mu\theta)(\bar{\theta}\gamma_5\gamma_\nu\theta)\partial^\mu\phi_n^*\partial^\nu\phi_n = \frac{1}{4}(\bar{\theta}\gamma_5\theta)^2\partial^\mu\phi_n^*\partial_\mu\phi_n. \tag{15.328}$$

By putting these things together we get

$$[\Phi^+\Phi]_{\theta^4} = -\frac{1}{4}(\bar{\theta}\gamma_5\theta)^2\Big(\frac{1}{2}\partial^2\phi_n^*\cdot\phi_n + \frac{1}{2}\phi_n^*\cdot\partial^2\phi_n - \partial^\mu\phi_n^*\partial_\mu\phi_n - \bar{\psi}_{nL}\gamma^\mu\partial_\mu\psi_{nL} - \partial_\mu\bar{\psi}_{nL}\gamma_5\gamma^\mu\psi_{nL}$$
$$+ 2\mathcal{F}_n^*\mathcal{F}_n\Big). \tag{15.329}$$

The second contribution (the second term in (15.322)) is computed as follows

$$[\Phi_n^*\Phi_m V^A]_{\theta^4} = \frac{i}{2}V_\mu^A[\Phi_n^*\Phi_m\bar{\theta}\gamma_5\gamma^\mu\theta]_{\theta^4} - i[\Phi_n^*\Phi_m(\bar{\theta}\gamma_5\theta)(\bar{\theta}\lambda^A)]_{\theta^4} - \frac{1}{4}[\Phi_n^*\Phi_m(\bar{\theta}\gamma_5\theta)^2 D^A]_{\theta^4}$$
$$= \Big(\frac{i}{2}V_\mu^A\Big)\Phi_n^*\Big(\frac{1}{2}\bar{\theta}\gamma_5\gamma_\nu\theta\cdot\partial^\nu\phi_m\Big)(\bar{\theta}\gamma_5\gamma^\mu\theta) + \Big(\frac{i}{2}V_\mu^A\Big)(-\sqrt{2}\bar{\psi}_{nL}\theta)(-\sqrt{2}\bar{\theta}\psi_{mL})(\bar{\theta}\gamma_5\gamma^\mu\theta)$$
$$+ \Big(\frac{i}{2}V_\mu^A\Big)\Big(-\frac{1}{2}\bar{\theta}\gamma_5\gamma_\nu\theta\cdot\partial^\nu\phi_n^*\Big)\phi_m(\bar{\theta}\gamma_5\gamma^\mu\theta) - i\phi_n^*(-\sqrt{2}\bar{\theta}\psi_{mL})(\bar{\theta}\gamma_5\theta)(\bar{\theta}\lambda^A)$$
$$- i(-\sqrt{2}\bar{\psi}_{nL}\theta)\phi_m(\bar{\theta}\gamma_5\theta)(\bar{\theta}\lambda^A) - \frac{1}{4}\phi_n^*\phi_m(\bar{\theta}\gamma_5\theta)^2 D^A. \tag{15.330}$$

We use now

$$\Big(\frac{i}{2}V_\mu^A\Big)\Phi_n^*\Big(\frac{1}{2}\bar{\theta}\gamma_5\gamma_\nu\theta\cdot\partial^\nu\phi_m\Big)(\bar{\theta}\gamma_5\gamma^\mu\theta) = -\frac{i}{4}(\bar{\theta}\gamma_5\theta)^2 V_\mu^A\phi_n^*\partial^\mu\phi_m. \tag{15.331}$$

$$\Big(\frac{i}{2}V_\mu^A\Big)(-\sqrt{2}\bar{\psi}_{nL}\theta)(-\sqrt{2}\bar{\theta}\psi_{mL})(\bar{\theta}\gamma_5\gamma^\mu\theta) = \frac{i}{2}(\bar{\theta}\gamma_5\theta)^2 V_\mu^A(\bar{\psi}_{nL}\gamma^\mu\psi_{mL}). \tag{15.332}$$

$$\Big(\frac{i}{2}V_\mu^A\Big)\Big(-\frac{1}{2}\bar{\theta}\gamma_5\gamma_\nu\theta\cdot\partial^\nu\phi_n^*\Big)\phi_m(\bar{\theta}\gamma_5\gamma^\mu\theta) = \frac{i}{4}(\bar{\theta}\gamma_5\theta)^2 V_\mu^A\partial^\mu\phi_n^*\cdot\phi_m. \tag{15.333}$$

$$-i\phi_n^*(-\sqrt{2}\bar{\theta}\psi_{mL})(\bar{\theta}\gamma_5\theta)(\bar{\theta}\lambda^A) = -i\frac{\sqrt{2}}{4}(\bar{\theta}\gamma_5\theta)^2\phi_n^*\bar{\lambda}^A\psi_{mL}. \tag{15.334}$$

$$-i(-\sqrt{2}\bar{\psi}_{nL}\theta)\phi_m(\bar{\theta}\gamma_5\theta)(\bar{\theta}\lambda^A) = i\frac{\sqrt{2}}{4}(\bar{\theta}\gamma_5\theta)^2\phi_m\bar{\psi}_{nL}\lambda^A. \tag{15.335}$$

Thus the second contribution takes the final form

$$[\Phi_n^*\Phi_m V^A]_{\theta^4} = -\frac{1}{4}(\bar{\theta}\gamma_5\theta)^2\Big(iV_\mu^A\phi_n^*\partial^\mu\phi_m - iV_\mu^A\bar{\psi}_{nL}\gamma^\mu\psi_{mL} - iV_\mu^A\partial^\mu\phi_n^* \cdot \phi_m + i\sqrt{2}\,\phi_n^*(\bar{\lambda}^A\psi_{mL})$$
$$- i\sqrt{2}\,\phi_m\bar{\psi}_{nL}\lambda^A + \phi_n^*\phi_m D^A\Big). \tag{15.336}$$

The final result from (15.322), (15.329) and (15.336) is given by

$$[\Phi^+\Gamma\Phi]_{\theta^4} = -\frac{1}{4}(\bar{\theta}\gamma_5\theta)^2\Big(\frac{1}{2}\partial^2\phi_n^* \cdot \phi_n + \frac{1}{2}\phi_n^* \cdot \partial^2\phi_n - \partial^\mu\phi_n^*\partial_\mu\phi_n - \bar{\psi}_{nL}\gamma^\mu\partial_\mu\psi_{nL} + \partial_\mu\bar{\psi}_{nL}\gamma^\mu\psi_{nL}$$
$$+ 2\mathcal{F}_n^*\mathcal{F}_n - 2i(t_A)_{nm}V_\mu^A\phi_n^*\partial^\mu\phi_m + 2i(t_A)_{nm}V_\mu^A\partial^\mu\phi_n^* \cdot \phi_m - 2i\sqrt{2}\,(t_A)_{nm}\phi_n^*(\bar{\lambda}^A\psi_{mL})$$
$$+ 2i\sqrt{2}\,(t_A)_{nm}\phi_m(\bar{\psi}_{nL}\lambda^A) + 2i(t_A)_{nm}V_\mu^A(\bar{\psi}_{nL}\gamma^\mu\psi_{mL}) - 2(t_A)_{nm}\phi_n^*\phi_m D^A$$
$$- 2\phi_n^*(t_A t_B)_{nm}\phi_m V_\mu^A V^{\mu B}\Big). \tag{15.337}$$

The d-term of $\Phi^+\Gamma\Phi$ is the coefficient of $-(\bar{\theta}\gamma_5\theta)^2/4$ in the expansion of $\Phi^+\Gamma\Phi$ minus $\partial^2/2$ acting on $[\Phi^+\Gamma\Phi]_{\theta^0} = \phi_n^*\phi_n$. We use the result

$$-2\partial^\mu\phi_n^*\partial_\mu\phi_n = \partial^2\phi_n^* \cdot \phi_n + \phi_n^* \cdot \partial^2\phi_n - \partial^2(\phi_n^*\phi_n). \tag{15.338}$$

We find the d-term

$$[\Phi^+\Gamma\Phi]_D = -2\partial^\mu\phi_n^*\partial_\mu\phi_n - \bar{\psi}_{nL}\gamma^\mu\partial_\mu\psi_{nL} + \partial_\mu\bar{\psi}_{nL}\gamma^\mu\psi_{nL} + 2\mathcal{F}_n^*\mathcal{F}_n - 2i(t_A)_{nm}V_\mu^A\phi_n^*\partial^\mu\phi_m$$
$$+ 2i(t_A)_{nm}V_\mu^A\partial^\mu\phi_n^* \cdot \phi_m + 2i(t_A)_{nm}V_\mu^A(\bar{\psi}_{nL}\gamma^\mu\psi_{mL}) - 2i\sqrt{2}\,(t_A)_{nm}\phi_n^*(\bar{\lambda}^A\psi_{mL}) \tag{15.339}$$
$$+ 2i\sqrt{2}\,(t_A)_{nm}\phi_m(\bar{\psi}_{nL}\lambda^A) - 2(t_A)_{nm}\phi_n^*\phi_m D^A - 2\phi_n^*(t_A t_B)_{nm}\phi_m V_\mu^A V^{\mu B}.$$

This term can be put in the equivalent but more elegant form

$$[\Phi^+\Gamma\Phi]_D = -2(D_\mu\phi_n)^*(D^\mu\phi_n) - \bar{\psi}_{nL}\gamma^\mu D_\mu\psi_{nL} + \bar{D}_\mu\bar{\psi}_{nL}\gamma^\mu\psi_{nL} + 2\mathcal{F}_n^*\mathcal{F}_n$$
$$- 2i\sqrt{2}\,(t_A)_{nm}\phi_n^*(\bar{\lambda}^A\psi_{mL}) + 2i\sqrt{2}\,(t_A)_{nm}\phi_m(\bar{\psi}_{nL}\lambda^A) - 2(t_A)_{nm}\phi_n^*\phi_m D^A. \tag{15.340}$$

The covariant derivatives are defined by

$$D_\mu\phi_n = \partial_\mu\phi_n - i(t_A)_{nm}V_\mu^A\phi_m$$
$$(D_\mu\phi_n)^* = \partial_\mu\phi_n^* + i(t_A)_{mn}V_\mu^A\phi_m^*. \tag{15.341}$$

Also

$$D_\mu\psi_{nL} = \partial_\mu\psi_{nL} - i(t_A)_{nm}V_\mu^A\psi_{mL}$$
$$(D_\mu\psi_{nL})^+ = \partial_\mu\psi_{nL}^+ + i(t_A)_{mn}V_\mu^A\psi_{mL}^+ \quad \Leftrightarrow \quad \bar{D}_\mu\bar{\psi}_{nL} \equiv \partial_\mu\bar{\psi}_{nL} + i(t_A)_{mn}V_\mu^A\bar{\psi}_{mL}. \tag{15.342}$$

15.6 Exercises

Exercise 1: Show that P^2 and W^2 are Casimirs by computing their commutators with all generators. What are their eigenvalues for massive and massless particles.

Exercise 2: Verify that if a and b are two 4-vectors such that $a.\,b = a.\,a = b.\,b = 0$ then $a^\mu = hb^\mu$.

Exercise 3:

- Check explicitly the summation convention

$$\psi\chi = \psi^\alpha\chi_\alpha = \varepsilon^{\alpha\beta}\psi_\beta\chi_\alpha = -\varepsilon^{\alpha\beta}\psi_\alpha\chi_\beta = -\psi_\alpha\chi^\alpha = \chi^\alpha\psi_\alpha = \chi\psi. \tag{15.343}$$

- Check explicitly the Fierz identities (15.33)–(15.41).

Exercise 4:

- Verify that the supersymmetry generators commute with the generators of translations, viz

$$[P_\mu,\, Q_\alpha^A] = [P_\mu,\, \bar{Q}_A^{\dot\alpha}] = 0. \tag{15.344}$$

- Show that the central charges X^{AB} commute with $\bar{Q}_{\dot\alpha}^A$ and $\bar{Q}_{\dot\beta}^B$ by using the Jacobi identity.
- Verify explicitly that P^2 and C^2 are Casimirs of the supersymmetry algebra.

Exercise 5:

- Show that

$$\{D_\alpha,\, \bar{D}_{\dot\beta}\} = 2i(\sigma^\mu)_{\alpha\dot\beta}\partial_\mu. \tag{15.345}$$

- Show that the condition for a chiral superfield, which is given by $\bar{D}_{\dot\alpha}\Phi = 0$ is equivalent to

$$\bar{\chi}_{\dot\alpha} = 0, \quad n = 0, \quad v_\mu = i\partial_\mu f, \quad \bar\psi = 0, \quad \bar{\lambda}_{\dot\alpha} = -\frac{i}{2}\partial_\mu\phi^\beta(\sigma^\mu)_{\beta\dot\alpha}, \quad d = -\frac{1}{4}\partial_\mu\partial^\mu f. \tag{15.346}$$

- Show that the Wess–Zumino gauge given by equation (15.128) is not preserved by supersymmetry transformations.
- Show that chiral spinor superfield satisfies $\bar{D}_{\dot\alpha} W_\beta = 0$.
- Show that chiral and antichiral spinor superfields satisfy

$$\bar{D}_{\dot\alpha}\bar{W}^{\dot\alpha} = D^\alpha W_\alpha. \tag{15.347}$$

Exercise 6: Verify explicitly that by using the infinitesimal non-Abelian gauge transformation (15.159) we can bring the vector superfield V into the Wess–Zumino form, viz

$$V_{\mathrm{WZ}} = (\theta\sigma^\mu\bar\theta)v_\mu(x) + (\theta\theta)(\bar\theta\bar\lambda(x)) + (\bar\theta\bar\theta)(\theta\lambda(x)) + (\theta\theta)(\bar\theta\bar\theta)d(x). \tag{15.348}$$

Exercise 7: Show that the Wess–Zumino gauge does not fix the non-Abelian gauge freedom generated, as in the Abelian case, by the scalar component of $\Phi - \Phi^+$.

Exercise 8: Show that the the component field F of the chiral superfield Φ is auxiliary and therefore it can be set equal to its classical value in the Wess–Zumino model. Verify then that the bosonic part of the Wess–Zumino model will only depend on the component field A and it reads

$$
(S_{\text{WZ}})_{\text{bosonic}} = \int d^4x [\partial^\mu A^+ \partial_\mu A - V_F(A^+, A)], \quad V_F(A^+, A)
$$
$$
= F^+ F = (mA + gA^2)^+ (mA + gA^2). \tag{15.349}
$$

Exercise 9: Show that the total potential of the field components A and A^+ in $N = 1$ super Yang–Mills action with matter given by a chiral superfield is given by

$$
V(A^+, A) = V_F(A^+, A) + V_D(A^+, A) = F^+ F + \frac{2}{g^2} d^2. \tag{15.350}
$$

References

[1] Lykken J D 1996 Introduction to supersymmetry arXiv:hep-th/9612114
[2] Weinberg S 2005 *The Quantum Theory of Fields, Volume III: Supersymmetry* (Cambridge: Cambridge Univ. Press)
[3] Wess J and Bagger J 1992 *Supersymmetry and Supergravity* (Princeton: University Press) 259
[4] West P C 1990 *Introduction to Supersymmetry and Supergravity* (Singapore: World Scientific) 425
[5] Bilal A 2001 Introduction to supersymmetry arXiv:hep-th/0101055
[6] Coleman S R and Mandula J 1967 All possible symmetries of the s matrix *Phys. Rev.* **159** 1251
[7] Haag R, Lopuszanski J T and Sohnius M 1975 All possible generators of supersymmetries of the s matrix *Nucl. Phys.* B **88** 257

IOP Publishing

A Modern Course in Quantum Field Theory, Volume 2
(Second Edition)
Advanced topics
Badis Ydri

Chapter 16

The AdS/CFT correspondence

The goal in this chapter is to provide a pedagogical presentation of the celebrated AdS/CFT correspondence adhering mostly to the language of quantum field theory (QFT). This is certainly possible, and perhaps even natural, if we recall that in this correspondence we are positing that quantum gravity in an anti-de Sitter spacetime AdS_{d+1} is nothing else but a conformal field theory (CFT_d) at the boundary of AdS spacetime. Some of the reviews of the AdS/CFT correspondence, which emphasize the QFT aspects and language include Kaplan [11], Zaffaroni [10] and Ramallo [6].

This chapter contains therefore a thorough introductions to conformal symmetries, anti-de Sitter spacetimes, conformal field theories and the AdS/CFT correspondence. The primary goal however in this chapter is the holographic entanglement entropy. In other words, how spacetime geometry as encoded in Einstein's equations in the bulk of AdS spacetime can emerge from the quantum entanglement entropy of the CFT living on the boundary of AdS.

A sample of the original literature for the holographic entanglement entropy is [39–41, 43]. However, a very good concise and pedagogical review of the formalism relating spacetime geometry to quantum entanglement due to Van Raamsdonk and collaborators is found in [41, 42].

16.1 Conformal symmetry

In this section we mostly follow [9]

16.1.1 The conformal groups $SO(p + 1, q + 1)$

We assume a spacetime with signature $(-1, \ldots, +1, \ldots)$ with q minus signs and p plus signs where $q + p = d$ is the dimension of spacetime.

We have then $q = 0$ and $p = d$ for Euclidean, and $q = 1$ and $p = d - 1$ for Lorentzian.

doi:10.1088/978-0-7503-5834-7ch16

We start with the group of diffeomorphisms, i.e., the group of general coordinate transformations. The conformal group is a subgroup of the diffeomorphism group, which preserves the conformal flatness of the metric. The corresponding transformations preserve the angles but change the lengths.

A manifold is called conformally flat if the metric takes the following form

$$ds^2 = \exp(\omega(x))dx^\mu dx_\mu. \tag{16.1}$$

Conformal transformations are given by

$$x^\mu \longrightarrow x'^\mu(x), \quad dx'^\mu dx'_\mu = \Omega^2(x)dx_\mu dx^\mu. \tag{16.2}$$

Thus if the manifold is conformally flat, it will remain so under conformal transformations. These conformal transformations are generalization of the scale transformation

$$x^\mu \longrightarrow x'^\mu = \alpha x^\mu, \quad dx'^\mu dx'_\mu = \alpha^2 dx_\mu dx^\mu. \tag{16.3}$$

We consider infinitesimal conformal transformations $x'_\mu = x_\mu + v_\mu$, $\Omega = 1 + \omega/2$. We get immediately from (16.2) the conditions

$$\omega = \frac{2}{d}\partial_\mu v^\mu. \tag{16.4}$$

$$\partial_\nu v^\mu + \partial^\mu v_\nu - \frac{2}{d}\partial_\rho v^\rho \eta^\mu_\nu = 0. \tag{16.5}$$

In $d = 2$ this equation admits an infinite number of solutions and thus the conformal group in two dimensions is infinite dimensional.

In $d \neq 2$ there is a finite number of solutions given precisely by

$$v_\mu = \delta x^\mu = a^\mu + \omega^\mu{}_\nu x^\nu + \lambda x^\mu + b^\mu x^2 - 2x^\mu bx. \tag{16.6}$$

The conformal group contains therefore:
1. Lorentz transformations $\Lambda \in SO(p, q)$, i.e., rotations and boosts, with parameters $\omega^\mu{}_\nu$. The finite transformations are $x \longrightarrow x' = \Lambda x$. There are $d(d-1)/2$ generators denoted by $M_{\mu\nu}$. They satisfy

$$[M_{\mu\nu}, M_{\rho\sigma}] = i(\eta_{\nu\rho} M_{\mu\sigma} + \eta_{\mu\sigma} M_{\nu\rho} - \eta_{\mu\rho} M_{\nu\sigma} - \eta_{\nu\sigma} M_{\mu\rho}). \tag{16.7}$$

2. Translations with parameters a^μ. There are d generators denoted by P^μ.
3. The scale (dilatation) transformation $x^\mu \longrightarrow \lambda x^\mu$ where λ is a constant. The generator of dilatation is denoted by D. A field theory, which is invariant under scale transformations, will also, under mild conditions, be invariant under all conformal transformations.
4. The special conformal transformations with parameters b^μ given by

$$\delta x^\mu = b^\mu x^2 - 2x^\mu bx. \tag{16.8}$$

A special conformal transformation is obtained from the composition of an inversion, a translation by a vector b^μ and another inversion. The inversion is an element in the conformal group, which is not connected to the identity given by

$$x^\mu \longrightarrow \frac{x^\mu}{x^2}. \tag{16.9}$$

There are d generators of special conformal transformations denoted by K^μ. The finite form of special conformal transformations is

$$x^\mu \longrightarrow x'^\mu = \frac{x^\mu + b^\mu x^2}{1 + 2bx + b^2 x^2} \Rightarrow x \longrightarrow x'^2 = \frac{x^2}{1 + 2bx + b^2 x^2}. \tag{16.10}$$

Altogether we have then $(d+1)(d+2)/2$ conformal transformations. This is exactly the number of generators of the rotation group $SO(d+2)$. However, the conformal group must be a non-compact group. It is therefore given by $SO(p+1, q+1)$. In Lorentzian signature the conformal group is $SO(d, 2)$ whereas in Euclidean signature the conformal group is $SO(d+1, 1)$. For $d > 2$ the conformal group is actually given by $O(p+1, q+1)$ and consists of two disconnected components since the inversion element is not infinitesimally generated.

16.1.2 Differential representation of the conformal algebra

The conformal group in Lorentzian signature $\eta = (-1, +1, \ldots, +1)$ is $O(d, 2)$ with generators P_μ, K_μ, $M_{\mu\nu}$ and D, which satisfy the algebra:

1. $M_{\mu\nu}$ generate the algebra of the Lorentz group $SO(d-1, 1)$, viz

$$[M_{\mu\nu}, M_{\rho\sigma}] = i(\eta_{\mu\sigma} M_{\rho\nu} - \eta_{\mu\rho} M_{\sigma\nu} + \eta_{\sigma\nu} M_{\mu\rho} - \eta_{\rho\nu} M_{\mu\sigma}). \tag{16.11}$$

2. D is a scalar under the Lorentz group, viz

$$[M_{\mu\nu}, D] = 0. \tag{16.12}$$

3. P_μ, K_μ are vectors under the Lorentz group, viz

$$[M_{\mu\nu}, P_\rho] = i(\eta_{\mu\rho} P_\nu - \eta_{\nu\rho} P_\mu), \quad [M_{\mu\nu}, K_\rho] = i(\eta_{\mu\rho} K_\nu - \eta_{\nu\rho} K_\mu). \tag{16.13}$$

4. D is the Hamiltonian and P_μ, K_μ are the raising and lowering operators since

$$[D, P_\mu] = P_\mu, \quad [D, K_\mu] = -K_\mu. \tag{16.14}$$

5. P_μ, K_μ close on a dilatation and a Lorentz transformation, viz

$$[P_\mu, K_\nu] = 2(\eta_{\mu\nu} D + iM_{\mu\nu}). \tag{16.15}$$

We define the new generators $\mathcal{M}_{AB}$ by the relations

$$\mathcal{M}_{AB} = M_{\mu\nu}, \quad A = \mu = 0, \ldots, d-1, \quad B = \nu = 0, \ldots, d-1. \tag{16.16}$$

$$\mathcal{M}_{\mu d} = -\mathcal{M}_{d\mu} = i\frac{K_\mu - P_\mu}{2}, \quad \mu = 0, \ldots, d-1. \tag{16.17}$$

$$\mathcal{M}_{\mu d+1} = -\mathcal{M}_{d+1\mu} = -i\frac{K_\mu + P_\mu}{2}, \quad \mu = 0, \ldots, d-1. \tag{16.18}$$

$$\mathcal{M}_{d d+1} = -\mathcal{M}_{d+1 d} = iD. \tag{16.19}$$

The algebra becomes

$$[\mathcal{M}_{AB}, \mathcal{M}_{CD}] = i(\eta_{AD}\mathcal{M}_{CB} - \eta_{AC}\mathcal{M}_{DB} + \eta_{DB}\mathcal{M}_{AC} - \eta_{CB}\mathcal{M}_{AD}). \tag{16.20}$$

The metric η_{AB} is a flat $(d+2)$-dimensional with signature $-1+1+1\ldots-1$. By going from Lorentzian to the Euclidean spacetime the conformal group $SO(d, 2)$ becomes the group $SO(d+1, 1)$ while the Lorentz group $SO(d-1, 1)$ becomes $SO(d)$ and the metric η_{AB} becomes of signature $-1+1+1\ldots+1$. The Poincaré and dilatation generators form together a subgroup of the conformal group.

The mass operator $P_\mu P^\mu$ is a Casimir of the Poincaré group but it is not a Casimir of the conformal group. States in a conformal field theory are therefore not classified by their mass, as in the Poincaré group, but they are instead classified by their scaling dimension, i.e., by the eigenvalue of the dilatation operator

$$D|\Delta\rangle = i\Delta|\Delta\rangle. \tag{16.21}$$

The representation of the dilatation operator (which is a Hermitian operator) on classical fields is not unitary and hence the factor i. Indeed, dilatations are not bounded transformations and as a consequence a finite dimensional representation of a non-compact Lie algebra must be necessarily non-unitary. This happens also for example with boosts in the Lorentz group [10, 14, 15].

The scaling dimension Δ of a field Φ is defined by the action of dilatations on the field given by the equation

$$\Phi(\lambda x) = \lambda^{-\Delta}\Phi(x). \tag{16.22}$$

If we take for example a free scalar field theory in d dimensions given by the action

$$\begin{aligned}
S &= \int d^d x\, \partial_\mu\Phi(x)\partial^\mu\Phi(x) \\
&= \lambda^{-d}\lambda^{2\Delta}\int d^d(\lambda x)\, \partial_\mu\Phi(\lambda x)\partial^\mu\Phi(\lambda x) \tag{16.23} \\
&= \lambda^{-d+2\Delta+2}\int d^d x'\, \partial'_\mu\Phi(x')\partial^{\mu'}\Phi(x').
\end{aligned}$$

Hence this invariant under dilatation if and only if

$$-d + 2\Delta + 2 = 0. \tag{16.24}$$

The differential representation of the dilatation operator on scalar fields with scaling dimension Δ is given by

$$D\Phi = -i(x^\mu \partial_\mu + \Delta)\Phi. \tag{16.25}$$

This is the analogue of the differential representation of the momentum and Lorentz generators given by

$$P_\mu \Phi = -i\partial_\mu \Phi. \tag{16.26}$$

$$M_{\mu\nu}\Phi = i(x_\mu \partial_\nu - x_\nu \partial_\mu)\Phi + S_{\mu\nu}\Phi. \tag{16.27}$$

The special conformal generator is represented differentially by

$$K_\mu \Phi = (-2i\Delta x_\mu - x^\nu S_{\mu\nu} - 2ix_\mu x^\nu \partial_\nu + ix^2 \partial_\mu)\Phi. \tag{16.28}$$

By using (16.25) we can rewrite the quantum analogue of the scaling transformation $\Phi(x) \longrightarrow \lambda^\Delta \phi(\lambda x)$ as

$$[D, \Phi] = -i(x^\mu \partial_\mu + \Delta)\Phi. \tag{16.29}$$

Under a coordinate transformation $x \longrightarrow x'$ the metric tensor transforms as $g_{\alpha\beta}(x) \longrightarrow g'_{\alpha\beta}(x') = (\partial x^\mu / \partial x'^\alpha)(\partial x^\nu / \partial x'^\beta)g_{\mu\nu}(x)$. The conformal group is the subgroup, which satisfies $g_{\alpha\beta}(x) \longrightarrow g'_{\alpha\beta}(x') = \Omega^2(x)g_{\mu\nu}(x)$. From the invariance of the volume element we can see that the Jacobian of the conformal transformation $x \longrightarrow x'$, $dx'^\mu dx'_\mu = \Omega^2(x)dx_\mu dx^\mu$ is given by

$$\left|\frac{\partial x'}{\partial x}\right| = \frac{1}{\sqrt{\det g'_{\mu\nu}}} = \Omega^{-d}. \tag{16.30}$$

The action of the generators of the conformal group on the scalar field Φ with scaling dimension Δ given by the transformations (16.25), (16.26), (16.27), (16.28) translates into the transformation law

$$\Phi(x) \longrightarrow \Phi'(x') = \left|\frac{\partial x'}{\partial x}\right|^{-\Delta/d}\Phi(x). \tag{16.31}$$

The scalar field Φ is called quasi-primary operator. The covariance of the theory on conformal transformations is then given by the behavior of the correlation functions under conformal transformations given by

$$\langle \Phi_1(x_1) \ldots \Phi_N(x_N)\rangle \longrightarrow \langle \Phi'(x_1') \ldots \Phi'(x_N')\rangle = \left|\frac{\partial x'}{\partial x}\right|_{x=x_1}^{-\Delta_1/d} \ldots \left|\frac{\partial x'}{\partial x}\right|_{x=x_N}^{-\Delta_N/d} \langle \Phi_1(x_1) \ldots \Phi_N(x_N)\rangle. \tag{16.32}$$

For a scalar field we have obviously $\Phi' = \Phi$.

16.1.3 Constraints of conformal symmetry

Conformal invariance fully constrains the two- and three-point functions of the conformal field theory. For the two-point function of two quasi-primary operators ϕ_1 and ϕ_2 we have

$$\langle \Phi_1(x_1)\Phi_2(x_2)\rangle \longrightarrow \langle \Phi(x_1')\Phi(x_2')\rangle = \left|\frac{\partial x'}{\partial x}\right|_{x=x_1}^{-\Delta_1/d}\left|\frac{\partial x'}{\partial x}\right|_{x=x_2}^{-\Delta_2/d} \langle \Phi_1(x_1)\Phi_2(x_2)\rangle. \tag{16.33}$$

In other words, we have

$$\langle \Phi_1(x_1)\Phi_2(x_2)\rangle = |\frac{\partial x'}{\partial x}|_{x=x_1}^{\Delta_1/d}|\frac{\partial x'}{\partial x}|_{x=x_2}^{\Delta_2/d}\langle \Phi(x_1')\Phi(x_2')\rangle. \tag{16.34}$$

Invariance under translations and rotations (for which the Jacobian is equal 1) yields a dependence only on the combination $r_{12} = |x_1 - x_2|$ (the difference is due to translations and the modulus is due to rotations). Invariance under the scaling transformations $x \longrightarrow x' = \lambda x$ leads to

$$\langle \Phi_1(x_1)\Phi_2(x_2)\rangle = \lambda^{\Delta_1 + \Delta_2}\langle \Phi(x_1')\Phi(x_2')\rangle. \tag{16.35}$$

This gives immediately the behavior

$$\langle \Phi_1(x_1)\Phi_2(x_2)\rangle = \frac{C_{12}}{r_{12}^{\Delta_1 + \Delta_2}}. \tag{16.36}$$

Under special conformal transformation we have

$$x^\mu \longrightarrow x'^\mu = \frac{x^\mu + b^\mu x^2}{1 + 2bx + b^2x^2} \Rightarrow r_{12}'^2 = \frac{r_{12}^2}{(1 + 2bx_1 + b^2x_1^2)(1 + 2bx_2 + b^2x_2^2)}. \tag{16.37}$$

$$|\frac{\partial x'}{\partial x}| = \frac{1}{(1 + 2bx + b^2x^2)^d}. \tag{16.38}$$

We can then easily verify that we must have $\Delta_1 = \Delta_2 = \Delta$. Hence the two-point function of a conformal field theory must be constrained such that

$$\langle \Phi_1(x_1)\Phi_2(x_2)\rangle = \frac{C_{12}}{r_{12}^{2\Delta}}. \tag{16.39}$$

However, if $\Delta_1 \neq \Delta_2$ then we must have $\langle \Phi_1(x_1)\Phi_2(x_2)\rangle = 0$.

Similarly, the three-point function of a conformal field theory must be constrained by the invariance under translations, rotations, scalings and special conformal transformations such that

$$\langle \Phi_1(x_1)\Phi_2(x_2)\Phi_3(x_3)\rangle = \frac{C_{123}}{r_{12}^{\Delta_1 + \Delta_2 - \Delta_3} r_{13}^{\Delta_1 + \Delta_3 - \Delta_2} r_{23}^{\Delta_2 + \Delta_3 - \Delta_1}}. \tag{16.40}$$

16.1.4 Conformal algebra in two dimensions

We consider a free boson Φ in two dimensions with the Euclidean action (with $z = \sigma^1 + i\sigma^2$)

$$S = \frac{1}{4\pi} \int d^2\sigma \partial_\mu \Phi \partial^\mu \Phi = \frac{1}{2\pi} \int dz d\bar{z} \partial \Phi \bar{\partial} \Phi. \tag{16.41}$$

The symmetries of this action are given by the conformal mappings

$$z \longrightarrow f(z), \quad \bar{z} \longrightarrow \bar{z} = \bar{f}(\bar{z}). \tag{16.42}$$

These are angle-preserving transformations when f and its inverse are both holomorphic, i.e., f is biholomorphic. For example, $z \longrightarrow z + a$ is a translation, $z \longrightarrow \zeta z$ where $|\zeta| = 1$ is a rotation, and $z \longrightarrow \zeta z$ where ζ is real not equal to 1 is a scale transformation called also dilatation.

We work with the complex coordinates

$$z = \sigma^1 + i\sigma^2 \longrightarrow z = e^{2(\sigma^2 - i\sigma^1)}, \quad \bar{z} = \sigma^1 - i\sigma^2 \longrightarrow \bar{z} = e^{2(\sigma^2 + i\sigma^1)}. \tag{16.43}$$

The factor of 2 in the exponent is included for consistency with the periodicity of the closed string given by $\sigma^1 \longrightarrow \sigma^1 + \pi$. The worldsheet for a closed string is a cylinder, which is topologically an $\mathbf{R}^2$. It can also be regarded as a Riemann surface, i.e., as a deformation of the complex plane. The Euclidean time σ^2 on the worldsheet corresponds to the radial distance $r = \exp(2\sigma^2)$ on the complex plane, with the infinite past $\sigma^2 = -\infty$ at $r = 0$, and the infinite future $\sigma^2 = +\infty$ is a circle at $r = \infty$.

The generators of conformal mappings are given by the infinitesimal transformations

$$z \longrightarrow z' = z - \epsilon_n z^{n+1}, \quad \bar{z} \longrightarrow \bar{z}' = \bar{z} - \bar{\epsilon}_n \bar{z}^{n+1}, \quad n \in Z. \tag{16.44}$$

The generators are immediately given by

$$l_n = -z^{n+1}\partial, \quad \bar{l}_n = -\bar{z}^{n+1}\bar{\partial}, \quad n \in Z. \tag{16.45}$$

The generators with $n < -1$ are defined on the punctured complex plane whereas the generators with $n > 1$ are defined on the complex plane with the point at infinity removed. The generators l_{-1}, l_0, l_1 are defined on the whole Riemann sphere, i.e., the complex plane+the point at infinity. They satisfy the classical Virasoro algebra

$$[l_m, l_n] = (m - n)l_{m+n}, \quad [\bar{l}_m, \bar{l}_n] = (m - n)\bar{l}_{m+n}, \quad [l_m, \bar{l}_n] = 0. \tag{16.46}$$

It is also easily seen that the Virasoro algebra is the same as the algebra of infinitesimal diffeomorphisms of the circle $\mathbf{S}^1$.

The group $SO(3, 1)$ is called the restricted conformal group. The full conformal group in two dimensions is infinite dimensional. The finite dimensional subgroup $SO(3, 1)$ is generated by $l_0, l_{\pm 1}, \bar{l}_0, \bar{l}_{\pm 1}$. These are given explicitly by

$$
\begin{aligned}
l_{-1}: & \quad z \longrightarrow z - \epsilon \\
l_0: & \quad z \longrightarrow z - \epsilon z \\
l_1: & \quad z \longrightarrow z - \epsilon z^2.
\end{aligned} \tag{16.47}
$$

Similarly for $\bar{l}_{-1}, \bar{l}_0, \bar{l}_1$. We have the following interpretation:

$$
\begin{aligned}
l_{-1}, \; \bar{l}_{-1} & \quad \text{translations} \\
l_0 - \bar{l}_0 & \quad \text{rotations} \\
l_0 + \bar{l}_0 & \quad \text{scalings} \\
l_1, \; \bar{l}_1 & \quad \text{special conformal transformations.}
\end{aligned} \tag{16.48}
$$

The finite or global form of these transformations are:

$$\begin{aligned}
l_{-1}:\ &z \longrightarrow z + \alpha \\
l_0:\ &z \longrightarrow \lambda z \\
l_1:\ &z \longrightarrow \frac{z}{1 - \beta z}.
\end{aligned} \qquad (16.49)$$

By combining these transformations we obtain

$$z \longrightarrow \frac{az + b}{cz + d}, \quad ad - bc = 1. \qquad (16.50)$$

For infinitesimal z we obtain

$$z \longrightarrow \frac{b}{d} + \frac{1}{d^2}z - \frac{c}{d^3}z^2 \qquad (16.51)$$

Only three parameters are independent as it should be. We obtain a linear combination of the transformations (16.47).

The group given by the relation (16.50) is $SO(3, 1) = SL(2, C)/\mathbf{Z}_2$. The division by $\mathbf{Z}_2$ is to take into account the property that the above transformations remain unchanged if $a, b, c, d \longrightarrow -a, -b, -c, -d$. The Lorentzian analogue is the group $SO(2, 2) = SL(2, R) \times SL(2, R)$ where one factor of $SL(2, R)$ stands for left-movers and the other factor stands for right-movers.

16.2 The AdS spacetime

For an extensive discussion of anti-de Sitter spacetimes and their uses, see [3, 4].

16.2.1 Maximally symmetric spaces

The maximally symmetric manifolds of dimension d are those spaces with the maximum number $d(d + 1)/2$ of Killing vector fields generating isometries, i.e., symmetries, consisting of d translations and $d(d - 1)/2$ rotations/boosts. A space is a maximally symmetric manifold if and only if its Riemann curvature tensor is given by

$$R_{\mu\nu\alpha\beta} = \frac{R}{d(d - 1)}(g_{\mu\alpha}g_{\nu\beta} - g_{\mu\beta}g_{\nu\alpha}), \qquad (16.52)$$

where g is of course the metric tensor. This means that in a maximally symmetric space the curvature tensor looks the same everywhere and in every direction and thus it is fully specified locally by the Ricci scalar curvature R. Hence, there are only three possible maximally symmetric spaces locally specified by the sign of the Ricci scalar R (since the scale of R specifies only the size of the space), which are given by:

1. **Positive Curvature:** The spheres $\mathbf{S}^d$ in Euclidean signature and de Sitter spacetimes dS_d in Lorentzian signature. The de Sitter spacetime plays a crucial role in cosmology in the early Universe during inflation (when the cosmological constant was very large and the expansion was exponential) as

well as in the final state of the Universe, which is seen to be dominated by a very small cosmological constant and an accelerated expansion.

2. **Zero Curvature:** The Euclidean spaces $\mathbf{R}^d$ in Euclidean signature and Minkowski spacetimes $\mathbf{M}^d$ in Lorentzian signature.

3. **Negative Curvature:** The hyperboloids $\mathbf{H}^d$ in Euclidean signature and anti-de Sitter spacetimes AdS_d in Lorentzian signature. The anti-de Sitter spacetime is crucial for the holographic principle and the AdS/CFT correspondence. This is the primary point of interest to us here in this chapter.

For more detail see chapters 3 and 8 in [2].

16.2.2 Global and Poincaré coordinates

16.2.2.1 Global coordinates

We are interested mostly in AdS_5. Thus, we start from a six-dimensional flat spacetime $\mathbf{R}^{2,4}$ with metric

$$ds_5^2 = -dX_0^2 - dX_5^2 + dX_1^2 + dX_2^2 + dX_3^2 + dX_4^2. \tag{16.53}$$

The isometry group of this metric is obviously $SO(2,4)$, which is precisely the conformal group in 4 dimensions. For AdS_{d+1} we need to start from $\mathbf{R}^{2,d}$ with isometry group $SO(2,d)$. We Lobachevski-like embed in this Minkowski spacetime $\mathbf{R}^{2,4}$ the following hyperboloid

$$-X_0^2 - X_5^2 + X_1^2 + X_2^2 + X_3^2 + X_4^2 = -L^2. \tag{16.54}$$

This hyperboloid is obviously five-dimensional. Now we can induce global coordinates $(\tau, \rho, \hat{x}_i)$ on this hyperboloid where $\hat{x}_i$ define an $\mathbf{S}^3$, i.e., $\sum_{i=1}^4 \hat{x}_i^2 = 1$, by the relations

$$\begin{aligned}
X_0 &= L \cos\tau \cosh\rho \\
X_5 &= L \sin\tau \cosh\rho \\
X_i &= L \sinh\rho \hat{x}_i.
\end{aligned} \tag{16.55}$$

The metric becomes (with $d\Omega_3$ is the solid angle on $\mathbf{S}^3$)

$$ds_5^2 = L^2(-\cosh^2\rho d\tau^2 + d\rho^2 + \sinh^2\rho d\Omega_3). \tag{16.56}$$

The coordinate ρ plays the role of a radius in AdS_5 since $\rho \in [0, +\infty[$ whereas the coordinate τ is timelike with $\tau \in [0, 2\pi]$. These coordinates cover the whole of AdS_5, which is the reason why they are called global. On the other hand, τ is periodic, which signals the existence of closed timelike curves. However, this property is not intrinsic to the AdS_5 space but it is an artifact of this system of coordinates. Instead of the above space we will then take its universal cover space, in which we allow τ to run over the unrestricted range $]-\infty, +\infty[$, as the definition of anti-de Sitter spacetime AdS_5. The isometry group becomes a cover of $SO(2,4)$.

The anti-de Sitter spacetime, as opposed to de Sitter spacetime, is not a globally hyperbolic spacetime. This means that AdS does not admit a well-defined time evolution starting from suitable initial data on a spacelike (Cauchy) hypersurface. Indeed, specifying the initial data on a spacelike hypersurface in *AdS* (together with the knowledge of the equations of motion) is not sufficient to determine the future evolution uniquely and deterministically. This is due to the existence of a boundary at timelike infinity (see below) and thus information can flow in from infinity. See [1] for a systematic discussion of this issue.

The other difference between *dS* and *AdS* is with regard to topology. AdS_5 spacetime is topologically equivalent to $\mathbf{R}^4 \times \mathbf{S}^1$ whereas dS_5 spacetime is topologically equivalent to $\mathbf{R} \times \mathbf{S}^4$. This can be fleshed out using Penrose diagrams. See [2, 4] for explicit discussion.

16.2.2.2 Poincaré coordinates

Next set of coordinates is much more important for quantum field theory. We introduce the following so-called Poincaré coordinates given by a Minkowski four-vector x_μ and a radial coordinate u defined by the relations

$$
\begin{aligned}
X_0 &= \frac{1}{2u}(1 + u^2(L^2 + \vec{x}^2 - t^2)) \\
X_i &= Lux_i \\
X_4 &= \frac{1}{2u}(1 - u^2(L^2 - \vec{x}^2 + t^2)) \\
X_5 &= Lut.
\end{aligned}
\tag{16.57}
$$

We get immediately the metric

$$
ds_5^2 = L^2\left(\frac{du^2}{u^2} + u^2(dx_i^2 - dt^2)\right) = L^2\left(\frac{du^2}{u^2} + u^2(dx_\mu dx^\mu)\right).
\tag{16.58}
$$

Thus, for each fixed value of the radial coordinate u we have a four-dimensional ordinary Minkowski spacetime, i.e., the ordinary spacetime is foliated over the radial coordinate u which takes the value from 0 to infinity. However, because of the overall conformal factor R^2u^2 multiplying the Minkowski metric all distances in the four-dimensional theory on the Minkowski slices or branes are rescaled by a factor of Ru.

The point $u = \infty$ is a conformal boundary since it is the conformally equivalent metric ds_5^2/u^2, which is seen to have a boundary $\mathbf{M}^4$ at $u = \infty$. However, the point $u = 0$ is a horizon since the 00 component of the metric vanishes there and hence the Killing vector $\partial/\partial t$ becomes of zero norm at this point. Furthermore, the metric can be extended beyond the horizon $u = 0$, which is only a coordinate singularity and the Poincaré coordinates cover only half of the hyperboloid.

Another form of the metric can be obtained by the substitution $u = 1/z$ and thus the conformal boundary becomes located at $z = 0$ whereas the horizon becomes located at $z = \infty$. The metric takes the form

$$
ds_5^2 = \frac{L^2}{z^2}(dz^2 + dx_\mu dx^\mu). \; 4
\tag{16.59}
$$

Another system of coordinates can be obtained by the substitution $\sinh \rho = 2r/(1 - r^2)$ in the system of coordinates, i.e., r runs from 0 to 1. We obtain the metric

$$ds_5^2 = \frac{L^2}{(1 - r^2)^2}(-(1 + r^2)^2 d\tau^2 + 4dr^2 + 4r^2 d\Omega_3). \tag{16.60}$$

The metric (16.59) should be thought of as an approximation of the metric (16.60) in the vicinity of a point on the boundary located now at $r = 1$. In this approximation the 6jsphere is replaced with the flat coordinates x_i, t is replaced with τ, and the radial coordinate z is replaced with $z = 1 - r$. In this system of coordinates $r = 0$ is the center of anti-de Sitter and there is no horizon since the metric (16.60) is geodesically complete. However, the metric (16.59) is geodesically incomplete because we can reach $z = \infty$ along a timelike geodesic in a finite proper time, i.e., $z = \infty$ is a horizon.

Furthermore, we can check that we can travel from the center $r = 0$ of anti-de Sitter to the boundary $r = 1$ (which is an infinite proper distance) and back along a null curve satisfying $(1 + r^2)d\tau = 2dr$ in a finite proper time given by $\tau = \pi$. This means that anti-de Sitter space is causally finite and it behaves as a finite box of size L. See [5] for more discussion on the relation between the metrics (16.59) and (16.60).

16.2.2.3 Generalization

We generalize to anti-de Sitter in $d + 1$ dimensions, i.e., AdS_{d+1}, with isometry group $SO(2, d)$ which is precisely the conformal group in d dimensions. AdS_{d+1} spacetime is topologically equivalent to $\mathbf{R}^d \times \mathbf{S}^1$ (to be contrasted with the topology of dS_{d+1} spacetime given by $\mathbf{R} \times \mathbf{S}^d$). We embed AdS_{d+1} in the Minkowski spacetime $\mathbf{R}^{2, d}$ by

$$-X_0^2 - X_{d+1}^2 + X_1^2 + X_2^2 + X_3^2 + \cdots + X_d^2 = -L^2. \tag{16.61}$$

Now we can induce global coordinates $(\tau, \rho, \hat{x}_i)$, or the slightly different ones $(\tau, r, \hat{x}_i)$, on this hyperboloid by the relations

$$X_0 = L \cos \tau \cosh \rho = L\frac{\cos \tau}{\cos r}$$

$$X_{d+1} = L \sin \tau \cosh \rho = L\frac{\sin \tau}{\cos r} \tag{16.62}$$

$$X_i = L \sinh \rho \hat{x}_i = L \tan r \hat{x}_i.$$

The range is $\rho \in [0, +\infty[$ or correspondingly $r \in [0, \pi/2[$, and $\tau \in [0, 2\pi] \longrightarrow \tau \in [-\infty, \infty]$ as we unwrap to the universal cover, and $\hat{x}_i$ defines a $(d - 1)$-dimensional sphere $\mathbf{S}^{d-1}$, i.e, $\sum_{i=1}^4 \hat{x}_i^2 = 1$. The metric reads

$$ds_{d+1}^2 = L^2(-\cosh^2 \rho d\tau^2 + d\rho^2 + \sinh^2 \rho d\Omega_{d-1}) = \frac{L^2}{\cos^2 r}(-d\tau^2 + dr^2 + \sin^2 r d\Omega_{d-1}). \tag{16.63}$$

Thus, AdS_{d+1} can be viewed as a cylinder with bases at $\tau = -\infty$ and $\tau = +\infty$, a center at $\rho = r = 0$ whereas the spatial infinity $\rho = \infty$ is at $r = \pi/2$, while going around the cylinder is given by the angular variables Ω_{d-1}. The anti-de Sitter space is therefore not compact both in time and space yet it behaves as a box as we will discuss.

The symmetry group of AdS_{d+1} is given by the conformal group $SO(2, d)$. We have $SO(d)$ rotations among the X_i, $SO(2)$ rotation in the plane $X_0 X_{d+1}$, d boosts in the planes $X_0 X_i$ and d boosts in the planes $X_{d+1} X_i$. In total we have $d(d-1)/2 + 1 + 2d = (d+1)(d+2)/2$ generators. These generators can be represented in terms of the coordinates X_A as

$$L_B^A = X^A \frac{\partial}{\partial X^B} - X_B \frac{\partial}{\partial X_A}. \tag{16.64}$$

We will also need later

$$L^{AB} = X^A \frac{\partial}{\partial X_B} - X^B \frac{\partial}{\partial X_A}, \quad L_{AB} = X_A \frac{\partial}{\partial X^B} - X_B \frac{\partial}{\partial X^A}. \tag{16.65}$$

For example, the rotation in the timelike plane $X_0 X_{d+1}$ is given by (recall the metric signature $-++\cdots-$)

$$L_{d+1}^{0} = X_0 \frac{\partial}{\partial X_{d+1}} - X_{d+1} \frac{\partial}{\partial X_0}. \tag{16.66}$$

On the other hand, we have

$$\frac{\partial}{\partial t} = \frac{\partial X_0}{\partial t} \frac{\partial}{\partial X_0} + \frac{\partial X_{d+1}}{\partial t} \frac{\partial}{\partial X_{d+1}} = -X_{d+1} \frac{\partial}{\partial X_0} + X_0 \frac{\partial}{\partial X_{d+1}}. \tag{16.67}$$

Hence the Hamiltonian in anti-de Sitter spacetime is given by

$$\text{Hamiltonian} \equiv -i \frac{\partial}{\partial t} = -i L_{d+1}^{0}. \tag{16.68}$$

16.2.3 Euclidean Poincaré patch and RG equation

16.2.3.1 Poincaré coordinates revisited
Let us start with the most general metric in $(d+1)$-dimension which enjoys Poincaré invariance in d dimensions given by

$$ds_{d+1}^2 = \Omega^2(z)(dz^2 + dx_\mu dx^\mu) = \Omega^2(z)(dz^2 + d\vec{x}^2 - dt^2). \tag{16.69}$$

As we will see the extra coordinate z corresponds to an energy scale of a conformally invariant theory. Hence the above metric must be invariant under $z \longrightarrow z, \vec{x} \longrightarrow \lambda \vec{x}$ and $t \longrightarrow \lambda t$. This leads immediately to the requirement

$$ds_{d+1}^2 = \Omega^2(z)(dz^2 + d\vec{x}^2 - dt^2) \longrightarrow \lambda^2 \Omega^2(\lambda z)(dz^2 + d\vec{x}^2 - dt^2) = \Omega^2(z)(dz^2 + d\vec{x}^2 - dt^2). \tag{16.70}$$

The function Ω must then satisfy

$$\Omega(z) = \frac{L}{z}. \tag{16.71}$$

We get then

$$ds_{d+1}^2 = \frac{L^2}{z^2}(dz^2 + dx_\mu dx^\mu) = \frac{L^2}{z^2}(dz^2 + d\vec{x}^2 - dt^2). \tag{16.72}$$

This is precisely anti-de Sitter spacetime AdS_{d+1} in Poincaré coordinates with boundary at $z = 0$ and horizon at $z = \infty$. The constant L is the radius of anti-de Sitter. This metric solves Einstein equations with cosmological constant Λ, viz

$$R_{\mu\nu} - \frac{1}{2}g_{\mu\nu}R = -\Lambda g_{\mu\nu}. \tag{16.73}$$

The Ricci tensor and Ricci scalar of the metric (16.165) are given by

$$R_{\mu\nu} = -\frac{d}{L^2}g_{\mu\nu} \Rightarrow R = -\frac{d(d+1)}{L^2}. \tag{16.74}$$

Thus, the metric (16.165) defines an Einstein space. However, by contracting both sides of the Einstein equations (16.73) we obtain the Ricci scalar

$$R = 2\frac{d+1}{d-1}\Lambda. \tag{16.75}$$

By comparing the above two last equations we can determine the cosmological constant in terms of the radius of anti-de Sitter by the relation

$$\Lambda = -\frac{d(d-1)}{2L^2}. \tag{16.76}$$

16.2.3.2 Euclidean Poincaré coordinates
The relationship between the global coordinates τ, ρ and $\hat{x}_i$ (with a manifest $SO(2) \times SO(d)$ symmetry subgroup) and the Poincaré coordinates t, z, $\vec{x}$ (with the d-dimensional Poincaré symmetry subgroup manifest) is given by

$$X_0 = L\cos\tau\cosh\rho = \frac{z}{2}\left(1 + \frac{L^2 + \vec{x}^2 - t^2}{z^2}\right)$$

$$X_i = L\sinh\rho\hat{x}_i = L\frac{x_i}{z}, \quad i < d$$

$$X_d = L\sinh\rho\hat{x}_d = \frac{z}{2}\left(1 - \frac{L^2 - \vec{x}^2 + t^2}{z^2}\right) \tag{16.77}$$

$$X_{d+1} = L\sin\tau\cosh\rho = L\frac{t}{z}.$$

Although these Poincaré coordinates cover only half of the AdS_{d+1} spacetime (with conformal group $SO(2, d)$) their Euclidean analogues cover all of the Euclidean AdS_{d+1} space (with conformal group $SO(1, d + 1)$). The Euclidean AdS_{d+1} space is obtained by the Wick rotation $X_{d+1} \longrightarrow - iX_{d+1}$, i.e., it is given by embedding

$$-X_0^2 + X_{d+1}^2 + \sum_{i=1}^{d} X_i^2 = L^2. \tag{16.78}$$

This corresponds to the Wick rotations $t \longrightarrow - it$ and $\tau \longrightarrow - i\tau$. The metric becomes

$$ds_{d+1}^2 = R^2(\cosh^2 \rho d\tau^2 + d\rho^2 + \sinh^2 \rho d\Omega_{d-1}) = \Omega^2(z)(dz^2 + d\vec{x}^2 + dt^2). \tag{16.79}$$

The map between these systems is then given by

$$X_0 = L \cosh \tau \cosh \rho = \frac{z}{2}\left(1 + \frac{L^2 + \vec{x}^2 + t^2}{z^2}\right)$$

$$X_i = L \sinh \rho \hat{x}_i = L\frac{x_i}{z}, \quad i < d + 1 \tag{16.80}$$

$$X_{d+1} - L \sinh \tau \cosh \rho - \frac{z}{2}\left(1 - \frac{L^2 - \vec{x}^2 - t^2}{z^2}\right).$$

The boundary $z = 0$ in the Minkowski metric becomes $\mathbf{R}^d$ in the Euclidean metric whereas the horizon $z = \infty$ in the Minkowski metric shrinks to a point in the Euclidean metric. By adding the point $z = \infty$ to the boundary $\mathbf{R}^d$ we obtain a sphere $\mathbf{S}^d$. This compactified Euclidean AdS_{d+1} is thus the solid $(d + 1)$-dimensional ball.

16.2.3.3 Renormalization group equation
As we said the extra coordinate z corresponds to an energy scale of a conformal field theory, i.e., it defines a lattice spacing a. The d-dimensional slices or branes, defined by the points $x = (x_0, x_1, \ldots, x_d)$ in the higher dimensional AdS_{d+1} spacetime, should then be regarded as lattices of increasing size in a Kadanoff–Wilson renormalization group approach [7, 8].

To exhibit this crucial point in some detail we start with a field theory Hamiltonian on a lattice a, with coupling constants or sources $J_i(x, a)$ and field operators $\mathcal{O}_i$, given by [6]

$$H = \sum_{i,x} J_i(x, a)\mathcal{O}_i(x). \tag{16.81}$$

Then under the Kadanoff–Wilson renormalization group approach the lattice is coarse grained, i.e., we increase the lattice spacing successively as $\mu = a \longrightarrow 2a \longrightarrow 4a\ldots$ and replace at each step the spins (fields) by block spins (averages of fields) [7, 8]. As a result the Hamiltonian will change in such a way that its form remains invariant but only the coupling constants $J_i(x, \mu)$ change or flow, i.e., the weightings of the field operators $\mathcal{O}_i$ flow, according to the renormalization group equation

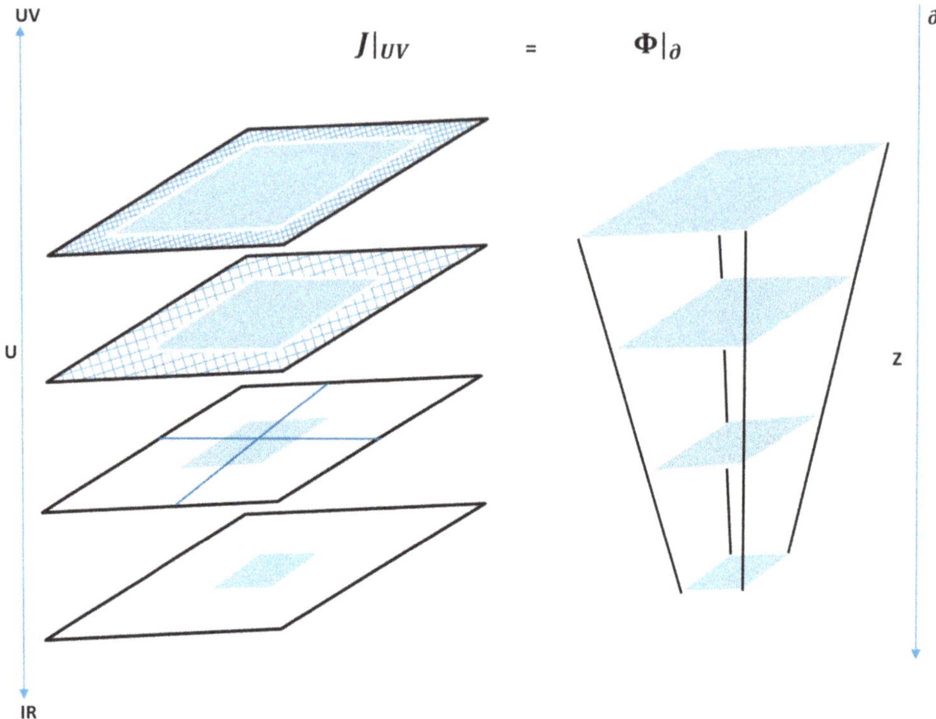

Figure 16.1. AdS_{d+1} as a foliation of M_d over the energy scale.

$$\mu\frac{\partial}{\partial\mu}J_i(x,\mu) = \beta_i(J_j(x,\mu),\mu). \tag{16.82}$$

In the AdS/CFT proposal we view the lattice scale μ as an extra dimension and as a consequence the collection of the lattices $\mu = a, 2a, 4a, \ldots$ should be viewed as slices of a higher dimensional space. The coupling constants $J_i(x,\mu)$ should then be reinterpreted as fields $\Phi_i(x,\mu)$ in this new higher dimensional space with the equation of motion in the extra dimension given by the above renormalization group equation. The dynamics of these bulk fields $\Phi_i(x,\mu)$ is determined by a gravitational action with their boundary values equated with the microscopic/continuum values of the coupling constants $J_i(x,\mu)$ of the field theory (corresponding to the operators $\mathcal{O}_i$) in the UV [6]. See figure 16.1.

16.3 Scalar field in AdS_{d+1}

16.3.1 The Klein–Gordon equation

Recall that AdS_{d+1} in this system can be viewed as a cylinder with bases at $\tau = -\infty$ and $\tau = +\infty$, a center at $\rho = r = 0$ whereas the spatial infinity is at $r = \pi/2$, while going around the cylinder is given by the angular variables Ω_{d-1}. The embedding is given explicitly by (the radius is denoted here by R)

$$X_0 = R\frac{\cos \tau}{\cos r}$$

$$X_{d+1} = R\frac{\sin \tau}{\cos r} \tag{16.83}$$

$$X_\mu = R \tan r \hat{x}_\mu.$$

Recall also the ranges $r \in [0, \pi/2[$, and $\tau \in [-\infty, \infty]$ as we unwrap to the universal cover and $\hat{x}_\mu$ defines a $(d-1)$-dimensional sphere $\mathbf{S}^{d-1}$. We will consider global coordinates on AdS_{d+1} given by

$$ds_{d+1}^2 = \frac{R^2}{\cos^2 r}(-d\tau^2 + dr^2 + \sin^2 r\, d\Omega_{d-1}). \tag{16.84}$$

We start by writing the action of a scalar field ϕ in AdS_{d+1} with $SO(2, d)$ invariance given by

$$S = \int d^{d+1}x \sqrt{g}\left[-\frac{1}{2}g^{MN}\partial_M\phi\partial_N\phi - \frac{1}{2}m^2\phi^2 \right]. \tag{16.85}$$

The Klein–Gordon equation is the Euler–Lagrange equation derived from this action given by

$$\frac{1}{\sqrt{g}}\partial_M(\sqrt{g}\,g^{MN}\partial_N\phi) - m^2\phi = 0. \tag{16.86}$$

16.3.2 Example of AdS_2

For simplicity we consider AdS_2 with isometry given by $SO(2, 1)$ and metric $ds_2^2 = R^2(-d\tau^2 + dr^2)/\cos^2 r$. We obtain the equation

$$-\partial_\tau^2\phi = \left(-\partial_r^2 + \frac{R^2m^2}{\cos^2 r}\right)\phi. \tag{16.87}$$

We pull the time dependence as $\phi = \exp(i\Delta\tau)\chi(r)$ where E is the energy. We get immediately

$$\Delta^2\chi = \left(-\partial_r^2 + \frac{R^2m^2}{\cos^2 r}\right)\chi. \tag{16.88}$$

A solution is given by the ansatz

$$\chi(r) = \cos^\alpha r. \tag{16.89}$$

The equation of motion becomes algebraic given by

$$(\Delta^2 - \alpha^2)\chi = (-\alpha(\alpha - 1) + R^2m^2)\chi \Rightarrow \Delta^2 - \alpha^2 = -\alpha(\alpha - 1) + R^2m^2 = 0. \tag{16.90}$$

In other words, $\chi(r) = \cos^\alpha r$ is a solution if the mass m of the scalar field is related to its energy Δ (in the ground state) by the relation

$$R^2m^2 = \Delta(\Delta - 1). \tag{16.91}$$

This crucial result can also be found using group theoretic method as follows.

Recall the signature $-++\cdots-$ and that the number of generators of $SO(2, d)$ is $(d + 1)(d + 2)/2$. Thus, the number of generators of $SO(2, 1)$ is 3 and they are given by L_2^0, L_1^0 and L_2^1. Explicitly we have

$$L_2^0 = X^0\frac{\partial}{\partial X^2} - X_2\frac{\partial}{\partial X_0} = \partial_\tau. \tag{16.92}$$

$$L_1^0 = X^0\frac{\partial}{\partial X^1} - X_1\frac{\partial}{\partial X_0} = \sin\tau\,\sin r\partial_\tau - \cos\tau\,\cos r\partial_r. \tag{16.93}$$

$$L_2^1 = X^1\frac{\partial}{\partial X^2} - X_2\frac{\partial}{\partial X_1} = -\cos\tau\,\sin r\partial_\tau - \sin\tau\,\cos r\partial_r. \tag{16.94}$$

They satisfy the algebra

$$[L_2^0, L_1^0] = -L_2^1, \quad [L_2^0, L_2^1] = L_1^0, \quad [L_1^0, L_2^1] = L_2^0. \tag{16.95}$$

Or equivalently

$$[D, P] = P, \quad [D, K] = -K, \quad [K, P] = -2D. \tag{16.96}$$

The operators D, P and K are given in terms of the operators L_B^A by the relations:

- The dilatation generator or Hamiltonian operator:

$$D = -iL_2^0. \tag{16.97}$$

  The representations of the $SO(2, 1)$ algebra are then labeled by the eigenvalues Δ of D, viz

$$D|\psi\rangle = \Delta|\psi\rangle. \tag{16.98}$$

- The special conformal generator or lowering operator

$$K = L_1^0 - iL_2^1. \tag{16.99}$$

  Thus, by acting on the ground state $|\psi_0\rangle$ with the lowering operator K, we must have

$$K|\psi_0\rangle = 0. \tag{16.100}$$

- The momentum generator or raising operator

$$P = L_1^0 + iL_2^1. \tag{16.101}$$

  All other states above the ground state $|\psi_0\rangle$ are obtained by acting successively with the raising operator P on $|\psi_0\rangle$.

The conditions $D|\psi_0\rangle = \Delta|\psi_0\rangle$ and $K|\psi_0\rangle = 0$ read in the global coordinates τ, r as follows

$$-i\partial_\tau \psi_0(t, r) = \Delta\psi_0(t, r) \Rightarrow \psi_0(t, r) = \exp(i\Delta t)\chi(r). \tag{16.102}$$

$$(i \sin r\partial_\tau - \cos r\partial_r)\psi_0(t, r) = 0 \Rightarrow \chi(r) = \cos^\Delta r. \tag{16.103}$$

All other states $|\psi_n\rangle$ can now be obtained by the action of the raising operator P on this ground state $|\psi_0\rangle$, i.e., $|\psi_n\rangle = P^n|\psi_0\rangle$. The energy is found to be quantized as

$$E_n = \Delta + n. \tag{16.104}$$

In other words, the energy levels are integer spaced analogously to the harmonic oscillator motion and hence all AdS_2 orbits have the same period with respect to the AdS_2 time τ.

16.3.3 Generalization

The generators of the conformal algebra $SO(d, 2)$ are:

- The dilatation generator $D = -iL^0_{d+1}$, which plays the role of the Hamiltonian.
- The momentum generators $P_\mu = L^0_\mu + iL^{d+1}_\mu$, which play the role of raising operators. Note that $L^{d+1}_\mu = L^\mu_{d+1}$.
- The special conformal generators $K_\mu = L^0_\mu - iL^{d+1}_\mu$, which play the role of lowering operators.
- The rotation generators $M_{\mu\nu} = iL^\nu_\mu$, which generate the Lie algebra $SO(d)$.

The generators L^B_A satisfy the algebra

$$[L^B_A, L^D_C] = \eta^D_A L^B_C - \eta_{AC}L^{BD} - \eta^{DB}L_{AC} - \eta^B_C L^D_A. \tag{16.105}$$

The rotation generators were absent in $SO(1, 2)$. Also, we have d momentum generators and d special conformal generators in the case of $SO(d, 2)$, i.e., P_μ and K_μ transform as vectors under $SO(d)$, viz

$$[M_{\mu\nu}, P_\rho] = i(\eta_{\mu\rho}P_\nu - \eta_{\nu\rho}P_\mu), \quad [M_{\mu\nu}, K_\rho] = i(\eta_{\mu\rho}K_\nu - \eta_{\nu\rho}K_\mu). \tag{16.106}$$

The generators P_μ and K_μ are indeed the raising and lowering operators, respectively, viz

$$[D, P_\mu] = P_\mu, \quad [D, K_\mu] = -K_\mu. \tag{16.107}$$

Also we have

$$[P_\mu, K_\nu] = 2(\eta_{\mu\nu}D + iM_{\mu\nu}). \tag{16.108}$$

The $SO(d)$ rotation generators satisfy

$$[M_{\mu\nu}, M_{\rho\sigma}] = i(\eta_{\mu\sigma}M_{\rho\nu} - \eta_{\mu\rho}M_{\sigma\nu} + \eta_{\sigma\nu}M_{\mu\rho} - \eta_{\rho\nu}M_{\mu\sigma}). \tag{16.109}$$

The dilatation generator D is a scalar under $SO(d)$ rotations, viz

$$[M_{\mu\nu}, D] = 0. \tag{16.110}$$

This means that the Hamiltonian and the angular momentum operators can be diagonalized simultaneously.

The ground state $|\psi_0\rangle$ will correspond to the smallest possible eigenvalue Δ of the Hamiltonian D and it must be annihilated by all the lowering operators K_μ, viz

$$D|\psi_0\rangle = \Delta|\psi_0\rangle, \quad K_\mu|\psi_0\rangle = 0. \tag{16.111}$$

The ground state or highest weight state $|\psi_0\rangle$ is called a primary state. All other states (descendant states) can be obtained by acting successively with the raising operators P_μ (which commute among themselves), viz

$$|\psi_{nlm}\rangle \sim (P_\mu^2)^n P_{\mu_1} \cdots P_{\mu_l}|\psi_0\rangle. \tag{16.112}$$

The energy of this state is clearly (since each P_μ raises the energy by a single unit) is

$$E_{nl} = \Delta + 2n + l. \tag{16.113}$$

These states carry $SO(d)$ indices and hence they are characterized by an angular momentum quantum number equal exactly to the integer l, i.e., to the number of uncontracted vector indices. If $l = 0$ then the state does not carry a free $SO(d)$ index, i.e., it is a scalar, and as consequence its spin is given directly by $l = 0$. Whereas if $l = 1$ the state carries a single free vector index, i.e., it transforms as a vector under $SO(d)$, and as a consequence its spin must be $l = 1$. The quantum number l is therefore the angular momentum or spin quantum number.

The quantum numbers m^1 appearing in (16.112) relate to the spherical harmonics $Y_{lm}(\Omega)$ on AdS_{d+1}, which, by rotational symmetry, carry the angular dependence of the wave functions $\psi_{nlm}(\tau, r, \Omega) = \langle \tau, r, \Omega|\psi_{nlm}\rangle$. For example for $d = 3$ we have a single (integer) number m, which is the usual magnetic quantum number and $Y_{lm}(\Omega) = Y_{lm}(\theta, \phi)$ are the usual spherical harmonics. Thus, the wave functions $\psi_{nlm}(\tau, r, \Omega)$ are of the general form [11]

$$\psi_{nlm}(\tau, r, \Omega) = \exp(iE_{nl}\tau)\, Y_{lm}(\Omega)\psi_{nl}(r). \tag{16.114}$$

The radial part $\psi_{nl}(r)$ is proportional to a hypergeometric function, which for the ground state reduces to

$$\psi_{00}(r) \sim \cos^\Delta r. \tag{16.115}$$

This can be shown as follows. We start from the identity

$$
\begin{aligned}
K_\mu =& -(X_0 - iX_{d+1})\frac{\partial}{\partial X_\mu} + X_\mu\left(\frac{\partial}{\partial X_0} - i\frac{\partial}{\partial X_{d+1}}\right) \\
=& -\frac{\exp(-i\tau)\cos^2 r\hat{x}_\mu}{r}\frac{\partial}{\partial \hat{x}_\mu} + X_\mu\left(\frac{\partial}{\partial X_0} - i\frac{\partial}{\partial X_{d+1}}\right).
\end{aligned}
\tag{16.116}
$$

---

[1] It should not be confused with the mass m. But it should also be clear from the context which one is which.

The ground state $|\psi_0\rangle$ is characterized by zero angular momentum, i.e., $l = 0$, and hence it does not depend on the angles $\hat{x}_\mu$ or equivalently Ω. The condition $K_\mu|\psi_0\rangle = 0$ reduces then to

$$\left(\frac{\partial}{\partial X_0} - i\frac{\partial}{\partial X_{d+1}}\right)|\psi_0\rangle = 0. \tag{16.117}$$

We compute (by dropping the derivative with respect to $\hat{x}_\mu$ in $\partial/\partial r$)

$$\frac{\partial}{\partial X_0} = \frac{\cos r}{R}\left(-\sin\tau\frac{\partial}{\partial\tau} + \tan^{-1}r\,\cos\tau\frac{\partial}{\partial r}\right). \tag{16.118}$$

$$\frac{\partial}{\partial X_{d+1}} = \frac{\cos r}{R}\left(\cos\tau\frac{\partial}{\partial\tau} + \tan^{-1}r\,\sin\tau\frac{\partial}{\partial r}\right). \tag{16.119}$$

The above condition becomes then

$$(-i\sin r\frac{\partial}{\partial\tau} + \cos r\frac{\partial}{\partial r})|\psi_0\rangle = 0. \tag{16.120}$$

This is immediately solved by

$$\psi_0 \sim \exp(i\Delta t)\cos^\Delta r. \tag{16.121}$$

This fundamental result can be found from another route. The generalization of the Klein–Gordon equation (16.87) to AdS_{d+1} is given by

$$-\partial_\tau^2\phi = \left(-\partial_r^2 + \frac{1-d}{\cos r\,\sin r}\partial_r + \frac{l(l+d-2)}{\sin^2\rho} + \frac{R^2m^2}{\cos^2 r}\right)\phi. \tag{16.122}$$

In this equation $l(l+d-2)$ are the eigenvalues of the Laplacian on the sphere $\mathbf{S}^{d-1}$ with corresponding eigenfunctions given by the d-dimensional spherical harmonics Y_{lm} where m is the corresponding set of magnetic quantum numbers on AdS_{d+1}. In other words, the complete scalar field on AdS_{d+1} is actually $Y_{lm}(\Omega)\phi(\tau, r)$. We separate the remaining variables as $\phi(\tau, r) = \exp(iE\tau)\chi(\rho)$. For the ground state we must have $E = \Delta$ and $l = 0$ and the Klein–Gordon equation reduces to

$$-\Delta^2\chi = \left(\partial_r^2 + \frac{d-1}{\cos r\,\sin r}\partial_r - \frac{R^2m^2}{\cos^2 r}\right)\chi. \tag{16.123}$$

We make the change of variables

$$\chi = \sin^\alpha r\,\cos^\beta r\hat{\chi}. \tag{16.124}$$

The exponents α and β are determined from the requirement that the linear derivative vanishes. We have

$$\sin^{\alpha-1}r\,\cos^{\beta-1}r\partial_r\hat{\chi}(-2\beta + d - 1 + (2\beta + 2\alpha)\cos^2 r) = 0 \Rightarrow \beta = -\alpha = \frac{d-1}{2}. \tag{16.125}$$

Hence

$$\partial_r^2 + \frac{d-1}{\cos r \sin r}\partial_r = \sin^\alpha r \cos^\beta r \partial_r^2 \hat\chi + \sin^\alpha r \cos^\beta r \hat\chi\left[\frac{\beta(\beta-d)}{\cos^2 r} + \frac{\beta(\beta-d+2)}{\sin^2 r}\right]. \quad (16.126)$$

The equation of motion becomes

$$-\Delta^2\hat\chi = \partial_r^2\hat\chi - \frac{1}{4}\left[\frac{d^2-1+4R^2m^2}{\cos^2 r} + \frac{(d-1)(d-3)}{\sin^2 r}\right]\hat\chi. \quad (16.127)$$

We propose the solution

$$\hat\chi = \sin^{\alpha_1} r \cos^{\beta_1} r. \quad (16.128)$$

We compute immediately the second derivative

$$-(\alpha_1 + \beta_1)^2\hat\chi = \partial_r^2\hat\chi - \left[\frac{\beta_1(\beta_1-1)}{\cos^2 r} + \frac{\alpha_1(\alpha_1-1)}{\sin^2 r}\right]\hat\chi. \quad (16.129)$$

By comparing the above two final equations we obtain

$$\alpha_1(\alpha_1-1) = \frac{1}{4}(d-1)(d-3) \Rightarrow \alpha_1 = \frac{d-1}{2}. \quad (16.130)$$

$$\beta_1(\beta_1-1) = \frac{1}{4}(d^2-1+4R^2m^2) \Rightarrow \beta_1 = \frac{1}{2} + \frac{1}{2}\sqrt{d^2+4R^2m^2}. \quad (16.131)$$

$$\alpha_1 + \beta_1 = \Delta \Rightarrow \frac{d-1}{2} + \frac{1}{2} + \frac{1}{2}\sqrt{d^2+4R^2m^2} = \Delta \Rightarrow R^2m^2 = \Delta(\Delta-d). \quad (16.132)$$

16.4 Representation theory of the conformal group

16.4.1 More on dilatation operator and primary/descendant operators

Hamiltonian quantization of ordinary quantum field theory involves foliation of the d-dimensional spacetime by equal time $(d-1)$-dimensional surfaces characterized by the same Hilbert space. The unitary evolution operator $U = \exp(iH(t_2 - t_1))$ allows us to advance from the surface $t = t_1$ to the surface $t = t_2$. The theory in this case is covariant under the Poincaré group and the states of the Hilbert space are specified by two quantum numbers: mass m and spin s.

In conformal field theory the dilatation operator is what plays the role of the Hamiltonian and the scaling dimension Δ is what plays the role of the momentum. In this case Euclidean spacetime is foliated using spheres $\mathbf{S}^{d-1}$ characterized by the same Hilbert space. This is called radial quantization. By the action of the dilatation operator we move from one sphere to another. States of the Hilbert space are specified now by the scaling dimension Δ and the spin s, viz

$$D|\Delta\rangle = i\Delta|\Delta\rangle. \quad (16.133)$$

$$M_{\mu\nu}|\Delta, s\rangle = \Sigma_{\mu\nu}|\Delta, s\rangle. \tag{16.134}$$

The metric reads (with Ω the solid angle on $\mathbf{S}^{d-1}$ and $\tau = \log r$)

$$\begin{aligned} ds^2 &= dr^2 + r^2 d\Omega^2 \\ &= e^{2\tau}(d\tau^2 + d\Omega^2). \end{aligned} \tag{16.135}$$

This metric is conformally equivalent to the metric on the cylinder. In other words, the transformation $r \longrightarrow \tau = \log r$ maps $\mathbf{R} \times \mathbf{S}^{d-1}$ to $\mathbf{R}^d$. The parameter τ plays then the role of the time parameter. The lower base of the cylinder is at the infinite past $\tau \longrightarrow -\infty$ ($r = 0$) whereas the upper base of the cylinder is at the infinite future $\tau \longrightarrow +\infty$ ($r = +\infty$). Going around the cylinder is given by the solid angle Ω. The evolution operator is given by

$$U|\Delta\rangle = \exp(i\tau D)|\Delta\rangle = r^{-\Delta}|\Delta\rangle. \tag{16.136}$$

There is a unique vacuum state $|0\rangle$ which is invariant under the global conformal group. This corresponds to no operator insertion in the cylinder, which would create a state at a given time τ (corresponding to a given radius r).

Let $\mathcal{O}_\Delta(x)$ be some operator with scaling dimension Δ. The insertion of this operator at the origin $r = 0$ (or infinite past $\tau = -\infty$) creates the state $|\Delta\rangle = \mathcal{O}_\Delta(0)|0\rangle$ with scaling dimension Δ. By inserting the operator $\mathcal{O}_\Delta(x)$ at an arbitrary point x will create the state

$$\begin{aligned} |\chi\rangle &= \mathcal{O}_\Delta(x)|0\rangle \\ &= \exp(iPx)\mathcal{O}_\Delta(0)\exp(-iPx)|0\rangle \\ &= \exp(iPx)|\Delta\rangle. \end{aligned} \tag{16.137}$$

The momentum operator P_μ is a raising operator with respect to the eigenvalues of the dilatation operator, i.e., it raises the scaling dimension Δ by unity. Thus, by expanding the exponential $\exp(iPx)$ and acting with the momentum operator we obtain a linear superposition of states with different eigenvalues Δ.

Similarly, the special conformal generator K_μ is a lowering operator with respect to the eigenvalues of the dilatation operator, i.e., it lowers the scaling dimension Δ by unity. An operator annihilated by K_μ is called a primary operator. By acting on this primary operator with P_μ we obtain the so-called descendant operators. The primary operator and its descendant operators form a conformal family.

Each state then corresponds to an operator and vice versa. This state-operator correspondence is one-to-one. For example, by inserting a primary operator with scaling dimension Δ at the origin we obtain a state with scaling dimension Δ annihilated by K_μ. Conversely, given a state with a scaling dimension Δ annihilated by K_μ we can construct a local primary operator at the origin by constructing its correlators with other operators, viz

$$\langle \phi(x_1)\phi(x_2) \ldots \mathcal{O}_\Delta(0)\rangle = \langle 0|\phi(x_1)\phi(x_2) \ldots |\Delta\rangle \tag{16.138}$$

16.4.2 Representation theory of $O(4, 2)$

A concise description of this topic can be found in [10, 15] and references therein. See also [16,17].

The conformal group in four dimensions in Lorentzian signature is given by $SO(4, 2)$ or more precisely $O(4, 2)$. There are 15 generators $\mathcal{M}_{AB} = -\mathcal{M}_{BA}$. The irreducible representations of the conformal group are infinite dimensional. They are characterized by the eigenvalues of the three Casimir (quadratic, cubic and quartic) operators

$$C_I = \mathcal{M}_{AB}\mathcal{M}^{AB}. \tag{16.139}$$

$$C_{II} = \epsilon_{ABCDEF}\mathcal{M}^{AB}\mathcal{M}^{CD}\mathcal{M}^{EF}. \tag{16.140}$$

$$C_{III} = \mathcal{M}^B_A\mathcal{M}^C_B\mathcal{M}^D_C\mathcal{M}^A_D. \tag{16.141}$$

An infinite dimensional irreducible representation of the conformal group is determined by an irreducible representation of the Lorentz group with definite conformal dimension and annihilated by the special conformal operators K_μ. The stability algebra at the origin consists of the generators D, K_μ and $M_{\mu\nu}$. A primary conformal operator O (the lowest weight state) in a given representation of the Lorentz group is defined by

$$[D, O(0)] = i\Delta O(0). \tag{16.142}$$

$$[K, O(0)] = 0. \tag{16.143}$$

The descendants $\partial \cdots \partial O(0)$ are obtained by the repeated action of the momentum operators P_μ. The eigenvalues of the Lorentz operators $M_{\mu\nu}$ on the primary operator O are spin quantum numbers denoted for example by j_L and j_R. This defines an irreducible representation of the conformal group characterized by Δ, j_L and j_R.

The three Casimirs of the stability algebra are

$$\mathcal{D} = \Delta. \tag{16.144}$$

$$\frac{1}{2}M_{\mu\nu}M^{\mu\nu} = j_L(j_L + 1) + j_R(j_R + 1). \tag{16.145}$$

$$\frac{1}{2}\epsilon_{\mu\nu\alpha\beta}M_{\mu\nu}M^{\alpha\beta} = j_L(j_L + 1) - j_R(j_R + 1). \tag{16.146}$$

The eigenvalues of the conformal group $O(4, 2)$ are then given by

$$C_I = \Delta(\Delta - 4) + 2j_L(j_L + 1) + 2j_R(j_R + 1). \tag{16.147}$$

$$C_{II} = (\Delta - 2)(j_L(j_L + 1) - j_R(j_R + 1)). \tag{16.148}$$

$$C_{III} = (\Delta - 2)^4 - 4(\Delta - 2)^2(j_L(j_L + 1) + j_R(j_R + 1) + 1) + 16j_Lj_R(j_L + 1)(j_R + 1). \tag{16.149}$$

In particular for tensor representations of spin s associated with the quantum numbers $(\Delta, j_L = s/2, j_R = s/2)$ we get

$$C_I = \Delta(\Delta - 4) + s(s + 2). \tag{16.150}$$

$$C_{II} = 0. \tag{16.151}$$

$$C_{III} = (\Delta(\Delta - 2) - s(s + 2))((\Delta - 2)(\Delta - 4) - s(s + 2)). \tag{16.152}$$

The requirement of unitarity imposes the following constraints on the possible values of Δ, j_L and j_R. We have

$$\Delta \geqslant j + 1, \quad j_L j_R = 0. \tag{16.153}$$

$$\Delta \geqslant j_L + j_R + 2, \quad j_L j_R \neq 0. \tag{16.154}$$

The first constraint (16.153) is saturated by massless fields satisfying $\partial^2 \Phi_{(0,j)} = 0$. Indeed, this wave equation is conformally covariant only if $\Delta = j + 1$. Similarly, the second constraint (16.154) is saturated by conserved tensor fields satisfying $\partial^{\alpha_1 \dot\alpha_1} O_{\alpha_1 \ldots \alpha_{2j_L}, \dot\alpha_1 \ldots \dot\alpha_{2j_R}} = 0$, which is a conformally covariant equation only if $\Delta = 2 + j_L + j_R$.

16.4.3 $O(4, 2)$ and isometries of AdS_5

The conformal group $O(d, 2)$ is the isometry group of AdS_{d+1}. By the AdS/CFT correspondence there is a d-dimensional conformal field theory (CFT$_d$) living on the boundary of AdS_{d+1} where gauge invariant composite operators in the CFT$_d$ are associated with fields in AdS_{d+1}.

This means in particular that the scaling dimension Δ can be reinterpreted as the energy of a particle moving in anti-de Sitter space. Indeed, particles in AdS_5 are characterized by the quantum numbers (E, j_L, j_R) where the energy E is identified with the scaling dimension of the corresponding conformal primary operator living on the boundary.

Indeed, the covariant wave equation of a particle in AdS_5 can be re-expressed in terms of the Casimir of the conformal group and hence the mass of the particle can be determined in terms of the quantum numbers (E, j_L, j_R). For example, for a scalar field with $j_L = j_R = 0$ the Laplacian operator in AdS_5 is precisely the Casimir operator C_I and hence from (16.150) we obtain the mass squared $m^2 = \Delta(\Delta - 4)$, which will be obtained more directly in due course. By using also equation (16.150) we obtain for fermions of spin 1/2 with $j_L = 1/2$, $j_R = 0$ or $j_L = 0$, $j_R = 1/2$ the mass $m = \Delta - 2$, and for vector fields of spin 1 with $j_L = j_R = 1/2$ we get the mass squared $m^2 = \Delta(\Delta - 4) + 3 = (\Delta - 1)(\Delta - 3)$, whereas for symmetric tensor fields of spin 2 with $j_L = j_R = 1$ we obtain the mass squared $m^2 = \Delta(\Delta - 4)$. See [15] for more detail.

The generators D and $M_{\mu\nu}$, giving the quantum numbers (Δ, j_L, j_R), correspond to the non-compact subgroup $O(3, 1)$ of the conformal group $O(4, 2)$. However, we observe that the operators in the representations (Δ, j_L, j_R) yields, when applied to the vacuum, non-normalizable states. This is because these states cannot furnish a finite dimensional unitary representation of a non-compact group.

Another set of good quantum numbers corresponds to the maximal compact subgroup $O(2) \times O(4)$ of the conformal group $O(4, 2)$. This compact group $O(2) \times O(4)$ allows us to obtain finite dimensional unitary representations of the conformal group using states with finite norm. Indeed, the quantum numbers (E, j_L, j_R) can now be viewed as the eigenvalues of the Cartan generators of $O(2) \times O(4)$. In particular, the quantum number E is associated with the $O(2)$ generator $H = (K_0 + P_0)/2$, which is called the conformal energy. This can also be seen by going to the Euclidean spacetime $\mathbf{R}^4$, which can be mapped via a conformal transformation to $\mathbf{R} \times \mathbf{S}^3$ (radial quantization). In this case the group $O(2)$ is seen acting on $\mathbf{R}$ and hence H is the Hamiltonian corresponding to translations in this direction whereas the factor $O(4)$ acts on $\mathbf{S}^3$.

Furthermore, we remark that since P_0 is a raising operator and K_0 is a lowering operator the eigenvalue E seems to be integer-valued. But in the quantum theory it is the covering space of the conformal group that is being realized and is obtained by unwinding the factor $O(2)$ giving rise to a continuous spectrum of E. Hence the identification of E with Δ.

16.4.4 The fields and operators in CFT_d

Although the scaling dimension Δ in the CFT_4 can be identified with the conformal energy E in the AdS_5 we strictly speaking do not have particle states in a conformal field theory. The first obvious reason is that the mass operator $P_\mu P^\mu$ is not a Casimir of the conformal group, i.e., it does not commute with the dilatation operator. Hence if a state in a given representation of the conformal group has an energy E_0, then by the action of the dilatation operator we can obtain states in this representation with any other value of the energy between 0 and ∞.

This can be understood more precisely by means of the Kallen–Lehmann spectral representation of the two-point function of a general interacting QFT_d given in terms of the free propagator $\Delta(p, \mu^2)$ and the spectral density $\rho(\mu^2)$ by the relation [18, 19]

$$\Delta(p) = \int_0^\infty \rho(\mu^2) \Delta_0(p, \mu^2) d\mu^2, \quad \Delta(p, \mu^2) = \frac{1}{p^2 - \mu^2 + i\epsilon}. \qquad (16.155)$$

The spectral density $\rho(\mu^2)$ encodes the contributions of the states $|s\rangle$ with momenta p_s to the two-point function and it is given explicitly by

$$\rho(p^2) = \sum_s \delta^d(p - p_s)|\langle s|\phi\rangle|^2. \qquad (16.156)$$

For a free scalar field ϕ we have

$$\rho(p^2) = \delta^d(p^2 - m^2). \qquad (16.157)$$

This corresponds to a single massive excitation. If ϕ overlaps with heavier states then there will be other terms with $p^2 > m^2$.

On the other hand, for a conformal field theory in four-dimension we know that the two-point function should behave as

$$\Delta(x) \sim \frac{1}{x^2} \;\Rightarrow\; \rho(p^2) = \delta^4(p^2). \tag{16.158}$$

This also corresponds to a single massless excitation. But for a generic scalar operator in CFT_4 characterized by a non-trivial scaling dimension (anomalous dimension) the behavior of the two-point function is altered as

$$\Delta(x) \sim \frac{1}{x^{2+2\delta}} \;\Rightarrow\; \rho(p^2) = (p^2)^{\delta-1}, \quad \delta > 0. \tag{16.159}$$

This is a continuous power-law spectrum characterized by δ. In other words, there is no mass scale nor a discrete set of particles but the operator just creates a scale-invariant continuous set of states.

Hence, fields in an ordinary QFT_d are local operators which furnish a representation of the Lorentz group, generate the Hilbert space, and create particle states. But in a CFT_d the basic objects are operators which are not necessarily fields since they do not create particle-like excitations. Thus, the formalism of scattering theory and the S-matrix does not apply for conformal field theory.

16.4.5 Unitary bounds revisited

The states that saturate the unitarity bound (16.153) are called singletons and they are topological configurations living at the boundary of AdS_5 associated with fundamental fields (and not gauge invariant operators) of the CFT_4.

On the other hand, the constraint equation (16.154) enjoys a profound physical meaning. Recall that this inequality is saturated by conserved tensor fields satisfying $\partial^{\alpha_1 \dot\alpha_1} O_{\alpha_1 \dots \alpha_{2j_L}, \dot\alpha_1 \dots \dot\alpha_{2j_R}} = 0$ which is a conformally covariant equation only if $\Delta = 2 + j_L + j_R$. These conserved tensor fields are precisely the conserved currents in the CFT_4 which are associated with massless fields in AdS_5 with local gauge invariance. In other words, global symmetries in CFT_4 corresponds to local symmetries in AdS_5. For example, the energy–momentum tensor $T_{\mu\nu}$ is associated with the graviton field $g_{\mu\nu}$ and the global current J_μ is associated with the gauge field A_μ, etc. The equation $\partial^{\alpha_1 \dot\alpha_1} O_{\alpha_1 \dots \alpha_{2j_L}, \dot\alpha_1 \dots \dot\alpha_{2j_R}} = 0$ satisfied by these conserved currents means that the number of degrees of freedom contained in the conserved tensor in CFT_4 is precisely the number of degrees of freedom contained in the massless field in AdS_5. Obviously, the tensor contains $(2j_L + 1)(2j_R + 1) - (2j_L)(2j_R) = 2(j_L + j_R) + 1$ degrees of freedom, i.e., a massless particle of spin $j_L + j_R$. For $\Delta > 2 + j_L + j_R$ the fields in AdS_5 are massive and the corresponding tensor fields are not conserved. See [15] for more detail.

16.5 Holography

The number of degrees of freedom, or equivalently the amount of information, contained in a quantum system is measured as we know by thermodynamic entropy. In quantum mechanics the entropy is an extensive quantity and thus the entropy of a d-dimensional spatial region $\mathbf{R}^d$ is proportional to its volume V_d.

However, in quantum gravity the entropy is sub-extensive, i.e., the entropy of a d-dimensional spatial region $\mathbf{R}^d$ is actually proportional to the surface area $S_{d-1} = \partial V_d$ which bounds its volume V_d and not proportional to the volume V_d itself. In other words, the entropy of a d-dimensional spatial region, in a gravitational theory, is bounded by the entropy of the black hole, which fits inside that spatial region. This is essentially what is called the holographic principle introduced first by 't Hooft [20] and then extended to string theory by Susskind [21] (see also [22]). As we can see this principle is largely inspired by the Bekenstein–Hawking formula, which states that the entropy of a black hole $\mathcal{S}_{BH}$ is proportional to the surface area A_H of the black hole horizon with the constant of proportionality equal $1/4G_N$ where G_N is Newton's constant, viz

$$\mathcal{S}_{BH} = \frac{A_H}{4G_N}. \tag{16.160}$$

The holographic principle provides therefore a partial answer to the question of how could a higher dimensional gravity theory (AdS_5) contain the same number of degrees of freedom, the same amount information, and have the same entropy as a lower dimensional quantum field theory (CFT_4), i.e., it lies at the heart of the celebrated AdS/CFT correspondence [23].

This can be seen more explicitly as follows. By the AdS/CFT correspondence, the AdS space AdS_{d+1} is the gravity dual of a d-dimensional conformal field theory CFT_d living on the boundary $z = \epsilon$ of AdS space. The radial coordinate z should be thought of as a lattice spacing, i.e., as a UV cutoff. Thus, the boundary theory is a quantum field theory on a d-dimensional lattice with lattice spacing ϵ. At any given instant of time, the boundary theory is also regulated by placing it in a spatial box of size R (IR cutoff). Hence, the number of cells in the box is given by $(R/L)^{d-1}$.

The central charge c_{QFT} of the CFT is by definition equal to the number of degrees of freedom per lattice site. Thus the total number of degrees of freedom contained in the box is given by

$$N_{\text{QFT}} = \left(\frac{R}{\epsilon}\right)^{d-1} c_{\text{QFT}}. \tag{16.161}$$

From the AdS space side the estimation of the degrees of freedom can be carried out as follows. The metric at $z = \epsilon$ is

$$ds_d^2 = \frac{L^2}{\epsilon^2} dx_\mu dx^\mu. \tag{16.162}$$

By using the holographic principle, i.e., the Bekenstein–Hawking formula, the number of degrees of freedom at a given instant of time contained in the spatial volume of AdS space is given by the maximum entropy given by

$$N_{\text{AdS}} = \frac{A_H}{4G_N}. \tag{16.163}$$

Here A_H is the area of the spatial boundary of AdS_{d+1} delimiting the spatial volume of AdS space. We use the metric to compute this area as follows:

$$A_H = \int_{z=\epsilon} d^{d-1}x \sqrt{g} = \left(\frac{L}{\epsilon}\right)^{d-1} \int d^{d-1}x = \left(\frac{L}{\epsilon}\right)^{d-1} R^{d-1}. \tag{16.164}$$

By using also the fact that for gravity in $d+1$ dimensions the Newton constant is given by $G_N = l_P^{d-1} = 1/M_P^{d-1}$ we arrive at the result

$$N_{\text{AdS}} = \frac{1}{4}\left(\frac{R}{\epsilon}\right)^{d-1}\left(\frac{L}{l_P}\right)^{d-1}. \tag{16.165}$$

By comparing (16.161) and (16.165) we obtain the central charge

$$c_{\text{QFT}} = \frac{1}{4}\left(\frac{L}{l_P}\right)^{d-1}. \tag{16.166}$$

Hence semi-classical gravity corresponding to $L \gg l_P$ is dual to a CFT with a large central charge. If the conformal field theory is an $SU(N)$ gauge theory then the central charge is proportional to N^2 and as a consequence semi-classical gravity is dual in this case to a large N gauge theory.

16.6 The AdS/CFT correspondence

In this section we follow the presentation of [6].

16.6.1 Approaching the AdS boundary

We go back to Euclidean AdS_{d+1} in the Poincaré patch:

$$ds_{d+1}^2 = \frac{L^2}{z^2}(dz^2 + d\vec{x}^2 + dt^2). \tag{16.167}$$

The action of a scalar field in AdS_{d+1} is given by

$$S = \int d^{d+1}x \sqrt{g}\left[-\frac{1}{2}g^{MN}\partial_M\phi\partial_N\phi - \frac{1}{2}m^2\phi^2\right]. \tag{16.168}$$

The Klein–Gordon equation in the AdS_{d+1} background reads explicitly

$$z^{d+1}\partial_z(z^{1-d}\partial_z\phi) + z^2\partial^2\phi - m^2L^2\phi = 0. \tag{16.169}$$

We perform Fourier transform in the x-space, viz

$$\phi(z, x) = \int \frac{d^dk}{(2\pi)^d} \exp(ikx)f_k(z). \tag{16.170}$$

The Klein–Gordon equation reduces to

$$z^{d+1}\partial_z(z^{1-d}\partial_z f_k) - k^2z^2 f_k - m^2L^2 f_k = 0. \tag{16.171}$$

Near the conformal boundary $z = 0$ we may expect $f_k \sim z^\beta$. This gives immediately

$$\beta(\beta - d) - m^2 L^2 = 0 \Rightarrow \beta = \frac{d}{2} \pm \sqrt{\frac{d^2}{4} + m^2 L^2}. \tag{16.172}$$

The solution near the boundary is therefore of the general form

$$f_k(z) \longrightarrow A(k) z^{d-\Delta} + B(k) z^\Delta, \quad z \longrightarrow 0. \tag{16.173}$$

The exponent Δ is the so-called scaling dimension of the field and it is given by

$$\Delta = \frac{d}{2} + \sqrt{\frac{d^2}{4} + m^2 L^2}. \tag{16.174}$$

The scaling dimension Δ is real if the mass m satisfies the Breitenlohner-Freedman (BF) bound

$$m^2 > -\frac{d^2}{4L^2}. \tag{16.175}$$

In this case $d - \Delta \leqslant \Delta$ and hence $z^{d-\Delta}$ is the dominant term as $z \longrightarrow 0$. We place the boundary at $z = \epsilon \longrightarrow 0$. Then the behavior of the scalar field on the boundary is given by

$$\phi(z = \epsilon, x) = A(x) \epsilon^{d-\Delta}. \tag{16.176}$$

For $m^2 > 0$ the exponent $d - \Delta$ is negative and hence this field is divergent. The quantum field theory source $\varphi(x)$, i.e., the scalar field living on the boundary, is then identified with $A(x)$, viz

$$\varphi(x) = \lim_{\epsilon \longrightarrow 0} \epsilon^{\Delta-d} \phi(\epsilon, x). \tag{16.177}$$

This is the scalar field representing (or dual to) the anti-de Sitter scalar field $\phi(z, x)$ at the boundary $z = 0$. The field φ is the holographic dual of the AdS field ϕ. The scaling dimension of the source φ is given by $d - \Delta$.

Let $\mathcal{O}(z, x)$ and $\mathcal{O}(x)$ be the dual operators to the scalar fields $\phi(z, x)$ and $\varphi(x)$, respectively. Their coupling is a boundary term of the form

$$\begin{aligned}
S_{\text{bound}} &= \int d^d x \sqrt{\gamma} \, \phi(\epsilon, x) \mathcal{O}(\epsilon, x) \\
&= \int d^d x \frac{L^d}{\epsilon^d} \epsilon^{d-\Delta} \varphi(x) \mathcal{O}(\epsilon, x) \\
&= L^d \int d^d x \, \varphi(x) \mathcal{O}(x),
\end{aligned} \tag{16.178}$$

where

$$\mathcal{O}(\epsilon, x) = \epsilon^\Delta \mathcal{O}(x). \tag{16.179}$$

This is the wave function renormalization of the operator $\mathcal{O}$ as we move into the bulk. This also shows explicitly that Δ is the scaling dimension of the dual operator

$\mathcal{O}$ since going from $z = 0$ to $z = \epsilon$ is a dilatation operation in the quantum field theory.

In summary, the scaling dimensions of the source φ and its dual operator $\mathcal{O}$ are given by $d - \Delta$ and Δ, respectively. Any deformation of the conformal field theory with the operator $\mathcal{O}$ takes then the form

$$\Delta S = \int d^d x M^{d-\Delta} \mathcal{O}. \tag{16.180}$$

We distinguish the usual three cases:

- For $m^2 > 0$ we have $\Delta > d$ and thus corrections to various amplitudes are of the form $(E/M)^\alpha$ for some positive α. In other words, these corrections are negligible for low energies and the operator $\mathcal{O}$ does not change the IR behavior of the theory. It is an irrelevant operator.
- For $m^2 = 0$ we have $\Delta = d$ and the operator $\mathcal{O}$ is marginal.
- For $m^2 < 0$ we have $\Delta < d$ and the operator $\mathcal{O}$ is relevant since it changes the IR behavior of the theory.

16.6.2 Correlation functions

We have now an operator $\mathcal{O}(x)$ living on the AdS boundary. We are interested in computing the correlation functions

$$\langle \mathcal{O}(x_1) \dots \mathcal{O}(x_n) \rangle. \tag{16.181}$$

These will determine the CFT living on the boundary completely. On the conformal field theory side with Lagrangian $\mathcal{L}$ the calculation of these correlation functions proceeds as usual by introducing the generating functional

$$Z_{\text{CFT}}[J] = \int \exp(\mathcal{L} + \int J(x)\mathcal{O}(x)) = \langle \exp(\int J(x)\mathcal{O}(x)) \rangle. \tag{16.182}$$

Then we have immediately

$$\langle \mathcal{O}(x_1) \dots \mathcal{O}(x_n) \rangle = \frac{\delta^n \log Z_{CFT}[J]}{\delta J(x_1) \dots \delta J(x_n)} |_{J=0}. \tag{16.183}$$

The operator $\mathcal{O}(x)$ is sourced by a scalar field $\varphi(x)$ living on the boundary, which is related to the boundary value of the bulk scalar field $\phi(z, x)$ by the relation

$$\varphi(x) = \lim_{z \to 0} z^{\Delta - d} \phi(z, x). \tag{16.184}$$

The boundary scalar field $\phi_0(x)$ is actually divergent and it is simply defined by

$$\phi_0(x) = \phi(0, x). \tag{16.185}$$

The AdS/CFT correspondence states then that the CFT generating functional with source $J = \phi_0$ is equal to the path integral on the gravity side evaluated over a bulk field, which has the value ϕ_0 at the boundary of AdS [24, 25]. We write

$$Z_{\text{CFT}}[\phi_0] = \langle \exp(\int \phi_0(x)\mathcal{O}(x)) \rangle = \int_{\phi \to \phi_0} \mathcal{D}\phi \exp(S_{\text{grav}}) = Z_{\text{grav}}[\phi \longrightarrow \phi_0]. \tag{16.186}$$

In the limit in which classical gravity is a good approximation the gravity path integral can be replaced by the classical amplitude given by the classical on-shell gravity action, i.e.,

$$Z_{\text{CFT}}[\phi_0] = \exp(S_{\text{grav}}^{\text{on-shell}}[\phi \longrightarrow \phi_0]). \tag{16.187}$$

Typically the on-shell gravity action is divergent requiring thus regularization and renormalization using the so-called holographic renormalization [26,27]. The on-shell action gets renormalized and the above prescription becomes

$$Z_{\text{CFT}}[\phi_0] = \exp(S_{\text{grav}}^{\text{renor}}[\phi \longrightarrow \phi_0]). \tag{16.188}$$

The correlation functions are then renormalized as

$$\langle \mathcal{O}(x_1) \ldots \mathcal{O}(x_n) \rangle = \frac{\delta^n S_{\text{grav}}^{\text{renor}}[\phi \longrightarrow \phi_0]}{\delta\varphi(x_1) \ldots \delta\varphi(x_n)} \Big|_{\varphi=0}. \tag{16.189}$$

Let us compute the one-point correlation function, viz

$$\langle \mathcal{O}(x) \rangle_\varphi = \frac{\delta S_{\text{grav}}^{\text{renor}}[\phi \longrightarrow \phi_0]}{\delta\varphi(x)} = \lim_{z \to 0} z^{d-\Delta} \frac{\delta S_{\text{grav}}^{\text{renor}}[\phi \longrightarrow \phi_0]}{\delta\phi(z, x)}. \tag{16.190}$$

We can rewrite this in terms of an action of the form

$$S_{\text{grav}} = \int_{\mathcal{M}} dz d^d x \mathcal{L}(\phi, \partial\phi). \tag{16.191}$$

Under $\phi \longrightarrow \phi + \delta\phi$, and assuming the Euler–Lagrange equations of motion, and also by assuming that the variation $\delta\phi$ vanishes whenever $x \longrightarrow \pm\infty$ or $z \longrightarrow \infty$, we obtain the on-shell variation

$$\delta S_{\text{grav}}^{\text{on-shell}} = \int_{\partial\mathcal{M}} d^d x \Pi. \, \delta\phi \, |_{z=\epsilon}. \tag{16.192}$$

The canonical momentum with respect to z is given by

$$\Pi = \frac{\delta S_{\text{grav}}^{\text{on-shell}}}{\delta\phi(z, x)} = -\frac{\delta\mathcal{L}}{\delta\partial_z\phi(z, x)}. \tag{16.193}$$

The renormalized action is obtained by adding counter terms, viz

$$S_{\text{grav}}^{\text{renor}} = S_{\text{grav}}^{\text{on-shell}} + S_{\text{ct}}. \tag{16.194}$$

We define in this case the renormalized canonical momentum as

$$\Pi^{\text{renor}} = \frac{\delta S_{\text{grav}}^{\text{renor}}}{\delta\phi(z, x)}. \tag{16.195}$$

Hence

$$\langle \mathcal{O}(x) \rangle_\varphi = \lim_{z \to 0} z^{d-\Delta} \Pi^{\text{renor}}(z, x). \tag{16.196}$$

From the one-point function we can obtain the two-point function as follows. In the quantum field theory with fields ψ the one-point function is given by the path integral

$$\langle \mathcal{O}(x) \rangle_\varphi = \int [d\psi] \mathcal{O}(x) \exp(S_E[\psi] + \int d^d y \, \varphi(y) \mathcal{O}(y)). \tag{16.197}$$

By expanding in power series of the source φ we obtain

$$\langle \mathcal{O}(x) \rangle_\varphi = \langle \mathcal{O}(x) \rangle_{\varphi=0} + \int d^d y \, G_E(x - y) \varphi(y) + \cdot s \tag{16.198}$$

$G_E(x - y)$ is obviously the two-point function defined by

$$G_E(x - y) = \langle \mathcal{O}(x) \mathcal{O}(y) \rangle_{\varphi=0}. \tag{16.199}$$

By assuming that the observable $\mathcal{O}(x)$ has been normal-ordered we have $\langle \mathcal{O}(x) \rangle_{\varphi=0} = 0$. In other words, $\langle \mathcal{O}(x) \rangle_\varphi$ measures really fluctuations of the observable away from its expectation value without source. Hence

$$\langle \mathcal{O}(x) \rangle_\varphi = \int d^d y \, G_E(x - y) \varphi(y) + \cdots \tag{16.200}$$

In momentum space this reads

$$\langle \mathcal{O}(k) \rangle_\varphi = G_E(k) \varphi(k). \tag{16.201}$$

Therefore the two-point function in momentum space is given immediately by

$$G_E(k) = \frac{\langle \mathcal{O}(k) \rangle_\varphi}{\varphi(k)}. \tag{16.202}$$

16.6.3 The two-point function

We return to the action of a scalar field in Euclidean AdS_{d+1} given by (including an overall constant η)

$$S_{\text{grav}} = \eta \int d^{d+1} x \sqrt{g} \left[-\frac{1}{2} g^{MN} \partial_M \phi \partial_N \phi - \frac{1}{2} m^2 \phi^2 \right]. \tag{16.203}$$

By using the Euler–Lagrange equations of motion we obtain the on-shell action and variation as

$$S_{\text{grav}}^{\text{on-shell}} = \frac{\eta}{2} \int d^d x (\sqrt{g} \, \phi g^{zz} \partial_z \phi)_{z=\epsilon}. \tag{16.204}$$

$$\delta S_{\text{grav}}^{\text{on-shell}} = \eta \int d^d x (\sqrt{g} \, \phi g^{zz} \partial_z \phi)_{z=\epsilon}. \tag{16.205}$$

The canonical momentum with respect to z is then given by

$$\Pi = \frac{\delta S_{\text{grav}}^{\text{on-shell}}}{\delta \phi(z, x)} = -\frac{\delta \mathcal{L}}{\delta \partial_z \phi(z, x)} = \eta \sqrt{g} \, g^{zz} \partial_z \phi. \tag{16.206}$$

Thus the on-shell action is of the form

$$
\begin{aligned}
S_{\text{grav}}^{\text{on-shell}} &= \frac{1}{2} \int_{z=\epsilon} d^d x \, \Pi(z, x)\phi(z, x) \\
&= \frac{1}{2} \int_{z=\epsilon} \frac{d^d k}{(2\pi)^d} \pi_{-k}(z) f_k(z).
\end{aligned}
\tag{16.207}
$$

The field $f_k(z)$ satisfies the equation of motion (16.171). We introduce another function $g_k(z)$ by $f_k(z) = z^{d/2} g_k(z)$ and substitute in (16.171) to obtain the differential equation

$$
z^2 \partial_z^2 g_k + z \partial_z g_k - (\nu^2 + k^2 z^2) g_k = 0, \quad \nu^2 = \frac{d^2}{4} + m^2 L^2 = (\Delta - \frac{d}{2})^2. \tag{16.208}
$$

This is the modified Bessel equation. The solutions are therefore the modified Bessel functions $g_k(z) = I_{\pm\nu}(kz)$, viz $f_k(z) = z^{d/2} I_{\pm\nu}(kz)$. Recall the small and large z limits of the modified Bessel functions

$$
I_{\pm\nu}(kz) \sim \frac{1}{\Gamma(1 \pm \nu)}\left(\frac{kz}{2}\right)^{\pm\nu}, \quad z \longrightarrow 0. \tag{16.209}
$$

$$
I_{\pm\nu}(kz) \sim \frac{e^{kz}}{\sqrt{2\pi kz}}, \quad z \longrightarrow \infty. \tag{16.210}
$$

The two independent solutions of (16.171) are taken to be given by

$$
\phi_1(z, k) = \Gamma(1 - \nu)\left(\frac{k}{2}\right)^{\nu} z^{d/2} I_{-\nu}(kz) \longrightarrow z^{d-\Delta}, \quad z \longrightarrow 0 \tag{16.211}
$$

$$
\phi_2(z, k) = \Gamma(1 + \nu)\left(\frac{k}{2}\right)^{-\nu} z^{d/2} I_{\nu}(kz) \longrightarrow z^{\Delta}, \quad z \longrightarrow 0 \tag{16.212}
$$

Remark that the small z behavior agrees with the one previously found in (16.173). Equation (16.173) should then be generalized as

$$
\begin{aligned}
f_k(z) &= A(k)\phi_1(z, k) + B(k)\phi_2(z, k) \\
&= z^{d/2}\left[\Gamma(1 - \nu)\left(\frac{k}{2}\right)^{\nu} A(k) I_{-\nu}(kz) + \Gamma(1 + \nu)\left(\frac{k}{2}\right)^{-\nu} B(k) I_{\nu}(kz)\right].
\end{aligned}
\tag{16.213}
$$

The large z limit of this solution is given by

$$
f_k(z) = z^{d/2}\frac{e^{kz}}{\sqrt{2\pi kz}}\left[\Gamma(1 - \nu)\left(\frac{k}{2}\right)^{\nu} A(k) + \Gamma(1 + \nu)\left(\frac{k}{2}\right)^{-\nu} B(k)\right], \quad z \longrightarrow \infty. \tag{16.214}
$$

This diverges in the limit $z \longrightarrow \infty$ unless the term between brackets vanishes. Thus we obtain a regular field in the IR limit $z \longrightarrow \infty$ if

$$
\frac{B(k)}{A(k)} = -\frac{\Gamma(1 - \nu)}{\Gamma(1 + \nu)}\left(\frac{k}{2}\right)^{2\nu}. \tag{16.215}
$$

Now in the small z limit the field behaves as

$$f_k(z) = A(k)z^{d-\Delta} + B(k)z^{\Delta}, \quad z \longrightarrow 0. \tag{16.216}$$

We compute immediately the small z limit of the conjugate field as

$$\pi_{-k}(z) = \eta L^{d-1}[(d - \Delta)A(-k)z^{-\Delta} + \Delta B(-k)z^{\Delta-d}], \quad z \longrightarrow 0. \tag{16.217}$$

The on-shell action becomes

$$S_{\text{grav}}^{\text{on-shell}} = \frac{\eta}{2}L^{d-1} \int \frac{d^d k}{(2\pi)^d}((d - \Delta)A(k)A(-k)\epsilon^{-2\nu} + dA(k)B(-k)). \,) \tag{16.218}$$

The first term is divergent and thus this action requires a renormalization. The required counter term is a quadratic local term living on the boundary of AdS space given by [6]

$$S_{\text{ct}} = \frac{\eta}{2}\eta_1 \int \sqrt{\gamma}d^d x\phi^2. \tag{16.219}$$

The metric γ is the induced metric on the boundary, viz $ds^2 = L^2 dx_\mu dx_\mu / \epsilon^2$. Indeed, we compute

$$S_{\text{ct}} = \frac{\eta}{2}\eta_1 L^d \int \frac{d^d k}{(2\pi)^d}(A(k)A(-k)\epsilon^{-2\nu} + 2A(k)B(-k)). \tag{16.220}$$

The divergence is canceled if $\eta_1 = -(d - \Delta)/L$. The renormalized action is therefore

$$\begin{aligned}
S_{\text{grav}}^{\text{renor}} &= S_{\text{grav}}^{\text{on-shell}} + S_{\text{ct}} \\
&= \frac{\eta}{2}L^{d-1}(2\Delta - d)\int \frac{d^d k}{(2\pi)^d}A(k)B(-k) \\
&= -\frac{\eta}{2}L^{d-1}(2\Delta - d)\frac{\Gamma(1 - \nu)}{\Gamma(1 + \nu)} \int \frac{d^d k}{(2\pi)^d}A(k)\left(\frac{k}{2}\right)^{2\nu}A(-k) \\
&= -\frac{\eta}{2}L^{d-1}(2\Delta - d)\frac{\Gamma(1 - \nu)}{\Gamma(1 + \nu)} \int \frac{d^d k}{(2\pi)^d}\varphi(k)\left(\frac{k}{2}\right)^{2\nu}\varphi(-k),
\end{aligned} \tag{16.221}$$

where we have used (16.215) and

$$\varphi(x) = \lim_{z \longrightarrow 0}z^{\Delta-d}\phi(z, x) = A(x). \tag{16.222}$$

The one-point correlator of the dual operator on the boundary is given by (16.190) or equivalently

$$\langle \mathcal{O}(k)\rangle_\varphi = (2\pi)^d \frac{\delta S_{\text{grav}}^{\text{renor}}[\phi \longrightarrow \phi_0]}{\delta\varphi(-k)}. \tag{16.223}$$

We get

$$\langle \mathcal{O}(k)\rangle_\varphi = -\eta L^{d-1}(2\Delta - d)\frac{\Gamma(1 - \nu)}{\Gamma(1 + \nu)}\left(\frac{k}{2}\right)^{2\nu}\varphi(k). \tag{16.224}$$

The two-point function is then given in terms of the one-point function by the formula (16.202), viz

$$G_E(k) = \frac{\langle \mathcal{O}(k) \rangle_\varphi}{\varphi(k)} = -\eta L^{d-1}(2\Delta - d)\frac{\Gamma(1 - \nu)}{\Gamma(1 + \nu)}\left(\frac{k}{2}\right)^{2\nu}. \tag{16.225}$$

In position space this two-point function reads

$$\langle \mathcal{O}(x)\mathcal{O}(0) \rangle = \frac{2\nu L^{d-1}\eta}{\pi^{d/2}}\frac{\Gamma\left(\dfrac{d}{2} + \nu\right)}{\Gamma(-\nu)}\frac{1}{|x|^{2\Delta}}. \tag{16.226}$$

This is the correct behavior of a conformal field of scaling dimension Δ, i.e., the exponent Δ is indeed the scaling dimension of the boundary operator $\mathcal{O}(x)$.

16.7 Conformal field theory on the torus

16.7.1 Modular invariance

We will follow [9, 12].

We go from the complex plane z to the cylinder w via the conformal exponential mapping, viz $w \longrightarrow z = \exp(w)$ or equivalently $r\exp(i\theta) = \exp(\sigma^2)\exp(-i\sigma^1)$. Thus, the Euclidean time $\sigma^2 = -\infty$ on the worldsheet corresponds to $r = 0$ on the plane while $\sigma^2 = +\infty$ corresponds to $r = +\infty$. The spacelike worldsheet coordinate σ^1 is identified with the angle $-\theta$ on the plane, i.e., it is periodic.

A primary field $\Phi(z, \bar{z})$ of conformal dimension $(h, \bar{h})$ transforms under the conformal mapping $z \longrightarrow w$ covariantly as

$$\Phi(z, \bar{z}) \longrightarrow \left(\frac{\partial w}{\partial z}\right)^h\left(\frac{\partial \bar{w}}{\partial \bar{z}}\right)^{\bar{h}}\Phi(w, \bar{w}). \tag{16.227}$$

Thus, $\Phi(z\bar{z})dz^h d\bar{z}^{\bar{h}}$ is invariant under the conformal mapping. On the cylinder the field is given by

$$\Phi_{\text{cyl}}(w, \bar{w}) = \left(\frac{\partial z}{\partial w}\right)^h\left(\frac{\partial \bar{z}}{\partial \bar{w}}\right)^{\bar{h}}\Phi(z, \bar{z}) = z^h\bar{z}^{\bar{h}}\Phi(z, \bar{z}). \tag{16.228}$$

Laurent expansion on the complex plane becomes Fourier expansion on the cylinder under the conformal mapping. Indeed, we have (for holomorphic fields $\Phi(z)$ and $\Phi_{\text{cyl}}(w)$)

$$\Phi(z) = \sum_n \frac{\phi_n}{z^{n+h}} \Rightarrow \Phi_{\text{cyl}}(w) = z^h\Phi(z) = \sum_n \phi_n \exp(-nw). \tag{16.229}$$

It is obvious that under a full rotation in the plane $z \longrightarrow \exp(2\pi i)z, \bar{z} \longrightarrow \exp(-2\pi i)\bar{z}$ the field on the cylinder Φ_{cyl} rotates as $\Phi_{\text{cyl}}(w, \bar{w}) \longrightarrow \exp(2\pi i(h - \bar{h}))\Phi_{\text{cycl}}(w, \bar{w})$. The spin of the field Φ is precisely $s = h - \bar{h}$ if you recall that $l_0 - \bar{l}_0$ generates rotation. Thus a bosonic field Φ will satisfy the same boundary condition on the plane and on the cylinder whereas for a fermionic field on the plane Φ with periodic (anti-periodic) boundary condition the field on the cylinder Φ_{cyl} satisfies anti-periodic (periodic) boundary condition.

Next we construct the torus via discrete identification on the cylinder. Let H and P be the energy and momentum operators, which generate translation in the time $\mathrm{Re}\ w = \sigma^2$ and space $\mathrm{Im}\ w = -\sigma^1$ directions, respectively. On the plane we know that $l_0 + \bar{l}_0$ and $l_0 - \bar{l}_0$ are the generators of dilatations and rotations, respectively, and thus on the cylinder we must have $H = (l_0)_{\mathrm{cyl}} + (\bar{l}_0)_{\mathrm{cyl}}$ and $P = (l_0)_{\mathrm{cyl}} - (\bar{l}_0)_{\mathrm{cyl}}$.

The torus is defined by two periods in w. First we redefine w as $w \longrightarrow iw$, i.e., $z = \exp(iw)$ and $\mathrm{Re}\ w = \sigma^1$, $\mathrm{Im}\ w = \sigma^2$. The first period is then 1 since $w \longrightarrow w + 2\pi$ leaves z invariant. This corresponds to an identification along the real axis and rolling up the complex plane to a cylinder. The second period is taken to be equal to $\tau = \tau_1 + i\tau_2$ where τ is called the modular parameter. Thus the identification $w \equiv w + 2\pi\tau$ along the vector τ in the complex plane rolls up the cylinder to a torus. See figure 16.2. This torus is then effectively a lattice described by the complex number τ. Any field on this torus must then satisfy

$$\Phi(w + 1) = \Phi(w + \tau) = \Phi(w). \tag{16.230}$$

The complex number w can be rewritten conveniently in terms of two real numbers $\tilde{\sigma}^i \in [0, 1[$ each of period 1 as

$$w = \tilde{\sigma}^1 - \tau\tilde{\sigma}^2. \tag{16.231}$$

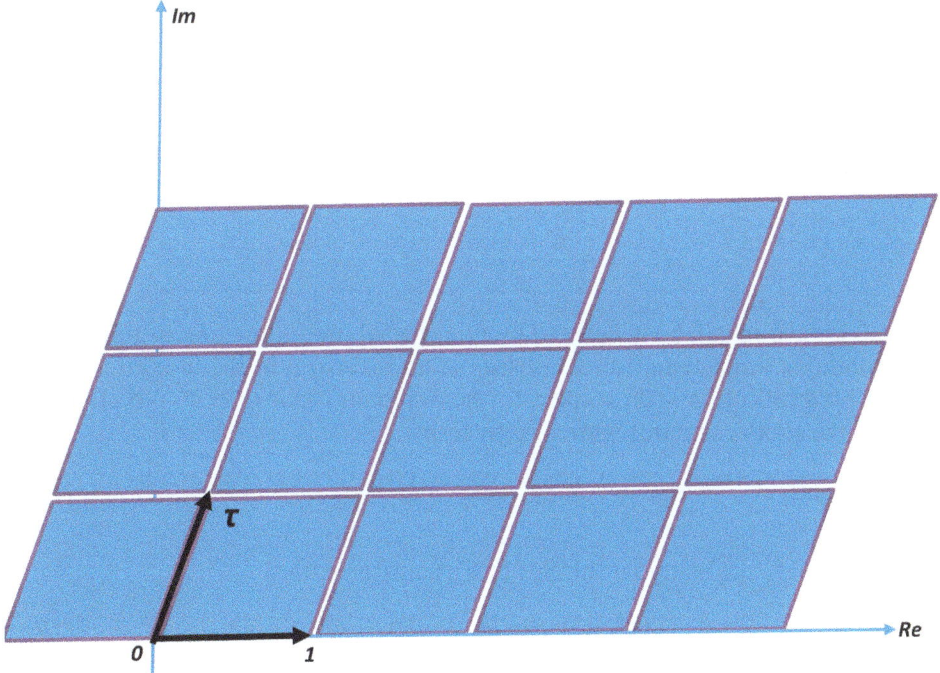

Figure 16.2. The torus.

The torus is then defined by the more transparent equivalence relation

$$(\tilde{\sigma}^1, \tilde{\sigma}^2) \sim (\tilde{\sigma}^1 + n, \tilde{\sigma}^2 + m), \quad n, m \in \mathbf{Z}. \tag{16.232}$$

This torus admits diffeomorphisms, which are not connected continuously to the identity given by [13]

$$(\tilde{\sigma}^1, \tilde{\sigma}^2) \longrightarrow (a\tilde{\sigma}^1 + b\tilde{\sigma}^2, c\tilde{\sigma}^1 + d\tilde{\sigma}^2). \tag{16.233}$$

The points that are equivalent under (16.232) are mapped under the transformations (16.233) to other points, which are also equivalent under (16.232) if and only of a, b, c and d are integers. The transformation (16.233) is invertible and one-to-one if and only if

$$ad - bc = 1. \tag{16.234}$$

This means that the transformation (16.233) defines an element of $SL(2, Z)$, which is the group of reparametrizations of the torus. This is what is called the modular group of the torus. These transformations cannot be reached continuously from the identity by exponentiating infinitesimal reparametrizations. This $SL(2, Z)$ is the part of the reparametrization invariance that is not taken into account in the Fadeev-Popov procedure [13].

Under the modular transformation (16.233) we also have, similarly to (16.50), the transformation law

$$\tau \longrightarrow \tau' = \frac{a\tau + b}{c\tau + d} \tag{16.235}$$

and

$$\begin{aligned}
w \longrightarrow w' &= \tilde{\sigma}^{1\prime} - \tau'\tilde{\sigma}^{2\prime} \\
&= (a + c\tau')\tilde{\sigma}^1 + (b + d\tau')\tilde{\sigma}^2 \\
&= \frac{\tilde{\sigma}^1 - \tau\tilde{\sigma}^2}{c\tau + d} \\
&= \frac{w}{c\tau + d}.
\end{aligned} \tag{16.236}$$

The modular group can be generated by two special modular transformations T and S. The modular transformation S consists of an inversion in the unit circle $r' = 1/r$ followed by a reflection with respect to the imaginary axis $\theta' = \pi - \theta$. Indeed, the action on the modular parameter τ is given by

$$S: \quad \tau \longrightarrow \tau' = -\frac{1}{\tau}. \tag{16.237}$$

This corresponds explicitly to the $SL(2, Z)$ transformation

$$(\tilde{\sigma}^1, \tilde{\sigma}^2) \longrightarrow (\tilde{\sigma}^2, -\tilde{\sigma}^1). \tag{16.238}$$

This transformation interchanges then the two basis vectors with a change of sign.

The modular transformation T is a translation given explicitly by

$$T: \quad \tau \longrightarrow \tau' = \tau + 1. \tag{16.239}$$

This corresponds explicitly to the $SL(2, Z)$ transformation

$$(\tilde{\sigma}^1, \tilde{\sigma}^2) \longrightarrow (\tilde{\sigma}^1 + \tilde{\sigma}^2, \tilde{\sigma}^2). \tag{16.240}$$

The transformations S and T generate the whole modular group. They satisfy $S^2 = (ST)^3 = 1$. The modular group is actually isomorphic to $SL(2, Z)/\mathbf{Z}_2$ since the two elements 1 and -1 are indistinguishable.

16.7.2 Free bosons on a torus

Let us then consider the action

$$S = \frac{1}{2\pi} \int \partial\Phi\bar{\partial}\Phi. \tag{16.241}$$

The measure is normalized such that (with $z = \sigma^1 + \tau\sigma^2$ and $\sigma^{1,2} \in [0, 1[$)

$$\int 1 = \int 2i dz \wedge d\bar{z} = \int 4\tau_2 d\sigma^1 \wedge d\sigma^2 = 4\tau_2. \tag{16.242}$$

We consider compactification on a circle of radius r, viz

$$\Phi = \Phi + 2\pi r. \tag{16.243}$$

The torus has actually two periods τ and 1. Thus the periodic boundary conditions are given explicitly by

$$\Phi_0(z + \tau, \bar{z} + \bar{\tau}) = \Phi_0(z, \bar{z}) + 2\pi r n', \quad \Phi_0(z + 1, \bar{z} + 1) = \Phi_0(z, \bar{z}) + 2\pi r n. \tag{16.244}$$

The field space decomposes then into instanton sectors characterized by the pair (n', n). The solution of the equation of motion $\partial\bar{\partial}\Phi_0 = 0$ in the instanton sector (n', n) is given explicitly by

$$\Phi_0(z, \bar{z}) = \frac{2\pi r}{2i\tau_2}[n'(z - \bar{z}) + n(\tau\bar{z} - \bar{\tau}z)]. \tag{16.245}$$

We expand the field as $\Phi = \Phi_0 + Y$ where the fluctuation Y is assumed to vanish at infinity. The action becomes

$$S[\Phi] = S[\Phi_0] + \frac{1}{2\pi} \int Y \square Y, \tag{16.246}$$

where $\square = -\partial\bar{\partial}$. The relevant path integral is

$$\int \mathcal{D}\Phi \exp(-S[\Phi]) = \exp(-S[\Phi_0]) \int \mathcal{D}Y \exp\left(-\frac{1}{2\pi} \int Y \square Y\right). \tag{16.247}$$

The fluctuation Y is split into a constant part $\tilde{Y}$ and a fluctuation Y' containing no zero mode. Thus, we have

$$\int \mathcal{D}\Phi \exp(-S[\Phi]) = \exp(-S[\Phi_0]) \int d\tilde{Y}\mathcal{D}Y' \exp\left(-\frac{1}{2\pi}\int Y' \,\Box\, Y'\right)$$

$$= 2\pi r \exp(-S[\Phi_0]) \int \mathcal{D}Y' \exp\left(-\frac{1}{2\pi}\int Y' \,\Box\, Y'\right). \tag{16.248}$$

The path integral is normalized such that

$$1 = \int \mathcal{D}X \exp\left(-\frac{1}{2\pi}\int X^2\right)$$

$$= \int d\tilde{X}\exp\left(-\frac{1}{2\pi}(4\tau_2)\tilde{X}^2\right)\int \mathcal{D}X' \exp\left(-\frac{1}{2\pi}\int X'^2\right) \tag{16.249}$$

$$= \frac{\pi}{\sqrt{2\tau_2}} \int \mathcal{D}X' \exp\left(-\frac{1}{2\pi}\int X'^2\right)$$

Therefore the partition function (16.248) on the torus becomes (by taking also the sum over all instanton sectors)

$$\int \mathcal{D}\Phi \exp(-S[\Phi]) = 2\pi r\frac{\sqrt{2\tau_2}}{\pi}\frac{1}{\det{}'^{1/2}\Box}\sum_{n',n=-\infty}^{+\infty} \exp(-S[\Phi_0^{(n',n)}])$$

$$= 2\pi r\frac{\sqrt{2\tau_2}}{\pi}\frac{1}{\det{}'^{1/2}\Box}\sum_{n',n=-\infty}^{+\infty} \exp\left(4\tau_2\frac{1}{2\pi}\left(\frac{2\pi r}{2i\tau_2}\right)^2(n' - \bar{\tau}n)(n' - \tau n)\right). \tag{16.250}$$

The crucial piece is the determinant. We use the single-valued (under both $z \longrightarrow z + \tau$ and $z \longrightarrow z + 1$) eigenfunctions

$$\psi_{n,m}(z, \bar{z}) = \exp\left(\frac{2\pi r}{2i\tau_2}[n(z - \bar{z}) + m(\tau\bar{z} - \bar{\tau}z)]\right). \tag{16.251}$$

Indeed, we have

$$\Box\psi_{n,m}(z, \bar{z}) = \frac{\pi^2}{\tau_2^2}(n - m\bar{\tau})(n - m\tau)\psi_{n,m}(z, \bar{z}). \tag{16.252}$$

The determinant $\det{}'\Box$, which does not involve the zero mode, is then given by

$$\det{}'\Box = \prod_{(m,n)\neq(0,0)} \frac{\pi^2}{\tau_2^2}(n - m\bar{\tau})(n - m\tau)$$

$$= \prod_{(m,n)\neq(0,0)} \frac{\pi^2}{\tau_2^2}\prod_{n\neq0} n^2 \prod_{m\neq0,n\in Z} (n - m\bar{\tau})(n - m\tau). \tag{16.253}$$

We use zet-function regularization $\zeta(s) = \sum_{n=1}^{\infty}1/n^s$, $\zeta(-1) = -1/12$, $\zeta(0) = -1/2$, $\zeta'(0) = -\ln\sqrt{2\pi}$. Then

$$\prod_{n=1}^{\infty} a = a^{\zeta(0)} = a^{-1/2}. \tag{16.254}$$

$$\prod_{n=1}^{\infty} n^{\alpha} = e^{\alpha \sum_{n=1}^{\infty} \ln n} = e^{-\alpha \zeta'(0)} = (2\pi)^{\alpha/2}. \tag{16.255}$$

The determinant becomes

$$\begin{aligned}
\det{}'\square &= \frac{\tau_2^2}{\pi^2}(2\pi)^2 \prod_{m>0,n\in Z} (n - m\bar{\tau})(n + m\bar{\tau})(n - m\tau)(n + m\tau) \\
&= \frac{\tau_2^2}{\pi^2}(2\pi)^2 \prod_{m>0} (e^{i\pi m\tau} - e^{-i\pi m\tau})^2 (e^{i\pi m\bar{\tau}} - e^{-i\pi m\bar{\tau}})^2,
\end{aligned} \tag{16.256}$$

where we have used the identity

$$\prod_{n\in Z} (n + a) = a \prod_{n=1}^{\infty} (-n^2)(1 - a^2/n^2) = 2i \sin \pi a. \tag{16.257}$$

We get finally

$$\begin{aligned}
\det{}'\square &= \frac{\tau_2^2}{\pi^2}(2\pi)^2 \prod_{m>0} (q\bar{q})^{-m}(1 - q^m)^2(1 - \bar{q}^m)^2 \\
&= 4\tau_2^2(q\bar{q})^{1/12} \prod_{m>0} (1 - q^m)^2(1 - \bar{q}^m)^2 \\
&= 4\tau_2^2 \eta^2 \bar{\eta}^2.
\end{aligned} \tag{16.258}$$

We have defined

$$\eta = q^{1/24} \prod_{m>0} (1 - q^m), \quad \bar{\eta} = \bar{q}^{1/24} \prod_{m>0} (1 - \bar{q}^m), \quad q = \exp(2i\pi\tau), \quad \bar{q} = \exp(-2i\pi\bar{\tau}). \tag{16.259}$$

The partition function on the torus becomes

$$\int \mathcal{D}\Phi \exp(-S[\Phi]) = r\sqrt{\frac{2}{\tau_2}} \frac{1}{\eta\bar{\eta}} \sum_{n',n=-\infty}^{+\infty} \exp\left(-2\pi\left[\frac{1}{\tau_2}(n'r - \tau_1 nr)^2 + r^2 n^2 \tau_2\right]\right). \tag{16.260}$$

The summation over the winding n' in the time direction (corresponding to the period τ) will be converted into a summation over a conjugate momentum by means of Poisson resummation formula

$$\sum_{n'=-\infty}^{+\infty} f(n'r) = \frac{1}{r} \sum_{m=-\infty}^{+\infty} \tilde{f}\left(\frac{m}{r}\right), \quad \tilde{f}(p) = \int_{-\infty}^{+\infty} dx \, e^{2\pi i p x} f(x). \tag{16.261}$$

We take the function f and its Fourier transform $\tilde{f}$ to be

$$f(n'r) = \exp\left(-2\pi\frac{1}{\tau_2}(n'r - \tau_1 nr)^2\right) \Rightarrow \tilde{f}(p) = \sqrt{\frac{\tau_2}{2}} \exp\left(2\pi i \tau_1 nrp - \frac{1}{2}\pi\tau_2 p^2\right). \tag{16.262}$$

The partition function becomes

$$\int \mathcal{D}\Phi \exp(-S[\Phi]) = \frac{1}{\eta\bar{\eta}} \sum_{m,n=-\infty}^{+\infty} \exp\left(-2\pi r^2 n^2 \tau_2 + 2\pi i \tau_1 nm - \frac{1}{2}\pi\tau_2\left(\frac{m}{r}\right)^2\right)$$

$$= \frac{1}{\eta\bar{\eta}} \sum_{m,n=-\infty}^{+\infty} q^{\frac{1}{2}\left(\frac{m}{2r}+nr\right)^2} \bar{q}^{\frac{1}{2}\left(\frac{m}{2r}-nr\right)^2}. \tag{16.263}$$

As we know compactification on a circle leads to a momentum and a winding defined by $p = m/r$ and $w = nr$. The zero modes are given by

$$\alpha_0 = p_L = \frac{p}{2} + w, \quad \bar{\alpha}_0 = p_R = \frac{p}{2} - w. \tag{16.264}$$

Hence we get the partition function

$$\int \mathcal{D}\Phi \exp(-S[\Phi]) = \frac{1}{\eta\bar{\eta}} \sum_{m,n=-\infty}^{+\infty} q^{\frac{1}{2}p_L^2} \bar{q}^{\frac{1}{2}p_R^2}. \tag{16.265}$$

We recall that in the quantum theory the generators L_0 and $\bar{L}_0$ are given by

$$L_0 = \sum_{l=1} \alpha_{-l}\alpha_l + \frac{1}{2}\alpha_0^2, \quad \bar{L}_0 = \sum_{l=1} \bar{\alpha}_{-l}\bar{\alpha}_l + \frac{1}{2}\bar{\alpha}_0^2. \tag{16.266}$$

The string ground states $|n, m\rangle$ are characterized by the fact that the oscillators α_l and $\bar{\alpha}_l$ are in their ground state, viz $\alpha_l|n, m\rangle = \bar{\alpha}_l|n, m\rangle = 0$, whereas the string center of mass has precisely a non-zero momentum equal m and non-zero winding equal n. In other words, we have

$$L_0|n, m\rangle = \frac{1}{2}\left(\frac{m}{2r} + nr\right)^2 |n, m\rangle, \quad \bar{L}_0|n, m\rangle = \frac{1}{2}\left(\frac{m}{2r} - nr\right)^2 |n, m\rangle. \tag{16.267}$$

The Hamiltonian and momentum eigenvalues of the string ground states $|n, m\rangle$ are then given by

$$H|n, m\rangle = (L_0 + \overline{L_0})|n, m\rangle = \left(\frac{m^2}{4r^2} + n^2r^2\right)|n, m\rangle, \quad P|n, m\rangle = (L_0 - \overline{L_0})|n, m\rangle = nm|n, m\rangle. \tag{16.268}$$

The partition function (16.265) can be rewritten in terms of the generators L_0 and $\bar{L}_0$ as

$$\int \mathcal{D}\Phi \exp(-S[\Phi]) = (q\bar{q})^{-c/24} tr q^{L_0} \bar{q}^{\bar{L}_0}. \tag{16.269}$$

This is the most general form of the partition function of the conformal field theory of free bosons on the torus. We can verify modular invariance by studying the effect of the modular transformation $\tau \longrightarrow -1/\tau$ on η.

16.8 Holographic entanglement entropy

16.8.1 Entanglement entropy

In quantum mechanics it is shown by the EPR experiment for example that entanglement is at odd with locality. The action (due to a measurement) seems to propagate with an infinite velocity and although it cannot carry any energy we are left in an uncomfortable position. Entanglement as opposed to energy is not conserved and there are degrees of entanglement. Mathematically, entanglement means that the vector state is not separable, i.e., it cannot be written as a tensor product.

Quantum entanglement is measured by entropy or more precisely by entanglement entropy. However, entropy has actually two sources: statistical and quantum.

1. **Statistical/Thermal Entropy**: The thermal or Boltzmann entropy of a macroscopic state is the logarithm of the number n of microscopic states consistent with this state. Thus this entropy measures the lack of resolution, i.e., the fact that a large number of microscopic configurations correspond to the same macroscopic thermodynamical state. The thermal entropy is defined in terms of the Boltzmann density matrix $\rho_{\text{ther}} = \exp(-\beta E)/Z$ by

$$\begin{aligned} S_{\text{ther}} &= \ tr\rho_{\text{ther}} \log \rho_{\text{ther}} \\ &= \log n. \end{aligned} \tag{16.270}$$

   The second equality holds if the microstates are equally probable.

2. **Entanglement Entropy:**
   - **Measurement**: In quantum mechanics, there is another source of entropy associated with the restriction of observers, who are performing the experiments, to finite volume. Indeed, a typical observer performing an experiment on a closed system, which is supposed to be in a pure ground state $|\Psi\rangle$, will only be able to access a particular subsystem, i.e., a partial set of the relevant observables such as those with support in a restricted volume.

     We will denote the accessible subsystem by A (where the observers are restricted) and the inaccessible subsystem is B. The total system $\Sigma = A \cup B$ is in a pure ground state $|\Psi\rangle$. See figure 16.3.

   - **Reduced Density Matrix**: The state of the system will be given by a mixed density matrix ρ and the entropy will measure the correlation between the inaccessible subsystem B and the accessible part A of the closed system. The total Hilbert space is $\mathcal{H}_{\text{Tot}} = \mathcal{H}_A \otimes \mathcal{H}_B$.

     The observer who cannot access the subsystem B will describe the total system by the reduced density matrix (obtained by tracing over the inaccessible degrees of freedom)

$$\rho_{\text{Red}} \equiv \rho_A = tr_B \rho_{\text{Tot}}. \tag{16.271}$$

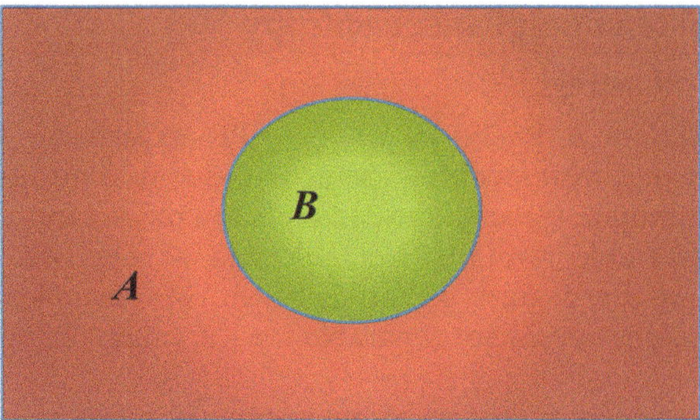

Figure 16.3. The accessible and inaccessible regions.

In other words, we trace (integrate) over the inaccessible subsystem B, i.e., we take average over the inaccessible degrees of freedom.

- **Mixed versus Pure:** The reduced density matrix is an incoherent (mixed) superposition (statistical ensemble, classical probabilities, no interference terms, random relative phases). It is not an idempotent and it satisfies $Tr\rho^2 < 1$.

  In contrast, a pure state is a vector in the Hilbert space, which is a coherent superposition (interference terms, coherent relative phases) represented by a projector.

  Mixed states are relevant if the exact initial state vector is unknown.

- **Entanglement Entropy:** The entropy of the subsystem A, which measures the correlation between the inaccessible subsystem and the accessible part of the closed system, is defined by the von Neumann entropy of this reduced density matrix, viz

$$S_{\text{Red}} \equiv S_A = -tr_A \rho_A \log \rho_A = -\sum_i \rho_i \log \rho_i. \qquad (16.272)$$

Thus, entanglement entropy is the logarithm of the number of microscopic states of the inaccessible subsystem B, which are consistent with observations restricted to the accessible subsystem A, together with the assumption that the total system is in a pure state. It measures the degree of entanglement between A and B. This is different from the thermodynamic Boltzmann entropy.

- **Examples:** For a pure (separable) state, i.e., when all eigenvalues with the exception of one vanish, we get $S = 0$. For mixed states we have $S > 0$.

  In the case of a totally incoherent mixed density matrix in which all the eigenvalues are equal to $1/N$ where N is the dimension of the Hilbert space we get the maximum value of the Von Neumann entropy given by

$$S_{\text{Red}} = S_{\text{max}} = \log N. \qquad (16.273)$$

In the case that ρ is proportional to a projection operator onto a subspace of dimension n we find

$$S_{\text{Red}} = \log n. \tag{16.274}$$

In other words, the Von Neumann entropy measures the number of important states in the statistical ensemble, i.e., those states, which have an appreciable probability. This entropy is also a measure of the degree of entanglement between subsystems A and B and hence its other name entanglement entropy.

- **Information:** The von Neumann entropy $S_V \equiv S_A$ is not additive as opposed to the thermal entropy S_B defined with respect to Boltzmann distribution. We have $S_B \geqslant S_V$, i.e., the Boltzmann thermal entropy (coarse grained, macroscopic) is always greater or equal to von Neumann entanglement (fine grained, microscopic) entropy.

  The amount of information is the difference:

$$I = S_B - S_V. \tag{16.275}$$

If $\Sigma = A$ then there is no entanglement and the amount of information is maximal, i.e., $S_V = 0 \Rightarrow I = S_B$. If $A \ll B$ then in this case the amount of information is zero, i.e., $S_V = S_B$. Equivalently, if $A \ll B$ then the entanglement entropy becomes maximal equal to the thermal entanglement.

Remark that the von Neumann entropy of the total system is zero, viz $S_{\text{Tot}} = tr\rho_{\text{Tot}} \log \rho_{\text{Tot}} = 0$ since there is no inaccessible part here.

16.8.2 Entanglement entropy in quantum mechanics

For detail of the formalism used here we refer to [28]. We will consider a Hamiltonian of the form

$$H = \frac{1}{2}\sum_{A,B}(\delta_{A,B}\pi^A\pi^B + V_{AB}\varphi^A\varphi^B). \tag{16.276}$$

In this equation V is a real symmetric matrix with positive definite eigenvalues. The normalized ground state of this model is given in the Schrodinger representation by

$$\langle \varphi^A | 0 \rangle = \left[\det \frac{W}{\pi}\right]^{1/4} \exp\left[-\frac{1}{2}W_{AB}\varphi^A\varphi^B\right]. \tag{16.277}$$

W is the square root of the matrix V. The corresponding density matrix is

$$\rho(\varphi, \varphi') = \left[\det \frac{W}{\pi}\right]^{1/2} \exp\left[-\frac{1}{2}W_{AB}(\varphi^A\varphi^B + \varphi'^A\varphi'^B)\right]. \tag{16.278}$$

If we suppose that the field degrees of freedom φ^α, $\alpha = \overline{1, n}$, are inaccessible then the correct description of the state of the system will be given by the reduced density matrix in which we integrate out these inaccessible degrees of freedom, viz

$$\rho_{\text{red}}(\varphi^{n+1}, \varphi^{n+2}, \ldots, \varphi'^{n+1}, \varphi'^{n+2}, \ldots) = \int \prod_{\alpha=1}^{n} d\varphi^{\alpha} \rho(\varphi, \varphi') \tag{16.279}$$

The entanglement entropy is the associated Von Newman entropy of ρ_{red} defined by $S = -Tr\rho_{\text{red}} \log \rho_{\text{red}}$. The entanglement entropy for any Hamiltonian of the form (16.276) can be shown to be given by [28]

$$S_{\text{ent}} = \sum_i \left[\log\left(\frac{1}{2}\sqrt{\lambda_i}\right) + \sqrt{1 + \lambda_i} \, \log\left(\frac{1}{\sqrt{\lambda_i}} + \sqrt{1 + \frac{1}{\lambda_i}}\right) \right]. \tag{16.280}$$

The λ_i are the eigenvalues of the following matrix

$$\Lambda_{i,j} = -\sum_{\alpha=1}^{n} W_{i\alpha}^{-1} W_{\alpha j} \tag{16.281}$$

$W_{\alpha j}$ and $W_{i\alpha}^{-1}$ are elements of W and W^{-1}, respectively, with i, j running from $n + 1$ to $\mathcal{N}$ and α from 1 to n, i.e., Λ is an $(\mathcal{N} - n) \times (\mathcal{N} - n)$ matrix and i, j run from $n + 1$ to $\mathcal{N}$.

16.8.3 Entanglement entropy in conformal field theory

In this section we follow mostly [29, 31]. See also [30].

The entropy of a macroscopic state, in statistical mechanics, is defined by the logarithm of the number of microscopic states, which are consistent with it. Thus this entropy measures the lack of resolution, i.e., the fact that a large number of microscopic configurations correspond to the same macroscopic thermodynamical state.

However, in quantum mechanics, there is another source of entropy associated with the restriction of observers, who are performing the experiments, to finite volume. Indeed, a typical observer performing an experiment on a closed system, which is supposed to be in a pure ground state $|\Psi\rangle$, will only be able to access a particular subsystem, i.e., a partial set of the relevant observables such as those with support in a restricted volume.

We will denote the accessible subsystem by A and the inaccessible subsystem by B (see (16.3)). In this case the state of the system will be given by a mixed density matrix ρ and the entropy will measure the correlation between the inaccessible subsystem and the accessible part of the closed system. The total Hilbert space is clearly given by $\mathcal{H}_{\text{Tot}} = \mathcal{H}_A \otimes \mathcal{H}_B$. The observer who cannot access the subsystem B will describe the total system not by the ground state $|\Psi\rangle$ (or its corresponding density matrix $\rho_{\text{Tot}} = |\Psi\rangle\langle\Psi|$) but by the reduced density matrix

$$\rho_{\text{Red}} \equiv \rho_A = tr_B \rho_{\text{Tot}}. \tag{16.282}$$

In other words, we trace (integrate) over the inaccessible subsystem B, i.e., we take average over the inaccessible degrees of freedom. The entanglement entropy of the subsystem A is defined by the von Neumann entropy of this reduced density matrix, viz

$$S_{\text{Red}} \equiv S_A = -tr_A \rho_A \log \rho_A. \tag{16.283}$$

The entanglement entropy is then essentially the logarithm of the number of microscopic states of the inaccessible part of the system, which are consistent with the observations restricted to the accessible subsystem, together with the assumption that the total system is in a pure state. It measures as we said the degree of correlation (entanglement) between the accessible subsystem and the inaccessible part of the total system.

Remark that the von Neumann entropy of the total system is zero, viz $S_{\text{Tot}} = tr\rho_{\text{Tot}} \log \rho_{\text{Tot}} = 0$ since there is no inaccessible part here.

We assume now a conformal field theory in two dimensions with complex coordinate $z = \sigma + i\tau$. The spatial dimension σ is given by $C = [0, L[$ where L is an infrared cutoff and we will assume periodic boundary condition, i.e., $L \equiv 0$. The subsystem playing the role of the accessible region (where measurements are performed) is $A = [0, \Sigma[$ whereas the unavailable region (to be traced over) is $B = [\Sigma, L[$. The ultraviolet cutoffs are introduced by considering instead the intervals $A = [\epsilon_1, \Sigma - \epsilon_2[$ and $B = [\Sigma + \epsilon_2, L - \epsilon_1[$. We perform the conformal mapping

$$z \longrightarrow w = -\frac{\sin \dfrac{\pi}{L}(z - \Sigma)}{\sin \dfrac{\pi}{L}z}. \tag{16.284}$$

The regularized intervals A and B become (with the assumption $\Sigma \ll L$) the positive half-axis and the negative half-axis, respectively, viz

$$A =]\epsilon_2/\Sigma, \Sigma/\epsilon_1], \quad B =]-\Sigma/\epsilon_1, -\epsilon_2/\Sigma]. \tag{16.285}$$

This is spatial section at $\tau = 0$. Extrapolation into the past $\tau \longrightarrow -\infty$ corresponds to extrapolation to the lower half-plane with inner and outer radii ϵ_2/Σ and Σ/ϵ_1, respectively.

Lastly we perform the conformal mapping

$$w \longrightarrow y = \frac{1}{\kappa} \ln w. \tag{16.286}$$

Our points from $\tau = -\infty$ to $\tau = 0$ are given by $w = R \exp(i\theta)$ where R ranges from ϵ_2/Σ to Σ/ϵ_1 and θ ranges from $-\pi$ to 0. Hence $y = \ln R/\kappa + i\theta/\kappa$. We have then

$$\theta = \text{fixed}, \quad \Delta y = \frac{1}{\kappa} \ln \frac{\Sigma}{\epsilon_1} - \frac{1}{\kappa} \ln \frac{\epsilon_2}{\Sigma} = \frac{2}{\kappa} \ln \frac{\Sigma}{\sqrt{\epsilon_1 \epsilon_2}} \equiv L. \tag{16.287}$$

$$R = \text{fixed}, \quad \Delta y = i\frac{\Delta\theta}{\kappa} = i\frac{\pi}{\kappa} \equiv ih. \tag{16.288}$$

This is a finite strip of length L and width h. The accessible region A corresponds to fixed θ between $-\pi/2$ and 0 and thus corresponds to the upper side of the strip whereas the inaccessible region B corresponds to fixed θ between $-\pi$ and $-\pi/2$ and thus to the lower side of the strip. The 'upper' and 'lower' are with respect to the width direction h.

The ground state wave functional $\Psi(\phi_0(x))$ can be defined via a path integral with an appropriate boundary conditions specifying the field on the Cauchy surface $C = A \cup B$. Explicitly we have

$$\Psi(\phi_0(x)) = \int_{\tau=-\infty}^{\tau=0} \mathcal{D}\phi \, \exp(-S(\phi)), \quad \phi_0(x) = \phi(0, x). \tag{16.289}$$

The field is also assumed to vanish in the limit $\tau \longrightarrow -\infty$. The complex conjugate wave functional $\bar{\Psi}(\phi_0'(x))$ is given similarly by

$$\bar{\Psi}(\phi_0'(x)) = \int_{\tau=0}^{\tau=+\infty} \mathcal{D}\phi \, \exp(-S(\phi)), \quad \phi_0'(x) = \phi'(0, x). \tag{16.290}$$

We can write $\phi_0 = XY$ where $\phi_0 = X$ on A (upper side of the first copy of strip) and $\phi_0 = Y$ on B (lower side of this strip) and similarly we can write $\phi_0' = X'Y'$ where $\phi_0' = X'$ on A (lower side of another copy of the strip) and $\phi_0' = Y'$ on B (upper side of the second copy of the strip).

The total density matrix is then given by

$$\rho_{\phi_0 \phi_0'} = \Psi(\phi_0(x))\bar{\Psi}(\phi_0'(x)). \tag{16.291}$$

However, the reduced density matrix ρ_A, which describes the observations from the subsystem A, is obtained by tracing ρ over the inaccessible subsystem B. Thus we need to integrate ϕ_0 on the region B with the condition $\phi_0(x) = \phi_0'(x)$ when $x \in B$. Then the reduced density matrix is given by

$$(\rho_A)_{XX'} = \int \mathcal{D}Y \Psi(X, Y)\bar{\Psi}(Y, X'). \tag{16.292}$$

This is generally a mixed density matrix as opposed to the total density matrix ρ, which is a pure density matrix. This integral involves pasting together two copies of the strip along the inaccessible region B. Thus it corresponds to a functional integral over a strip of height $2\pi/\kappa$ with boundary conditions given by $\phi_0 = X$ on the upper side of the strip and $\phi_0 = X'$ on the lower side of the strip given by

$$(\rho_A)_{XX'} = \frac{1}{Z_1} \int \mathcal{D}\phi \, \exp(-S(\phi)). \tag{16.293}$$

The path integral Z_1 is determined by the normalization condition $tr\rho_A = 1$. This is clearly periodic in the height direction h since the fields are identified by the tracing. If we also impose a periodic boundary condition in the length direction L then Z_1 is nothing else but the partition function on a torus.

The entropy is actually going to be calculated using the so-called replica trick given by the relation

$$S_A = -tr\rho_A \log \rho_A = -\frac{\partial}{\partial n} tr_A \rho_A^n \big|_{n=1}. \tag{16.294}$$

The trace $tr_A \rho_A^n$ is computed by pasting together n copies of the strip along the inaccessible region B. We start thus from n copies of the reduced density matrix, viz

$$(\rho_A)_{X_{1+}X_{1-}}(\rho_A)_{X_{2+}X_{2-}} \cdots (\rho_A)_{X_{n+}X_{n-}}. \tag{16.295}$$

The pasting or gluing is done by the conditions $X_{i-}(x) = X_{i+1+}(x)$, $\forall\, i = 1, \ldots, n-1$, and then integrating over X_{i+}. If we choose $X_{1+} = X$ and $X_{n-} = X'$ then we obtain the matrix element

$$(\rho_A^n)_{XX'} = \frac{1}{Z_1^n} \int \mathcal{D}\phi \, \exp(-S(\phi)). \tag{16.296}$$

The functional integral is over a strip of width $2\pi n/\kappa$ with boundary conditions given by $\phi_0 = X$ on the upper side of the strip and $\phi_0 = X'$ on the lower side of the strip. By setting $X = X'$ and integrating over X we obtain the desired trace as

$$tr_A \rho_A^n = \frac{1}{Z_1^n} \int \mathcal{D}X \int \mathcal{D}\phi \, \exp(-S(\phi)) = \frac{Z_n}{Z_1^n}. \tag{16.297}$$

The Z_n is the partition function on the torus of lengths $2\pi n/\kappa$ and L around its two cycles. The entanglement entropy S_A finally takes the form

$$S_A = \left(\left[1 - n\frac{\partial}{\partial n} \right] \ln Z_n \right)\Big|_{n=1}. \tag{16.298}$$

16.8.4 Ryu–Takayanagi formula

In this section we follow mainly [31].

The Bekenstein–Hawking formula states that the entropy of a black hole S_{BH} is proportional to the surface area of the event horizon A_H, viz

$$S_{BH} = \frac{A_H}{4G_N}. \tag{16.299}$$

On the other hand, the entanglement entropy S_A for observers accessible to a subsystem A (outside event horizon) who cannot receive any signals from the subsystem B (inside the black hole) is given by

$$S_A = -tr_A \rho_A \log \rho_A. \tag{16.300}$$

The entanglement entropy satisfies the following properties:
1. For three subsystems A, B and C, which do not intersect each other, we have the so-called strong subadditivity relations

$$\begin{aligned} S_{A+B+C} + S_B &\leqslant S_{A+B} + S_{B+C} \\ S_A + S_C &\leqslant S_{A+B} + S_{B+C}. \end{aligned} \tag{16.301}$$

2. By choosing B empty in the above relations we obtain

$$S_{A+B} \leqslant S_A + S_B. \tag{16.302}$$

The mutual information is defined by

$$I(A, B) = S_A + S_B - S_{A+B} \geqslant 0. \tag{16.303}$$

3. If we choose B to be the complement of A then

$$S_A = S_B \Rightarrow S_{A+B} \leqslant 2S_A. \tag{16.304}$$

Hence the entanglement entropy is not an extensive quantity.

In a QFT on a $(d + 1)$-dimensional manifold $\mathbf{R} \times \mathbf{N}$ where $d \geqslant 2$ and $\mathbf{N} = A \cup B$ it is found that the entanglement entropy (1) depends only on the geometry of A (this is why entanglement entropy is also called geometric entropy), (2) is UV divergent and hence the continuum theory should be regularized by a lattice a, and (3) it is proportional to the area of the boundary ∂A of A since the entanglement between A and B occurs strongly obviously on the boundary. We have explicitly [28, 32]

$$S_A = \gamma \cdot \frac{\text{Area}(\partial A)}{a^{d-1}} + \text{subleading terms.} \tag{16.305}$$

This entanglement entropy formula (includes UV divergences, proportional to the number of matter fields) is very similar to the Bekenstein–Hawking formula (does not include UV divergences, is not proportional to the number of matter fields). In fact the quantum corrections to the Bekenstein–Hawking black hole entropy in the presence of matter fields is given by the entanglement entropy [33,34].

The Ryu–Takayanagi formula is a generalization of the Bekenstein–Hawking formula, based on the $\text{AdS}_{d+2}/\text{CFT}_{d+1}$ correspondence, in which we identify the entanglement entropy in $(d + 1)$-dimensional QFT with a geometric quantity in $(d + 2)$-dimensional gravity.

We consider the metric in AdS_{d+2} in Poincaré patch given by

$$ds^2 = g^{\mu\nu}dx_\mu dx_\nu = \frac{R^2}{z^2}\left(-dt^2 + \sum_{i=1}^{d}dx_i^2 + dz^2\right). \tag{16.306}$$

The dual $(d + 1)$-dimensional CFT lives on the boundary located at $z = 0$. The radial coordinate z as we have seen is a lattice spacing and the theory on the boundary should be properly understood as the continuum limit (in the sense of RG) of a regularize CFT, i.e., with a cutoff Λ. This regularized CFT lives on a surface $z = a$ where $a = 1/\Lambda$ and $a \longrightarrow 0$.

Our observers live on the boundary $z = a$. The accessible region A and the inaccessible region B are both on this boundary $z = a$. The entanglement entropy in the CFT_{d+1}, which lives on this boundary can be compute from the gravity theory, which lives in the bulk AdS_{d+2} as follows. We extend $\mathbf{N} = A \cup B$ to the time slice M of the bulk spacetime and we extend ∂A to a d-dimensional surface $\Gamma_A \in M$ such that $\partial\Gamma_A = \partial A$ (figure 16.4). The time slice M is the $(d + 1)$-dimensional hyperbolic space $\mathbf{H}_{d+1}$, which is an Euclidean manifold whereas Γ_A is a minimal area surface. The entanglement entropy in the CFT_{d+1} is then given by the formula [35, 36]

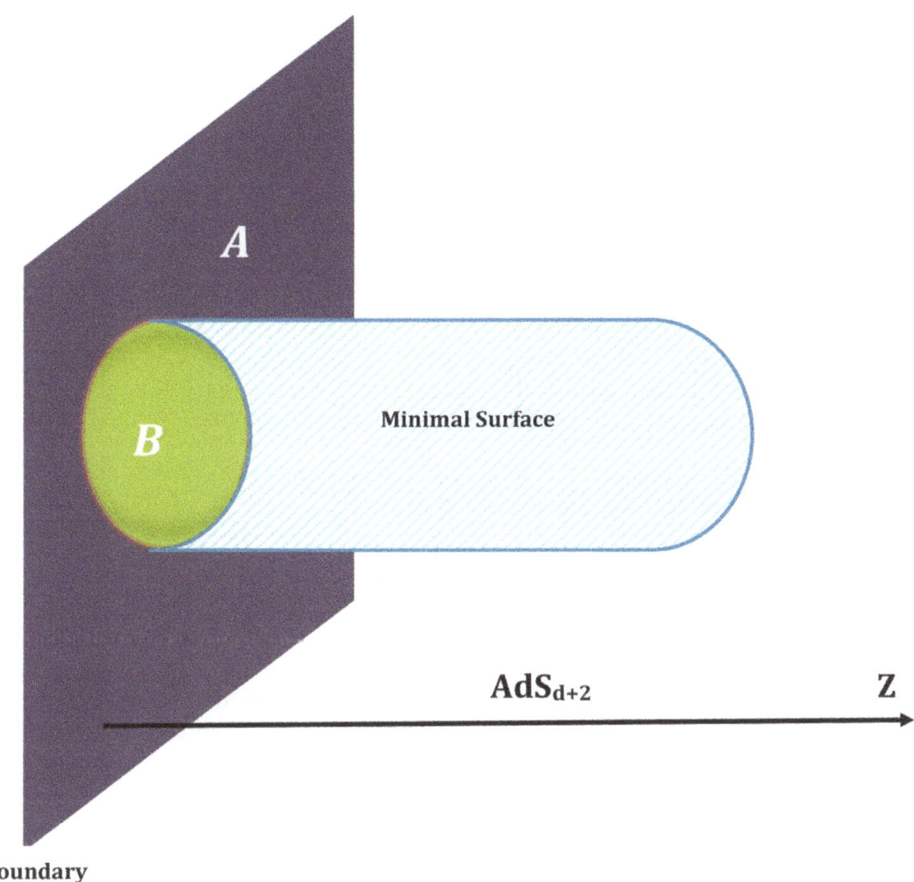

Figure 16.4. The accessible and inaccessible regions in AdS.

$$S_A = \frac{\text{Area}(\Gamma_A)}{4G_N^{(d+2)}}.$$

(16.307)

As an example we consider the case of AdS_3. The dual field theory is a two-dimensional conformal field theory with central charge c. The corresponding partition function is given by the formula (16.269). This is the partition function of the conformal field theory of free bosons on the torus with periods 1 (real axis) and τ (imaginary axis). We have then

$$Z(\tau, \bar{\tau}) = (q\bar{q})^{-c/24} tr q^{L_0} \bar{q}^{\bar{L}_0}.$$

(16.308)

The variables q and $\bar{q}$ are given in terms of the modular parameter τ by the relations

$$q = \exp(2i\pi\tau), \quad \bar{q} = \exp(-2i\pi\bar{\tau}).$$

(16.309)

On the other hand, the entanglement entropy is given in terms of the partition function Z_n on the torus of lengths $2\pi n/\kappa$ and L around its two cycles by the relation

$$S_A = \left(\left[1 - n\frac{\partial}{\partial n} \right] \ln Z_n \right)\Big|_{n=1}. \tag{16.310}$$

We can take the modular parameter τ to be $\tau = (i2\pi n/\kappa)/L = 2i\pi n/\kappa L$. But the partition function is invariant under the modular transformation $\tau \longrightarrow -1/\tau$ and hence we can take τ to be $\tau = i\kappa L/2\pi n$. In other words, $q = \bar{q} = \exp(-\kappa L/n)$. The entanglement entropy becomes

$$
\begin{aligned}
S_A &= \left(1 + \ln q\frac{\partial}{\partial \ln q} \right) \ln Z(\tau, \bar{\tau}) \\
&= \left(1 + \ln q\frac{\partial}{\partial \ln q} \right)\left(-\frac{c}{12}\ln q + \ln trq^{L_0 + \bar{L}_0} \right) \\
&= -\frac{c}{6}\ln q + \left(1 + \ln q\frac{\partial}{\partial \ln q} \right) \ln trq^{L_0 + \bar{L}_0} \\
&= -\frac{c}{6}(-\kappa L) \\
&= \frac{c}{3}\ln \frac{\Sigma}{\epsilon}.
\end{aligned}
\tag{16.311}
$$

In going from the third to the fourth lines we have assumed that $\ln trq^{L_0 + \bar{L}_0}$ is exponentially suppressed whereas in the last line we have used the result (16.287).

This result can be re-derived from the gravity side as follows.

The metric in AdS_3 can be given by (Poincaré coordinates)

$$ds^2 = g^{\mu\nu}dx_\mu dx_\nu = \frac{R^2}{z^2}(-dt^2 + dx^2 + dz^2). \tag{16.312}$$

The boundary lies at $z = 0$. We will regularize by taking the restriction $z \geqslant a$. On the boundary $z = a$ we are interested in the line segment $x \in [-l/2, l/2]$. This segment is extended in the bulk (the plane xz) to a line of minimal length, i.e., a spacelike geodesic, which can be found as follows. The length is written as (fixed time)

$$L = \int \frac{R}{z}\sqrt{\dot{z}^2 + \dot{x}^2}\, ds. \tag{16.313}$$

Then we write Euler–Lagrange equations for x and z as (with c a constant)

$$
\begin{aligned}
x &\longrightarrow \frac{\dot{x}}{z} = c\sqrt{\dot{z}^2 + \dot{x}^2} \\
z &\longrightarrow \frac{\dot{x}^2}{z^2} = -\frac{\sqrt{\dot{z}^2 + \dot{x}^2}}{z}\frac{d}{ds}\left(\frac{\dot{z}}{\sqrt{\dot{z}^2 + \dot{x}^2}} \right).
\end{aligned}
\tag{16.314}
$$

The solution is given by the circle

$$z = A \sin s, \quad x = A \cos s. \tag{16.315}$$

By imposing the boundary conditions that the circle start at $(a, +l/2)$ and terminates at $(a, -l/2)$ we obtain

$$z = \frac{l}{2}\sin s, \quad x = \frac{l}{2}\cos s, \quad \epsilon \leqslant s \leqslant \pi - \epsilon, \quad \epsilon = \frac{2a}{l} \longrightarrow 0. \tag{16.316}$$

We compute now the actual length of the minimal line in the bulk as

$$L = \int_{\epsilon}^{\pi - \epsilon} \frac{R}{\sin s} ds = 2R \ln \frac{l}{a}. \tag{16.317}$$

The central charge c of CFT_2 is related to the radius R of AdS_3 by the relation (16.166). The proportionality factor is given precisely by [37]

$$c = \frac{3R}{2G_N^{(3)}}. \tag{16.318}$$

The entanglement entropy becomes

$$S_A = \frac{L}{4G_N^{(3)}} = \frac{c}{3} \ln \frac{l}{a}. \tag{16.319}$$

16.9 Einstein's gravity from quantum entanglement

A sample of the original literature for this section is [39–41, 43]. However, a very good concise and pedagogical review of the formalism relating spacetime geometry to quantum entanglement due to Van Raamsdonk and collaborators is found in [42].

16.9.1 The CFT/black hole correspondence

The starting point is the statement that Einstein's theory of general relativity on anti-de Sitter spacetime AdS_{d+1} is dual to a conformal field theory CFT_d on its boundary M^d. Let $|\psi(0)\rangle$ be the vacuum state of the CFT_d. This state is dual to the Poincaré patch of the pure AdS spacetime given by the metric

$$ds^2 = \frac{L^2}{z^2}(dz^2 + dx_\mu dx^\mu). \tag{16.320}$$

Let $|\psi(\zeta)\rangle$ be a one-parameter family of CFT_d excited states, which are dual to the perturbed metrics

$$ds^2 = \frac{L^2}{z^2}(dz^2 + \Gamma_{\mu\nu}(z, x)dx^\mu dx^\nu). \tag{16.321}$$

This corresponds to a spacetime M_ζ with boundary at $z \longrightarrow 0$ denoted by ∂M_ζ where the state $|\psi(\zeta)\rangle$ is living. Thus, $\Gamma_{\mu\nu} \longrightarrow \eta_{\mu\nu}$ and $M_\zeta \longrightarrow AdS_{d+1}$ when $\zeta \longrightarrow 0$. For small z (near the boundary) the metric behaves as

$$\Gamma_{\mu\nu}(z, x) = \eta_{\mu\nu} + z^d \bar{h}_{\mu\nu}(z, x) \Rightarrow ds^2 = \frac{L^2}{z^2}(dz^2 + dx_\mu dx^\mu + z^d \bar{h}_{\mu\nu}(z, x)dx^\mu dx^\nu). \tag{16.322}$$

This is called the Fefferman–Graham coordinates.

However, this metric can also be understood as corresponding to a spacetime M_ζ, which is a perturbation of pure AdS, dual to a small perturbation $|\psi(\zeta)\rangle_{\zeta \longrightarrow 0}$ of the CFT$_d$ vacuum $|\psi(0)\rangle$. This is an asymptotically AdS spacetime.

For higher excited states $|\psi(\zeta)\rangle$ we cannot assume classical supergravity solution since l_s is no longer much less than L and as a consequence stringy corrections of the order l_s^2 and higher become important. The geometry (and even the topology) of AdS_{d+1} becomes therefore very different.

An example of a non-trivial dual spacetime is the Schwarzschild-AdS black hole in $d + 1$ dimensions given by the metric

$$ds^2 = -f_M(r)dt^2 + \frac{dr^2}{f_M(r)} + r^2 d\Omega_{d-2}. \tag{16.323}$$

The function f_M is given by the difference of two pieces (the first being the usual Schwarzschild term)

$$f_M(r) = 1 - \frac{2\mu}{r^{d-3}} + \frac{r^2}{L^2}, \quad \mu = \frac{8\pi G_N M}{(d-1)\mathrm{Vol}(\mathbf{S}^{d-1})}, \quad \mu_{d=4} = G_N M. \tag{16.324}$$

This is an eternal black hole in $d + 1$ dimensions, which is asymptotically a pure AdS in contrast to the eternal Schwarzschild black hole, which is asymptotically a flat Minkowski spacetime.

Indeed, if we set $L = \infty$ we obtain Schwarzschild black hole. This is characterized by the Penrose diagram (figure 16.5), which summarizes the causal structure of the maximally extended Schwarzschild spacetime in the Kruskal–Szekeres coordinates $-\infty < R < +\infty$ and $-\infty < T < +\infty$. The two light-like infinities $\mathcal{J}^\pm$ (where light rays begin and end) and the spacelike infinity i^0 ($r = \infty$) are the same as those of Minkowski spacetime and hence the eternal Schwarzschild black hole is asymptotically a flat Minkowski spacetime. The timelike trajectories begin and end on the two timelike infinities i^- and $i+$, respectively, which are distinct surfaces from the singularity at $r = 0$. The horizon is located at $r_h = (2\mu)^{1/d-3}$. The region II is the interior of the black hole whereas the region I is the exterior. The region III lies also outside the black hole but it is spatially separated and therefore causally disconnected from region I. Region IV is the interior of a white hole, i.e., we can never go there but things can emerge from that region towards region I.

The Penrose diagram of the Schwarzschild-AdS spacetime is shown in figure 16.6. The two light-like infinities $\mathcal{J}^\pm$ and the spacelike infinity i^0 are replaced with the universal covering of global AdS spacetime in both regions I and III. These asymptotic regions are denoted A and B and they are the conformal boundary of AdS spacetime, i.e., $\mathbf{R} \times \mathbf{S}^{d-1}$. We have two possible situations:

- For $r_h \ll L$ we can neglect the term r^2/L^2 and we end up with an ordinary Schwarzschild spacetime. The Schwarzschild black hole at the center will then evaporate due to Hawking emission.
- For $r_h \gg L$ the Schwarzschild-AdS spacetime is in equilibrium with its Hawking radiation, i.e., emission and absorption rates are equal and the black hole will not evaporate. This is due to the fact that radiation can reach the asymptotic AdS boundary and return in a finite time.

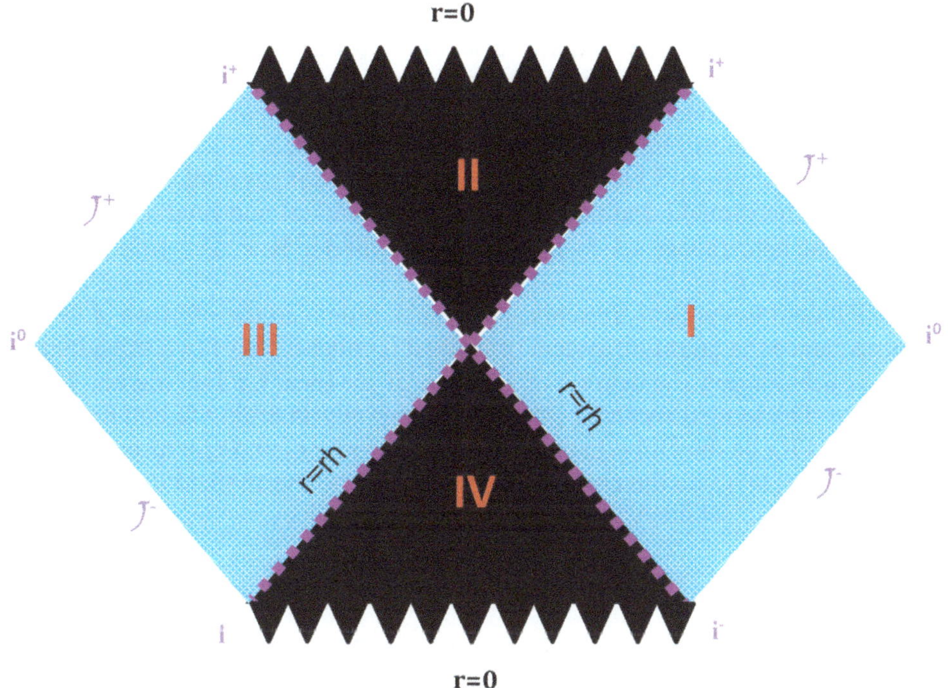

Figure 16.5. The Schwarzschild spacetime.

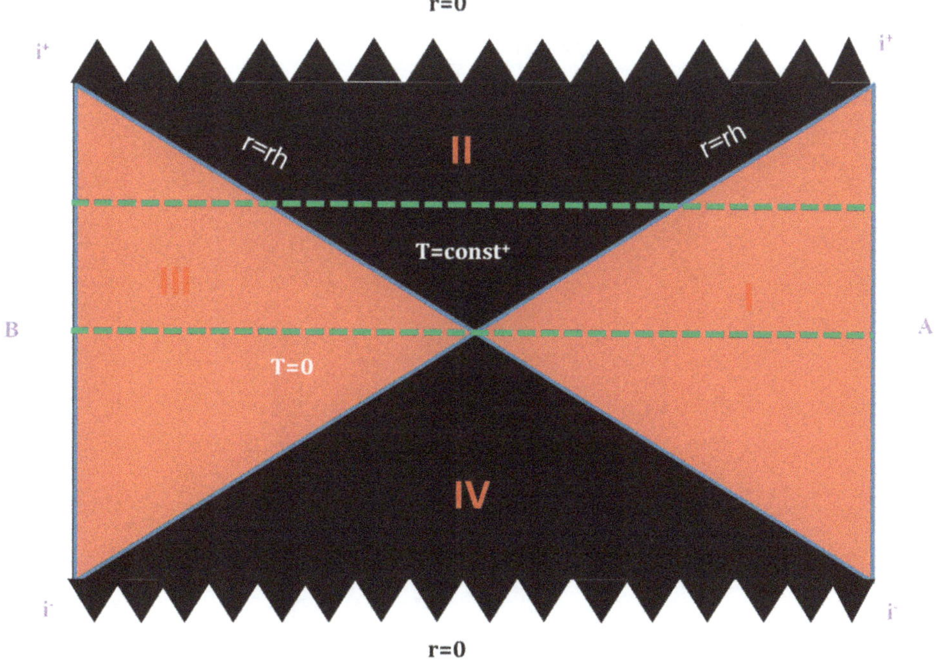

Figure 16.6. The Schwarzschild-AdS spacetime.

The regions I and III are causally disconnected but signals from region I can intersect with signals from region III behind the horizon. This is then a two-sided black hole, i.e., with two different exterior regions, which can also be viewed as a wormhole as depicted in figure 16.7 for $T = 0$ and $T > 0$.

The Schwarzschild-AdS black hole in $d + 1$ dimensions is conjectured in [38] to be dual to the thermofield double state $|\Psi\rangle$ of two identical non-interacting copies of the conformal field theory CFT_d living on the cylinder $\mathbf{R} \times \mathbf{S}^{d-1}$ (the asymptotic boundaries A and B).

The CFT dual to the Schwarzschild-AdS black is defined therefore on $A \cup B$ and as a consequence the quantum system of interest is constituted of two subsystems Q_A and Q_B, which contain the degrees of freedom of the local CFT living on A and B, respectively. The two subsystems Q_A and Q_B can only interact via entanglement.

Let $\mathcal{H}_A$ and $\mathcal{H}_B$ be the Hilbert spaces associated with Q_A and Q_B, respectively, and let $\{|E_i^A\rangle\}$ and $\{|E_i^B\rangle\}$ be the corresponding bases. The thermofield double state $|\Psi\rangle$ dual to the Schwarzschild-AdS black hole is then given by

$$|\Psi\rangle = \frac{1}{\sqrt{Z(\beta)}} \sum_i \exp(-\beta E_i/2)|E_i^A\rangle \otimes |E_i^B\rangle. \tag{16.325}$$

The partition function at inverse temperature β is given by

$$Z(\beta) = \sum_i \exp(-\beta E_i). \tag{16.326}$$

The reduced density matrix for the subsystem Q_A is immediately given by integrating out the degrees of freedom associated with the subsystem Q_B, viz

$$\rho_A = tr_B |\Psi\rangle\langle\Psi| = \frac{1}{\sqrt{Z(\beta)}} \sum_i \exp(-\beta E_i)|E_i^A\rangle\langle E_i^A|. \tag{16.327}$$

The corresponding entanglement entropy is given by

$$S_A = - tr_A \rho_A \log \rho_A$$
$$= \frac{1}{\sum_i e^{-\beta E_i}} \sum_i e^{-\beta E_i}\beta E_i + \log(\sum_i e^{-\beta E_i}). \tag{16.328}$$

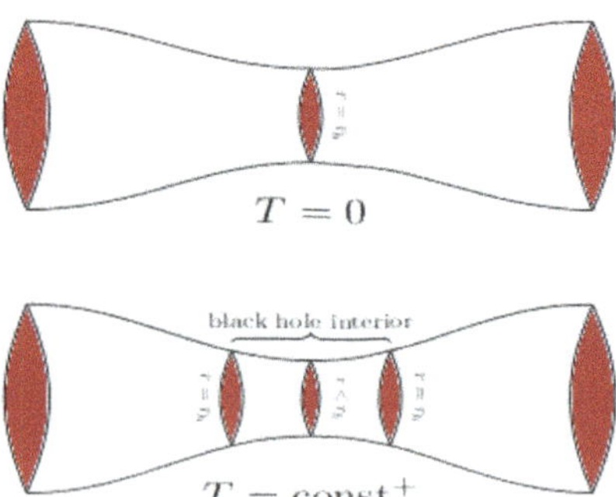

Figure 16.7. The two-sided Schwarzschild-AdS black hole as a wormhole.

This entanglement entropy is always non-zero except in the limit $\beta \longrightarrow \infty$. Indeed, in the limit of zero temperature the reduced density matrix ρ_A approaches $\rho_A = |E_0^A\rangle\langle E_0^A|$ where E_0 denotes the energy of the ground state and $E_0 = 0$. Hence, in this limit only the ground state is occupied and as a consequence the entanglement entropy vanishes. We have then, in the limit of zero temperature, the behavior

$$|\Psi\rangle = \frac{1}{\sqrt{Z(\beta)}}\sum_i \exp(-\beta E_i/2)|E_i^A\rangle \otimes |E_i^B\rangle \longrightarrow |\Phi\rangle = |E_0^A\rangle \otimes |E_0^B\rangle. \quad (16.329)$$

This is a product state with zero entanglement entropy describing two completely uncorrelated subsystems Q_A and Q_B.

The thermofield double state $|\Psi\rangle$ where Q_A and Q_B are entangled is dual to the Schwarzschild-AdS black hole, which is a connected spacetime in which light signals traveling from A and B can intersect. Similarly, the product state $|\Phi\rangle$ where Q_A and Q_B are now disentangled or uncorrelated is dual to a spacetime consisting of the product of two pieces corresponding to two disconnected regions A and B, i.e., light signals traveling from A and B cannot intersect. Thus, in the limit $\beta \longrightarrow \infty$ entanglement between Q_A and Q_B is removed and correspondingly connectivity between regions A and B is reduced until they become disconnected in the limit of zero temperature. This shows clearly that entanglement between Q_A and Q_B is a necessary condition for classical connectivity between A and B.

This picture can also be confirmed from the Ryu–Takayanagi formula $S_{\text{BH}} = A_{\text{BH}}/4G$. In the limit $\beta \longrightarrow \infty$ in the thermofield double state the entanglement between Q_A and Q_B is removed and thus $S_{\text{BH}} \longrightarrow 0$ or equivalently $A_{\text{BH}} \longrightarrow 0$. In other words, the limit $\beta \longrightarrow \infty$ in the thermofield double state decreases the connectivity in the dual black hole spacetime as seen on figure 16.8. Indeed, the proper distance between any two points in A and B (encoded in the mutual information between S_A and S_B) goes to ∞ as the entanglement between Q_A and Q_B goes to zero in the limit $\beta \longrightarrow \infty$.

This relation between quantum entanglement and spacetime connectivity generalizes to any quantum state with a classical spacetime dual.

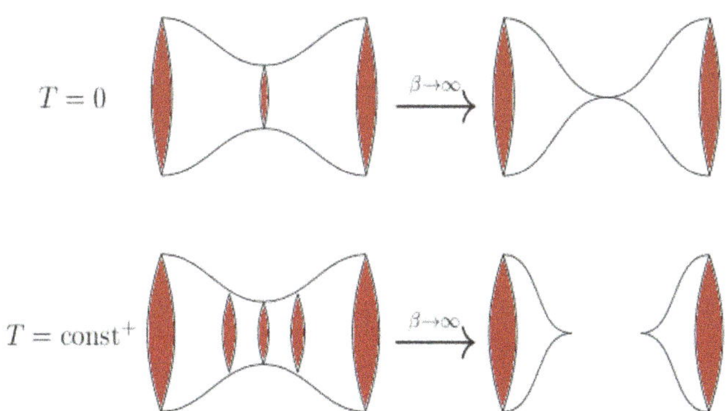

Figure 16.8. The limit $\beta \longrightarrow \infty$ in the Schwarzschild-AdS black hole.

16.9.2 The first law of thermodynamics and CFT

As before we will consider a one-parameter family of CFT states $|\psi(\zeta)\rangle$ living on the boundary of AdS spacetime AdS_{d+1}. The dual spacetime of the vacuum state $|\psi(0)\rangle$ of the CFT is assumed to be given by the Poincaré patch of the anti-de Sitter spacetime defined by the metric

$$ds^2 = \frac{L^2}{z^2}(dz^2 + dx_\mu dx^\mu). \tag{16.330}$$

The dual spacetime of the perturbed state $|\psi(\zeta)\rangle$, i.e., with $\zeta \longrightarrow 0$, is denoted by $\mathcal{M}_\psi$ and is a small perturbation of anti-de Sitter spacetime with boundary $\partial\mathcal{M}_\psi$, which is itself a small perturbation of Minkowski spacetime.

Let B some spatial region on the boundary $\partial\mathcal{M}_\psi$ and let $\bar{B}$ its complement. The accessible region B is a ball-shaped region of radius R on the Cauchy surface $\Sigma_{\partial\mathcal{M}_\psi} \in \partial\mathcal{M}_\psi$. The entanglement entropy of this spatial region in the CFT is equal to the von Neumann entropy of the reduced density matrix

$$\rho_B = tr_{\bar{B}}|\psi(\zeta)\rangle\langle\psi(\zeta)|. \tag{16.331}$$

The entanglement entropy is thus given by

$$S_B = -tr_B\,\rho_B \log \rho_B. \tag{16.332}$$

We can immediately compute

$$\frac{d}{d\zeta}S_B = -tr_B \frac{d}{d\zeta}\rho_B \cdot \log \rho_B. \tag{16.333}$$

We define the so-called modular Hamiltonian by the relation

$$H_B = -\log \rho_B(\zeta = 0). \tag{16.334}$$

Thus

$$\begin{aligned}
\frac{d}{d\zeta}S_B &= tr_B H_B \frac{d}{d\zeta}\rho_B + O(\zeta) \\
&= \frac{d}{d\zeta}tr_B H_B \rho_B + O(\zeta) \\
&= \frac{d}{d\zeta}\langle H_B \rangle + O(\zeta).
\end{aligned} \tag{16.335}$$

The expectation value of the modular Hamiltonian is what we call the hyperbolic energy, viz $\langle H_B \rangle = E_B^{\text{Hyp}}$. We get then

$$\frac{d}{d\zeta}S_B = \frac{d}{d\zeta}E_B^{\text{Hyp}}. \tag{16.336}$$

This is effectively the first law of thermodynamics $dE = dS$, which holds in the CFT for arbitrary perturbations ζ of the vacuum state $|\psi(0)\rangle$ and not only for thermal and equilibrium configurations as it is usually the case for the first law of thermodynamics.

16.9.3 Holographic or AdS entanglement entropy

The variation in the entanglement entropy in the CFT is given by $dS_B/d\zeta$. In the AdS the variation in the entanglement entropy is given by the Ryu–Takayanagi formula, viz

$$\delta S_B = \frac{\delta A_B}{4G_N} \equiv \frac{d}{d\zeta} S_B. \tag{16.337}$$

A_B is the area of the extremal surface $\tilde{B}$ in the bulk such that $\partial\tilde{B} = \partial B$ where the accessible region B is a ball-shaped region of radius R. This is the holographic interpretation of the entanglement entropy.

Thus, the variation in the entanglement entropy $dS_B/d\zeta$ of the CFT state $|\psi\rangle$ with a dual spacetime given by pure AdS corresponding to a ball-shaped region B is proportional to the variation δA_B in the area of the extremal surface $\tilde{B}$ due to the corresponding perturbation of the AdS space.

Let g^0_{ab} be the metric of pure AdS corresponding to the vacuum state $|\psi(0)\rangle$ and g_{ab} be the metric of the perturbed AdS corresponding to the excited state $|\psi(\zeta)\rangle$. We have

$$g_{ab} = g^0_{ab} + \delta g_{ab}. \tag{16.338}$$

The perturbation is given by Fefferman–Graham coordinates

$$\delta g_{ab} = L^2 z^{d-2}\bar{h}_{ab}(z, x) = z^{d-2}h_{ab}(z, x). \tag{16.339}$$

The surface $\tilde{B}$ is an extension of the spatial region B into the bulk and thus it is a co-dimension two surface characterized by some embedding functions $X^a(\sigma)$ with area A_B given by

$$A_B(g, X) = \int d^{d-1}\sigma \sqrt{\det \gamma_{ij}}. \tag{16.340}$$

The induced metric γ_{ij} on this surface $\tilde{B}$ is given by

$$\gamma_{ij} = g_{ab}\frac{\partial X^a}{\partial\sigma^i}\frac{\partial X^b}{\partial\sigma^j}. \tag{16.341}$$

In pure AdS this area is extremized by some embedding functions X^0_{ext} whereas in perturbed AdS it is extremized by some other embedding functions X_{ext}. We have then the variation in the area of the extremal surface given by

$$\begin{aligned}
\delta A_B(g, X_{\text{ext}}) &= \frac{\delta A_B(g, X^0_{\text{ext}})}{\delta g}\delta g + \frac{\delta A_B(g^0, X_{\text{ext}})}{\delta X^a}\delta X^a \\
&= \frac{\delta A_B(g, X^0_{\text{ext}})}{\delta g}\delta g + O(\delta g^2).
\end{aligned} \tag{16.342}$$

In this equation we have used the fact that δX is of order δg and as a consequence the second term is subleading of order δg^2. The variation in the area of the extremal surface is thus obtained by holding the embedding functions fixed and varying the spacetime metric.

Hence, the variation in the induced metric, which in fact directly controls the variation in $A_B(g, X_{\text{ext}})$ should be obtained by holding the embedding functions fixed and varying the spacetime metric, viz

$$\delta\gamma_{ij} = \delta g_{ab}\frac{\partial X_{\text{ext}}^a}{\partial\sigma^i}\frac{\partial X_{\text{ext}}^b}{\partial\sigma^j}.$$
(16.343)

In this equation $X_{\text{ext}} = X_{\text{ext}}^0$ and obviously

$$\delta\gamma_{ij} = \frac{d}{d\zeta}\gamma_{ij}\,|_{\zeta=0}.$$
(16.344)

The variation in the area of the extremal surface is given explicitly in terms of the variation of the induced metric by

$$\delta A_B(g, X_{\text{ext}}) = \int d^{d-1}\sigma\sqrt{\det\gamma_{ij}^0}\left(\frac{1}{2}\gamma_0^{kl}\delta\gamma_{kl}\right).$$
(16.345)

The embedding functions X_{ext}^a are mappings from the spatial surface $\tilde{B}$ to the bulk whose boundary at $z = 0$ is the spatial surface B, which is a ball-shaped region of radius R. Both B and $\tilde{B}$ are co-dimension two surfaces.

For pure AdS the bulk surface $\tilde{B}$, which turns out to be extremal, is given by the sphere in d dimensions of radius R, i.e.,

$$\vec{x}^2 + z^2 = R^2.$$
(16.346)

This has been shown explicitly for AdS_3 and the argument for higher dimensional AdS spacetimes is similar.

We can now parameterize the extremal surface $\tilde{B}$ for pure AdS by $\sigma^i = x^i$ and we choose the embedding functions X_{ext}^a: $\mathbf{R}^{d-1} \longrightarrow \mathbf{R}^{d+1}$ as follows

$$X_{\text{ext}}^0 = t_0, \quad X_{\text{ext}}^1 = x^1, \quad \dots, \quad X_{\text{ext}}^{d-1} = x^{d-1}, \quad X_{\text{ext}}^d = z = \sqrt{R^2 - \vec{x}^2}.$$
(16.347)

We compute in the radial gauge (defined by $h_{zz} = h_{z\mu} = 0$) the variation in the induced metric given by

$$\delta\gamma_{ij} = z^{d-2}h_{ij}.$$
(16.348)

By using the above embedding functions we can easily compute the unperturbed induced metric

$$\gamma_{ij}^0 = g_{ab}^0\frac{\partial X^a}{\partial x^i}\frac{\partial X^b}{\partial x^j}$$
$$= \frac{L^2}{z^2}(\delta_{ij} + \frac{x_i x_j}{z^2}).$$
(16.349)

By using these two last results the variation in the area of the extremal surface can now be computed explicitly to obtain

$$\delta A_B = \frac{L^{d-3}R}{2} \int_{\tilde{B}} d^{d-1}x \left(\delta^{ij} - \frac{1}{R^2}(x - x_0)^i(x - x_0)^j \right) h_{ij}. \tag{16.350}$$

The point $\vec{x}_0$ is the center of the ball-shaped region B. The variation in the entanglement entropy in the AdS is thus given by

$$\delta S_B = \frac{L^{d-3}R}{8G_N} \int_{\tilde{B}} d^{d-1}x \left(\delta^{ij} - \frac{1}{R^2}(x - x_0)^i(x - x_0)^j \right) h_{ij}. \tag{16.351}$$

16.9.4 Modular Hamiltonian and hyperbolic energy

In this section we will follow closely [43]. First we will define the so-called modular Hamiltonian in the CFT without holography, then we will construct the AdS expression.

16.9.4.1 Modular Hamiltonian

We start by defining the causal development (also called domain of dependence) $\mathcal{D}$ of the region B by the set of all points p for which every inextensible causal (timelike or lightlike) curve through p necessarily intersect B. Classically, the field values on $\mathcal{D}$ are completely determined in terms of the initial field values on B. Quantum mechanically, any operator in $\mathcal{D}$ is determined in terms of the fields in B alone and hence it can be calculated using the reduced density matrix ρ_B. The domain of dependence $\mathcal{D}$ is also the causal development of any other spacelike surface whose boundary is the same as the boundary of B.

The reduced density matrix ρ_B is Hermitian and positive semi-definite and thus it can be rewritten in terms of a Hermitian operator H_B as

$$\rho_B = \exp(-H_B). \tag{16.352}$$

This operator H_B is called the modular Hamiltonian or entanglement Hamiltonian. Generically, this Hamiltonian is not local since it cannot be represented by a local expression in terms of fields defined on B. This Hamiltonian generates the modular group given by the transformations

$$tr\rho\mathcal{O} = tr\rho U(s)\mathcal{O}U(-s), \quad U(s) = \rho^{is} = \exp(-isH). \tag{16.353}$$

The operator $U(s)$, although it looks like a time evolution transformation along s, does not generate a local flow on $\mathcal{D}$ since H_B is not local. Indeed, the operator defined by $\mathcal{O}(s) = U(s)\mathcal{O}U(-s)$ is not local, i.e., it is not defined at a point x, even if we start with a local operator $\mathcal{O}$ defined at a point x, e.g. $\mathcal{O} = \phi(x)$.

16.9.4.2 Rindler wedge

However, there are special circumstances where the modular Hamiltonian is local. The first example is Rindler space $\mathcal{R}$ [43]. The metric is given by ($T = \rho \sinh \omega$ and $Z = \rho \cosh \omega$)

$$ds^2 = \rho^2 d\omega^2 - d\rho^2 - dX^2 - dY^2$$
$$= dT^2 - dZ^2 - dX^2 - dY^2.$$

(16.354)

The Rindler wedge is region I, which is the part of Minkowski spacetime accessible to a uniformly accelerating observer. Quadrants II and III have no causal relations with quadrant I. Quadrant IV provides initial data for the Rindler wedge. This is the Rindler decomposition of Minkowski spacetime. See figure 16.9.

A translation in the Rindler time $\omega \longrightarrow \omega + c$ corresponds to a Lorentz boost along the Z direction in Minkowski spacetime. The corresponding generator is the Rindler Hamiltonian H_ω, which is precisely the Hamiltonian in quadrant I, given by

$$H_\omega = \int_{\rho=0}^{\rho=\infty} \rho \, d\rho \, dX \, dY \, T^{00}(\rho, X, Y).$$

(16.355)

T^{00} is the Hamiltonian density with respect to the Minkowski observer. Note also that in the $T - Z$ plane the lines of constant Rindler time ω are straight lines through the origin. The proper time separation between these lines is $\delta\tau = \rho\delta\omega$. This is the origin of the ρ factor multiplying T^{00}.

The surface $T = 0$ is divided into two halves. The first half in region I and the second half in region III. The fields in region I ($Z > 0$) act in the Hilbert space $\mathcal{H}_R$ and those in region III ($Z < 0$) act in the Hilbert space $\mathcal{H}_L$. We have then

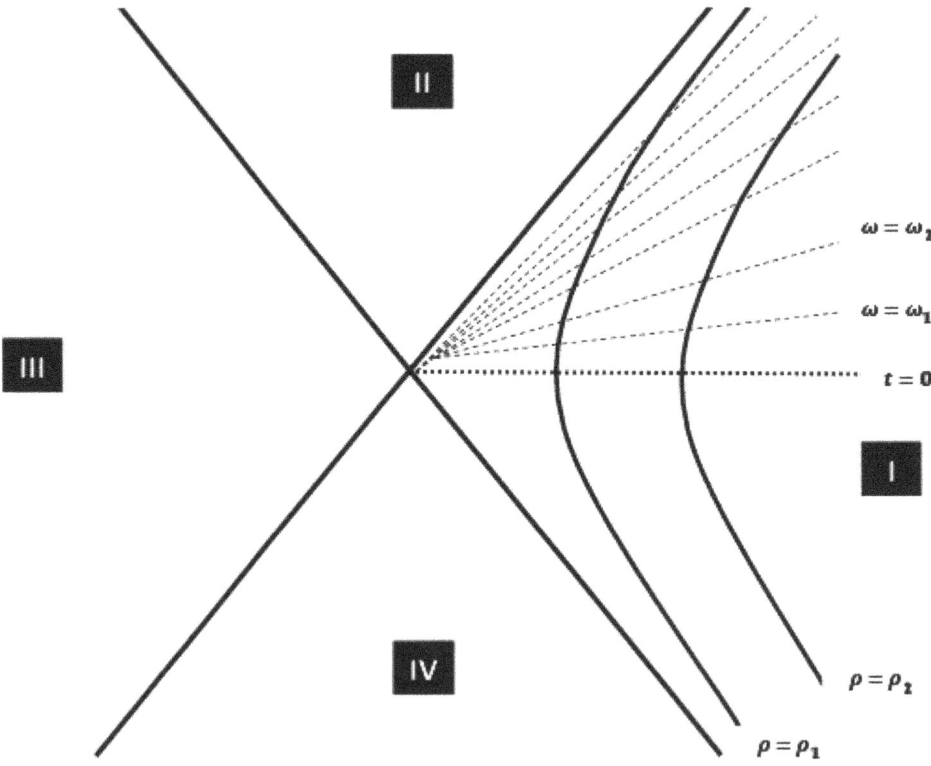

Figure 16.9. Rindler decomposition.

$$\phi(X, Y, Z) = \phi_R(X, Y, Z), \quad Z > 0. \tag{16.356}$$

$$\phi(X, Y, Z) = \phi_L(X, Y, Z), \quad Z < 0. \tag{16.357}$$

The general wave functional of interest is the pure state

$$\Psi = \Psi(\phi_L, \phi_R). \tag{16.358}$$

This is the ground state of the Minkowski Hamiltonian, which can be computed using Euclidean path integrals. We get after some calculation the transition matrix element [44]

$$\begin{aligned} \Psi(\phi_L, \phi_R) &= \langle \phi_R | \langle \phi_L | \Omega \rangle \\ &\propto \sum_i e^{-\pi E_i} \langle \phi_R | i_R \rangle \langle \phi_L | i_L^* \rangle. \end{aligned} \tag{16.359}$$

In other words, we get the ground state

$$|\Omega\rangle = \frac{1}{\sqrt{Z}} \sum_i e^{-\pi E_i} |i_R\rangle |i_L^*\rangle. \tag{16.360}$$

The entanglement between the left and right wedges is now fully manifest.

However, we want to compute the reduced density matrix ρ_A used by observers in the Rindler quadrant (quadrant I) to describe the Minkowski vacuum. This is the density matrix associated with the spatial half-space $Z > 0$, $X = Y = 0$ corresponding to the time slice $T = 0$, which is denoted by A. In other words, the Rindler wedge $\mathcal{R}$ is the causal development of the half-space A. We can define immediately the reduced matrix ρ_A by the relation

$$\begin{aligned} \rho_A(\phi_R, \phi_R') &= \int \Psi^*(\phi_L, \phi_R)\Psi(\phi_L, \phi_R')d\phi_L \\ &= \frac{1}{Z}\sum_i e^{-2\pi E_i} \langle i_R | \phi_R \rangle \langle \phi_R' | i_R \rangle \\ &= \frac{1}{Z} \langle \phi_R' | e^{-2\pi H_\omega} | \phi_R \rangle. \end{aligned} \tag{16.361}$$

We get then the reduced density matrix

$$\rho_A = \frac{1}{Z} e^{-2\pi H_\omega}. \tag{16.362}$$

Thus the fiducial observers in the Rindler wedge $\mathcal{R}$ see the vacuum as a thermal ensemble with a Maxwell–Boltzmann distribution at a temperature

$$T_A = \frac{1}{2\pi}. \tag{16.363}$$

This is the Unruh effect.

The modular flow on the Rindler wedge $\mathcal{R}$ corresponds therefore to the Rindler time translations $\omega \longrightarrow \omega + c$ and the modular Hamiltonian is given immediately by

$$H_A = 2\pi H_\omega + \log Z. \tag{16.364}$$

16.9.4.3 The ball-shaped region B

Another very important example in which the modular flow and the modular Hamiltonian can be shown to be local is a conformal field theory defined on the causal development $\mathcal{D}$ of a ball-shaped region B of radius R. It was shown in [43] that this causal development $\mathcal{D}$ is conformally related to the Rindler wedge by a special conformal transformation.

Let us denote now the coordinates on the Rindler wedge by X^μ with metric $ds^2 = \eta_{\mu\nu} dX^\mu dX^\mu$. The Rindler Wedge is the causal development or domain of dependence of the half-space A corresponding to $X^0 = 0$, $X^1 > 0$ and $X^2 = X^3 = 0$. This causal development can be given by $X^1 \geqslant |X^0|$ or equivalently in terms of the null coordinates $X^\pm = X^1 \pm X^0$ by $\{X^+ \geqslant 0\} \cap \{X^- \geqslant 0\}$.

The causal development $\mathcal{D}$ with coordinates x^μ and metric $ds^2 = \Omega^{-2}\eta_{\mu\nu} dx^\mu dx^\mu$ is obtained from the Rindler edge via the special conformal transformation [43]

$$x^\mu = \frac{X^\mu - (X.X)C^\mu}{\Omega} + 2R^2 C^\mu, \quad C^\mu = (0, \frac{1}{2R}, 0, \ldots, 0), \quad \Omega = 1 - 2(X.C) + (X.X)(C.C). \tag{16.365}$$

This maps the domain of dependence $\{X^+ \geqslant 0\} \cap \{X^- \geqslant 0\}$ to the domain of dependence $\{x^+ \leqslant R\} \cap \{x^- \leqslant R\}$ where the null coordinates $x^\pm$ are defined by $x^\pm = r \pm t$ where $r = \sqrt{(x^1)^2 + \cdots + (x^{d-1})^2}$.

Let us introduce the hyperbolic coordinates on the Rindler wedge by $X^1 = \rho \cosh \omega$ and $X^0 = \rho \sinh \omega$. We have then $X^\pm = \rho \exp(\pm\omega)$. The modular flow on the Rindler wedge is given by the time translations $\omega \longrightarrow \omega' = \omega + c$. We will write $c = 2\pi s$ since the period is given precisely by the inverse temperature $1/T = 2\pi$. Under this modular flow the coordinates transform as $X^\pm(0) = X^\pm \longrightarrow X^\pm(s) = X^\pm \exp(\pm 2\pi s)$. Under the special conformal transformation (16.365) this modular flow on the Rindler wedge transforms to the modular flow on $\mathcal{D}$ given explicitly by

$$x^\pm(s) = R\frac{R + x^\pm - e^{\mp 2\pi s}(R - x^\pm)}{R + x^\pm + e^{\mp 2\pi s}(R - x^\pm)}. \tag{16.366}$$

We compute for the Rindler wedge $\mathcal{R}$ the variation of the modular flow

$$\delta X^\pm(s) = \pm 2\pi \delta s X^\pm(s) \Leftrightarrow \delta X^1(s) = 2\pi \delta s X^0(s), \quad \delta X^0(s) = 2\pi \delta s X^1(s). \tag{16.367}$$

We focus on the time slice $s = \omega = 0$. We immediately get the equations

$$\delta X^1 = \delta\rho = 0, \quad \delta X^0 = \rho\delta\omega = 2\pi\rho\delta s. \tag{16.368}$$

The corresponding generator H_A is then given by

$$H_A = 2\pi \int d\rho dX dY \rho T^{00}(\rho, X, Y) + \text{constant}. \tag{16.369}$$

We go through the same steps for the causal development $\mathcal{D}$. We compute immediately for $\mathcal{D}$ the variation of the modular flow

$$\delta x^{\pm}(s) = \frac{\pm 4\pi R \delta s e^{\mp 2\pi s}(R^2 - (x^{\pm})^2)}{(R + x^{\pm} + e^{\mp 2\pi s}(R - x^{\pm}))^2}. \tag{16.370}$$

We focus on the ball B of radius R corresponding to the time slice $s = t = 0$ and hence $x^{\pm} = r$. The variation of the mdoular flow becomes

$$\delta x^{\pm} = \pm \frac{\pi \delta s(R^2 - r^2)}{R}. \tag{16.371}$$

This gives immediately the shift

$$\delta r = 0, \quad \delta t = \frac{2\pi \delta s(R^2 - r^2)}{2R}. \tag{16.372}$$

The corresponding generator H_B is then given by

$$H_B = 2\pi \int_B d^{d-1}x \frac{R^2 - r^2}{2R} T^{00}(x) + \text{constant}. \tag{16.373}$$

A careful derivation of H_B is given in [43]. The special conformal transformation (16.365), which is a symmetry of Minkowski spacetime, is associated with a unitary operator U_0 in the CFT, which leaves the vacuum invariant, viz $U_0|0\rangle = |0\rangle$. This unitary operator acts on primary operators as

$$\phi(x) = \Omega^{\Delta} U_0 \phi(X) U_0^{-1}. \tag{16.374}$$

This unitary operator must then act on the density matrices as follows:

$$\begin{aligned}
\rho_B &= U_0 \rho_A U_0^{-1} \\
&= \frac{1}{Z} e^{-2\pi U_0 H_\omega U_0^{-1}} \\
&= \frac{1}{Z} e^{-H_B}.
\end{aligned} \tag{16.375}$$

Next, if $U_{\mathcal{R}}$ is the quantum operator, which generates the modular flow on the Rindler wedge $\mathcal{R}$, then the quantum operator, which generates the modular flow on $\mathcal{D}$, must be given by

$$U_{\mathcal{D}}(s) = U_0 U_{\mathcal{R}}(s) U_0^{-1}. \tag{16.376}$$

We can then act on primary operators as follows:

$$\begin{aligned}
U_{\mathcal{D}}(s)\phi(x[s_0])U_{\mathcal{D}}(-s) &= U_0 U_{\mathcal{R}}(s) U_0^{-1}\phi(x[s_0])U_0 U_{\mathcal{R}}(-s)U_0^{-1} \\
&= \Omega^{\Delta}(x[s_0]) U_0 U_{\mathcal{R}}(s)\phi(X[s_0])U_{\mathcal{R}}(-s)U_0^{-1} \\
&= \Omega^{\Delta}(x[s_0])U_0\phi(X[s_0 + s])U_0^{-1} \\
&= \Omega^{\Delta}(x[s_0])\Omega^{-\Delta}(x[s_0 + s])\phi(x[s_0 + s]).
\end{aligned} \tag{16.377}$$

The pre-factor $\Omega^\Delta(x[s_0])\Omega^{-\Delta}(x[s_0 + s])$ evaluated on the surface $t = 0$ is equal to 1 at the order δs and hence it does not contribute to the infinitesimal shift (16.372). The modular Hamiltonian is therefore given indeed by (16.373).

16.9.4.4 Hyperbolic energy

The hyperbolic energy is the expectation value of the modular Hamiltonian. We are interested in the first order variation in the ζ of the hyperbolic energy.

If we assume that the ball-shaped region B of radius R is centered at $\vec{x}_0$ then the hyperbolic energy must be given by

$$E_B^{\text{Hyp}} = 2\pi \int_B d^{d-1}x \frac{R^2 - (\vec{x} - \vec{x}_0)^2}{2R} \langle T^{00}(x) \rangle. \tag{16.378}$$

For an infinitesimal ball B we can safely replace $\langle T^{00}(x) \rangle$ with $\langle T^{00}(x_0) \rangle$. We can then easily compute

$$\begin{aligned}
\delta E_B &= \frac{d}{d\zeta}\Big|_{\zeta=0} E_{B_{\text{inf}}}^{\text{Hyp}} \\
&= 2\pi \int_B d^{d-1}x \frac{R^2 - (\vec{x} - \vec{x}_0)^2}{2R} \frac{d}{d\zeta}\Big|_{\zeta=0} \langle T^{00}(x) \rangle \\
&= 2\pi \frac{d}{d\zeta}\Big|_{\zeta=0} \langle T^{00}(x_0) \rangle \int_B d^{d-1}x \frac{R^2 - (\vec{x} - \vec{x}_0)^2}{2R} \\
&= \frac{2\pi R^d S_{d-2}}{d^2 - 1} \frac{d}{d\zeta}\Big|_{\zeta=0} \langle T^{00}(x_0) \rangle,
\end{aligned} \tag{16.379}$$

where $S_{d-2} = \int d\Omega_{d-2}$ is the area of the unit sphere $\mathbf{S}^{d-2}$.

However, by using the first law of thermodynamics, the holographic interpretation of the entanglement entropy and the Ryu–Takayanagi formula we obtain

$$\begin{aligned}
\frac{d}{d\zeta}\Big|_{\zeta=0} E_{B_{\text{inf}}}^{\text{Hyp}} &= \frac{d}{d\zeta}\Big|_{\zeta=0} S_{B_{\text{inf}}} \\
&= \lim_{R \to 0} \delta S_B \\
&= \lim_{R \to 0} \frac{\delta A_B}{4G_N}.
\end{aligned} \tag{16.380}$$

We employ now the holographic result (16.350) to obtain

$$\frac{d}{d\zeta}\Big|_{\zeta=0} \langle T^{00}(x_0) \rangle = \frac{L^{d-3}d}{16\pi G_N} \delta^{ij} h_{ij}(x_0, z = 0). \tag{16.381}$$

We have thus another crucial holographic result given by

$$\frac{d}{d\zeta}\Big|_{\zeta=0} \langle T^{00}(x) \rangle = \frac{L^{d-3}d}{16\pi G_N} \delta^{ij} h_{ij}(x, z = 0). \tag{16.382}$$

By substituting back into the variation of the hyperbolic energy for a ball-shaped region B of radius R we obtain the holographic interpretation of the second side of the first law of thermodynamics, viz

$$\delta E_B = \frac{L^{d-3}d}{16G_N} \int_B d^{d-1}x \frac{R^2 - (\vec{x} - \vec{x}_0)^2}{R} \delta^{ij} h_{ij}(x, z = 0). \tag{16.383}$$

16.9.5 Einstein's equations

In the remainder of this chapter we will consider a ball-shaped region B of radius R centered around the origin, i.e., $\vec{x}_0 = 0$. This is a spatial surface at the time slice $t = 0$.

The first law of thermodynamics in the CFT is given by

$$\frac{d}{d\zeta} S_B = \frac{d}{d\zeta} E_B^{\text{Hyp}}. \tag{16.384}$$

The holographic interpretation of the two sides of this equation is given by the two equations (16.351) and (16.383), viz

$$\frac{d}{d\zeta} S_B = \delta S_B = \frac{L^{d-3}R}{8G_N} \int_{\tilde{B}} d^{d-1}x \left(\delta^{ij} - \frac{1}{R^2} x^i x^j \right) h_{ij}(x, z), \tag{16.385}$$

where $\tilde{B}$ is the extremal surface in the bulk such that $\partial\tilde{B} = \partial B$ and z should be understood as $z = \sqrt{R^2 - \vec{x}^2}$, and

$$\frac{d}{d\zeta} E_B^{\text{Hyp}} \equiv \delta E_B = \frac{L^{d-3}d}{16G_N} \int_B d^{d-1}x \frac{R^2 - r^2}{R} \delta^{ij} h_{ij}(x, z = 0). \tag{16.386}$$

The first law of thermodynamics in the AdS is then given by the non-local constraints

$$\int_{\tilde{B}} d^{d-1}x (R^2 \delta^{ij} - x^i x^j) h_{ij}(x, z) = \frac{d}{2} \int_B d^{d-1}x (R^2 - r^2) \delta^{ij} h_{ij}(x, z = 0). \tag{16.387}$$

Remark that this equation relates the metric perturbation on the boundary $z = 0$ to the metric perturbation in the bulk $z \neq 0$. The central claim is that these non-local equations are precisely the Einstein's equations linearized around AdS [39, 40].

Also we have seen that the modular Hamiltonian of a ball-shaped region B of radius R evaluated at time $t = 0$ is given by the following expression

$$H_B = 2\pi \int_B d^{d-1}x \frac{R^2 - r^2}{2R} T^{00}(x) + \text{constant}. \tag{16.388}$$

The constant is chosen such that the trace of the density matrix is equal to 1. The corresponding modular flow (16.366) is a symmetry that is generated by the conformal timelike Killing vector of Minkowski spacetime given by

$$\xi = \frac{\pi}{R}((R^2 - t^2 - r^2)\partial_t - 2tx^i\partial_i). \tag{16.389}$$

We associate to this a bulk Killing vector ξ_B, which obviously approaches ξ as we approach the AdS boundary $z = 0$. This is given explicitly by

$$\xi_B = \frac{\pi}{R}((R^2 - z^2 - t^2 - r^2)\partial_t - 2t(z\partial_z + x^i\partial_i)). \tag{16.390}$$

This is indeed a symmetry that preserves the metric, i.e., an isometry, since the Lie derivative of the metric with respect to ξ_B is zero, viz $\mathcal{L}_{\xi_B} g_{AdS} = 0$. The corresponding conserved quantity is the so-called canonical energy associated with the Killing vector ξ_B. This Killing vector ξ_B also vanishes on the Rindler horizon, i.e., on the extremal surface $\tilde{B}$ in the unperturbed AdS given by $\vec{x}^2 + z^2 = R^2$.

Clearly the function ξ_B is defined in the spatial region Σ between B and $\tilde{B}$, i.e., Σ is the volume bounded by $B \cup \tilde{B} = \partial\Sigma$. In other words, Σ is a co-dimension d-dimensional surface in the bulk. See figure 16.10. The volume form on this region Σ is given by

$$\epsilon_{a_1} = \frac{1}{d!}\epsilon_{a_1 a_2 \ldots a_{d+1}}dx^{a_2} \wedge \ldots \wedge dx^{a_{d+1}}, \tag{16.391}$$

where $\epsilon_{a_1 \ldots a_{d+1}}$ is the anti-symmetric tensor in $d + 1$ dimensions normalized such that $\epsilon_{ztx^1 \ldots x^{d-1}} = \sqrt{-g}$. Since the region Σ is a spatial surface all components but ϵ^t vanish and as a consequence the volume form is simply given by

$$\mathrm{vol}_\Sigma = \epsilon^t. \tag{16.392}$$

The approach to convert the above non-local constraints (16.387) to local equations is similar to the way we convert the integral form of Maxwell's equations to differential forms where we mainly use Stokes' theorem. Indeed, the non-local constraints (16.387) can be converted into the local Einstein's equations if for every choice of B there exists a differential form χ_B such that

$$\delta S_B = \int_{\tilde{B}} \chi_B. \tag{16.393}$$

$$\delta E_B = \int_{B} \chi_B. \tag{16.394}$$

The non-local constraints $\delta E_B = \delta S_B$ become therefore

$$\int_{B} \chi_B = \int_{\tilde{B}} \chi_B \Rightarrow \int_{B} \chi_B - \int_{\tilde{B}} \chi_B = 0$$

$$\Rightarrow \int_{\partial\Sigma} \chi_B = 0 \tag{16.395}$$

$$\Rightarrow \int_{\Sigma} d\chi_B = 0.$$

The form χ_B is obtained by means of the Iyer-Wald formalism [45]. Explicitly the form χ_B can be shown to satisfy

$$d\chi_B = 2\xi_B^0 \delta E_{00}\mathrm{vol}_\Sigma. \tag{16.396}$$

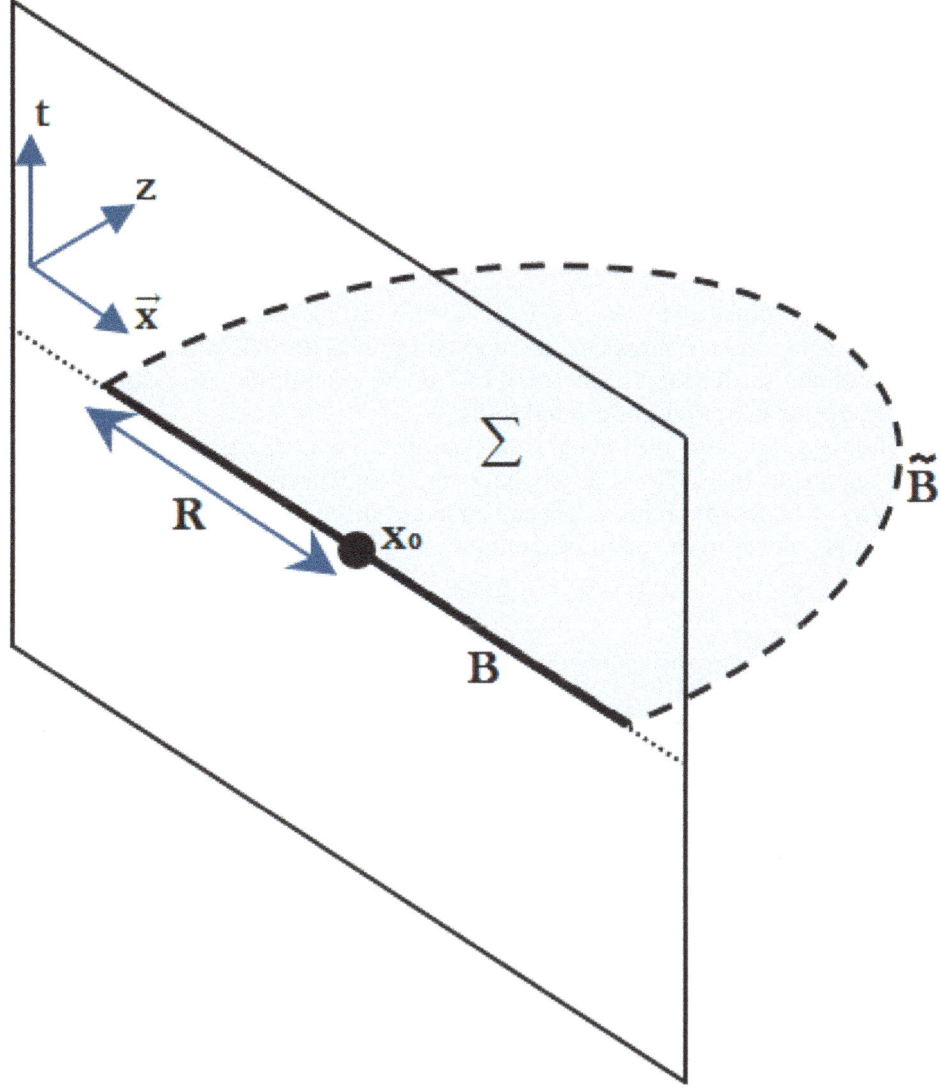

Figure 16.10. The volume Σ bounded by $B \cup \tilde{B} = \partial\Sigma$.

In this equation δE_{00} is the time-time component of the Einstein equations linearized about AdS spacetime given explicitly by

$$\delta E_{00} = \frac{z^d}{2L^2}\left(\partial_z^2 h_i^i + \frac{d+1}{z}\partial_z h_i^i + \partial_j \partial^j h_i^i - \partial^i \partial^j h_{ij}\right). \tag{16.397}$$

Hence, the first law of thermodynamics in the CFT, in which the energy is understood as hyperbolic energy and the entropy is understood as entanglement entropy, yields under holographic extension, for ball-shaped regions in slices of constant time, a set of non-local constraints in the bulk, which are precisely the time-time component of the Einstein equations linearized about AdS spacetime, i.e.

$$\delta E_{00} = 0. \tag{16.398}$$

To obtain the other components of the Einstein equations we repeat the same steps for ball-shaped regions in general frames of reference. We obtain then

$$\delta E_{\mu\nu} = 0. \tag{16.399}$$

The remaining equations $\delta E_{zz} = \delta E_{z\mu} = 0$ are in fact constraints in the dual geometry, which are equivalent to the conservation and tracelessness of the energy–momentum tensor $T_{\mu\nu}$ in the CFT at the boundary. This can be verified explicitly by considering infinitesimal balls.

In summary, we have thus shown that the first law of thermodynamics for ball-shaped regions in the CFT is exactly equivalent to linearized Einstein equations in the gravity dual theory. This is a very concrete mechanism for the emergence of spacetime geometry from quantum entanglement.

16.10 Exercises

Exercise 1: Show that in anti-de Sitter the proper time to go from the center $r = 0$ to the boundary $r = 1$ and back is finite equal to $\tau = \pi$. What do you conclude?

Exercise 2: Show that the boundary $z = 0$ of AdS_{d+1} in the Poincaré patch becomes $\mathbf{R}^d$ in the Euclidean rotation whereas the horizon $z = \infty$ shrinks to a point. What do you get by adding the point $z = \infty$ to the boundary $\mathbf{R}^d$ and what the resulting compactified Euclidean AdS_{d+1}?.

Exercise 3:
1. Show that the generators of $SO(2, 1)$ in the global coordinates τ, r are given by the differential operators

$$L_2^0 = \partial_\tau. \tag{16.400}$$

$$L_1^0 = \sin \tau \sin r \partial_\tau - \cos \tau \cos r \partial_r. \tag{16.401}$$

$$L_2^1 = -\cos \tau \sin r \partial_\tau - \sin \tau \cos r \partial_r. \tag{16.402}$$

2. Verify that they satisfy the algebra

$$[L_2^0, L_1^0] = -L_2^1, \quad [L_2^0, L_2^1] = L_1^0, \quad [L_1^0, L_2^1] = L_2^0. \tag{16.403}$$

3. Show that the lowering and raising operators (special conformal and momentum generators) K and P take in the global coordinates τ, r the form

$$K = \exp(-i\tau)(i \sin r \partial_\tau - \cos r \partial_r). \tag{16.404}$$

$$P = -\exp(i\tau)(i \sin r \partial_\tau + \cos r \partial_r). \tag{16.405}$$

4. Show that the Casimir operator for the conformal group $SO(2, 1)$ which is precisely the Klein–Gordon Laplacian is given by

$$\nabla = D^2 + \frac{1}{2}(PK + KP)$$
$$= \cos^2 r(-\partial_\tau^2 + \partial_r^2). \tag{16.406}$$

Exercise 4: Show that the diffeomorphisms

$$(\tilde{\sigma}^1, \tilde{\sigma}^2) \longrightarrow (a\tilde{\sigma}^1 + b\tilde{\sigma}^2, c\tilde{\sigma}^1 + d\tilde{\sigma}^2). \tag{16.407}$$

are compatible with the equivalence relation

$$(\tilde{\sigma}^1, \tilde{\sigma}^2) \sim (\tilde{\sigma}^1 + n, \tilde{\sigma}^2 + m), \quad n, m \in \mathbb{Z}. \tag{16.408}$$

Determine the domain of the coefficients a, b, c and d. By requiring invertibility and one-to-one determine the group structure corresponding to the above transformations.

Exercise 5:
- Show that

$$\int 1 = \int 2idz \wedge d\bar{z} = \int 4\tau_2 d\sigma^1 \wedge d\sigma^2 = 4\tau_2. \tag{16.409}$$

- Show that the Fourier transform of the function

$$f(n'r) = \exp\left(-2\pi\frac{1}{\tau_2}(n'r - \tau_1 nr)^2\right) \tag{16.410}$$

fis

$$\tilde{f}(p) = \sqrt{\frac{\tau_2}{2}} \exp\left(2\pi i \tau_1 nrp - \frac{1}{2}\pi\tau_2 p^2\right). \tag{16.411}$$

- Show that the partition function of the conformal field theory of free bosons on the torus is given by

$$\int \mathcal{D}\Phi \exp(-S[\Phi]) = (q\bar{q})^{-c/24} tr q^{L_0} \bar{q}^{\bar{L}_0}. \tag{16.412}$$

Verify its modular invariance under the modular transformation $\tau \longrightarrow -1/\tau$.

Exercise 6: Show that the dilatation operator, the translation and Lorentz generators, and the special conformal generators are represented on scalar fields with scaling dimension Δ by the differential operators

$$D\Phi = -i(x^\mu \partial_\mu + \Delta)\Phi. \tag{16.413}$$

$$P_\mu \Phi = -i\partial_\mu \Phi. \tag{16.414}$$

$$M_{\mu\nu}\Phi = i(x_\mu \partial_\nu - x_\nu \partial_\mu)\Phi + S_{\mu\nu}\Phi. \tag{16.415}$$

$$K_\mu \Phi = (-2i\Delta x_\mu - x^\nu S_{\mu\nu} - 2ix_\mu x^\nu \partial_\nu + ix^2\partial_\mu)\Phi. \tag{16.416}$$

Work for simplicity in three dimensions. Show that the transformation law of the scalar field Φ under conformal transformations is given by

$$\Phi(x) \longrightarrow \Phi'(x') = |\frac{\partial x'}{\partial x}|^{-\Delta/d}\Phi(x). \tag{16.417}$$

Exercise 7: Show that under special conformal transformation we have the behavior

$$x^\mu \longrightarrow x'^\mu = \frac{x^\mu + b^\mu x^2}{1 + 2bx + b^2 x^2} \Rightarrow r_{12}'^2 = \frac{r_{12}^2}{(1 + 2bx_1 + b^2 x_1^2)(1 + 2bx_2 + b^2 x_2^2)}. \tag{16.418}$$

$$|\frac{\partial x'}{\partial x}| = \frac{1}{(1 + 2bx + b^2 x^2)^d}. \tag{16.419}$$

Exercise 8: Show that the requirement of invariance under translations, rotations, scalings and special conformal transformations constrains the three-point function of a conformal field theory such that

$$\langle \Phi_1(x_1)\Phi_2(x_2)\Phi_3(x_3)\rangle = \frac{C_{123}}{r_{12}^{\Delta_1 + \Delta_2 - \Delta_3} r_{13}^{\Delta_1 + \Delta_3 - \Delta_2} r_{23}^{\Delta_2 + \Delta_3 - \Delta_1}}. \tag{16.420}$$

Exercise 9: Show that for a scalar field in Euclidean AdS_{d+1} the canonical momentum with respect to z is given by

$$\Pi = \eta\sqrt{g}\,\phi g^{zz}\partial_z\phi. \tag{16.421}$$

Exercise 10: Show that the required counter term to renormalize the action (16.218) is given by

$$S_{\text{ct}} = \frac{\eta}{2}\eta_1 \int \sqrt{\gamma}\,d^d x \phi^2$$

$$= \frac{\eta}{2}\eta_1 L^d \int \frac{d^d k}{(2\pi)^d}(A(k)A(-k)\epsilon^{-2\nu} + 2A(k)B(-k)). \tag{16.422}$$

Determine the value of η_1.

Exercise 11: Show that the two-point function (16.225) takes in position space the form

$$\langle \mathcal{O}(x)\mathcal{O}(0)\rangle = \frac{2\nu L^{d-1}\eta}{\pi^{d/2}}\frac{\Gamma\left(\dfrac{d}{2}+\nu\right)}{\Gamma(-\nu)}\frac{1}{|x|^{2\Delta}}. \tag{16.423}$$

Use the identity

$$\int \frac{d^d k}{(2\pi)^d}\exp(ikx)k^n = \frac{2^n}{\pi^{d/2}}\frac{\Gamma\left(\dfrac{d+n}{2}\right)}{\Gamma\left(-\dfrac{n}{2}\right)}\frac{1}{|x|^{d+n}}. \tag{16.424}$$

Exercise 12: Show that $\ln tr q^{L_0+\bar{L}_0}$ admits an expansion in positive powers of q due to the requirement of positive dimensions of all fields (locality). Show then that the contributions from $\ln tr q^{L_0+\bar{L}_0}$ to the entanglement entropy are exponentially suppressed. Recall that $q = \exp(-\kappa L)$.

Exercise 13: The thermofield double state $|\psi\rangle$ dual to the Schwarzschild-AdS black hole is given by

$$|\Psi\rangle = \frac{1}{\sqrt{Z(\beta)}}\sum_i \exp(-\beta E_i/2)|E_i^A\rangle \otimes |E_i^B\rangle. \tag{16.425}$$

Compute the reduced density matrix for the subsystem Q_A and its entanglement entropy. Show that in the limit $\beta \longrightarrow \infty$ we have the behavior

$$|\Psi\rangle = \rangle \longrightarrow |\Phi\rangle = |E_0^A\rangle \otimes |E_0^B\rangle. \tag{16.426}$$

Exercise 14: The entanglement entropy of a spatial region B in the CFT is equal to the von Neumann entropy of the reduced density matrix

$$\rho_B = tr_{\bar{B}}|\psi(\zeta)\rangle\langle\psi(\zeta)|. \tag{16.427}$$

Show that

$$\frac{d}{d\zeta}S_B = -tr_B\frac{d}{d\zeta}\rho_B \cdot \log \rho_B. \tag{16.428}$$

Exercise 15:
  1. Show that

$$\frac{\delta A_B(g^0, X_{\text{ext}})}{\delta X^a}\delta X^a \tag{16.429}$$

  is of order δg^2.

2. Show that the variation in the area of the extremal surface is given in terms of the variation of the induced metric by

$$\delta A_B(g, X_{\text{ext}}) = \int d^{d-1}\sigma \sqrt{\det \gamma^0_{\mu\nu}} \left(\frac{1}{2}\gamma_0^{\rho\lambda}\delta\gamma_{\rho\lambda}\right). \tag{16.430}$$

3. Show that in the radial gauge the variation in the induced metric given by

$$\delta\gamma_{ij} = z^{d-2}h_{ij}. \tag{16.431}$$

4. Verify that the unperturbed induced metric is given by

$$\gamma^0_{ij} = \frac{L^2}{z^2}(\delta_{ij} + \frac{x_i x_j}{z^2}). \tag{16.432}$$

5. Show that the variation in the area of the extremal surface is given by

$$\delta A_B = \frac{L^{d-3}R}{2}\int_{\tilde B} d^{d-1}x\left(\delta^{ij} - \frac{1}{R^2}(x - x_0)^i(x - x_0)^j\right)h_{ij}. \tag{16.433}$$

Exercise 16:

1. Verify that under the special conformal transformation (16.365) the metric transforms as

$$ds^2 = \eta_{\mu\nu}dX^\mu dX^\mu = \Omega^{-2}\eta_{\mu\nu}dx^\mu dx^\mu. \tag{16.434}$$

2. Show that the domain of dependence $\{X^+ \geqslant 0\} \cap \{X^- \geqslant 0\}$ of the Rindler wedge maps under the special conformal transformation (16.365) to the domain of dependence $\{x^+ \leqslant R\} \cap \{x^- \leqslant R\}$ of $\mathcal{D}$ where the null coordinates $x^\pm$ are defined by $x^\pm = r \pm t$ where $r = \sqrt{(x^1)^2 + \cdots + (x^{d-1})^2}$.

3. Show that the modular flow on the Rindler wedge given by $X^\pm(s) = X^\pm \exp(\pm 2\pi s)$ transforms under the special conformal transformation (16.365) to the modular flow on $\mathcal{D}$ given explicitly by

$$x^\pm(s) = R\frac{R + x^\pm - e^{\mp 2\pi s}(R - x^\pm)}{R + x^\pm + e^{\mp 2\pi s}(R - x^\pm)}. \tag{16.435}$$

Exercise 17: By employ the holographic result (16.350) verify that

$$\frac{d}{d\zeta}\big|_{\zeta=0}\langle T^{00}(x)\rangle = \frac{L^{d-3}d}{16\pi G_N}\delta^{ij}h_{ij}(x, z = 0). \tag{16.436}$$

Exercise 18:

1. Show that the time-time component of the Einstein equations linearized about AdS spacetime is given by

$$\delta E_{00} = \frac{z^d}{2L^2}\left(\partial_z^2 h_i^i + \frac{d+1}{z}\partial_z h_i^i + \partial_j \partial^j h_i^i - \partial^i \partial^j h_{ij}\right). \qquad (16.437)$$

Use the Fefferman–Graham metric.

2. By employing the Iyer-Wald formalism show that the form χ_B satisfies

$$d\chi_B = 2\xi_B^0 \delta E_{00} \mathrm{vol}_\Sigma. \qquad (16.438)$$

3. Show that the form χ_B is given explicitly by

$$\chi_B \mid_\Sigma = \frac{z^d}{16\pi G_N}\left\{\epsilon_z^t\left(\frac{2\pi z}{R} + \frac{d}{z}\xi^t + \xi^t\partial_z\right)h_i^i + \epsilon_i^t\left[\left(\frac{2\pi x^i}{R} + \xi^t\partial^i\right)h_j^j - \left(\frac{2\pi x^j}{R} + \xi^t\partial^j\right)h_j^i\right]\right\}. \qquad (16.439)$$

4. Verify explicitly that

$$\delta S_B = \int_{\tilde{B}} \chi_B. \qquad (16.440)$$

$$\delta E_B = \int_B \chi_B. \qquad (16.441)$$

References

[1] Ishibashi A and Wald R M 2004 Dynamics in nonglobally hyperbolic static space-times. 3. Anti-de Sitter space-time *Class. Quant. Grav.* **21** 2981

[2] Carroll S 2004 *Spacetime and Geometry: An Introduction to General Relativity* (Reading, MA: Addison Wesley)

[3] Gibbons G W 2011 Anti-de-Sitter spacetime and its uses arXiv:1110.1206 [hep-th]

[4] Bengtsson I Anti-de Sitter space https://3dhouse.se/ingemar/relteori/Kurs.pdf

[5] Susskind L and Lindesay J 2005 *An Introduction to Black Holes, Information and the String Theory Revolution: The Holographic Universe Hackensack, USA* (Singapore: World Scientific) p 183

[6] Ramallo A V 2015 Introduction to the AdS/CFT correspondence Springer *Proc. Phys.* **161** 411 [arXiv:1310.4319 [hep-th]].

[7] Kadanoff L P 1966 Scaling laws for Ising models near T(c) *Physics* **2** 263

[8] Wilson K G and Kogut J B 1974 The renormalization group and the epsilon expansion *Phys. Rep.* **12** 75

[9] Schellekens A N 1996 *Fortsch. Phys.* **44** 605

[10] Zaffaroni A 2000 Introduction to the AdS-CFT correspondence *Class. Quant. Grav.* **17** 3571

[11] Kaplan J *Lectures on AdS/CFT from the Bottom Up* https://sites.krieger.jhu.edu/jared-kaplan/files/2016/05/AdSCFTCourseNotesCurrentPublic.pdf

[12] Ginsparg P H 1988 Applied conformal field theory arXiv:hep-th/9108028

[13] Green M B, Schwarz J H and Witten E 1987 *Superstring Theory. Vol. 2: Loop Amplitudes, Anomalies And Phenomenology* (Cambridge: Cambridge Univ. Press)

[14] Qualls J D 2015 Lectures on conformal field theory arXiv:1511.04074 [hep-th]

[15] Ferrara S and Zaffaroni A 1998 Bulk gauge fields in AdS supergravity and supersingletons arXiv:hep-th/9807090

[16] Mack G and Salam A 1969 Finite component field representations of the conformal group *Ann. Phys.* **53** 174

[17] Ferrara S and Fronsdal C 1998 Gauge fields as composite boundary excitations *Phys. Lett.* B **433** 19

[18] Kallen G 1952 On the definition of the renormalization constants in quantum electrodynamics *Helv. Phys. Acta* **25** 417

[19] Lehmann H 1954 Über Eigenschaften von Ausbreitungsfunktionen und Renormierungskonstanten quantisierter Felder Nuovo Cimento (in German) *Soc. Ital. Fis.* **11** 342–57

[20] 't Hooft G 1993 Dimensional reduction in quantum gravity *Conf. Proc.* C **930308** 284 arXiv: gr-qc/9310026

[21] Susskind L 1995 The World as a hologram *J. Math. Phys.* **36** 6377

[22] Bousso R 2002 The Holographic principle *Rev. Mod. Phys.* **74** 825

[23] Maldacena J M 1999 The large N limit of superconformal field theories and supergravity *Int. J. Theor. Phys.* **38** 1113–33

[24] Gubser S S, Klebanov I R and Polyakov A M 1998 Gauge theory correlators from noncritical string theory *Phys. Lett.* B **428** 105

[25] Witten E 1998 Anti-de Sitter space and holography *Adv. Theor. Math. Phys.* **2** 253

[26] Henningson M and Skenderis K 1998 The Holographic Weyl anomaly *JHEP* **9807** 023

[27] Skenderis K 2002 Lecture notes on holographic renormalization *Class. Quant. Grav.* **19** 5849

[28] Bombelli L, Koul R K, Lee J and Sorkin R D 1986 A quantum source of entropy for black holes *Phys. Rev.* D **34** 373

[29] Holzhey C, Larsen F and Wilczek F 1994 Geometric and renormalized entropy in conformal field theory *Nucl. Phys.* B **424** 443

[30] Calabrese P and Cardy J L 2004 Entanglement entropy and quantum field theory *J. Stat. Mech.* **0406** P06002

[31] Nishioka T, Ryu S and Takayanagi T 2009 Holographic entanglement entropy: an overview *J. Phys.* A **0905** 0932

[32] Srednicki M 1993 Entropy and area *Phys. Rev. Lett.* **71** 666

[33] Susskind L and Uglum J 1994 Black hole entropy in canonical quantum gravity and superstring theory *Phys. Rev.* D **50** 2700

[34] Solodukhin S N 1995 On 'nongeometric' contribution to the entropy of black hole due to quantum corrections *Phys. Rev.* D **51** 618

[35] Ryu S and Takayanagi T 2006 Holographic derivation of entanglement entropy from AdS/CFT *Phys. Rev. Lett.* **96** 181602

[36] Ryu S and Takayanagi T 2006 Aspects of holographic entanglement entropy *JHEP* **0608** 045

[37] Brown J D and Henneaux M 1986 Central charges in the canonical realization of asymptotic symmetries: an example from three-dimensional gravity *Commun. Math. Phys.* **104** 207

[38] Maldacena J M 2003 Eternal black holes in anti-de Sitter *JHEP* **0304** 021

[39] Lashkari N, McDermott M B and Van Raamsdonk M 2014 Gravitational dynamics from entanglement 'thermodynamics' *JHEP* **2014** 195

[40] Faulkner T, Guica M, Hartman T, Myers R C and Van Raamsdonk M 2014 Gravitation from entanglement in holographic CFTs *JHEP* **2014** 051

[41] Van Raamsdonk M 2016 Lectures on gravity and entanglement *New Frontiers in Fields and Strings* (Singapore: World Scientific)

[42] Jaksland R 2017 A review of the holographic relation between linearized gravity and the first law of entanglement entropy arXiv:1711.10854 [hep-th]

[43] Casini H, Huerta M and Myers R C 2011 Towards a derivation of holographic entanglement entropy *JHEP* **2011** 036

[44] Ydri B 2017 Quantum black holes arXiv:1708.00748 [hep-th]

[45] Iyer V and Wald R M 1994 *Phys. Rev.* D **50** 846

IOP Publishing

A Modern Course in Quantum Field Theory, Volume 2 (Second Edition)
Advanced topics

Badis Ydri

Chapter 17

Conformal field theory and emergence of conformal symmetry

In the first section of this chapter we give a brief overview of conformal field theory following [1] then we discuss, in the remainder of the chapter, the celebrated SYK model and the emergence of conformal invariance in its large N IR limit [2, 3].

17.1 An overview of conformal field theory

In this section we give a brief overview of conformal field theory following the lectures by David Tong [1].

17.1.1 Conformal invariance

Conformal symmetry is a crucial ingredient in: (i) statistical mechanics (second order phase transitions, critical exponents), (ii) quantum field theory (fixed points of the renormalization group equation) and (iii) string theory (Polyakov path integral, AdS/CFT correspondence). A conformal field theory is a quantum field theory, which enjoys conformal symmetry, which means essentially that it enjoys scale invariance, i.e., there is no preferred length scale.

Conformal transformations are given explicitly by

$$x^\alpha \longrightarrow \tilde{x}^\alpha(x) \;\; \Rightarrow \;\; g_{\alpha\beta}(x) \longrightarrow \tilde{g}_{\alpha\beta}(\tilde{x}) = \Omega^2(x)g_{\alpha\beta}(x). \tag{17.1}$$

Here, there are two possible interpretations:
1. The metric is dynamical. In this case conformal transformations are diffeomorphisms, i.e., local gauge transformations, which can be undone by Weyl transformations.
2. The metric is fixed. In this case conformal transformations are global transformations, which physically change the point x to the point $\tilde{x}$.

doi:10.1088/978-0-7503-5834-7ch17   17-1

In here we are interested in the second interpretation.

A typical example is two-dimensional massless scalar field ϕ. The Hamilton action in this case is given by

$$S = \frac{1}{4\pi\alpha'} \int d^2x \sqrt{g}\, \partial_\mu \phi \partial^\mu \phi. \tag{17.2}$$

The coordinates of the flat and Euclidean spacetime are alternatively given by the complex coordinates

$$z = x^1 + ix^2, \quad \bar{z} = x^1 - ix^2. \tag{17.3}$$

We have effectively a complex plane. The holomorphic derivatives are introduced by

$$\partial_z = \partial = \frac{1}{2}(\partial_1 - i\partial_2), \quad \partial_{\bar{z}} = \bar{\partial} = \frac{1}{2}(\partial_1 + i\partial_2). \tag{17.4}$$

Infinitesimal conformal transformations in a flat background (the metric is fixed) are given explicitly by

$$x^\mu \longrightarrow \tilde{x}^\mu(x) = x^\mu + \epsilon^\mu(x). \tag{17.5}$$

These transformations become in the complex plane conformal mappings given by

$$z \longrightarrow z' = f(z), \quad \bar{z} \longrightarrow \bar{z}' = \bar{f}(\bar{z}). \tag{17.6}$$

The most important conformal transformations are translations, rotations and dilatations given explicitly by

$$
\begin{aligned}
z &\longrightarrow z' = z + a, \quad \text{Translations} \\
z &\longrightarrow z' = \zeta z, \quad |\zeta| = 1, \quad \text{Rotations} \\
z &\longrightarrow z' = \zeta z, \quad |\zeta| \neq 1, \quad \text{Dilatations.}
\end{aligned}
\tag{17.7}
$$

17.1.2 The stress–energy–momentum tensor

Infinitesimal diffeomorphisms (used in the computation of the stress–energy–momentum tensor) are given by the transformations

$$x^\mu \longrightarrow x^\mu + \epsilon^\mu, \quad g_{\mu\nu} \longrightarrow g_{\mu\nu} + \partial_\mu \epsilon_\nu + \partial_\nu \epsilon_\mu. \tag{17.8}$$

The variation of the action (17.2) under the conformal transformations (17.5) is given, in terms of the stress–energy–momentum tensor $T^{\alpha\beta}$, by

$$\delta S = \frac{1}{2\pi} \int d^2x\, T^{\alpha\beta} \partial_\alpha \epsilon_\beta. \tag{17.9}$$

The stress–energy–momentum tensor itself $T^{\alpha\beta}$ is explicitly defined by

$$T_{\alpha\beta} = -\frac{4\pi}{\sqrt{g}}\left(\frac{\delta S}{\delta g^{\alpha\beta}}\right)_{g=\delta}$$

$$= -\frac{1}{\alpha'}(\partial_\alpha\phi\partial_\beta\phi - \frac{1}{2}\delta_{\alpha\beta}(\partial\phi)^2).$$

(17.10)

The stress–energy–momentum tensor is conserved (Noether's theorem). In other words, we have the identity

$$\partial^\alpha T_{\alpha\beta} = 0.$$

(17.11)

The stress–energy–momentum tensor is also traceless (which is due to the invariance under scale transformations), viz

$$g_{\mu\nu} \longrightarrow \epsilon g_{\mu\nu}, \quad S \longrightarrow S + \delta S: \quad \delta S = 0 \Rightarrow T_\alpha^\alpha = 0.$$

(17.12)

The components of the stress–energy–momentum tensor itself $T^{\alpha\beta}$ in the complex plane are given by

$$T(z) = T_{zz}(z) = -\frac{1}{\alpha'}\partial\phi\partial\phi, \quad \bar\partial T = 0$$

$$\bar T(\bar z) = T_{\bar z\bar z}(\bar z) = -\frac{1}{\alpha'}\bar\partial\phi\bar\partial\phi, \quad \partial\bar T = 0$$

$$T_{z\bar z} = 0.$$

(17.13)

Noether's currents (for generic conformal transformations $z \longrightarrow z + \epsilon(z)$, $\bar z \longrightarrow \bar z + \bar\epsilon(\bar z)$) are given by

$$2J_z = J^z = T(z)\epsilon(z), \quad \bar\partial J^z = 0$$

$$2\bar J_{\bar z} = \bar J^{\bar z} = \bar T(\bar z)\bar\epsilon(\bar z), \quad \partial\bar J^{\bar z} = 0.$$

(17.14)

17.1.3 Ward identities: the operator product expansion (OPE)

After quantization, the conservation of Noether's currents J^μ is replaced by Ward identities. Explicitly, we have

$$\partial_\mu J^\mu = 0, \quad \text{Classical}$$

$$\Rightarrow -\frac{1}{2\pi}\int_{\epsilon\neq0} \sqrt{g}\,d^2x\partial_\mu\langle 0|T(J^\mu(x)O_1(x_1) \ldots O_N(x_N))|0\rangle$$

$$= \langle 0|T(\delta O_1(x_1) \ldots O_N(x_N))|0\rangle, \quad x \longrightarrow x_1, \quad \text{Quantum.}$$

(17.15)

These so-called Ward identities are given explicitly by

$$-\frac{1}{2\pi}\int_{\epsilon\neq0} \sqrt{g}\,d^2x\partial_\mu\langle 0|T(J^\mu(x)O_1(x_1)\ldots)|0\rangle = \langle 0|T(\delta O_1(x_1)\ldots)|0\rangle, \quad x \longrightarrow x_1. \quad (17.16)$$

In deriving this equation we have assumed that the support of the conformal transformation $\epsilon(x)$ includes only the operator insertion $O_1(x_1)$.

Recall that the variation of the field ϕ and the operator insertions $O_i(x)$ under conformal transformations are defined by $\phi \longrightarrow \phi + \epsilon\delta\phi$ and $O_i(x) \longrightarrow O_i(x) + \epsilon\delta O_i(x)$, respectively.

The above result is true for any $\epsilon(x)$. Thus, we have actually the operator identity

$$-\frac{1}{2\pi}\int_{\epsilon \neq 0}\sqrt{g}\,d^2x\,\partial_\mu(J^\mu(x)O_1(x_1)...) = \delta O_1(x_1) \;...\;, \quad x \longrightarrow x_1. \tag{17.17}$$

However, the expectation values of time-ordered products of operators computed using the Feynman path integral are implicitly understood if and when needed.

In two dimensions, we can use Stokes' theorem, which allows us to simplify the calculation to a contour integral. Furthermore, for conformal transformations, the Noether's currents are either holomorphic or anti-holomorphic functions, i.e., we can use the residue theorem. We get then the result (for $z \longrightarrow w$)

$$\frac{i}{2\pi}\oint_{\partial\epsilon}dz J_z(z)O_1(w,\,\bar{w}) = -\mathrm{Res}(J_z O_1) \Rightarrow \delta O_1(w,\,\bar{w}) = -\frac{1}{2}\mathrm{Res}(\epsilon(z)T(z)O_1(w,\,\bar{w})). \tag{17.18}$$

We have then the behavior

$$J_z(z)O_1(w,\,\bar{w}) = \cdots + \frac{\mathrm{Res}(J_z(z)O_1(w,\,\bar{w}))}{z - w} + \cdots \tag{17.19}$$

This is an example of an operator product expansion (OPE).

We conclude that if we know the OPE of an operator with the stress–energy–momentum tensor we can determine the transformation law of this operator under conformal transformations.

17.1.4 Primary operators: the spin s and scaling dimension Δ

As we have just explained, Noether's theorem is replaced in quantum mechanics by Ward identities given by

$$\frac{i}{2\pi}\oint_{\partial\epsilon}dz J_z(z)O(w,\,\bar{w}) = -\mathrm{Res}(J_z O) = \delta O(w,\,\bar{w}). \tag{17.20}$$

These identities are equivalent to the operator product expansions:

$$\begin{aligned}
J_z(z)O(w,\,\bar{w}) &= \cdots + \frac{\mathrm{Res}(J_z(z)O(w,\,\bar{w}))}{z - w} + \cdots \\
&= \cdots - \frac{\delta O(w,\,\bar{w})}{z - w} + \cdots, \quad z \longrightarrow w.
\end{aligned} \tag{17.21}$$

In other words, the OPE between a given operator and Noether's currents (which represent the stress–energy–momentum tensor) determine the transformation law of that operator under conformal transformations.

The main examples are translations, rotations and scaling.

1. **Translations:** In this case

$$\delta z = \epsilon = \text{constant}$$
$$O(z - \epsilon) = O(z) - \epsilon \partial O(z). \tag{17.22}$$

In other words, we have $\delta O = -\partial O$ and $J_z(z) = T(z)$. Hence, we have the OPE give by

$$T(z)O(w, \bar{w}) = \cdots + \frac{\partial O(w, \bar{w})}{z - w} + \cdots, \quad z \longrightarrow w. \tag{17.23}$$

2. **Rotations and Scaling:** In this case we have

$$\delta z = \epsilon z, \quad \delta \bar{z} = \bar{\epsilon} \bar{z} \tag{17.24}$$

and

$$O(z - \epsilon z, \bar{z} - \bar{\epsilon} \bar{z}) = O(z, \bar{z}) - \epsilon(hO + z\partial O) - \bar{\epsilon}(\tilde{h}O + \bar{z}\bar{\partial}O). \tag{17.25}$$

The pair $(h, \tilde{h})$ is called the weight of the operator O. The spin s (eigenvalue under rotations) and the scaling dimension Δ (eigenvalue under scaling transformations, which plays the role of energy in conformal field theory) are defined by

$$s = h - \tilde{h}, \quad \Delta = h + \tilde{h}. \tag{17.26}$$

In the case of rotations and scaling transformations, we have the holomorphic transformations $\delta z = \epsilon z$, $\delta O = -hO - z\partial O$ while Noether's current is $J_z(z) = zT(z)$. The OPE between J and O will now determine the $1/z^2$ term in the OPE between T and O, viz

$$T(z)O(w, \bar{w}) = \cdots + h\frac{O(w, \bar{w})}{(z - w)^2} + \frac{\partial O(w, \bar{w})}{z - w} + \cdots, \quad z \longrightarrow w. \tag{17.27}$$

Similarly, we obtain for the anti-holomorphic transformations $\delta \bar{z} = \bar{\epsilon} \bar{z}$ the OPE given by

$$\bar{T}(\bar{z})O(w, \bar{w}) = \cdots + \tilde{h}\frac{O(w, \bar{w})}{(\bar{z} - \bar{w})^2} + \frac{\bar{\partial} O(w, \bar{w})}{\bar{z} - \bar{w}} + \cdots, \quad z \longrightarrow w. \tag{17.28}$$

Thus, translations determine the $1/z$ term in the OPE while rotations and scaling transformations determine the $1/z^2$ term in the OPE.

The operators whose OPEs with the stress–energy–momentum tensor truncates at the $1/z^2$ order are called primary operators. These operators transform covariantly under conformal transformations. Indeed, general holomorphic conformal transformations are given by

$$\delta z = \epsilon(z) = \epsilon(w) + \epsilon'(w)(z - w) + \cdots. \tag{17.29}$$

The corresponding transformation laws are given by

$$\delta O = - \operatorname{Res}\left[\epsilon(z)T(z)O(w, \bar{w})\right]$$
$$= - h\epsilon'(w)O(w, \bar{w}) - \epsilon(w)\partial O(w, \bar{w}). \tag{17.30}$$

Similarly, for anti-holomorphic transformations.

The corresponding finite transformation law of a primary operator under the conformal transformations $z \longrightarrow \tilde{z}(z)$ and $\bar{z} \longrightarrow \tilde{\bar{z}}(\bar{z})$ reads

$$O(z, \bar{z}) \longrightarrow \tilde{O}(\tilde{z}, \tilde{\bar{z}}) = \left(\frac{\partial \tilde{z}}{\partial z}\right)^{-h}\left(\frac{\partial \tilde{\bar{z}}}{\partial \bar{z}}\right)^{-\tilde{h}} O(z, \bar{z}). \tag{17.31}$$

The goal is to compute the spectrum of weights $(h, \tilde{h})$.

17.1.5 The conformal anomaly

In two dimensions the propagator is given by

$$\langle 0| T(\phi(x)\phi(x'))|0\rangle = -\frac{\alpha'}{2}\ln(x - x')^2. \tag{17.32}$$

We can now include operator insertions $O_i(x_i)$ in the Feynman path integral defining this expectation value (assuming also that $x_i \neq x$ and $x_i \neq x'$) to obtain the operator product expansion

$$\phi(x)\phi(x') = -\frac{\alpha'}{2}\ln(x - x')^2 + \cdots. \tag{17.33}$$

By using the fact that the field ϕ splits into holomorphic (left-moving) and anti-holomorphic (right-moving) parts, which do not communicate, we can rewrite the above OPE in the complex plane as

$$\phi(z)\phi(w) = -\frac{\alpha'}{2}\ln(z - w) + \cdots. \tag{17.34}$$

We conclude immediately that ϕ is not a primary operator. In fact ϕ is not even a good conformal field theory operator since it has no good transformation properties under conformal transformations.

By taking the derivatives with respect to z and w we obtain the good OPE:

$$\partial\phi(z)\partial\phi(w) = -\frac{\alpha'}{2}\frac{1}{(z - w)^2} + \cdots. \tag{17.35}$$

Indeed, $\partial\phi$ is a primary operator of weights $h = 1$ and $\tilde{h} = 0$.

The stress–energy–momentum tensor in the classical theory is given by

$$T = -\frac{1}{\alpha'}\partial\phi\partial\phi. \tag{17.36}$$

In the quantum theory this is defined using the normal-ordering prescription:

$$T = -\frac{1}{\alpha'} : \partial\phi\partial\phi := -\frac{1}{\alpha'}(\partial\phi(z)\partial\phi(w) - \langle 0|T(\partial\phi(z)\partial\phi(w))|0\rangle), \quad z \longrightarrow w. \quad (17.37)$$

Physically this means that the vacuum energy is identically zero, i.e., $\langle T \rangle = 0$.

Let us recall that the OPEs involve time-ordered products of operators. Here, we want to compute the time-ordered product $T(z)\partial\phi(w)$. To this end, we use Wick's theorem, which allows us to convert the normal-ordered product to a time-ordered product minus all contractions. We write this schematically as

$$\text{normalordered} = \text{timeordered} - \sum\text{contractions}. \quad (17.38)$$

In our case, this reads explicitly

$$:T(z)\partial\phi(w) := T(z)\partial\phi(z) + \frac{1}{\alpha'}\partial\phi(z)\langle\partial\phi(z)\partial\phi(w)\rangle + \frac{1}{\alpha'}\langle\partial\phi(z)\partial\phi(w)\rangle\partial\phi(z). \quad (17.39)$$

We get immediately the OPE:

$$\begin{aligned}
T(z)\partial\phi(w) &= -\frac{2}{\alpha'}\left(-\frac{\alpha'}{2}\frac{1}{(z-w)^2}\right)\partial\phi(z) \\
&= \frac{\partial\phi(z)}{(z-w)^2} + \cdots \\
&= \frac{\partial\phi(w)}{(z-w)^2} + \frac{\partial^2\phi(w)}{z-w} + \cdots
\end{aligned} \quad (17.40)$$

Thus, $\partial\phi$ is a primary operator of weights $h = 1$ and $\tilde{h} = 0$.

As a second example, we compute the OPE between the stress–energy–momentum tensor and itself. We have

$$\begin{aligned}
(w) :=\ &T(z)T(w) - \frac{2}{\alpha'^2}\langle\partial\phi(z)\partial\phi(w)\rangle\langle\partial\phi(z)\partial\phi(w)\rangle \\
&- \frac{4}{\alpha'^2}\langle\partial\phi(z)\partial\phi(w)\rangle : \partial\phi(z)\partial\phi(w): .
\end{aligned} \quad (17.41)$$

The second term contains the two ways in which we can perform two contractions. The third term contains the four ways in which we can perform a single contraction. (In both cases we exclude those contractions involving operators at the same point which were already taken into account in the definition (17.37) of the stress–energy–momentum tensor T.)

We get then the OPE:

$$\begin{aligned}
T(z)T(w) &= \frac{1/2}{(z-w)^4} - \frac{1}{\alpha'}\frac{2}{(z-w)^2} : \partial\phi(z)\partial\phi(w): \\
&= \frac{1/2}{(z-w)^4} + \frac{2T(w)}{(z-w)^2} - \frac{1}{\alpha'}\frac{2}{z-w} : \partial^2\phi(w)\partial\phi(w): \\
&= \frac{1/2}{(z-w)^4} + \frac{2T(w)}{(z-w)^2} + \frac{\partial T(w)}{z-w}.
\end{aligned} \quad (17.42)$$

We conclude immediately that the stress–energy–momentum tensor $T(z)$ is a conformal operator of weights $h = 2$ and $\tilde{h} = 0$ and hence it is of spin $s = 2$ (it is a symmetric 2-tensor which couples to the metric) and scaling dimension $\Delta = 2$ (it involves two derivatives each giving a mass dimension $+1$ whereas the inverse of the string tension α' is of mass dimension -2 and the filed ϕ is of mass dimension -1).

However, the stress–energy–momentum tensor $T(z)$ is not a primary operator because of the first term, which corresponds to the so-called conformal anomaly and is proportional to the so-called central charge.

For a general conformal field theory the above equation changes to

$$T(z)T(w) = \frac{c/2}{(z-w)^4} + \frac{2T(w)}{(z-w)^2} + \frac{\partial T(w)}{z-w} + \cdots. \tag{17.43}$$

The number c is called the central charge and it is the most important number characterizing a conformal field theory. For example, in the case of N non-interacting free massless scalar fields in two dimensions we have $c = N$. In general, the central charge is a measure of the number of the degrees of freedom (Cardy's formula and Zamalodchikov's c-theorem).

We are ready now to compute the transformation law of the stress–energy–momentum tensor under infinitesimal conformal transformations $z \longrightarrow \tilde{z} + \epsilon(z)$ with Noether's current $J_z(z) = \epsilon(z)T(z)$. We get

$$\begin{aligned}
\delta T(w) = {}& - \operatorname{Res}\left[\epsilon(z)T(z) \cdot T(w)\right] \\
= {}& - \operatorname{Res}\left[\left(\epsilon(w) + \epsilon'(w)(z-w) + \frac{1}{2}\epsilon''(w)(z-w)^2 + \frac{1}{6}\epsilon'''(w)(z-w)^3 + \cdots\right)\right. \\
& \left. \times \left(\frac{c/2}{(z-w)^4} + \frac{2T(w)}{(z-w)^2} + \frac{\partial T(w)}{z-w} + \cdots\right)\right] \\
= {}& - \epsilon(w)\partial T(w) - 2\epsilon'(w)T(w) - \frac{c}{12}\epsilon'''(w).
\end{aligned} \tag{17.44}$$

This can be integrated to obtain the transformation law:

$$\left(\frac{\partial \tilde{z}}{\partial z}\right)^2 \tilde{T}(\tilde{z}) = T(z) - \frac{c}{12}\{\tilde{z}, z\}. \tag{17.45}$$

The bracket $\{\tilde{z}, z\}$ is called the Schwarzian derivative of $\tilde{z}$ with respect to z, denoted also by $S(\tilde{z}, z)$, and it is given by

$$\begin{aligned}
\{\tilde{z}, z\} = S(\tilde{z}, z) &= \frac{\tilde{z}'''}{z'} - \frac{3}{2}\frac{\tilde{z}''}{z'^2} \\
&= \left(\frac{\tilde{z}''}{z'}\right)' - \frac{1}{2}\left(\frac{\tilde{z}''}{z'}\right)^2.
\end{aligned} \tag{17.46}$$

This derivative is invariant under $SL(2, R)$ or Mobius transformations:

$$z \longrightarrow f(z) = \frac{az+b}{cz+d}, \quad ad - bc = 1. \tag{17.47}$$

Thus, if $\langle T(z) \rangle = 0$ then $\langle \tilde{T}(\tilde{z}) \rangle \neq 0$ and it is completely determined by the conformal anomaly and the central charge. This is the conformal anomaly.

The central charge c determines therefore the Casimir energy.

The conformal anomaly is intimately related to the Weyl anomaly and to the gravitational anomaly.

The Weyl anomaly is the statement that the tracelessness property $T^\alpha_\alpha = 0$ of the stress–energy–momentum tensor is lost in the quantum theory. Indeed, we compute for a conformally flat spacetime the expectation value

$$\langle T^\alpha_\alpha \rangle = -\frac{c}{12} R. \tag{17.48}$$

Here, R is the Ricci scalar. Thus, the symmetry under Weyl transformations is lost since R takes different values for the different metric related by Weyl transformations.

Of course c is the central charge associated with the holomorphic (left-moving) sector. There is also a central charge $\tilde{c}$ associated with the anti-holomorphic (right-moving) sector. For the anti-holomorphic (right-moving) sector we obtain a similar formula to (17.48), viz

$$\langle T^\alpha_\alpha \rangle = -\frac{\tilde{c}}{12} R. \tag{17.49}$$

Thus, in curved background we must have $c = \tilde{c}$. This is a gravitational anomaly.

17.1.6 Summary

We conclude this section by summarizing the main points we have discussed so far.

1. Conformal systems have no preferred length scale. Example: A two-dimensional massless free scalar field ϕ with action

$$S = \frac{1}{4\pi\alpha'} \int d^2x \sqrt{g}\, \partial_\mu \phi \partial^\mu \phi. \tag{17.50}$$

2. This action is invariant under the conformal transformations:

$$x^\alpha \longrightarrow \tilde{x}^\alpha(x) \;\; \Rightarrow \;\; g_{\alpha\beta}(x) \longrightarrow \tilde{g}_{\alpha\beta}(\tilde{x}) = \Omega^2(x) g_{\alpha\beta}(x). \tag{17.51}$$

A conformally flat spacetime remains therefore conformal.

3. Euclidean flat spacetime is the complex plane with coordinates z and $\bar{z}$. Conformal transformations are conformal mapping in the complex plane:

$$z \longrightarrow z' = f(z), \;\; \bar{z} \longrightarrow \bar{z}' = \bar{f}(\bar{z}). \tag{17.52}$$

4. These are global transformations, which are symmetries of the action and thus they must be associated with conserved quantities called Noether's currents. This is Noether's theorem. The transformations, the currents and the conservation laws are given by

$$\delta z = \epsilon(z), \quad J_z = \frac{1}{2}T(z)\epsilon(z), \quad \bar{\partial}J_z = 0: \quad \text{holomorphic}$$

$$\delta\bar{z} = \bar{\epsilon}(\bar{z}), \quad \bar{J}_{\bar{z}} = \frac{1}{2}\bar{T}(\bar{z})\bar{\epsilon}(\bar{z}), \quad \partial\bar{J}_{\bar{z}} = 0: \quad \text{anti-holomorphic.}$$

$$(17.53)$$

5. The stress–energy–momentum tensor, with components $T(z)$ and $\bar{T}(\bar{z})$, is conserved (under translations) and traceless (under Weyl transformations) and it is the Noether's current associated with translations.
6. In the case of rotations and scaling transformations we have $\delta z = \epsilon z$ while Noether's current is $J_z(z) = zT(z)$. Similarly, for the anti-holomorphic transformations.
7. Noether's theorem is replaced in the quantum theory by Ward identities with arbitrary number of operator insertions in the complex plane. This can be turned into the operator identities:

$$-\frac{1}{2\pi}\int_{\epsilon \neq 0} \sqrt{g}\,d^2x\,\partial_\mu(J^\mu(x)O_1(x_1)\,\ldots\,O_N(x_N)) = \delta O_1(x_1)\,\ldots\,O_N(x_N). \quad (17.54)$$

The conformal transformation $\epsilon(x)$ is non-zero only around the operator insertion $O_1(x_1)$ and $\delta O_1(x_1)$ is the variation of the operator $O_1(x_1)$ under conformal transformations.
8. In two dimensions this result reduces to the contour integral:

$$\frac{i}{2\pi}\oint_{\partial\epsilon} dz J_z(z)O(w, \bar{w}) = -\mathrm{Res}(J_z O) = \delta O(w, \bar{w}). \quad (17.55)$$

These identities are equivalent to the operator product expansions:

$$J_z(z)O(w, \bar{w}) = \cdots + \frac{\mathrm{Res}(J_z(z)O(w, \bar{w}))}{z - w} + \cdots$$

$$= \cdots - \frac{\delta O(w, \bar{w})}{z - w} + \cdots, \quad z \longrightarrow w.$$

$$(17.56)$$

In other words, the OPE between a given operator and Noether's currents (stress–energy–momentum tensor) determine the transformation law of the operator under conformal transformations.
9. The weights h and $\tilde{h}$ of an operator O are defined by

$$O(z - \epsilon z, \bar{z} - \bar{\epsilon}\bar{z}) = O(z, \bar{z}) - \epsilon(hO + z\partial O) - \bar{\epsilon}(\tilde{h}O + \bar{z}\bar{\partial}O). \quad (17.57)$$

The spin s (eigenvalue under rotations) and the scaling dimension Δ (eigenvalue under scaling transformations, which plays the role of energy) are defined by:

$$s = h - \tilde{h}, \quad \Delta = h + \tilde{h}. \quad (17.58)$$

10. Thus, translations determine the $1/z$ term in the OPE while rotations and scaling transformations determine the $1/z^2$ term in the OPE, viz

$$T(z)O(w, \bar{w}) = \cdots + h\frac{O(w, \bar{w})}{(z - w)^2} + \frac{\partial O(w, \bar{w})}{z - w} + \cdots, \quad z \longrightarrow w. \qquad (17.59)$$

$$\bar{T}(\bar{z})O(w, \bar{w}) = \cdots + \tilde{h}\frac{O(w, \bar{w})}{(\bar{z} - \bar{w})^2} + \frac{\bar{\partial}O(w, \bar{w})}{\bar{z} - \bar{w}} + \cdots, \quad z \longrightarrow w. \qquad (17.60)$$

The operators whose OPE's with the stress–energy–momentum tensor truncates at the $1/z^2$ order are called primary operators. These operators transform covariantly under conformal transformations.

11. The stress–energy–momentum tensor $T(z)$ is a conformal operator of weights $h = 2$ and $\tilde{h} = 0$ and hence it is of spin $s = 2$ and scaling dimension $\Delta = 2$. However, the stress–energy–momentum tensor is not a primary operator due to the conformal anomaly.

17.2 Overview of the SYK model and its connection to the Jackiw–Teitelboim (JT) gravity

The SYK (Sachdev–Ye–Kitaev) model is a remarkable quantum mechanical model, which originates in condensed matter physics [2]. Its interest to us lies in its connection to holography as outlined first by Kitaev [3]. In particular, this model garnered significant attention in the high energy theoretical physics community due to its being the only known solvable holographic model [4–6].

The SYK model is/was a strong candidate for the field theory which lives at the boundary of $\mathbf{AdS}_2$ black hole, which is described by the Jackiw–Teitelboim gravity. Indeed, both the SYK model and the Jackiw–Teitelboim gravity are described in the IR by the so-called Schwarzian action with an $SL(2, \mathbf{R})$ symmetry. They share similar two-point and four-point functions. The time-ordered correlation (TOC) and the out-of-time-ordered correlation (OTOC) of the four-point function in both theories are the same exhibiting maximal chaos characterized by the Lyapunov exponent $\kappa = 2\pi/\beta$, i.e., it saturates the Maldacena–Shenker–Stanford bound on the Lyapunov exponent.

Recall that maximal chaos is equivalent to the fastest scrambling which is measured by the out-of-time-ordered correlators (OTOCs) which in turn measure how two initially commuting operators fail to commute after evolving over time due to interactions providing therefore insight into how information spreads through the system.

Thus, the SYK model and the $\mathbf{AdS}_2$ black hole of JT gravity are both fastest scrambler of information.

The SYK model consists of N Majorana fermions with random all-to-all interactions described by coupling constants drawn from a Gaussian distribution.

The SYK model is intimately connected to tensor models (fields with more than 2 indices, exactly solvable in the large N limit where melon diagrams dominate). This should be contrasted with vectors models (fields with a single index, exactly solvable in the large N limit where bubble diagrams dominate and can be summed up). Thus, the SYK model (fields with a single index but includes disorder) is essentially as easy as vector models but far much easier than matrix models (fields with two indices, generally not exactly solvable in the large N limit where planar diagrams dominate but can not be summed up).

More precisely, the SYK model enjoys the three remarkable properties [4]:

- Solvability at strong coupling in the large N limit.
- Emergent conformal symmetry in the IR limit implying the existence of a holographic gravity dual [8, 9].
- Maximally chaotic behavior similar to black holes in Einstein gravity [10–12].

For a more detailed pedagogical presentation of this model we refer to the review [13, 14]. See also the Sachdev–Ye–Kitaev model in nlab [15].

17.3 Large N quantization of the SYK model

17.3.1 Effective action

We work in Euclidean signature, i.e., the Euclidean time τ is related to the Lorentzian time t by $\tau = it$. The Euclidean time is compactified on a circle of circumference β. In other words, $\tau \in [-\beta/2, +\beta/2]$ and τ is periodic with period β, i.e., $\tau + \beta = \tau$.

The degrees of freedom in the SYK model consist of N Majorana fermions denoted by χ_i where $i = 1, 2,..., N$. These fermions are real, i.e., $\chi_i^\dagger = \chi_i$ and they satisfy the anti-commutation relations

$$\{\chi_i(\tau), \chi_j(\tau')\} = \delta_{ij}\delta(\tau - \tau'). \tag{17.61}$$

In the SYK model these Majorana fermions are interacting with random all-to-all interactions. Explicitly, we have:

- The couplings J_{ijkl} are drawn from a Gaussian distribution with zero mean, viz

$$P(J_{ijkl}) = \sqrt{\frac{N^3}{12\pi J^2}} \exp\left(-\frac{N^3 J_{ijkl}^2}{12 J^2}\right), \quad \langle J_{ijkl} \rangle = 0. \tag{17.62}$$

- The variance of the couplings is given by:

$$\left\langle J_{ijkl}^2 \right\rangle = \frac{3! J^2}{N^3}. \tag{17.63}$$

Here, J is a parameter that sets the strength of the interactions.

- The couplings J_{ijkl} are antisymmetric with respect to any permutation of their indices, i.e.,

$$J_{ijkl} = -J_{jikl} = -J_{ijlk} = \cdots. \tag{17.64}$$

The action, Lagrangian and Hamiltonian of the SYK model for N Majorana fermions χ_i are given by

$$S_{\text{SYK}} = \int d\tau \left[\frac{1}{2}\sum_{i=1}^{N} \chi_i(\tau)\frac{d\chi_i(\tau)}{d\tau} - \frac{1}{4!}\sum_{i,j,k,l=1}^{N} J_{ijkl}\chi_i(\tau)\chi_j(\tau)\chi_k(\tau)\chi_l(\tau) \right], \tag{17.65}$$

$$L_{\text{SYK}} = \frac{1}{2}\sum_{i=1}^{N} \chi_i(\tau)\frac{d\chi_i(\tau)}{d\tau} - \frac{1}{4!}\sum_{i,j,k,l=1}^{N} J_{ijkl}\chi_i(\tau)\chi_j(\tau)\chi_k(\tau)\chi_l(\tau). \tag{17.66}$$

$$H_{\text{SYK}} = \frac{1}{4!}\sum_{i,j,k,l=1}^{N} J_{ijkl}\chi_i(\tau)\chi_j(\tau)\chi_k(\tau)\chi_l(\tau). \tag{17.67}$$

The effective action is computed as follows. We start from the partition function and then we average over the disorder. We obtain a Gaussian integral over disorder, which can be done trivially and we end up with a theory involving only the Majorana fermions χ_i. Explicitly, we have

$$Z = \int \mathcal{D}\chi \exp\left(-S_{\text{SYK}}[\chi]\right)$$

$$\langle Z \rangle_J = \int \mathcal{D}J_{ijkl}\, P(J_{ijkl}) \int \mathcal{D}\chi \exp\left(-\int d\tau \left[\frac{1}{2}\sum_{i=1}^{N}\chi_i\frac{d\chi_i(\tau)}{d\tau} - \frac{1}{4!}\sum_{i,j,k,l} J_{ijkl}\chi_i\chi_j\chi_k\chi_l\right]\right)$$

$$= \int \mathcal{D}\chi \exp\left(-\int d\tau \frac{1}{2}\sum_{i=1}^{N}\chi_i\frac{d\chi_i(\tau)}{d\tau}\right) \times \exp\left(\frac{J^2}{8N^3}\sum_{i,j,k,l}\left(\int d\tau\chi_i\chi_j\chi_k\chi_l\right)^2\right) \tag{17.68}$$

$$= \int \mathcal{D}\chi \exp\left(-\int d\tau \frac{1}{2}\sum_{i=1}^{N}\chi_i\frac{d\chi_i(\tau)}{d\tau}\right) \times \exp\left(\frac{J^2}{8N^3}\int d\tau d\tau'\left(\sum_i \chi_i(\tau)\chi_i(\tau')\right)^4\right).$$

We have also used the fact that

$$\sum_{i,j,k,l}\left(\int d\tau\chi_i\chi_j\chi_k\chi_l\right)^2 = 4!\sum_{i<j<k<l}\left(\int d\tau\chi_i\chi_j\chi_k\chi_l\right)^2$$

$$= \int d\tau d\tau'\left(\sum_i \chi_i(\tau)\chi_i(\tau')\right)^4. \tag{17.69}$$

This is because the fermions totally anti-commute because of the constraint $i < j < k < l$.

We define now the mean-field variable Θ by

$$\Theta(\tau, \tau') = \frac{1}{N}\sum_i \chi_i(\tau)\chi_i(\tau'). \tag{17.70}$$

The kinetic term can be rewritten in terms of this variable as follows

$$
\begin{aligned}
-\frac{N}{2}\int d\tau d\tau' \delta(\tau - \tau')\partial_\tau \Theta(\tau, \tau') &= -\frac{N}{2}\int d\tau d\tau' \delta(\tau - \tau')\partial_\tau\left(\frac{1}{N}\sum_i \chi_i(\tau)\chi_i(\tau')\right) \\
&= -\frac{1}{2}\int d\tau d\tau' \delta(\tau - \tau')\left(\sum_i \partial_\tau \chi_i(\tau). \chi_i(\tau')\right) \quad (17.71) \\
&= \frac{1}{2}\int d\tau \sum_i \chi_i \partial_\tau \chi_i(\tau).
\end{aligned}
$$

The partition function becomes

$$\langle Z \rangle_J = \int \mathcal{D}\chi \exp\left(\frac{N}{2}\int d\tau d\tau' G_0^{-1}(\tau, \tau')\Theta(\tau, \tau') + \frac{J^2 N}{8}\int d\tau d\tau' \Theta^4(\tau, \tau')\right). \tag{17.72}$$

The tree-level inverse propagator is defined by

$$G_0^{-1}(\tau, \tau') = \delta(\tau - \tau')\partial_\tau. \tag{17.73}$$

Next, we introduce bilocal fields $G(\tau, \tau')$ and $\Sigma(\tau, \tau')$ as follows. We start from the identity

$$
\begin{aligned}
f(z) &= \int dx f(x)\delta(x - z) \\
&= \frac{N}{4\pi}\int dx dy f(x)\exp(i\frac{N}{2}y(x - z)).
\end{aligned}
\tag{17.74}
$$

We choose:

$$
\begin{aligned}
z &\longrightarrow \Theta(\tau, \tau') \\
x &\longrightarrow G(\tau, \tau') \\
y &\longrightarrow i\Sigma(\tau, \tau') \\
f(z) &\longrightarrow \exp\left(\frac{J^2 N}{8}\int d\tau d\tau' \Theta^4(\tau, \tau')\right).
\end{aligned}
\tag{17.75}
$$

We have then

$$
\begin{aligned}
&\exp\left(\frac{J^2 N}{8}\int d\tau d\tau' \Theta^4(\tau, \tau')\right) \\
&= \int \mathcal{D}G\mathcal{D}\Sigma \exp\left(\frac{J^2 N}{8}\int d\tau d\tau' G^4(\tau, \tau') - \frac{N}{2}\int d\tau d\tau' \Sigma(\tau, \tau')(G(\tau, \tau') - \Theta(\tau, \tau'))\right).
\end{aligned}
\tag{17.76}
$$

The path integral is normalized such that

$$1 = \int \mathcal{D}G\mathcal{D}\Sigma \exp\left(\frac{J^2 N}{8}\int d\tau d\tau' G^4(\tau, \tau') - \frac{N}{2}\int d\tau d\tau'\Sigma(\tau, \tau')G(\tau, \tau')\right). \quad (17.77)$$

This step is something which resembles a Hubbard–Stratonovich transformation as it will allow us to integrate the fermions. Indeed, we can rewrite the partition function as follows

$$\langle Z \rangle_J = \int \mathcal{D}G\mathcal{D}\Sigma \int \mathcal{D}\chi \exp\left(\frac{N}{2}\int d\tau d\tau'\left(G_0^{-1}(\tau, \tau') + \Sigma(\tau, \tau')\right)\Theta(\tau, \tau')\right)$$

$$\times \exp\left(\frac{J^2 N}{8}\int d\tau d\tau' G^4(\tau, \tau') - \frac{N}{2}\int d\tau d\tau'\Sigma(\tau, \tau')G(\tau, \tau')\right)$$

$$= \int \mathcal{D}G\mathcal{D}\Sigma \int \mathcal{D}\chi \exp\left(\frac{1}{2}\int d\tau d\tau'\sum_i \chi_i(\tau')(-\delta(\tau - \tau')\partial_\tau + \Sigma(\tau, \tau'))\chi_i(\tau)\right) \quad (17.78)$$

$$\times \exp\left(\frac{J^2 N}{8}\int d\tau d\tau' G^4(\tau, \tau') - \frac{N}{2}\int d\tau d\tau'\Sigma(\tau, \tau')G(\tau, \tau')\right)$$

$$= \int \mathcal{D}G\mathcal{D}\Sigma \exp\left(-NS_{\text{eff}}\right).$$

The effective action reads

$$S_{\text{eff}} = -\frac{1}{2}\text{logdet}_{(}G_0^{-1} + \Sigma) - \frac{J^2}{8}\int d\tau d\tau' G^4(\tau, \tau') + \frac{1}{2}\int d\tau d\tau'\Sigma(\tau, \tau')G(\tau, \tau')$$

$$= -\frac{1}{2}\text{Trlog}_{(}G_0^{-1} + \Sigma) - \frac{J^2}{8}\int d\tau d\tau' G^4(\tau, \tau') + \frac{1}{2}\int d\tau d\tau'\Sigma(\tau, \tau')G(\tau, \tau'). \quad (17.79)$$

In the above equations we have used the fact that path integration over a real fermion θ with a Gaussian weight gives the square root of the determinant, viz

$$\int \mathcal{D}\theta e^{-\theta M\theta} = \sqrt{\det M}. \quad (17.80)$$

In our case we have N independent real fermions.

17.3.2 Dyson–Schwinger saddle point equations

The saddle point equations are obtained by varying the effective action with respect to G and Σ:

$$\partial_G S_{\text{eff}} = 0 \Rightarrow \Sigma(\tau, \tau') = J^2 G^3(\tau, \tau'), \quad (17.81)$$

$$\partial_\Sigma S_{\text{eff}} = 0 \Rightarrow \left(G_0^{-1} + \Sigma\right)^{-1} = G$$

$$\Rightarrow \left(G_0^{-1} + \Sigma\right)G = 1 \quad (17.82)$$

$$\Rightarrow \int d\tau'(\delta(\tau - \tau')\partial_\tau + \Sigma(\tau, \tau'))G(\tau', \tau'') = \delta(\tau - \tau'').$$

The solution of the second equation, in fact from the first line of the above equation, is explicitly given by

$$G^{-1} = G_0^{-1} + \Sigma. \tag{17.83}$$

Alternatively, from the third line of the above equation, the solution is more explicitly given by

$$G(\tau_1, \tau_2) = G_0(\tau_1, \tau_2) - \int d\tau_3 d\tau_4 \, G_0(\tau_1, \tau_3)\Sigma(\tau_3, \tau_4)G(\tau_4, \tau_2). \tag{17.84}$$

This means that G is the full two-point function or propagator and Σ is the self-energy.

Thus, by combining these equations we obtain the Dyson–Schwinger equation:

$$G(\tau_1, \tau_2) = G_0(\tau_1, \tau_2) - J^2 \int d\tau_3 d\tau_4 \, G_0(\tau_1, \tau_3)G^3(\tau_3, \tau_4)G(\tau_4, \tau_2). \tag{17.85}$$

The full two-point function G starts with the free propagator G_0 then it is dominated, in the large N limit, by the so-called melonic diagrams (see figure 17.1). The first correction is given by the melonic diagram:

$$-J^2 \int d\tau_3 d\tau_4 \, G_0(\tau_1, \tau_3)G_0^3(\tau_3, \tau_4)G_0(\tau_4, \tau_2). \tag{17.86}$$

This is given by the first diagram in figure 17.1. The three internal lines are associated with the internal propagator $-G_0(\tau_3, \tau_4)$, the two external lines are associated with the external propagators $-G_0(\tau_1, \tau_3)$ and $-G_0(\tau_4, \tau_2)$ while the vertices are associated with the disorder strength J.

In fact the second correction is obtained by inserting this melonic diagram in place of the free propagator in all possible ways. We get the third and fourth melonic diagrams in figure 17.1. Already at these orders we have many other Feynman diagrams but they are all subleading in the large N limit.

We derive the free propagator as follows. We have the definitions

$$G_0(\tau, \tau') = \frac{1}{N}\sum_i \langle 0| T(\chi_i(\tau)\chi_i(\tau'))|0\rangle \tag{17.87}$$

and

$$G_0^{-1}(\tau, \tau') = \delta(\tau - \tau')\partial_\tau \Rightarrow \partial_\tau G_0(\tau, \tau') = \delta(\tau - \tau'). \tag{17.88}$$

By using invariance under translation and Fourier transforming to frequency space we have

$$G_0(\tau, \tau') \equiv G_0(\tau - \tau') = \int \frac{d\omega}{2\pi} G_0(\omega)e^{i\omega(\tau-\tau')} \Rightarrow G_0(\omega) = \frac{1}{i\omega}. \tag{17.89}$$

Thus, we obtain the propagator

$$G_0(\tau, \tau') = \int \frac{d\omega}{2\pi} \frac{1}{i\omega} e^{i\omega(\tau-\tau')} = \frac{1}{2}\text{sign}(\tau - \tau'). \tag{17.90}$$

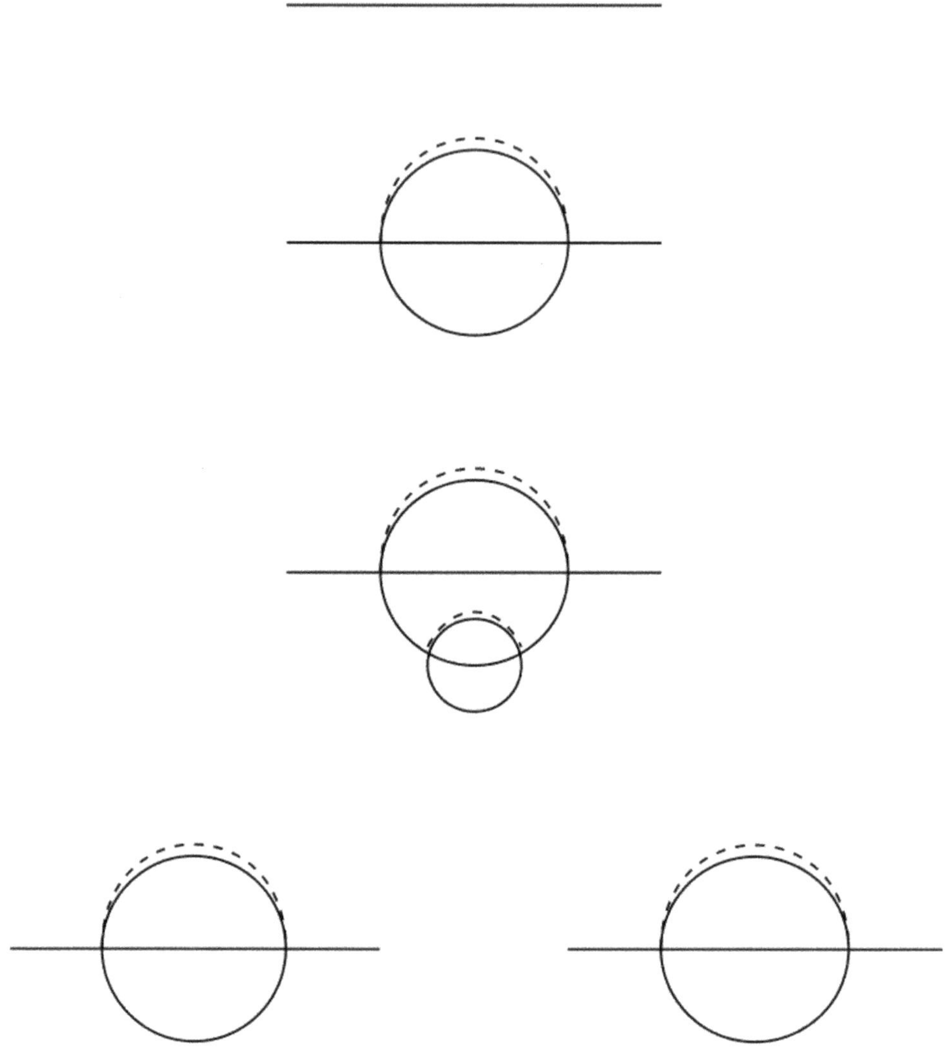

Figure 17.1. The first three melonic diagrams (second, third and fourth lines) correcting the free propagator (first line) of the SYK model.

17.4 Emergence of conformal symmetry in the SYK model

17.4.1 The IR propagator

We started with the Hamiltonian of the SYK model which involves a random all-to-all interactions among N Majorana fermions. First, the effective action in terms of the collective bilocal fields G and Σ is given by

$$S_{\text{eff}} = -\frac{1}{2}\text{Trlog}(G_0^{-1} + \Sigma) - \frac{J^2}{8} \int d\tau d\tau' G^4(\tau, \tau') + \frac{1}{2} \int d\tau d\tau' \Sigma(\tau, \tau') G(\tau, \tau'). \quad (17.91)$$

We also have the Dyson–Schwinger equations at large N for the two-point Green's function $G(\tau)$ and the self-energy $\Sigma(\tau)$, which are given by

$$\partial_G S_{\text{eff}} = 0 \Rightarrow \Sigma(\tau, \tau') = J^2 G^3(\tau, \tau'). \tag{17.92}$$

$$\partial_\Sigma S_{\text{eff}} = 0 \Rightarrow \frac{\partial \Sigma}{G_0^{-1} + \Sigma}$$
$$= G \Rightarrow G(\tau_1, \tau_2) = G_0(\tau_1, \tau_2) - \int d\tau_3 d\tau_4\, G_0(\tau_1, \tau_3)\Sigma(\tau_3, \tau_4)G(\tau_4, \tau_2). \tag{17.93}$$

The saddle point solution is a maximum of the effective action in G but is a minimum of it in Σ, and solves these large N Dyson–Schwinger equations. As we have seen, the large N limit is dominated by melonic diagrams.

The free propagator, the exact propagator and the self-energy are related by the equations

$$G^{-1} = G_0^{-1} + \Sigma. \tag{17.94}$$

$$G_0^{-1}(\tau, \tau') = \delta(\tau - \tau')\partial_\tau, \quad G_0^{-1} = i\omega. \tag{17.95}$$

Here, $i\omega$ is the Matsubara frequency in the imaginary time formalism.

The second Dyson–Schwinger equation can also be put in the form (multiplying by $G_0^{-1}(\tau_0, \tau_1)$ and integrating over τ_1)

$$\int d\tau_1\, G_0^{-1}(\tau_0, \tau_1)G(\tau_1, \tau_2) = \int d\tau_1\, G_0^{-1}(\tau_0, \tau_1)G_0(\tau_1, \tau_2)$$
$$- \int d\tau_3 d\tau_4 \int d\tau_1\, G_0^{-1}(\tau_0, \tau_1)G_0(\tau_1, \tau_3)\Sigma(\tau_3, \tau_4)G(\tau_4, \tau_2)$$
$$\partial_{\tau_0} G(\tau_0, \tau_2) = \delta(\tau_0 - \tau_2) - \int d\tau_3 d\tau_4 \delta(\tau_0 - \tau_3)\Sigma(\tau_3, \tau_4)G(\tau_4, \tau_2)$$
$$\partial_{\tau_0} G(\tau_0, \tau_2) = \delta(\tau_0 - \tau_2) - \int d\tau_4 \Sigma(\tau_0, \tau_4)G(\tau_4, \tau_2). \tag{17.96}$$

By using again invariance under translation we obtain (with $\tau_0 \neq \tau_2$)

$$\partial_\tau G(\tau) = -\int d\tau'\Sigma(\tau - \tau')G(\tau'). \tag{17.97}$$

The second Dyson–Schwinger equation takes the form

$$\Sigma(\tau) = J^2 G^3(\tau). \tag{17.98}$$

The full response of the fermion system is the sum of a free propagation (the derivative term) and an interaction term given by a convolution of the Green's function with the self-energy, which in the SYK model is a function of the Green's function itself.

We consider now the IR or low-energy limit of the SYK model when $N \longrightarrow \infty$ and the energy scale is much smaller than the characteristic interaction energy, i.e., $E \ll J$ or $\tau \gg 1/J$ (since the disorder J sets the energy scale).

We will also consider strong coupling defined by $\beta J \gg 1$. In fact, in the IR limit the system reaches the strong coupling regime.

In this limit the kinetic term $\partial_\tau G(\tau)$ of the free propagation becomes negligible compared to the interaction term leading to the simplified Dyson–Schwinger equations

$$0 = \int d\tau' \Sigma(\tau - \tau') G(\tau'), \quad \Sigma(\tau) = J^2 G^3(\tau). \tag{17.99}$$

More carefully, we must have in the IR limit the following simplified Dyson–Schwinger equation (from the last line of equation (17.96))

$$\delta(\tau_0 - \tau_2) = \int d\tau_4 \Sigma(\tau_0, \tau_4) G(\tau_4, \tau_2). \tag{17.100}$$

Equivalently, we have

$$\delta(\tau) = \int d\tau' \Sigma(\tau - \tau') G(\tau'), \quad \Sigma(\tau) = J^2 G^3(\tau). \tag{17.101}$$

In the IR limit, this equation is dominated by long-time correlations, and thus we expect the Green's function to exhibit a power-law behavior, viz

$$G(\tau) \sim \frac{1}{|\tau|^{2\Delta}}. \tag{17.102}$$

Here, Δ is the scaling dimension of the fermion field. The above form reflects the expected scale invariance of the theory in the IR limit. Indeed, in this limit, the system exhibits emergent conformal symmetry, and the Green's function takes on a power-law form that is characteristic of conformally invariant systems.

To see that this is indeed a solution we use the result that the convolution of two power laws $|\tau|^{-a}$ and $|\tau|^{-b}$ in one dimension results in a power law $|\tau|^{-(a+b-1)}$. Thus, we have immediately the integral

$$\int d\tau' \Sigma(\tau - \tau') G(\tau') = J^2 \int d\tau' G^3(\tau - \tau') G(\tau')$$
$$\sim J^2 \int d\tau' \frac{1}{|\tau - \tau'|^{6\Delta}} \frac{1}{|\tau'|^{2\Delta}} \tag{17.103}$$
$$\sim J^2 |\tau|^{-(8\Delta - 1)}.$$

This corresponds to a decaying behavior if $8\Delta - 1 > 0$. This is still larger than the decaying behavior of the kinetic term $\partial_\tau G(\tau)$, which is $|\tau|^{-(2\Delta+1)}$.

The above simplified Dyson–Schwinger equation is invariant under the time reparametrizations, $\tau \longrightarrow \bar{\tau} = f(\tau)$, $f'(\tau) > 0$:

$$G(\tau_1, \tau_2) \to \bar{G}(\bar{\tau}_1, \bar{\tau}_2) = G(\tau_1, \tau_2) f'(\tau_1)^{-\Delta} f'(\tau_2)^{-\Delta}$$
$$\Sigma(\tau_1, \tau_2) \to \bar{\Sigma}(\bar{\tau}_1, \bar{\tau}_2) = \Sigma(\tau_1, \tau_2) f'(\tau_1)^{-3\Delta} f'(\tau_2)^{-3\Delta}. \tag{17.104}$$

Indeed, we have

$$\int d\bar{\tau}' \bar{\Sigma}(\bar{\tau} - \bar{\tau}')G(\bar{\tau}') = (f'(\tau))^{-3\Delta}(f'(0))^{-\Delta}\int d\tau'(f(\tau'))^{-4\Delta+1}\Sigma(\tau - \tau')G(\tau')$$

$$\delta(\bar{\tau}) = \delta(f(\tau)) = \frac{\delta(\tau)}{f'(\tau)} = \frac{\delta(\tau)}{f'(0)} \tag{17.105}$$

$$\Rightarrow \int df(\tau')(f(\tau'))^{-4\Delta+1}\Sigma(\tau - \tau')G(\tau') = \frac{\delta(\tau)}{(f'(0))^{1-4\Delta}}.$$

This shows that we must have the scaling dimension (anomalous conformal dimension):

$$\Delta = \frac{1}{4}. \tag{17.106}$$

Thus, we have on the one hand

$$\int d\tau'\Sigma(\tau - \tau')G(\tau') \sim J^2|\tau|^{-(8\Delta-1)} \longrightarrow J^2|f(\tau)|^{-(8\Delta-1)} \equiv J^2|\bar{\tau}|^{-1}. \tag{17.107}$$

On the other hand, we have

$$\int d\tau'\Sigma(\tau - \tau')G(\tau') \longrightarrow (f'(\tau))^{\frac{3}{4}}(f'(0))^{\frac{1}{4}}\int d\bar{\tau}'\Sigma(\bar{\tau} - \bar{\tau}')\bar{G}(\bar{\tau}') \sim (f'(\tau))^{\frac{3}{4}}(f'(0))^{\frac{1}{4}}J^2|\bar{\tau}|^{-1}$$

$$\delta(\tau) \longrightarrow f'(0)\delta(\bar{\tau}) \tag{17.108}$$

$$\Rightarrow \int d\bar{\tau}'\Sigma(\bar{\tau} - \bar{\tau}')G(\bar{\tau}') = \delta(\bar{\tau}) \sim J^2|\bar{\tau}|^{-1}.$$

The exact propagator in the IR or low-energy limit (low energy $E \ll J$, strong coupling $J \gg 1/\beta$, large times $\tau \gg 1/J$, zero temperature $\beta = \infty$) is approximately given by the conformal propagator [2]

$$G_c(\tau) = \frac{B}{2}\frac{\text{sign}(\tau)}{|J\tau|^{2\Delta}}, \quad \Delta = \frac{1}{4}. \tag{17.109}$$

Here, B is some dimensionless coefficient. This propagator should be compared with the UV or high-energy limit (high energy $E \gg J$, weak coupling $J \ll 1/\beta$, short times $\tau \gg 1/J$, zero temperature $\beta = \infty$), which is approximately given by the free propagator

$$G_0(\tau) = \frac{1}{2}\text{sign}(\tau). \tag{17.110}$$

The actual exact propagator interpolates between these two results. The large N limit approaches the conformal propagator in the same way that perturbation theory approaches the free propagator.

Also, it is not difficult to show that the Dyson–Schwinger equation in the IR limit given by equation (17.100) is equivalent to the original Dyson–Schwinger equation (17.93) after dropping from it the exact propagator from the left-hand side, viz

$$0 = G_0(\tau_1, \tau_2) - \int d\tau_3 d\tau_4\, G_0(\tau_1, \tau_3)\Sigma(\tau_3, \tau_4)G(\tau_4, \tau_2). \tag{17.111}$$

In order to complete our construction we prove here explicitly that the convolution of two power-law functions in one dimension follows the rule used in equation (17.103). In other words, that the convolution of two power laws $|\tau|^{-a}$ and $|\tau|^{-b}$ in one dimension results in a power law $|\tau|^{-(a+b-1)}$, viz

$$\int_{-\infty}^{\infty} d\tau' \, |\tau - \tau'|^{-a} |\tau'|^{-b} \sim |\tau|^{-(a+b-1)}. \tag{17.112}$$

First, and without any loss of generality, we assume that $\tau > 0$ (since the integrand only depends on the absolute values of τ and τ', the behavior for negative τ will follow similarly). Then, we split the integral into two parts:

$$\begin{aligned}
I(\tau) &= \int_{-\infty}^{\infty} d\tau' \, |\tau - \tau'|^{-a} |\tau'|^{-b} \\
&= \int_{-\infty}^{0} d\tau' |\tau - \tau'|^{-a} |\tau'|^{-b} + \int_{0}^{\infty} d\tau' |\tau - \tau'|^{-a} |\tau'|^{-b}.
\end{aligned} \tag{17.113}$$

Since both terms behave similarly, we can focus on the second term (the first term is symmetric and can be evaluated similarly):

$$I_1(\tau) = \int_{0}^{\infty} d\tau' (\tau - \tau')^{-a} (\tau')^{-b}. \tag{17.114}$$

Next, we perform the substitution $\tau' = u\tau$, which rescales the integration variable to make it dimensionless:

$$I_1(\tau) = \tau^{-(a+b-1)} \int_{0}^{1} du \, (1 - u)^{-a} u^{-b} = \tau^{-(a+b-1)} B(1 - b, 1 - a). \tag{17.115}$$

Here, $B(x, y)$ is the standard Beta function. Thus, the convolution of two power laws $|\tau|^{-a}$ and $|\tau|^{-b}$ indeed results in a power law $|\tau|^{-(a+b-1)}$. This result holds for any a and b such that the Beta function $B(1 - b, 1 - a)$ converges, meaning that we must have $a, b < 1$. The convolution leads to a shift in the power-law exponent by 1, consistent with dimensional analysis of the integral.

17.4.2 The Schwarzian action

The basic result is the effective action given by

$$\begin{aligned}
S_{\text{eff}} = &-\frac{1}{2}\text{Trlog}\big(G_0^{-1} + \Sigma\big) - \frac{J^2}{8} \int d\tau d\tau' G^4(\tau, \tau') \\
&+ \frac{1}{2} \int d\tau d\tau' \Sigma(\tau, \tau') G(\tau, \tau').
\end{aligned} \tag{17.116}$$

Now the variable Σ is shifted as follows $\Sigma' = G_0^{-1} + \Sigma$. The effective action becomes then

$$\begin{aligned}
S_{\text{eff}} = &-\frac{1}{2}\text{Trlog}\Sigma' - \frac{J^2}{8} \int d\tau d\tau' G^4(\tau, \tau') \\
&+ \frac{1}{2} \int d\tau d\tau' \big(\Sigma'(\tau, \tau') - G_0^{-1}(\tau, \tau')\big) G(\tau, \tau').
\end{aligned} \tag{17.117}$$

This effective action can be separated into a conformally invariant part and a conformally-non-invariant part as follows

$$S_{\text{eff}} = S_{\text{CFT}} + S_I. \tag{17.118}$$

$$S_{\text{CFT}} = -\frac{1}{2}\text{Trlog}\Sigma' - \frac{J^2}{8}\int d\tau d\tau' G^4(\tau, \tau') + \frac{1}{2}\int d\tau d\tau' \Sigma'(\tau, \tau')G(\tau, \tau'). \tag{17.119}$$

$$S_I = -\frac{1}{2}\int d\tau d\tau' G_0^{-1}(\tau, \tau')G(\tau, \tau'). \tag{17.120}$$

In other words, it is sufficient to drop the free propagator to obtain the conformal theory of the low energy IR limit. Indeed, the conformally invariant action S_{CFT} leads to the Dyson–Schwinger equations

$$\partial_G S_{\text{eff}} = 0 \Rightarrow \Sigma'(\tau, \tau') = J^2 G^3(\tau, \tau'). \tag{17.121}$$

$$\partial_\Sigma S_{\text{eff}} = 0 \Rightarrow \frac{\partial \Sigma'}{\Sigma'} = G \Rightarrow 0 = G_0(\tau_1, \tau_2)$$
$$- \int d\tau_3 d\tau_4 G_0(\tau_1, \tau_3)\Sigma'(\tau_3, \tau_4)G(\tau_4, \tau_2). \tag{17.122}$$

Thus, the first equation (which is a constraint representing the fact that the self-energy is a Lagrange multiplier) is the same as before whereas the second equation is precisely equation (17.111), i.e., this Dyson–Schwinger equation is equivalent to the original Dyson–Schwinger equation (17.93) after dropping from it the exact propagator from the left-hand side. In the IR limit, the SYK model therefore exhibits reparametrization invariance, which is captured by the conformal action S_{CFT}.

In the following we will drop, for simplicity, the prime over the shifted self-energy.

The addition of the conformally-non-invariant action S_I (which corresponds to the inclusion of the kinetic term $\partial_\tau G$ from (17.97)) will lead to the spontaneous breaking of the conformal symmetry of the action S_{CFT}, which induces as a consequence Goldstone modes given in terms of a scalar field $f(\tau)$, which is precisely the time reparametrization function of the conformal symmetry. These soft reparametrization modes are governed by the Schwarzian action, which captures the low energy dynamics of the SYK model.

The Schwarzian action also captures the chaotic behavior of the system, as it controls the exponential growth of out-of-time-ordered correlators (OTOCs), which are used to diagnose chaos in quantum systems. It has also an important interpretation in 2D gravity, in particular the Jackiw–Teitelboim (JT) theory, where it describes the dynamics of the boundary mode of $\mathbf{AdS}^2$ spacetimes.

To construct the Schwarzian action we start from the observation that the solutions of the above Dyson–Schwinger equations are given by a manifold of vacuum states obtained from the conformal propagator G_c by the action of the reparametrization transformation $\tau \longrightarrow f(\tau)$. In other words, there is an infinite number of solutions G_f of the above Dyson–Schwinger equations, which are parameterized by the function $f(\tau)$ and which all have the same action S_{CFT}.

We should then consider fluctuations around the conformal solution, which are captured by the reparametrization function $f(\tau)$. The variation of the effective action around these fluctuations is then given precisely by the variation of the conformally-non-invariant action S_I, which in turn is given by the variation of the propagator G_f with respect to f. More explicitly, we have

$$S_{\text{CFT}}[G_c] = S_{\text{CFT}}[G_f] \rightarrow \delta S_{\text{CFT}}[G_f] = 0 \Rightarrow \delta S_{\text{eff}}[G_f]$$
$$= \delta S_I[G_f] = -\frac{1}{2}\int d\tau d\tau' G_0^{-1}(\tau, \tau')\delta G_f(\tau, \tau'). \tag{17.123}$$

The computation of δG_f leads to the so-called Schwarzian derivative of the reparametrization function $f(\tau)$. This then gives us the Schwarzian action.

This calculation goes explicitly as follows. The conformal propagator, which is a solution of the Dyson–Schwinger equations (17.121) and (17.122), is given by

$$G_c(\tau, \tau_2) = \frac{B}{2}\frac{\text{sign}(\tau_{12})}{|J\tau_{12}|^{2\Delta}}, \quad \tau_{12} = \tau_1 - \tau_2. \tag{17.124}$$

We consider now the reparametrization transformations

$$\tau \longrightarrow \bar{\tau} = f(\tau)$$
$$G(\tau_1, \tau_2) \rightarrow \bar{G}(\bar{\tau}_1, \bar{\tau}_2) = G(\tau_1, \tau_2)f'(\tau_1)^{-\Delta}f'(\tau_2)^{-\Delta} \tag{17.125}$$
$$\Sigma(\tau_1, \tau_2) \rightarrow \bar{\Sigma}(\bar{\tau}_1, \bar{\tau}_2) = \Sigma(\tau_1, \tau_2)f'(\tau_1)^{-3\Delta}f'(\tau_2)^{-3\Delta}.$$

The reparametrization symmetry in one dimension is the full conformal invariance. This symmetry corresponds also to the group of diffeomorphisms of the circle, **Diff(S¹)**, reflecting the infinite-dimensional conformal symmetry in one-dimensional systems.

The IR conformal action S_{CFT} is in fact invariant under these conformal reparametrization transformations. As a consequence, the IR propagator G_c is conformally invariant in the sense that we must have

$$\bar{G}_c(\bar{\tau}_1, \bar{\tau}_2) = G_c(\bar{\tau}_1, \bar{\tau}_2). \tag{17.126}$$

This means that in addition to the conformal propagator G_c we also have other solutions of the Dyson–Schwinger equation, which have the same value of S_{CFT}, and which are obtained by the action of the above reparametrization transformations as follows

$$G_f(\tau_1, \tau_2) = \bar{G}_c(\bar{\tau}_1, \bar{\tau}_2)f'(\tau_1)^{\Delta}f'(\tau_2)^{\Delta}$$
$$= \frac{B}{2}\frac{\text{sign}(\tau_{12})}{|J|^{2\Delta}}\frac{f'(\tau_1)^{\Delta}f'(\tau_2)^{\Delta}}{|f(\tau_1) - f(\tau_2)|^{2\Delta}}. \tag{17.127}$$

In the second line of this equation we have used the fact that f is a monotonically increasing function of τ, i.e., $f'(\tau) > 0$ which means that $\text{sign}(\tau_{12}) = \text{sign}(\bar{\tau}_{12})$. Clearly, we have

$$G_f(\tau_1, \tau_2)|_{f(\tau)=\tau} = G_c(\tau_1, \tau_2) \tag{17.128}$$

Either we work with $\bar{G}_c$ with the time variable $\bar{\tau}$ or we work with G_f with the original time variable τ, viz

$$\begin{aligned}
&S_{\text{CFT}}[\bar{G}_c(\bar{\tau}_1, \bar{\tau}_2)] = S_{\text{CFT}}[G_f(\tau_1, \tau_2)] = S_{\text{CFT}}[\bar{G}_c(\tau_1, \tau_2)] \\
&\Rightarrow \delta S_{\text{CFT}}[G_f(\tau_1, \tau_2)] = S_{\text{CFT}}[G_f(\tau_1, \tau_2)] - S_{\text{CFT}}[\bar{G}_c(\tau_1, \tau_2)].
\end{aligned} \tag{17.129}$$

We must have

$$\delta S_{\text{CFT}}[G_f(\tau_1, \tau_2)] = 0. \tag{17.130}$$

The invariance of the conformally invariant action can be shown explicitly as follows.

First, the above conformally invariant fluctuations are then of the form

$$G = G_c + \delta G, \quad \Sigma = J^2 G_c^3 + 3J^2 G_c^2 \delta G. \tag{17.131}$$

The Dyson–Schwinger equation $\int d\tau \Sigma(\tau_1, \tau) G(\tau, \tau_2) = \delta(\tau_1 - \tau_2)$ takes the form

$$\int d\tau \left[\Sigma_c(\tau_1, \tau) \delta G(\tau, \tau_2) + 3J^2 G_c^2(\tau_1, \tau) G_c(\tau, \tau_2) \delta G(\tau_1, \tau) \right] = 0. \tag{17.132}$$

Then, we multiply by $G_c(\tau_1, \tau_3)$ and integrate over τ_1 to get (using again the above Dyson–Schwinger equation)

$$\int d\tau d\tau_1 K(\tau_3, \tau_2; \tau_1, \tau) \delta G(\tau_1, \tau) = \delta G(\tau_3, \tau_2). \tag{17.133}$$

The kernel K_c is defined by

$$K(\tau_3, \tau_2; \tau_1, \tau) = -3J^2 G_c(\tau_3, \tau_1) G_c(\tau_2, \tau) G_c^2(\tau_1, \tau). \tag{17.134}$$

The fluctuations δG are then the zero modes of the operator $K - 1$.

Let us now show explicitly that fluctuation of the conformally invariant action S_{CFT} around the **IR** conformal solution is zero.

We parameterize the configurations around the saddle point $\tilde{G}$ of the effective action (which is approximately equal to the conformal solution G_c) as follows: $G = \tilde{G} + \delta G = \tilde{G} + \delta\bar{G}/|\tilde{G}|$ and $\Sigma = \tilde{\Sigma} + \delta\Sigma = \tilde{\Sigma} + |\tilde{G}|\delta\bar{\Sigma}$.

We then compute (the linear term vanishes by the equations of motion and we keep terms up to quadratic in δG and $\delta\Sigma$)

$$\begin{aligned}
&-\frac{J^2}{8} \int d\tau d\tau' G^4(\tau, \tau') = -\frac{J^2}{8} \int d\tau d\tau' \tilde{G}^4(\tau, \tau') - \frac{3J^2}{4} \int d\tau d\tau' \delta\bar{G}^2(\tau, \tau') \\
&\frac{1}{2} \int d\tau d\tau' \Sigma(\tau, \tau') G(\tau, \tau') = \frac{1}{2} \int d\tau d\tau' \tilde{\Sigma}(\tau, \tau') \tilde{G}(\tau, \tau') + \frac{1}{2} \int d\tau d\tau' \delta\bar{\Sigma}(\tau, \tau') \delta\bar{G}(\tau, \tau')
\end{aligned} \tag{17.135}$$

and

$$-\frac{1}{2}\mathrm{Trlog}\Sigma = \frac{1}{2}\mathrm{Trlog}\tilde{\Sigma} + \frac{1}{4}\,\mathrm{Tr}\left(\frac{\delta\Sigma}{\tilde{\Sigma}}\right)^2$$

$$= -\frac{1}{2}\mathrm{Trlog}\tilde{\Sigma} + \frac{1}{4}\,\mathrm{Tr}\,(\tilde{G}\delta\Sigma)^2 \tag{17.136}$$

$$= -\frac{1}{2}\mathrm{Trlog}\tilde{\Sigma} - \frac{1}{12J^2}\int\,d\tau_1 d\tau_2 d\tau_3 d\tau_4\,\delta\tilde{\Sigma}(\tau_1,\,\tau_2)\bar{K}(\tau_1,\,\tau_2;\,\tau_3,\,\tau_4)\delta\tilde{\Sigma}(\tau_3,\,\tau_4)$$

$$\bar{K}(\tau_1,\,\tau_2;\,\tau_3,\,\tau_4) = -3J^2\tilde{G}(\tau_1,\,\tau_3)\tilde{G}(\tau_2,\,\tau_4)|\tilde{G}(\tau_1,\,\tau_2)||\tilde{G}(\tau_3,\,\tau_4)|.$$

Thus, we have

$$S_{\mathrm{CFT}} = -\frac{1}{2}\mathrm{Trlog}\Sigma - \frac{J^2}{8}\int\,d\tau d\tau' G^4(\tau,\,\tau') + \frac{1}{2}\int\,d\tau d\tau'\Sigma(\tau,\,\tau')G(\tau,\,\tau')$$

$$= \tilde{S}_{\mathrm{CFT}} - \frac{1}{12J^2}(\delta\tilde{\Sigma},\,\bar{K}\delta\tilde{\Sigma}) - \frac{3J^2}{4}(\delta\bar{G},\,\delta\bar{G}) + \frac{1}{2}(\delta\tilde{\Sigma},\,\delta\bar{G})$$

$$\delta S_{\mathrm{CFT}} = -\frac{1}{12J^2}(\delta\tilde{\Sigma},\,\bar{K}\delta\tilde{\Sigma}) - \frac{3J^2}{4}(\delta\bar{G},\,\delta\bar{G}) + \frac{1}{2}(\delta\tilde{\Sigma},\,\delta\bar{G}) \tag{17.137}$$

$$= -\frac{1}{12J^2}(\delta\tilde{\Sigma},\,\bar{K}\delta\tilde{\Sigma}) + \frac{3J^2}{4}(\delta\bar{G},\,(\bar{K}^{-1} - 1)\delta\bar{G}),\quad \delta\tilde{\Sigma} = \delta\bar{\Sigma} - 3J^2\bar{K}^{-1}\delta G.$$

Note that $\bar{K}$ is an antisymmetric matrix. Thus, by integrating over $\delta\Sigma$ we obtain

$$\delta S_{\mathrm{CFT}} = \frac{3J^2}{4}(\delta\bar{G},\,(\bar{K}^{-1} - 1)\delta\bar{G}) = 0. \tag{17.138}$$

This is zero since $\delta\bar{G}$ is a zero mode of $\bar{K} - 1$. Indeed, equation (17.133) holds also for $\bar{K}$ if we substitute the scaled fluctuation $\delta\bar{G}$ in place of the fluctuation δG.

We are left with the computation of the variation of the kinetic action S_I (thought of as a perturbation to the kinetic action) around the conformal solution. This is given by

$$\delta S_{\mathrm{eff}}[G_f] = \delta S_I[G_f] = -\frac{1}{2}\int\,d\tau_1 d\tau_2\,G_0^{-1}(\tau_1,\,\tau_2)\delta G_f(\tau_1,\,\tau_2). \tag{17.139}$$

We have (with $\tau = \frac{1}{2}(\tau_1 + \tau_2)$ and $\tau_{12} = \tau_1 - \tau_2$)

$$\delta S_{\mathrm{eff}}[G_f] = \delta S_I[G_f] = -\frac{1}{2}\int\,d\tau d\tau_{12}\delta(\tau_{12})\left[\frac{1}{2}\partial_\tau\delta G_f(\tau,\,\tau_{12}) + \partial_{\tau_{12}}\delta G_f(\tau,\,\tau_{12})\right]. \tag{17.140}$$

Thus, the delta function in the free propagator G_0^{-1} picks out the $\tau_{12} \ll 1$ regime. We compute then (with $\tau_{12} = \tau_1 - \tau_2 \longrightarrow 0$)

$$\frac{f'(\tau_1)f'(\tau_2)}{(f(\tau_1) - f(\tau_2))^2} = \frac{f'\left(\tau + \frac{1}{2}\tau_{12}\right)f'\left(\tau - \frac{1}{2}\tau_{12}\right)}{\left(f\left(\tau + \frac{1}{2}\tau_{12}\right) - f\left(\tau - \frac{1}{2}\tau_{12}\right)\right)^2}$$

$$= \frac{\left(f'(\tau) + \frac{1}{2}\tau_{12}f''(\tau) + \frac{1}{8}\tau_{12}^2 f'''(\tau) + O_3\right)\left(f'(\tau) - \frac{1}{2}\tau_{12}f''(\tau) + \frac{1}{8}\tau_{12}^2 f'''(\tau) + O_3\right)}{\left(\tau_{12}f'(\tau) + \frac{1}{24}\tau_{12}^3 f'''(\tau) + O_5\right)^2}$$

$$= \frac{\left(f'(\tau) + \frac{1}{8}\tau_{12}^2 f'''(\tau)\right)^2 - \left(\frac{1}{2}\tau_{12}f''(\tau)\right)^2}{\left(\tau_{12}f'(\tau) + \frac{1}{24}\tau_{12}^3 f'''(\tau) + O_5\right)^2} \qquad (17.141)$$

$$= \frac{(f'(\tau))^2 + \frac{1}{4}\tau_{12}^2 f'(\tau)f'''(\tau) - \frac{1}{4}\tau_{12}^2 (f''(\tau))^2 + O_3}{\tau_{12}^2 (f'(\tau))^2 + \frac{1}{12}\tau_{12}^4 f'(\tau)f'''(\tau) + O_5}$$

$$= \frac{1}{\tau_{12}^2} \times \frac{(f'(\tau))^2 + \frac{1}{4}\tau_{12}^2 f'(\tau)f'''(\tau) - \frac{1}{4}\tau_{12}^2 (f''(\tau))^2 + O_3}{(f'(\tau))^2 + \frac{1}{12}\tau_{12}^2 f'(\tau)f'''(\tau) + O_3}$$

$$= \frac{1}{\tau_{12}^2} \times \frac{1 + \frac{1}{4}\tau_{12}^2 \frac{f'''(\tau)}{f'(\tau)} - \frac{1}{4}\tau_{12}^2 \left(\frac{f''(\tau)}{f'(\tau)}\right)^2 + O_3}{1 + \frac{1}{12}\tau_{12}^2 \frac{f'''(\tau)}{f'(\tau)} + O_3}.$$

We have then

$$\frac{f'(\tau_1)f'(\tau_2)}{(f(\tau_1) - f(\tau_2))^2} = \frac{1}{\tau_{12}^2} \times \left(1 + \frac{1}{6}\tau_{12}^2 \frac{f'''(\tau)}{f'(\tau)} - \frac{1}{4}\tau_{12}^2 \left(\frac{f''(\tau)}{f'(\tau)}\right)^2 + O_3\right). \qquad (17.142)$$

Thus, in summary we obtain the expansion of the conformal propagator in terms of the so-called Schwarzian derivative, viz

$$\left[\frac{f'(\tau_1)f'(\tau_2)}{(f(\tau_1) - f(\tau_2))^2}\right]^\Delta = \frac{1}{\tau_{12}^{2\Delta}}\left(1 + \frac{\Delta\tau_{12}^2}{6}\left[\frac{f'''(\tau)}{f'(\tau)} - \frac{3}{2}\left(\frac{f''(\tau)}{f'(\tau)}\right)^2\right] + O_3\right)$$

$$= \frac{1}{\tau_{12}^{2\Delta}}\left(1 + \frac{\Delta\tau_{12}^2}{6}\mathrm{Sch}(f(\tau), \tau) + O_3\right). \qquad (17.143)$$

The Schwarzian derivative is explicitly defined by

$$\mathrm{Sch}(f(\tau), \tau) = \frac{f'''(\tau)}{f'(\tau)} - \frac{3}{2}\left(\frac{f''(\tau)}{f'(\tau)}\right)^2. \qquad (17.144)$$

We compute then that the variation of the kinetic action S_I is given by

$$\delta G_f(\tau_1, \tau_2) = \frac{B}{2} \frac{\mathrm{sign}(\tau_{12})}{|J|^{2\Delta}} \frac{\Delta \tau_{12}^{2-2\Delta}}{6} \mathrm{Sch}(f(\tau), \tau)$$

$$\delta S_{\mathrm{eff}}[G_f] = \delta S_I[G_f] = -\frac{1}{2} \int d\tau d\tau_{12} \, \delta(\tau_{12})[\partial_{\eta_2} \delta G_f(\tau, \tau_{12})]$$

$$= -\frac{B\Delta}{24|J|^{2\Delta}} \int d\tau_{12} \, \delta(\tau_{12}) \partial_{\eta_2}\left(\mathrm{sign}(\tau_{12})\tau_{12}^{2-2\Delta}\right) \int d\tau \, \mathrm{Sch}(f(\tau), \tau) \quad (17.145)$$

$$= -\frac{B\Delta}{24J} \int d\eta \delta(\eta) \partial_{\eta}(\mathrm{sign}(\eta)\eta^{2-2\Delta}) \int d\tau \, \mathrm{Sch}(f(\tau), \tau), \quad \eta = J\tau_{12}$$

$$= -\frac{C}{J} \int d\tau \, \mathrm{Sch}(f(\tau), \tau).$$

This is the celebrated Schwarzian action.

In the above calculation, the coefficient C comes out in fact to be undefined. This coefficient is determined numerically to be given by [5–7]

$$C = 0.48 \frac{\Delta}{12}. \qquad (17.146)$$

By choosing a particular reparametrization function $f(\tau)$ at the level of the Schwarzian action the original conformal symmetry of the theory is explicitly spontaneously broken down to an $SL(2, \mathbf{R})$ symmetry given by

$$\tau \longrightarrow \frac{a\tau + b}{c\tau + d}, \quad ad - bc = 1. \qquad (17.147)$$

17.4.3 On finite temperature

- By using reparametrization invariance of the IR theory we know that the zero-temperature conformal propagator $G_c(\tau_1, \tau_2)$ is related to the finite-temperature conformal propagator $G_c^{\beta}(\tau_1, \tau_2)$ by the conformal map

$$\tau_{\mathrm{line}} = f(\tau_{\mathrm{circle}}) = \tan \frac{\pi \tau_{\mathrm{circle}}}{\beta}. \qquad (17.148)$$

  Obviously, the infinite line corresponds to the zero-temperature while the compact circle corresponds to finite-temperature (in fact β is the circumference of the circle).

- The same recipe applies of course to the free propagator. We have

$$G_0(\tau, \tau') = \frac{1}{2}\mathrm{sign}(\tau - \tau') \longrightarrow G_0^{\beta}(\tau, \tau') = \frac{1}{2}\mathrm{sign}(\tan \frac{\pi}{\beta}(\tau - \tau')) = \frac{1}{2}\mathrm{sign}(\tau - \tau'). \quad (17.149)$$

  Remark that the finite-temperature and zero-temperature propagators coincide for $\tau \in [-\frac{\beta}{2}, +\frac{\beta}{2}]$. We have also used the fact that $\mathrm{sign}(\tan \frac{\pi\tau}{\beta}) = \mathrm{sign}(\tau)$.

- We have then (we must include also the conformal factor $f'(\tau)^{\Delta} f'(0)^{\Delta}$)

$$G_c(\tau) = \frac{B}{2} \frac{\mathrm{sign}(\tau)}{|J\tau|^{2\Delta}} \Rightarrow G_c^\beta(\tau) = \frac{B'}{2} \frac{\mathrm{sign}(\tau)}{|J\sin\frac{\pi\tau}{\beta}|^{2\Delta}}, \quad \tau = \left[-\frac{\beta}{2}, +\frac{\beta}{2}\right[. \tag{17.150}$$

The coefficients B and B' are given by

$$\frac{B}{2} = \frac{1}{(4\pi)^{1/4}}, \quad \frac{B'}{2} = \frac{B}{2} \frac{\pi^{1/2}}{\beta^{1/2}} \tag{17.151}$$

- Similarly, the finite-temperature Schwarzian action is related to the finite-temperature Schwarzian action by the same conformal map. Explicitly, we have

$$S_I = -\frac{C}{J} \int_{-\infty}^{+\infty} d\tau \, \mathrm{Sch}(f(\tau), \tau) \Rightarrow S_I^\beta = -\frac{C}{J} \int_{-\frac{\beta}{2}}^{+\frac{\beta}{2}} d\tau \, \mathrm{Sch}(\tan\frac{\pi f(\tau)}{\beta}, \tau). \tag{17.152}$$

$f(\phi)$ is now a map, which takes the circle to itself preserving the orientation. We will also need the finite-temperature Schwarzian action with the rescaling $\bar\tau = \frac{2\pi}{\beta}\tau$ and $\bar f = \frac{2\pi}{\beta}f$, i.e.,

$$S_I^\beta = -\frac{2\pi C}{J\beta} \int_{-\pi}^{+\pi} d\bar\tau \, \mathrm{Sch}(\tan\frac{\bar f(\bar\tau)}{2}, \bar\tau). \tag{17.153}$$

17.5 The four-point function of the SYK model

17.5.1 Contribution of the Schwarzian soft modes and OTOC

In order to probe the chaotic properties of the SYK model we discuss in some detail the four-point function of the model [4, 5].

The four-point function is expressed as a sum over indices i and j of the time-ordered product of Majorana fermions χ_i and χ_j at different times τ_1, τ_2, τ_3, τ_4:

$$F(\tau_1, \tau_2, \tau_3, \tau_4) = \frac{1}{N^2}\sum_{i,j}\langle T\chi_i(\tau_1)\chi_i(\tau_2)\chi_j(\tau_3)\chi_j(\tau_4)\rangle. \tag{17.154}$$

The indices of the fermions are contracted in pairs because this is the only possibility to get non-zero contribution, and the division by N^2 is due to the averaging over i and j.

We will consider the IR or low-energy limit when $N \longrightarrow \infty$ and $\tau \gg 1/J$ (since the disorder J sets the energy scale). In this IR limit the system reaches the strong coupling regime $\beta J \gg 1$ and the theory is controlled by the Schwarzian action around the conformal theory.

We will restrict ourselves to $\tau_1 > \tau_2$, $\tau_1 > \tau_3$ and $\tau_3 > \tau_4$.

We will also work at finite temperature and thus the conformal propagator is given by $G_c^\beta(\tau_1, \tau_2)$. The saddle point solution $\tilde G_c^\beta(\tau_1, \tau_2)$ of the Dyson–Swinger equations is approximately equal to the conformal propagator in the strong coupling limit. The saddle point involves also the self-energy $\tilde{\Sigma}_c^\beta(\tau_1, \tau_2)$, which is approximately equal to the conformal self-energy $\Sigma_c^\beta(\tau_1, \tau_2)$.

The first contribution to the four-point function is the tree-level or classical contribution given by the disconnected four-point function, viz

$$
\begin{aligned}
F(\tau_1, \tau_2, \tau_3, \tau_4) &= \frac{1}{N^2} \sum_{i,j} \langle T \chi_i(\tau_1)\chi_i(\tau_2)\chi_j(\tau_3)\chi_j(\tau_4) \rangle \\
&= \frac{1}{Z} \int \mathcal{D}G \mathcal{D}\Sigma \left(G(\tau_{12})G(\tau_{34}) + \frac{1}{N}G(\tau_{14})G(\tau_{23}) \right. \\
&\quad \left. - \frac{1}{N}G(\tau_{13})G(\tau_{24}) \right) \exp(-S_{\text{eff}}) \\
&= G_c^\beta(\tau_{12})G_c^\beta(\tau_{34}) \equiv F_{\text{disc}}(\tau_1, \tau_2, \tau_3, \tau_4).
\end{aligned}
\tag{17.155}
$$

The leading quantum contribution to the four-point function is of order $1/N$ and it is given by the soft modes of the Schwarzian in the Gaussian approximation.

Again, we will consider the effect of conformally invariant fluctuations around the conformal propagator. The fluctuations are denoted here by $\delta G_f^\parallel$ as they are parallel to the conformal propagator and they are given by

$$
\begin{aligned}
\delta G_f^\parallel(\tau_1, \tau_2) &= G_f^\beta(\tau_1, \tau_2) - G_c^\beta(\tau_1, \tau_2) \\
&= G_c^\beta(f(\tau_1), f(\tau_2))f'(\tau_1)^{1/4}f'(\tau_2)^{1/4} - G_c^\beta(\tau_1, \tau_2) \\
&= \left(\delta f(\tau_1)\partial_{\tau_1} + \frac{1}{4}\partial_{\tau_1}\delta f(\tau_1) + \delta f(\tau_2)\partial_{\tau_2} + \frac{1}{4}\partial_{\tau_2}\delta f(\tau_2) \right) G_c^\beta(\tau_1, \tau_2).
\end{aligned}
\tag{17.156}
$$

The fluctuation δf is defined by

$$
\delta f(\tau) = f(\tau) - \tau = \frac{1}{\sqrt{2\pi}} \sum_{m \in \mathbf{Z}} \delta f(m)e^{im\tau}.
\tag{17.157}
$$

In these fluctuations we compute the Schwarzian action in the Gaussian approximation as follows (we include an appropriate re-scaling)

$$
\begin{aligned}
L_f(\tau) &= \text{Sch}(\tan \frac{f(\tau)}{2}, \tau) \\
&= \frac{(f'(\tau))^2}{2} + \text{Sch}(f(\tau), \tau) \\
&= \frac{1}{2} + \delta f' + \frac{1}{2}(\delta f')^2 + \delta f''' - \delta f'''.\delta f' - \frac{3}{2}(\delta f'')^2 + O_3 \\
&= \frac{1 + (\delta f')^2 - (\delta f'')^2}{2} + \partial_\tau(\delta f'' - \delta f''.\delta f' + \delta f) + O_3
\end{aligned}
\tag{17.158}
$$

$$
\begin{aligned}
S_I^\beta &= -\frac{2\pi C}{\beta J} \int_{-\pi}^{+\pi} d\tau L_f(\tau) \\
&= -\frac{2\pi^2 C}{\beta J} + \frac{1}{2}\frac{2\pi C}{\beta J}\sum_m \delta f(m)(m^4 - m^2)\delta f(-m).
\end{aligned}
$$

From this Gaussian approximation we can obtain immediately the propagator of the fluctuation field $\delta f(m)$. This is given by

$$\langle \delta f(m)\delta f(n)\rangle_I = \frac{1}{N}\frac{\beta J}{2\pi C}\frac{1}{m^4 - m^2}\delta_{m,-n}, \quad m \neq 0, +1, -1. \tag{17.159}$$

The coefficient $1/N$ is due to the factor of N in front of the action S_I. The excluded modes $m = 0, \pm 1$ are non-physical due to the residual $SL(2, \mathbf{R})$ gauge symmetry of the Schwarzian action.

The propagator of the Schwarzian field $\delta L_f(\tau) = L_f(\tau) - \frac{1}{2}$ can then be computed in terms of the fluctuation fields $\delta f(m)$ as follows

$$\delta L_f(\tau) = L_f(\tau) - \frac{1}{2} = = \frac{i}{\sqrt{2\pi}}\sum_{m\in\mathbf{Z}} \delta f(m)(m - m^3)e^{im\tau} + O((\delta f)^2)$$

$$\Rightarrow \langle \delta L_f(\tau)\delta L_f(\tau')\rangle_I = -\frac{1}{2\pi}\sum_{m,n}(m - m^3)(n - n^3)e^{im\tau + in\tau'}\langle \delta f(m)\delta f(n)\rangle_I$$

$$= \frac{1}{2\pi}\frac{1}{N}\frac{\beta J}{2\pi C}\sum_{m\neq 0}(m^2 - 1)e^{im(\tau - \tau')} \tag{17.160}$$

$$= \frac{1}{2\pi}\frac{1}{N}\frac{\beta J}{2\pi C}[1 - 2\pi\delta(\tau - \tau') - 2\pi\delta''(\tau - \tau')].$$

The contribution to the four-point function due to the soft modes is then given by (recall that S_{CFT} takes the same value in the conformally invariant fluctuations)

$$\begin{aligned}
F(\tau_1, \tau_2, \tau_3, \tau_4) &= \frac{1}{N^2}\sum_{i,j}\langle T\chi_i(\tau_1)\chi_i(\tau_2)\chi_j(\tau_3)\chi_j(\tau_4)\rangle \\
&= \frac{1}{Z}\int \mathcal{D}G\mathcal{D}\Sigma \left(G(\tau_{12})G(\tau_{34}) + \frac{1}{N}G(\tau_{14})G(\tau_{23})\right. \\
&\quad \left. - \frac{1}{N}G(\tau_{13})G(\tau_{24})\right)\exp(-NS_{\mathrm{eff}}) \\
&= F_{\mathrm{disc}}(\tau_1, \tau_2, \tau_3, \tau_4) + \frac{1}{Z}\int \mathcal{D}G_f^\beta\mathcal{D}\Sigma \left(G_f^\beta(\tau_{12})\right. \\
&\quad \left. G_f^\beta(\tau_{34}) - G_c^\beta(\tau_{12})G_c^\beta(\tau_{34})\right)\exp(-NS_{\mathrm{eff}}) \\
&= F_{\mathrm{disc}}(\tau_1, \tau_2, \tau_3, \tau_4) + \frac{1}{Z}\int \mathcal{D}\delta G_f^\parallel\mathcal{D}\Sigma \left(\delta G_f^\parallel(\tau_{12})\right. \\
&\quad \left. \delta G_f^\parallel(\tau_{34})\right)\exp(-NS_{\mathrm{CFT}} - NS_I) \\
&= F_{\mathrm{disc}}(\tau_1, \tau_2, \tau_3, \tau_4) + \frac{1}{Z_I}\int \mathcal{D}f \left(\delta G_f^\parallel(\tau_{12})\right. \\
&\quad \left. \delta G_f^\parallel(\tau_{34})\right)e^{-NS_I[f]} \\
&= F_{\mathrm{disc}}(\tau_1, \tau_2, \tau_3, \tau_4) + \langle \delta G_f^\parallel(\tau_{12})\delta G_f^\parallel(\tau_{34})\rangle_I.
\end{aligned} \tag{17.161}$$

The leading connected contribution to the four-point function is thus given by the propagator of the conformally invariant fluctuation field $\delta G_f^{\parallel}$ which, as it turns out, can be expressed in terms of the propagator of the Schwarzian field $\delta L_f(\tau) = L_f(\tau) - \frac{1}{2}$ in the Gaussian approximation given by (17.160). This is done in great pedagogical detail in [13]. More explicitly, we have (with $s_{ij} = 2\sin(\tau_i - \tau_j)/2$)

$$
\begin{aligned}
\langle \delta G_f^{\parallel}(\tau_{12}) \delta G_f^{\parallel}(\tau_{34}) \rangle_I &= \frac{1}{16} G_c^{\beta}(\tau_{12}) G_c^{\beta}(\tau_{34}) \int_{\tau_2}^{\tau_1} d\tau_5 \frac{s_{15}s_{52}}{s_{12}} \int_{\tau_4}^{\tau_3} d\tau_6 \\
&\quad \frac{s_{36}s_{64}}{s_{32}} \langle \delta L_f(\tau_5) \delta L_f(\tau_6) \rangle_I \\
&= \frac{1}{16} \frac{1}{2\pi} \frac{1}{N} \frac{\beta J}{2\pi C} G_c^{\beta}(\tau_{12}) G_c^{\beta}(\tau_{34}) \int_{\tau_2}^{\tau_1} d\tau_5 \frac{s_{15}s_{52}}{s_{12}} \qquad (17.162) \\
&\quad \int_{\tau_4}^{\tau_3} d\tau_6 \frac{s_{36}s_{64}}{s_{32}} \Big[1 - 2\pi\delta(\tau_5 - \tau_6) \\
&\quad - 2\pi\delta''(\tau_5 - \tau_6) \Big].
\end{aligned}
$$

Here, we have two cases of interests.

Operator Product Expansion: The first case of interest is the operator product expansion corresponding to the time ordering

$$\tau_1 > \tau_2 > \tau_3 > \tau_4. \qquad (17.163)$$

In this case the delta functions in (17.162) give zero contribution since the integration intervals over τ_5 and τ_6 decouple. In this case we get the correction (with $\mathcal{N} = \frac{1}{16} \frac{1}{2\pi} \frac{1}{N} \frac{\beta J}{2\pi C}$)

$$
\begin{aligned}
\frac{\langle \delta G_f^{\parallel}(\tau_{12}) \delta G_f^{\parallel}(\tau_{34}) \rangle_I}{G_c^{\beta}(\tau_{12}) G_c^{\beta}(\tau_{34})} &= \mathcal{N} \int_{\tau_2}^{\tau_1} d\tau_5 \frac{s_{15}s_{52}}{s_{12}} \int_{\tau_4}^{\tau_3} d\tau_6 \frac{s_{36}s_{64}}{s_{32}} \\
&= \mathcal{N} \int_{\tau_2}^{\tau_1} d\tau_5 \frac{\cos\left(\dfrac{\tau_1 + \tau_2}{2} - \tau_5\right) - \cos\dfrac{\tau_1 - \tau_2}{2}}{\sin\dfrac{\tau_1 - \tau_2}{2}} \\
&\quad \int_{\tau_4}^{\tau_3} d\tau_6 \frac{\cos\left(\dfrac{\tau_3 + \tau_4}{2} - \tau_6\right) - \cos\dfrac{\tau_3 - \tau_4}{2}}{\sin\dfrac{\tau_3 - \tau_4}{2}} \qquad (17.164) \\
&= 4\mathcal{N} \left(\frac{\tau_{12}}{2\tan\dfrac{\tau_{12}}{2}} - 1 \right) \left(\frac{\tau_{34}}{2\tan\dfrac{\tau_{34}}{2}} - 1 \right).
\end{aligned}
$$

The time-ordered correlation (TOC) function is then given by

$$F(\tau_1, \tau_2, \tau_3, \tau_4) = \left[1 + 4\mathcal{N}\left(\frac{\tau_{12}}{2\tan\frac{\tau_{12}}{2}} - 1\right)\left(\frac{\tau_{34}}{2\tan\frac{\tau_{34}}{2}} - 1\right)\right]$$
$$G_c^\beta(\tau_{12})G_c^\beta(\tau_{34}) + F_0(\tau_1, \tau_2, \tau_3, \tau_4)$$
$$F_0(\tau_1, \tau_2, \tau_3, \tau_4) = \frac{1}{N}G_c^\beta(\tau_{14})G_c^\beta(\tau_{23}) - \frac{1}{N}G_c^\beta(\tau_{13})G_c^\beta(\tau_{24}).$$

$$(17.165)$$

Recall that the times appearing in this equation are actually $\bar{\tau} = \frac{2\pi}{\beta}\tau$, i.e., we have to re-scale back the temperature on the time circle by $\tau \longrightarrow \frac{2\pi}{\beta}\tau$. We choose now the times such that $\tau_{12} = \pi \longrightarrow \frac{\beta}{2}$ and $\tau_{34} = \pi \longrightarrow \frac{\beta}{2}$. More explicitly, we choose

$$\tau_1 = \pi + \Delta\tau \longrightarrow \frac{\beta}{2} + \Delta\tau = \frac{\beta}{2} + it$$
$$\tau_2 = \Delta\tau \longrightarrow it$$
$$\tau_3 = 0 \longrightarrow 0$$
$$\tau_4 = -\pi \longrightarrow -\frac{\beta}{2}.$$

$$(17.166)$$

We have also Wick-rotated back to Lorentzian signature, viz $t = -i\Delta\tau$. The time-ordered correlation (TOC) function becomes

$$\mathrm{TOC}(t) \equiv F\left(\frac{\beta}{2} + it, it, 0, -\frac{\beta}{2}\right) = (1 + 4\mathcal{N})G_c^\beta\left(\frac{\beta}{2}\right)G_c^\beta\left(\frac{\beta}{2}\right) + F_0(\tau_1, \tau_2, \tau_3, \tau_4). \quad (17.167)$$

We need also to Wick-rotate the conformal projector to Lorentzian times $t = -i\tau$ ($\tau > 0$) as follows

$$G_c^\beta(\tau) = \frac{\pi^{1/4}}{(2\beta J)^{1/2}}\frac{\mathrm{sign}(\tau)}{|\sin\frac{\pi\tau}{\beta}|^{2\Delta}} \longrightarrow \frac{\pi^{1/4}}{(2\beta J)^{1/2}}\frac{1}{|\sinh\frac{\pi t}{\beta}|^{2\Delta}} = \frac{\pi^{1/4}}{(2\beta J)^{1/2}}e^{-\frac{2\pi\Delta}{\beta}t}. \quad (17.168)$$

We get the final result for the time-ordered correlation (TOC) function to be given by

$$\mathrm{TOC}(t) \equiv F\left(\frac{\beta}{2} + it, it, 0, -\frac{\beta}{2}\right) = \left(1 + \frac{c_0}{N}\right)\frac{\sqrt{\pi}}{2\beta J}. \quad (17.169)$$

Out-of-time Ordering: There is also the possibility of the out-of-time ordering corresponding to

$$\tau_1 > \tau_3 > \tau_2 > \tau_4. \quad (17.170)$$

In this case the delta functions in (17.162) give non-zero contribution since the integration intervals over τ_5 and τ_6 overlap. Indeed, there is an extra contribution compared to the previous case, which comes from the first delta function and which is given for $\tau_1 - \tau_2 = \pi$ and $\tau_3 - \tau_4 = \pi$ by the formula (with $\Delta\tau = \frac{\tau_1+\tau_2}{2} - \frac{\tau_3-\tau_4}{2}$)

$$\mathcal{N}(-2\pi)\int_{\tau_2}^{\tau_1} d\tau_5 \frac{\cos\left(\dfrac{\tau_1+\tau_2}{2} - \tau_5\right) - \cos\dfrac{\tau_1-\tau_2}{2}}{\sin\dfrac{\tau_1-\tau_2}{2}}$$

$$\frac{\cos\left(\dfrac{\tau_3+\tau_4}{2} - \tau_5\right) - \cos\dfrac{\tau_3-\tau_4}{2}}{\sin\dfrac{\tau_3-\tau_4}{2}} = \tag{17.171}$$

$$\mathcal{N}(-2\pi)\int_{\tau_2}^{\tau_1} d\tau_5 \cos\left(\frac{\tau_1+\tau_2}{2} - \tau_5\right)\cos\left(\frac{\tau_3+\tau_4}{2} - \tau_5\right) =$$

$$\mathcal{N}(-2\pi)\left(\frac{\pi}{2}\cos\Delta\tau\right).$$

In this case we get then correction

$$\begin{aligned}
\frac{\langle \delta G_f^{\parallel}(\tau_{12})\delta G_f^{\parallel}(\tau_{34})\rangle_I}{G_c^{\beta}(\tau_{12})G_c^{\beta}(\tau_{34})} &= 4\mathcal{N} + \mathcal{N}(-2\pi)\left(\frac{\pi}{2}\cos\Delta\tau\right) \\
&= 4\mathcal{N}\left(1 - \left(\frac{\pi}{2}\right)^2 \cos\Delta\tau\right) \\
&\longrightarrow 4\mathcal{N}\left(1 - \left(\frac{\pi}{2}\right)^2 \cos\frac{2\pi\Delta\tau}{\beta}\right) \\
&= 4\mathcal{N}\left(1 - \left(\frac{\pi}{2}\right)^2 \cos\frac{2\pi it}{\beta}\right) \\
&= -2\mathcal{N}\left(\frac{\pi}{2}\right)^2 e^{\frac{2\pi t}{\beta}} \\
&= -\frac{\Delta^2\beta J}{8NC}e^{\frac{2\pi t}{\beta}}.
\end{aligned} \tag{17.172}$$

We choose the times

$$\begin{aligned}
\tau_1 &= \frac{\pi}{2} + \Delta\tau \longrightarrow \frac{\beta}{4} + \Delta\tau = \frac{\beta}{4} + it \\
\tau_2 &= -\frac{\pi}{2} + \Delta\tau \longrightarrow -\frac{\beta}{4} + \Delta\tau = -\frac{\beta}{4} + it \\
\tau_3 &= 0 \longrightarrow 0 \\
\tau_4 &= -\pi \longrightarrow -\frac{\beta}{2}.
\end{aligned} \tag{17.173}$$

We get then the final result for the out-of-time-ordered correlation (OTOC) function to be given by

$$\text{OTOC}(t) \equiv F\left(\frac{\beta}{4} + it, -\frac{\beta}{4} + it, 0, -\frac{\beta}{2}\right) = \left(1 - \frac{\Delta^2 \beta J}{8NC} e^{\frac{2\pi t}{\beta}}\right)\frac{\sqrt{\pi}}{2\beta J}. \quad (17.174)$$

17.5.2 Conformal spectrum and bound on chaos

The other correction to the four-point function which is of order $1/N$ will correspond to fluctuations which are perpendicular to the conformally invariant space of configurations. We consider then

$$G_f^\beta(\tau_1, \tau_2) = G_c^\beta(\tau_1, \tau_2) + \delta G_f^\parallel(\tau_1, \tau_2) + \delta G_f^\perp(\tau_1, \tau_2). \quad (17.175)$$

As before, $\delta G_f^\parallel$ is a conformally invariant fluctuation, i.e., these fluctuations are the zero modes of the operator $K - 1$, and hence the variation of the effective action around the above fluctuations δS_{eff}, and thus the effective action S_{eff} itself, will only depend on $\delta G_f^\perp$. The operator K is given by equation (17.134).

The four-point function will then take the form

$$
\begin{aligned}
F(\tau_1, \tau_2, \tau_3, \tau_4) &= \frac{1}{N^2}\sum_{i,j}\langle T\chi_i(\tau_1)\chi_i(\tau_2)\chi_j(\tau_3)\chi_j(\tau_4)\rangle \\
&= \frac{1}{Z}\int \mathcal{D}G\mathcal{D}\Sigma \left(G(\tau_{12})G(\tau_{34}) + \frac{1}{N}G(\tau_{14})G(\tau_{23})\right. \\
&\qquad\left. - \frac{1}{N}G(\tau_{13})G(\tau_{24})\right)\exp(-NS_{\text{eff}}) \\
&= G_c^\beta(\tau_{12})G_c^\beta(\tau_{34}) + \langle\delta G_f^\parallel(\tau_{12})\delta G_f^\parallel(\tau_{34})\rangle_I \\
&\qquad + \frac{1}{N}\mathcal{F}(\tau_1, \tau_2, \tau_3, \tau_4) + \cdots
\end{aligned}
\quad (17.176)
$$

where

$$
\begin{aligned}
\frac{1}{N}\mathcal{F}(\tau_1, \tau_2, \tau_3, \tau_4) &= \frac{1}{N}\sum_{n=0}\mathcal{F}_n(\tau_1, \tau_2, \tau_3, \tau_4) \\
&= \frac{1}{Z_\perp}\int \mathcal{D}\delta G_f^\perp \left(\delta G_f^\perp(\tau_{12})\delta G_f^\perp(\tau_{34})\right)\exp(-NS_{\text{eff}}[\delta G_f^\perp]).
\end{aligned}
\quad (17.177)
$$

The first term in $\mathcal{F}$ is the disconnected piece given by

$$\mathcal{F}_0(\tau_1, \tau_2, \tau_3, \tau_4) = G_c^\beta(\tau_{14})G_c^\beta(\tau_{23}) - G_c^\beta(\tau_{13})G_c^\beta(\tau_{24}). \quad (17.178)$$

The next contributions are connected pieces given by the so-called ladder diagrams shown in figure 17.2. These diagrams can be summed up in a geometric series to give the very simple result [5]

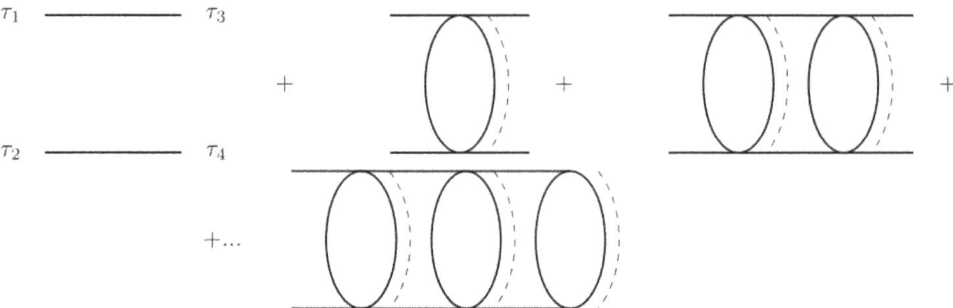

Figure 17.2. The ladder diagrams contributing to the four-point function of the SYK model.

$$\mathcal{F}(\tau_1, \tau_2, \tau_3, \tau_4) = \sum_{n=0} \mathcal{F}_n(\tau_1, \tau_2, \tau_3, \tau_4)$$

$$= \sum_{n=0} K^n \mathcal{F}_0(\tau_1, \tau_2, \tau_3, \tau_4) \tag{17.179}$$

$$= \frac{1}{1 - K} \mathcal{F}_0(\tau_1, \tau_2, \tau_3, \tau_4).$$

Here, K is precisely given by the conformal kernel (17.134).

This correction to the four-point function is thus determined in terms of the conformal spectrum of the operator K. By conformal invariance, the conformal kernel K commutes with the Casimir operator C of $SL(2, R)$. Thus, we can use the complete set of common eigenstates of C and K as a basis. These common eigenstates are denoted by $|\Psi_h\rangle$. We have then

$$C|\Psi_h\rangle = h(h - 1)|\Psi_h\rangle, \quad K|\Psi_h\rangle = k(h)|\Psi_h\rangle. \tag{17.180}$$

Explicitly, we find [4]

$$k(h) = -\frac{3}{2} \frac{\tan \frac{\pi}{2}(h - \frac{1}{2})}{h - \frac{1}{2}}. \tag{17.181}$$

Clearly, the mode of K with eigenvalue equal $k(h_0) = 1$ or $h_0 = 2$ does not contribute as ladder diagrams to the four-point function. These modes are precisely the soft mode captured by the Schwarzian action which come with a conformal enhancement. However, there are other solutions of the equation $k(h_m) = 1$ with $m > 0$ which are not eigenvalues of K. These are given explicitly by

$$k(h_m) = 1 \Rightarrow h_m = 2\Delta + 2m + 1 + \epsilon_m, \quad \epsilon_m \simeq \frac{3}{2\pi m}, \quad m \gg 1. \tag{17.182}$$

In summary, the modes h_m with $m > 0$ (which are not part of the spectrum of K) appear as simple poles in the expression of $\mathcal{F}$ as opposed to the zero mode $h_0 = 2$ of

K which is excluded. Indeed, the conformal correction to the four-point function is explicitly given by

$$\mathcal{F}(\tau_1, \tau_2, \tau_3, \tau_4) = \frac{4}{3} G_c^\beta(\tau_{12}) G_c^\beta(\tau_{34}) \tilde{F}(\chi)$$

$$\tilde{F}(\chi) = \int_C \frac{dh}{2\pi i} \left[\frac{k(h)}{1 - k(h)} \frac{h - \frac{1}{2}}{\tan \frac{h\pi}{2}} \Psi_h(\chi) \right]_{h \neq 2} . \tag{17.183}$$

The $SL(2, R)$-harmonics $\psi_h(\chi)$ depend only on the finite-temperature cross ratio χ defined by

$$\chi = \frac{\sin \frac{\pi \tau_{12}}{\beta} \sin \frac{\pi \tau_{34}}{\beta}}{\sin \frac{\pi \tau_{13}}{\beta} \sin \frac{\pi \tau_{24}}{\beta}} \longrightarrow \chi = \frac{\tau_{12} \tau_{34}}{\tau_{13} \tau_{24}}, \quad \beta \longrightarrow \infty. \tag{17.184}$$

Thus, this conformal correction to the four-point function is dominated by the modes h_m, with $m > 0$, given by equation (17.182). In fact, h_m furnish the conformal weights (dimensions) of the operators appearing in the operator product expansion (OPE) of the four-point function in the limit $\chi \longrightarrow 0$. This can be viewed as two fermion fields with some interaction.

The contribution of the conformal correction to the time-ordered and out-of-time-ordered four-point functions can be computed by following the same steps as in the previous section. We find similar expressions to (17.169) and (17.174) as shown simply and clearly in [13]. However, these conformal contributions, although they are at the same order in $1/N$, are in fact suppressed by the factor $1/\beta J$ compared to the contributions of the Schwarzian soft modes.

The time-ordered and out-of-time-ordered correlation functions are then given by

$$\text{TOC}(t) \equiv F\left(\frac{\beta}{2} + it, it, 0, -\frac{\beta}{2}\right) = \frac{\sqrt{\pi}}{2\beta J}\left(1 + \frac{c_0'}{N}\right). \tag{17.185}$$

$$\text{OTOC}(t) \equiv F\left(\frac{\beta}{4} + it, -\frac{\beta}{4} + it, 0, -\frac{\beta}{2}\right) = \frac{\sqrt{\pi}}{2\beta J}\left(1 - \frac{c_1 \beta J}{N} e^{\kappa t}\right). \tag{17.186}$$

Where (recalling also that $C = 0.48\Delta/12$ from (17.146))

$$c_1 = \frac{\Delta^2}{8C}\left(1 - O\left(\frac{1}{\beta J}\right)\right) \tag{17.187}$$

$$\kappa = \frac{2\pi}{\beta}\left(1 - \frac{6.05}{\beta J} + \cdots\right) < \frac{2\pi}{\beta}. \tag{17.188}$$

The above results are only valid for larger times compared to the dissipation time $t_d = \beta$, which are much smaller than the scrambling time $t_* = \beta \log \frac{N}{\beta J}$, viz $\beta \ll t \ll t_*$. In this regime the out-of-time-ordered four-point function decays approaching 0 at greater times. In the above equations the number κ is the quantum Lyapunov exponent and it satisfies the bound on chaos [12] given in equation (17.188). Here, scrambling (which is the exponential decay of OTOC) is considered the same as chaos.

References

[1] Tong D Introducing conformal field theory https://www.damtp.cam.ac.uk/user/tong/string/string4.pdf

[2] Sachdev S and Ye J 1993 Gapless spin-fluid ground state in a random quantum Heisenberg magnet *Phys. Rev. Lett.* **70** 3339

[3] Kitaev A 2015 A simple model of quantum holography KITP strings seminar and Entanglement 2015 program (Feb.12, April 7, and May 27). http://online.kitp.ucsb.edu/online/entangled15/

[4] Polchinski J and Rosenhaus V 2016 The spectrum in the Sachdev-Ye-Kitaev model *JHEP* **04** 001

[5] Maldacena J and Stanford D 2016 Remarks on the Sachdev-Ye-Kitaev model *Phys. Rev. D* **94** 106002

[6] Kitaev A and Suh S J 2018 The soft mode in the Sachdev-Ye-Kitaev model and its gravity dual *JHEP* **05** 183

[7] Jevicki A and Suzuki K 2016 Bi-local holography in the SYK model: perturbations *JHEP* **11** 046

[8] Sachdev S 2010 Holographic metals and the fractionalized fermi liquid *Phys. Rev. Lett.* **105** 151602

[9] Sachdev S 2010 Strange metals and the AdS/CFT correspondence *J. Stat. Mech.* **1011** P11022

[10] Shenker S H and Stanford D 2014 Black holes and the butterfly effect *JHEP* **03** 067

[11] Shenker S H and Stanford D 2015 Stringy effects in scrambling *JHEP* **05** 132

[12] Maldacena J, Shenker S H and Stanford D 2016 A bound on chaos *JHEP* **08** 106

[13] Trunin D A 2021 Pedagogical introduction to the Sachdev-Ye-Kitaev model and two-dimensional dilaton gravity *Usp. Fiz. Nauk.* **191** 225–61

[14] Rosenhaus V 2019 An introduction to the SYK model *J. Phys. A* **52** 323001

[15] nlab, Sachdev-Ye-Kitaev model in nlab https://ncatlab.org/nlab/show/Sachdev-Ye-Kitaev+model

IOP Publishing

A Modern Course in Quantum Field Theory, Volume 2 (Second Edition)
Advanced topics
Badis Ydri

Chapter 18

Novel phenomena in noncommutative field theory: emergent geometry

Noncommutative field theory (NCFT) [1, 2] is an extension of quantum field theory (QFT) that redefines spacetime, replacing commuting coordinates with a noncommutative structure. This shift fundamentally alters the way fields, interactions, and symmetries are understood. NCFT uniquely integrates with supersymmetry, making it a natural framework for unifying quantum mechanics and gravity. It also provides a consistent mechanism for spontaneous supersymmetry breaking.

Unlike conventional QFT, which quantizes fields on a fixed spacetime, NCFT begins by quantizing spacetime itself. This perspective reveals novel phenomena, such as ultraviolet-infrared mixing and a natural transition from discrete to continuous geometries. It offers insights into quantum gravity at the Planck scale.

Mathematically, NCFT bridges quantum mechanics and geometry through operator algebras, enabling the exploration of new theories unattainable in traditional frameworks. This dual role cements NCFT as a cornerstone of modern theoretical physics.

18.1 Introduction

Noncommutative field theory is quantum field theory defined on a noncommutative spacetime which is primarily characterized by two fundamental features:

1. A Planck-like length scale (noncommutativity parameter).
2. A discrete geometry which becomes continuous at large distances (commutative limit).

Noncommutative field theory and the underlying noncommutative spacetime require for their description the language of noncommutative geometry [3, 4].

doi:10.1088/978-0-7503-5834-7ch18

In this context the geometry is in a precise sense emergent (here from operator algebras) given in terms of a spectral triple $(\mathcal{A}, \Delta, \mathcal{H})$ rather than in terms of a set of points. The algebra $\mathcal{A}$ is typically an operator operator represented on the Hilbert space $\mathcal{H}$ whereas Δ is the Laplace operator, which defines the metric aspects of the geometry. We really should think of the algebra $\mathcal{A}$ as defining the topology whereas the Laplacian Δ defines the metric. In the non-perturbative formulation of the spectral triple $(\mathcal{A}, \Delta, \mathcal{H})$ as a path or functional integral we require the language of matrix models.

A noncommutative scalar field theory is given by an action functional, which can be split in the usual fashion into a kinetic part and a potential term. This action functional is defined non-perturbatively by means of a matrix/operator model (after proper regularization and Euclidean rotation). The potential term represents the algebraic structure, i.e., the algebra, the inner product, the Hilbert space, etc., and as such it defines in some sense the topological aspects of the noncommutative space. The kinetic term on the other hand, as it involves a Laplacian operator Δ, plays a pivotal role in defining the metric aspects of the geometry following Fröhlich and Gawędzki [5] (or Connes [3] for spin geometry).

Thus, a noncommutative scalar field theory on some underlying noncommutative spacetime (assumed to be a Groenewold–Moyal–Weyl spacetime [6–8] for concreteness) is given generically by a matrix model of the form (with a single Hermitian scalar field $\Phi = \Phi^\dagger \in \mathcal{A}$)

$$S = a \operatorname{Tr} \Phi \Delta \Phi + \operatorname{Tr} V(\Phi), \quad \Delta(\dots) = \alpha[X_a, [X_a,\dots]]. \tag{18.1}$$

The trace Tr is over an infinite dimensional Hilbert space $\mathcal{H}$. We can generally regularize the Moyal–Weyl spacetime using noncommutative tori [9, 10] or fuzzy spheres [11, 12].

The operators X_a entering the definition of the Laplacian operator Δ are the coordinate operators, which define operationally the noncommutative spacetime. For example, the Moyal–Weyl spacetime is precisely defined in terms of the operators X_a by the Heisenberg relations

$$[X_a, X_b] = i\Theta_{ab}, \quad \Theta_{ab} = \text{constant}. \tag{18.2}$$

The Groenewold–Moyal–Weyl spacetime is the most important noncommutative spacetime as it plays for Poisson manifolds (via Darboux theorem) the same role played by flat Minkowski spacetime for curved manifolds, i.e. Darboux theorem is effectively the equivalence principle in this case as noted for example in [13, 14].

The most important interaction term is given by a quartic phi-four potential, viz

$$V(\phi) = b\Phi^2 + c\Phi^4. \tag{18.3}$$

The phase structure of the matrix model (18.1), with the potential (18.3), was calculated non-perturbatively mostly in two dimensions on the fuzzy sphere (but also on the noncommutative torus) using the Monte Carlo method [15–18]. It is believed that the phase diagram of noncommutative phi-four theory in any dimension and on any noncommutative background is identical in features to the phase diagram of the potential (18.3) on the fuzzy sphere.

The first result here is the emergence of a novel phase, termed the stripe or non-uniform-ordered phase, in addition to the usual commutative phases (disordered and uniform-ordered phases), and hence the emergence of a triple point in the phase diagram where the three vacuum solutions coexist. The non-uniform-ordered phase might also be called the matrix phase as it is generated in the real quartic matrix model (18.3) alone.

The rich phase structure characterizing a noncommutative scalar field theory is of great implications for both particle physics, e.g., spontaneous symmetry breaking and condensed matter systems, e.g., exotic phases of matter and quantum phase transitions.

A far more important result is the emergence of geometry itself in the multitrace matrix models, which were originally proposed as approximations of noncommutative scalar field theory on the fuzzy sphere [19, 20]. These models fall in the 'universality class' of the real quartic matrix model (18.3) and as such their phase structure is only characterized by disordered and non-uniform-ordered phases.

A priori, a generic multitrace matrix model does not require for its definition a spectral triple and thus it does not entail any geometric content. However, by tuning the parameters of the model appropriately, a uniform-ordered phase might emerge signaling the emergence of an underlying geometry. The associated critical exponents can then be used to determine its dimension while the behavior of the Wigner's semicircle law near the origin can be used to determine its metric. Furthermore, by expanding around the uniform-ordered phase, we get a noncommutative gauge theory on the emergent space.

Thus, in the context of noncommutative scalar field theories, the emergence of the non-uniform-ordered stripe phase is important from the perspective of quantum field theory but the emergence of the uniform-ordered phase, in their multitrace matrix models, is important from the perspective of quantum gravity.

The phenomena of emergent geometry does also occur in the Yang–Mills matrix models associated with noncommutative gauge theories. The noncommutative Moyal–Weyl space (18.2) should be though of as a background solution of some Yang–Mills matrix model. First, we start from the operator model given by

$$S = \frac{\sqrt{\Theta^D \det(\pi B)}}{2g^2} \operatorname{Tr}_{\mathcal{H}} \sum_{a,b=1}^{D} \left(i[D_a, D_b] - \frac{1}{\Theta} B_{ab}^{-1} \right)^2. \tag{18.4}$$

Here, the degrees of freedom are given by the connections or covariant derivatives D_a, which are Hermitian operators. The noncommutativity parameter Θ is of dimension length-squared so that the connection operators D_a are of dimension (length)$^{-1}$. The gauge coupling constant g is of dimension (mass)$^{2-\frac{D}{2}}$ while B^{-1} is a dimensionless invertible tensor, which in two dimensions is given by $B_{ij}^{-1} = \varepsilon_{ij}^{-1} = -\varepsilon_{ij}$ while in higher even dimensions is given by (using Darboux theorem)

$$B_{ij}^{-1} = \begin{pmatrix} -\varepsilon_{ij} & & \\ & \ddots & \\ & & -\varepsilon_{ij} \end{pmatrix}. \tag{18.5}$$

The Moyal–Weyl space (18.2) correspond to the classical configurations $D_a = -B_{ab}^{-1}X_b$ with $\Theta_{ab} = \Theta B_{ab}$. Then, by computing the effective potential in the configuration $D_a = -\varphi B_{ab}^{-1}X_b$ we verify that the Moyal–Weyl space itself, i.e., the Heisenberg algebra (18.2) ceases to exist above a certain value g_* of the gauge coupling constant. Conversely, by moving in the phase diagram from strong coupling to weak coupling the geometry of the Moyal–Weyl spaces (star product and the Weyl map) emerges at the critical point g_*.

This discussion of emergent geometry can be made more precise by considering a natural regularization of the Moyal–Weyl space, which involves the matrix algebra Mat_N of $N \times N$ Hermitian matrices. In other words, we must regularize the action (18.4) by i) replacing the operators D_a by $N \times N$ Hermitian matrices and ii) replacing the trace $\mathrm{Tr}_{\mathcal{H}}$ by the ordinary trace Tr on the finite-dimensional Hilbert space associated with the matrix algebra Mat_N.

The regularization of choice in this work is given by fuzzy projective spaces [21]. In particular, the fuzzy sphere will regularize effectively all noncommutative Moyal–Weyl spaces as it is evident from the symplectic structure (18.5), i.e., each two-dimensional block/plane is regularized by a fuzzy sphere and we get therefore a Cartesian product of fuzzy spheres approximating even-dimensional Moyal–Weyl spaces.

In this regularization, the Heisenberg algebra (18.2) in two dimensions will be regularized by the $SU(2)$ angular momentum algebra in the spin $s \equiv (N-1)/2$ irreducible representation given by $[L_i, L_j] = i\varepsilon_{ijk}L_k, \sum_i L_i^2 = c_2, c_2 = (N^2 - 1)/2$. This gives immediately the fuzzy sphere, which is defined by three $N \times N$ matrices $\hat{x}_i = RL_i/\sqrt{c_2}$, playing the role of coordinates operators and satisfying commutation relations and embedding constraint given by

$$[\hat{x}_i, \hat{x}_j] = i\Theta \varepsilon_{ijk} \hat{x}_k, \quad \sum_i \hat{x}_i^2 = R^2, \quad \Theta = \frac{R}{\sqrt{c_2}}. \tag{18.6}$$

The Yang–Mills action (18.4) in two dimensions will then be regularized by a genuine $D = 3$ Yang–Mills matrix model of the IKKT-type [22] given explicitly by

$$S = \frac{1}{g^2 N} \mathrm{Tr} \sum_{i,j=1}^{3} \left(i[D_i, D_j] + \sum_{k=1}^{3} \varepsilon_{ijk} D_k \right)^2. \tag{18.7}$$

This is the basic model whose degrees of freedom are $N \times N$ Hermitian matrices D_i, which should be interpreted as covariant matrix coordinates. This model can be immediately generalized by adding mass deformation terms in the matrices D_i. It can also be generalized to a supersymmetric $D = 3$ Yang–Mills matrix model, which is the fundamental $SO(3)$symmetric theory with Euclidean action given by

$$S = \frac{1}{g^2} \, \mathrm{Tr} \left[-\frac{1}{4} [X_A, X_B]^2 - \frac{1}{2} \bar{\Psi} \gamma^A [X_A, \Psi] \right]. \tag{18.8}$$

This supersymmetric model can also be defined in $D = 10$, $D = 6$ and $D = 4$. In fact, the partition functions of these models are convergent only in these dimensions $D = 4, 6, 10$ [23–26, 27–29]. In $D = 3$ the partition function can be made finite by adding appropriate mass deformation, which consists of a positive quadratic term in the matrices X_A, which damps flat directions.

However, Yang–Mills matrix models in $D = 10$ and $D = 6$ have their $SO(D)$ rotational symmetry group spontaneously broken in the large N limit to $SO(3)$ [30] while the $D = 4$ case is equivalent to the $D = 3$ model coupled to a scalar field Φ [31]. In other words, we are very interested in the $D = 3$ Yang–Mills matrix model (18.7), its supersymmetric extension, generic mass deformation terms and Cartesian products thereof.

The emergent geometry transition manifests itself in the $D = 3$ Yang–Mills matrix model (18.7) as an exotic phase transition where the geometry disappears as the temperature is increased. The fuzzy sphere phase (low temperature phase where $D_i = \varphi L_i$ and $\varphi \sim 1$) has the background geometry of a two-dimensional spherical noncommutative manifold, which macroscopically becomes a standard commutative sphere for $N \longrightarrow \infty$. The fluctuations are then of a noncommutative gauge theory, which mixes with a normal scalar field on this background. In the Yang–Mills matrix phase (high temperature phase) the order parameter φ is not well defined and the fluctuations are around diagonal matrices so the model is a pure matrix one corresponding to a zero-dimensional Yang–Mills theory in the large N limit. The fuzzy sphere phase occurs for $\tilde{\alpha} > \tilde{\alpha}_*$ while the matrix phase occurs for $\tilde{\alpha} < \tilde{\alpha}_*$ where $\tilde{\alpha}^4 = 1/g^2$.

The programme of emergent geometry from Yang–Mills matrix models revolves therefore around the two hypotheses:

1. The existence of a noncommutative structure with a Planck-like length scale corresponding to a discrete geometry. In other words, noncommutative geometry is thought of as 'first quantization' of the geometry.
2. The existence of a Yang–Mills matrix action functional where the non-commutative structure appears as its classical background solution. The Feynman path integral associated with this action defines therefore 'second quantization' of the geometry where the noncommutative structure becomes dynamical, i.e., this structure can emerge/evaporates from/into a pure matrix model as some parameters in the model are varied.

Emergent geometry in this programme is therefore defined as 'quantum non-commutative geometry'.

As it turns out, emergent geometry in the scalar multitrace matrix models is more fundamental than emergent geometry in the gauge Yang–Mills matrix model. This is simply because a full-blown $D = 3$ Yang–Mills matrix model can emerge in the uniform-ordered phase of a cleverly-engineered multitrace matrix model.

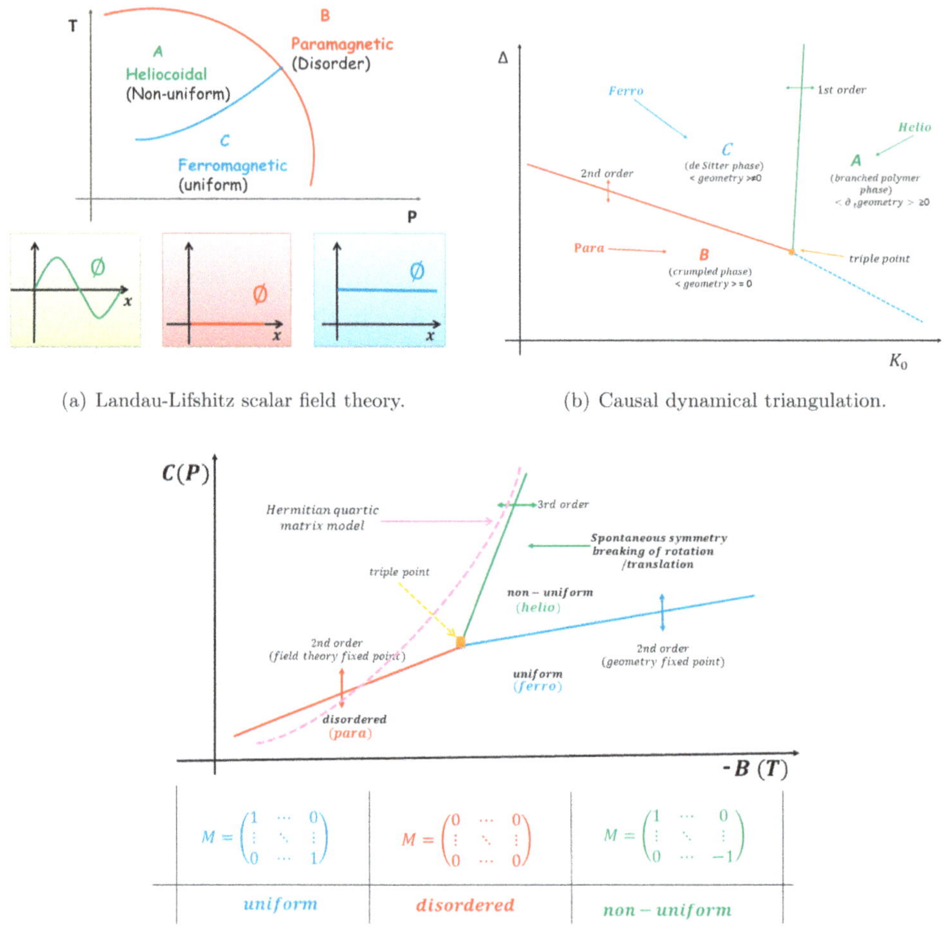

(a) Landau-Lifshitz scalar field theory.

(b) Causal dynamical triangulation.

$M = \begin{pmatrix} 1 & \cdots & 0 \\ \vdots & \ddots & \vdots \\ 0 & \cdots & 1 \end{pmatrix}$	$M = \begin{pmatrix} 0 & \cdots & 0 \\ \vdots & \ddots & \vdots \\ 0 & \cdots & 0 \end{pmatrix}$	$M = \begin{pmatrix} 1 & \cdots & 0 \\ \vdots & \ddots & \vdots \\ 0 & \cdots & -1 \end{pmatrix}$
uniform	*disordered*	*non − uniform*

(c) Multitrace matrix model.

Figure 18.1. The phase diagrams of causal dynamical triangulation, Lifshitz scalar field theory and multitrace matrix model.

This approach to emergent geometry should also be contrasted with other approaches to quantum geometry. For example, see [32–37]. In particular, causal dynamical triangulation (CDT) [38–42] and Horava-Lifshitz (HL) gravity [43–45] are very similar in structure to the multitrace matrix model approach. See figure 18.1.

The remainder of this work is organized as follows. In section 2, we discuss aspects of noncommutative phi-four theory on the Moyal–Weyl space. In section 3, we discuss aspects of noncommutative phi-four theory on the fuzzy sphere. In section 4, we discuss emergent geometry from multitrace matrix models. In section 5, we discuss emergent geometry from Yang–Mills matrix models.

18.2 Aspects of noncommutative phi-four theory on the Moyal–Weyl space

18.2.1 UV–IR mixing and stripe phase on noncommutative $\mathbf{R}_\theta^d$

Noncommutativity is naturally described in terms of operator algebras and operators can be intuitively viewed (albeit in a rough sense) as matrix algebras. It is believed that noncommutativity is the only non-trivial extension of supersymmetry and thus noncommutative field theory provides a natural framework for spontaneous supersymmetry breaking, which is central to most particle physics beyond the standard model.

The simplest (at least formally) noncommutative spacetime is the Moyal–Weyl space $\mathbf{R}_\theta^d$ [6–8]. Two nonperturbative regularizations of the Moyal–Weyl space exist. The noncommutative torus obtained via Eguchi–Kawai construction [9, 10] and the fuzzy sphere obtained via co-adjoint orbit quantization [8, 12].

The coordinates on $\mathbf{R}_\theta^d$ are operators which satisfy

$$[x_\mu, x_\nu] = i\theta_{\mu\nu}. \tag{18.9}$$

The Weyl map allows us to work with a C^*-algebra of functions where the pointwise multiplication of function is replaced by an appropriate deformed (star) product. The Moyal–Weyl star product is given by

$$f*g(x) = e^{\frac{i}{2}\sum_{i,j=1}^{d}\theta_{ij}\frac{\partial}{\partial\xi_i}\frac{\partial}{\partial\eta_j}} f(x + \xi)g(x + \eta)|_{\xi=\eta=0}. \tag{18.10}$$

The noncommutative phi-four theory is then given by the action

$$S = \int d^d x\left[(\partial_\mu\phi)^2 + \mu^2\phi^2 + \frac{\lambda}{4!}(\phi^*\phi)^2\right]. \tag{18.11}$$

This theory suffers from the UV–IR mixing problem which hampers perturbative renormalization of the theory [46]. There seems to exist a unique renormalizable solution in which the propagator is altered by the addition of a harmonic oscillator term [47]. A self-consistent Hartree–Fock analysis of the above field theory [48] reveals a complicated phase structure beyond the single critical line present in the commutative theory.

18.2.1.1 UV–IR mixing
The two-point function is given explicitly by

$$\Gamma^{(2)}(p) = p^2 + \mu^2 + 2\frac{\lambda}{4!}I_2(0, \Lambda) + \frac{\lambda}{4!}I_2(p, \Lambda). \tag{18.12}$$

The second and third terms correspond to the planar and non-planar diagrams, respectively. See figure 18.2. These contributions are given by

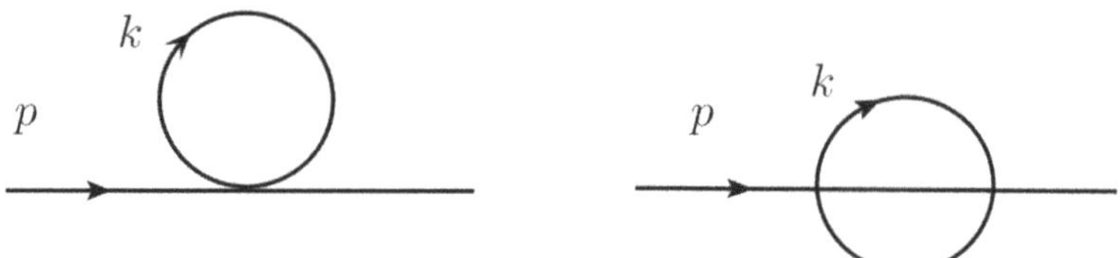

Figure 18.2. The one-loop planar and non-planar contributions.

$$I_2(p, \Lambda) = \int \frac{d^d k}{(2\pi)^d} \frac{1}{k^2 + \mu^2} \, e^{-i\theta_{ij}k_i p_j}$$

$$= \frac{1}{(4\pi)^{\frac{d}{2}}} \int_0^\infty \frac{d\alpha}{\alpha^{\frac{d}{2}}} \, e^{-\frac{1}{\alpha\Lambda^2}} \, e^{-\frac{(\theta_{ij}p_j)^2}{4\alpha} - \alpha\mu^2}$$

$$= (2\pi)^{-\frac{d}{2}} \mu^{\frac{d-2}{2}} \left(\frac{4}{\Lambda_{\text{eff}}^2}\right)^{\frac{2-d}{4}} K_{\frac{d-2}{2}}\left(\frac{2\mu}{\Lambda_{\text{eff}}}\right). \tag{18.13}$$

Here, $K_{\frac{d-2}{2}}$ is the modified Bessel function. The effective cutoff is defined by the equation

$$\frac{4}{\Lambda_{\text{eff}}^2} = \frac{4}{\Lambda^2} + (\theta_{ij}p_j)^2. \tag{18.14}$$

Four dimensions:

$$I_2(p, \Lambda) = \frac{\mu}{4\pi^2} \frac{\Lambda_{\text{eff}}}{2} K_1\left(\frac{2\mu}{\Lambda_{\text{eff}}}\right), \quad K_1(z) = \frac{1}{z} + \frac{z}{2}\ln\frac{z}{2} + \cdots \Rightarrow I_2(p, \Lambda)$$

$$= \frac{1}{16\pi^2}\left(\Lambda_{\text{eff}}^2 - \mu^2 \ln\frac{\Lambda_{\text{eff}}^2}{\mu^2} + \cdots\right). \tag{18.15}$$

Two dimensions:

$$I_2(p, \Lambda) = \frac{1}{2\pi} K_0\left(\frac{2\mu}{\Lambda_{\text{eff}}}\right), \quad K_0(z) = -\ln\frac{z}{2} + \cdots \Rightarrow I_2(p, \Lambda) = \frac{1}{4\pi}\ln\frac{\Lambda_{\text{eff}}^2}{\mu^2} + \cdots \tag{18.16}$$

In the limit $\Lambda \longrightarrow \infty$ the non-planar one-loop contribution remains finite whereas the planar one-loop contribution diverges quadratically/logarithmically as usual. However, the two-point function $\Gamma^{(2)}(p)$, which can be made finite in the limit $\Lambda \longrightarrow \infty$ through the introduction of the renormalized mass $m^2 = \mu^2 + 2\frac{\lambda}{4!}I_2(0, \Lambda)$ is singular in the limit $p \longrightarrow 0$ or $\theta \longrightarrow 0$. This is because the effective cutoff $\Lambda_{\text{eff}} = 2/|\theta_{ij}p_j|$ diverges as $p \longrightarrow 0$ or $\theta \longrightarrow 0$. This is the celebrated UV–IR mixing problem discussed originally in [46].

18.2.1.2 Stripe phase

The self-consistent 2-point proper vertex in the disordered phase is found to be given by (with m^2 being the renormalized mass and $g^2 = \lambda/4$!)

$$\Gamma^{(2)}(p) = p^2 + m^2 + g^2 \int \frac{d^d k}{(2\pi)^d} \frac{e^{-i\theta_{ij} k_i p_j}}{\Gamma^{(2)}(k)}. \tag{18.17}$$

We can infer the existence of a minimum p_c in $\Gamma^{(2)}(p)$ from the behavior at $p \longrightarrow 0$ and at $p \longrightarrow \infty$ of $\Gamma^{(2)}(p)$ given by

$$\Gamma^{(2)}(p) = p^2, \text{ for } p \text{ large.} \tag{18.18}$$

$$\Gamma^{(2)}(p) \propto \frac{g^2}{\theta^2 p^2}, \text{ for } p \text{ small.} \tag{18.19}$$

Around the minimum p_c we can write $\Gamma^{(2)}(p)$ as

$$\Gamma^{(2)}(p) = \xi_0^2 (p^2 - p_c^2)^2 + r, \text{ for } p \simeq p_c. \tag{18.20}$$

Explicitly, we find in the disordered phase the result (with $\alpha \propto g^{7/2}/\theta^{3/2}$, $\tau - 2p_c^2 + m^2$)

$$p_c = \sqrt{\frac{g}{2\pi\theta}}, \quad \Gamma^{(2)}(p_c) \equiv r = \tau + \frac{\alpha}{\sqrt{r}}, \quad \xi_0 = \frac{1}{2p_c}. \tag{18.21}$$

Next, we consider the ordered phase. Here, we must perform perturbation around a stripe configuration ϕ_0, which breaks translational invariance, i.e., we expand the scalar field as $\phi = \phi_0 + \tilde{\phi} + X$ where X is the quantum fluctuation field. The stripe configuration ϕ_0 is given explicitly by

$$\phi_0 = A \cos p_c x. \tag{18.22}$$

The scalar field is then expanded as $\phi = \phi_0 + \varphi$. The condition of vanishing tadpole graphs gives then either $A = 0$ (disordered phase) or

$$\text{Tadpole} = 0 \Leftrightarrow r = 4A^2 g^2, \quad \text{ordered phase.} \tag{18.23}$$

In this case the 2-point proper vertex is given by

$$\Gamma^{(2)}(p_c) \equiv r = \tau + \frac{\alpha}{\sqrt{r}} + 6g^2 A^2. \tag{18.24}$$

Equations (18.23) and (18.24) imply that there is an ordered solution if

$$\tau + \frac{\alpha}{\sqrt{r}} + 6g^2 A^2 = 4A^2 g^2 \Rightarrow 2\tau + \frac{2\alpha}{\sqrt{r_0}} + r_0$$

$$= 0 \Rightarrow \varepsilon \equiv -\frac{1}{\sqrt{3}} \frac{\tau^3}{\alpha^2} > \varepsilon_{*1} = \sqrt{3}\frac{9}{8}. \tag{18.25}$$

This should be contrasted with the corresponding equation in the disordered phase given by

$$\tau + \frac{\alpha}{\sqrt{r}} = r \Rightarrow \tau + \frac{\alpha}{\sqrt{r_d}} - r_d = 0. \tag{18.26}$$

However, the transition between the two phases is given by the condition that the free energy difference must vanish, i.e., this is a first order transition. We have

$$\frac{dF}{dA} = \frac{d}{d\tilde{\phi}_c}\text{Tadople} = 2A(r - 4g^2A^2)$$

$$A = \frac{1}{2g}\sqrt{\frac{2}{3}}\sqrt{r - \tau - \frac{\alpha}{\sqrt{r}}} \tag{18.27}$$

$$\Rightarrow F_d - F_o = \frac{1}{12g^2}(\sqrt{r_d} - \sqrt{r_o})(3\alpha + \tau(\sqrt{r_d} + \sqrt{r_o})).$$

We find then the transition point

$$3r_o^2 + 6r_d^2 - 4\tau(r_d - r_o) \Rightarrow \varepsilon_{*2} = c\sqrt{3}\frac{9}{8}. \tag{18.28}$$

The coefficient $c > 1$ must be determined from numeric.

The region $\varepsilon_{*1} < \varepsilon < \varepsilon_{*2}$ corresponds to the disordered phase whereas $\varepsilon > \varepsilon_{*2}$ corresponds to the ordered phase.

The transition point reads in terms of m and g as follows

$$m_{*2}^2 = -\frac{g}{\pi\theta} - c'\frac{g^{\frac{7}{3}}}{\theta}. \tag{18.29}$$

The coefficient c' depends on c and on the proportionality factor between α and $g^{7/2}/\theta^{3/2}$. The system avoids the second-order behavior by precisely an amount proportional to $g^{\frac{7}{3}}/\theta^{3/2}$. It is important to compare the second-order behavior $m_{*2}^2 = -2g/\theta$ with the critical point of real quartic matrix models. The phenomena of a phase transition of an isotropic system to a non-uniform phase was in fact realized a long time ago by Brazovkii [49].

For more detail see [48]. See also figure 18.3.

18.2.1.3 Dispersion relations on the noncommutative torus

The phase diagram and the dispersion relation of noncommutative phi-four theory on the noncommutative torus in $d=3$ is also discussed using the Monte Carlo method in [50]. Here, the noncommutative torus should be thought of as a regularization of the Moyal–Weyl space. See also [51, 52]. The phase diagram, with exactly the same qualitative features conjectured by Gubser and Sondhi [48], is shown on figure 5 of [50]. Three phases, which meet at a triple point, are identified. The Ising (disordered-to-uniform) transition exists for small θ whereas transitions to

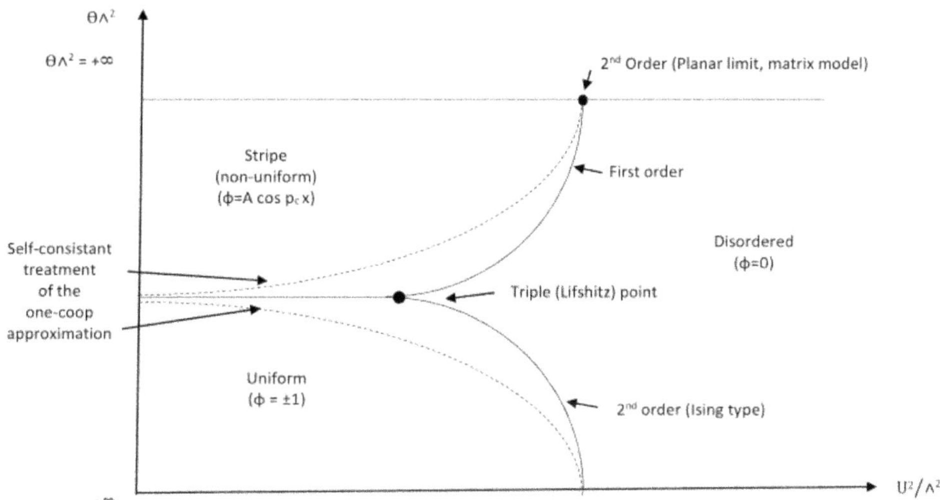

Figure 18.3. The phase diagram of noncommutative phi-four in $d = 4$ at fixed $\lambda \sim g^2$.

the stripe phase (disordered-to-stripe and uniform-to-stripe) are favored at large θ. The collapsed parameters in this case are found to be given by

$$N^2 m^2, \quad N^2 \lambda. \tag{18.30}$$

On the other hand, the dispersion relations are computed as usual from the exponential decay of the correlation function

$$\frac{1}{T}\sum_t \langle \tilde{\Phi}^*(\vec{p}, t)\tilde{\Phi}(\vec{p}, t + \tau)\rangle. \tag{18.31}$$

This behaves as $\exp(-E(\vec{p})\tau)$ for large τ and thus we can extract the energy $E(\vec{p})$ by computing the above correlator as a function of τ. In the disordered phase, i.e., for small λ near the uniform phase, we find the usual linear behavior $E(\vec{p}) = a\vec{p}^2$ and thus in this region the model looks like its commutative counterpart. As we increase λ we observe that the rest energy $E_0 \equiv E(\vec{0})$ increases, followed by a sharp dip at some small value of the momentum $\vec{p}^2$, then the energy rises again with $\vec{p}^2$ and approaches asymptotically the linear behavior $E(\vec{p}) = a\vec{p}^2$ for $\vec{p}^2 \longrightarrow \infty$. An example of a dispersion relation near the stripe phase, for $m^2 = -15$ and $\lambda = 50$, is shown on figure 14 of [50] with a fit given by

$$E^2(\vec{p}) = c_0\vec{p}^2 + m^2 + \frac{c_1}{\sqrt{\vec{p}^2 + \bar{m}^2}} \exp(-c_2\sqrt{\vec{p}^2 + \bar{m}^2}). \tag{18.32}$$

The parameters c_i and $\bar{m}^2$ are given by equation (6.2) of [50]. The minimum in this case occurs around the cases $k = N|\vec{p}|/2\pi = \sqrt{2}, 2, \sqrt{5}$ so this corresponds actually to a multi-stripe pattern.

The above behavior of the dispersion relations stabilizes in the continuum limit defined by the double scaling limit $N \longrightarrow \infty$ (planar limit), $a \longrightarrow 0$

$(m^2 \longrightarrow m_c^2 = -15.01(8))$ keeping λ and $\theta = Na^2/\pi$ fixed. In this limit the rest energy E_0 is found to be divergent, linearly with $\sqrt{N} \propto 1/a$, in full agreement with the UV–IR mixing.

The shifting of the energy minimum to a finite non-vanishing value of the momentum in this limit indicates the formation of a stable stripe phase in the continuum noncommutative theory. The existence of a continuum limit is also a strong indication that the theory is non-perturbatively renormalizable.

18.2.1.4 The Grosse–Wulkenhaar model

For simplicity, we consider a two-dimensional noncommutative space $\mathbf{R}_\theta^2$. We introduce noncommutativity in momentum space by introducing a minimal coupling to a constant background magnetic field B_{ij}. The derivation operators become

$$\hat{D}_i = \hat{\partial}_i - iB_{ij}X_j, \quad \hat{C}_i = \hat{\partial}_i + iB_{ij}X_j, \quad X_i = \frac{\hat{x}_i + \hat{x}_i^R}{2}. \tag{18.33}$$

Instead of the conventional Laplacian $\Delta = (-\hat{\partial}_i^2 + \mu^2)/2$ we will consider the generalized Laplacians

$$\Delta = -\sigma\hat{D}_i^2 - \tilde{\sigma}\hat{C}_i^2 + \frac{\mu^2}{2}$$

$$= -(\sigma + \tilde{\sigma})\hat{\partial}_i^2 + (\sigma - \tilde{\sigma})iB_{ij}\{X_j, \hat{\partial}_i\} - (\sigma + \tilde{\sigma})(B^2)_{ij}X_iX_j + \frac{\mu^2}{2}. \tag{18.34}$$

The case $\sigma = \tilde{\sigma}$ corresponds to the Grosse–Wulkenhaar model [47, 53, 54], while the model $\sigma = 1$, $\tilde{\sigma} = 0$ corresponds to Langmann–Szabo–Zarembo model considered in [55].

We introduce two sets of creation and annihilation operators $(\hat{a}, \hat{a}^\dagger)$ and $(\hat{b}, \hat{b}^\dagger)$ by the relations (with $\theta_0 = \theta/2$)

$$\hat{a} = \frac{1}{2}(\sqrt{\theta_0}\hat{\partial} + \frac{1}{\sqrt{\theta_0}}Z^\dagger), \quad \hat{a}^\dagger = \frac{1}{2}(\sqrt{\theta_0}\hat{\partial}^\dagger + \frac{1}{\sqrt{\theta_0}}Z). \tag{18.35}$$

$$\hat{b} = \frac{1}{2}(-\sqrt{\theta_0}\hat{\partial}^\dagger + \frac{1}{\sqrt{\theta_0}}Z), \quad \hat{b}^\dagger = \frac{1}{2}(-\sqrt{\theta_0}\hat{\partial} + \frac{1}{\sqrt{\theta_0}}Z^\dagger). \tag{18.36}$$

In these equations we have also used the definitions

$$Z = X_1 + iX_2, \quad Z^\dagger = X_1 - iX_2, \quad \hat{\partial} = \hat{\partial}_1 - i\hat{\partial}_2, \quad \hat{\partial}^\dagger = -\hat{\partial}_1 - i\hat{\partial}_2. \tag{18.37}$$

We expand the scalar field operator $\hat{\Phi}$ as follows

$$\hat{\Phi} = \sum_{l,m=1}^{\infty} M_{lm}\hat{\phi}_{l,m}, \quad \hat{\phi}_{l,m} = |l\rangle\langle m|. \tag{18.38}$$

The infinite dimensional matrix M should be thought of as a compact operator acting on some separable Hilbert space $\mathbf{H}_1$ of Schwartz sequences with sufficiently rapid decrease [56]. This in particular will guarantee the convergence of the expansion of the scalar operator $\hat{\Phi}$.

In the operators $\hat{\phi}_{l,m} = |l\rangle\langle m|$ we can identify the kets $|l\rangle$ with the states of the harmonic oscillator operators $\hat{a}$ and $\hat{a}^\dagger$ whereas the bras $\langle m|$ can be identified with the states of the harmonic oscillator operators $\hat{b}$ and $\hat{b}^\dagger$. More precisely, the operators $\hat{\phi}_{l,m}$ are in one-to-one correspondence with the wave functions $\phi_{l,m}(x) = \langle x|l, m\rangle$, which are known as the Landau states [57, 58].

The field/operator Weyl map is then given by

$$\sqrt{2\pi\theta}\ \phi_{l_1,m_1} \leftrightarrow \hat{\phi}_{l_1,m_1}. \tag{18.39}$$

$$\int d^2x \leftrightarrow \sqrt{\det(2\pi\theta)}\,\mathrm{Tr}_{\mathcal{H}}. \tag{18.40}$$

$$\Phi = \sqrt{2\pi\theta}\sum_{l,m=1}^{\infty} M_{lm}\phi_{l,m} \leftrightarrow \hat{\Phi}. \tag{18.41}$$

$$\Phi*\Phi' = \sqrt{2\pi\theta}\sum_{l,m=1}^{\infty} (MM')_{lm}\phi_{l,m} \leftrightarrow \hat{\Phi}\hat{\Phi}'. \tag{18.42}$$

In other words, the star product is mapped to the operator product as it should be. Furthermore, the differential operators $\hat{D}_i^2$ and $\hat{C}_i^2$ will be represented in the star picture by the differential operators

$$D_i = \partial_i - iB_{ij}x_j, \quad C_i = \partial_i + iB_{ij}x_j. \tag{18.43}$$

The Landau states are actually eigenstates of the Laplacians D_i^2, and C_i^2 at the special point

$$B^2\theta_0^2 \equiv \frac{B^2\theta^2}{4} = 1. \tag{18.44}$$

The most general single-trace action with a phi-four interaction on a noncommutative $\mathbf{R}_\theta^d$ under the effect of a magnetic field is then given by

$$\begin{aligned}
S &= \int d^2x\left[\Phi^\dagger\left(-\sigma D_i^2 - \tilde{\sigma}C_i^2 + \frac{\mu^2}{2}\right)\Phi + \frac{\lambda}{4!}\Phi^\dagger*\Phi*\Phi^\dagger*\Phi\right] \\
&= \sqrt{\det(2\pi\theta)}\,\mathrm{Tr}_{\mathcal{H}}\left[\hat{\Phi}^\dagger\left(-\sigma\hat{D}_i^2 - \tilde{\sigma}\hat{C}_i^2 + \frac{\mu^2}{2}\right)\hat{\Phi} + \frac{\lambda}{4!}\hat{\Phi}^\dagger\hat{\Phi}\ \hat{\Phi}^\dagger\hat{\Phi}\right].
\end{aligned} \tag{18.45}$$

This action enjoys a remarkable symmetry under a duality transformation, which exchanges positions and momenta. Explicitly, this duality transformation under which the action retains the same form reads [55, 56, 59]

$$x_i \leftrightarrow \tilde{k}_i = B_{ij}^{-1}k_j, \quad \Phi(x) \leftrightarrow \bar{\Phi}(\tilde{k}) = \sqrt{\left|\det\frac{B}{2\pi}\right|}\,\hat{\Phi}(B\tilde{k}). \tag{18.46}$$

$$\theta \leftrightarrow \bar{\theta} = -B^{-1}\theta^{-1}B^{-1}, \quad \lambda \leftrightarrow \bar{\lambda} = \frac{\lambda}{|\det B\theta|}. \tag{18.47}$$

Furthermore, we can re-express the above action in terms of the compact operators M and $M^{\dagger}$ as follows (with $\alpha = 1 + B\theta_0$, $\beta = 1 - B\theta_0$)

$$
\begin{aligned}
S = \frac{\sqrt{\det(2\pi\theta)}}{\theta_0}\Big[&-(\sigma + \tilde{\sigma})\alpha\beta\, \mathrm{Tr}_{\mathbf{H}_1}(\Gamma^{+}M^{+}\Gamma M + M^{+}\Gamma^{+}M\Gamma) + (\sigma\alpha^2 + \tilde{\sigma}\beta^2)\mathrm{Tr}_{\mathbf{H}_1}M^{+}EM \\
&+ (\sigma\beta^2 + \tilde{\sigma}\alpha^2)\mathrm{Tr}_{\mathbf{H}_1}MEM^{+} + \frac{\mu^2\theta_0}{2}\mathrm{Tr}_{\mathbf{H}_1}M^{+}M + \frac{\lambda\theta_0}{4!}Tr_{\mathbf{H}_1}M^{+}MM^{+}M \Big].
\end{aligned}
\tag{18.48}
$$

The infinite dimensional matrices Γ and E are defined by

$$(\Gamma)_{lm} = \sqrt{m-1}\,\delta_{lm-1}, \quad (E)_{lm} = (l - \tfrac{1}{2})\delta_{lm}. \tag{18.49}$$

Now, we regularize the theory by taking M to be an $N \times N$ matrix. The states $\phi_{l,m}(x)$, with $l, m < N$, where N is some large integer, correspond to simultaneous cut-offs in position and momentum spaces [47]. The infrared cutoff is found to be proportional to $R = \sqrt{2\theta N}$ while the UV cutoff is found to be proportional to $\Lambda = \sqrt{8N/\theta}$.

The so-called Grosse–Wulkenhaar model corresponds to the values $\sigma = \tilde{\sigma} \neq 0$ so that the mixing term in (18.34) cancels. This contains, compared with the usual case, a harmonic oscillator term in the Laplacian, which modifies and thus allows us to control the IR behavior of the theory. This model is perturbatively renormalizable, which makes it the more interesting. We consider, without any loss of generality, $\sigma = \tilde{\sigma} = 1/4$. We obtain therefore the action

$$
\begin{aligned}
S &= \int d^2x \left[\Phi^{\dagger}\left(-\frac{1}{2}\partial_i^2 + \frac{1}{2}\Omega^2\tilde{x}_i^2 + \frac{\mu^2}{2}\right)\Phi + \frac{\lambda}{4!}\Phi^{\dagger}*\Phi*\Phi^{\dagger}*\Phi \right] \\
&= \sqrt{\det(2\pi\theta)}\,\mathrm{Tr}_{\mathcal{H}}\left[\hat{\Phi}^{\dagger}\left(-\frac{1}{2}\hat{\partial}_i^2 + \frac{1}{2}\Omega^2\hat{X}_i^2 + \frac{\mu^2}{2}\right)\hat{\Phi} + \frac{\lambda}{4!}\hat{\Phi}^{\dagger}\hat{\Phi}\,\hat{\Phi}^{\dagger}\hat{\Phi} \right].
\end{aligned}
\tag{18.50}
$$

Here, $\tilde{x}_i = 2(\theta^{-1})_{ij}x_j$ and the parameter Ω is defined by

$$B\theta = 2\Omega. \tag{18.51}$$

The above action is also covariant under a duality transformation which exchanges positions and momenta as $x_i \leftrightarrow \tilde{p}_i = B_{ij}^{-1}p_j$. The value $\Omega^2 = 1$ gives an action, which is actually invariant under this duality transformation.

In the Landau basis, the above action reads

$$
\begin{aligned}
S = \frac{\nu_2}{\theta}\Big[&(\Omega^2 - 1)Tr_H(\Gamma^{+}M^{+}\Gamma M + M^{+}\Gamma^{+}M\Gamma) + (\Omega^2 + 1)Tr_H(M^{+}EM + MEM^{+}) \\
&+ \frac{\mu^2\theta}{2}Tr_H M^{+}M + \frac{\lambda\theta}{4!}Tr_H M^{+}MM^{+}M \Big].
\end{aligned}
\tag{18.52}
$$

This is a special case of (18.48). Equivalently

$$S = \nu_2 \sum_{m,n,k,l} \left(\frac{1}{2}(M^+)_{mn} G_{mn,kl} M_{kl} + \frac{\lambda}{4!}(M^+)_{mn} M_{nk}(M^+)_{kl} M_{lm} \right). \tag{18.53}$$

$$G_{mn,kl} = \left(\mu^2 + \mu_1^2(m + n - 1) \right)\delta_{n,k}\delta_{m,l} - \mu_1^2 \sqrt{\omega(m-1)(n-1)}\ \delta_{n-1,k}\delta_{m-1,l}$$
$$- \mu_1^2 \sqrt{\omega mn}\ \delta_{n+1,k}\delta_{m+1,l}. \tag{18.54}$$

The parameters of the model are μ^2, λ and

$$\nu_2 = \sqrt{\det(2\pi\theta)}, \quad \mu_1^2 = 2(\Omega^2 + 1)/\theta, \quad \sqrt{\omega} = (\Omega^2 - 1)/(\Omega^2 + 1). \tag{18.55}$$

However, there are only three independent coupling constants in this theory, which we can take to be μ^2, λ, and Ω^2.

18.2.2 Wilson renormalization group recursion formula

18.2.2.1 The Wilson–Fisher fixed point in noncommutative phi-four
In this section we will apply the renormalization group recursion formula of Wilson [60] to noncommutative phi-four theory on the Moyal–Weyl space $\mathbf{R}_\theta^d$. This formula was applied in [61, 62] to vector and Hermitian matrix models in the large N limit. Their method can be summarized as follows:

(1) We split the field into a background and a fluctuation and then integrate the fluctuation obtaining therefore an effective action for the background field alone.

(2) We keep, following Wilson, only induced corrections to the terms that are already present in the classical action. Thus, we will only need to calculate quantum corrections to the 2- and 4-point functions.

(3) We perform the so-called Wilson contraction which consists in estimating momentum loop integrals using the following three approximations or rules:

  - **Rule** 1: All external momenta, which are wedged with internal momenta, will be set to zero.
  - **Rule** 2: We approximate every internal propagator $\Delta(k)$ by $\Delta(\lambda)$ where λ is a typical momentum in the range $\rho\Lambda \leqslant \lambda \leqslant \Lambda$.
  - **Rule** 3: We replace every internal momentum loop integral $\int_k$ by a typical volume.

The two last approximations are equivalent to the reduction of all loop integrals to their zero-dimensional counterparts. These two approximations are quite natural in the limit $\rho \longrightarrow 1$.

As it turns out we do not need to use the first approximation in estimating the 2-point function. In fact rule 1 was proposed first in the context of a noncommutative Φ^4 theory in [63] in order to simply the calculation of the 4-point function. In some sense the first approximation is equivalent to taking the limit $\bar{\theta} = \theta\Lambda^2 \longrightarrow 0$.

(4) The last step in the renormalization group program of Wilson consists in rescaling the momenta so that the cutoff is restored to its original value. We can then obtain renormalization group recursion equations, which relate the new values of the coupling constants to the old values.

The action we will study is given by

$$S[\Phi] = \int d^d x \left[\Phi(-\partial_i^2 + \mu^2)\Phi + \frac{\lambda}{4!}\Phi_*^4 \right] = S_0 + S_4. \tag{18.56}$$

We introduce a sharp momentum cutoff Λ. We introduce the modes with low and high momenta by (with $0 \leqslant b \leqslant 1$)

$$\phi(k) \equiv \phi_L(k), \quad k \leqslant b\Lambda. \tag{18.57}$$

$$\phi(k) \equiv \phi_H(k), \quad b\Lambda \leqslant k \leqslant \Lambda. \tag{18.58}$$

By integrating over the modes $\Phi_H(k)$ in the path integral we obtain

$$Z = \int d\phi \; e^{-S_0[\phi]-S_4[\phi]} = \int d\phi_L \; d\phi_H \; e^{-S_0[\phi_L]-S_0[\phi_H]-S_4[\phi_L, \, \phi_H]}$$

$$= \int d\phi_L \; e^{-S_0[\phi_L]} \; e^{-S_4'[\phi_L]} \tag{18.59}$$

$$e^{-S_4'[\phi_L]} = \int d\phi_H \; e^{-S_0[\phi_H]-S_4[\phi_L, \, \phi_H]} \propto \left\langle e^{-S_4[\phi_L, \, \phi_H]} \right\rangle_{0H}.$$

The expectation value $\langle \dots \rangle_{0H}$ is taken with respect to the probability distribution $e^{-S_0[\phi_H]}$. We verify the identity (cumulant expansion)

$$\left\langle e^{-S_4[\phi_L, \, \phi_H]} \right\rangle_{0H} = \exp\left[-\langle S_4[\phi_L, \phi_H] \rangle_{0H} + \frac{1}{2}\left(\langle S_4^2[\phi_L, \phi_H] \rangle_{0H} - \langle S_4[\phi_L, \phi_H] \rangle_{0H}^2 \right) \right]$$

$$= \exp\left[-S_4[\phi_L] - \langle \delta S[\phi_L, \phi_H] \rangle_{0H} + \frac{1}{2}\left(\langle \delta S^2[\phi_L, \phi_H] \rangle_{0H} - \langle \delta S[\phi_L, \phi_H] \rangle_{0H}^2 \right) \right]. \tag{18.60}$$

The action δS is defined by $\delta S[\phi_L, \phi_H] = \delta S_1 + \delta S_2 + \delta S_3 + \delta S_4$ where δS_i involves i fields ϕ_H and $4 - i$ fields ϕ_L.

The correction to the 2-point function is found to be given by (with $n = (d-2)/2$, $K_d = S_d/(2\pi)^d$, $\mu^2 = r\Lambda^2$ and $g = \lambda\Lambda^{d-4}$)

$$\Delta\Gamma_2(p) = \frac{gK_d\Lambda^2}{12(1+r)}(1-b)\left[1 + 2^{n-1}n!\frac{J_n(\theta\Lambda p)}{(\theta\Lambda p)^n} \right]. \tag{18.61}$$

The corrected mass parameter and its renormalization group equation are then given by (with the mass parameter r assumed to be near 0 whereas b is by construction near 1)

$$r' = \frac{r}{b^2} - \frac{gK_d}{12}\frac{1}{1+r}\frac{\ln b}{b^2}\left[1 + 2^{n-1}n!\frac{J_n(b\theta\Lambda p)}{(b\theta\Lambda p)^n}\right]_{p=0}$$

$$= \frac{r}{b^2} - \frac{gK_d}{8}(1-r)\frac{\ln b}{b^2} \tag{18.62}$$

$$\Rightarrow b\frac{dr}{db} = (-2 + \frac{gK_d}{8})r' - \frac{gK_d}{8}$$

In this equation we have also employed the renormalization group transformations of the action of the background field ϕ_L, which are given by

$$p \longrightarrow p' = \frac{p}{b}$$

$$\phi_L(p) \longrightarrow \phi'_L(p') = b^{\frac{d}{2}+1}\phi_L(p)$$

$$\mu'^2 = \frac{\mu^2}{b^2} \Leftrightarrow r'^2 = \frac{r^2}{b^2} \tag{18.63}$$

$$\lambda' = b^{-4+d}\lambda \Leftrightarrow g' = b^{-4+d}g$$

$$\theta' = b^2\theta.$$

The above result (18.61) gives also the wave function renormalization, which is contained in the p-dependent part of the quadratic term. Explicitly, we have

$$\left(1 + \frac{gK_d(\theta\Lambda^2)^2}{192}\ln b\right)\int_{p\leqslant b\Lambda}\phi_L(p)\phi_L(-p)p^2 = \int_{p'\leqslant\Lambda}\phi'_L(p')\phi'_L(-p')p'^2, \quad \phi'_L(p')$$

$$= b^{\frac{d+2-\gamma}{2}}\phi_L(p). \tag{18.64}$$

Thus, we have obtained a negative wave function renormalization, which signals an instability in the theory. Indeed, the anomalous dimension γ is negative given explicitly by

$$\gamma = -\frac{gK_d(\theta\Lambda^2)^2}{192} < 0. \tag{18.65}$$

This is also θ-dependent. This result also implies a novel behavior for the 2-point function which must behave as

$$\langle\phi(x)\phi(0)\rangle \sim \frac{1}{|x|^{d-2+\gamma}}. \tag{18.66}$$

This should vanish for large distances as it should be as long as $d - 2 + \gamma > 0$. Thus, for large values of θ we get an instability because $d - 2 + \gamma$ becomes negative. The critical value of θ is precisely given by

$$d - 2 + \gamma_c = 0 \Rightarrow (\theta_c\Lambda^2)^2 = \frac{196(d-2)}{gK_d}. \tag{18.67}$$

The corrected quartic coupling and its renormalization group equation are found to be given by (by assuming the external momenta to be very small compared to the cutoff)

$$g' = b^{d-4}\left(g + g^2\frac{3K_d}{8(1+r)^2}\ln b\right)$$

$$= b^{d-4}\left(g + g^2\frac{3K_d}{8}(1 - 2r)\ln b\right) \tag{18.68}$$

$$\Rightarrow b\frac{dg'}{db} = (d-4)g' + \frac{3g'^2 K_d}{8}.$$

The fixed points are then given by the equation

$$0 = (-2 + \frac{g_*K_d}{8})r_* - \frac{g_*K_d}{8}. \tag{18.69}$$

$$0 = (d-4)g_* + \frac{3g_*^2 K_d}{8}. \tag{18.70}$$

We get immediately the two solutions (in dimension $d < 4$ with small $\varepsilon = 4 - d$)

$$r_* = g_* = 0, \quad \text{trivial(Gaussian)fixed point,} \tag{18.71}$$

$$r_* = -\frac{\varepsilon}{6}, \quad g_* = \frac{64\pi^2\varepsilon}{3}, \quad \text{Wilson–Fisher fixed point.} \tag{18.72}$$

The critical exponent ν is given by the usual value whereas the critical exponent η is now θ-dependent given by

$$\eta = \gamma\,|_* = -\frac{g_*K_d(\theta\Lambda^2)^2}{192} = -\frac{(\theta\Lambda^2)^2\varepsilon}{72}. \tag{18.73}$$

This is proportional to ε (and not ε^2) and is negative. The behavior of the 2-point function is now given by

$$\langle\phi(x)\phi(0)\rangle \sim \frac{1}{|x|^{2-\varepsilon(1+(\theta\Lambda^2)^2/72)}}. \tag{18.74}$$

We obtain now the critical point

$$\theta_c\Lambda^2 = \frac{12}{\sqrt{\varepsilon}}. \tag{18.75}$$

The noncommutative Wilson–Fisher fixed point is only stable for $\theta < \theta_c$.

The above negative anomalous dimension, which is due to the non-locality of the theory, leads immediately to the existence of a first order transition to a modulated phase via the Lifshitz scenario [64]. Indeed, we can show that below the critical value θ_c the coefficient of k^2 is positive whereas above θ_c the coefficient of k^2 becomes negative and thus one requires, in order to maintain stability, the inclusion of the term proportional to k^4, which turns out to have a positive coefficient as opposed to the commutative theory. We can show explicitly that at $\theta = \theta_c$ the dispersion relation changes from k^2 to k^4. Thus, the effective action is necessarily of the form (with positive a and b)

$$\int \frac{d^d k}{(2\pi)^d} \phi(k)\phi(-k)[(1 - ag\theta\Lambda^2)k^2 + bk^4] + \text{interaction}. \tag{18.76}$$

The Lifshitz point is a tri-critical point in the phase diagram where the coefficient of k^2 vanishes exactly and that of k^4 is positive. In this case, this point is given precisely by the value $\theta = \theta_c$, and the transition is a first order transition because it is not related to a change of symmetry. In this transition the system develops a soft mode associated with the minimum of the kinetic energy and as a consequence the ordering above θ_c is given by a modulating order parameter. A more thorough discussion of this point can be found in [63].

18.2.2.2 The noncommutative $O(N)$ Wilson–Fisher fixed point

A nonperturbative study of the fixed point in noncommutative $O(N)$ model can be carried out along the above lines [65, 66]. In this case the analysis is exact in $1/N$. It is found that the Wilson–Fisher fixed point makes good sense only for sufficiently small values of θ up to a certain maximal noncommutativity. This fixed point describes the transition from the disordered phase to the uniform ordered phase, i.e., the Ising universality class.

As it turns out, there is a θ-dependent fixed point, termed the noncommutative Wilson–Fisher fixed point, which interpolates between the commutative Wilson–Fisher fixed point of the Ising universality class, which is found to lie at $t = 1$ and $a_* = 0$, and a novel strongly interacting fixed point, which lies at $t = 2$ and $a_* \longrightarrow \infty$. Here, t is the effective noncommutativity parameter and a is the coupling constant of the zero-dimensional reduction of the theory. Thus, the point $t = 2$ corresponds to maximal noncommutativity and the corresponding fixed point is identified with the transition between non-uniform and uniform orders.

Indeed, the noncommutative Wilson–Fisher fixed point does not scale to zero when we send the dilation parameter ρ to zero, which corresponds to a single step of the renormalization group transformation, in contrast with the commutative Wilson–Fisher fixed point which still scales to zero in the limit $\rho \longrightarrow 1$. The main obstacle comes from the fact that the critical coupling constant a_* which starts from 0 at $\rho^\varepsilon = 1$, and then increases to ∞ at $\rho^\varepsilon = 1 - t/2$, does not return to zero as we decrease ρ^ε back from $\rho^\varepsilon = 1 - t/2$ to 0. In other words, the non-perturbative sheet shrinks as we increase the non-commutativity until it disappears at $t = 2$.

18.2.2.3 The $\theta = \infty$ matrix model fixed point

As discussed above, in the Wilson recursion formula we perform the usual truncation but also we perform a reduction to zero dimension, which allows explicit calculation, or more precisely estimation, of Feynman diagrams. This method was also applied in [66, 67] to noncommutative phi-four theory, with a harmonic oscillator term at the self-dual point, on a degenerate Moyal–Weyl space with two strongly noncommuting coordinates, viz $\mathbf{R}_\theta^d = \mathbf{R}^D \times \mathbf{R}_\theta^2$. In the matrix basis this theory becomes, after appropriate non-perturbative definition, an $N \times N$ matrix model where N is a regulator in the noncommutative directions, which is directly connected to the noncommutativity parameter θ itself. The action is explicitly given by

$$S[M] = \int d^D x \, \mathrm{Tr}_N \left[\frac{1}{2}(\partial_\mu M)^2 + \frac{1}{2}\mu^2 M^2 + r^2 E M^2 + \frac{u}{N} M^4 \right]. \qquad (18.77)$$

The external (diagonal) matrix E originates from the harmonic oscillator term and $r^2 = 4/\theta$.

Thus, in order to compute the effective action we employ, following [61, 62, 68], a combination of two methods given by

    (1) The Wilson approximate renormalization group recursion formula.

    (2) The solution of the zero-dimensional large N counting problem (given in this case by the Penner matrix model $\mathrm{Tr}_N[\frac{1}{2}\mu^2 M^2 + r^2 E M^2 + \frac{u}{N} M^4]$ which can be turned into a multitrace matrix model for large values of θ).

As discussed neatly in [68] the virtue and power of combining these two methods lies in the crucial fact that all leading Feynman diagrams in $1/N$ will be counted correctly in this scheme including the so-called 'setting sun' diagrams. Here, we choose to take the limit $\theta \longrightarrow \infty$ first and then $N \longrightarrow \infty$ so that the $1/N$ expansion of the theory is still given by that of the D-dimensional hermitian matrix model, i.e., the action with $r^2 = 0$.

The obtained matrix model fixed point describes the transition from the one-cut (disordered) phase to the two-cut (non-uniform ordered, stripe) phase in the same way that the noncommutative Wilson–Fisher fixed point describes transition from the disordered phase to the uniform ordered phase.

Thus, the analysis of phi-four theory on noncommutative spaces using a combination of the Wilson renormalization group recursion formula and the solution to the zero-dimensional vector/matrix models at large N suggests the existence of three fixed points. The matrix model $\theta = \infty$ fixed point, which describes the disordered-to-non-uniform-ordered transition, the Wilson–Fisher fixed point at $\theta = 0$, which describes the disordered-to-uniform-ordered transition, and a non-commutative Wilson–Fisher fixed point at a maximum value of θ, which is associated with the transition between non-uniform-order and uniform-order phases.

18.3 Noncommutative phi-four on the fuzzy sphere $\mathbf{S}_N^2$

18.3.1 UV–IR mixing and renormalizability

A real scalar field Φ on the fuzzy sphere $\mathbf{S}_N^2$ is an $N \times N$ Hermitian matrix where $N = L + 1$. Let L_a be the generators of $SU(2)$ in the spin $s = L/2$ irreducible representation. The noncommutative phi-four theory on the fuzzy sphere $\mathbf{S}_N^2$ is then given by

$$
\begin{aligned}
S &= \frac{1}{N}\,\mathrm{Tr}\,[\Phi[L_a,\,[L_a,\,\Phi]] + m^2\Phi^2 + \lambda\Phi^4] \\
&= \frac{1}{N}\,\mathrm{Tr}\,(\Phi\Delta\Phi + m^2\Phi^2 + \lambda\Phi^4), \quad \Delta = \mathcal{L}_a^2.
\end{aligned}
\tag{18.78}
$$

This model has the correct commutative large N limit, viz

$$
S = \int \frac{d\Omega}{4\pi}\Big[\Phi\mathcal{L}_a^2\Phi + m^2\Phi^2 + \lambda\Phi^4\Big].
\tag{18.79}
$$

Quantum field theories on the fuzzy sphere were proposed originally in [69, 70]. In perturbation theory of the matrix model (18.78) only the tadpole diagram can diverge in the limit $N \longrightarrow \infty$ [71, 72]. See also [73, 74]. On the fuzzy sphere, and similarly to the Moyal–Weyl plane, the planar and non-planar tadpole graphs are different and their difference is finite in the limit. This is the infamous UV–IR mixing. This problem can be removed in this case by standard normal ordering of the interaction [75].

We use the background field method to quantize this model. We write $\Phi = \Phi_0 + \Phi_1$ where Φ_0 is a background field, which satisfies the classical equation of motion and Φ_1 is a fluctuation. By integrating over Φ_1 in the path integral we obtain the effective action

$$
S_{\mathrm{eff}}[\Phi_0] = S[\Phi_0] + \frac{1}{2}\mathrm{TR}\,\log\,\Omega, \quad \Omega = \Delta + m^2 + 4\lambda\Phi_0^2 + 2\lambda\Phi_0\Phi_0^R.
\tag{18.80}
$$

The 2-point function is deduced from

$$
S_{\mathrm{eff}}^{\mathrm{quad}} = \frac{1}{N}\,\mathrm{Tr}\,\Phi_0(\Delta + m^2)\Phi_0 + \lambda\mathrm{TR}\left(\frac{2}{\Delta + m^2}\Phi_0^2 + \frac{1}{\Delta + m^2}\Phi_0\Phi_0^R\right).
\tag{18.81}
$$

The free propagator is given explicitly by

$$
\left(\frac{1}{\Delta + m^2}\right)^{AB,\,CD} = \sum_{k,k_3}\frac{1}{\Delta(k) + m^2}T_{kk_3}^{AB}(T_{kk_3}^+)^{DC}.
\tag{18.82}
$$

The planar and non-planar contributions are found explicitly to be given by

$$
\mathrm{TR}\frac{2}{\Delta + m^2}\Phi_0^2 = 2\sum_{p,p_3}|\phi(pp_3)|^2\Pi^P, \quad \Pi^P = \frac{1}{N}\sum_k\frac{2k+1}{k(k+1)+m^2}.
\tag{18.83}
$$

$$
\begin{aligned}
\mathrm{TR}\frac{1}{\Delta + m^2}\Phi_0\Phi_0^R &= \sum_{p,p_3}|\phi(pp_3)|^2\Pi^{N-P}(p), \quad \Pi^{N-P}(p) \\
&= \sum_k\frac{2k+1}{k(k+1)+m^2}(-1)^{p+k+2s}\begin{Bmatrix} p & s & s \\ k & s & s \end{Bmatrix}.
\end{aligned}
\tag{18.84}
$$

The UV–IR mixing is measured by the difference of the two contributions, viz

$$\Pi^{N-P} - \Pi^P = \frac{1}{N}\sum_k \frac{2k+1}{k(k+1)+m}\left[N(-1)^{p+k+2s}\left\{\begin{matrix} p & s & s \\ k & s & s \end{matrix}\right\} - 1\right]. \tag{18.85}$$

Remark that the non-planar contribution depends on the external momentum p. When p is small compared to $2s = N - 1$ one is justified to use the following approximation for the $6j$ symbols [76]:

$$\left\{\begin{matrix} p & s & s \\ k & s & s \end{matrix}\right\} \approx \frac{(-1)^{p+k+2s}}{N}P_p\left(1 - \frac{2k^2}{N^2}\right), \quad s \to \infty, \quad p \ll 2s, \quad 0 \leqslant k \leqslant 2s. \tag{18.86}$$

Since $P_p(1) = 1$ for all p, only $k \gg 1$ contribute in the above sum, and therefore it can be approximated by an integral as follows

$$\begin{aligned}
\Pi^{N-P} - \Pi^P &= \frac{1}{N}\sum_k \frac{2k+1}{k(k+1)+m^2}\left[P_p\left(1 - \frac{2k^2}{N^2}\right) - 1\right] \\
&= \frac{1}{N}h_p, \quad h_p = \int_{-1}^{+1} \frac{dx}{1 - x + \frac{2m^2}{N^2}}[P_p(x) - 1] \\
&= -\frac{2}{N}\sum_{n=1}^{p}\frac{1}{n}.
\end{aligned} \tag{18.87}$$

This is non-zero in the continuum limit. It has also the correct planar limit on the Moyal–Weyl plane. See for example [1].

The planar contribution Π^P is given explicitly by $\frac{1}{N}\log\frac{N^2}{m^2}$ (if we replace the sum in (18.83) by an integral). Thus, the total quadratic effective action is given by

$$S_{\text{eff}}^{\text{quad}} = \frac{1}{N}Tr\Phi_0\left(\Delta + m^2 + 3\lambda\log\frac{N^2}{m^2} - 2\lambda Q\right)\Phi_0. \tag{18.88}$$

The operator $Q = Q(\mathcal{L}^2)$ is defined by its eigenvalues $Q(p)$ given by

$$Q\hat{Y}_{pm} = Q(p)\hat{Y}_{pm}, \quad Q(p) = \sum_{n=1}^{p}\frac{1}{n}. \tag{18.89}$$

The quartic effective action is also deduced from (18.80). Formally, we obtain

$$\begin{aligned}
S_{\text{eff}}^{\text{quart}} = \frac{\lambda}{N}\,\text{Tr}\,\Phi_0^4 - \lambda^2\text{TR}&\left[2\left(\frac{1}{\Delta + m^2}\Phi_0^2\right)^2\right. \\
&+ 2\left(\frac{1}{\Delta + m^2}\Phi_0^2\right)\left(\frac{1}{\Delta + m^2}(\Phi_0^R)^2\right) + \left(\frac{1}{\Delta + m^2}\Phi_0\Phi_0^R\right)^2 \\
&+ \left.4\left(\frac{1}{\Delta + m^2}\Phi_0^2\right)\left(\frac{1}{\Delta + m^2}\Phi_0\Phi_0^R\right)\right].
\end{aligned} \tag{18.90}$$

Explicitly, we have

$$S_{\text{eff}}^{\text{quart}} = \frac{\lambda}{N}\,\text{Tr}\,\Phi_0^4 - \lambda^2 \sum_{jj_3}\sum_{ll_3}\sum_{qq_3}\sum_{tt_3} \phi(jj_3)\phi(ll_3)\phi(qq_3)\phi(tt_3)$$

$$[2V_{P,P} + 2\bar{V}_{P,P} + V_{NP,NP} + 4V_{P,NP}]. \tag{18.91}$$

We introduce the interaction vertex (with the notation $\vec{k} = (kk_3)$)

$$v(\vec{k},\vec{j},\vec{p},\vec{q}) = \text{Tr}\,T_{kk_3}^{+}T_{jj_3}\,T_{pp_3}\,T_{qq_3}. \tag{18.92}$$

The two planar-planar contributions are given by

$$V_{P,P}(\vec{j},\vec{l},\vec{q},\vec{t}) = \sum_{kk_3}\sum_{pp_3}\frac{v(\vec{k},\vec{j},\vec{l},\vec{p})}{k(k+1)+m^2}\frac{v(\vec{p},\vec{q},\vec{t},\vec{k})}{p(p+1)+m^2}$$

$$\bar{V}_{P,P}(\vec{j},\vec{l},\vec{q},\vec{t}) = \sum_{kk_3}\sum_{pp_3}\frac{v(\vec{k},\vec{j},\vec{l},\vec{p})}{k(k+1)+m^2}\frac{v(\vec{p},\vec{k},\vec{q},\vec{t})}{p(p+1)+m^2}. \tag{18.93}$$

The non-planar-non-planar and planar-non-planar contributions are given by

$$V_{NP,NP}(\vec{j},\vec{l},\vec{q},\vec{t}) = \sum_{kk_3}\sum_{pp_3}\frac{v(\vec{k},\vec{j},\vec{p},\vec{l})}{k(k+1)+m^2}\frac{v(\vec{p},\vec{q},\vec{k},\vec{t})}{p(p+1)+m^2}. \tag{18.94}$$

$$V_{P,NP}(\vec{j},\vec{l},\vec{q},\vec{q},\vec{t}) = \sum_{kk_3}\sum_{pp_3}\frac{v(\vec{k},\vec{j},\vec{l},\vec{p})}{k(k+1)+m^2}\frac{v(\vec{p},\vec{q},\vec{k},\vec{t})}{p(p+1)+m^2}. \tag{18.95}$$

In the commutative limit the planar-planar contribution $V_{P,P}$ remains finite and tends in the commutative large N limit to the result

$$V_{P,P}(\vec{j},\vec{l},\vec{q},\vec{t}) = \frac{1}{N}\sum_{kk_3}\sum_{pp_3}\frac{w(\vec{k},\vec{j},\vec{l},\vec{p})}{k(k+1)+m^2}\frac{w(\vec{p},\vec{q},\vec{t},\vec{k})}{p(p+1)+m^2}, \quad w(\vec{k},\vec{j},\vec{p},\vec{q})$$

$$= \int \frac{d\Omega}{4\pi}\, Y_{kk_3}^{+}Y_{jj_3}\,Y_{pp_3}\,Y_{qq_3}. \tag{18.96}$$

Furthermore, it can be shown that all other contributions become equal in the commutative large N limit to the above result [75]. In other words, there is no difference between planar and non-planar graphs and the UV–IR mixing is absent in this case.

Hence, to remove the UV–IR mixing from this model a standard prescription of normal ordering, which amounts to the substraction of tadpoloe contributions, will be sufficient. We consider therefore the action

$$S = \frac{1}{N}Tr[\Phi(\Delta + m^2 - 3N\lambda\Pi^P + 2\lambda Q)\Phi + \lambda\Phi^4]. \tag{18.97}$$

In above $Q = Q(\mathcal{L}_a^2)$ is given for any N by the expression

$$Q\hat{Y}_{pp_3} = Q(p)\hat{Y}_{pp_3}, \quad Q(p)$$
$$= -\frac{1}{2}\sum_k \frac{2k+1}{k(k+1)+m}\left[N(-1)^{p+k+2s}\begin{Bmatrix} p & s & s \\ k & s & s \end{Bmatrix} - 1\right]. \tag{18.98}$$

The first substraction is the usual tadpole substraction, which renders the limiting commutative theory finite. The second substraction is to cancel the UV–IR mixing. Although this action does not have the correct commutative limit (due to the non-local substraction) the corresponding quantum theory is standard phi-four in two dimensions.

18.3.2 The phase diagram

The coordinates operators $\hat{x}_a$ on the fuzzy sphere are $N \times N$ matrices defined in terms of the angular momentum generators by

$$\hat{x}_a = \frac{\theta}{R}L_a, \quad \theta = \frac{2R^2}{\sqrt{N^2 - 1}}. \tag{18.99}$$

Here, L_a are the angular momentum generators in the spin $s \equiv (N-1)/2$ irreducible representation of $SU(2)$. We have then

$$\hat{x}_1^2 + \hat{x}_2^2 + \hat{x}_3^2 = R^2, \quad [\hat{x}_a, \hat{x}_b] = \frac{\theta}{R}i\varepsilon_{abc}\hat{x}_c. \tag{18.100}$$

In the limit $N \longrightarrow \infty$ we recover the commuting sphere.

In the limit $N \longrightarrow \infty$ and $R \longrightarrow \infty$ keeping θ fixed we get the Moyal–Weyl plane.

We can construct a Weyl map and a star product, which allows us to work with the C^*-algebra of functions on the sphere $C^\infty(\mathbf{S}^2)$ and thus functions can be expanded in the basis of spherical harmonics. However, functions on the fuzzy sphere are $N \times N$ matrices, which can be expanded in $SU(2)$ polarization tensors. The Weyl map of the polarization tensors are precisely the spherical harmonics.

A scalar field on the fuzzy sphere is then an $N \times N$ matrix, which can be expanded in $SU(2)$ polarization tensors. The action and partition function of a phi-four theory on the fuzzy sphere is given by

$$S = -aTr[L_a, \Phi]^2 + Tr(b\Phi^2 + c\Phi^4). \tag{18.101}$$

$$Z[J] = \int d\Phi \ exp(-S(\Phi) - TrJ\Phi). \tag{18.102}$$

We can choose

$$a = \frac{2\pi}{N}, \quad b = \frac{2\pi r R^2}{N}, \quad c = \frac{\pi \lambda R^2}{6N}. \tag{18.103}$$

In the commutative limit $N \longrightarrow \infty$ we obtain the phi-four theory on the ordinary sphere $\mathbf{S}^2$, viz

$$S = \int_{S^2} d\Omega (\mathcal{L}_a \Phi)^2 + \int_{S^2} d\Omega \left(\frac{rR^2}{2} \Phi^2 + \frac{\lambda R^2}{4!} \Phi^4 \right). \tag{18.104}$$

However, in the limit $N \longrightarrow \infty$ and $R \longrightarrow \infty$ keeping θ fixed we get noncommutative phi-four on the Moyal–Weyl plane.

A Monte Carlo study of the $d=2$ noncommutative phi-four theory on the fuzzy sphere established the existence of an extra phase (known variously as the non-uniform phase, the stripe phase, the matrix phase) together with the two usual phases observed in the commutative theory [15–17]. Similarly, the existence of the stripe phase in $d=2$ noncommutative phi-four theory on the noncommutative torus is shown in [52]. This structure is expected to hold true in all dimensions, e.g., a Monte Carlo study of the $d=3$ phi-four theory on the noncommutative torus revealed the existence of a minimum in the dispersion relation corresponding to a stripe phase [50]. This phase structure is also observed in condensed matter physics (Lifshitz triple points) [77].

The collapsed (scaled) variables of the model are given by

$$\tilde{b} = \frac{b}{aN^{3/2}}, \quad \tilde{c} = \frac{c}{a^2 N^2}. \tag{18.105}$$

The Ising (disordered-to-uniform) transition is found to lie at the critical point

$$\tilde{c} = -0.23\tilde{b}. \tag{18.106}$$

This is a second-order phase transition, which can be located at the peaks of the specific heat and susceptibility defined by

$$C_v = \langle S^2 \rangle - \langle S \rangle^2, \quad \chi = \langle |\mathrm{Tr}\,\Phi\,|^2 \rangle - \langle |\mathrm{Tr}\,\Phi| \rangle^2. \tag{18.107}$$

The non-uniform-to-uniform transition is found to lie at the critical point

$$\tilde{c} = -0.2\tilde{b} + 0.07. \tag{18.108}$$

This seems also to be a second-order phase transition, which is in fact a continuation of the Ising transition to larger negative values of the mass parameter.

The disorder-to-non-uniform transition for small $\tilde{b}$ is located at the critical point

$$\tilde{c} = -2.29\tilde{b} - 4.74. \tag{18.109}$$

The behavior for large $\tilde{b}$ must be of the form (prediction of the real quartic matrix model)

$$\tilde{c} = \tilde{b}^2 / 4. \tag{18.110}$$

This is a third-order transition, which can be located at the point where the eigenvalue distribution of Φ splits into two disjoint sets. At this point the first derivative of C_v has a finite discontinuity.

The magnetization is defined by

$$M = \langle |Tr\Phi| \rangle. \tag{18.111}$$

The magnetization seems to change smoothly across the disorder-to-uniform transition point but it jumps at the uniform-to-uniform transition point. The magnetization seems to go through a maximum across the disorder-to-non-uniform transition.

The intersection of the above three lines yields an estimation of the tripe point of the noncommutative phi-four theory on the fuzzy sphere $\mathbf{S}_N^2$. We get

$$(\tilde{b}_T,\, \tilde{c}_T) = (-2.3,\, 0.52). \tag{18.112}$$

In summary, the noncommutative phi-four theory on the fuzzy sphere $\mathbf{S}_N^2$ exhibits the following three known phases:

- The usual second-order Ising phase transition between disordered $\langle \Phi \rangle = 0$ and uniform-ordered $\langle \Phi \rangle \sim 1_N$ phases. This appears for small values of c. This is the only transition observed in commutative phi-four, and thus, it can be accessed in a small noncommutativity parameter expansion.

  Thus, the uniform-ordered phase $\langle \Phi \rangle \sim 1_N$, is stable in this theory. This is in contrast with the case of the real quartic matrix model given by

$$V = \mathrm{Tr}(bM^2 + cM^4). \tag{18.113}$$

  Indeed, in this case the uniform-ordered phase becomes unstable for all values of the couplings. The source of this stability is obviously the addition of the kinetic term (geometry) to the action.

- A matrix transition between disordered $\langle \Phi \rangle = 0$ and non-uniform-ordered $\langle \Phi \rangle \sim \gamma$ phases with $\gamma^2 = 1_N$. This transition coincides, for very large values of c, with the third-order transition of the real quartic matrix model, i.e., the model with $a = 0$, which occurs at

$$b = -2\sqrt{Nc}. \tag{18.114}$$

  This is therefore a transition from a one-cut (disc) phase to a two-cut (annulus) phase.

- A transition between uniform-ordered $\langle \Phi \rangle \sim 1_N$ and non-uniform-ordered $\langle \Phi \rangle \sim \gamma$ phases. Some of the properties of the non-uniform phase are:
  - The non-uniform phase, in which translational/rotational invariance is spontaneously broken, is absent in the commutative theory.
  - The non-uniform phase is essentially the stripe phase observed originally on Moyal–Weyl spaces.
  - The non-uniform ordered phase is a full-blown nonperturbative manifestation of the perturbative UV–IR mixing effect, which is due to the underlying highly non-local matrix degrees of freedom of the non-commutative scalar field.

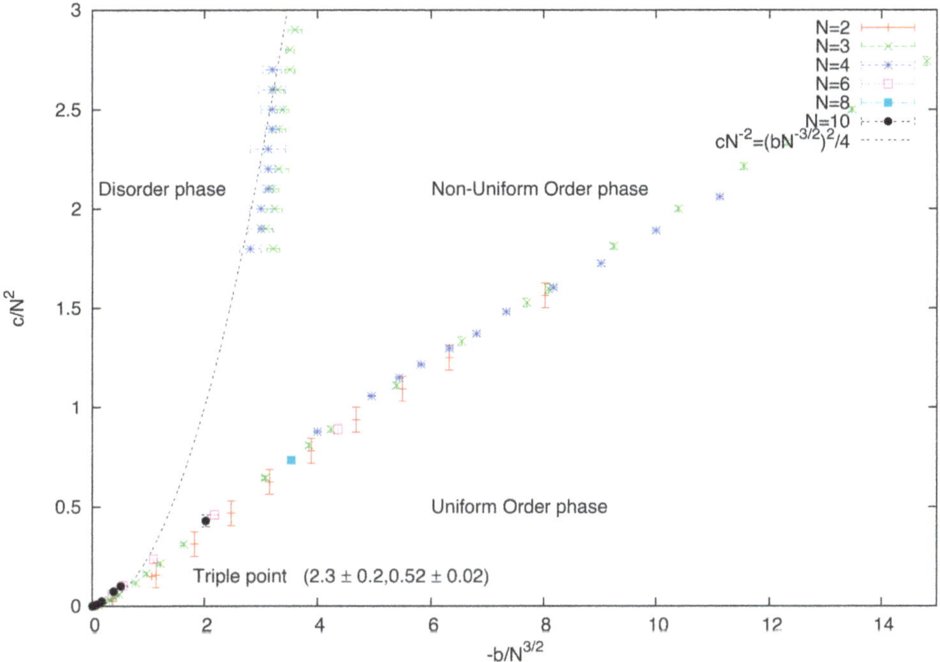

Figure 18.4. The phase diagram of phi-four theory on the fuzzy sphere.

The phase diagram is shown in figure 18.4.

18.3.3 On the phase diagram of the real quartic matrix model

The basic model is given by

$$V = B \operatorname{Tr} \Phi^2 + C \operatorname{Tr} \Phi^4, \quad \Phi^\dagger = \Phi. \tag{18.115}$$

The ground state configurations of this classical pure potential model are given by the matrices:

$$\Phi_0 = 0. \tag{18.116}$$

$$\Phi_\gamma = \sqrt{-\frac{B}{2C}}\, U\gamma U^+, \quad \gamma^2 = 1_N, \quad UU^+ = U^+U = 1_N. \tag{18.117}$$

The first configuration corresponds to the disordered (one-cut) phase whereas the second solution, valid only for $B \leqslant 0$, corresponds to the ordered (two-cut) phase. The idempotent γ can be always chosen such that $\gamma = \gamma_k = \operatorname{diag}(1_k, -1_{N-k})$.

From energy consideration the configurations Φ_γ are degenerate. If we include the effect of the kinetic energy then the configurations corresponding to $k = 0$ and $k = N$ become favorable. This is the Ising configuration $\Phi_\gamma \propto \pm 1$.

The orbit of γ_k is the Grassmannian manifold $U(N)/(U(k) \times U(N-k))$ whose dimension is $d_k = 2kN - 2k^2$. It is not difficult to show that this dimension is maximum at $k = N/2$ (assuming that N is even). Thus, from entropy consideration the most important two-cut solution is the so-called stripe configuration given by $\gamma = \mathrm{diag}(1_{N/2}, -1_{N/2})$.

In summary, we have therefore three possible phases:

$$\langle \Phi \rangle = 0: \quad \text{disordered phase.} \tag{18.118}$$

$$\langle \Phi \rangle = \pm\sqrt{-\frac{B}{2C}}\, 1_N: \quad \text{uniform phase.} \tag{18.119}$$

$$\langle \Phi \rangle = \pm\sqrt{-\frac{B}{2C}}\,\gamma: \quad \text{non-uniform phase.} \tag{18.120}$$

However, quantum mechanically, there are only two stable phases in this model [78, 79]

- **Disordered phase (one-cut) for $B \geqslant B_c$:**

$$\rho(\lambda) = \frac{1}{N\pi}(2C\lambda^2 + B + C\delta^2)\sqrt{\delta^2 - \lambda^2}, \quad -\delta \leqslant \lambda \leqslant \delta. \tag{18.121}$$

$$\delta^2 = \frac{1}{3C}(-B + \sqrt{B^2 + 12NC}). \tag{18.122}$$

- **Non-uniform ordered phase (two-cut) for $B \leqslant B_c$:**

$$\rho(\lambda) = \frac{2C|\lambda|}{N\pi}\sqrt{(\lambda^2 - r_-^2)(r_+^2 - \lambda^2)}, \quad r_- \leqslant |\lambda| \leqslant r_+. \tag{18.123}$$

$$r_\mp^2 = \frac{1}{2C}(-B \mp 2\sqrt{NC}). \tag{18.124}$$

Here, $\rho(\lambda)$ is the eigenvalue distribution of the matrix Φ.

Remark:

- **Critical point:** A third-order transition between the above two phases occurs at the critical point

$$B_c^2 = 4NC \leftrightarrow B_c = -2\sqrt{NC}. \tag{18.125}$$

- **Specific heat:** The behavior of the specific heat across the matrix transition is given by (with $\bar{B} = B/B_c$)

$$\frac{C_v}{N^2} = \begin{cases} \dfrac{1}{4} & \bar{B} < -1 \\[2ex] \dfrac{1}{4} + \dfrac{2\bar{B}^4}{27} - \dfrac{\bar{B}}{27}(2\bar{B}^2 - 3)\sqrt{\bar{B}^2 + 3} & \bar{B} > -1 \end{cases}. \tag{18.126}$$

- **Uniform ordered phase:** The real quartic matrix model admits also a solution with $\operatorname{Tr} M \neq 0$ corresponding to a possible uniform-ordered (Ising) phase. This $U(N)$-like solution can appear only for negative values of the mass parameter.

  The density of eigenvalues in this case is given by

$$\rho(z) = \frac{1}{\pi N}(2Cz^2 + 2\sigma Cz + B + 2C\sigma^2 + C\tau^2) \qquad (18.127)$$
$$\sqrt{((\sigma + \tau) - z)(z - (\sigma - \tau))}\,.$$

This is a one-cut solution centered around τ in the interval $[\sigma - \tau, \sigma + \tau]$ where σ and τ are given by

$$\sigma^2 = \frac{1}{10C}(-3B + 2\sqrt{B^2 - 15NC}), \quad \tau^2 = \frac{1}{15C}(-2B - 2\sqrt{B^2 - 15NC}). \qquad (18.128)$$

This solution makes sense only for

$$B \leqslant B_c = -\sqrt{15}\,\sqrt{NC}\,. \qquad (18.129)$$

18.3.4 A brief outline of the multitrace approach

We consider now phi-four theory on the fuzzy sphere $\mathbf{S}^2_N$ given by the action

$$S = -a\,\operatorname{Tr}[L_a, \Phi]^2 + \operatorname{Tr}(b\Phi^2 + c\Phi^4). \qquad (18.130)$$

By diagonalizing the scalar field Φ as $\Phi = U\Lambda U^\dagger$ we can put the partition function in the form

$$Z = \int d\Lambda \Delta(\Lambda) \exp\left(-\operatorname{Tr}(b\Lambda^2 + c\Lambda^4)\right) \int dU \exp\left(a\,\operatorname{Tr}[U^\dagger L_a U, \Lambda]^2\right). \qquad (18.131)$$

Expanding in powers of a and integrating over $U(N)$ leads to a multitrace matrix model of the noncommutative phi-four on the fuzzy sphere [19, 20]. For example, we find up to order a^2 the multitrace matrix model [80, 81]

$$S_{\text{eff}} = \sum_i (b\lambda_i^2 + c\lambda_i^4) - \frac{1}{2}\sum_{i\neq j}\ln(\lambda_i - \lambda_j)^2$$
$$+ \left[\frac{aN}{4}v_{2,1}\sum_{i\neq j}(\lambda_i - \lambda_j)^2 + \frac{a^2N^2}{12}v_{4,1}\sum_{i\neq j}(\lambda_i - \lambda_j)^4 - \frac{a^2}{6}v_{2,2}\left[\sum_{i\neq j}(\lambda_i - \lambda_j)^2\right]^2 + \cdots\right]. \qquad (18.132)$$

The logarithmic potential arises from the Vandermonde determinant $\Delta(\Lambda)$, i.e., from diagonalization. The coefficients $v_{2,1}$, $v_{4,1}$ and $v_{2,2}$ are given by $v_{2,1} = +1$, $v_{4,1} = 0$, $v_{2,2} = 1/8$. It is not difficult to convince ourselves that the above action is indeed a multitrace matrix model since it can be expressed in terms of various moments $m_n = \operatorname{Tr} M^n$ of the matrix $M \equiv \Phi$. Indeed, the above multitrace matrix model takes the form

$$V = B \operatorname{Tr} M^2 + C \operatorname{Tr} M^4 + D(\operatorname{Tr} M^2)^2$$
$$+ B^{'}(\operatorname{Tr} M)^2 + C^{'} \operatorname{Tr} M \operatorname{Tr} M^3 + D^{'}(\operatorname{Tr} M)^4 + A^{'} \operatorname{Tr} M^2(\operatorname{Tr} M)^2 + \cdots. \tag{18.133}$$

$$B = b + \frac{a^2 N}{2}, \quad C = c, \quad D = -\frac{a^2 N^2}{12}. \tag{18.134}$$

$$B^{'} = -\frac{aN}{2}, \quad C^{'} = 0, \quad D^{'} = -\frac{a^2}{12}, \quad A^{'} = \frac{a^2 N}{6}. \tag{18.135}$$

The saddle point method can be applied to the above multitrace matrix model to give us an estimation of the non-uniform-to-uniform transition line and the triple point. Furthermore, the whole phase diagram of the noncommutative phi-four on the fuzzy sphere can be constructed from Monte Carlo simulation of the multitrace matrix model [81, 82].

Let us consider the example of the $(\operatorname{Tr} M^2)^2$ multitrace matrix model. Remark that in the above model all multitrace terms depend on the odd moment $\operatorname{Tr} M$ with the exception of the doubletrace term $(\operatorname{Tr} M^2)^2$. The $(\operatorname{Tr} M^2)^2$ multitrace matrix model is given by the model (18.133) with all odd moments set to zero, i.e., by imposing the symmetry $M \longrightarrow - M$. We obtain a doubletrace matrix model given by

$$V = B \operatorname{Tr} M^2 + C \operatorname{Tr} M^4 + D(\operatorname{Tr} M^2)^2. \tag{18.136}$$

The two stable phases are given by the disordered (one-cut) phase and the non-uniform-ordered (two-cut) phase separated by a deformation of the line $\tilde{B}_* = -2\sqrt{\tilde{C}}$ given by [80, 83]

$$\tilde{B}_* = -2\sqrt{\tilde{C}} - \frac{2\tilde{D}}{\sqrt{\tilde{C}}}. \tag{18.137}$$

Here, $\tilde{D} = D/N = -2v_{2,2}\tilde{a}^2/3$. There exists a termination point in this model since the critical line does not extend to zero [80, 82]. Indeed, in order for the critical value $\tilde{B}_*$ to be negative one must have $\tilde{C}$ in the range

$$\tilde{C} \geqslant \tilde{C}_* = \frac{\tilde{a}^2}{12}. \tag{18.138}$$

Thus, the termination point is located at (for $\tilde{a} \equiv a\sqrt{N} = 1$)

$$(\tilde{B}, \tilde{C}) = (0, 1/12). \tag{18.139}$$

The triple point is identified as a termination point of the one-cut-to-two-cut transition line and is located in the Monte Carlo data of the full model, i.e., with the odd terms included, at $(-1.05, 0.4)$, which compares favorably with previous Monte Carlo estimate. See figure 18.5.

Thus, the phase diagram of the multitrace matrix model (18.133) does not contain the uniform-ordered phase. In fact, this phase diagram will not contain the uniform-

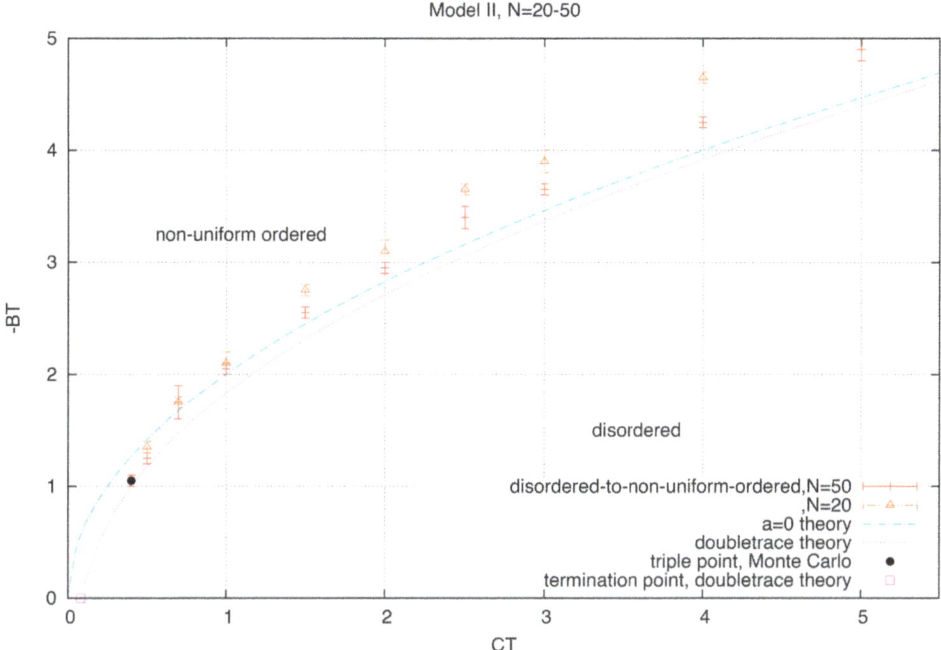

Figure 18.5. The termination point of the doubletrace matrix model as the triple point.

ordered phase even after including the terms depending on the odd moment $\operatorname{Tr} M$. This means in particular that the above multitrace matrix model (18.133), with the parameters (18.134) and (18.135), lies in the universality class of the real quartic matrix model. It is expected that the correct phase diagram of the noncommutative phi-four on the fuzzy sphere $\mathbf{S}_N^2$ can be reproduced by considering higher order multitrace corrections. The inclusion of the terms depending on the odd moment $\operatorname{Tr} M$ is absolutely necessary in order to correctly reproduce the uniform-ordered phase. For example, in [81, 82] the multitrace matrix model (18.132) is considered with the parameters [19]

$$v_{2,1} = -1, v_{4,\,1} = \frac{3}{2}, v_{2,2} = 0. \tag{18.140}$$

Explicitly, we have the multitrace action

$$V = B \operatorname{Tr} M^2 + C \operatorname{Tr} M^4 + D[\operatorname{Tr} M^2]^2 + B^{'}(\operatorname{Tr} M)^2 + C^{'} \operatorname{Tr} M \operatorname{Tr} M^3. \tag{18.141}$$

$$B = b - \frac{aN^2}{2}, \quad C = c + \frac{a^2N^3}{4}, \quad D = \frac{3a^2N^2}{4}, \quad B^{'} = \frac{aN}{2}, \quad C^{'} = -a^2N^2. \tag{18.142}$$

The phase diagram of this model is shown to contain the three phases of the noncommutative phi-four on the fuzzy sphere $\mathbf{S}_N^2$. Indeed, the uniform-ordered phase exists in this model only with the terms depending on the odd moment $\operatorname{Tr} M$

included. If we assume the symmetry $M \longrightarrow -M$ then the second line of (18.133) becomes identically zero and the uniform ordered phase disappears.

In [84] an even simpler multitrace matrix model is shown to exhibit the correct phase structure of the noncommutative phi-four on the fuzzy sphere $\mathbf{S}_N^2$. This model includes a single term depending on the odd moment $\mathrm{Tr}\,M$ given by $\mathrm{Tr}\,M\,\mathrm{Tr}\,M^3$. Explicitly, the action of this multitrace matrix model is given by

$$V = B\,\mathrm{Tr}\,M^2 + C\,\mathrm{Tr}\,M^4 + C'\,\mathrm{Tr}\,M\,\mathrm{Tr}\,M^3. \tag{18.143}$$

$$B = b, \quad C = c, \quad C' = -a^2 N^2. \tag{18.144}$$

18.3.5 Critical exponents

The critical exponents of the noncommutative phi-four theory on the fuzzy sphere $\mathbf{S}_N^2$ can be computed from its first multitrace approximation, which reproduces the correct phase diagram, i.e., a phase diagram, which includes the Ising phase and a triple point. A priori, the emergence of the Ising phase in any matrix model is rather surprising but in these multitrace matrix models it is guaranteed that the Ising phase appears at some higher order of the multitrace approximation.

A simpler approach is to consider a simpler multitrace matrix model, which involves the odd moment $\mathrm{Tr}\,M$ and hence is likely to include in its phase diagram an Ising phase and a triple point representing therefore the universality class of the noncommutative phi-four theory on the fuzzy sphere $\mathbf{S}_N^2$. This is the route taken in [81, 82] with the simple multitrace matrix model (18.141)+(18.142) and also taken in [84] with the even much simpler multitrace matrix model (18.144)+(18.144).

In the remainder we discuss in some detail the phase structure and the critical exponents of the multitrace matrix model (18.141)+(18.142) discussed in [81, 82]. We claim that this model contains three stable phases: disorder, uniform order and non-uniform order. See figure 18.6.

- The Ising and the matrix phase transitions:
  - **Ising**: The critical point is measured at the peak of the susceptibility $\chi = \langle\,|TrM\,|^2\rangle - \langle\,|TrM\,|\rangle^2$. The fit for the extrapolated critical value is given by

$$\tilde{C} = 0.291(0).\,(-\tilde{B}) + 0.104(1). \tag{18.145}$$

  - **Matrix**: The critical point is determined at the point where the eigenvalue distributions go from one-cut, in the disorder phase, to two-cut, in the non-uniform phase. The splitting of the distribution is considered to have been occurred when the height of the distribution at $\lambda = 0$ is less than some tolerance. The fit for the extrapolated critical value is given by

$$\tilde{C} = 2.206(67).\,(-\tilde{B}) - 7.039(301). \tag{18.146}$$

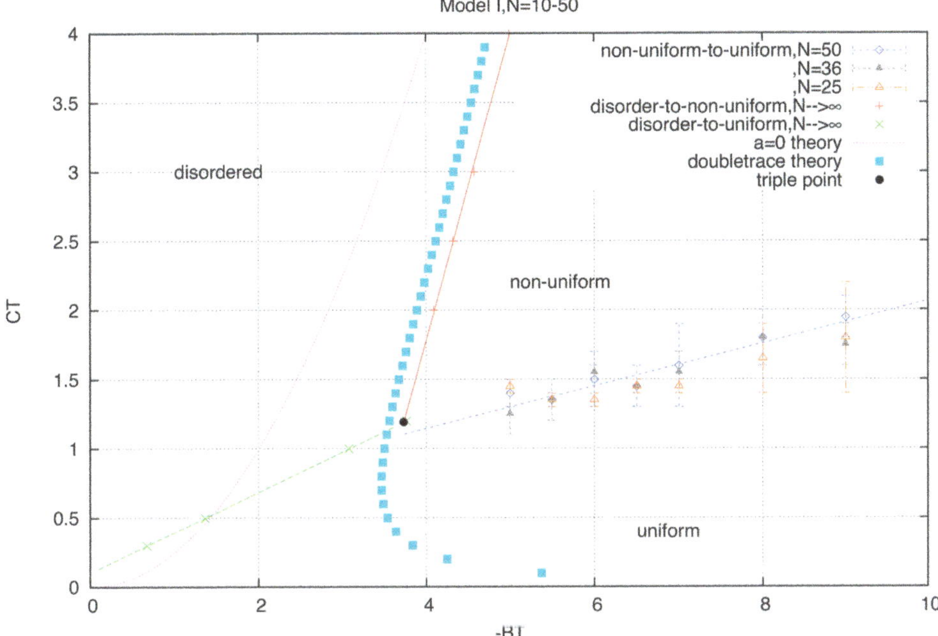

Figure 18.6. The phase diagram of the multitrace model (18.141)+(18.142).

The behavior of the specific heat across this transition is effectively that of the pure quartic matrix model $a = 0$. However, the critical line is better approximated with the doubletrace matrix model prediction.

- The stripe phase transition:
  - We approach the critical boundary by fixing $\tilde{B}$ and changing $\tilde{C}$ starting from small values, i.e., inside the uniform ordered phase, until the curves for the total and zero powers start to diverge, marking the transition to the non-uniform ordered phase.
  - Since this is a very delicate transition we do not perform any extrapolation of the critical point and the critical boundary is given by the fit of the largest value of N. In any case we observe no strong dependence on N of the measured critical value $\tilde{C}$. The stripe critical line is approximated by the fit for $N = 50$ given by

$$\tilde{C} = 0.154(22).\,(-\tilde{B}) + 0.530(131), \quad N = 50. \tag{18.147}$$

The critical exponents in the context of the multitrace model (18.141)+(18.142) are computed using the Monte Carlo method in [82].

- **The critical exponent ν:**
  - The correlation length should behave as

$$\xi \sim |B - B_c|^{-\nu} \sim N, \quad \nu_{\text{Onsgaer}} = 1. \tag{18.148}$$

- This is very delicate to check explicitly in Monte Carlo. Indeed, since we must necessarily deal with the critical region we must face the two famous problems: (1) finite size effects and (2) critical slowing down. The critical slowing down problem can be shown to start appearing in Monte Carlo simulations around $N > 60$ so we will keep below this value and employ very large statistics of the order of 2^{20}. The problem of finite size effects is also very serious for the measurement of the critical exponents since the above behavior (18.183) is supposed to hold only for large N. This problem can be avoided by including values of N larger or equal than 20.
- We choose $\tilde{C} = 1.0$, which is relatively large, but well within the Ising phase, before the appearance of the transition between the disordered and non-uniform-ordered phases around $\tilde{C} = 1.5$.
- We plot the critical point $\tilde{B}_c$ versus N. We get the $N = \infty$ critical point $\tilde{B}_*$ and the critical exponent ν:

$$\tilde{B}_c = -1.061(168). \; N^{-0.926(83)} - 3.074(6) \Rightarrow , \quad \nu = 0.926(83). \tag{18.149}$$

Also, we obtain

$$\tilde{B}_* = -3.074(6). \tag{18.150}$$

This prediction for ν agrees reasonably well with the Onsager calculation $\nu = 1$.

- **The critical exponent β:**
  - The magnetization is defined by

$$m = \langle |\mathrm{Tr}\, M| \rangle. \tag{18.151}$$

  - The critical behavior is

$$m/N = \langle |\mathrm{Tr}\, M| \rangle / N \sim (B_c - B)^\beta \sim N^{-\beta/\nu}, \quad \beta_{\mathrm{Onsager}} = \frac{1}{8}. \tag{18.152}$$

  - Measurements of the magnetization m/N were performed near the extrapolated critical point, $\tilde{B} = -3.07$, for $\tilde{C} = 1.0$, but inside the uniform-ordered phase. These are then used to compute the critical exponent β by searching for a power law behavior.
  - We measure $\ln(m/N)$ versus $\ln N$, for each value of $\tilde{B}$ very near and around $\tilde{B} = -3.10$, fit to a straight line in the range $20 \leqslant N \leqslant 60$, and compute the slope β, then search for the flattest line, i.e., the smallest slope β.
  - Deep inside the Ising phase the slope should approach the mean field value $-1/4$, which can be shown from the scaling behavior of the dominant configuration.

- After determining the critical value we then consider the value of $\tilde{B}$ nearest to it but within the Ising phase and take the slope there to be the value of the critical exponent β.
- In our example, here, with $\tilde{C} = 1.0$, the flattest line occurs at $\tilde{B} = -3.13$, with slope $-0.088(10)$, thus the measurement of the critical value from the magnetization is

$$\tilde{B}_* = -3.13. \tag{18.153}$$

- After this value the slope becomes $-0.109(11)$ at $\tilde{B} = -3.14$. The slope goes fast to the mean field value -0.25 as we keep decreasing $\tilde{B}$.
- Our measured value, for $\tilde{C} = 1.0$, of the critical exponent β is:

$$\ln \frac{m}{N} = -0.109(11). \ln N - 1.423(43) \Rightarrow \beta = -0.109(11). \tag{18.154}$$

- **The critical exponent γ:**
  - The critical behavior of the susceptibility is given by

$$\chi = \langle |\mathrm{Tr}\, M\,|^2 \rangle - \langle |\mathrm{Tr}\, M| \rangle^2 \sim (B - B_c)^{-\gamma} \sim N^{\gamma/\nu} \sim N^{2-\eta}. \tag{18.155}$$

  - If we try to fit the values of the susceptibility at its maximum, i.e., at the peak, which keeps slowly moving with $\tilde{B}$, then we will obtain a very bad underestimate of the critical exponent γ given by

$$\ln \chi_{\mathrm{max}} = 0.515(08). \ln N - 0.652(30) \Rightarrow \gamma = 0.515(08). \tag{18.156}$$

    This is due in part to the dependence of $\tilde{B}_c$ on N, and in another part, is an indication of the critical slowing down problem showing up in the measurement of this second moment, i.e., the size of the fluctuations is observed to grow with N at the critical point but not at the correct rate indicated by the independent measurements of the zero moment and the magnetization.
  - The measurement of the critical exponent γ is thus, quite delicate, and will be done indirectly as follows. We rewrite the susceptibility in terms of the zero power and magnetization as

$$\begin{aligned}
\chi &= \langle |\mathrm{Tr}\, M\,|^2 \rangle - \langle |\mathrm{Tr}\, M| \rangle^2 \\
&= N^2 P_0 - m^2.
\end{aligned} \tag{18.157}$$

$$P_0 = \left\langle \left(\frac{1}{N} \mathrm{Tr}\, M \right)^2 \right\rangle. \tag{18.158}$$

The critical exponent γ in terms of the critical exponent γ' of P_0 is then given by

$$\gamma = 2 + \gamma'. \tag{18.159}$$

- We can check that the second term in the susceptibility behaves using the result (18.154) as

$$\ln m^2 = 1.782(22). \ln N - 2.846(86) \Rightarrow \gamma = 1.782(22). \tag{18.160}$$

This measurements of the critical exponent γ agree reasonably well with the Onsager values.
- From the results, at $\tilde{B} = -3.14$, we obtain the exponent

$$\ln N^2 P_0 = 1.648(10). \ln N - 2.289(36) \Rightarrow \gamma = 1.648(10). \tag{18.161}$$

- **The critical exponent α:**
  - The sepcific heat is defined by

$$C_v = \langle S^2 \rangle - \langle S \rangle^2. \tag{18.162}$$

The critical point $\tilde{B}_*$ as measured from the specific heat is identified by the intersection point of the various curves with different N. We get

$$\tilde{B}_* = -3.08. \tag{18.163}$$

This measurement is contrasted very favorably with the independent measurement obtained from the extrapolated value of $\tilde{B}_c$ shown in equation (18.150) but should also be contrasted with the measurement obtained from the magnetization shown in equation (18.153).
  - The critical behavior is

$$C_v/N^2 \sim (B - B_c)^{-\alpha} \sim N^{\alpha/\nu}, \quad \alpha_{\text{Onsager}} = 0. \tag{18.164}$$

  - From the results, at the critical point $\tilde{B} = -3.08$, we obtain

$$\ln \frac{C_v}{N^2} = 0.024(9). \ln N - 0.623(31) \Rightarrow \alpha = 0.024(9). \tag{18.165}$$

18.3.6 Coupling to a Yang–Mills theory

It is very hard to observe the phase transition from the non-uniform-ordered to the uniform-ordered phases using the ordinary Metropolis algorithm due to the existence of an infinite number of non-uniform-ordered vacuum states that can be occupied by the system. In fact, there is a tendency for the system to become stuck in one of these non-uniform-ordered vacuum states, and as a consequence, tunneling to other vacuum states is very hard to observe in practice.

The correct sampling of the tunneling between vacuum states is essential for any Monte Carlo method to work properly. The only known method, which was quite successful in overcoming this problem is due to Garcia Flores [15, 85], which uses together with the Metropolis algorithm elements from the annealing algorithm. His algorithm does however break detail balance.

We claim here that an exact Metropolis algorithm using gauge invariance is sufficient to probe the transition from the non-uniform-ordered to the uniform-ordered phases.

The construction of this algorithm goes as follows [31]. By coupling the scalar field to a $U(1)$ gauge field, in a particular way, we can use gauge symmetry to completely diagonalize the scalar sector and reduce it thus to an eigenvalue problem.

In more detail, coupling the noncommutative scalar field Φ on the fuzzy sphere $\mathbf{S}_N^2$ to a Yang–Mills theory is done through the substitution $L_a \longrightarrow X_a/\alpha$. The parameter α plays the role of the gauge coupling constant. The matter action, i.e., the action of the noncommutative scalar field becomes

$$S_m = -\frac{a}{\alpha^2} Tr[X_a, \Phi]^2 + Tr(b\Phi^2 + c\Phi^4). \tag{18.166}$$

The dynamics of the covariant matrix coordinates X_a is given by a Yang–Mills matrix model in $D=3$ dimensions. In order for the underlying geometry to be a fuzzy sphere it is absolutely necessary to include also a Chern–Simons term. The gauge action reads then

$$S_g = NTr\left(-\frac{1}{4}[X_a, X_b]^2 + \frac{2i\alpha}{3}\varepsilon_{abc} X_a X_b X_c\right). \tag{18.167}$$

For $b = c = 0$ the relevant solutions of the classical equations of motion are $X_a = \alpha L_a$ and $\Phi = 0$. In fact, for $b = c = 0$ the action $S_g + S_m$ describes a Yang–Mills matrix model in $D=4$ dimensions where the fourth covariant matrix coordinate is proportional to the scalar field, viz $\Phi = \alpha\sqrt{N/2a}\,X_4$ [86].

The gauge field A_a is introduced as $X_a = \alpha(L_a + A_a)$. The scalar field Φ is in the adjoint representation of the $U(1)$ gauge group. Thus, in the commutative limit $N \longrightarrow \infty$ the scalar and gauge fields decouple and the Φ-dynamics of the action $S_m + S_g$ reduces to the dynamics of phi-four theory on the sphere.

It is well established that the theory with $b = c = 0$ suffers from an emergent geometry transition. The fuzzy sphere is only stable for [86]

$$\tilde{\alpha} = \sqrt{N}\alpha \geqslant \tilde{\alpha}_* = 2.55. \tag{18.168}$$

We must choose always $\tilde{\alpha} \gg \tilde{\alpha}_*$ to avoid this transition.

We diagonalize the scalar field as before, viz $\Phi = U\Lambda U^+$. However, we can now use the available gauge invariance to perform the integral over U. The resulting partition function is

$$Z = \int \prod_a dX_a \int d\Lambda \, \exp\left[NTr\left(\frac{1}{4}[X_a, X_b]^2 - \frac{2i\alpha}{3}\varepsilon_{abc} X_a X_b X_c\right) + S[\Lambda]\right]. \tag{18.169}$$

The scalar action is given by

$$S[\Lambda] = N\sum_i (X_a^2)_{ii}\lambda_i^2 - N\sum_{ij}(X_a)_{ij}(X_a)_{ji}\lambda_i\lambda_j + \sum_i (r\lambda_i^2 + u\lambda_i^4) - \sum_{i\neq j}\ln|\lambda_i - \lambda_j|. \tag{18.170}$$

We use an exact Metropolis algorithm to generate statistically independent matrices X_a and eigenvalues λ_i. This is done for a large value of $\tilde{\alpha} = \sqrt{N}\alpha$ where we know that the scalar field propagates on a fuzzy sphere.

In summary, instead of sampling the unitary matrix U in the noncommutative phi-four theory on the fuzzy sphere (which is very inefficient) we have now to sample the covariant matrix coordinates X_a (which is known to be very efficient).

18.4 Emergent geometry from multitrace matrix models

18.4.1 The $\mathrm{Tr}\, M\, \mathrm{Tr}\, M^3$ multitrace matrix model

The multitrace approach to noncommutative scalar field theories was initiated in [19, 20] on the fuzzy sphere. For an earlier approach see [87] and for a similar non-perturbative approach see [83, 88–90] and [91–93].

In this approach, as we have seen, we diagonalize the Hermitian scalar matrix Φ as $\Phi = UMU^\dagger$ and then integrate over the unitary matrix U using the group theoretic structure and properties of $SU(2)$ and $SU(N)$ extensively. In effect, in this approach the kinetic term is expanded while the potential term is treated exactly. This is in fact a hopping-parameter-like expansion. The end result is to convert the kinetic term into a multitrace matrix model, which to the lowest non-trivial order, is of the form

$$\int dU \exp\left(a\,\mathrm{Tr}[L_a,\, UMU^\dagger]^2 - \mathrm{Tr}\, V(M)\right)$$
$$= \exp\big(B\,\mathrm{Tr}\, M^2 + C\,\mathrm{Tr}\, M^4 + D(\mathrm{Tr}\, M^2)^2 + B'(\mathrm{Tr}\, M)^2 \tag{18.171}$$
$$+ C'\,\mathrm{Tr}\, M\,\mathrm{Tr}\, M^3 + D'(\mathrm{Tr}\, M)^4 + A'\,\mathrm{Tr}\, M^2(\mathrm{Tr}\, M))^2 + \cdots\big).$$

The parameters B and C have shifted values with respect to the original parameters of the potential b and c, respectively, while the primed parameters and D have purely quantum values coming from the hopping-parameter-like expansion of the kinetic term.

The basic statement is that the Laplacian operator on the fuzzy sphere (which is the example considered here) is exactly equivalent to this multitrace matrix model with these particular value of the coefficients. Each Laplacian operator, i.e., any other noncommutative space comes with its own set of multitrace coefficients.

It has been shown in [81, 84] that the essential features of the phase diagram of noncommutative phi-four theory in two dimensions can be captured by a truncated multitrace matrix model depending on the cubic moment $\mathrm{Tr}\, M^3$, i.e., a multitrace matrix model given simply by the potential

$$V_{\mathrm{trunc}} = B\,\mathrm{Tr}\, M^2 + C\,\mathrm{Tr}\, M^4 + C'\,\mathrm{Tr}\, M\,\mathrm{Tr}\, M^3. \tag{18.172}$$

Naturally, stability requirement of this multitrace matrix model constrains the range of the quartic coupling C in a particular way. The statistical physics of this multitrace matrix model interpolates between the statistical physics of the noncommutative field theory (18.1) and (18.3) and the statistical physics of the real quartic matrix model given by

$$V_{\text{pure}} = B \operatorname{Tr} M^2 + C \operatorname{Tr} M^4. \tag{18.173}$$

Indeed, the random multitrace matrix model (18.172) lies in the universality class of noncommutative phi-four theory for C' negative whereas it lies in the universality class of the real quartic matrix model for C' positive.

The above random multitrace matrix model should be thought of as a generalization of the discretization of random Riemannian surfaces with regular polygons such as dynamical triangulation [94, 95]. Indeed, and as we will discuss shortly, random multitrace matrix models can sustain emergent geometry as well as growing dimensions and topology change. For example, the multitrace matrix model (18.172) works in two dimensions but also it works away from two dimensions where it can generate a large class of spaces, such as fuzzy projective space $\mathbf{CP}^n_N$ [21], which admit finite spectral triples that can be captured by the multitrace term $\operatorname{Tr} M \operatorname{Tr} M^3$.

In terms of the collapsed or scaled parameters $\tilde{B} = B/N^{3/2}$, $\tilde{C} = C/N^2$ and $\tilde{C}' = C'/N$ the multitrace matrix model (18.172) is rewritten (with the scaled field $\tilde{M} = N^{1/4}M$) as

$$V_{\text{trunc}} = N\tilde{B}\operatorname{Tr} M^2 + N\tilde{C}\operatorname{Tr} M^4 + \tilde{C}' \operatorname{Tr} M \operatorname{Tr} M^3. \tag{18.174}$$

In the large N limit the saddle point equation controls the statistical quantum physics of the theory. The saddle point equation and the corresponding free energy read explicitly

$$2\tilde{B}\lambda + 4\tilde{C}\lambda^3 + \tilde{C}'m_3 + 3\tilde{C}'m_1\lambda^2 - 2\int d\lambda'\rho(\lambda')\frac{1}{\lambda - \lambda'} = 0. \tag{18.175}$$

$$\frac{E}{N^2} = \int d\lambda\rho(\lambda)(\tilde{B}\lambda^2 + \tilde{C}\lambda^4) + \tilde{C}' \int d\lambda\rho(\lambda) \int d\lambda'\rho(\lambda')\lambda'^3$$
$$- \frac{1}{2} \int d\lambda\rho(\lambda) \int d\lambda'\rho(\lambda')\log(\lambda - \lambda')^2. \tag{18.176}$$

The last term in both equations comes from the Vandermonde determinant, which causes the eigenvalues to repel each other and become spread evenly around zero. In the saddle point equation (18.175) there appears also the moments $m_q = \int d\lambda\rho(\lambda)\lambda^q$, which modify the behavior of the real quartic matrix model through the multitrace coupling C'.

The matrix model (18.1)+(18.3) without kinetic term ($a = 0$) can be solved exactly to obtain a phase structure consisting of
- A) a disordered (symmmetric, one-cut, disk) phase, and
- B) a non-uniform ordered (stripe, two-cut, anuulus) phase,

separated by a third-order transition line [78, 79].

This is the so-called matrix line and it corresponds to the matrix model fixed point of noncommutative field theory at infinite noncommutativity $\theta = \infty$ [47, 67, 96, 97].

The matrix model (18.1)+(18.3) with a kinetic term ($a \neq 0$) is much harder to solve but much more interesting. The phase diagram, in addition to the above two phases A) and B), involves

- a C) uniform (asymmetric, one-cut, Ising) ordered phase.

The three phases meet at a triple point where the three coexistence curves intersect. This phase diagram is also observed in the truncated multitrace matrix model (18.172).

The main difference between the real quartic matrix model (18.173) from the one hand, and the noncommutative phi-four theory (18.1)+(18.3) on the other hand, lies in the fact that the uniform-ordered phase is stable in the latter.

The three phases A), B) and C) in the real quartic matrix model are given explicitly by the following density eigenvalues [79]

$$A): \quad \rho(z) = \frac{1}{N\pi}(2Cz^2 + B + C\delta^2)\sqrt{\delta^2 - z^2},$$
$$- B \leqslant -B_c = 2\sqrt{NC}. \tag{18.177}$$

$$B): \quad \rho(z) = \frac{2C|z|}{N\pi}\sqrt{(z^2 - \delta_1^2)(\delta_2^2 - z^2)},$$
$$- B \geqslant -B_c = 2\sqrt{NC}. \tag{18.178}$$

$$C): \quad \rho(z) = \frac{1}{\pi N}(2Cz^2 + 2\sigma Cz + B + 2C\sigma^2 + C\tau^2)$$
$$\times \sqrt{((\sigma + \tau) - z)(z - (\sigma - \tau))},$$
$$- B \geqslant -B_c = \sqrt{15}\sqrt{NC}. \tag{18.179}$$

The expressions of the various cuts δ, δ_1, δ_2, σ and τ in terms of the parameters of the model can be found in [79]. Following Shimamune it is then not very difficult to show that the free energy E_C in the uniform-ordered phase is always higher than the free energies E_A and E_B in the disordered and non-uniform-ordered phases and hence the uniform-ordered phase is metastable in this model. In particular, the energies E_B and E_C are given explicitly by [79]

$$E_B = -\frac{B^2}{4NC} + \frac{1}{4}\ln\frac{NC}{4B^2} - \frac{3}{8}$$
$$E_C = -\frac{B^2}{4NC} + O(C^0), \quad C \longrightarrow 0. \tag{18.180}$$

In the limit $C \longrightarrow 0$ we have $E_B < E_C$ because of the logarithmic contribution. See figure 18.7.

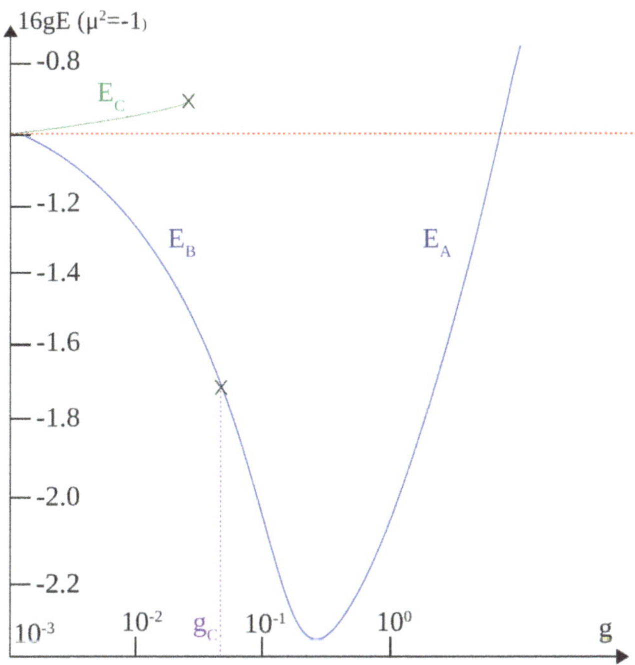

Figure 18.7. The free energy of the real quartic matrix model (18.173) in the disordered, uniform-ordered non-uniform-ordered phases for $\mu^2 = -1$. The coupling constants are identified as $g \equiv NC$ and $\mu^2 \equiv 2B$. The uniform-ordered phase is metastable in this model [79].

Similarly, the main difference between the real quartic matrix model (18.173) from the one hand, and the multitrace matrix model (18.172) on the other hand, lies in the fact that the uniform-ordered phase is stable in the latter. This important fact can be seen in both the saddle point equation and the Monte Carlo algorithm. For example, the saddle point equation (18.175) of the multitrace matrix model (18.172) can be solved exactly to determine the boundaries between the three phases A), B) and C) and the location of the triple point by following the remarkable work of [20] who solved a much larger class of multitrace matrix models.

The important point to stress here is the fact that the uniform-ordered phase becomes stable within the multitrace matrix model (18.172), in contrast to the case of the pure matrix model (18.173), which can be shown using a mean-field-like analysis as follows.

First, by using the invariance of the partition function of the multitrace matrix model (18.172) under $M \longrightarrow (1 + \varepsilon)M$ we can derive the Schwinger–Dyson identity

$$\frac{\langle V \rangle}{N^2} = \frac{1}{4} + \frac{\tilde{B}}{2}\langle \frac{1}{N} \operatorname{Tr} M^2 \rangle. \tag{18.181}$$

Next we can use the classical configurations $M = 0$, $M = m_B \gamma$ and $M = m_C 1$ deep inside the disordered, non-uniform-ordered and uniform-ordered phases to estimate the energies E_A, E_B and E_C, respectively.

Thus, we calculate $m_C^2 = -\tilde{B}/2(\tilde{C}+\tilde{C}')$ which indicates that the model for $\tilde{B}<0$ is only stable if $\tilde{C}>-\tilde{C}'$. The energy E_C in the uniform configuration is then estimated to be given by $E_C = (\tilde{C}+\tilde{C}' - \tilde{B}^2)/4(\tilde{C}+\tilde{C}')$. Similarly, we calculate $m_B^2 = -\tilde{B}/2\tilde{C}$ using the fact that $\gamma^2 = 1$. The energy E_B in the non-uniform configuration is then estimated to be given by $E_B = (\tilde{C}-\tilde{B}^2)/4\tilde{C}$. It is not difficult to check that $E_B > E_C$ (and hence the uniform-ordered phase is much more stable than the non-uniform-ordered phase) if and only if $\tilde{C}' < 0$ and vice versa.

Remark that the effect of the multitrace term in the Ising phase is then only to shift the quartic coupling as $\tilde{C} \longrightarrow \tilde{C}+\tilde{C}'$ whereas there is no effect in the stripe phase from the multitrace term. Hence, the energies E_B and E_C in this mean-field-like approximation are estimated as follows

$$E_B = -\frac{B^2}{4NC} + \frac{1}{4}\ln\frac{NC}{4B^2} - \frac{3}{8}$$

$$E_C = -\frac{B^2}{4N(C + C')} + O((C + C')^0), \quad C + C' \longrightarrow 0. \tag{18.182}$$

Since we cannot reach the point $C=0$ (the model becomes unstable there) we can only take the limit $C \longrightarrow - C'$. In this case we have instead $E_C < E_B$.

18.4.2 The phase diagram

As we have said, the noncommutative phi-four theory (18.1)+(18.3) and the random multitrace matrix model (18.172) share similar phase diagrams up to and including the triple point. Indeed, the multitrace matrix model (18.172) captures, in fact surprisingly very well, the phase structure of noncommutative phi-four theory. This phase structure can be probed non-perturbatively using the Monte Carlo algorithm.

A direct simulation of the noncommutative phi-four theory (18.1)+(18.3) is very involved for various physical and technical reasons [15–18, 31]. Thus, the numerical results reported here are simply obtained by applying the Monte Carlo method or more precisely the Metropolis algorithm to the multitrace matrix model (18.172) with the value $C' = -N$, which is the value obtained for the multitrace interaction $\mathrm{Tr}\, M\, \mathrm{Tr}\, M^3$ in the multitrace expansion of the kinetic term in the noncommutative phi-four theory (18.1)+(18.3) on the fuzzy sphere [19, 20].

We have for both cases, i.e., for the noncommutative phi-four theory (18.1)+(18.3) and for the random multitrace matrix model (18.172), the following phase structure:

1. A second-order phase transition between disordered ($\Phi \sim 0$) and uniform-ordered ($\Phi \sim 1$) phases at small values of the quartic coupling. This is the usual Ising phase transition [98]. The eigenvalue distribution $\rho(\lambda)$ as it transits between the disordered and uniform-ordered phases in the multitrace matrix model (18.172) is shown in figure 18.8.

2. A second-order phase transition between non-uniform-ordered ($\Phi \sim \gamma$ with $\gamma^2 = 1$) and uniform-ordered ($\Phi \sim 1$) phases, which is a continuation of the Ising transition to large values of the quartic coupling [49]. The eigenvalue

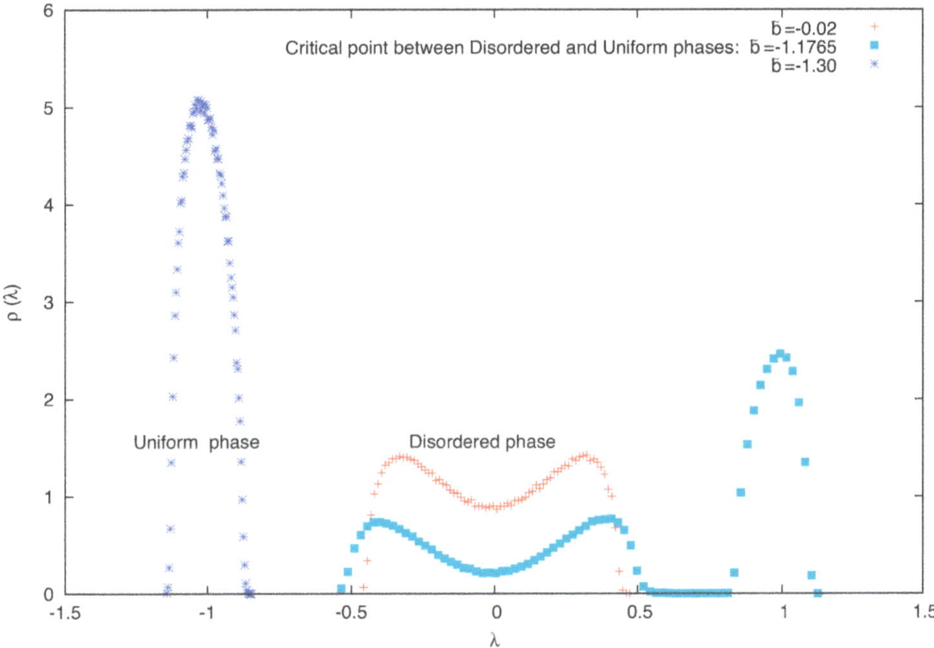

Figure 18.8. The eigenvalue distribution as it transits between the disordered and uniform-ordered phases in the multitrace matrix model (18.172).

distribution $\rho(\lambda)$ as it transits between the uniform-ordered and non-uniform-ordered phases in the multitrace matrix model (18.172) is shown in figure 18.9.

3. A third-order phase transition between disordered and non-uniform ordered phases. This is the matrix phase transition observed previously in the real quartic matrix model (18.173) (model without kinetic term). The eigenvalue distribution $\rho(\lambda)$ as it transits between the disordered and non-uniform-ordered phases in the multitrace matrix model (18.172) is shown in figure 18.10.

4. The three coexistence curves intersect at a triple point. The full phase diagram of the multitrace matrix model (18.172) in the plane $(-\tilde{B}, \tilde{C})$, as measured by the Monte Carlo method, is shown in figure 18.11 where the coexistence line $\tilde{C} = \tilde{B}_c^2/4$ of the real quartic matrix model (18.173) is also included for comparison. The agreement with the phase diagram of non-commutative phi-four (18.1)+(18.3) on the fuzzy sphere is extremely favorable.

The observables used in the Monte Carlo measurement of the phase diagram include, in addition to the eigenvalue (EV) distribution $\rho(\lambda)$, the specific heat $C_v = \langle S^2 \rangle - \langle S \rangle^2$ where $S \equiv V_{\text{trunc}}$, the susceptibility (sus) defined by $\chi = \langle |\text{Tr}\, M|^2 \rangle - \langle |\text{Tr}\, M| \rangle^2$ and the power in the zero mode $P_0 = \langle (\text{Tr}\, M)^2 / N^2 \rangle$. We also measure the magnetization $m = \langle |\text{Tr}\, M| \rangle$, the total power $P_T = \langle \text{Tr}\, M^2 \rangle / N$ and the quartic coupling $\langle \text{Tr}\, M^4 \rangle$.

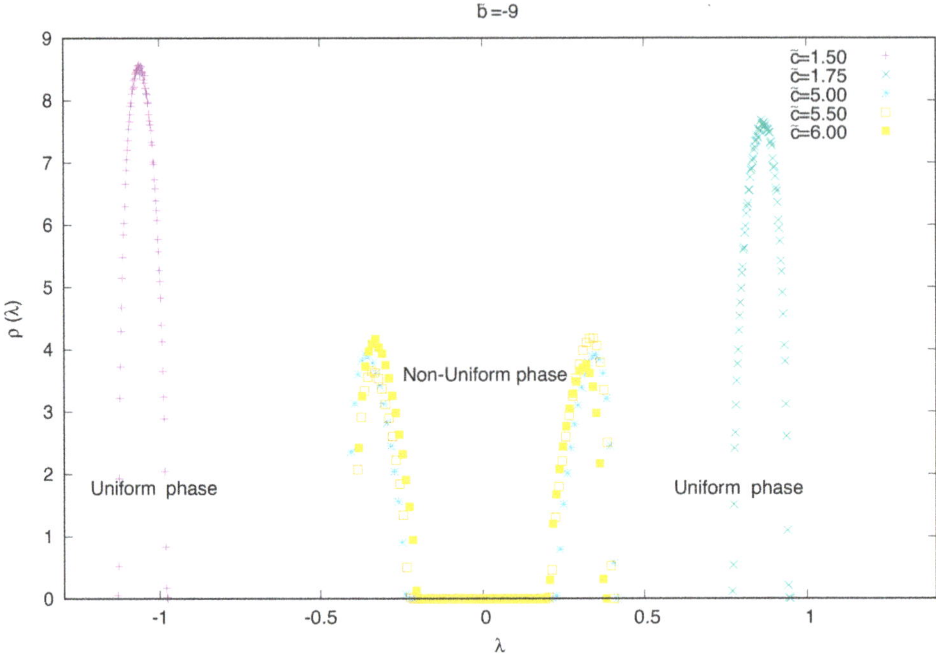

Figure 18.9. The eigenvalue distribution as it transits between the uniform-ordered and non-uniform-ordered phases in the multitrace matrix model (18.172).

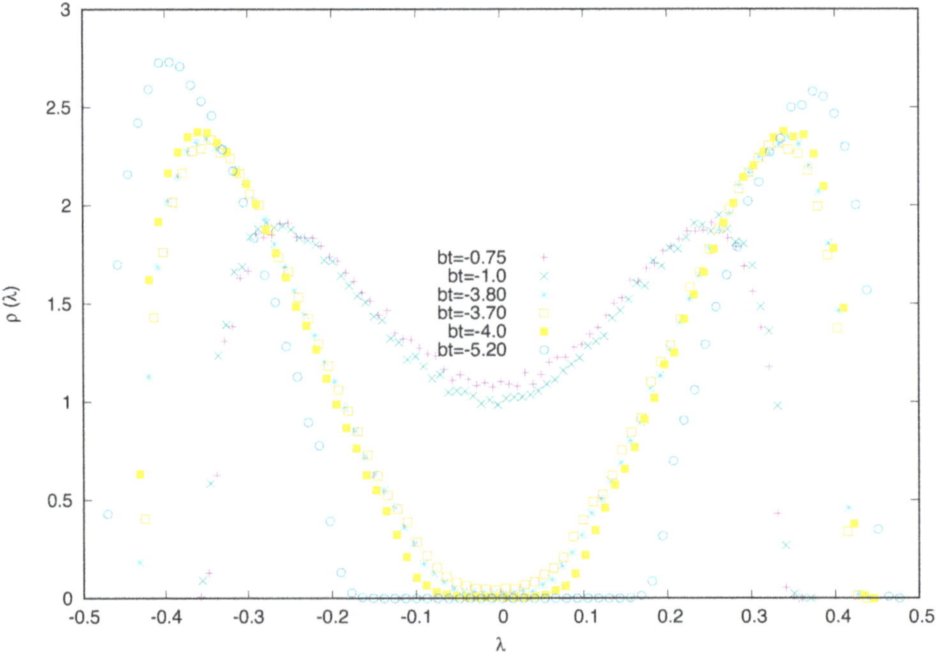

Figure 18.10. The eigenvalue distribution as it transits between the disordered and non-uniform-ordered phases in the multitrace matrix model (18.172).

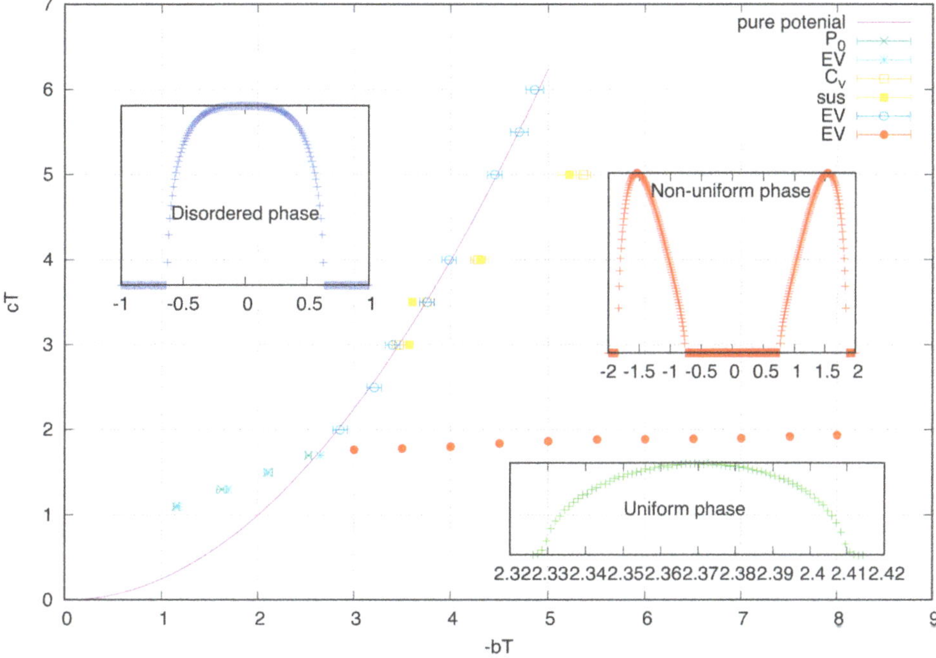

Figure 18.11. The phase diagram of the noncommutative-phi four theory (18.1)+(18.3) on the fuzzy sphere as approximated by the multitrace matrix model (18.172).

18.4.3 The uniform-ordered phase $\Rightarrow$ geometry

Now, the existence of the uniform-ordered phase and the Ising phase transition between this uniform-ordered phase and disordered phase signal as usual the spontaneous symmetry breaking of the discrete symmetry $\Phi \longrightarrow -\Phi$. The fundamental underlying hypothesis in this chapter is the statement that the existence of an Ising uniform-ordered phase in a 'pure matrix model' (such as the multitrace matrix model (18.172), which does not come with a pre-defined Laplacian operator) is an unambiguous signal for a lurking underlying geometry. In other words, an emergent geometry transition in this scenario (obtained by allowing the value of the coefficient C' of the multitrace term to change) is identified with the existence of a stable uniform-ordered phase and a corresponding Ising transition for some values of the coefficient C'.

On the other hand, the existence of the non-uniform-ordered phase signals the spontaneous symmetry breaking of translation symmetry (see for example [51]), which is quite remarkable for two reasons. First, this breaking is possible even in two dimensions in contrast to the Coleman–Mermin–Wagner theorem [99, 100] and it is due to the fact that noncommutative field theories are non-local by construction. Second, this breaking can be extended to supersymmetric models allowing us to obtain spontaneous supersymmetry breaking, which is usually quite hard to obtain otherwise (see [101] for a courageous attempt). The non-uniform-ordered phase is precisely the so-called stripe phase and it is the non-perturbative manifestation of the celebrated phenomena of the UV–IR mixing in noncommutative field theories [46].

18.4.4 Critical exponents $\Rightarrow$ dimension

The uniform-ordered phase is called the Ising phase precisely because we believe that the corresponding transition to the disordered phase is characterized by the universal critical exponents of the Ising model in two dimensions given by the Onsager solution. These critical exponents are defined as usual by the behavior near the critical point, viz

$$
\begin{aligned}
&m/N = \langle |TrM| \rangle / N \sim (B_c - B)^\beta \sim N^{-\beta/\nu} \\
&C_v/N^2 \sim (B - B_c)^{-\alpha} \sim N^{\alpha/\nu} \\
&\chi = \langle |TrM|^2 \rangle - \langle |TrM| \rangle^2 \sim (B - B_c)^{-\gamma} \sim N^{\gamma/\nu} \sim N^{2-\eta} \\
&\xi \sim |B - B_c|^{-\nu} \sim N.
\end{aligned}
\tag{18.183}
$$

There are in total six critical exponents, the above five plus the critical exponent δ, which controls the equation of state, but only two are truly in dependent because of the so-called scaling laws. The Onsager solution of the Ising model in two dimensions gives the following celebrated values [102]

$$
\nu = 1, \quad \beta = 1/8, \quad \gamma = 7/4, \quad \alpha = 0, \quad \eta = 1/4, \quad \delta = 15.
\tag{18.184}
$$

This fundamental result can be established for the multitrace matrix model (18.172) following the same steps used in deriving the critical exponents of the multitrace model (18.141)+(18.142). This is a very delicate exercise, which is essential in order to confirm the existence of the uniform-ordered phase beyond any doubt and thus confirm the fact that we are really dealing with a two-dimensional space geometry.

18.4.5 Wigner's semicircle law $\Rightarrow$ free propagator (metric)

As it turns out, a free noncommutative scalar field theory with mass parameter m^2 (the theory given by (18.1)+(18.3) with zero interaction) is characterized by a Wigner's semicircle law in stark contrast to free commutative scalar field theory. This stems from the fact that planar diagrams dominate over the non-planar ones in the limit of infinite cutoff [87, 89].

Explicitly, a noncommutative phi-four on a d-dimensional noncommutative spacetime $\mathbf{R}_\theta^d$ reads in position representation

$$
S = \int d^d x \left(\frac{1}{2} \partial_i \Phi \partial_i \Phi + \frac{1}{2} m^2 \Phi^2 + \frac{g}{4} \Phi_*^4 \right).
\tag{18.185}
$$

The first step is to regularize this theory in terms of a finite N_0-dimensional matrix Φ and rewrite the theory in matrix representation. Then, we diagonalize the matrix Φ, i.e. we write $\Phi = U^\dagger \Lambda U$ where $U \in U(N_0)$ and Λ is the matrix of eigenvalues λ_i. The measure becomes $\int \prod_i d\lambda_i \Delta^2(\Lambda) \int dU$ where dU is the Haar measure and $\Delta^2(\Lambda) = \prod_{i<j} (\lambda_i - \lambda_j)^2$ is the Vandermonde determinant.

The effective probability distribution of the eigenvalues λ_i can be determined uniquely from the behavior of the expectation values $\langle \int d^d x \Phi_*^{2n}(x) \rangle$. These objects clearly depend only on the eigenvalues λ_i and are computed using a sharp UV cutoff Λ.

If we are only interested in the eigenvalues of the scalar matrix Φ then the free theory $g = 0$ can be replaced by the effective matrix model [87]

$$S = \frac{2N}{\delta^2} Tr\Phi^2. \tag{18.186}$$

This result can be traced to the fact that planar diagrams dominates over the non-planar ones in the limit $\Lambda \longrightarrow \infty$. This means in particular that the eigenvalues λ_i are distributed according to the famous Wigner semicircle law with δ being the largest eigenvalue, viz

$$\rho(x) = \frac{2}{\pi\delta^2}\sqrt{\delta^2 - x^2}, \quad -\delta \leqslant x \leqslant +\delta. \tag{18.187}$$

The most important case is $d = 2$ since the case $d = 4$ is eliminated by the results of the critical exponents. In this case we find the radius [87]

$$\delta(m, \Lambda) = \frac{1}{\pi} \ln\left(1 + \frac{\Lambda^2}{m^2}\right), \quad d = 2. \tag{18.188}$$

In two dimensions the regulator Λ originates either from the noncommutative torus or from the fuzzy sphere.

The behavior of the squared-radius δ^2 on the noncommutative torus $\mathbf{T}_\theta^2$ (with cutoff $\Lambda = \sqrt{N\pi/\theta}$) is found to be different from the above sharp UV cutoff result (18.188) due to the different behavior of the propagator for large momenta. This behavior can also be excluded in our Monte Carlo data and by hindsight we know that this should be indeed the case since the original multitrace approximation was obtained from noncommutative phi-four theory on the fuzzy sphere.

We are left therefore with the case of the fuzzy sphere $\mathbf{S}_N^2$ where we have $N_0 \equiv N$ since the scalar field Φ is an $N \times N$ matrix ϕ given by $\phi = \sqrt{2\pi/Na}\,\Phi$. In this case the cutoff Λ is given in terms of the matrix size N and the radius R of the sphere by the relation $\Lambda = N/R$ and the mass parameters B and m^2 are related by $m^2 = B/aR^2$. By using $\tilde{B} = B/N^{3/2}$, and choosing $a = 2\pi/N$ so that $\Phi = \phi$, we obtain

$$\frac{\Lambda^2}{m^2} = \frac{2\pi N}{B} = \frac{2\pi}{\sqrt{N}\tilde{B}}. \tag{18.189}$$

In the limit $B \longrightarrow \infty$ we get the squared-radius $\delta^2 = 2N/B$, which corresponds to the Gaussian matrix model $B\,Tr\,M^2$, viz $B = 2N/\delta^2$. This can also be obtained by taking the limit $B \longrightarrow \infty$ of the one-cut solution. The Wigner's semicircle law is thus also obtained for the free quadratic matrix model $B\,Tr\,M^2$ with a squared-radius given simply by.

$$\delta^2 = \frac{2N}{B}. \tag{18.190}$$

In fact, the eigenvalues distribution of a free scalar field theory on the fuzzy sphere $\mathbf{S}_N^2$ with an arbitrary kinetic term, viz $S = \mathrm{Tr}(M\mathcal{K}M + BM^2)/2$, where $\mathcal{K}(0) = 0$ and $\mathcal{K}$ is diagonal in the basis of polarization tensors T_l^m, is always given by a Wigner semicircle law [89]. In this case the radius-squared is given by

$$\delta^2 = \frac{4f(B)}{N}, \quad f(B) = \sum_{l=0}^{N-1} \frac{2l+1}{\mathcal{K}(l) + B}. \tag{18.191}$$

The Wigner's semicircle law computed near the origin using the multitrace matrix model (18.172) should then be compared with the noncommutative field theory prediction (18.187)+(18.188) for negative values of the multitrace coupling C'. But for positive values of C' the behavior should be compared with the prediction of pure matrix model (18.190). This will allow us to confirm directly that the underlying space or emergent geometry is indeed a fuzzy sphere $\mathbf{S}_N^2$. See figure 18.12, but for more detail see [81].

18.4.6 Emergent Yang–Mills matrix model $\Rightarrow$ noncommutative gauge theory

The emergent geometry of the model (18.172) can be exhibited in a drastic way by exhibiting an emergent noncommutative gauge theory on a fuzzy sphere $\mathbf{S}_N^2$ in the uniform-ordered phase [103]. This gauge theory is an $SO(3)$-symmetric Yang–Mills matrix model of the IKKT-type, i.e., it is a matrix model which involves three covariant Hermitian matrix coordinates X_a with $SO(3)$ rotational symmetry and $U(N)$ gauge invariance.

In order to see this result we need to assume that we are in the uniform-ordered phase and thus the matrix M can be expanded around the identity matrix 1. For simplicity, we further assume that the dimension is even, i.e., the matrix M is $2N \times 2N$. Then, without any loss of generality, we can expand the matrix M as

$$M = M_0 \mathbf{1}_{2N} + M_1, \quad TrM_1 = 0. \tag{18.192}$$

Hence

$$M_1 = \sigma_a X_a, \quad M_0 = a + m, \tag{18.193}$$

where σ_a are the standard Pauli matrices, m is the fluctuation in the zero mode, and X_a are three Hermitian $N \times N$ matrices. By substitution, we obtain immediately the model

$$Z = \int \mathcal{D}X_a \exp(-V[X])) \int dm \exp(-f[m]). \tag{18.194}$$

The potential V is given by an $SO(3)$-symmetric three-matrix model, which is precisely equal to the $D = 3$ Yang–Mills matrix model defined by

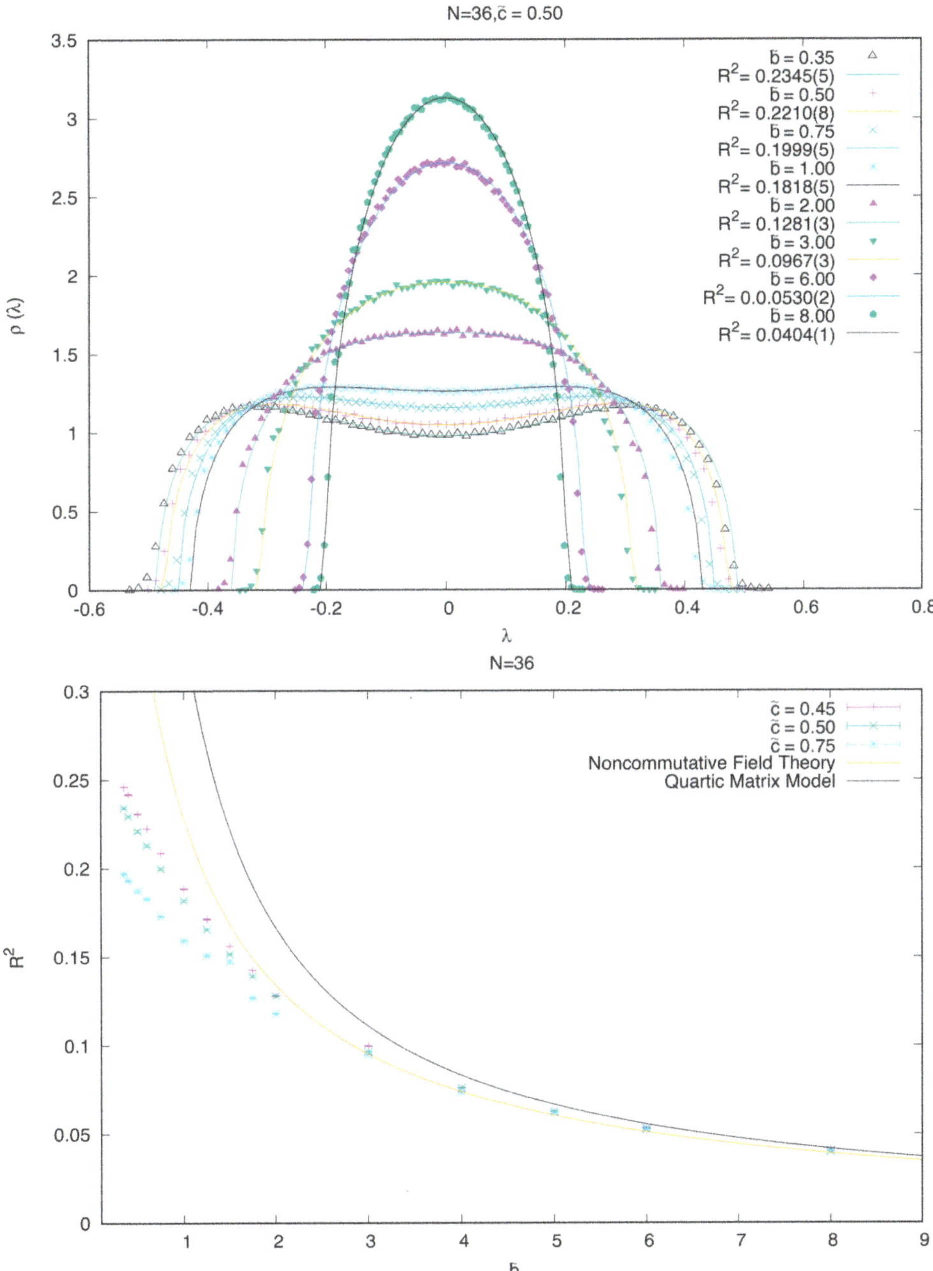

Figure 18.12. Wigner's semicircle law.

$$V = - CTr[X_a, X_b]^2 + 2CTr(X_a^2)^2$$
$$+2(B + 6Na^2C')TrX_a^2 + 4iNaC'\varepsilon_{abc}TrX_aX_bX_c. \tag{18.195}$$

The integration over m can be done and the result consists of some function of TrX_a^2 and $i\varepsilon_{abc}TrX_aX_bX_c$. This next-to-leading contribution (in $1/N$) is essentially the one-loop result and it is by construction subleading compared to V.

The Chern–Simons term is proportional to the value a of the order parameter. Thus, it is non-zero only in the Ising phase, and as a consequence, by tuning the parameters appropriately to the region in the phase diagram where the Ising phase exists, we will induce a non-zero value for the Chern–Simons. This Chern–Simons term is effectively the Myers term responsible for the condensation of the geometry [104, 105].

The above three-matrix model is then precisely a random matrix theory describing noncommutative gauge theory on the fuzzy sphere, where the first term is the Yang–Mills piece, whereas the second and third terms combine to give mass and linear terms for the normal scalar field on the sphere (recall that the index a runs from 1 to 3). This is essentially the random matrix theory describing noncommutative gauge theory on the fuzzy sphere found in [106]. However, we should emphasis here that we have obtained dynamically this gauge theory on the fuzzy sphere by going to the phase where a non-zero uniform order persists and thus securing a non-zero Chern–Simons term crucial for the condensation of the fuzzy sphere geometry. In [106], this was achieved by constraining the matrix M directly by hand in a particular way.

In summary, by expanding the above three-matrix model (18.195) around its background solution (which is a fuzzy sphere solution) we obtain a noncommutative $U(1)$ gauge theory on the fuzzy sphere [107, 108]. Thus, this Yang–Mills matrix model sustains itself emergent geometry [109, 110]. Furthermore, we know that noncommutative $U(1)$ gauge theory is effectively a gravity theory. Thus, this Yang–Mills matrix model sustains also emergent gravity [111–115].

18.4.7 Emergent geometry

In summary, the non-uniform-ordered phase is important for noncommutative geometry and matrix models whereas the uniform-ordered phase is essential to commutative geometry. Indeed, the central proposal of this chapter is to turn the original logic of noncommutative geometry and its matrix models on its head by starting from matrix models and attempt to reach noncommutative geometry and not the other way around. This will be precisely/explicitly done by searching in the phase diagram for a uniform-ordered phase.

Hence, we will take as first principle the random multitrace matrix model (18.172), or a generalization thereof, then attempt to reach the noncommutative field theory (18.1)+(18.3) by searching for a uniform-ordered phase in the phase diagram of the former. This relies on three facts:

- First, multitrace matrix models such as (18.172) do not involve in their definition a spectral triple specifying the geometry like those implicitly defining noncommutative field theories such as (18.1)+(18.3).
- Second, the uniform-ordered phase is a commutative order requiring the existence of an underlying space, which is *a priori* commutative.
- Third, the underlying geometry turns out to be quantized, i.e., noncommutative geometry because of the existence of a triple point in which a commutative order (uniform/Ising), a noncommutative order (non-uniform/stripe) and a matrix order coexist.

Thus, from the existence of a uniform-ordered phase and a corresponding Ising phase transition, for some values of the coefficient of the multitrace coupling, we can infer the existence of an underlying space and an emergent geometry transition. The dimension of this emergent space can be determined from the critical exponents of the uniform-to-disordered phase transition by virtue of scaling and universality properties of second-order phase transitions [60]. This exercise was done in great detail in [81].

Furthermore, the existence of a uniform-ordered phase and an Ising transition from a disordered phase to this uniform-ordered phase in a pure matrix model such as the multitrace matrix model (18.172) is indicative that this multitrace matrix model falls (for negative values of the multitrace coupling C') in the universality class of the noncommutative phi-four theory (18.1)+(18.3). In other words, the multitrace matrix model (18.172) captures the same geometry as the geometry specified by the spectral triple, which went implicitly into the definition of the noncommutative phi-four theory (18.1)+(18.3). For positive values of the multitrace coupling C' the multitrace matrix model (18.172) does not sustain a uniform-ordered phase and hence it falls in the universality class of the real quartic matrix model (18.173). We have then an emergent geometry as we vary C' from positive to negative values. The dimension of the underlying space, or more precisely this emergent geometry, are determined from the measurement/calculation of the critical exponents characterizing the Ising transition.

Hence, it is the statistical physics of the commutative (uniform/Ising) phase of the multitrace matrix model that captures the geometry. And, as it turns out, this geometry is quantum in the sense that it is emergent (reminiscent of second quantization of geometry) but it is also quantized in the sense that it is non-commutative (first quantization of geometry).

Indeed, this emergent quantum geometry is necessarily noncommutative since the multitrace matrix model (18.172) includes necessarily a stripe or non-uniform-ordered phase, which is a noncommutative order by excellence (a transition from a disordered phase to a non-uniform-ordered phase is necessarily generated by the single trace terms of the multitrace matrix model (18.172)). This important fact can be further confirmed by studying the transition from non-uniform-ordered to uniform-ordered and verifying that this transition is also second order (similarly to the Ising transition) and it is in fact a continuation of the Ising transition to larger values of the quartic coupling.

The precise metric on the emergent geometry can be fixed by studying the Wigner's semicircle law near the origin, i.e., by studying the eigenvalue distribution of the matrix M for vanishingly small values of the quartic coupling. This eigenvalue distribution captures clearly the properties of the free propagator.

This fourth ingredient in our proposal allows us, for example if the dimension is determined to be two from the critical exponents, to discriminate between the noncommutative torus $\mathbf{T}_\theta^2$ and the fuzzy sphere $\mathbf{S}_N^2$, which can be both used to regularize non-perturbatively the Moyal–Weyl plane. In fact, these two spaces lead to different behavior of the radius δ of the Wigner's semi-circle law as a function of the matrix size N and the mass parameter B. This issue was discussed in the previous section.

The fifth and final ingredient in our emergent geometry proposal consists in verifying explicitly the geometrical content of the multitrace matrix model (18.172) by expanding the matrix M around the uniform-ordered configuration $M_0 = m1$. We obtain an $SO(3)$-symmetric three-matrix model with a Chern–Simons term proportional to the value m of the order parameter in the uniform-ordered phase, i.e., to the magnetization. This three-matrix model describes a noncommutative gauge theory on the fuzzy sphere [106]. The Chern–Simons term is precisely Myers term responsible for the condensation of the geometry [104, 105].

In summary, our proposal for 'emergent geometry from random multitrace matrix models' consists of five ingredients [81, 84, 103, 116]:

1. The random multitrace matrix model (18.172) is taken as 'first principle'. There is no Laplacian entering the definition of this action and thus there is no geometry *a priori*. First, we need to determine whether or not this pure matrix model contains in its phase diagram a uniform-ordered phase and an Ising transition from this uniform-ordered phase to disordered phase.

2. The existence of a uniform-ordered phase in a multitrace matrix model such as (18.172) signals the existence of an underlying space and an emergent geometry as we vary the multitrace coupling C' from positive to negative values. By using scaling and universality of second-order phase transitions we can determine the dimension of the underlying space or emergent geometry from the critical exponents of the Ising line.

3. From the existence of a stripe or non-uniform-ordered phase with a transition line between uniform-ordered and non-uniform-ordered, which is second order and a continuation of the Ising line we can infer that the underlying space or emergent geometry is in fact noncommutative.

4. The study of the Wigner's semi-circle law of the multitrace matrix model (18.172) at the origin will allow us to discriminate decisively between the behavior of the noncommutative field theory and the real quartic matrix model.

5. In some interesting cases the expansion around the uniform-ordered phase exhibits the geometrical content of the multitrace matrix model as a gauge theory on a noncommutative background.

18.4.8 Wilsonian renormalization group equation

The partition function of the multitrace matrix model $\mathrm{Tr}\, M\, \mathrm{Tr}\, M^3$ can be rewritten in the usual form

$$Z = \int \mathcal{D}M \ \exp(-NTr_N V(M))$$

$$V(M) = \frac{g_2}{2}M^2 + \frac{g_4}{4}M^4 + \frac{g}{3N}(Tr_N M)M^3.$$

(18.196)

We have

$$g_2 = \pm 1, \quad g_4 = \frac{\tilde{C}}{\tilde{B}^2}, \quad g = \frac{3}{4}\frac{\tilde{D}}{\tilde{B}^2}.$$

(18.197)

We choose $g_2 = -1$ since B is taken to be negative, which is the region of interest for noncommutative scalar phi-four theories. Note also that the multitrace term is of the same order as the quartic term and stability requires that $\tilde{C} > -\tilde{D}$, which was noted previously on several occasions.

Now, by employing repeatedly large N factorization of multi-point functions of $U(N)$-invariant objects into product of one-point functions [117–121] we can effectively convert multitrace terms into single trace terms In addition, by expanding the multitrace and quartic terms and using the Schwinger–Dyson identities in the cubic potential we can rewrite the above multitrace matrix model as a cubic potential of the form [116]

$$V(M) = \frac{g_2'}{2}M^2 + \frac{g_3}{3}M^3$$

$$g_2' = \pm\sqrt{(g_2 + \frac{g_4}{2})^2 - 2g_4}$$

$$g_3 = ga_1 = \frac{g}{N}\langle \mathrm{Tr}\, M\rangle.$$

(18.198)

This is essentially a mean-field-approximation, which seems to be exact in the large N limit. Remark that the cubic coupling is proportional to the magnetization, i.e., $a_1 \equiv m$.

This cubic potential admits the standard Liouville quantum gravity fixed point given by [117–120]

$$\frac{g_3}{(g_{2*}')^{\frac{3}{2}}} = \varepsilon \equiv \frac{1}{432^{1/4}}.$$

(18.199)

The solution, which goes through the two standard fixed points $(0, 0)$ and $(0, \varepsilon)$ of the pure cubic potential, is given explicitly by [116]

$$g_{4*} = g_3^2\left[g_3^2 - \sqrt{g_3^4 - 2g_2 g_3^2 + \left(\frac{g_3}{\varepsilon}\right)^{4/3}} - g_2\right].$$

(18.200)

The behavior for $g_3^2 \longrightarrow \infty$ and $g_3^2 \longrightarrow 0$ is given explicitly by

$$g_{4*} = -\frac{1}{2}\left(\frac{g_3}{\varepsilon}\right)^{4/3}, \quad g_3^2 \longrightarrow \infty \tag{18.201}$$

and

$$g_{4*} = g_3^2\left(-g_2 - \left(\frac{g_3}{\varepsilon}\right)^{2/3}\right), \quad g_3^2 \longrightarrow 0. \tag{18.202}$$

We have then a critical line of fixed points (18.200), which interpolate smoothly between the fixed point $(0, 0)$ (interpreted as the fixed point of the third-order matrix coexistence line) and the fixed point $(0, \varepsilon)$ (interpreted as the fixed point of the second-order commutative/Ising coexistence line) going through a maximum in between. This critical line, restricted to the positive quadrant since the behavior $g_{4*} \longrightarrow -\infty$ as $g_3 \longrightarrow +\infty$ is unphysical for noncommutative field theory, is interpreted as the critical line of fixed points associated with the noncommutative/stripe coexistence line. The RG fixed points are sketched in figure 18.13.

18.5 Emergent geometry from Yang–Mills matrix models

18.5.1 IKKT model and other Yang–Mills matrix models

The one-dimensional reduction of $\mathcal{N} = 1$ supersymmetric Yang–Mills theory in $D = 10$ dimensions gives the so-called BFSS matrix model known also as M-(atrix) theory [122]. This is a Yang–Mills matrix quantum, which describes the low energy dynamics of a system of N type IIA D0-branes [123]. The BFSS matrix model is also conjectured to describe the discrete light-cone quantization (DLCQ) of M-theory. This is therefore our most fundamental matrix model as it gives a quantum mechanical theory of gravity.

The zero-dimensional reduction of the BFSS matrix model is precisely the IKKT matrix model [22]. This model is supposed to give a non-perturbative regularization of type IIB superstring theory in the Schild gauge. The IKKT model is the $D = 10$ Yang–Mills matrix model.

The IKKT model itself is equivalent to Connes' approach to geometry.

Recall that a commutative/noncommutative space in Connes' approach to geometry is given in terms of a spectral triple $(\mathcal{A}, \Delta, \mathcal{H})$ [124]. Here, $\mathcal{A}$ is the algebra of functions or bounded operators on the space, Δ is the Laplace operator which encodes the metric aspects, and $\mathcal{H}$ is the Hilbert space on which the algebra of bounded operators and the differential operator Δ are represented. The Laplacian Δ should be replaced with the Dirac operator $\mathcal{D}$ in the case of a spin structure.

In the IKKT matrix model we have, in a precise sense, an emergent matrix geometry, which makes it essentially equivalent to Connes' noncommutative geometry. Here, the algebra $\mathcal{A}$ is given, in the large N limit, by Hermitian matrices with smooth eigenvalue distributions and bounded square traces [125]. The Laplacian/Dirac operator is given in terms of the background solutions while the

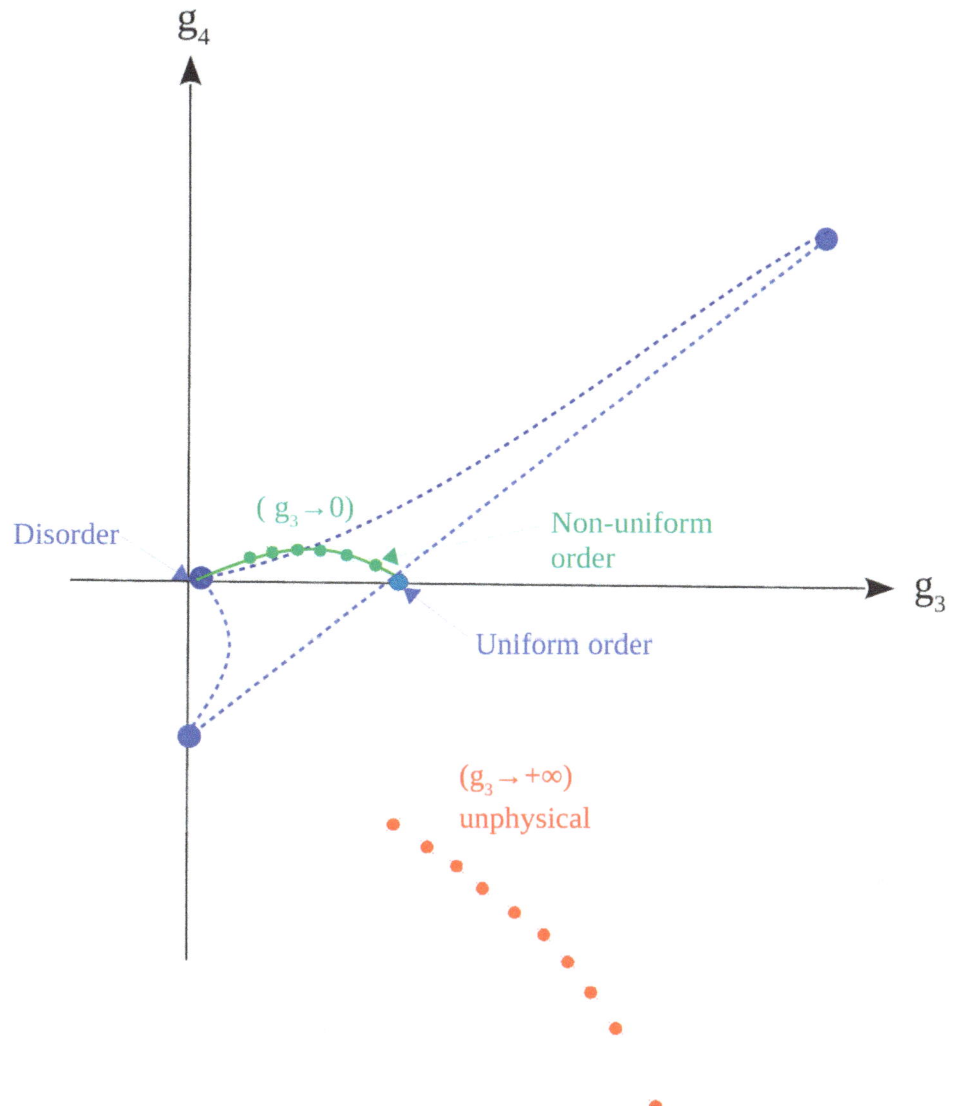

Figure 18.13. The RG fixed points of the multitrace matrix model (18.172) (green and red lines) compared with the RG fixed points of the single trace cubic-quartic matrix model (blue line) [116].

Hilbert space $\mathcal{H}$ is given by the adjoint representation of the gauge group $U(N)$ of the IKKT matrix model.

The BMN Yang–Mills matrix quantum mechanics is the plane-wave deformation of the BFSS matrix model [126]. In fact, it is the unique one-parameter deformation of the BFSS Yang–Mills matrix quantum mechanics, which preserves all $\mathcal{N} = 16$ supersymmetries. This deformation is given in terms of a mass parameter μ. It is conjectured that this matrix model describes the light-cone quantization of superparticles and supermembranes in the maximally supersymmetric pp-wave background of 11-dimensional M-theory and 11-dimensional supergravity.

The zero-dimensional reduction of the BMN matrix model gives again the IKKT matrix model but with additional mass deformation terms.

In the following, we will derive the IKKT-type IIB matrix model, with and without mass deformation terms, from the dimensional reduction of the BFSS/BMN matrix model.

- First, the action of the BFSS matrix model is written in terms of nine $N \times N$ bosonic Hermitian matrices X_a ($a = 1,..., 9$), a $U(N)$ gauge field A_t and sixteen $N \times N$ fermionic Majorana matrices ψ_α ($\alpha = 1,..., 16$). The bosonic and fermionic fields X_a and ψ_α transform in the adjoint representation of the gauge group, viz $D_t X_a = \partial_t X_a - i[A_t, X_a]$ and $D_t \psi_\alpha = \partial_t \psi_\alpha - i[A_t, \psi_\alpha]$. The Euclidean BFSS action, on a circle of circumference β, is given explicitly by

$$S_{\text{BFSS}} = N \int_0^\beta dt \, \text{Tr} \left[\frac{1}{2}(D_t X_a)^2 - \frac{1}{4}[X_a, X_b]^2 + \frac{1}{2}\psi^T C_9 D_t \psi - \frac{1}{2}\psi^T C_9 \gamma^a [X_a, \psi] \right]. \quad (18.203)$$

- The Euclidean BMN action, on a circle of circumference β, is a one-parameter deformation of the BFSS model which reads

$$S_{\text{BMN}} = S_{\text{BFSS}} + S_{\text{MASS}}. \quad (18.204)$$

$$S_{\text{MASS}} = N \int_0^\beta dt \, \text{Tr} \left[-i\frac{\mu}{6}\varepsilon_{ijk} X_i[X_j, X_k] + \frac{\mu^2}{18}X_i^2 + \frac{\mu^2}{72}X_a^2 - i\frac{\mu}{8}\psi^T C_9 \gamma^{789}\psi \right]. \quad (18.205)$$

- The local minimum of this model is a maximally supersymmetric config-uration preserving all dynamical supersymmetries given by a static fuzzy sphere spanning the 7, 8, 9 directions, viz

$$[X_i, X_j] = i\alpha\varepsilon_{ijk} X_k, \quad X_{a \neq i} = 0, \quad D_t X_a = 0, \quad \alpha = -\frac{\mu}{3}. \quad (18.206)$$

The IKKT-type IIB matrix model is obtained from the actions (18.203), (18.204) and (18.205) by considering their zero-dimensional reduction.

The BFSS model, in Lorentzian signature, can also be rewritten in terms of a 32-component Majorana–Weyl spinor Ψ. The corresponding action is given explicitly by

$$S_{\text{BFSS}} = N \int dt \, \text{Tr} \left[\frac{1}{2}(D_t X_a)^2 + \frac{1}{4}[X_a, X_b]^2 + \frac{i}{2}\bar{\Psi}\Gamma^0 D_t \Psi + \frac{1}{2}\bar{\Psi}\Gamma^a[X_a, \Psi] \right]. \quad (18.207)$$

Here, Γ^a are the Dirac matrices in D dimensions in the Weyl representation and $\bar{\Psi} = \Psi^T C_{10}$. The zero-dimensional reduction of this action is immediately given by (with $A_t \equiv X_0$)

$$S_{\text{IKKT}} = \frac{1}{g^2} \text{Tr} \left[\frac{1}{4}[X_A, X_B][X^A, X^B] + \frac{1}{2}\bar{\Psi}\gamma^A[X_A, \Psi] \right]. \quad (18.208)$$

The IIB superstring action can be obtained from this matrix model in the double scaling limit $N \longrightarrow \infty$ and $g^2 \longrightarrow 0$ keeping Ng^2 is kept fixed. In Euclidean signature this action reads

$$S_{\mathrm{IKKT}} = \frac{1}{g^2} \mathrm{Tr} \left[-\frac{1}{4} [X_A, X_B]^2 - \frac{1}{2} \bar{\Psi} \gamma^A [X_A, \Psi] \right]. \tag{18.209}$$

Equations (18.208) and (18.209) define the IKKT matrix model, without mass deformation, in Lorentzian and Euclidean signatures, respectively. Equation (18.209) defines in fact a supersymmetric matrix model in dimensions $D = 3, 4, 6$ where, similarly to the case $D = 10$, can be obtained from the zero-dimensional reduction of $\mathcal{N} = 1$ supersymmetric Yang–Mills theory in D dimensions. These are the basic $SO(D)$-symmetric Yang–Mills matrix models of the IKKT-type. The convergence properties of their corresponding partition functions are studied in [27–29]. The partition function exists only in $D = 10, 6, 4$. The $D = 3$ supersymmetric partition function is divergent while the bosonic truncation is convergent.

The most generic mass deformation of the Yang–Mills matrix model (18.209) is of the general form

$$S_{\mathrm{MASS}} = \frac{1}{g^2} \mathrm{Tr} \left[-i \frac{\mu_1}{6} \varepsilon_{ijk} X_i [X_j, X_k] + \frac{\mu_2^2}{18} X_i^2 + \frac{\mu_3^2}{72} X_a^2 - i \frac{\mu_4}{8} \bar{\Psi} \Gamma \Psi \right]. \tag{18.210}$$

For example, the mass deformation for $D = 10$ is obtained from the zero-dimensional reduction of the action (18.205) with $\Gamma = \Gamma^{789}$. Clearly, the Dirac matrix Γ depends on the dimension D. We have then obtained a four-parameter action where the parameters μ_i are expected to be considerably constrained by the requirement of supersymmetry, e.g., in $D = 10$ we have $\mu_i = \mu$.

The fermionic determinant in $D = 4$ is real positive while the determinants in $D = 6$ and $D = 10$ are complex, which means in particular that in $D = 10$ and $D = 6$ we can have spontaneous symmetry breaking of rotational invariance [127–130] while in $D = 4$ there is no spontaneous symmetry breaking [131–134].

The cases $D = 10$ and $D = 6$ will then enjoy $SO(D)$ rotational symmetry, which can be shown, by means of the so-called Gaussian expansion method [30], to become in the large N limit spontaneously broken down to $SO(3)$ due precisely to the phase of the Pfaffian [135, 136]. This spontaneous symmetry breaking corresponds effectively to a dynamical compactification to three dimensions.

The $D = 3$ Yang–Mills matrix model seems therefore to be the most fundamental Yang–Mills matrix model while the case $D = 4$, as we have already discussed albeit briefly, is actually a $D = 3$ model coupled to a scalar fluctuation.

18.5.2 The effective potential of the $D = 3$ Yang–Mills matrix model

We are therefore led to consider the bosonic truncation of the $D = 3$ Yang–Mills matrix model. Thus, the fundamental dynamical variables are three $N \times N$ Hermitian matrices X_i, which can be rotated into each other by $SO(3)$

transformations. The most general action (up to quartic power in the matrices X_i) is then given by

$$S[X] = N \, \text{Tr} \left[-\frac{1}{4}[X_i, X_j]^2 + \frac{2i\alpha}{3}\varepsilon_{ijk} X_i X_j X_k + BX_i^2 + C(X_i^2)^2 \right]. \quad (18.211)$$

The last term is not induced by maximally supersymmetric mass deformation of the BFSS/IKKT matrix models. However, it is a central ingredient of noncommutative gauge theory on the fuzzy sphere as we will discuss shortly. The inclusion of this quartic term may even improve the convergence properties of the partition function of the supersymmetric $D = 3$ Yang–Mills matrix model.

The above action can also be rewritten as

$$S[D] = \frac{1}{g^2 N} \, \text{Tr} \left[-\frac{1}{4}[D_i, D_j]^2 + \frac{2i}{3}\varepsilon_{ijk} D_i D_j D_k + B\alpha^2 D_i^2 + C(D_i^2)^2 \right], \quad X_i = \alpha D_i$$

$$= \frac{1}{g^2 N} \, \text{Tr} \left[-\frac{1}{4}[D_i, D_j]^2 + \frac{2i}{3}\varepsilon_{ijk} D_i D_j D_k + b(D_i^2 - c_2) + C(D_i^2 - c_2)^2 \right] \quad (18.212)$$

$$\frac{1}{g^2} = \alpha^4 N^2 \equiv \tilde{\alpha}^4, \quad c_2 = \frac{N^2 - 1}{4}, \quad b = B\alpha^2 + 2c_2 C.$$

This action enjoys a $U(N)$ gauge invariance. Indeed, gauge transformations are implemented here by the unitary transformations $U \in U(N)$ as follows: $D_i \rightarrow D_i' = U D_i U^{-1}$, i.e., the $N \times N$ Hermitian matrices D_i play the role of the covariant derivatives of this gauge symmetry. The field strength or curvature tensor F_{ij} associated with these covariant matrices D_i is then explicitly defined by

$$F_{ij} = [D_i, D_j] - i\varepsilon_{ijk} D_k. \quad (18.213)$$

The equations of motion derived from the action (18.212) read in terms of the gauge-covariant current J_i as follows

$$[D_j, F_{ij}] = 2C[D_i, D_j^2 - c_2]_+ + 2bD_i + J_i \, . \quad (18.214)$$

The local minimum of the model is given by the generators L_i of the spin $s \equiv (N - 1)/2$ irreducible representation of $SU(2)$ from which the fuzzy sphere configuration can be constructed. Indeed, for $J_i = 0$, we have the solution

$$b = 0: \quad D_i = L_i \text{ where } [L_i, L_j] = i\varepsilon_{ijk} L_k, \quad L_i^2 = c_2. \quad (18.215)$$

$$b \neq 0: \quad D_i = \varphi L_i \text{ where } (1 + m^2)\varphi^2 - \varphi - \mu = 0,$$

$$C = \frac{m^2}{2c_2}, \quad -\mu = b - m^2 = B\alpha^2. \quad (18.216)$$

The matrix coordinates $\hat{x}_i$ on the fuzzy sphere can then be constructed from the generators L_i in the usual way, viz

$$\hat{x}_i = \frac{RL_i}{\sqrt{c_2}} \text{ where } [\hat{x}_i, \hat{x}_j] = i\theta\varepsilon_{ijk}\hat{x}_k, \quad \hat{x}_i^2 = R^2, \quad \theta = \frac{R}{\sqrt{c_2}}. \quad (18.217)$$

This is the origin of the phenomena of emergent geometry from Yang–Mills matrix models. Thus, the covariant derivatives/matrices D_i act really as covariant matrix coordinates, which is their correct interpretation.

The partition function of the theory is given by

$$Z_N[J] = \int \prod_{i=1}^{3} [dD_i] e^{-S[D] - \frac{1}{Ng^2} \operatorname{Tr} J_i D_i}. \tag{18.218}$$

Now we adopt the background field method to the problem of quantization of this theory [107]. This method consists in making a perturbation of the field around the classical solution and then quantizing the fluctuation. Towards this end we first separate the field as $D_i = B_i + Q_a$ and write the action in the form

$$S[D_i] = S[B_i] + \frac{1}{g^2 N} \operatorname{Tr}(\hat{J}_i - J_i)Q_i - \frac{1}{2g^2 N} \operatorname{Tr}[B_i, Q_j]^2 + \frac{1}{2g^2 N}(1 + 2C)\operatorname{Tr}[B_i, Q_i]^2$$

$$+ \frac{1}{g^2 N} \operatorname{Tr} Q_i[F_{ij}^B, Q_j] + \frac{1}{2g^2 N} \operatorname{Tr} Q_i\left[2b\delta_{ij} + 4C(B_k^2 - c_2)\delta_{ij} + 8CB_i B_j\right]Q_j + \cdots \tag{18.219}$$

$$\hat{J}_a = -[B_j, F_{ij}^B] + 2C[B_i, B_j^2 - c_2]_+ + 2bB_i + J_i.$$

Here, $F_{ij}^B = [B_i, B_j] - i\varepsilon_{ijk} B_k$. This action is invariant under the gauge transformations $B_i \longrightarrow B_i, Q_i \longrightarrow UQ_i U^\dagger + U[B_i, U^\dagger]$. This means in particular that in order to fix the gauge in a consistent way the gauge fixing term should be covariant with respect to the background field. We impose here the covariant Lorentz gauge $[B_i, Q_i] = 0$. The gauge fixing term and the Faddeev–Popov term are therefore given by

$$S_{g.f} + S_{gh} = -\frac{1}{2g^2 N} \operatorname{Tr} \frac{[B_i, Q_i]^2}{\xi} + \frac{1}{g^2 N} \operatorname{Tr} b^\dagger[B_i, [B_i, b]]. \tag{18.220}$$

We will choose now for simplicity the gauge $\xi^{-1} = 1 + 2C$, which will cancel the fourth term in (18.219). This gauge becomes the Feynman gauge $\xi = 1$ in the limit $N \longrightarrow \infty$ and the Landau gauge $\xi = 0$ in the limit $C \longrightarrow \infty$. Furthermore, we will assume that the background field B_i satisfies the classical equations of motion (18.214) and hence the second term of (18.219) also vanishes. The partition function then becomes

$$Z_N[J] = e^{-S[B_i] - \frac{1}{g^2 N} \operatorname{Tr} B_i J_i} \det\left(\mathcal{B}_i^2\right) \int \prod_{i=1}^{3} [dQ_i]\, e^{-\frac{1}{2g^2 N} \operatorname{Tr} Q_i \Omega_{ij} Q_j + \cdots}. \tag{18.221}$$

Here, $\det(\mathcal{B}_i^2)$ comes from the integration over the ghost field whereas the gauge field Laplacian Ω_{ab} is defined by

$$\Omega_{ij} = 2b\delta_{ij} + \mathcal{B}_k^2 \delta_{ij} + 2\mathcal{F}_{ij}^B + 4C(\mathcal{B}_k^2 - c_2)\delta_{ij} + 8C\mathcal{B}_i \mathcal{B}_j. \tag{18.222}$$

In this equation the notation $\mathcal{B}_i$ and $\mathcal{F}_{ij}^B$ means that the covariant derivatives B_i and their curvature F_{ij}^B act by commutators, viz $\mathcal{B}_i(M) = [B_i, M]$ and $\mathcal{F}_{ij}^B(M) = [F_{ij}^B, M]$ for any $N \times N$ Hermitian matrix M. Similarly,

$\mathcal{B}_i^2(M) = [B_i, [B_i, M]]$. By performing the Gaussian path integral we obtain the one-loop effective action

$$\Gamma[B_i] = S[B_i] + \frac{1}{2}\,\mathrm{Tr}_3\mathrm{TR}\log\Omega - \mathrm{TR}\log\mathcal{B}_i^2. \tag{18.223}$$

Here, TR is a trace over four indices corresponding to the left and right actions of operators on matrices and Tr_3 is the trace associated with three-dimensional rotations.

We can immediately use the above expression to compute the effective potential in the fuzzy sphere configuration (18.217), i.e., in the covariant matrix coordinates $B_i = \varphi L_i$. We find (by assuming that $|b|$, $C \ll 1$ and $\varphi \sim 1$) [107]

$$\begin{aligned}
\frac{V_{\mathrm{eff}}(\varphi)}{2c_2} &\equiv \frac{\Gamma[B_i = \varphi L_i]}{2c_2}\\
&= \tilde{\alpha}^4\left[\frac{1}{4}\varphi^4 - \frac{1}{3}\varphi^3 + \frac{1}{4}m^2\varphi^4 - \frac{1}{2}\mu\varphi^2\right] + \frac{1}{4c_2}\,\mathrm{Tr}_3\mathrm{TR}\log\left[2b\delta_{ij} + \varphi^2\mathcal{L}_k^2\delta_{ij} + (\varphi^2 - \varphi)i\varepsilon_{ijk}\mathcal{L}_k\right.\\
&\quad \left. + 4Cc_2(\varphi^2 - 1) + 8C\varphi^2 L_i L_j\right] - \frac{1}{2c_2}\mathrm{TR}\log\varphi^2\mathcal{L}_i^2\\
&= \tilde{\alpha}^4\left[\frac{1}{4}\varphi^4 - \frac{1}{3}\varphi^3 + \frac{1}{4}m^2\varphi^4 - \frac{1}{2}\mu\varphi^2\right] + \frac{1}{4c_2}\,\mathrm{Tr}_3\mathrm{TR}\log\left[\varphi^2\mathcal{L}_k^2\delta_{ij}\right] - \frac{1}{2c_2}\mathrm{TR}\log\varphi^2\mathcal{L}_i^2\\
&= \tilde{\alpha}^4\left[\frac{1}{4}\varphi^4 - \frac{1}{3}\varphi^3 + \frac{1}{4}m^2\varphi^4 - \frac{1}{2}\mu\varphi^2\right] + \log\varphi^2.
\end{aligned} \tag{18.224}$$

18.5.3 Emergent geometry: the matrix-to-sphere phase transition

The basic result summarizing the physics of emergent geometry from Yang–Mills matrix models is the statement that the classical global minimum of the underlying matrix model, which realizes the corresponding noncommutative space, becomes unstable under quantum fluctuations, as we increase the gauge coupling constant, and then at some critical value evaporates into a pure matrix configurations without any geometric content.

The existence of a classical background of the Yang–Mills matrix model defines 'first quantization' of the geometry whereas the stability of this background under quantum fluctuations defines 'second quantization' of the geometry. In the first instance we have noncommutative geometry whereas in the second instance we have emergent geometry.

For example, in the context of the $D = 3$ Yang–Mills matrix model (18.211) or (18.212) the fuzzy sphere configuration (18.217), although completely stable classically, becomes unstable under the effect of quantum fluctuations as we decrease the inverse gauge coupling constant $1/g^2 = \tilde{\alpha}^4$. We will illustrate this effect first in the case of the so-called Alekseev–Recknagel–Schomerus matrix model [137, 138], which is associated with the parameters $b = C = 0$. Explicitly, the action is given by

$$S[D] = \frac{1}{g^2 N}\,\mathrm{Tr}\left[-\frac{1}{4}[D_i, D_j]^2 + \frac{2i}{3}\varepsilon_{ijk}D_iD_jD_k\right]. \tag{18.225}$$

The classical potential and the global minimum of the model are then given by

$$\frac{V(\varphi)}{2c_2} = \tilde{\alpha}^4\left[\frac{1}{4}\varphi^4 - \frac{1}{3}\varphi^3\right]$$

$$\frac{V'(\varphi)}{2c_2} = \tilde{\alpha}^4[\varphi^3 - \varphi^2] = 0 \Rightarrow \varphi$$

$$= 0 \ \text{(matrix)} \ \text{or} \ \varphi = 1 \ \text{fuzzy sphere}$$

$$\frac{V(\varphi = 1)}{2c_2} = -\frac{1}{12}\tilde{\alpha}^4 < \frac{V(\varphi = 0)}{2c_2} = 0 \quad : \quad \text{fuzzy sphere is the global minimum.}$$

(18.226)

The quantum potential is, on the other hand, given by

$$\frac{V_{\text{eff}}(\varphi)}{2c_2} = \tilde{\alpha}^4\left[\frac{1}{4}\varphi^4 - \frac{1}{3}\varphi^3\right] + \log \varphi^2. \tag{18.227}$$

By taking the first and second derivatives of this potential we obtain

$$\frac{V'_{\text{eff}}}{2c_2} = \frac{1}{g^2}(\varphi^3 - \varphi^2) + \frac{2}{\varphi}$$

$$\frac{V''_{\text{eff}}}{2c_2} = \frac{1}{g^2}(3\varphi^2 - 2\varphi) - \frac{2}{\varphi^2}. \tag{18.228}$$

The condition $V'(\varphi) = 0$, which also reads $\varphi^4 - \varphi^3 + 2g^2 = 0$, will give us extrema of the model. These extrema are minima and thus stable if the condition $V''(\varphi) > 0$ (or equivalently $3\varphi^4 - 2\varphi^3 - 2g^2 > 0$) is satisfied whereas they are maxima if $3\varphi^4 - 2\varphi^3 - 2g^2 < 0$. The equation that tells us therefore when we go from a bounded potential to an unbounded potential is given by (see figure 18.14)

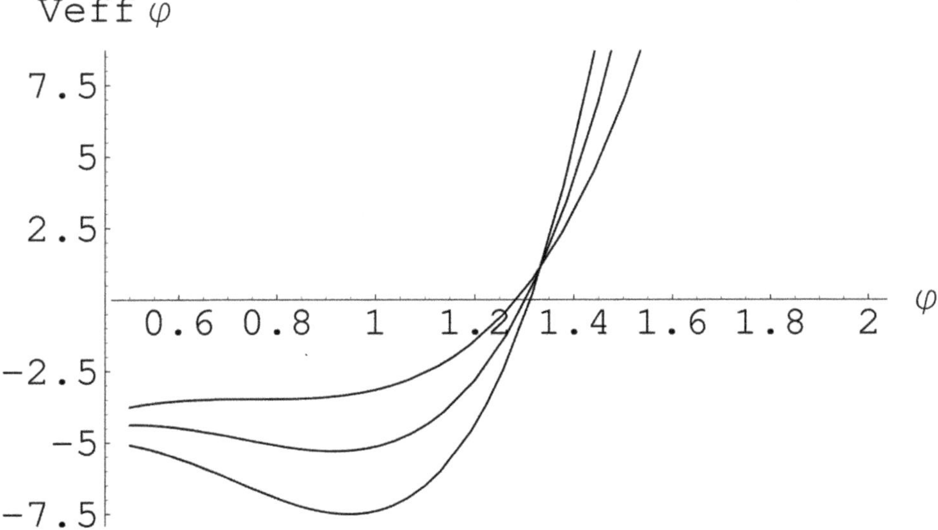

Figure 18.14. Effective potential for different values of g for the Alekseev–Recknagel–Schomerus.

$$3\varphi^4 - 2\varphi^3 - 2g^2 = 0. \tag{18.229}$$

Solving the above two equations yields immediately the critical values

$$\varphi \equiv \varphi_* = \frac{3}{4}, \quad g^2 \equiv g_*^2 = \left(\frac{3}{8}\right)^3, \quad \tilde{\alpha} \equiv \tilde{\alpha}_* = \left(\frac{8}{3}\right)^{3/4} = 2.08677944. \tag{18.230}$$

This one-loop calculation agrees nicely with the exact result obtained in [105] by means of Monte Carlo simulation of the $D = 3$ Yang–Mills matrix model (18.211) with $B = C = 0$.

It is not difficult to convince ourselves that the trace part of the covariant coordinate matrices X_i does not enter the Alekseev–Recknagel–Schomerus action (18.225) and thus it can be removed from the partition function. As it turns out, the removal of this trace part is crucial if we want the Monte Carlo method to thermalize. Hence, we should consider in this case the partition function given by

$$Z = \int \prod_{i=1}^{3} [dX_i] \, \exp(-S)\delta(\mathrm{Tr}\, X_i). \tag{18.231}$$

The order parameter in this problem is given by the extent of space $\langle \text{radius} \rangle$, or equivalently by the radius r, defined by

$$\langle \text{radius} \rangle = \langle \mathrm{Tr}\, X_i^2 \rangle, \quad \frac{1}{r} = \frac{1}{\tilde{\alpha}^2 c_2}\langle \text{radius} \rangle. \tag{18.232}$$

A more powerful set of order parameters is given by the eigenvalues distributions of the matrices X_3, $i[X_1, X_2]$ and X_i^2. We also measure the Yang–Mills and Chern–Simons terms given by

$$\mathrm{YM} = -\frac{N}{4} \mathrm{Tr}[X_i, X_j]^2, \quad \mathrm{CS} = \frac{2iN\alpha}{3}\varepsilon_{ijk} \, \mathrm{Tr}\, X_i X_j X_k. \tag{18.233}$$

The total action (energy) and the specific heat are given by

$$S = \mathrm{YM} + \mathrm{CS}, \quad C_v = \langle S^2 \rangle - \langle S \rangle^2. \tag{18.234}$$

An exact Schwinger–Dyson identity is given by

$$\text{identity:} \quad 4\langle \mathrm{YM} \rangle + 3\langle \mathrm{CS} \rangle \equiv 3(N^2 - 1). \tag{18.235}$$

The Monte Carlo results reported here are derived using the Metropolis algorithm in [139].

The Alekseev–Recknagel–Schomerus model is characterized by two phases: the fuzzy sphere phase and the Yang–Mills phase. Some of the fundamental results are:

1. **The Fuzzy Sphere (Geometric) Phase:**
   - This appears for large values of $\tilde{\alpha}$. It corresponds to the class of solutions of the equations of motion given by

$$[X_i, X_j] = i\alpha\varphi\varepsilon_{ijk} X_k, \quad \varphi = 1. \tag{18.236}$$

The global minimum is given by the largest irreducible representation of $SU(2)$, which fits in $N \times N$ matrices. This corresponds to the spin $s = (N - 1)/2$ irreducible representation, viz

$$X_i = \varphi \alpha L_i: \quad [L_i, L_j] = i\varepsilon_{ijk}L_c, \quad c_2 = L_i^2 = s(s + 1)1_N, \quad s = \frac{N - 1}{2}. \quad (18.237)$$

The values of the various observables in these configurations are

$$\mathrm{YM} = \frac{\varphi^4 \tilde{\alpha}^4 c_2}{2}, \quad \mathrm{CS} = -\frac{2\varphi^3 \tilde{\alpha}^4 c_2}{3},$$

$$S = \varphi^3 \tilde{\alpha}^4 c_2 \left(\frac{\varphi}{2} - \frac{2}{3} \right), \quad \langle \mathrm{radius} \rangle = \varphi^2 \tilde{\alpha}^2 c_2. \quad (18.238)$$

- The eigenvalues of $D_3 = X_3/\alpha$ and $i[D_1, D_2] = i[X_1, X_2]/\alpha^2$ are given by

$$\lambda_i = -s, -s + 1,\ldots, s - 1, +s, \quad s = \frac{N - 1}{2}. \quad (18.239)$$

The spectrum of $i[D_1, D_2]$ is a better measurement of the geometry since quantum fluctuations around L_3 are more suppressed. Some illustrative data for $\tilde{\alpha} = 3$ and $N = 4$ is shown in figure 18.15.

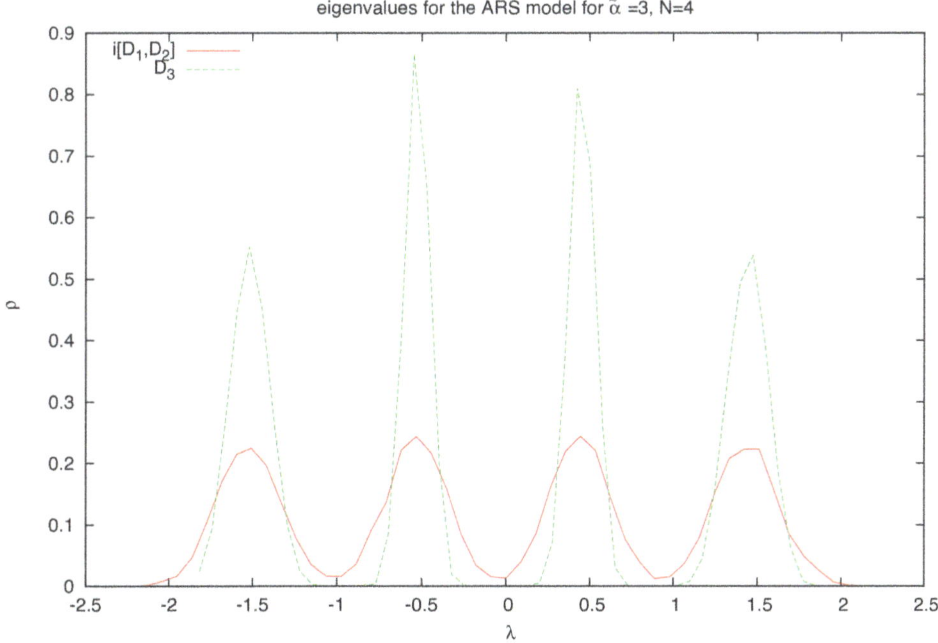

Figure 18.15. The eigenvalue distribution in the fuzzy sphere phase.

2. **The Yang–Mills (Matrix) Phase:**
   - This appears for small values of $\tilde{\alpha}$. It corresponds to the class of solutions of the equations of motion given by

$$[X_i, X_j] = 0. \tag{18.240}$$

   This is the phase of almost commuting matrices. It is characterized by the eigenvalues distribution

$$\rho(\lambda) = \frac{3}{4R^3}(R^2 - \lambda^2). \tag{18.241}$$

   It is believed that $R = 2$. We compute

$$\begin{aligned}
\langle \text{radius} \rangle &= 3\langle \text{Tr } X_3^2 \rangle \\
&= 3N \int_{-R}^{R} d\lambda \rho(\lambda)\lambda^2 \\
&= \frac{3}{5}R^2 N.
\end{aligned} \tag{18.242}$$

   - The above eigenvalues distribution can be derived by assuming that the joint eigenvalues distribution of the three commuting matrices X_1, X_2 and X_3 is uniform inside a solid ball of radius R. This can actually be proven by quantizing the system in the Yang–Mills phase around commuting matrices [140].
   - The value of the radius R is determined numerically as follows:
     - The first measurement R_1 is obtained by comparing the numerical result for $\langle \text{radius} \rangle$, for the biggest value of N, with the formula (18.242).
     - We use R_1 to restrict the range of the eigenvalues of X_3.
     - We fit the numerical result for the density of eigenvalues of X_3, for the biggest value of N, to the parabola (18.241) in order to get a second measurement R_2.
     - We may take the average of R_1 and R_2.

Sample data for $\tilde{\alpha} = 0$ with $N = 6$, 8 and 10 is shown in figure 18.16.
   - It is found that the eigenvalues distribution, in the Yang–Mills phase, is independent of $\tilde{\alpha}$. Sample data for $\tilde{\alpha} = 0 - 2$ and $N = 10$ is shown in figure 18.17.

3. **Critical Fluctuations:** The transition between the two phases occur at $\tilde{\alpha} = 2.1$. The specific heat diverges at this point from the Yang–Mills side while it remains constant from the fuzzy sphere side. This indicates a second-order behavior with critical fluctuations only from one side of the transition. The Yang–Mills and Chern–Simons actions, and as a consequence the total action, as well as the extent of space $\langle \text{radius} \rangle$ suffer a discontinuity at this point reminiscent of a first order behavior. In particular, the sphere expands then evaporates at the critical point, i.e., its radius r diverges at the transition

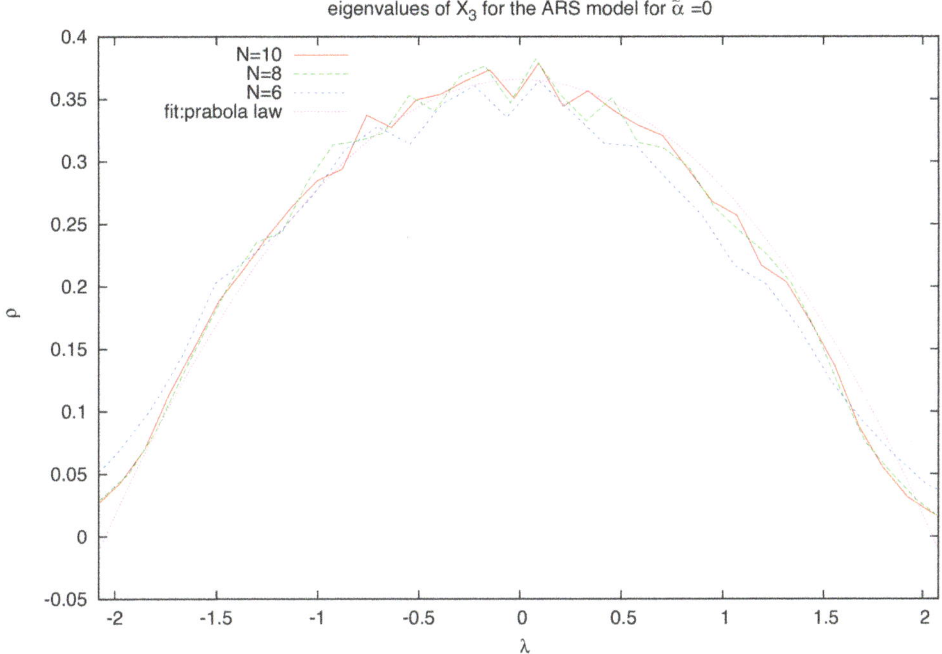

Figure 18.16. The eigenvalue distribution in the Yang–Mills matrix phase as a function of N.

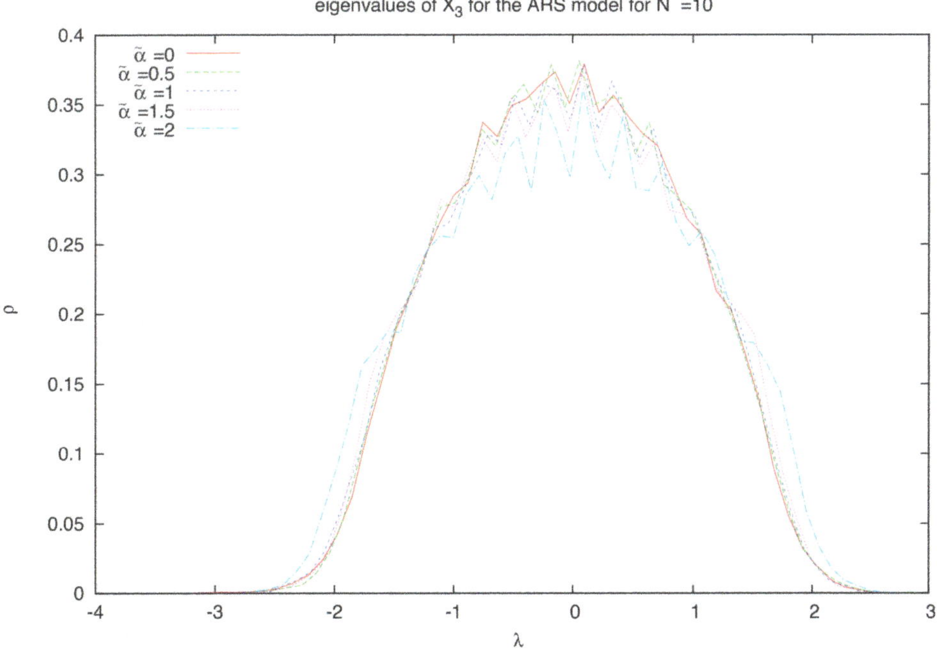

Figure 18.17. The eigenvalue distribution in the Yang–Mills matrix phase as a function of $\tilde{\alpha}$.

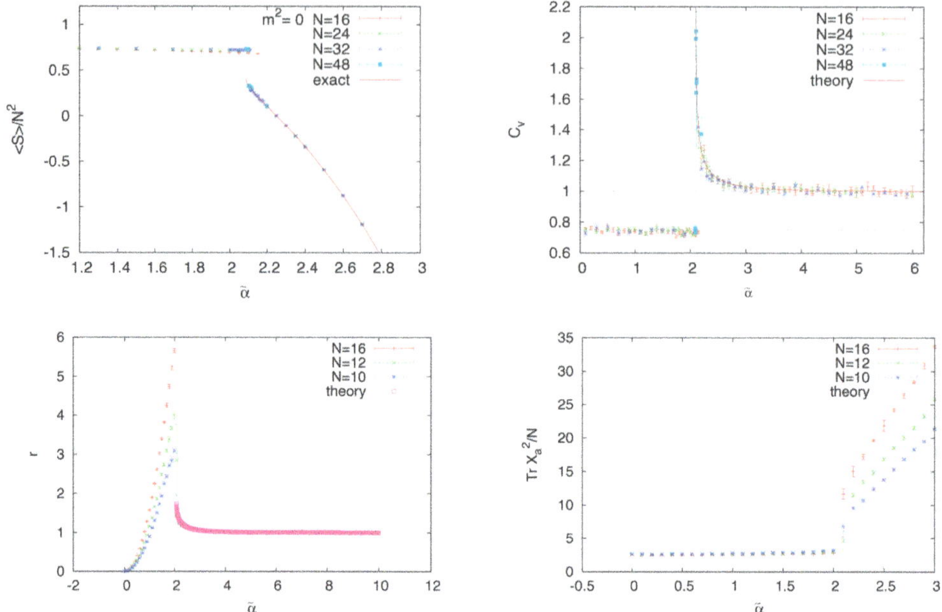

Figure 18.18. The behavior of various observables across the phase transition between the fuzzy sphere and the Yang–Mills phase.

point then it starts decreasing fast in the matrix phase until it reaches the value $r = 0$. See figure 18.18.

The different phases of the model are characterized by:

fuzzy sphere ($\tilde{\alpha} > \tilde{\alpha}_*$)	matrix phase ($\tilde{\alpha} < \tilde{\alpha}_*$)
$r = 1$	$r = 0$
$C_v = 1$	$C_v = 0.75$

18.5.4 Noncommutative gauge theory: UV–IR mixing on the fuzzy sphere

The second basic result summarizing the physics of emergent geometry from Yang–Mills matrix models is the statement that by expanding around the quantum background corresponding to the geometric phase we obtain a noncommutative gauge theory on the corresponding noncommutative space.

For example, it was shown in [137, 138] that, to leading order in the string tension, the dynamics of open strings moving in a curved space with S^3 metric, in the presence of a non-vanishing Neveu-Schwarz B-field and with Dp-branes, is equivalent to a noncommutative gauge theory on the fuzzy sphere with a Chern–Simons term given precisely by the Alekseev–Recknagel–Schomerus matrix model (18.225). Indeed, this action (18.225) can be rewritten as a sum of a Yang–Mills term S_{YM} and a Chern–Simons term S_{CS} as follows

$$S[D_i] = S_{\mathrm{YM}}[D_i] + S_{\mathrm{CS}}[D_i]$$

$$S_{\mathrm{YM}}[D_i] = -\frac{1}{4g^2 N}\,\mathrm{Tr}\,F_{ij}^2$$

$$S_{\mathrm{CS}}[D_i] = -\frac{1}{6g^2 N}\,\mathrm{Tr}\left[i\varepsilon_{ijk}F_{ij}D_k + (D_i^2 - c_2)\right].$$

(18.243)

By expanding now the covariant matrix coordinates D_i around the fuzzy sphere configuration given by (18.217) as $D_i = L_i + A_i$ we can rewrite the field strength $F_{ij} = [D_i, D_j] - i\varepsilon_{ijk}D_k$ as follows

$$D_i = L_i + A_i \Rightarrow F_{ij} = F_{ij}^{(0)} + [A_i, A_j], \quad F_{ij}^{(0)} = [L_i, A_j] - [L_j, A_i] - i\varepsilon_{ijk}A_k. \quad (18.244)$$

Clearly, we have in the commutative limit the behavior

$$F_{ij} \longrightarrow F_{ij}^{(0)}, \quad F_{ij}^{(0)} \longrightarrow \mathcal{L}_i A_j - \mathcal{L}_j A_i - i\varepsilon_{ijk}, \quad N \longrightarrow \infty. \quad (18.245)$$

In other words, A_i is identified as a noncommutative gauge field on the fuzzy sphere with a field strength given exactly by F_{ij}. In fact, the $U(N)$ gauge symmetry $D_i \longrightarrow U D_i U^\dagger$ is equivalent to the $U(N)$ transformations: $A_i \longrightarrow U A_i U^\dagger + U[L_i, U^\dagger]$. This gauge field, which has three components, describes therefore a two-dimensional noncommutative gauge field tangent to the sphere plus a scalar component normal to the sphere defined by the gauge-covariant prescription [141]

$$\Phi = R\frac{D_i^2 - c_2}{2\sqrt{c_2}} = \frac{1}{2}\left(\hat{x}_i A_i + A_i \hat{x}_i + \frac{R A_i^2}{\sqrt{c_2}}\right) \longrightarrow x_i A_i, \quad N \longrightarrow \infty. \quad (18.246)$$

Thus, the second term in the Chern–Simons action S_{CS} is a linear term in the normal scalar field Φ. In fact, the whole Alekseev–Recknagel–Schomerus action (18.243) can be rewritten in the form

$$S_N[A_i] = -\frac{1}{4g^2 N}\,\mathrm{Tr}\left[F_{ij}^{(0)} + [A_i, A_j]\right]^2$$
$$-\frac{i}{2g^2 N}\varepsilon_{ijk}\,\mathrm{Tr}\left[\frac{1}{2}F_{ij}^{(0)}A_k + \frac{1}{3}[A_i, A_j]A_k\right].$$

(18.247)

This is indeed a noncommutative gauge theory on the fuzzy sphere $\mathbf{S}_N^2$ given by the sum of a Yang–Mills term and a Chern–Simons term. In the commutative limit $N \longrightarrow \infty$, where all commutators vanish, we get the action

$$S_\infty[A_i] = -\frac{1}{4g^2}\int_{S^2}\frac{d\Omega}{4\pi}(F_{ij}^{(0)})^2 - \frac{i}{2g^2}\varepsilon_{ijk}\int_{S^2}\frac{d\Omega}{4\pi}\frac{1}{2}F_{ij}^{(0)}A_k. \quad (18.248)$$

The commutative action S_∞ is at most quadratic in the field A_i and as a consequence the corresponding effective action will be essentially given by S_∞ itself. However, quantization of the noncommutative action S_N is much more involved and yields a non-trivial effective action. As it turns out, the commutative limit of this noncommutative effective action does not tend to S_∞, which is the signature of the UV–IR mixing in this model. In [107] the quadratic effective action was computed explicitly and was found to be given in the commutative limit $N \longrightarrow \infty$ by the expression

$$\Gamma_2 = -\frac{1}{4g^2} \int \frac{d\Omega}{4\pi} F_{ij}^{(0)} (1 + 2g^2 \Delta_3) F_{ij}^{(0)}$$

$$- \frac{i}{4g^2} \varepsilon_{ijk} \int \frac{d\Omega}{4\pi} F_{ij}^{(0)} (1 + 2g^2 \Delta_3) A_k + 4\sqrt{c_2} \int \frac{d\Omega}{4\pi} \Phi \qquad (18.249)$$

$$+ \text{non}-\text{local quadratic terms}.$$

The derivation of this result starts from the effective action (18.223). First, we should expand the background field as $B_i = L_i + A_i$. The quadratic tree-level action S_2, as derives from (18.247), explicitly reads

$$S_2[A_i] = -\frac{1}{4g^2 N} \text{Tr} \left[F_{ij}^{(0)} \right]^2 - \frac{i}{2g^2 N} \varepsilon_{ijk} \text{Tr} \left[\frac{1}{2} F_{ij}^{(0)} A_k \right]. \qquad (18.250)$$

Second, we will apply directly the result (18.223) to find quantum corrections to this quadratic action. This will of course capture all quantum corrections to the vacuum polarization tensor as well as tadpole corrections. The quadratic effective action is given explicitly by

$$\Gamma_2 = S_2 + \frac{1}{2} \text{TR} \left(\Delta^{(1)} + \Delta^{(2)} - \frac{1}{2} (\Delta^{(1)})^2 \right) - \frac{1}{4} \text{TR}(\Delta^{(3)})_{ii}^2. \qquad (18.251)$$

Here, $\Delta^{(1)}$, $\Delta^{(2)}$ and $\Delta^{(3)}$ are defined by

$$\Delta^{(1)} = \frac{1}{\mathcal{L}^2} (\mathcal{L}\mathcal{A} + \mathcal{A}\mathcal{L}), \quad \Delta^{(2)} = \frac{1}{\mathcal{L}^2} \mathcal{A}^2, \quad \Delta_{ij}^{(3)} = \frac{2}{\mathcal{L}^2} \mathcal{F}_{ij}^{(0)}. \qquad (18.252)$$

It is obvious, from these expressions, that the propagators of the fluctuation field Q_i and the ghost fields $b^\dagger$ and b are given by the inverse of the Laplacian $\mathcal{L}^2$. In fact, it is a property of the Alekseev–Recknagel–Schomerus action given by either (18.225) or (18.243) or (18.247) that the propagator in the Feynman gauge $\xi = 1$ is given simply by $1/\mathcal{L}^2$.

We start with the tadpole contribution. Quantum correction to the tree-level linear term, which is in fact identically zero, is given by the combination of the two tadpole diagrams of figure 18.19. These diagrams are also equal to the second term in the expansion (18.251), viz $\Gamma_1 = \frac{1}{2} \text{TR} \Delta^{(1)}$. We find

$$\Gamma_1 = 4 \frac{\sqrt{c_2}}{N} \text{Tr} \, \Phi - \frac{2}{N} \text{Tr} \, A_i^2. \qquad (18.253)$$

This identity is exact and as it turns out it is crucial in establishing gauge invariance of the quantum noncommutative gauge theory on the fuzzy sphere.

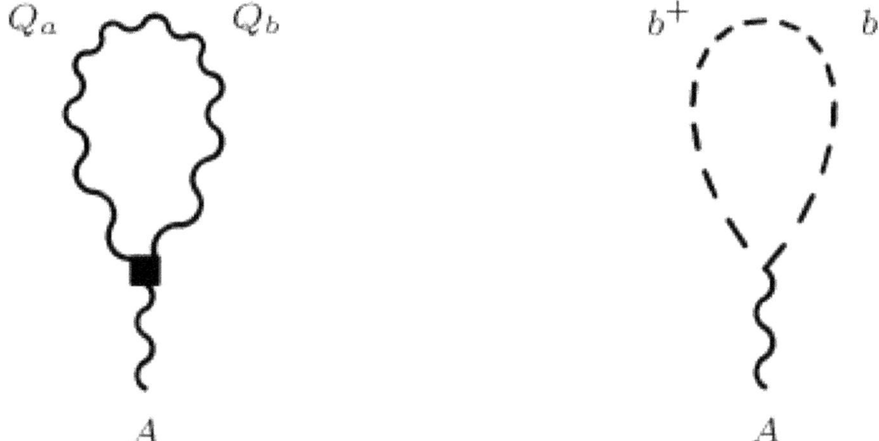

Figure 18.19. Tadpole diagrams.

Figure 18.20. Vacuum polarization diagrams (4-vertices).

Next, we compute the vacuum polarization tensor. The 4-vertex contribution to the vacuum polarization tensor is given by the diagrams of figure 18.20, which are also equal to the third term in the expansion (18.251), viz $\Gamma_2^{(4)} = \frac{1}{2}\mathrm{TR}\,\Delta^{(2)}$. After some calculation we get the explicit answer

$$\Gamma_2^{(4)} = \frac{1}{N}\,\mathrm{Tr}\,A_i \mathcal{L}^2 \Delta_4 A_i. \tag{18.254}$$

The operator $\Delta_4 \equiv \Delta_4(\mathcal{L}^2)$ is defined by its eigenvalues on polarization tensor $\hat{Y}_{p_1 n_1}$ given by (with $L \equiv N - 1$)

$$\Delta_4(p_1) = \sum_{l_1, l_2} \frac{2l_1 + 1}{l_1(l_1 + 1)} \frac{2l_2 + 1}{l_2(l_2 + 1)} (1 - (-1)^{l_1 + l_2 + p_1})(L + 1) \left\{ \begin{matrix} p_1 & l_1 & l_2 \\ \frac{L}{2} & \frac{L}{2} & \frac{L}{2} \end{matrix} \right\}^2 \frac{l_2(l_2 + 1)}{p_1(p_1 + 1)}. \tag{18.255}$$

The 3-vertex contribution comes from three different diagrams. The contribution of the $\mathcal{F}$ term is given by the diagram of figure 18.21 and it corresponds to the last term in expansion (18.251), namely $\Gamma_2^{(3F)} = -\frac{1}{4}\mathrm{TR}(\Delta^{(3)})_{ii}^2$ whereas the fourth term in the expansion (18.251), i.e., $\Gamma_2^{(3A)} = -\frac{1}{4}\mathrm{TR}(\Delta^{(1)})^2$ corresponds to the combination of the diagrams displayed in figure 18.22.

Figure 18.21. Vacuum polarization diagrams (3-vertices with the F-field).

Figure 18.22. Vacuum polarization diagrams (3-vertices with the A-field).

The diagram of figure 18.21 is found to be given by

$$\Gamma_2^{(3F)} = \frac{1}{N}\,\mathrm{Tr}\,F_{ij}^{(0)}\Delta_F F_{ij}^{(0)}.$$
(18.256)

The operator $\Delta_F \equiv \Delta_F(\mathcal{L}^2)$ is defined by its spectrum

$$\Delta_F(p_1) = 2\sum_{l_1,l_2}\frac{2l_1+1}{l_1(l_1+1)}\frac{2l_2+1}{l_2(l_2+1)}(1-(-1)^{l_1+l_2+p_1})(L+1)\left\{\begin{matrix} l_1 & l_2 & p_1 \\ \frac{L}{2} & \frac{L}{2} & \frac{L}{2} \end{matrix}\right\}^2.$$
(18.257)

Similarly, the diagrams of figure 18.22 are given by

$$\Gamma_2^{(3A)} = -\frac{1}{N}\,\mathrm{Tr}\,A_i \mathcal{L}_i \Delta_3 \mathcal{L}_j A_j + \Gamma_2^{(3A)}.$$
(18.258)

The second term is given explicitly by

$$\begin{aligned}
\Gamma_2^{(3A)} &= \sum_{p_1 n_1}\sum_{p_2 n_2} A_{-\mu}(p_1 n_1)A_{-\nu}(p_2 n_2)(-1)^{n_1+\nu} \\
&\quad \left[C^{p_1-1m}_{p_1 n_1 1\mu} C^{p_1-1-m}_{p_2 n_2 1\nu}\left(\Lambda^{(-)}(p_1,p_2)+\Sigma^{(-)}(p_1,p_2)\right)\right. \\
&\quad \left. + C^{p_1+1m}_{p_1 n_1 1\mu} C^{p_1+1-m}_{p_2 n_2 1\nu}\left(\Lambda^{(+)}(p_1,p_2)+\Sigma^{(+)}(p_1,p_2)\right)\right].
\end{aligned}$$
(18.259)

Again the operator $\Delta_3 \equiv \Delta_3(\mathcal{L}^2)$ is defined by its spectrum

$$\Delta_3(p_1) = \sum_{l_1,l_2} \frac{2l_1 + 1}{l_1(l_1 + 1)} \frac{2l_2 + 1}{l_2(l_2 + 1)}(1 - (-1)^{l_1 + l_2 + p_1})(L + 1)\begin{Bmatrix} p_1 & l_1 & l_2 \\ \frac{L}{2} & \frac{L}{2} & \frac{L}{2} \end{Bmatrix}^2$$

$$\times \frac{l_2(l_2 + 1)}{p_1^2(p_1 + 1)^2}(l_2(l_2 + 1) - l_1(l_1 + 1)). \tag{18.260}$$

We remark that all quantum corrections to the vacuum polarization tensor given by the equations (18.254), (18.256) and (18.258) are written in terms of the operator

$$\Delta(p_1, p_2) = \sum_{l_1 l_2} \frac{2l_1 + 1}{l_1(l_1 + 1)} \frac{2l_2 + 1}{l_2(l_2 + 1)}(L + 1)$$

$$\begin{Bmatrix} p_1 & l_1 & l_2 \\ \frac{L}{2} & \frac{L}{2} & \frac{L}{2} \end{Bmatrix}\begin{Bmatrix} p_2 & l_1 & l_2 \\ \frac{L}{2} & \frac{L}{2} & \frac{L}{2} \end{Bmatrix}X(l_1, l_2, p_1, p_2), \tag{18.261}$$

X is of the form $X(l_1, l_2, p_1, p_2) = \delta_{p_1 p_2} \bar{X}(l_1, l_2, p_1)$ and where the sums are always over l_1 and l_2 such that $l_1 + l_2 + p_1$ is an odd number.

Similarly, the functions $\Lambda^{(\pm)}(p_1, p_2)$ and $\Sigma^{(\pm)}(p_1, p_2)$ in the definition of $\Gamma_2^{(3A)}$ (equation (18.259)) are of the form (18.261) with some X such that $\Lambda^{(\pm)}(p_1, p_2) = \delta_{p_1 p_2} \bar{\Lambda}^{(\pm)}(p_1)$ and $\Sigma^{(\pm)}(p_1, p_2) = \delta_{p_1 \pm 2, p_2} \bar{\Sigma}^{(\pm)}(p_1)$ respectively. Furthermore, we can see by inspection that the Clebsch–Gordan coefficients appearing in the action $\Gamma_2^{(3A_2)}$ are exactly those which appear in the scalar mass term

$$\frac{1}{4N} \text{Tr}[\hat{x}_i, A_i]_+^2 = \sum_{p_1 n_1 p_2 n_2} A_{-\mu}(p_1 n_1)A_{-\nu}(p_2 n_2)(-1)^{n_1 + \nu}\left[C_{p_1 n_1 \mu}^{p_1 - 1m} C_{p_2 n_2 1\nu}^{p_1 - 1 - m}\right.$$

$$\times (\lambda^{(-)}(p_1, p_2) + \sigma^{(-)}(p_1, p_2)) + C_{p_1 n_1 \mu}^{p_1 + 1m} C_{p_2 n_2 1\nu}^{p_1 + 1 - m}(\lambda^{(+)}(p_1, p_2) + \sigma^{(+)}(p_1, p_2))\Big]. \tag{18.262}$$

Here, $\lambda^{(\pm)}(p_1, p_2)$ and $\sigma^{(\pm)}(p_1, p_2)$ are some other functions which are such that $\lambda^{(\pm)}(p_1, p_2) = \delta_{p_1 p_2} \bar{\lambda}^{(\pm)}(p_1), \sigma^{(\pm)}(p_1, p_2) = \delta_{p_1 \pm 2, p_2} \bar{\sigma}^{(\pm)}(p_1)$.

By comparing (18.259) and (18.262) we can immediately deduce that the action $\Gamma_2^{(3A_2)}$, in position space, must involve anticommutators of $\hat{x}_i$ and A_i instead of commutators and hence it is a scalar-like contribution. As it turns out, this action contains (in the commutative limit) gauge-invariant terms, which describe non-local interactions between the normal scalar and the tangent gauge fields on the sphere [107].

By putting together equations (18.253), (18.254), (18.256) and (18.258) we obtain the full quadratic effective action of noncommutative gauge theory on the fuzzy sphere $\mathbf{S}_N^2$. Explicitly, we have

$$\Gamma_2 = S_2 + \frac{4\sqrt{c_2}}{N} \text{Tr}\,\Phi + \frac{1}{N} \text{Tr}\,A_i(\mathcal{L}^2\Delta_4 - 2)A_i$$

$$- \frac{1}{N} \text{Tr}\,A_i\mathcal{L}_i\Delta_3\mathcal{L}_j A_j + \frac{1}{N} \text{Tr}\,F_{ij}^{(0)}\Delta_F F_{ij}^{(0)} + \Gamma_2^{(3A_2)}. \tag{18.263}$$

In is rather clear that the first two terms and the last two terms are gauge invariant (in the commutative limit $N \longrightarrow \infty$). Naturally, we also expect that the third and fourth terms in (18.263) to become gauge invariant in the limit. To check this property explicitly we rewrite these two terms as follows

$$
\mathrm{Tr}\, A_i (\mathcal{L}^2 \Delta_4 - 2) A_i - \mathrm{Tr}\, A_i \mathcal{L}_i \Delta_3 \mathcal{L}_j A_j = -\frac{1}{2}\, \mathrm{Tr}\, F_{ij}^{(0)} \Delta_3 F_{ij}^{(0)}
$$
$$
-\frac{i}{2}\varepsilon_{ijk}\, \mathrm{Tr}\, F_{ij}^{(0)}(\Delta_3 + \mathcal{L}^2(\Delta_3 - \Delta_4) + 2) A_k \qquad (18.264)
$$
$$
+ i\varepsilon_{ijk}\, \mathrm{Tr}\, \mathcal{L}_i A_j (\mathcal{L}^2(\Delta_3 - \Delta_4) + 2) A_k .
$$

The first two terms in this expression are now exactly gauge invariant in the commutative limit whereas the third term can not be gauge invariant unless it vanishes identically. We expect therefore by the requirement of gauge invariance alone that we have the asymptotic behavior $\mathcal{L}^2(\Delta_3 - \Delta_4) + 2 \longrightarrow 0$, $N \longrightarrow \infty$. As it turns out, this behavior is true for all finite values of N. In other words, we have in the identity [107]

$$
\mathcal{L}^2(\Delta_3 - \Delta_4) + 2 = 0. \qquad (18.265)
$$

Thus, the quadratic effective action of noncommutative gauge theory on the fuzzy sphere $\mathbf{S}_N^2$ reads

$$
\Gamma_2 = S_2 + \frac{4\sqrt{c_2}}{N}\,\mathrm{Tr}\,\Phi + \frac{1}{N}\,\mathrm{Tr}\, F_{ij}^{(0)}\left(\Delta_F - \frac{1}{2}\Delta_3\right)F_{ij}^{(0)}
$$
$$
-\frac{i}{2N}\varepsilon_{ijk}\,\mathrm{Tr}\, F_{ij}^{(0)}\Delta_3 A_k + \Gamma_2^{(3A_2)}. \qquad (18.266)
$$

In summary, the third and four4th terms of the effective action (18.266) give rise to a non-trivial quantum contribution to the noncommutative gauge theory on the fuzzy sphere reflecting the existence of a gauge invariant UV–IR mixing, which survives the commutative limit. Indeed, this effective action (18.266) goes, in the commutative limit $N \longrightarrow \infty$, to the action (18.249) where we have also to use the fact that the eigenvalues Δ_3 and Δ_F are given in the limit by

$$
\Delta_F(p_1) \longrightarrow 0, \quad \Delta_3(p_1) \longrightarrow \frac{2\displaystyle\sum_{l=2}^{p_1}\frac{1}{l}}{p_1(p_1 + 1)} \neq 0. \qquad (18.267)
$$

18.5.5 The UV–IR mixing in the large mass limit

The normal scalar fluctuation Φ can be suppressed, and thus reduce the non-commutative gauge theory to a purely two-dimensional model on the fuzzy sphere $\mathbf{S}_N^2$, by simply adding a large mass term to the Alekseev–Recknagel–Schomerus action (18.247). In other words, we should consider instead the action

$$S_N[A_i] = -\frac{1}{4g^2N} \operatorname{Tr}\left[F_{ij}^{(0)} + [A_i, A_j]\right]^2$$
$$-\frac{i}{2g^2N}\varepsilon_{ijk}\operatorname{Tr}\left[\frac{1}{2}F_{ij}^{(0)}A_k + \frac{1}{3}[A_i, A_j]A_k\right] + \frac{2m^2}{g^2N}\operatorname{Tr}\Phi^2. \tag{18.268}$$

In the presence of this mass term the quadratic tree-level action (18.250) and the quadratic effective action (18.251) become now given by

$$S_2 = -\frac{1}{2g^2N}\operatorname{Tr}[L_i, A_j]^2 + \frac{1}{2g^2N}\operatorname{Tr}[L_i, A_i]^2 + \frac{m^2}{2g^2N}\operatorname{Tr}(\hat{x}_i A_i + A_i \hat{x}_i)^2. \tag{18.269}$$

$$\Gamma_2 = S_2 + \operatorname{TR}\left[\frac{1}{2}\left(\frac{1}{\Delta}\right)_{ij}\omega_{ji}^{(1)} - \frac{1}{\mathcal{L}^2}(\mathcal{L}\mathcal{A} + \mathcal{A}\mathcal{L})\right]$$
$$+ \operatorname{TR}\left[\frac{1}{2}\left(\frac{1}{\Delta}\right)_{ij}\omega_{ji}^{(2)} - \frac{1}{4}\left(\frac{1}{\Delta}\right)_{ik}\omega_{kl}^{(1)}\left(\frac{1}{\Delta}\right)_{lj}\omega_{ja}^{(1)} - \frac{1}{\mathcal{L}^2}\mathcal{A}^2 + \frac{1}{2}\frac{1}{\mathcal{L}^2}(\mathcal{L}\mathcal{A} + \mathcal{A}\mathcal{L})\frac{1}{\mathcal{L}^2}(\mathcal{L}\mathcal{A} + \mathcal{A}\mathcal{L})\right]. \tag{18.270}$$

The operators $\omega^{(1)}$ and $\omega^{(2)}$ contain the linear and quadratic vertices, respectively, and they are given explicitly by

$$\omega_{ij}^{(1)} = (\mathcal{L}\mathcal{A} + \mathcal{A}\mathcal{L})\delta_{ij} + 2\mathcal{F}_{ij}^{(0)} + \frac{2m^2}{c_2}(LA + AL)\delta_{ij} + \frac{4m^2}{c_2}(L_i A_j + A_i L_j)$$
$$\omega_{ij}^{(2)} = \mathcal{A}^2\delta_{ij} + 2[\mathcal{A}_i, \mathcal{A}_j] + \frac{2m^2}{c_2}A^2\delta_{ij} + \frac{4m^2}{c_2}A_i A_j. \tag{18.271}$$

The Laplacian Δ is given, on the other hand, by $\Delta_{ij} = \mathcal{L}^2\delta_{ij} + 4m^2 P_{ij}^N$ where $P_{ij}^N = \hat{x}_i \hat{x}_j$ is the normal projector on the fuzzy sphere $\mathbf{S}_N^2$. The tangent projector on the fuzzy sphere is then obviously given by the orthogonal complement defined by $P_{ij}^T = \delta_{ij} - \hat{x}_i \hat{x}_j$. These two projectors correspond to the normal and tangent projective modules on the fuzzy sphere, respectively. These projective modules play in noncommutative geometry the same role played by fiber bundles in differential geometry.

The propagator, in the large mass limit $m^2 \longrightarrow \infty$, takes the relatively simple form

$$\frac{1}{\Delta} = P^T \frac{1}{\mathcal{L}^2} P^T + O\left(\frac{1}{m^2}\right). \tag{18.272}$$

It is then straightforward to see that the combined contribution of the tadpole diagrams and the 4-vertex corrections to the vacuum polarization tensor vanishes in this limit, viz

$$\Gamma_1 + \Gamma_2^{(4)} = \operatorname{TR}\frac{1}{2}\left(\frac{1}{\Delta}\right)_{ij}(\mathcal{L}\mathcal{A} + \mathcal{A}\mathcal{L} + \mathcal{A}^2)\delta_{ij} - \operatorname{TR}\frac{1}{2}\left(\frac{1}{\mathcal{L}^2}\right)(\mathcal{L}\mathcal{A} + \mathcal{A}\mathcal{L} + \mathcal{A}^2) \equiv 0. \tag{18.273}$$

This result should be compared with the commutative limit of the sum of the two actions (18.253) and (18.254), which as we have shown does depend in the limit on the two-dimensional gauge field. In other words, suppressing the normal component of the gauge field, by giving it a large mass, allowed us to suppress in the limit the contribution of the tangent gauge field to the tadpole and to the 4-vertex correction of the vacuum polarization tensor. By the requirement of gauge invariance this suppression will also occur in the other contributions to the vacuum polarization tensor and as consequence the large mass of the scalar field regulates effectively the UV–IR mixing.

18.5.6 Phase diagram in the large mass limit and stability of the geometric phase

The massive Alekseev–Recknagel–Schomerus action (18.268) can also be rewritten as a $D = 3$ Yang–Mills matrix model given by

$$S[D] = S_{\mathrm{ARS}}[D] + V[D]$$

$$S_{\mathrm{ARS}} = \frac{1}{g^2 N} \mathrm{Tr} \left[-\frac{1}{4}[D_i, D_j]^2 + \frac{2i}{3}\varepsilon_{ijk} D_i D_j D_k \right]$$

$$V[D] = \frac{1}{g^2 N} \mathrm{Tr} \left[-\mu D_i^2 + \frac{m^2}{2c_2}(D_i^2)^2 \right], \quad \mu = m^2. \tag{18.274}$$

The case $m^2 = 0$ is studied in [105, 142] whereas the case $m^2 \neq 0$ is studied in [108–110].

By following the standard Faddeev–Popov procedure [143] and taking the background field configuration to be $D_a = \varphi L_a$ one finds that the free energy, the effective potential and the equation of motion of the order parameter φ to be given by

$$F = -\log Z, \quad \frac{F}{N^2} = \frac{3}{4}\log \tilde{\alpha}^4 + \frac{\tilde{\alpha}^4}{2}\left[\frac{\varphi^4}{4} - \frac{\varphi^3}{3} + m^2\frac{\varphi^4}{4} - \mu\frac{\varphi^2}{2} \right] + \log \tilde{\alpha}\varphi. \tag{18.275}$$

$$\frac{V_{\mathrm{eff}}}{2c_2} = \tilde{\alpha}^4\left[\frac{\varphi^4}{4} - \frac{\varphi^3}{3} + m^2\frac{\varphi^4}{4} - \mu\frac{\varphi^2}{2} \right] + \log \varphi^2. \tag{18.276}$$

$$\frac{\tilde{\alpha}^4}{2}[\varphi^4 - \varphi^3 + m^2\varphi^4 - \mu\varphi^2] + 1$$
$$= 0 \Rightarrow \bar{\varphi}^4 - \bar{\varphi}^3 - t\bar{\varphi}^2 + \frac{2}{a^4} = 0, \quad \bar{\varphi} = (1 + m^2)\varphi. \tag{18.277}$$

Here, t and a are the actual parameters of the model defined explicitly by

$$t = \mu(1 + m^2), \quad a^4 = \tilde{\alpha}^4/(1 + m^2)^3. \tag{18.278}$$

Thus, the critical point found for the Alekseev–Recknagel–Schomerus model with $\mu = m^2 = 0$ will be replaced by a critical line in the $a - t$ plane.

First, we compute the expected radius of the sphere (extent of space) as follows

$$\frac{1}{r} = \frac{\langle TrD_i^2 \rangle}{Nc_2} = -\frac{2}{\tilde{\alpha}^4}\frac{dF}{d\mu} = \varphi^2. \tag{18.279}$$

We also compute the average value of the total action (energy) by

$$S = \langle S \rangle / N^2 = \tilde{\alpha}^4 \frac{d}{d\tilde{\alpha}^4}\left(\frac{F}{N^2}\right) = \frac{3}{4} - \frac{\tilde{\alpha}^4 \varphi^3}{24} - \frac{\tilde{\alpha}^4 \mu \varphi^2}{8}. \tag{18.280}$$

Scaling the covariant matrix coordinates D_i as $D_i \longrightarrow (1 - \varepsilon)D_i$ in both the total action $S[D]$ and the measure $\int \prod_{i=1}^{3} [dD_i]$ leaves the partition function $Z = \int \prod_{i=1}^{3} [dD_i] \exp(-S[D])$ invariant, which in turn leads to a non-trivial Ward identity given by

$$\frac{\tilde{\alpha}^4}{N}\langle K_m \rangle = 3N^2, \quad K_m = \mathrm{Tr}\left(-[D_i, D_j]^2 + 2i\varepsilon_{ijk} D_i D_j D_k - 2m^2 D_i^2 + \frac{2m^2}{c_2}(D_i^2)^2\right). \tag{18.281}$$

Using this identity we can express the energy as

$$S = \frac{3}{4} + \frac{\tilde{\alpha}^4}{6N^3}\langle i\varepsilon_{ijk}\, \mathrm{Tr}\, D_i D_j D_k \rangle - \frac{\tilde{\alpha}^4}{2N^3}m^2\langle \mathrm{Tr}\, D_i^2 \rangle. \tag{18.282}$$

Thus, we can immediately compute the specific heat to be given by

$$C_v = \frac{\langle S^2 \rangle - \langle S \rangle^2}{N^2} = \frac{\langle S \rangle}{N^2} - \tilde{\alpha}^4 \frac{d}{d\tilde{\alpha}^4}\left(\frac{\langle S \rangle}{N^2}\right) = \frac{3}{4} + \frac{\tilde{\alpha}^5 \varphi}{32}(\varphi + 2m^2)\frac{d\varphi}{d\tilde{\alpha}}. \tag{18.283}$$

We also define the Yang–Mills and Chern–Simons term by

$$4\mathrm{YM} = -\frac{\langle \mathrm{Tr}[D_i, D_j]^2 \rangle}{2Nc_2} = \varphi^4 + \frac{8}{\tilde{\alpha}^4}, \quad \text{and} \quad 3\mathrm{CS} = \frac{\langle i\varepsilon_{ijk}\, \mathrm{Tr}\, D_i D_j D_k \rangle}{Nc_2} = -\varphi^3. \tag{18.284}$$

Let us now examine the predictions for the quantum transition as determined by the effective potential given in (18.276).

The extrema of the classical potential occur at

$$\varphi = \frac{1}{1 + m^2}\left\{0, \quad \varphi_\pm = \frac{1 \pm \sqrt{1 + 4t}}{2}\right\}. \tag{18.285}$$

For μ positive the global minimum is φ_+. The solution $\varphi = 0$ is a local maximum and φ_- is a local minimum. In particular for $\mu = m^2$ we obtain the global minimum $\varphi_+ = 1$. For μ negative the global minimum is still φ_+ but 0 becomes a local minimum and φ_- a local maximum. If μ is sent more negative then the global minimum $\varphi_+ = 1$ becomes degenerate with $\varphi = 0$ at $t = -\frac{2}{9}$ and the maximum height of the barrier is given by $V_- = \tilde{\beta}^4/324$ which occurs at $\varphi_- = \frac{1}{3}$. The model has a first order transition at $t = -2/9$ where the classical ground states switches from φ_+ for $t > -2/9$ to 0 for $t < 2/9$.

Let us now consider the effect of quantum fluctuations. The condition $V'_{\mathrm{eff}} = 0$ gives us extrema of the model. For large enough $\tilde{\alpha}$, and large enough m and μ, it admits two positive solutions. The largest solution can be identified with the ground state of the system. It will determine the radius of the sphere. The second solution is

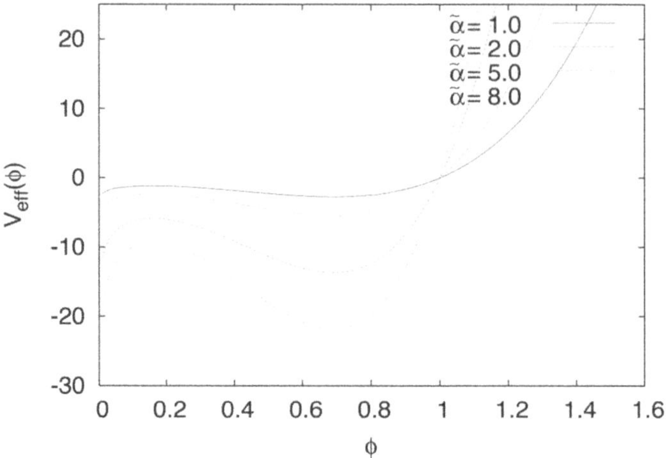

Figure 18.23. The effective potential for $m^2 = 20$.

the local maximum of V_{eff} and will determine the height of the barrier (figure 18.23). As the coupling is decreased these two solutions merge and the barrier disappears. This is the critical point of the model. For smaller couplings than the critical value $\tilde{\alpha}_*$ the fuzzy sphere solution $D_i = \varphi L_i$ no longer exists. Therefore, the classical transition described above is significantly affected by quantum fluctuations.

The condition when the barrier disappears is $V_{\text{eff}}'' = 0$. At this point the local minimum merges with the local maximum. Solving the two equations $V_{\text{eff}}' = V_{\text{eff}}'' = 0$ yield the critical value

$$g_*^2 = \frac{1}{\tilde{\alpha}_*^4} = \frac{\varphi_*^2(\varphi_* + 2\mu)}{8}, \quad \varphi_* = \frac{3}{8(1 + m^2)}\left[1 + \sqrt{1 + \frac{32\mu(1 + m^2)}{9}}\right]. \quad (18.286)$$

If we take μ negative we see that g_* goes to zero at $\mu(1 + m^2) = -1/4$ and the critical coupling $\tilde{\alpha}_*$ is sent to infinity and therefore for $\mu(1 + m^2) < -\frac{1}{4}$ the model has no fuzzy sphere phase. However, in the region $-\frac{1}{4} < \mu(1 + m^2) < -\frac{2}{9}$ the action $S_{\text{ARS}} + V$ is completely positive. It is therefore not sufficient to consider only the configuration $D_i = \varphi L_i$, but rather all $SU(2)$ representations must be considered. Furthermore, for large $\tilde{\alpha}$ the ground state will be dominated by those representations with the smallest Casimir. This means that there is no fuzzy sphere solution for $\mu(1 + m^2) < -\frac{2}{9}$.

The limits of interest are the limit $\mu = m^2 \longrightarrow 0$ and the limit $\mu = m^2 \longrightarrow \infty$. In these cases, we have the critical values

$$\varphi_* = \frac{3}{4}, \quad \tilde{\alpha}_*^4 = \left(\frac{8}{3}\right)^3, \quad \mu = m^2 \longrightarrow 0. \quad (18.287)$$

$$\varphi_* = \frac{1}{\sqrt{2}}, \quad \tilde{\alpha}_*^4 = \frac{8}{m^2}, \quad \mu = m^2 \longrightarrow \infty. \tag{18.288}$$

This means that the phase transition, in the limit of large mass, is located at a smaller value of the coupling constant $\tilde{\alpha}$ as m is increased. In other words, the region where the fuzzy sphere is stable is extended to lower values of the gauge coupling.

The critical behavior for the model $\mu = m^2 = 0$ is given explicitly by

$$\varphi = \frac{1}{4}\left[3 + \sqrt{6}\sqrt{\frac{\tilde{\alpha}-\tilde{\alpha}^*}{\tilde{\alpha}^*}} - \frac{4}{3}\frac{\tilde{\alpha}-\tilde{\alpha}^*}{\tilde{\alpha}^*} + O(\varepsilon^{\frac{3}{2}})\right], \quad \varepsilon = \frac{\tilde{\alpha}-\tilde{\alpha}^*}{\tilde{\alpha}^*}. \tag{18.289}$$

$$S = \frac{5}{12} - \frac{1}{3^{\frac{1}{8}}2^{\frac{5}{8}}}\sqrt{\tilde{\alpha}-\tilde{\alpha}_*} - \frac{7}{3^{\frac{5}{4}}2^{\frac{5}{4}}}(\tilde{\alpha}-\tilde{\alpha}_*) + O((\tilde{\alpha}-\tilde{\alpha}_*)^{\frac{3}{2}}). \tag{18.290}$$

$$C_v = \frac{29}{36} + \frac{1}{2^{\frac{11}{8}}3^{\frac{7}{8}}}\frac{1}{\sqrt{\tilde{\alpha}-\tilde{\alpha}_*}} + O((\tilde{\alpha}-\tilde{\alpha}_*)^{\frac{1}{2}}). \tag{18.291}$$

This gives immediately a divergent specific heat with critical exponent equal to 1/2.

The generic model with $\mu = m^2 \neq 0$ is given by the critical behavior

$$\bar{\phi} = \bar{\phi}_* + \frac{4}{a_*^{\frac{5}{2}}}\frac{1}{\sqrt{3\bar{\phi}_* + 4t}}\sqrt{a - a_*} + \cdots \tag{18.292}$$

$$S = S_* - a_c^4 \frac{\bar{\phi}_*(\bar{\phi}_* + 2t)}{\sqrt{3\bar{\phi}_* + 4t}}\sqrt{\frac{a - a_c}{a_c}} + \cdots, \quad S_* = \frac{3}{4} - \frac{(\bar{\phi}_* + 3t)}{3(\bar{\phi}_* + 2t)}. \tag{18.293}$$

$$C_v = C_v^B + \frac{1}{8\sqrt{1 + \frac{\tilde{\alpha}_*^4\bar{\phi}_*^3}{16}}}\frac{\sqrt{\tilde{\alpha}_*}}{\sqrt{\tilde{\alpha}-\tilde{\alpha}_*}} + \cdots, \quad C_v^B = \frac{3}{4} + \frac{(3 + 4t)\bar{\phi}_* + 2t}{8(3\bar{\phi} + 4t)^2}. \tag{18.294}$$

The basic prediction here is that the critical exponent of the specific heat for this model is given precisely by

$$\alpha = \frac{1}{2}. \tag{18.295}$$

If we extrapolate these results to large m^2 where we know that $\phi_* \longrightarrow 1/\sqrt{2}$ and $\tilde{\alpha}_*^4 \longrightarrow 8/m^2$ we get the specific heat in the fuzzy sphere phase to be given by

$$C_v = \frac{3}{4} + \frac{1}{32\sqrt{2}\,m^2} + \frac{1}{2^{\frac{21}{8}}m^{\frac{1}{4}}}\frac{1}{\sqrt{\tilde{\alpha}-\tilde{\alpha}_*}} + \cdots \tag{18.296}$$

The massive Alekseev–Recknagel–Schomerus action (18.274) is studied by means of the Monte Carlo method (Metropolis algorithm) in [110].

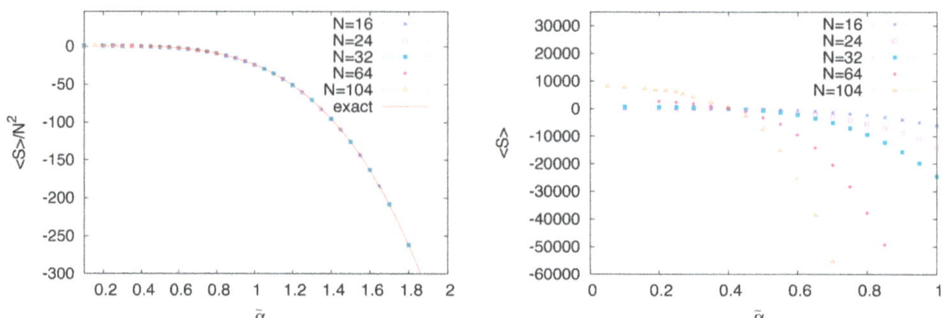

Figure 18.24. The critical value $\tilde{\alpha}_s$ is where the curves $\langle S \rangle$ for different values of N intersect.

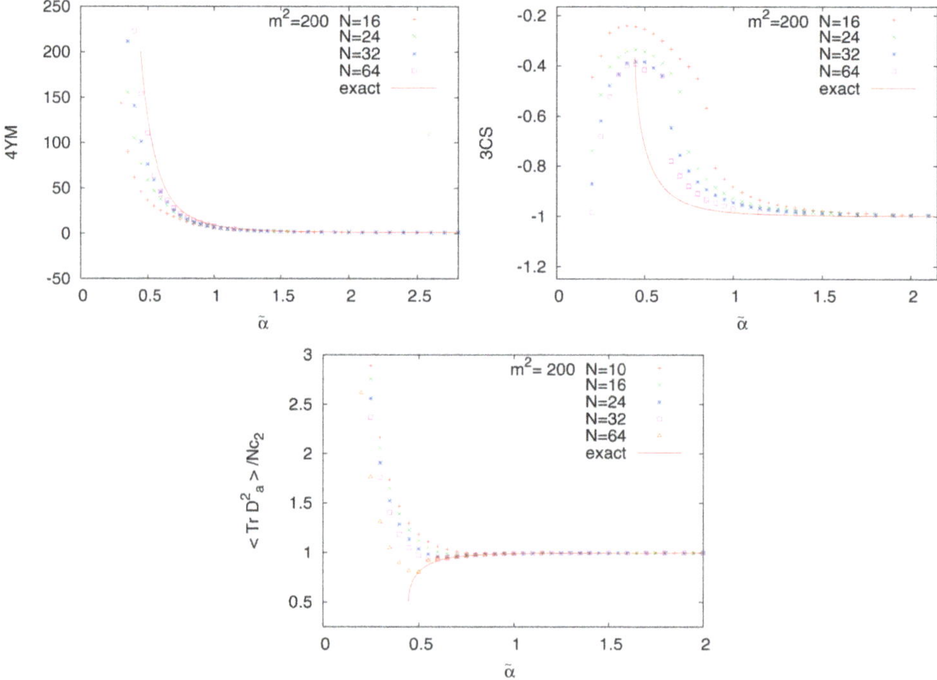

Figure 18.25. The observed data approaches the theoretical prediction as N is increased.

For example, the critical value $\tilde{\alpha}_s$ is measured as follows. We observe that different actions $\langle S \rangle$, which correspond to different values of N (for some fixed value of m^2), intersect at some value of the coupling constant $\tilde{\alpha}$, which we define to be the critical point $\tilde{\alpha}_s$ (figure 18.24). The quantities S, YM, CS and the radius $1/r$ are all continuous across the transition point in this regime. We observe that the Monte Carlo data tend to approach the theoretical prediction as N is increased (figure 18.25).

The specific heat in the fuzzy sphere phase is constant equal to 1, it starts to decrease at $\tilde{\alpha}_{\max}$, goes through a minimum at $\tilde{\alpha}_{\min}$, and then goes up again to the value 0.75 when $\tilde{\alpha} \longrightarrow 0$ (figure 18.26). Extrapolating $\tilde{\alpha}_{\max}$ and $\tilde{\alpha}_{\min}$ to $N = \infty$

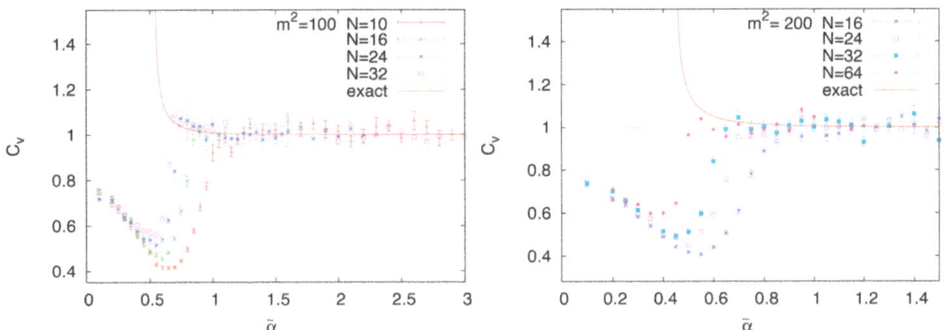

Figure 18.26. The specific heat of the matrix model (18.274) at large mass.

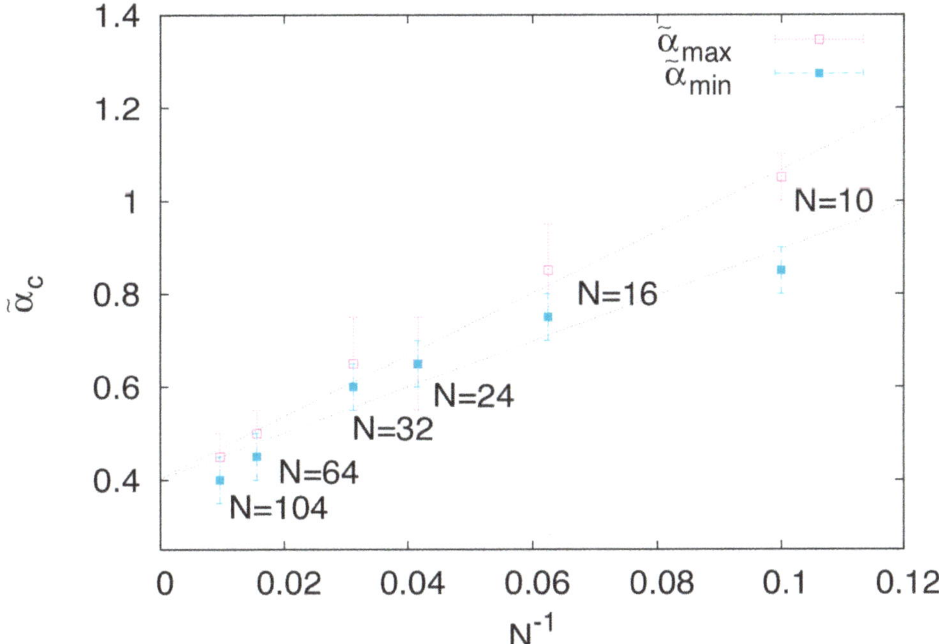

Figure 18.27. By extrapolating the measured values of $\tilde{\alpha}_{max}$ and $\tilde{\alpha}_{min}$ to $N = \infty$ we obtain the critical value $\tilde{\alpha}_c$.

(figure 18.27) we obtain our estimate for the critical coupling $\tilde{\alpha}_c$, which agree with $\tilde{\alpha}_s$ within statistical errors.

Thus, it seems that the specific heat in the regime of large values of m^2 becomes constant in the matrix phase equal to $C_v = 0.75$. There remains the question of whether or not the specific heat has critical fluctuations and diverges with a critical exponent $\alpha = 1/2$ as predicted by equation (18.294).

In fact, it seems that the matrix-to-sphere phase transition is third order in the limit of large mass, which is a behavior reminiscent of pure matrix models.

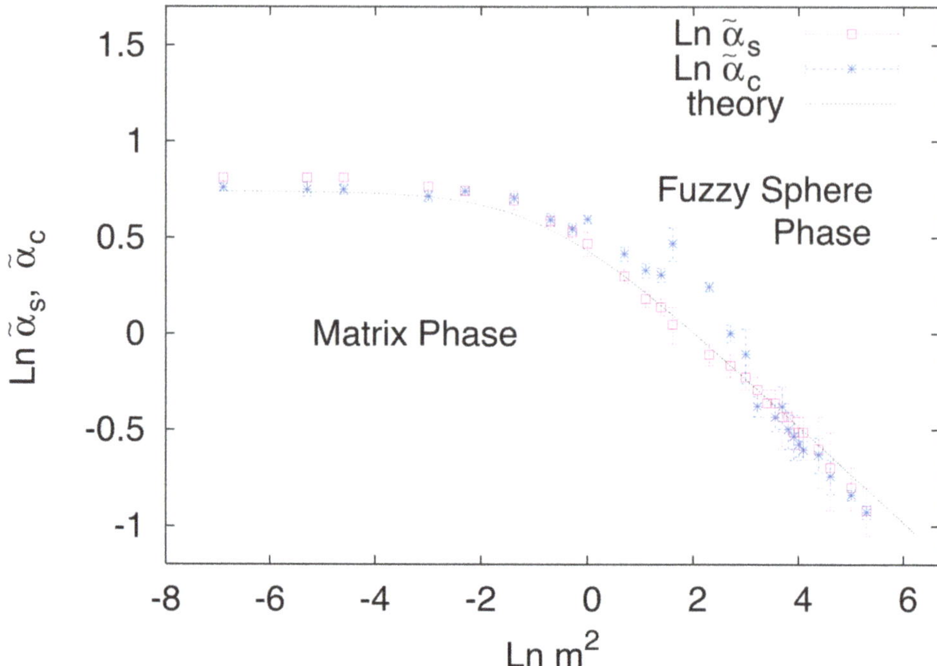

Figure 18.28. Phase diagram showing the measured critical line (theoretical prediction given by equation (18.286)) separating the geometrical and matrix phases of the model (18.274).

Indeed, the essential ingredient in producing this geometric transition is clearly the Chern–Simons term, which is due to the Myers effect [104]. For example, the SO (3) matrix model corresponding to the potential V (given in the last line of equation (18.274)) does not have any transition but when the Chern–Simons term is added we reproduce the one-cut-to-the two-cut transition of the real quartic matrix model. By adding then the Yang–Mills terms, i.e., by considering the full model we should get a generalization of the one-cut-to-the two-cut transition. As it turns out, the matrix-to-sphere transition is in a clear sense a one-cut to N-cut transition as shown explicitly for the model with $\mu = m^2 = 0$.

The phase digaram is given in figure 18.28. The Yang–Mills matrix model (18.274) is characterized by an exotic line of discontinuous transitions with a jump in the entropy, characteristic of a first order transition, yet with divergent critical fluctuations and a divergent specific heat with critical exponent $\alpha = 1/2$. The low temperature phase (small values of the gauge coupling constant) is a geometrical one with gauge fields fluctuating on a round sphere. As the temperature is increased the sphere evaporates in a transition to a pure matrix phase with no background geometrical structure. This model presents an appealing picture of a geometrical phase emerging as the system cools and suggests a scenario for the emergence of geometry in the early Universe.

18.5.7 Emergent geometry and phase diagram with a cosmological term

The Alekseev–Recknagel–Schomerus model with a cosmological term is a $D=3$ Yang–Mills matrix model obtained by adding only a mass term $\operatorname{Tr} X_i^2$ to the basic action (18.225). The action is given explicitly by (with $b \equiv \tau = \beta/\alpha^2$)

$$S[D] = \frac{1}{g^2 N} \operatorname{Tr} \left[-\frac{1}{4}[D_i,\, D_j]^2 + \frac{2i}{3}\varepsilon_{ijk} D_i D_j D_k + \tau D_i^2 \right]. \tag{18.297}$$

The effective potential in this case is give by

$$\frac{V_{\text{eff}}(\varphi)}{2c_2} = \tilde{\alpha}^4 \left[\frac{1}{4}\varphi^4 - \frac{1}{3}\varphi^3 + \frac{1}{2}\tau\varphi^2 \right] + \log \varphi^2. \tag{18.298}$$

Here, the scalar field φ play the role of the order parameter characterizing the phase diagram while the role of the temperature T is played by the gauge coupling constants squared, i.e. $T \equiv 1/\tilde{\alpha}^4 = g^2$.

The model on the fuzzy sphere is extensively studied by analytical and Monte Carlo methods for both $\tau = 0$ and $\tau \neq 0$ in [105, 109, 110, 144, 145]. The phase structure in this case can be summarized as follows:

- We start by setting the logarithmic quantum correction to zero. The classical equation of motion admits three solutions:

$$\varphi_0 = 0, \quad \varphi_{\pm} = \frac{1 \pm \sqrt{1 - 4\tau}}{2}. \tag{18.299}$$

  The solution $\varphi_0 = 0$ (the Yang–Mills or matrix phase) is the global minimum (ground state) of the system in the regime $\tau > 1/4$. The solution φ_- (the geometric or fuzzy sphere phase) is the global minimum in the regime $0 < \tau < 1/4$. The model has no ground state for $\tau < 0$, i.e., $\beta < 0$. The two global minima $D_i = 0$ and $D_i = \varphi_- L_i$ are separated by a potential barrier whose maximum height is reached at the local maximum φ_+.
- We should also mention here that the configuration $D_i = \varphi J_i$ is also a local minimum of the system. The J_i are the generators of $SU(2)$ in a reducible representation characterized by the spin quantum numbers $j_i < s = (N-1)/2$ satisfying $\sum_i(2j_i + 1) = N$. More precisely, we find that the configuration $D_i = \varphi_- L_i$ has a negative energy and thus lower than the zero energy of the configuration $D_i = 0$ only in the regime $0 < \tau < 2/9$. This negative energy is minimized when $J_i = L_i$. In this regime the fuzzy sphere is indeed stable and the expansion of the matrix model around the fuzzy background $D_i = \varphi_- L_i$ gives a noncommutative gauge theory, which also includes coupling to a normal scalar field, i.e., a noncommutative Higgs system.
- At $\tau = 2/9$ the two configurations $D_i = \varphi_- L_i$ and $D_i = 0$ become degenerate. Thus, in the regime $2/9 < \tau < 1/4$ the fuzzy sphere becomes unstable. The coexistence curve between the geometric fuzzy sphere phase and the Yang–Mills matrix phase asymptotes therefore to the line $\tau = 2/9$ (and not to the line $\tau = 1/4$) where the energy functional becomes a complete square.

- If we include the logarithmic quantum correction the potential becomes unbounded from below near $\varphi = \varphi_0 = 0$, i.e., the effective potential (18.298) is really valid only in the fuzzy sphere phase $\varphi = \varphi_- \neq 0$. But the Yang–Mills phase can still be accessed by Monte Carlo simulation of the matrix model (18.297).
- In the quantum case the minimum φ_- (corresponding to the geometric fuzzy sphere phase) becomes a function of both τ and $\tilde{\alpha}^4$. The critical coexistence curve exists therefore in the $(\tau, \tilde{\alpha})$ plane where the local minimum φ_- disappears. The conditions determining this curve are obviously given by $V'_{\text{eff}} = 0$ and $V''_{\text{eff}} = 0$. Explicitly, we obtain the curve $\tilde{\alpha}_* = \tilde{\alpha}_*(\tau)$ defined by the equations

$$\frac{1}{\tilde{\alpha}_*^4} = \frac{\varphi_*^2(\varphi_* - 2\tau)}{8}, \quad \varphi_* = \frac{3}{8}\left(1 + \sqrt{1 - \frac{32\tau}{9}}\right). \tag{18.300}$$

Thus, as we increase τ from 0 to 1/4 the critical value $\tilde{\alpha}_*$ increases from around 2 to infinity. Thus, the critical temperature $T_* \equiv 1/\tilde{\alpha}_*^4 = g_{S*}^2$ decreases towards zero as we increase τ to 1/4. In other words, the geometric fuzzy sphere phase exists in the region of low temperatures T (or large $\tilde{\alpha}$) and $\tau < 1/4$.

- Hence, as the temperature is increased the fuzzy sphere phase evaporates to a pure matrix phase with no background geometrical structure. In this model the geometry condenses or emerges only as the system cools.
- These predictions, which are based on the one-loop effective potential (18.298), are confirmed by Monte Carlo simulation only for $\tau < 2/9$. It is observed (in Monte Carlo simulation) that the coexistence curve between the geometric fuzzy sphere phase (low temperatures) and the Yang–Mills matrix phase (high temperatures) for $2/9 < \tau < 1/4$ asymptotes very rapidly to the line $\tau = 2/9$ for $\tilde{\alpha} > \tilde{\alpha}_* = 4.02$ [145]. In other words, the region $2/9 < \tau < 1/4$ corresponds to the Yang–Mills matrix phase for all values of $\tilde{\alpha}$.
- In fact for $\tau > 2/9$ the geometric fuzzy sphere background is a metastable state with an observable decay to the Yang–Mills matrix background. This decay is not observable for $\tau = 2/9$ although the fuzzy sphere is not the true ground state even here.
- In the Yang–Mills matrix phase the ground state is given by $\varphi = \varphi_0 = 0$ and fluctuations are insensitive to the value of $\tilde{\alpha}$ and are dominated by commuting matrices. In fact, in this phase the matrix model (18.297) is dominated by the Yang–Mills term [146].
- More precisely, the Yang–Mills matrix phase is characterized by a joint eigenvalue distribution, for the three matrices D_1, D_2 and D_3, which is uniform inside a solid ball of some radius $R = 2.0$ in $\mathbf{R}^3$. The eigenvalue distribution of a single matrix is then given by the so-called parabolic law, viz [140, 146–148]

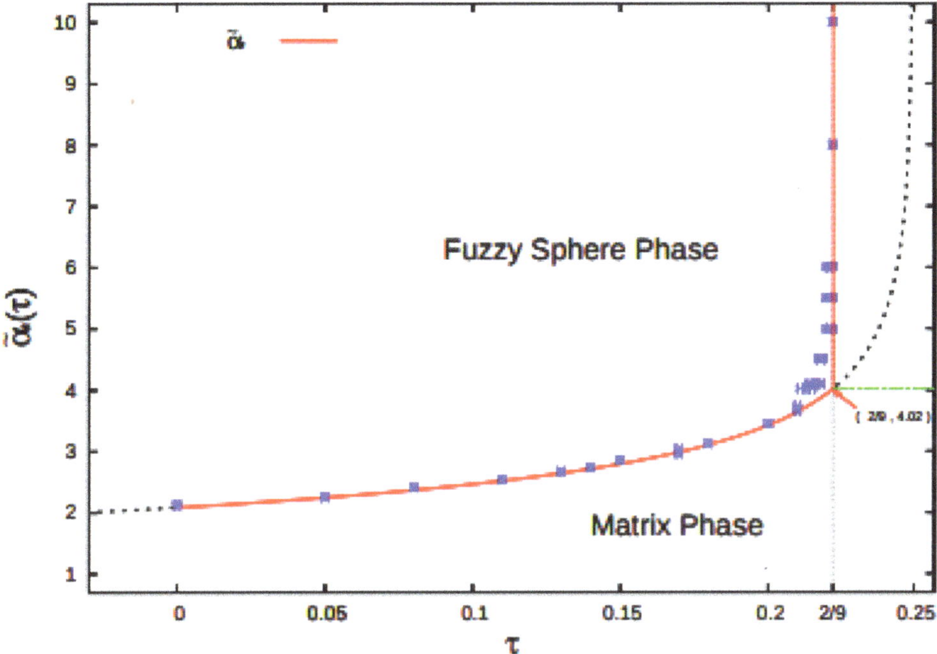

Figure 18.29. The phase diagram of the Yang–Mills matrix model (18.297).

$$\rho(x) = \frac{3}{4R^3}(R^2 - x^2). \tag{18.301}$$

- The transition from the geometric fuzzy sphere phase to the Yang–Mills matrix phase is of an exotic character in the sense that by crossing the coexistence curve at fixed τ from the fuzzy sphere side we encounter divergent specific heat with critical exponent equal 1/2. However, by crossing the coexistence curve at fixed $\tilde{\alpha} > \tilde{\alpha}_* = 4.02$ we find no critical fluctuations and the transition is associated with a continuous internal energy and discontinuous specific heat.

The phase diagram is shown in figure 18.29.

18.5.8 Generalization

Emergent geometry from Yang–Mills matrix models can be generalized to the following systems:
- Noncommutative scalar field coupled to a gauge theory on the fuzzy sphere can be described by $D = 4$ Yang–Mills matrix models [31, 86]. See also [149, 150].

- Fuzzy projective spaces $\mathbf{CP}_F^n$ are the most natural generalization of the fuzzy sphere $\mathbf{S}_N^2$ [21, 151]. Fuzzy $\mathbf{CP}_F^2$ is considered for example in [152–154].
- Fuzzy $\mathbf{S}_{N_1}^2 \times \mathbf{S}_{N_2}^2$ is described by $D = 6$ Yang–Mills matrix models [155, 156]. See also [157–159].
- Fuzzy $\mathbf{S}_F^4$ and $\mathbf{S}_F^3$ are obtained from fuzzy $\mathbf{CP}_N^3$ by squashing the unwanted modes [160, 161]. As a consequence, these three spaces can all be embedded in the same Yang–Mills matrix model.
- The noncommutative $\mathbf{AdS}_\theta^2$ space and noncommutative $\mathbf{AdS}_\theta^2$ black hole are described by Lorentzian $D = 3$ and $D = 4$ Yang–Mills matrix models respectively [102, 162, 163]. The Cartesian products of spheres and AdS spaces are then obtained by taking tensor products of their respective matrix models.
- Emergent gravity from Euclidean/Lorentzian Yang–Mills matrix models [111–113].
- Emergent time (cosmology) from Lorentzian $D = 10$ Yang–Mills matrix model [164–167].
- Emergent geometry, gravity and cosmology from supersymmetric Yang–Mills matrix quantum mechanics [168, 169].

References

[1] Ydri B 2017 Lectures on matrix field theory *Lect. Notes Phys.* (Berlin: Springer) **929** 1–352

[2] Ydri B 2018 *Matrix Models of String Theory* (Bristol: IOP Publishing)

[3] Connes A 1994 *Noncommutative Geometry* (London: Academic)

[4] Connes A 1996 Gravity coupled with matter and foundation of noncommutative geometry *Commun. Math. Phys.* **182** 155

[5] Fröhlich J and Gawędzki K 1993 Conformal field theory and geometry of strings [arXiv: hep-th/9310187]

[6] Weyl H 1931 *The Theory of Groups and Quantum Mechanics* (New York: Dover)

[7] Moyal J E 1949 Quantum mechanics as a statistical theory *Proc. Cambridge Phil. Soc* **45** 99

[8] Groenewold H J 1946 On the principles of elementary quantum mechanics *Physica* **12** 405

[9] Rieffel M A 1981 C*-algebras associated with irrational rotations *Pacific J. Math.* 93415–29

[10] Ambjorn J, Makeenko Y M, Nishimura J and Szabo R J 2000 Lattice gauge fields and discrete noncommutative Yang-Mills theory *JHEP* **05** 023

[11] Hoppe J 1982 Massachusetts Institute of Technology *PhD Thesis*

[12] Madore J 1992 The fuzzy sphere *Class. Quant. Grav.* **9** 69

[13] Lee J and Yang H S 2014 Quantum gravity from noncommutative spacetime *J. Korean Phys. Soc.* **65** 1754 [arXiv:1004.0745 [hep-th]]

[14] Blaschke D N and Steinacker H 2010 Schwarzschild geometry emerging from matrix models *Class. Quant. Grav.* **27** 185020

[15] Garcia Flores F, Martin X and O'Connor D 2009 Simulation of a scalar field on a fuzzy sphere *Int. J. Mod. Phys.* A **24** 3917–44

[16] Martin X 2004 A Matrix phase for the phi**4 scalar field on the fuzzy sphere *JHEP* **04** 077

[17] Panero M 2007 Numerical simulations of a non-commutative theory: the Scalar model on the fuzzy sphere *JHEP* **05** 082

[18] Das C R, Digal S and Govindarajan T R 2008 Finite temperature phase transition of a single scalar field on a fuzzy sphere *Mod. Phys. Lett.* A **23** 1781–91

[19] O'Connor D and Saemann C 2007 Fuzzy scalar field theory as a multitrace matrix model *JHEP* **0708** 066

[20] Saemann C 2010 The multitrace matrix model of scalar field theory on fuzzy CP**n *SIGMA* **6** 050

[21] Balachandran A P, Dolan B P, Lee J H, Martin X and O'Connor D 2002 Fuzzy complex projective spaces and their star products *J. Geom. Phys. J. Geom. Phys.* **43** 184–204

[22] Ishibashi N, Kawai H, Kitazawa Y and Tsuchiya A 1997 A large N reduced model as superstring *Nucl. Phys.* B **498** 467

[23] Krauth W and Staudacher M 1999 Eigenvalue distributions in Yang-Mills integrals *Phys. Lett.* B **453** 253

[24] Krauth W and Staudacher M 1998 Finite Yang-Mills integrals *Phys. Lett.* B **435** 350

[25] Austing P and Wheater J F 2003 Adding a Myers term to the IIB matrix model *JHEP* **0311** 009

[26] Austing P 2001 Yang–Mills matrix theory arXiv:hep-th/0108128

[27] Krauth W, Nicolai H and Staudacher M 1998 Monte Carlo approach to M theory *Phys. Lett.* B **431** 31–41

[28] Austing P and Wheater J F 2001 The convergence of Yang-Mills integrals *JHEP* **02** 028

[29] Austing P and Wheater J F 2001 Convergent Yang-Mills matrix theories *JHEP* **04** 019

[30] Nishimura J, Okubo T and Sugino F 2011 Systematic study of the SO(10) symmetry breaking vacua in the matrix model for type IIB superstrings *JHEP* **1110** 135

[31] Ydri B 2014 New algorithm and phase diagram of noncommutative ϕ^4 on the fuzzy sphere *JHEP* **03** 065

[32] Bombelli L, Lee J, Meyer D and Sorkin R 1987 Space-time as a causal set *Phys. Rev. Lett.* **59** 521

[33] Dowker F 2005 Causal sets and the deep structure of spacetime arXiv:gr-qc/0508109

[34] Thiemann T 2008 *Modern Canonical Quantum General Relativity* (Cambridge: Cambridge University Press)

[35] Rovelli C 2004 *Quantum Gravity* (Cambridge: Cambridge University Press)

[36] Maldacena J M 1999 The large N limit of superconformal field theories and supergravity *Int. J. Theor. Phys.* **38** 1113

[37] Seiberg N 2006 Emergent spacetime arXiv:0601234

[38] Ambjorn J, Jurkiewicz J and Loll R 2005 Reconstructing the universe *Phys. Rev.* D **72** 064014

[39] Ambjørn J, Janik R, Westra W and Zohren S 2006 The emergence of background geometry from quantum fluctuations *Phys. Lett.* B **641** 94

[40] Ambjorn J, Gorlich A, Jurkiewicz J and Loll R 2008 Planckian birth of the quantum de Sitter universe *Phys. Rev. Lett.* **100** 091304

[41] Ambjorn J, Gorlich A, Jordan S, Jurkiewicz J and Loll R 2010 CDT meets Horava-Lifshitz gravity *Phys. Lett.* B **690** 413

[42] Gorlich A 2011 Causal dynamical triangulations in four dimensions arXiv:1111.6938 [hep-th]

[43] Horava P 2009 Spectral dimension of the universe in quantum gravity at a lifshitz point *Phys. Rev. Lett.* **102** 161301

[44] Horava P 2009 Membranes at quantum criticality *JHEP* **0903** 020

[45] Horava P 2009 Quantum gravity at a Lifshitz point *Phys. Rev.* D **79** 084008
[46] Minwalla S, Van Raamsdonk M and Seiberg N 2000 Noncommutative perturbative dynamics *JHEP* **02** 020
[47] Grosse H and Wulkenhaar R 2003 Renormalization of phi**4 theory on noncommutative R**2 in the matrix base *JHEP* **12** 019
[48] Gubser S S and Sondhi S L 2001 Phase structure of noncommutative scalar field theories *Nucl. Phys.* B **605** 395–424
[49] Brazovkii S A 1975 Phase transition of an isotropic system to a nonuniform state *Zh. Eksp. Teor. Fiz* **68** 175–85
[50] Bietenholz W, Hofheinz F and Nishimura J 2004 Phase diagram and dispersion relation of the noncommutative lambda phi**4 model in d = 3 *JHEP* **0406** 042
[51] Mejía-Díaz H, Bietenholz W and Panero M 2014 The continuum phase diagram of the 2D non-commutative $\lambda\phi^4$ model *JHEP* **10** 056
[52] Ambjorn J and Catterall S 2002 Stripes from (noncommutative) stars *Phys. Lett.* B **549** 253–9
[53] Grosse H and Wulkenhaar R 2005 Renormalization of phi**4 theory on noncommutative R**4 in the matrix base *Commun. Math. Phys.* **256** 305–74
[54] Grosse H and Wulkenhaar R 2005 Power counting theorem for nonlocal matrix models and renormalization *Commun. Math. Phys.* **254** 91–127 [arXiv:hep-th/0305066 [hep-th]]
[55] Langmann E and Szabo R J 2002 Duality in scalar field theory on noncommutative phase spaces *Phys. Lett.* B **533** 168–77
[56] Langmann E, Szabo R J and Zarembo K 2004 Exact solution of quantum field theory on noncommutative phase spaces *JHEP* **01** 017
[57] Langmann E 2003 Interacting fermions on noncommutative spaces: exactly solvable quantum field theories in 2n+1 dimensions *Nucl. Phys.* B **654** 404
[58] Gracia-Bondia J M and Varilly J C 1988 Algebras of distributions suitable for phase space quantum mechanics. 1 *J. Math. Phys.* **29** 869–79
[59] Langmann E, Szabo R J and Zarembo K 2003 Exact solution of noncommutative field theory in background magnetic fields *Phys. Lett.* B **569** 95–101
[60] Wilson K G and Kogut J B 1974 The renormalization group and the epsilon expansion *Phys. Rep.* **12** 75–199
[61] Ferretti G 1995 On the large N limit of 3-D and 4-D Hermitian matrix models *Nucl. Phys.* B **450** 713
[62] Nishigaki S 1995 Wilsonian approximated renormalization group for matrix and vector models in $2 < d < 4$ *Phys. Lett.* B **376** 73
[63] Chen G-H and Wu Y-S 2002 Renormalization group equations and the Lifshitz point in noncommutative Landau-Ginsburg theory *Nucl. Phys.* B **622** 189
[64] Kleinert H and Nogueira F S 2003 Charged fixed point found in superconductor below T(c) *Nucl. Phys.* B **651** 361
[65] Ydri B and Bouchareb A 2012 The fate of the Wilson-Fisher fixed point in non-commutative ϕ^4 *J. Math. Phys.* **53** 102301
[66] Ydri B, Ahmim R and Bouchareb A 2015 Wilson RG of noncommutative Φ_4^4 *Int. J. Mod. Phys.* A **30** 1550195
[67] Ydri B and Ahmim R 2013 Matrix model fixed point of noncommutative ϕ^4 theory *Phys. Rev.* D **88** 106001
[68] Ferretti G 1997 The critical exponents of the matrix valued Gross-Neveu model *Nucl. Phys.* B **487** 739

[69] Grosse H, Klimcik C and Presnajder P 1996 Towards finite quantum field theory in noncommutative geometry *Int. J. Theor. Phys.* **35** 231

[70] Grosse H, Klimcik C and Presnajder P 1997 Field theory on a supersymmetric lattice *Commun. Math. Phys.* **185** 155

[71] Vaidya S 2001 Perturbative dynamics on the fuzzy S**2 and RP**2 *Phys. Lett.* B **512** 403

[72] Chu C S, Madore J and Steinacker H 2001 Scaling limits of the fuzzy sphere at one loop *JHEP* **0108** 038

[73] Vaidya S and Ydri B 2003 On the origin of the UV-IR mixing in noncommutative matrix geometry *Nucl. Phys.* B **671** 401

[74] Vaidya S and Ydri B 2002 New scaling limit for fuzzy spheres arXiv:hep-th/0209131

[75] Dolan B P, O'Connor D and Presnajder P 2002 Matrix phi**4 models on the fuzzy sphere and their continuum limits *JHEP* **0203** 013

[76] Varshalovich D A, Moskalev A N and Khersonsky V K 1988 *Quantum Theory Of Angular Momentum: Irreducible Tensors, Spherical Harmonics, Vector Coupling Coefficients, 3nj Symbols* (Singapore: World Scientific) 514

[77] Chaikin P M and Lubensky T C 1995 *Principles of Condensed Matter Physics* (Cambridge: Cambridge University Press)

[78] Brezin E, Itzykson C, Parisi G and Zuber J B 1978 Planar diagrams *Commun. Math. Phys.* **59** 35

[79] Shimamune Y 1982 On the phase structure of large N matrix models and gauge models *Phys. Lett.* B **108** 407

[80] Ydri B 2016 A multitrace approach to noncommutative Φ_2^4 *Phys. Rev.* D **93** 065041

[81] Ydri B, Rouag A and Ramda K 2016 Emergent geometry from random multitrace matrix models *Phys. Rev.* D **93** 065055

[82] Ydri B, Ramda K and Rouag A 2016 Phase diagrams of the multitrace quartic matrix models of noncommutative Φ^4 theory *Phys. Rev.* D **93** 065056

[83] Polychronakos A P 2013 Effective action and phase transitions of scalar field on the fuzzy sphere *Phys. Rev.* D **88** 065010

[84] Ydri B, Soudani C and Rouag A 2017 Quantum gravity as a multitrace matrix model *Int. J. Mod. Phys.* A **32** 1750180

[85] Garcia Flores F, O'Connor D and Martin X 2006 Simulating the scalar field on the fuzzy sphere *PoS LAT* **2005** 262 arXiv:hep-lat/0601012

[86] Ydri B 2012 Impact of supersymmetry on emergent geometry in Yang–Mills matrix models II *Int. J. Mod. Phys.* A **27** 1250088

[87] Steinacker H 2005 A non-perturbative approach to non-commutative scalar field theory *JHEP* **0503** 075

[88] Tekel J 2014 Uniform order phase and phase diagram of scalar field theory on fuzzy CP**n arXiv:1407.4061 [hep-th]

[89] Nair V P, Polychronakos A P and Tekel J 2012 Fuzzy spaces and new random matrix ensembles *Phys. Rev.* D **85** 045021

[90] Tekel J 2013 Random matrix approach to scalar fields on fuzzy spaces *Phys. Rev.* D **87** 085015

[91] Tekel J 2015 Phase strucutre of fuzzy field theories and multitrace matrix models *Acta Phys. Slov.* **65** 369 [arXiv:1512.00689 [hep-th]]

[92] Tekel J 2015 Matrix model approximations of fuzzy scalar field theories and their phase diagrams *JHEP* **1512** 176

[93] Subjakova M and Tekel J 2020 Multitrace matrix models of fuzzy field theories arXiv:2006.13577 [hep-th]

[94] Di Francesco P, Ginsparg P H and Zinn-Justin J 1995 2-D gravity and random matrices *Phys. Rep.* **254** 1

[95] Zarembo K L and Makeenko Y M 1998 An introduction to matrix superstring models *Phys. Usp.* **41** 1 [Usp. Fiz. Nauk **168** 3 (1998)]

[96] Bietenholz W, Hofheinz F and Nishimura J 2004 On the relation between non-commutative field theories at theta = infinity and large N matrix field theories *JHEP* **05** 047

[97] Becchi C, Giusto S and Imbimbo C 2003 The renormalization of noncommutative field theories in the limit of large noncommutativity *Nucl. Phys.* B **664** 371–99

[98] Onsager L 1944 Crystal statistics. 1. A two-dimensional model with an order disorder transition *Phys. Rev.* **65** 117

[99] Mermin N D and Wagner H 1966 Absence of ferromagnetism or antiferromagnetism in one-dimensional or two-dimensional isotropic Heisenberg models *Phys. Rev. Lett.* **17** 1133–6

[100] Coleman S R 1973 There are no goldstone bosons in two-dimensions *Commun. Math. Phys.* **31** 259–64

[101] Volkholz J and Bietenholz W 2007 Simulations of a supersymmetry inspired model on a fuzzy sphere *PoS* **LATTICE2007** 283 [arXiv:0808.2387 [hep-th]]

[102] Bouraiou L and Ydri B 2022 The $\mathbf{AdS}^2_\theta/\mathbf{CFT}_1$ correspondence and noncommutative geometry III: phase structure of the noncommutative $\mathbf{AdS}^2_\theta \times \mathbf{S}^2_N$ *Int. J. Mod. Phys.* A **37** 2250079

[103] Ydri B 2016 The multitrace matrix model: an alternative to Connes NCG and IKKT model in 2 dimensions *Phys. Lett.* B **763** 161–3 [arXiv:1608.02758 [hep-th]]

[104] Myers R C 1999 Dielectric branes *JHEP* **9912** 022

[105] Azuma T, Bal S, Nagao K and Nishimura J 2004 Nonperturbative studies of fuzzy spheres in a matrix model with the Chern-Simons term *JHEP* **0405** 005

[106] Steinacker H 2004 Quantized gauge theory on the fuzzy sphere as random matrix model *Nucl. Phys.* B **679** 66

[107] Castro-Villarreal P, Delgadillo-Blando R and Ydri B 2005 A gauge-invariant UV-IR mixing and the corresponding phase transition for U(1) fields on the fuzzy sphere *Nucl. Phys.* B **704** 111–53

[108] O'Connor D and Ydri B 2006 Monte Carlo simulation of a NC gauge theory on the fuzzy sphere *JHEP* **11** 016

[109] Delgadillo-Blando R, O'Connor D and Ydri B 2008 Geometry in transition: a model of emergent geometry *Phys. Rev. Lett.* **100** 100

[110] Delgadillo-Blando R, O'Connor D and Ydri B 2009 Matrix models, gauge theory and emergent geometry *JHEP* **05** 049

[111] Steinacker H 2010 Emergent geometry and gravity from matrix models: an introduction *Class. Quant. Grav.* **27** 133001

[112] Steinacker H C 2016 Emergent gravity on covariant quantum spaces in the IKKT model *JHEP* **1612** 156

[113] Steinacker H 2007 Emergent gravity from noncommutative gauge theory *JHEP* **0712** 049

[114] Yang H S 2009 Emergent spacetime and the origin of gravity *JHEP* **0905** 012

[115] Yang H S 2009 Emergent gravity from noncommutative spacetime *Int. J. Mod. Phys.* A **24** 4473

[116] Ydri B and Ahmim R 2020 Wilsonian matrix renormalization group arXiv:2008.09564 [hep-th]

[117] Higuchi S, Itoi C, Nishigaki S and Sakai N 1995 Renormalization group flow in one and two matrix models *Nucl. Phys.* B **434** 283 Erratum: [Nucl. Phys. B **441** 405 (1995)]

[118] Higuchi S, Itoi C, Nishigaki S and Sakai N 1993 Nonlinear renormalization group equation for matrix models *Phys. Lett.* B **318** 63

[119] Higuchi S, Itoi C, Nishigaki S and Sakai N 1993 Renormalization group approach to discretized gravity arXiv:hep-th/9307065

[120] Higuchi S, Itoi C, Nishigaki S and Sakai N 1994 Large N renormalization group approach to matrix models arXiv: hep-th/9307065

[121] Kawamoto S and Tomino D 2013 A renormalization group approach to a Yang-Mills two matrix model *Nucl. Phys.* B **877** 825–51

[122] Banks T, Fischler W, Shenker S H and Susskind L 1997 M theory as a matrix model: a conjecture *Phys. Rev.* D **55** 5112

[123] Witten E 1996 Bound states of strings and p-branes *Nucl. Phys.* B **460** 335

[124] Connes A 1996 Gravity coupled with matter and foundation of noncommutative geometry *Commun. Math. Phys.* **182** 155

[125] Sochichiu C 2000 M[any] vacua of IIB *JHEP* **0005** 026

[126] Berenstein D E, Maldacena J M and Nastase H S 2002 Strings in flat space and pp waves from N=4 superYang-Mills *JHEP* **0204** 013

[127] Anagnostopoulos K N and Nishimura J 2002 New approach to the complex action problem and its application to a nonperturbative study of superstring theory *Phys. Rev.* D **66** 106008

[128] Nishimura J 2002 Exactly solvable matrix models for the dynamical generation of space-time in superstring theory *Phys. Rev.* D **65** 105012

[129] Nishimura J, Okubo T and Sugino F 2005 Gaussian expansion analysis of a matrix model with the spontaneous breakdown of rotational symmetry *Prog. Theor. Phys.* **114** 487

[130] Anagnostopoulos K N, Azuma T and Nishimura J 2011 A practical solution to the sign problem in a matrix model for dynamical compactification *JHEP* **1110** 126

[131] Burda Z, Petersson B and Tabaczek J 2001 Geometry of reduced supersymmetric 4-D Yang-Mills integrals *Nucl. Phys.* B **602** 399

[132] Ambjorn J, Anagnostopoulos K N, Bietenholz W, Hofheinz F and Nishimura J 2002 On the spontaneous breakdown of Lorentz symmetry in matrix models of superstrings *Phys. Rev.* D **65** 086001

[133] Ambjorn J, Anagnostopoulos K N, Bietenholz W, Hotta T and Nishimura J 2000 Large N dynamics of dimensionally reduced 4-D SU(N) superYang-Mills theory *JHEP* **0007** 013

[134] Ambjorn J, Anagnostopoulos K N, Bietenholz W, Hotta T and Nishimura J 2000 Monte Carlo studies of the IIB matrix model at large N *JHEP* **0007** 011

[135] Nishimura J and Vernizzi G 2000 Spontaneous breakdown of Lorentz invariance in IIB matrix model *JHEP* **0004** 015

[136] Nishimura J and Vernizzi G 2000 Brane world from IIB matrices *Phys. Rev. Lett.* **85** 4664

[137] Alekseev A Y, Recknagel A and Schomerus V 2000 Brane dynamics in background fluxes and noncommutative geometry *JHEP* **05** 010

[138] Alekseev A Y and Schomerus V 1999 D-branes in the WZW model *Phys. Rev.* D **60** 061901

[139] Delgadillo-Blando R, O'Connor D and Ydri B 2008 Geometry in transition: a model of emergent geometry *Phys. Rev. Lett.* **100** 201601

[140] Filev V G and O'Connor D 2014 On the phase structure of commuting matrix models arXiv:1402.2476 [hep-th]

[141] Karabali D, Nair V P and Polychronakos A P 2002 Spectrum of Schrodinger field in a noncommutative magnetic monopole *Nucl. Phys.* B **627** 565–79

[142] Azuma T, Nagao K and Nishimura J 2005 Perturbative dynamics of fuzzy spheres at large N *JHEP* **06** 081

[143] Faddeev L D and Popov V N 1967 *Phys. Lett.* B **25** 29–30

[144] Azuma T, Bal S and Nishimura J 2005 Dynamical generation of gauge groups in the massive Yang-Mills-Chern-Simons matrix model *Phys. Rev.* D **72** 066005

[145] Delgadillo-Blando R and O'Connor D 2012 Matrix geometries and matrix models *JHEP* **11** 057

[146] O'Connor D and Filev V G 2013 Near commuting multi-matrix models *JHEP* **04** 144

[147] Berenstein D E, Hanada M and Hartnoll S A 2009 Multi-matrix models and emergent geometry *JHEP* **02** 010

[148] Filev V G and O'Connor D 2013 Multi-matrix models at general coupling *J. Phys.* A **46** 475403

[149] Ambjorn J, Anagnostopoulos K N, Bietenholz W, Hotta T and Nishimura J 2000 Large N dynamics of dimensionally reduced 4-D SU(N) superYang-Mills theory *JHEP* **07** 013

[150] Anagnostopoulos K N, Azuma T, Nagao K and Nishimura J 2005 Impact of supersymmetry on the nonperturbative dynamics of fuzzy spheres *JHEP* **09** 046

[151] Dolan B P, Huet I, Murray S and O'Connor D 2007 Noncommutative vector bundles over fuzzy CP^N and their covariant derivatives *JHEP* **0707** 007

[152] Grosse H and Steinacker H 2005 Finite gauge theory on fuzzy CP^2 *Nucl. Phys.* B **707** 145

[153] Dou D and Ydri B 2007 Topology change from quantum instability of gauge theory on fuzzy CP^2 *Nucl. Phys.* B **771** 167

[154] Azuma T, Bal S, Nagao K and Nishimura J 2006 Dynamical aspects of the fuzzy CP^2 in the large N reduced model with a cubic term *JHEP* **0605** 061

[155] Ydri B, Rouag A and Ramda K 2017 Emergent fuzzy geometry and fuzzy physics in four dimensions *Nucl. Phys.* B **916** 567–606

[156] Ydri B, Khaled R and Ahlam R 2016 Geometry in transition in four dimensions: a model of emergent geometry in the early universe *Phys. Rev.* D **94** 085020

[157] Delgadillo-Blando R and Ydri B 2007 Towards noncommutative fuzzy QED *JHEP* **0703** 056

[158] Behr W, Meyer F and Steinacker H 2005 Gauge theory on fuzzy $S^2 \times S^2$ and regularization on noncommutative R^4 *JHEP* **0507** 040

[159] Castro-Villarreal P, Delgadillo-Blando R and Ydri B 2005 Quantum effective potential for $U(1)$ fields on $S_N^2 \times S_N^2$ *JHEP* **0509** 066

[160] Dolan B P and O'Connor D 2003 A fuzzy three sphere and fuzzy tori *JHEP* **10** 060

[161] Medina J and O'Connor D 2003 Scalar field theory on fuzzy S**4 *JHEP* **11** 051

[162] Ydri B 2022 The $\mathbf{AdS}_\theta^2/\mathbf{CFT}_1$ correspondence and noncommutative geometry II: noncommutative quantum black holes *Int. J. Mod. Phys.* A **37** 2250078

[163] Ydri B 2022 The $\mathbf{AdS}_\theta^2/\mathbf{CFT}_1$ correspondence and noncommutative geometry I: A QM/NCG correspondence *Int. J. Mod. Phys.* A **37** 2250077

[164] Kim S W, Nishimura J and Tsuchiya A 2012 Expanding (3+1)-dimensional universe from a Lorentzian matrix model for superstring theory in (9+1)-dimensions *Phys. Rev. Lett.* **1108** 011601

[165] Ito Y, Nishimura J and Tsuchiya A 2015 Power-law expansion of the universe from the bosonic Lorentzian type IIB matrix model *JHEP* **1506** 070

[166] Kim S W, Nishimura J and Tsuchiya A 2012 Expanding universe as a classical solution in the Lorentzian matrix model for nonperturbative superstring theory *Phys. Rev.* D **86** 027901

[167] Kim S W, Nishimura J and Tsuchiya A 2012 Late time behaviors of the expanding universe in the IIB matrix model *JHEP* **10** 147

[168] Kim N and Park J H 2006 Massive super Yang-Mills quantum mechanics: classification and the relation to supermembrane *Nucl. Phys.* B **759** 249–82

[169] Park J H 2006 Noncritical *osp*(1|2, *R*) M-theory matrix model with an arbitrary time dependent cosmological constant *Nucl. Phys.* B **745** 123–41

IOP Publishing

A Modern Course in Quantum Field Theory, Volume 2 (Second Edition)
Advanced topics
Badis Ydri

Chapter 19

Noncommutative scalar field theory and its renormalizability

In this chapter, and after an efficient introduction to noncommutative scalar field theory, we will apply the Wilson–Polchinski renormalization group equation, discussed in the previous chapter, to the problem of renormalizing noncommutative phi-four theory in two and four dimensions with and without the harmonic oscillator term. Noncommutative field theory is discussed in great detail in our book [1] whereas we follow closely the original programme of Grosse and Wulkenhaar [2–4] in the very difficult problem of renormalization of noncommutative phi-four theory on Moyal–Weyl spaces.

19.1 Noncommutative Moyal–Weyl spaces

This section is taken from [1].

19.1.1 The Weyl map

A Groenewold–Moyal–Weyl space $\mathbf{R}^d_\theta$ is a deformation of ordinary d dimensional Euclidean space $\mathbf{R}^d$ in which the coordinates x_i are replaced with Hermitian operators $\hat{x}_i$ satisfying the Heisenberg-Weyl commutation relations

$$[\hat{x}_i, \hat{x}_j] = i\theta_{ij}. \tag{19.1}$$

The space $\mathbf{R}^d_\theta$ in general can be only partially noncommutative, i.e., the Poisson tensor θ_{ij} is of rank $2r \leqslant d$. This means in particular that we have only $2r$ noncommuting coordinates. The Poisson tensor, also known as the noncommutativity parameter, can thus be brought by means of an appropriate linear transformation of the coordinate operators to the canonical form

doi:10.1088/978-0-7503-5834-7ch19   19-1   © IOP Publishing Ltd 2025. All rights, including for text and data mining (TDM), artificial intelligence (AI) training, and similar technologies, are reserved.

$$\theta = \begin{pmatrix} 0 & \theta_1 & 0 & \cdot & \cdot & 0 & \cdot \\ -\theta_1 & 0 & 0 & \cdot & \cdot & 0 & \cdot \\ \cdot & \cdot & \cdot & \cdot & \cdot & \cdot & \cdot \\ 0 & 0 & \cdot & \cdot & 0 & \theta_r & \cdot \\ 0 & 0 & \cdot & \cdot & -\theta_r & 0 & \cdot \\ \cdot & \cdot & \cdot & \cdot & \cdot & \cdot & \cdot \end{pmatrix}. \tag{19.2}$$

In the above equation $\theta_r = \theta_{2r-12r}$. In the spirit of Connes' noncommutative geometry [5], we will describe the Groenewold–Moyal–Weyl spacetime $\mathbf{R}_\theta^d$ in terms of the algebra of functions on $\mathbf{R}^d$, endowed with an associative noncommutative product between elements f and g denoted by $f*g$. This star product is, precisely, the Groenewold–Moyal–Weyl star product [6–8], which we will derive shortly.

The algebra corresponding to the space $\mathbf{R}_\theta^d$ will be denoted $\mathcal{A}_\theta$. The algebra $\mathcal{A}_0$ corresponding to the commutative space $\mathbf{R}^d$ is clearly the algebra of functions on $\mathbf{R}^d$ with the usual pointwise multiplication of functions. Specification of the algebra will determine only topological properties of the space $\mathbf{R}_\theta^d$. In order to specify the metric aspects we must also define proper derivation operations on the algebra $\mathcal{A}_\theta$. The Weyl map will allow us to map the algebra $\mathcal{A}_\theta$ to the correct operator algebra generated by the coordinate operators $\hat{x}_i$.

The algebra of functions on $\mathbf{R}^d$, of interest to us here, is the algebra of Schwartz functions of sufficiently rapid decrease at infinity. These are functions with all their derivatives vanishing at infinity. Equivalently Schwartz functions are functions $f(x)$, which admit well defined Fourier transforms $\tilde{f}(k)$, viz

$$\tilde{f}(k) = \int d^d x f(x) e^{-ikx}. \tag{19.3}$$

The Fourier transforms $\tilde{f}(k)$ are also Schwartz functions, i.e., their derivatives to any order vanish at infinity in momentum space. The functions $f(x)$ are given by the inverse Fourier transforms, viz

$$f(x) = \int \frac{d^d k}{(2\pi)^d} \tilde{f}(k) e^{ikx}. \tag{19.4}$$

The Weyl operator $\hat{f}$ acting in some, infinite dimensional separable, Hilbert space $\mathcal{H}$, which corresponds to the function $f(x)$ is obtained by requiring that $f(x)$ is the Weyl symbol of $\hat{f}$, i.e.,

$$\hat{f} = \int \frac{d^d k}{(2\pi)^d} \tilde{f}(k) e^{ik\hat{x}}. \tag{19.5}$$

We have only replaced the coordinates x_i by the coordinate operators $\hat{x}_i$. This is a bounded operator, which is also compact.

For simplicity we will assume maximal noncommutativity, i.e., $d = 2r$. The Weyl map, or quantizer, is given by

$$\Delta(\hat{x}_i, x_i) = \int \frac{d^d k}{(2\pi)^d} e^{ik_i \hat{x}_i} e^{-ik_i x_i}. \tag{19.6}$$

This corresponds to the symmetric ordering of the operator. We have explicitly

$$\hat{f} = \int d^d x f(x_i) \Delta(\hat{x}_i, x_i). \tag{19.7}$$

It is obvious that if $f = f_k = \exp(ikx)$ then $\hat{f} = \hat{f}_k = \exp(ik\hat{x})$, viz

$$\exp(ik\hat{x}) = \int d^d x \, \exp(ikx) \Delta(\hat{x}_i, x_i). \tag{19.8}$$

The derivative operators on the noncommutative space $\mathbf{R}_\theta^d$ can be given by the inner derivations

$$\hat{\partial}_i = \frac{1}{i}(\theta^{-1})_{ij}(\hat{x}_j - \hat{x}_j^R). \tag{19.9}$$

The coordinate operators $\hat{x}_i^R$ act on the right of the algebra, viz $\hat{x}_i^R \hat{f} = \hat{f} \hat{x}_i$. These derivative operators satisfy the conditions

$$[\hat{\partial}_i, \hat{\partial}_j] = 0, \quad [\hat{\partial}_i, \hat{x}_j] = \delta_{ij}. \tag{19.10}$$

The derivative operators on $\mathbf{R}_\theta^d$ can also be given by any outer derivations satisfying the above two requirements.

We have the basic identity

$$\begin{aligned}
[\hat{\partial}_i, e^{ik\hat{x}}] &= \sum_{n=1}^{\infty} \frac{i^n}{n!}[\hat{\partial}_i, (k\hat{x})^n] \\
&= \sum_{n=1}^{\infty} \frac{i^n}{n!} n k_i (k\hat{x})^{n-1} \\
&= i k_i e^{ik\hat{x}}.
\end{aligned} \tag{19.11}$$

In the second line we have used the second equation of (19.10), which is valid for all derivations inner or outer. We have therefore the result

$$[\hat{\partial}_i, \hat{f}] = \int \frac{d^d k}{(2\pi)^d} i k_i \tilde{f}(k) e^{ik\hat{x}}. \tag{19.12}$$

This suggest that we associate the operator $[\hat{\partial}_i, \hat{f}]$ with the function $\partial_i f(x_i)$. The proof goes as follows. First we have

$$\begin{aligned}
[\hat{\partial}_i, \Delta(\hat{x}_i, x_i)] &= \int \frac{d^d k}{(2\pi)^d} i k_i e^{ik_i \hat{x}_i} e^{-ik_i x_i} \\
&= -\partial_i \Delta(\hat{x}_i, x_i).
\end{aligned} \tag{19.13}$$

By using the above result we have

$$\begin{aligned}
[\hat{\partial}_i, \hat{f}] &= \int d^d x f(x_i) [\hat{\partial}_i, \Delta(\hat{x}_i, x_i)] \\
&= -\int d^d x f(x_i) \partial_i \Delta(\hat{x}_i, x_i) \\
&= \int d^d x \partial_i f(x_i) \Delta(\hat{x}_i, x_i).
\end{aligned} \tag{19.14}$$

In other words the operator $[\hat{\partial}_i, \hat{f}]$ corresponds to the function $\partial_i f(x_i)$ as it should be.

In the commutative limit $\theta \longrightarrow 0$ the operator $\Delta(\hat{x}_i, x_i)$ reduces in an obvious way to the delta function $\delta^2(\hat{x} - x)$. This is in fact obvious from (19.8). For $\alpha_i \in \mathbf{R}$ we compute

$$
\begin{aligned}
e^{\alpha\hat{\partial}} e^{ik\hat{x}} e^{-\alpha\hat{\partial}} &= e^{\frac{i}{2}\alpha k} e^{\alpha\hat{\partial}+ik\hat{x}} e^{-\alpha\hat{\partial}} \\
&= e^{i\alpha k} e^{ik\hat{x}}.
\end{aligned}
\tag{19.15}
$$

The unitary operator $\exp(\alpha\hat{\partial})$ corresponds to a translation operator in space by a vector $\vec{\alpha}$. By using the above result we obtain

$$
e^{\alpha\hat{\partial}} \Delta(\hat{x}_i, x_i) e^{-\alpha\hat{\partial}} = \Delta(\hat{x}_i, x_i - \alpha_i).
\tag{19.16}
$$

We can then conclude that $Tr_{\mathcal{H}} \Delta(\hat{x}_i, x_i)$ is independent of x for any trace $Tr_{\mathcal{H}}$ on $\mathcal{H}$ since $Tr_{\mathcal{H}} \Delta(\hat{x}_i, x_i) = Tr_{\mathcal{H}} \Delta(\hat{x}_i, x_i - \alpha_i)$. In other words $Tr_{\mathcal{H}} \Delta(\hat{x}_i, x_i)$ is simply an overall normalization, which we can choose appropriately. We choose

$$
Tr_{\mathcal{H}} \Delta(\hat{x}_i, x_i) = \frac{1}{\sqrt{\det(2\pi\theta)}}.
\tag{19.17}
$$

In some sense $\sqrt{\det(2\pi\theta)}$ is the volume of an elementary cell in noncommutative spacetime if we think of $\mathbf{R}_\theta^d$ as a phase space. This can also be understood from the result

$$
\sqrt{\det(2\pi\theta)} \, Tr_{\mathcal{H}} e^{ik\hat{x}} = \int d^d x \, e^{ikx} = (2\pi)^d \delta^d(k).
\tag{19.18}
$$

Similarly we can compute

$$
\sqrt{\det(2\pi\theta)} \, Tr_{\mathcal{H}} \hat{f} = \int d^d x f(x_i).
\tag{19.19}
$$

The analogue of the identity (19.17) is the identity

$$
\int d^d x \Delta(\hat{x}_i, x_i) = 1.
\tag{19.20}
$$

We want now to show that the Weyl map is indeed one-to-one. The proof goes as follows. First we compute

$$
\begin{aligned}
\sqrt{\det(2\pi\theta)} \, Tr_{\mathcal{H}} e^{ik\hat{x}} e^{ip\hat{x}} &= \sqrt{\det(2\pi\theta)} \, Tr e^{-\frac{i}{2}\theta_{ij} k_i p_j} e^{i(k+p)\hat{x}} \\
&= (2\pi)^d \delta^d(k + p).
\end{aligned}
\tag{19.21}
$$

Hence

$$
\begin{aligned}
\sqrt{\det(2\pi\theta)} \, Tr_{\mathcal{H}} \Delta(\hat{x}_i, x_i) \Delta(\hat{x}_i, y_i) &= \int \frac{d^d k}{(2\pi)^d} e^{-ikx} \int \frac{d^d p}{(2\pi)^d} e^{-ipy} \sqrt{\det(2\pi\theta)} \, Tr_{\mathcal{H}} e^{ik\hat{x}} e^{ip\hat{x}} \\
&= \delta^2(x - y).
\end{aligned}
\tag{19.22}
$$

Using this last formula one can immediately deduce

$$f(x_i) = \sqrt{\det(2\pi\theta)}\, Tr_{\mathcal{H}}\hat{f}\ \Delta(\hat{x}_i, x_i). \tag{19.23}$$

This shows explicitly that the Weyl map Δ provides indeed a one-to-one correspondence between fields and operators.

19.1.2 Star product and scalar action

The most natural problem now is to determine the image under the Weyl map of the pointwise product $\hat{f}\hat{g}$ of the two operators $\hat{f}$ and $\hat{g}$.

First we compute the generalization of (19.21) given by

$$\sqrt{\det(2\pi\theta)}\, Tr_{\mathcal{H}} e^{ik\hat{x}} e^{ip\hat{x}} e^{iq\hat{x}} = e^{-\frac{i}{2}\theta_{ij}k_i p_j}(2\pi)^d \delta^d(k + p + q). \tag{19.24}$$

This leads immediately to

$$\sqrt{\det(2\pi\theta)}\, Tr_{\mathcal{H}}\, \Delta(\hat{x}, y)\Delta(\hat{x}, z)\Delta(\hat{x}, x) = \int \frac{d^d k}{(2\pi)^d} \frac{d^d p}{(2\pi)^d} e^{ik(x-y)} e^{ip(x-z)} e^{-\frac{i}{2}\theta_{ij}k_i p_j}. \tag{19.25}$$

Hence

$$\sqrt{\det(2\pi\theta)}\, Tr_{\mathcal{H}}\hat{f}\,\hat{g}\Delta(\hat{x}_i, x_i) = \int \frac{d^d k}{(2\pi)^d} \frac{d^d p}{(2\pi)^d} \tilde{f}(k)\tilde{g}(p) e^{-\frac{i}{2}\theta_{ij}k_i p_j} e^{i(k+p)x} \tag{19.26}$$

$$\equiv f^*g(x).$$

This star product can also be given by

$$f^*g(x) = e^{\frac{i}{2}\theta_{ij}\frac{\partial}{\partial\xi_i}\frac{\partial}{\partial\eta_j}} f(x + \xi)g(x + \eta)|_{\xi=\eta=0}. \tag{19.27}$$

The above result can also be put in the form

$$\hat{f}\hat{g} = \int d^d x f^*g(x_i)\Delta(\hat{x}_i, x_i). \tag{19.28}$$

This leads to the identity

$$\sqrt{\det(2\pi\theta)}\, Tr_{\mathcal{H}}\hat{f}\,\hat{g} \equiv \int d^d x\ f^*g(x) = \int d^d x f(x)g(x). \tag{19.29}$$

From the other hand we know that the operator $[\hat{\partial}_i, \hat{f}]$ corresponds to the function $\partial_i f(x_i)$. We can then also write down

$$\sqrt{\det(2\pi\theta)}\, Tr_{\mathcal{H}}[\hat{\partial}_i, \hat{f}]^2 \equiv \int d^d x\ \partial_i f^*\partial_i f(x) = \int d^d x(\partial_i f)^2. \tag{19.30}$$

We are now in a position to propose a free, i.e., quadratic scalar action. This will be given simply by

$$S_{\text{free}} = \int d^d x\ \Phi\left(-\frac{1}{2}\partial_i^2 + \frac{\mu^2}{2}\right)\Phi$$

$$= \sqrt{\det(2\pi\theta)}\, Tr_{\mathcal{H}}\hat{\Phi}\left(-\frac{1}{2}[\hat{\partial}_i, [\hat{\partial}_i, \dots]] + \frac{\mu^2}{2}\right)\hat{\Phi}. \tag{19.31}$$

Next we add a phi-four interaction as a typical example of noncommutative interacting field theory. First we note that the operators $\hat{\Phi}$ and $\hat{\Phi}^2$ correspond to the fields Φ and Φ^2, respectively. Indeed we have

$$\sqrt{\det(2\pi\theta)}\, Tr_{\mathcal{H}}\hat{\Phi}\Delta(\hat{x}_i,\, x_i) = \Phi(x). \tag{19.32}$$

$$\sqrt{\det(2\pi\theta)}\, Tr_{\mathcal{H}}\hat{\Phi}^2\Delta(\hat{x}_i,\, x_i) = \Phi*\Phi(x). \tag{19.33}$$

Hence we must have immediately

$$\sqrt{\det(2\pi\theta)}\, Tr_{\mathcal{H}}\hat{\Phi}^4\Delta(\hat{x}_i,\, x_i) = \Phi*\Phi*\Phi*\Phi\ (x). \tag{19.34}$$

The phi-four interaction term must therefore be of the form

$$S_{\text{interaction}} = \sqrt{\det(2\pi\theta)}\,\frac{\lambda}{4!}Tr_{\mathcal{H}}\hat{\Phi}^4 = \frac{\lambda}{4!}\int d^dx\, \Phi*\Phi*\Phi*\Phi\ (x). \tag{19.35}$$

19.1.3 The Langmann–Szabo–Zarembo models

In this section, we will write down the most general action, with a phi-four interaction, in a noncommutative $\mathbf{R}_\theta^d$, under the effect of a magnetic field, which induces noncommutivity also in momentum space. Then, we will regularize the partition function of the theory by replacing the field operator $\hat{\Phi}$ by an $N \times N$ matrix M, and also replacing the infinite dimensional trace $Tr_{\mathcal{H}}$ by a finite dimensional trace Tr_N. The resulting theory is a single-trace matrix model, with a matrix phi-four interaction, and a modified propagator. We will mostly follow [12].

For simplicity, we start by considering a two-dimensional Euclidean space, with noncommutativity given by θ_{ij}. Generalization to higher dimension will be sketched in section (19.1.6). We also introduce noncommutativity in momentum space, by introducing a minimal coupling to a constant background magnetic field B_{ij}. The derivation operators become

$$\hat{D}_i = \hat{\partial}_i - iB_{ij}X_j, \quad \hat{C}_i = \hat{\partial}_i + iB_{ij}X_j. \tag{19.36}$$

In the above equation, X_i is defined by

$$X_i = \frac{\hat{x}_i + \hat{x}_i^R}{2}. \tag{19.37}$$

Hence

$$\hat{D}_i = -i(\theta^{-1} + \frac{B}{2})_{ij}\hat{x}_j + i(\theta^{-1} - \frac{B}{2})_{ij}\hat{x}_j^R$$

$$\hat{C}_i = -i(\theta^{-1} - \frac{B}{2})_{ij}\hat{x}_j + i(\theta^{-1} + \frac{B}{2})_{ij}\hat{x}_j^R. \tag{19.38}$$

We also remark

$$[X_i,\, X_j] = 0. \tag{19.39}$$

$$[\hat{D}_i, \hat{D}_j] = 2iB_{ij}, \quad [\hat{C}_i, \hat{C}_j] = -2iB_{ij}. \tag{19.40}$$

$$[\hat{D}_i, X_j] = [\hat{C}_i, X_j] = [\hat{\partial}_i, X_j] = \delta_{ij}. \tag{19.41}$$

Instead of the conventional Laplacian $\Delta = (-\hat{\partial}_i^2 + \mu^2)/2$ we will consider the generalized Laplacians

$$\begin{aligned}
\Delta =& -\sigma \hat{D}_i^2 - \tilde{\sigma}\hat{C}_i^2 + \frac{\mu^2}{2} \\
=& -(\sigma + \tilde{\sigma})\hat{\partial}_i^2 + (\sigma - \tilde{\sigma})iB_{ij}\{X_j, \hat{\partial}_i\} - (\sigma + \tilde{\sigma})(B^2)_{ij}X_iX_j + \frac{\mu^2}{2}.
\end{aligned} \tag{19.42}$$

The case $\sigma = \tilde{\sigma}$ corresponds to the Grosse–Wulkenhaar model [2–4], while the model $\sigma = 1$, $\tilde{\sigma} = 0$ corresponds to Langmann–Szabo–Zarembo model considered in [9].

Let us introduce the operators $Z = X_1 + iX_2$, $\bar{Z} = Z^+ = X_1 - iX_2$, $\hat{\partial} = \hat{\partial}_1 - i\hat{\partial}_2$ and $\bar{\hat{\partial}} = -\hat{\partial}^+ = \hat{\partial}_1 + i\hat{\partial}_2$. Also introduce the creation and annihilation operators (with $\theta_0 = \theta/2$, $\theta = \theta_{12}$)

$$\hat{a} = \frac{1}{2}\left(\sqrt{\theta_0}\hat{\partial} + \frac{1}{\sqrt{\theta_0}}\bar{Z}\right), \quad \hat{a}^+ = \frac{1}{2}\left(-\sqrt{\theta_0}\bar{\hat{\partial}} + \frac{1}{\sqrt{\theta_0}}Z\right). \tag{19.43}$$

$$\hat{b} = \frac{1}{2}\left(\sqrt{\theta_0}\bar{\hat{\partial}} + \frac{1}{\sqrt{\theta_0}}Z\right), \quad \hat{b}^+ = \frac{1}{2}\left(-\sqrt{\theta_0}\hat{\partial} + \frac{1}{\sqrt{\theta_0}}\bar{Z}\right). \tag{19.44}$$

We have

$$\begin{aligned}
X_1 &= \frac{\sqrt{\theta_0}}{2}(\hat{a} + \hat{b}^+ + \hat{a}^+ + \hat{b}), \quad X_2 = \frac{i\sqrt{\theta_0}}{2}(\hat{a} + \hat{b}^+ - \hat{a}^+ - \hat{b}) \\
\hat{\partial}_1 &= \frac{1}{2\sqrt{\theta_0}}(\hat{a} - \hat{b}^+ - \hat{a}^+ + \hat{b}), \quad \hat{\partial}_2 = \frac{i}{2\sqrt{\theta_0}}(\hat{a} - \hat{b}^+ + \hat{a}^+ - \hat{b}).
\end{aligned} \tag{19.45}$$

We compute by using $[Z, \bar{Z}] = 0$, $[\hat{\partial}, Z] = [\bar{\hat{\partial}}, \bar{Z}] = 2$, $[\hat{\partial}, \bar{Z}] = [\bar{\hat{\partial}}, Z] = 0$, and $[\hat{\partial}, \bar{\hat{\partial}}] = 0$ the commutation relations

$$[\hat{a}, \hat{a}^+] = 1, \quad [\hat{b}, \hat{b}^+] = 1. \tag{19.46}$$

The rest are zero.

We consider, now, the rank-one Fock space operators

$$\hat{\phi}_{l,m} = |l\rangle\langle m|. \tag{19.47}$$

We have immediately

$$\hat{\phi}_{l,m}^{\,+} = \hat{\phi}_{m,l}. \tag{19.48}$$

$$\hat{\phi}_{l,m}\hat{\phi}_{l',m'} = \delta_{m,l'}\hat{\phi}_{l,m'}. \tag{19.49}$$

$$Tr_{\mathcal{H}}\hat{\phi}_{l,m} = \delta_{l,m}. \tag{19.50}$$

$$Tr_{\mathcal{H}}\hat{\phi}^+_{l,m}\hat{\phi}_{l',m'} = \delta_{l,l'}\delta_{m,m'}. \tag{19.51}$$

We are, therefore, led to consider expanding the arbitrary scalar operators $\hat{\Phi}$, and $\hat{\Phi}^+$, in terms of $\hat{\phi}_{l,m}$, as follows

$$\hat{\Phi} = \sum_{l,m=1}^{\infty} M_{lm}\hat{\phi}_{l,m}, \quad \hat{\Phi}^+ = \sum_{l,m=1}^{\infty} M^*_{lm}\hat{\phi}^+_{l,m}. \tag{19.52}$$

The infinite dimensional matrix M should be thought of, as a compact operator, acting on some separable Hilbert space $\mathbf{H}_1$ of Schwartz sequences, with sufficiently rapid decrease [12]. This, in particular, will guarantee the convergence of the expansions of the scalar operators $\hat{\Phi}$, and $\hat{\Phi}^+$.

Next, we compute

$$\hat{\Phi}\hat{\Phi}' = \sum_{l,m=1}^{\infty} (MM')_{lm}\hat{\phi}_{l,m}. \tag{19.53}$$

The representation of this operator product, in terms of the star product of functions, can be obtained as follows. In the operators $\hat{\phi}_{l,m} = |l\rangle\langle m|$, we can identify the kets $|l\rangle$ with the states of the harmonic oscillator operators $\hat{a}$, and $\hat{a}^+$, whereas the bras $\langle m|$ can be identified with the states of the harmonic oscillator operators $\hat{b}$, and $\hat{b}^+$. More precisely, the operators $\hat{\phi}_{l,m}$ are, in one-to-one correspondence, with the wave functions $\phi_{l,m}(x) = \langle x|l, m\rangle$ known as Landau states [10, 11].

Landau states are defined by

$$\hat{a}\hat{\phi}_{l,m} = \sqrt{l-1}\,\hat{\phi}_{l-1,m}, \quad \hat{a}^+\hat{\phi}_{l,m} = \sqrt{l}\,\hat{\phi}_{l+1,m}. \tag{19.54}$$

$$\hat{b}\hat{\phi}_{l,m} = \sqrt{m-1}\,\hat{\phi}_{l,m-1}, \quad \hat{b}\hat{\phi}_{l,m} = \sqrt{m}\,\hat{\phi}_{l,m+1}. \tag{19.55}$$

These states, as we will show, satisfy, among other things, the following properties

$$\phi^*_{l,m}(x) = \phi_{m,l}(x). \tag{19.56}$$

$$\phi_{l_1,m_1}{}^*\phi_{l_2,m_2}(x) = \frac{1}{\sqrt{4\pi\theta_0}}\delta_{m_1,l_2}\,\phi_{l_1,m_2}(x). \tag{19.57}$$

$$\int d^2x\,\phi_{l,m}(x) = \sqrt{4\pi\theta_0}\,\delta_{l,m}. \tag{19.58}$$

$$\int d^2x\,\phi^*_{l_1,m_1}{}^*\phi_{l_2,m_2}(x) = \delta_{l_1,l_2}\delta_{m_1,m_2}. \tag{19.59}$$

By comparing with (19.48), (19.49), (19.50), and (19.51) we conclude immediately that the field/operator (Weyl) map is given by

$$\sqrt{2\pi\theta}\ \phi_{l_1,m_1} \leftrightarrow \hat{\phi}_{l_1,m_1}. \tag{19.60}$$

$$\int d^2x \leftrightarrow \sqrt{\det(2\pi\theta)}\,\mathrm{Tr}_{\mathcal{H}}. \tag{19.61}$$

We are therefore led to consider scalar functions Φ and Φ^+ corresponding to the scalar operators $\hat{\Phi}$, $\hat{\Phi}^+$ given explicitly by

$$\Phi = \sqrt{2\pi\theta}\sum_{l,m=1}^{\infty} M_{lm}\phi_{l,m}, \quad \Phi^+ = \sqrt{2\pi\theta}\sum_{l,m=1}^{\infty} M_{lm}^*\phi_{l,m}^* \longrightarrow \hat{\Phi},\ \hat{\Phi}^+. \tag{19.62}$$

Indeed the star product is given by

$$\Phi*\Phi' = \sqrt{2\pi\theta}\sum_{l,m=1}^{\infty} (MM')_{lm}\phi_{l,m}. \tag{19.63}$$

In other words, the star product is mapped to the operator product as it should be, viz

$$\Phi*\Phi' \leftrightarrow \hat{\Phi}\hat{\Phi}'. \tag{19.64}$$

The operator $X_i\hat{\Phi} = \frac{1}{2}\hat{x}_i\hat{\Phi} + \frac{1}{2}\hat{\Phi}\hat{x}_i$ corresponds to the function $\frac{1}{2}x_i^*\Phi + \frac{1}{2}\Phi^*x_i = x_i\Phi$. As a consequence, the differential operators $\hat{D}_i^2$, and $\hat{C}_i^2$ will be represented in the star picture by the differential operators

$$D_i = \partial_i - iB_{ij}x_j, \quad C_i = \partial_i + iB_{ij}x_j. \tag{19.65}$$

The Landau states are actually eigenstates of the Laplacians D_i^2, and C_i^2 at the special point

$$B^2\theta_0^2 \equiv \frac{B^2\theta^2}{4} = 1. \tag{19.66}$$

Indeed, we can compute (with $\alpha = 1 + B\theta_0$, $\beta = 1 - B\theta_0$, and $\alpha\beta = 1 - B^2\theta_0^2$) the following

$$4\theta_0\hat{D}_i^2 = -4\alpha^2(\hat{a}^+\hat{a} + \frac{1}{2}) - 4\beta^2(\hat{b}^+\hat{b} + \frac{1}{2}) + 4\alpha\beta(\hat{a}\hat{b} + \hat{a}^+\hat{b}^+). \tag{19.67}$$

$$4\theta_0\hat{C}_i^2 = -4\beta^2(\hat{a}^+\hat{a} + \frac{1}{2}) - 4\alpha^2(\hat{b}^+\hat{b} + \frac{1}{2}) + 4\alpha\beta(\hat{a}\hat{b} + \hat{a}^+\hat{b}^+). \tag{19.68}$$

For $\beta = 0$, we observe that D_i^2, and C_i^2 depend only on the number operators $\hat{a}^+\hat{a}$, and $\hat{b}^+\hat{b}$, respectively. For $\alpha = 0$, the roles of D_i^2, and C_i^2, are reversed.

Next, we write down, the most general single-trace action with a phi-four interaction in a noncommutative $\mathbf{R}_\theta^d$ under the effect of a magnetic field, as follows

$$S = \sqrt{\det(2\pi\theta)}\, Tr_{\mathcal{H}}\left[\hat{\Phi}^+\left(-\sigma\hat{D}_i^2 - \check{\sigma}\hat{C}_i^2 + \frac{\mu^2}{2}\right)\hat{\Phi} + \frac{\lambda}{4!}\hat{\Phi}^+\hat{\Phi}\,\hat{\Phi}^+\hat{\Phi} + \frac{\lambda'}{4!}\hat{\Phi}^+\hat{\Phi}^+\,\hat{\Phi}\hat{\Phi}\right]. \quad (19.69)$$

By using the above results, as well as the results of the previous section, we can rewrite this action in terms of the star product as follows

$$S = \int d^2x\left[\Phi^+\left(-\sigma D_i^2 - \check{\sigma}C_i^2 + \frac{\mu^2}{2}\right)\Phi + \frac{\lambda}{4!}\Phi^+*\Phi*\Phi^+*\Phi + \frac{\lambda'}{4!}\Phi^+*\Phi^+*\Phi*\Phi\right]. \quad (19.70)$$

In most of the following we will assume $\lambda' = 0$.

19.1.4 Duality transformations

The action, of interest, reads

$$S = \sqrt{\det(2\pi\theta)}\, Tr_{\mathcal{H}}\left[\hat{\Phi}^+\left(-\sigma\hat{D}_i^2 - \check{\sigma}\hat{C}_i^2 + \frac{\mu^2}{2}\right)\hat{\Phi} + \frac{\lambda}{4!}\hat{\Phi}^+\hat{\Phi}\,\hat{\Phi}^+\hat{\Phi}\right]$$
$$= \int d^2x\left[\Phi^+\left(-\sigma D_i^2 - \check{\sigma}C_i^2 + \frac{\mu^2}{2}\right)\Phi + \frac{\lambda}{4!}\Phi^+*\Phi*\Phi^+*\Phi\right]. \quad (19.71)$$

This action enjoys, a remarkable, symmetry under certain duality transformations, which exchange, among other things, positions and momenta. See [9, 12, 13] for the original derivation. This property can be shown as follows. We start with the quadratic action given by

$$S_2[\Phi, B] = \int d^2x\left[\Phi^+\left(-\sigma D_i^2 - \check{\sigma}C_i^2 + \frac{\mu^2}{2}\right)\Phi\right]. \quad (19.72)$$

We define $\tilde{k}_i = B_{ij}^{-1}k_j$. The Fourier transform of $\Phi(x)$, and $D_i\Phi(x)$, are $\tilde{\Phi}(k)$, and $-\tilde{D}_i\tilde{\Phi}(k)$, where

$$\tilde{\Phi}(k) = \int d^2x\,\Phi(x)\, e^{-ik_ix_i}. \quad (19.73)$$

$$-\tilde{D}_i\tilde{\Phi}(k) = \int d^2x\, D_i\Phi(x)\, e^{-ik_ix_i}$$
$$= -\left(\frac{\partial}{\partial\tilde{k}_i} - iB_{ij}\tilde{k}_j\right)\tilde{\Phi}(k). \quad (19.74)$$

Then, we can immediately compute that

$$\int d^2x\,(D_i\Phi)^+(x)(D_i\Phi)(x) = \int d^2\tilde{k}\,(\tilde{D}_i\tilde{\Phi})^+(\tilde{k})(\tilde{D}_i\tilde{\Phi})(\tilde{k}). \quad (19.75)$$

The new field, $\bar{\Phi}$, is defined by

$$\bar{\Phi}(\tilde{k}) = \sqrt{\left|\det\frac{B}{2\pi}\right|}\,\tilde{\Phi}(B\tilde{k}). \quad (19.76)$$

A similar result holds for the other quadratic terms By renaming the variable as $\tilde{k} = x$, we can see that the resulting quadratic action has, therefore, the same form as the original quadratic action, viz

$$S_2[\Phi, B] = S_2[\bar{\Phi}, B].$$
(19.77)

Next, we consider the interaction term

$$S_{\text{int}}[\Phi, B] = \int d^2x \Phi^{+*}\Phi^*\Phi^{+*}\Phi$$
$$= \int \frac{d^2k_1}{(2\pi)^2} \cdots \int \frac{d^2k_1}{(2\pi)^2} \tilde{\Phi}^+(k_1)\tilde{\Phi}(k_2)\tilde{\Phi}^+(k_3)\tilde{\Phi}^+(k_4)\tilde{V}(k_1, k_2, k_3, k_4).$$
(19.78)

The vertex in momentum space is given by

$$\tilde{V}(k_1, k_2, k_3, k_4) = (2\pi)^2 \delta^2(k_1 - k_2 + k_3 - k_4)e^{-i\theta^{\mu\nu}(k_{1\mu}k_{2\nu} + k_{3\mu}k_{4\nu})}.$$
(19.79)

By substituting, $k = B\tilde{k}$, we obtain

$$S_{\text{int}}[\Phi, B] = \int d^2\tilde{k}_1 \cdots \int d^2\tilde{k}_4 \bar{\Phi}^+(\tilde{k}_1)\bar{\Phi}(\tilde{k}_2)\bar{\Phi}^+(\tilde{k}_3)\bar{\Phi}^+(\tilde{k}_4)\bar{V}(\tilde{k}_1, \tilde{k}_2, \tilde{k}_3, \tilde{k}_4).$$
(19.80)

The new vertex is given by

$$\bar{V}(\tilde{k}_1, \tilde{k}_2, \tilde{k}_3, \tilde{k}_4) = \frac{\det B}{(2\pi)^2}\delta^2(\tilde{k}_1 - \tilde{k}_2 + \tilde{k}_3 - \tilde{k}_4)e^{i(B\theta B)^{\mu\nu}(\tilde{k}_{1\mu}\tilde{k}_{2\nu} + \tilde{k}_{3\mu}\tilde{k}_{4\nu})}.$$
(19.81)

The interaction term, can also, be rewritten as

$$S_{\text{int}}[\Phi, B] = \int d^2x_1 \cdots \int d^2x_4 \Phi^+(x_1)\Phi(x_2)\Phi^+(x_3)\Phi^+(x_4)V(x_1, x_2, x_3, x_4).$$
(19.82)

The vertex in position space is given by

$$V(x_1, x_2, x_3, x_4) = \int \frac{d^2k_1}{(2\pi)^2} \cdots \int \frac{d^2k_4}{(2\pi)^2} \tilde{V}(k_1, k_2, k_3, k_4)e^{ik_1x_1 - ik_2x_2 + ik_3x_3 - ik_4x_4}$$
$$= \frac{1}{(2\pi)^2|\det\theta|}\delta^2(x_1 - x_2 + x_3 - x_4)e^{-i(\theta^{-1})_{\mu\nu}(x_1^\mu x_2^\mu + x_3^\mu x_4^\mu)}.$$
(19.83)

We can see immediately from comparing equations (19.80), and (19.81), to equations (19.82), and (19.83), that the interaction term in momentum space, has the same form as the interaction term in position space, provided that the new noncommutativity parameter, and the new coupling constant, are given by

$$\bar{\theta} = -B^{-1}\theta^{-1}B^{-1}.$$
(19.84)

$$\lambda \frac{\det B}{(2\pi)^2} = \frac{\bar{\lambda}}{(2\pi)^2}\frac{1}{\det\bar{\theta}} \Leftrightarrow \bar{\lambda} = \frac{\lambda}{|\det B\theta|}.$$
(19.85)

In summary, the duality transformations under which the action retains the same form, are given by

$$x_i \longrightarrow \tilde{k}_i = B_{ij}^{-1} k_j. \tag{19.86}$$

$$\Phi(x) \longrightarrow \tilde{\Phi}(\tilde{k}) = \sqrt{\left|\det \frac{B}{2\pi}\right|}\, \tilde{\Phi}(B\tilde{k}). \tag{19.87}$$

$$\theta \longrightarrow \bar{\theta} = -B^{-1}\theta^{-1}B^{-1}. \tag{19.88}$$

$$\lambda \longrightarrow \bar{\lambda} = \frac{\lambda}{|\det B\theta|}. \tag{19.89}$$

19.1.5 Matrix regularization

Now, we want to express the above action, which is given by equation (19.71), in terms of the compact operators M, and M^+. First, we compute

$$Tr_{\mathcal{H}}\Phi^+\Phi = Tr_{\mathrm{H}_1} M^+M. \tag{19.90}$$

$$Tr_{\mathcal{H}}\Phi^+(ab + a^+b^+)\Phi = Tr_{\mathrm{H}_1}(\Gamma^+M^+\Gamma M + M^+\Gamma^+M\Gamma). \tag{19.91}$$

$$Tr_{\mathcal{H}}\Phi^+(a^+a + \frac{1}{2})\Phi = Tr_{\mathrm{H}_1} M^+EM. \tag{19.92}$$

$$Tr_{\mathcal{H}}\Phi^+(b^+b + \frac{1}{2})\Phi = Tr_{\mathrm{H}_1} MEM^+. \tag{19.93}$$

$$Tr_{\mathcal{H}}\Phi^+\Phi \; \Phi^+\Phi = Tr_{\mathrm{H}_1} M^+MM^+M. \tag{19.94}$$

The infinite dimensional matrices Γ, and E are defined by

$$(\Gamma)_{lm} = \sqrt{m-1}\,\delta_{lm-1}, \quad (E)_{lm} = (l - \frac{1}{2})\delta_{lm}. \tag{19.95}$$

The action becomes

$$S = \frac{\sqrt{\det(2\pi\theta)}}{\theta_0}\left[-(\sigma + \tilde{\sigma})\alpha\beta Tr_{\mathrm{H}_1}(\Gamma^+M^+\Gamma M + M^+\Gamma^+M\Gamma) + (\sigma\alpha^2 + \tilde{\sigma}\beta^2)Tr_{\mathrm{H}_1} M^+EM \right.$$
$$\left. + (\sigma\beta^2 + \tilde{\sigma}\alpha^2)Tr_{\mathrm{H}_1} MEM^+ + \frac{\mu^2\theta_0}{2}Tr_{\mathrm{H}_1} M^+M + \frac{\lambda\theta_0}{4!}Tr_{\mathrm{H}_1} M^+MM^+M\right]. \tag{19.96}$$

We regularize the theory by taking M to be an $N \times N$ matrix. The states $\phi_{l,m}(x)$, with $l, m < N$, where N is some large integer, correspond to a cutoff in position, and momentum spaces [4]. The infrared cutoff is found to be proportional to $R = \sqrt{2\theta N}$, while the UV cutoff is found to be proportional to $\Lambda = \sqrt{8N/\theta}$. In [12], a double scaling strong noncommutativity limit, in which N/θ (and thus Λ) is kept fixed, was considered.

19.1.6 The Grosse–Wulkenhaar model

19.1.6.1 In two dimensions

We will be mostly interested in the so-called Grosse–Wulkenhaar model. This contains, compared with the usual case, a harmonic oscillator term in the Laplacian, which modifies, and thus allows us, to control the IR behavior of the theory. This model is perturbatively renormalizable, which makes it, the more interesting. The Grosse–Wulkenhaar model, corresponds to the values $\sigma = \tilde{\sigma} \neq 0$, so that the mixing term, in (19.42), cancels. We consider, without any loss of generality, $\sigma = \tilde{\sigma} = 1/4$. We obtain therefore the action

$$
\begin{aligned}
S &= \sqrt{\det(2\pi\theta)} \, Tr_{\mathcal{H}} \left[\hat{\Phi}^+ \left(-\frac{1}{2}\hat{\partial}_i^2 + \frac{1}{2}(B_{ij} X_j)^2 + \frac{\mu^2}{2} \right) \hat{\Phi} + \frac{\lambda}{4!} \hat{\Phi}^+ \hat{\Phi} \, \hat{\Phi}^+ \hat{\Phi} \right] \\
&= \int d^2x \left[\Phi^+ \left(-\frac{1}{2}\partial_i^2 + \frac{1}{2}(B_{ij}x_j)^2 + \frac{\mu^2}{2} \right) \Phi + \frac{\lambda}{4!} \Phi^{+}*\Phi*\Phi^{+}*\Phi(x) \right].
\end{aligned}
\tag{19.97}
$$

In two dimensions, we can show that $(B_{ij}x_j)^2 = \Omega^2 \tilde{x}_i^2$, where $\tilde{x}_i = 2(\theta^{-1})_{ij}x_j$, and Ω is defined by

$$
B\theta = 2\Omega.
\tag{19.98}
$$

We get therefore the action

$$
\begin{aligned}
S &= \sqrt{\det(2\pi\theta)} \, Tr_{\mathcal{H}} \left[\hat{\Phi}^+ \left(-\frac{1}{2}\hat{\partial}_i^2 + \frac{1}{2}\Omega^2 \tilde{X}_i^2 + \frac{\mu^2}{2} \right) \hat{\Phi} + \frac{\lambda}{4!} \hat{\Phi}^+ \hat{\Phi} \, \hat{\Phi}^+ \hat{\Phi} \right] \\
&= \int d^2x \left[\Phi^+ \left(-\frac{1}{2}\partial_i^2 + \frac{1}{2}\Omega^2 \tilde{x}_i^2 + \frac{\mu^2}{2} \right) \Phi + \frac{\lambda}{4!} \Phi^{+}*\Phi*\Phi^{+}*\Phi \right].
\end{aligned}
\tag{19.99}
$$

Similarly, to $\tilde{x}_i = 2(\theta^{-1})_{ij}x_j$, we have defined $\tilde{X}_i = 2(\theta^{-1})_{ij}X_j$. This action, is also found, to be covariant under a duality transformation, which exchanges, among other things, positions and momenta as $x_i \leftrightarrow \tilde{p}_i = B_{ij}^{-1}p_j$. Let us note here, that because of the properties of the star product, the phi-four interaction, is actually invariant under this duality transformation. The value $\Omega^2 = 1$, gives an action, which is invariant under this duality transformation, i.e., the kinetic term becomes invariant under this duality transformation for $\Omega^2 = 1$.

In the Landau basis, the above action, reads

$$
\begin{aligned}
S = \frac{\nu_2}{\theta}\Big[(\Omega^2 - 1) Tr_H \left(\Gamma^+ M^+ \Gamma M + M^+ \Gamma^+ M \Gamma \right) + (\Omega^2 + 1) Tr_H (M^+ E M + M E M^+) \\
+ \frac{\mu^2 \theta}{2} Tr_H M^+ M + \frac{\lambda \theta}{4!} Tr_H M^+ M M^+ M \Big].
\end{aligned}
\tag{19.100}
$$

This is a special case of (19.96). Equivalently

$$
S = \nu_2 \sum_{m,n,k,l} \left(\frac{1}{2}(M^+)_{mn} G_{mn,kl} M_{kl} + \frac{\lambda}{4!}(M^+)_{mn} M_{nk}(M^+)_{kl} M_{lm} \right).
\tag{19.101}
$$

$$G_{mn,kl} = \left(\mu^2 + \mu_1^2(m + n - 1)\right)\delta_{n,k}\delta_{m,l} - \mu_1^2\sqrt{\omega(m - 1)(n - 1)}\;\delta_{n-1,k}\delta_{m-1,l}$$
$$- \mu_1^2\sqrt{\omega mn}\;\delta_{n+1,k}\delta_{m+1,l}. \tag{19.102}$$

The parameters of the model are μ^2, λ, and

$$\nu_2 = \sqrt{\det(2\pi\theta)}, \quad \mu_1^2 = 2(\Omega^2 + 1)/\theta, \quad \sqrt{\omega} = (\Omega^2 - 1)/(\Omega^2 + 1). \tag{19.103}$$

There are only three independent coupling constants in this theory, which we can take to be μ^2, λ, and Ω^2.

19.1.6.2 Generalization

Generalization of the above results, to higher dimensions $d = 2n$, assuming maximal noncommutativity for simplicity, is straightforward. The action reads

$$S = \sqrt{\det(2\pi\theta)}\,Tr_{\mathcal{H}}\left[\hat{\Phi}^+\left(-\sigma\hat{D}_i^2 - \tilde{\sigma}\hat{C}_i^2 + \frac{\mu^2}{2}\right)\hat{\Phi} + \frac{\lambda}{4!}\hat{\Phi}^+\hat{\Phi}\;\hat{\Phi}^+\hat{\Phi}\right]$$
$$= \int d^dx\left[\Phi^+\left(-\sigma D_i^2 - \tilde{\sigma}\tilde{C}_i^2 + \frac{\mu^2}{2}\right)\Phi + \frac{\lambda}{4!}\Phi^{+*}\Phi^*\Phi^{+*}\Phi\right]. \tag{19.104}$$

In order to be able to proceed, we will assume that the noncommutativity tensor θ, and the magnetic tensor B, are simultaneously diagonalizable. In other words, θ and B, can be brought together, to the canonical form (19.2). For example, in four dimension, we will have

$$\theta = \begin{pmatrix} 0 & \theta_{12} & 0 & 0 \\ -\theta_{12} & 0 & 0 & 0 \\ 0 & 0 & 0 & \theta_{34} \\ 0 & 0 & -\theta_{34} & 0 \end{pmatrix}, \quad B = \begin{pmatrix} 0 & B_{12} & 0 & 0 \\ -B_{12} & 0 & 0 & 0 \\ 0 & 0 & 0 & B_{34} \\ 0 & 0 & -B_{34} & 0 \end{pmatrix}. \tag{19.105}$$

The d-dimensional problem will, thus, split into a direct sum, of n independent, and identical, two-dimensional problems.

The expansion of the scalar field operator is, now, given by

$$\hat{\Phi} = \sum_{\vec{l},\vec{m}}^{\infty} M_{\vec{l}\vec{m}}\hat{\phi}_{\vec{l},\vec{m}}, \quad \vec{l} = (l_1, \ldots, l_n), \quad \vec{m} = (m_1, \ldots, m_n). \tag{19.106}$$

Obviously

$$\hat{\phi}_{\vec{l},\vec{m}} = \prod_{i=1}^{n} \hat{\phi}_{l_i,m_i} \tag{19.107}$$

and

$$\hat{\phi}_{\vec{l},\vec{m}} \leftrightarrow \det(2\pi\theta)^{1/4}\;\phi_{\vec{l},\vec{m}}. \tag{19.108}$$

The quantum numbers l_i, and m_i correspond to the plane $x_{2i-1} - x_{2i}$. They correspond to the operators $\hat{x}_{2i-1}$, $\hat{x}_{2i}$, $\hat{\partial}_{2i-1}$, and $\hat{\partial}_{2i}$, or equivalently, to the creation, and annihilation operators $\hat{a}^{(i)}$, $\hat{a}^{(i)+}$, $\hat{b}^{(i)}$, $\hat{b}^{(i)+}$. Indeed, the full Hilbert space $\mathbf{H}_n$, in this case, is a direct sum of the individual Hilbert spaces $\mathbf{H}_1^{(i)}$, associated, with the individual planes.

The above action can be given, in terms of the compact operators M and M^+, by essentially equation (19.96).

The Grosse–Wulkenhaar model, in higher dimensions, corresponds, as before, to the values $\sigma = \tilde{\sigma} = 1/4$. However, in higher dimensions, we need also to choose the magnetic field B, such that

$$B\theta = 2\Omega 1. \tag{19.109}$$

The action reduces, then, to

$$S = \sqrt{\det(2\pi\theta)}\, Tr_{\mathcal{H}}\left[\hat{\Phi}^+\left(-\frac{1}{2}\hat{\partial}_i^2 + \frac{1}{2}\Omega^2\tilde{X}_i^2 + \frac{\mu^2}{2} \right)\hat{\Phi} + \frac{\lambda}{4!}\hat{\Phi}^+\hat{\Phi}\,\hat{\Phi}^+\hat{\Phi} \right]$$
$$= \int d^d x\left[\Phi^+\left(-\frac{1}{2}\partial_i^2 + \frac{1}{2}\Omega^2\tilde{x}_i^2 + \frac{\mu^2}{2} \right)\Phi + \frac{\lambda}{4!}\Phi^{+*}\Phi^*\Phi^{+*}\Phi \right]. \tag{19.110}$$

Again this action will be given, in terms of the compact operators M and M^+, by essentially the same equations (19.100), and (19.101).

For example we replace in (19.102) mn by $m_1 n_1 m_2 n_2$, kl by $k_1 l_1 k_2 l_2$, nk by $n_1 k_1 n_2 k_2$, kl by $k_1 l_1 k_2 l_2$ and lm by $l_1 m_1 l_2 m_2$ to obtain on $\mathbf{R}_\theta^2 \times \mathbf{R}_\theta^2$ the result

$$\begin{aligned}
G^{m_2 n_2, k_2 l_2}_{m_1 n_1,\, k_1 l_1} =\ & \left(\mu_0^2 + \mu_1^2(m_1 + n_1 + m_2 + n_2 - 2)\right)\delta_{n_1,k_1}\delta_{m_1,l_1}\delta_{n_2,k_2}\delta_{m_2,l_2} \\
& - \mu_1^2\sqrt{\omega}\left(\sqrt{(m_1 - 1)(n_1 - 1)}\ \delta_{n_1-1,k_1}\delta_{m_1-1,l_1} + \sqrt{m_1 n_1}\ \delta_{n_1+1,k_1}\delta_{m_1+1,l_1}\right) \\
& \times \delta_{n_2,k_2}\delta_{m_2,l_2} \\
& - \mu_1^2\sqrt{\omega}\left(\sqrt{(m_2 - 1)(n_2 - 1)}\ \delta_{n_2-1,k_2}\delta_{m_2-1,l_2} + \sqrt{m_2 n_2}\ \delta_{n_2+1,k_2}\delta_{m_2+1,l_2}\right) \\
& \times \delta_{n_1,k_1}\delta_{m_1,l_1}.
\end{aligned} \tag{19.111}$$

19.1.7 Noncommutative phi-four on $\mathbf{R}^D \times \mathbf{R}_\theta^2$

Let us consider a phi-four theory on a noncommutative Moyal–Weyl space with only two noncommuting coordinates, viz $\mathbf{R}_\theta^d = \mathbf{R}^D \times \mathbf{R}_\theta^2$ where $D = d - 2$. We also introduce noncommutativity in momentum space by introducing a minimal coupling to a constant background magnetic field B_{ij} as was done in [12]. This corresponds to the addition of a harmonic oscillator potential to the kinetic action, which modifies and thus allows us to control the IR behavior of the theory. A particular version of this theory was shown to be renormalizable by Grosse and Wulkenhaar in [3].

The action of interest is given by

$$S = \int d^d x\left[\Phi\left(-\frac{1}{2}\partial_i^2 + \frac{1}{2}(B_{ij}x_j)^2 - \frac{1}{2}\partial_\mu^2 + \frac{\mu^2}{2} \right)\Phi + \frac{\lambda}{4!}\Phi*\Phi*\Phi*\Phi \right]. \tag{19.112}$$

The index i runs over the noncommuting directions while the index μ runs over the commuting directions. We rewrite this action as

$$S = \int d^d x \left[\Phi\left(-\frac{1}{2}\partial_i^2 + \frac{1}{2}\Omega^2 \tilde{x}_i^2 - \frac{1}{2}\partial_\mu^2 + \frac{\mu^2}{2} \right)\Phi + \frac{\lambda}{4!}\Phi^*\Phi^*\Phi^*\Phi \right]. \tag{19.113}$$

The harmonic oscillator coupling constant Ω is defined by $\Omega^2 = B^2\theta^2/4$ whereas the coordinate $\tilde{x}_i$ is defined by $\tilde{x}_i = 2(\theta^{-1})_{ij}x_j$. It was shown in [9] that this action is covariant under a duality transformation, which exchanges among other things positions and momenta as $p_i \leftrightarrow \tilde{x}_i$. The value $\Omega^2 = 1$ in particular gives an action, which is invariant under this duality transformation. Under the field/operator Weyl map we can rewrite the action as

$$S = \nu_2 \int d^D x \; Tr_{\mathcal{H}}\left[\hat{\Phi}\left(-\frac{1}{2}\hat{\partial}_i^2 + \frac{1}{2}\Omega^2 \hat{X}_i^2 - \frac{1}{2}\partial_\mu^2 + \frac{\mu^2}{2} \right)\hat{\Phi} + \frac{\lambda}{4!}\hat{\Phi}^4 \right]. \tag{19.114}$$

The Planck volume ν_2 is defined by $\nu_2 = \sqrt{\det 2\pi\theta}$. We can expand the scalar fields in the Landau basis $\{\hat{\phi}_{m,n}\}$ as (with x standing for commuting coordinates)

$$\hat{\Phi} = \frac{1}{\sqrt{\nu_2}} \sum_{m,n=1}^{\infty} M_{mn}(x)\hat{\phi}_{m,n}. \tag{19.115}$$

The infinite dimensional matrix M should be thought of as a compact operator acting on the separable Hilbert space $H = \mathcal{S}(\mathbf{N})$ of Schwartz sequences $(a_m)_{m\geqslant 1}$ with sufficiently rapid decrease as $m \longrightarrow \infty$. In the Landau basis the action becomes

$$S = \int d^D x Tr_H \left[\frac{1}{2}(\partial_\mu M)^2 + \frac{1}{2}\mu^2 M^2 + \frac{1}{2}r^2 E M^2 + \frac{u}{N}M^4 + \text{remainder} \right]. \tag{19.116}$$

The coupling constants r^2 and u are defined by

$$r^2 = \frac{4\pi(\Omega^2 + 1)}{\nu_2}, \quad u = \frac{\lambda}{4!}\frac{N}{\nu_2}. \tag{19.117}$$

The remainder is given by

$$\text{remainder} = \frac{1}{2}r^2 \sqrt{\omega}(\Gamma^+ M^+ \Gamma M + M^+ \Gamma^+ M\Gamma), \quad \sqrt{\omega} = \frac{1 - \Omega^2}{1 + \Omega^2}. \tag{19.118}$$

The matrices Γ and E are given by

$$(\Gamma)_{lm} = \sqrt{m - 1}\,\delta_{lm-1}, \quad (E)_{lm} = (l - \frac{1}{2})\delta_{lm}. \tag{19.119}$$

The above action can also be put in the form

$$S = \int d^D x \sum_{m,n,k,l}\left(\frac{1}{2}M_{mn}G_{mn,kl}M_{kl} + \frac{\lambda}{4!}M_{mn}M_{nk}M_{kl}M_{lm} \right) \tag{19.120}$$

where

$$G_{mn,kl} = \left(-\partial_\mu^2 + \mu^2 + \mu_1^2(m + n - 1)\right)\delta_{n,k}\delta_{m,l} - \mu_1^2\sqrt{\omega(m-1)(n-1)}\ \delta_{n-1,k}\delta_{m-1,l}$$
$$- \mu_1^2\sqrt{\omega mn}\ \delta_{n+1,k}\delta_{m+1,l}. \tag{19.121}$$

In momentum space we have

$$S = \int \frac{d^D p}{(2\pi)^D} \sum_{m,n,k,l} \frac{1}{2}\tilde{M}_{mn}(-p)G_{mn,kl}(p)\tilde{M}_{kl}(p) + \int \frac{d^D p_1}{(2\pi)^D}\int \frac{d^D p_2}{(2\pi)^D}\int \frac{d^D p_3}{(2\pi)^D}$$
$$\times \sum_{m,n,k,l} \frac{\lambda}{4!}\tilde{M}_{mn}(p_1)\tilde{M}_{nk}(p_2)\tilde{M}_{kl}(p_3)\tilde{M}_{lm}(-p_1 - p_2 - p_3) \tag{19.122}$$

where

$$G_{mn,kl}(p) = \left(p_\mu^2 + \mu^2 + \mu_1^2(m + n - 1)\right)\delta_{n,k}\delta_{m,l} - \mu_1^2\sqrt{\omega(m-1)(n-1)}\ \delta_{n-1,k}\delta_{m-1,l}$$
$$- \mu_1^2\sqrt{\omega mn}\ \delta_{n+1,k}\delta_{m+1,l}. \tag{19.123}$$

19.2 Wilson–Polchinski renormalization group equation on $\mathbf{R}^D \times \mathbf{R}_\theta^2$

We are interested in the renormalizability of noncommutative phi-four theory with a harmonic oscillator term. Towards this end we need to develop the Wilson–Polchinski renormalization group equation for noncommutative scalar field theory in the matrix basis [2]. As an example, we consider in this section the theory on $\mathbf{R}^D \times \mathbf{R}_\theta^2$.

We consider the path integral

$$Z[J] = \int \prod_{m,n,p} d\tilde{M}_{mn}(p)\ \exp\left[-S[M] - \sqrt{\det(2\pi\tilde{\theta})}\int \frac{d^D p}{(2\pi)^D}\sum_{m,n}\tilde{M}_{mn}(p)J_{nm}(-p)\right]. \tag{19.124}$$

The total action S and the interaction term L are given by

$$S[M] = \sqrt{\det(2\pi\tilde{\theta})}\int \frac{d^D p}{(2\pi)^D}\sum_{m,n,k,l}\frac{1}{2}\tilde{M}_{mn}(-p)G_{mn,kl}(p)\tilde{M}_{kl}(p) + L[M]. \tag{19.125}$$

$$L[M] = \sqrt{\det(2\pi\tilde{\theta})}\int \frac{d^D p_1}{(2\pi)^D}\int \frac{d^D p_2}{(2\pi)^D}\int \frac{d^D p_3}{(2\pi)^D}\sum_{m,n,k,l}\frac{\lambda}{4!}\tilde{M}_{mn}(p_1)\tilde{M}_{nk}(p_2)\tilde{M}_{kl}(p_3)$$
$$\times \tilde{M}_{lm}(-p_1 - p_2 - p_3). \tag{19.126}$$

The volume of an elementary cell is $\nu_2 = \sqrt{\det 2\pi\tilde{\theta}} = \pi\theta = 2\pi(1 + \Omega^2)/\mu_1^2$ where we have used $\tilde{\theta} = \theta/2$ and $\theta = 2(1 + \Omega^2)/\mu_1^2$. We introduce a momentum space cutoff Λ_1 and a cutoff Λ_2 on discrete indices. The dimensionless noncommutativity is $\bar{\theta} = \theta\Lambda_2^2$. The regularized inverse propagator is

$$G_{mn,kl}(p, \Lambda) = G_{mn,kl}(p)K_1^{-1}\left(\frac{p^2}{\Lambda_1^2}\right)\prod_{i\in\{m,n,k,l\}}K_2^{-1}\left(\frac{\mu_1^2 i}{\Lambda_2^2}\right). \tag{19.127}$$

The ultraviolet cutoff function K_1 is such that it is equal to 1 for $p^2 \leqslant \Lambda_1^2$ and it vanishes rapidly for $p^2 > \Lambda_1^2$. Similarly the ultraviolet cutoff function $K_2(x)$ is such that it is equal to 1 for $0 \leqslant x \leqslant 1$ and it vanishes rapidly for $x > 1$. We will assume that $\Lambda_1 = \Lambda_2 = \Lambda$. We consider now the path integral

$$Z[J, \Lambda] = \int \prod_{m,n,p} d\tilde{M}_{mn}(p) \, \exp\left[-S[M, \Lambda]\right.$$

$$\left. - \sqrt{\det(2\pi\tilde{\theta})} \int \frac{d^D p}{(2\pi)^D} \sum_{m,n,k,l} \tilde{M}_{mn}(p) F_{mn,kl}[p, \Lambda] J_{kl}(-p) - S'[M, \Lambda] \right]. \tag{19.128}$$

The action $S[M, \Lambda]$ is given in terms of the inverse propagator $G_{mn,k,l}(p, \Lambda)$ and a general interaction action $L[M, \Lambda]$ as follows

$$S[M, \Lambda] = \sqrt{\det(2\pi\tilde{\theta})} \int \frac{d^D p}{(2\pi)^D} \sum_{m,n,k,l} \frac{1}{2} \tilde{M}_{mn}(-p) G_{mn,kl}(p, \Lambda) \tilde{M}_{kl}(p) + L[M, \Lambda]. \tag{19.129}$$

Clearly we must have

$$S[M, \infty] = S[M], \quad L[M, \infty] = L[M]. \tag{19.130}$$

$$G_{mn,kl}(p, \infty) = G_{mn,kl}(p), \quad F_{mn,kl}[p, \infty] = \delta_{ml}\delta_{nk}. \tag{19.131}$$

The extra piece $S'[M, \Lambda]$ is absent in the commutative theory. It is given by

$$S'[M, \Lambda] = C[\Lambda] + \frac{\nu_2}{2} \int \frac{d^D p}{(2\pi)^D} \sum_{m,n,k,l} J_{mn}(p) E_{mn,kl}[p, \Lambda] J_{kl}(-p). \tag{19.132}$$

We must have

$$C[\infty] = 0, \quad E_{mn,kl}[p, \infty] = 0. \tag{19.133}$$

The Polchinski exact renormalization group equation is derived from the requirement that the path integral $Z[J, \Lambda]$ is independent of the cutoff Λ. We compute

$$-\Lambda \frac{\partial Z[J, \Lambda]}{\partial \Lambda} = \int \prod_{m,n,p} d\tilde{M}_{mn}(p) \left[\Lambda \frac{\partial L[M, \Lambda]}{\partial \Lambda} \right.$$

$$+ \nu_2 \int_p \sum_{m,n,k,l} \frac{1}{2} \tilde{M}_{mn}(-p) \Lambda \frac{\partial G_{mn,kl}(p, \Lambda)}{\partial \Lambda} \tilde{M}_{kl}(p)$$

$$+ \nu_2 \int_p \sum_{m,n,k,l} \tilde{M}_{mn}(p) \Lambda \frac{\partial F_{mn,kl}[p, \Lambda]}{\partial \Lambda} J_{kl}(-p) + \Lambda \frac{\partial C[\Lambda]}{\partial \Lambda} \tag{19.134}$$

$$+ \frac{\nu_2}{2} \int_p \sum_{m,n,k,l} J_{mn}(p) \Lambda \frac{\partial E_{mn,kl}[p, \Lambda]}{\partial \Lambda} J_{kl}(-p) \bigg] \exp\left[-S[M, \Lambda]\right.$$

$$\left. - \nu_2 \int_p \sum_{m,n} \tilde{M}_{mn}(p) F_{mn,kl}[\Lambda] J_{kl}(-p) - S'[M, \Lambda] \right].$$

By using $\partial \tilde{M}_{mn}(p)/\partial \tilde{M}_{kl}(q) = \delta_{mk}\delta_{nl}(2\pi)^D \delta^D(p - q)$ and $G_{mn,kl}(p, \Lambda) = G_{kl,mn}(-p, \Lambda)$ we find the results

$$\tilde{M}_{mn}(-p) = \frac{1}{\nu_2} \Delta_{nm,lk}(-p, \Lambda) \frac{\partial}{\partial \tilde{M}_{kl}(p)} (S[M, \Lambda] - L[M, \Lambda]). \tag{19.135}$$

$$\frac{\partial S[M, \Lambda]}{\partial \tilde{M}_{mn}(p)} = \nu_2 \sum_{k,l} G_{mn,kl}(-p, \Lambda)\tilde{M}_{kl}(-p) + \frac{\partial L[M, \Lambda]}{\partial \tilde{M}_{mn}(p)}. \tag{19.136}$$

$$\frac{\partial S_J[M, \Lambda]}{\partial \tilde{M}_{mn}(p)} = \nu_2 \sum_{k,l} F_{mn,kl}[p, \Lambda]J_{kl}(-p). \tag{19.137}$$

The propagator is as usual defined by

$$\Delta_{nm,lk}(p, \Lambda)G_{kl,st}(p, \Lambda) = G_{nm,lk}(p, \Lambda)\Delta_{kl,st}(p, \Lambda) = \delta_{nt}\delta_{ms}. \tag{19.138}$$

We have the result

$$\begin{aligned}
\mathcal{R} &= \nu_2 \int_p \sum_{m,n,k,l} \frac{1}{2}\tilde{M}_{mn}(-p)\Lambda\frac{\partial G_{mn,kl}(p, \Lambda)}{\partial \Lambda}\tilde{M}_{kl}(p) \\
&= -\frac{1}{2\nu_2} \int_p \sum_{m,n,k,l} \Lambda\frac{\partial \Delta_{nm,lk}(-p, \Lambda)}{\partial \Lambda}\left[\frac{\partial S[M, \Lambda]}{\partial \tilde{M}_{kl}(p)}\frac{\partial S[M, \Lambda]}{\partial \tilde{M}_{mn}(-p)} + \frac{\partial L[M, \Lambda]}{\partial \tilde{M}_{kl}(p)}\frac{\partial L[M, \Lambda]}{\partial \tilde{M}_{mn}(-p)} \right. \\
&\quad \left. - 2\frac{\partial S[M, \Lambda]}{\partial \tilde{M}_{kl}(p)}\frac{\partial L[M, \Lambda]}{\partial \tilde{M}_{mn}(-p)} \right]
\end{aligned} \tag{19.139}$$

where

$$\begin{aligned}
\frac{\partial S[M, \Lambda]}{\partial \tilde{M}_{kl}(p)}\frac{\partial S[M, \Lambda]}{\partial \tilde{M}_{mn}(-p)} \, e^{\cdots} = &\left[\frac{\partial^2}{\partial \tilde{M}_{kl}(p)\partial \tilde{M}_{mn}(-p)} + \frac{\partial^2 S[M, \Lambda]}{\partial \tilde{M}_{kl}(p)\partial \tilde{M}_{mn}(-p)} \right. \\
&- \frac{\partial S_J[M, \Lambda]}{\partial \tilde{M}_{kl}(p)}\frac{\partial L[M, \Lambda]}{\partial \tilde{M}_{mn}(-p)} - \frac{\partial L[M, \Lambda]}{\partial \tilde{M}_{kl}(p)}\frac{\partial S_J[M, \Lambda]}{\partial \tilde{M}_{mn}(-p)} \\
&- \frac{\partial S_J[M, \Lambda]}{\partial \tilde{M}_{kl}(p)}\nu_2 G_{mn,st}(p, \Lambda)\tilde{M}_{st}(p) - \nu_2 G_{kl,st}(-p, \Lambda)\tilde{M}_{st}(-p)\frac{\partial S_J[M, \Lambda]}{\partial \tilde{M}_{mn}(-p)} \\
&\left. - \frac{\partial S_J[M, \Lambda]}{\partial \tilde{M}_{kl}(p)}\frac{\partial S_J[M, \Lambda]}{\partial \tilde{M}_{mn}(-p)} \right] e^{\cdots}
\end{aligned} \tag{19.140}$$

and

$$\frac{\partial^2 S[M, \Lambda]}{\partial \tilde{M}_{kl}(p)\partial \tilde{M}_{mn}(-p)} = \nu_2 G_{mn,kl}(p, \Lambda) + \frac{\partial^2 L[M, \Lambda]}{\partial \tilde{M}_{kl}(p)\partial \tilde{M}_{mn}(-p)}. \tag{19.141}$$

The first term in the above equation yields a total derivative in the path integral. The third and fourth terms lead to equal contributions. We have then

$$\begin{aligned}
\mathcal{R} &= \nu_2 \int_p \sum_{m,n,k,l} \frac{1}{2}\tilde{M}_{mn}(-p)\Lambda\frac{\partial G_{mn,kl}(p, \Lambda)}{\partial \Lambda}\tilde{M}_{kl}(p) \, e^{\cdots} \\
&= -\frac{1}{2\nu_2} \int_p \sum_{m,n,k,l} \Lambda\frac{\partial \Delta_{nm,lk}(-p, \Lambda)}{\partial \Lambda}\left[\nu_2 G_{mn,kl}(p, \Lambda) + \frac{\partial^2 L[M, \Lambda]}{\partial \tilde{M}_{kl}(p)\partial \tilde{M}_{mn}(-p)} \right. \\
&\quad \left. - 2\frac{\partial S_J[M, \Lambda]}{\partial \tilde{M}_{kl}(p)}\frac{\partial L[M, \Lambda]}{\partial \tilde{M}_{mn}(-p)} + \frac{\partial L[M, \Lambda]}{\partial \tilde{M}_{kl}(p)}\frac{\partial L[M, \Lambda]}{\partial \tilde{M}_{mn}(-p)} - 2\frac{\partial S[M, \Lambda]}{\partial \tilde{M}_{kl}(p)}\frac{\partial L[M, \Lambda]}{\partial \tilde{M}_{mn}(-p)} \right] e^{\cdots} \\
&= -\frac{1}{2\nu_2} \int_p \sum_{m,n,k,l} \Lambda\frac{\partial \Delta_{nm,lk}(-p, \Lambda)}{\partial \Lambda}\left[\nu_2 G_{mn,kl}(p, \Lambda) - \frac{\partial^2 L[M, \Lambda]}{\partial \tilde{M}_{kl}(p)\partial \tilde{M}_{mn}(-p)} \right. \\
&\quad \left. + \frac{\partial L[M, \Lambda]}{\partial \tilde{M}_{kl}(p)}\frac{\partial L[M, \Lambda]}{\partial \tilde{M}_{mn}(-p)} \right] e^{\cdots} + \cdots.
\end{aligned} \tag{19.142}$$

In the above equation we have used $\Delta_{nm,lk}(-p, \Lambda) = \Delta_{lk,nm}(p, \Lambda)$ then we have neglected a term, which will lead to a total derivative in the path integral. The last two terms can be cancelled if we choose $L[M, \Lambda]$ such that

$$\Lambda\frac{\partial L[M, \Lambda]}{\partial \Lambda} = \frac{1}{2\nu_2}\int_p \sum_{m,n,k,l}\Lambda\frac{\partial \Delta_{nm,lk}(-p, \Lambda)}{\partial \Lambda}\left[\frac{\partial L[M, \Lambda]}{\partial \tilde{M}_{kl}(p)}\frac{\partial L[M, \Lambda]}{\partial \tilde{M}_{mn}(-p)} - \frac{\partial^2 L[M, \Lambda]}{\partial \tilde{M}_{kl}(p)\partial \tilde{M}_{mn}(-p)}\right]. \quad (19.143)$$

The first term can be cancelled if we choose $C[\Lambda]$ (the vacuum energy) such that

$$\Lambda\frac{\partial C[\Lambda]}{\partial \Lambda} = \frac{1}{2}\int_p \sum_{m,n,k,l}\Lambda\frac{\partial \Delta_{lk,nm}(p, \Lambda)}{\partial \Lambda}G_{mn,kl}(p, \Lambda)$$

$$= \frac{1}{2}\int_p \sum_{m,n,k,l}\left(\Lambda\frac{\partial}{\partial \Lambda}\ln K_1\left(\frac{p^2}{\Lambda^2}\right) + \sum_i\Lambda\frac{\partial}{\partial \Lambda}\ln K_2\left(\frac{\mu_1^2 i}{\Lambda^2}\right)\right)\Delta_{lk,nm}(p, \Lambda)G_{mn,kl}(p, \Lambda) \quad (19.144)$$

$$= \frac{1}{2}\int_p \sum_{m,n}\Lambda\frac{\partial}{\partial \Lambda}\ln K_1\left(\frac{p^2}{\Lambda^2}\right) + \int_p \sum_{m,n}\Lambda\frac{\partial}{\partial \Lambda}\ln K_2\left(\frac{\mu_1^2 m}{\Lambda^2}\right)K_2\left(\frac{\mu_1^2 n}{\Lambda^2}\right).$$

We integrate this equation with the initial condition $C[\infty] = 0$ replaced with $C[\Lambda_0] = 0$. We obtain

$$C[\Lambda] = \frac{1}{2}\int_p \sum_{m,n}\ln K_1\left(\frac{p^2}{\Lambda^2}\right)K_1^{-1}\left(\frac{p^2}{\Lambda_0^2}\right) + 2\int_p \sum_n\ln \prod_m K_2\left(\frac{\mu_1^2 m}{\Lambda^2}\right)K_2^{-1}\left(\frac{\mu_1^2 m}{\Lambda_0^2}\right). \quad (19.145)$$

There remains the following terms

$$\frac{\partial S[M, \Lambda]}{\partial \tilde{M}_{kl}(p)}\frac{\partial S[M, \Lambda]}{\partial \tilde{M}_{mn}(-p)}e^{\cdots} = \left[-\frac{\partial S_J[M, \Lambda]}{\partial \tilde{M}_{kl}(p)}\nu_2 G_{mn,st}(p, \Lambda)\tilde{M}_{st}(p) - \nu_2 G_{kl,st}(-p, \Lambda)\tilde{M}_{st}(-p)\right.$$

$$\left.\times \frac{\partial S_J[M, \Lambda]}{\partial \tilde{M}_{mn}(-p)} - \frac{\partial S_J[M, \Lambda]}{\partial \tilde{M}_{kl}(p)}\frac{\partial S_J[M, \Lambda]}{\partial \tilde{M}_{mn}(-p)}\right]e^{\cdots}. \quad (19.146)$$

The first two terms lead to equal contributions. They give

$$\int_p \sum_{m,n,k,l}\sum_{s,t}\Lambda\frac{\partial \Delta_{nm,lk}(-p, \Lambda)}{\partial \Lambda}G_{kl,st}(-p, \Lambda)\tilde{M}_{st}(-p)\frac{\partial S_J[M, \Lambda]}{\partial \tilde{M}_{mn}(-p)} =$$

$$\nu_2 \int_p \sum_{m,n,k,l}\tilde{M}_{mn}(p)\Delta F_{mn,kl}(p)J_{kl}(-p) \quad (19.147)$$

where

$$\Delta F_{mn,kl}(p) = \sum_{s,t,s',t'}G_{mn,s't'}(-p, \Lambda)\Lambda\frac{\partial \Delta_{t's',ts}(-p, \Lambda)}{\partial \Lambda}F_{st,kl}[p, \Lambda]. \quad (19.148)$$

To cancel this term we choose $F_{mn,kl}[p, \Lambda]$ such that

$$\Lambda\frac{\partial F_{mn,kl}[p, \Lambda]}{\partial \Lambda} = -\Delta F_{mn,kl}(p) = -\sum_{s,t,s',t'}G_{mn,s't'}(-p, \Lambda)\Lambda\frac{\partial \Delta_{t's',ts}(-p, \Lambda)}{\partial \Lambda}F_{st,kl}[p, \Lambda]. \quad (19.149)$$

The last term is quadratic in the source and it leads to the contribution

$$\frac{1}{2\nu_2}\int_p \sum_{m,n,k,l}\Lambda\frac{\partial \Delta_{nm,lk}(-p, \Lambda)}{\partial \Lambda}\frac{\partial S_J[M, \Lambda]}{\partial \tilde{M}_{kl}(p)}\frac{\partial S_J[M, \Lambda]}{\partial \tilde{M}_{mn}(-p)} = \frac{\nu_2}{2}\int_p \sum_{m,n,k,l}J_{mn}(p)\Delta E_{mn,kl}(p)J_{kl}(-p) \quad (19.150)$$

where

$$\Delta E_{mn,kl}(p) = \sum_{s,t,s',t'} F_{st,mn}[-p, \Lambda]\Lambda\frac{\partial\Delta_{ts,t's'}(-p, \Lambda)}{\partial\Lambda}F_{s't',kl}[p, \Lambda]. \tag{19.151}$$

We choose $E_{mn,kl}[p, \Lambda]$ such that

$$\Lambda\frac{\partial E_{mn,kl}[p, \Lambda]}{\partial\Lambda} = -\Delta E_{mn,kl}(p) = -\sum_{s,t,s',t'} F_{st,mn}[-p, \Lambda]\Lambda\frac{\partial\Delta_{ts,t's'}(-p, \Lambda)}{\partial\Lambda}F_{s't',kl}[p, \Lambda]. \tag{19.152}$$

We integrate equation (19.149) with the initial condition $F_{mn,kl}[p, \infty] = \delta_{ml}\delta_{nk}$ replaced with $F_{mn,kl}[p, \Lambda_0] = \delta_{ml}\delta_{nk}$. We obtain immediately the solution

$$F_{mn,kl}[p, \Lambda] = \sum_{s',t'} G_{mn,s't'}(-p, \Lambda)\Delta_{t's',kl}(-p, \Lambda_0). \tag{19.153}$$

Thus equation (19.152) becomes

$$\Lambda\frac{\partial E_{mn,kl}[p, \Lambda]}{\partial\Lambda} = \sum_{s,t,s',t'} \Delta_{mn,st}[-p, \Lambda_0]\Lambda\frac{\partial G_{ts,t's'}(-p, \Lambda)}{\partial\Lambda}\Delta_{s't',kl}[-p, \Lambda_0]. \tag{19.154}$$

The initial condition $E_{mn,kl}[p, \infty] = 0$ will be replaced with $E_{mn,kl}[p, \Lambda_0] = 0$ and then we obtain immediately the solution

$$E_{mn,kl}[p, \Lambda] = \sum_{s,t,s',t'} \Delta_{mn,st}[-p, \Lambda_0](G_{ts,t's'}(-p, \Lambda) - G_{ts,t's'}(-p, \Lambda_0))\Delta_{s't',kl}[p, \Lambda_0]. \tag{19.155}$$

There remains finally to solve the Wilson–Polchinski equation

$$\Lambda\frac{\partial L[M, \Lambda]}{\partial\Lambda} = \frac{1}{2\nu_2}\int_p \sum_{m,n,k,l} \Lambda\frac{\partial\Delta_{nm,lk}(-p, \Lambda)}{\partial\Lambda}\left[\frac{\partial L[M, \Lambda]}{\partial\tilde{M}_{kl}(p)}\frac{\partial L[M, \Lambda]}{\partial\tilde{M}_{mn}(-p)} - \frac{\partial^2 L[M, \Lambda]}{\partial\tilde{M}_{kl}(p)\partial\tilde{M}_{mn}(-p)}\right]. \tag{19.156}$$

The initial condition is

$$L[M, \Lambda_0] = L[M] = \nu_2\int_{P_1}\int_{P_2}\int_{P_3}\sum_{m,n,k,l}\frac{\lambda}{4!}\tilde{M}_{mn}(p_1)\tilde{M}_{nk}(p_2)\tilde{M}_{kl}(p_3)\tilde{M}_{lm}(-p_1 - p_2 - p_3). \tag{19.157}$$

The vertices are given by

$$L_{m_1n_1\ldots m_Nn_N}[p_1, \ldots, p_N, \Lambda] = \frac{\partial^N L[M, \Lambda]}{\partial\tilde{M}_{m_1n_1}(p_1)\ldots\partial\tilde{M}_{m_Nn_N}(p_N)}\Big|_{M=0}. \tag{19.158}$$

In the solution of the theory given by Feynman graphs the vertices are connected to each other by internal lines $\Delta_{mn,kl}(p, \Lambda)$ while they are connected to sources $J_{mn}(p)$ by external lines $\Delta_{mn,kl}(p, \Lambda_0)$. For finite Λ loop summations are finite. Renormalizability is the requirement that the Wilson–Polchinski equation admits a regular solution, which depends on a finite number of initial conditions.

19.3 Renormalization of scalar ϕ^4 in two dimensions

19.3.1 Wilson–Polchinski renormalization group equation

We start with the model

$$S[M] = \nu_2 \sum_{m,n,k,l} \frac{1}{2}\tilde{M}_{mn}G_{mn,kl}\tilde{M}_{kl} + L[M]. \tag{19.159}$$

The Laplacian is

$$G_{mn,kl} = \left(\mu^2 + \mu_1^2(m+n-1)\right)\delta_{n,k}\delta_{m,l} - \mu_1^2\sqrt{\omega(m-1)(n-1)}\ \delta_{n-1,k}\delta_{m-1,l}$$
$$- \mu_1^2\sqrt{\omega mn}\ \delta_{n+1,k}\delta_{m+1,l}. \tag{19.160}$$

And the interaction term is

$$L[M] = \nu_2\frac{\lambda}{4!}\sum_{m,n,k,l}\tilde{M}_{mn}\tilde{M}_{nk}\tilde{M}_{kl}\tilde{M}_{lm}. \tag{19.161}$$

The parameters of the model are (with $\nu_2 = \sqrt{\det(2\pi\theta)} = \pi\theta, \sqrt{\omega} = (1-\Omega^2)/(1+\Omega^2)$)

$$\nu_2\mu^2 = \pi\theta\mu^2, \quad \nu_2\mu_1^2\sqrt{\omega} = 2\pi(1-\Omega^2), \quad \nu_2\lambda = \pi\theta\lambda. \tag{19.162}$$

We introduce a momentum space cutoff Λ, which will play the role of the running cutoff. The dimensionless noncommutativity is $\bar{\theta} = \theta\Lambda^2$. The regularized inverse propagator is

$$G_{mn,kl}(\Lambda) = G_{mn,kl}\prod_{i\in\{m,n,k,l\}}K^{-1}\left(\frac{\mu_1^2 i}{\Lambda^2}\right). \tag{19.163}$$

The ultraviolet cutoff function $K(x)$ is such that it is equal to 1 for $0 \leqslant x \leqslant 1$ and it vanishes rapidly for $x > 1$. The propagator is as usual defined by

$$\Delta_{nm,lk}(\Lambda)G_{kl,st}(\Lambda) = G_{nm,lk}(\Lambda)\Delta_{kl,st}(\Lambda) = \delta_{nt}\delta_{ms}. \tag{19.164}$$

We consider now the path integral

$$Z[J,\Lambda] = \int\prod_{m,n}d\tilde{M}_{mn}\ e^{-S[M,\Lambda]-\nu_2\sum_{m,n,k,l}\tilde{M}_{mn}F_{mn,kl}[\Lambda]J_{kl} - S'[M,\Lambda]}. \tag{19.165}$$

The action $S[M,\Lambda]$ is given in terms of the inverse propagator $G_{mn,k,l}(p,\Lambda)$ and a general interaction action $L[M,\Lambda]$ as follows

$$S[M,\Lambda] = \nu_2\sum_{m,n,k,l}\frac{1}{2}\tilde{M}_{mn}G_{mn,kl}(\Lambda)\tilde{M}_{kl} + L[M,\Lambda]. \tag{19.166}$$

Clearly we must have

$$S[M,\Lambda_0] = S[M], \quad L[M,\Lambda_0] = L[M]. \tag{19.167}$$

$$G_{mn,kl}(\Lambda_0) = G_{mn,kl}, \quad F_{mn,kl}[\Lambda_0] = \delta_{ml}\delta_{nk}. \tag{19.168}$$

In the above equation Λ_0 is the ultraviolet cutoff or initial cutoff, which must be sent to ∞ at the end. The extra piece $S'[M, \Lambda]$ is absent in the commutative theory. It is given by

$$S'[M, \Lambda] = C[\Lambda] + \frac{\nu_2}{2} \sum_{m,n,k,l} J_{mn} E_{mn,kl}[\Lambda] J_{kl}. \tag{19.169}$$

We must have

$$C[\Lambda_0] = 0, \quad E_{mn,kl}[\Lambda_0] = 0. \tag{19.170}$$

The Polchinski exact renormalization group equation is derived from the requirement that the path integral $Z[J, \Lambda]$ is independent of the cutoff Λ, viz

$$-\Lambda\frac{\partial Z[J, \Lambda]}{\partial \Lambda} = 0. \tag{19.171}$$

We must then choose $L[M, \Lambda]$ such that

$$\Lambda\frac{\partial L[M, \Lambda]}{\partial \Lambda} = \frac{1}{2\nu_2} \sum_{m,n,k,l} \Lambda\frac{\partial \Delta_{nm,lk}(\Lambda)}{\partial \Lambda}\left[\frac{\partial L[M, \Lambda]}{\partial \tilde{M}_{kl}} \frac{\partial L[M, \Lambda]}{\partial \tilde{M}_{mn}} - \frac{\partial^2 L[M, \Lambda]}{\partial \tilde{M}_{kl}\partial \tilde{M}_{mn}} \right]. \tag{19.172}$$

We must choose the vacuum energy such that

$$C[\Lambda] = 2\sum_{n} \ln \prod_{m} K_2\left(\frac{\mu_1^2 m}{\Lambda^2}\right) K_2^{-1}\left(\frac{\mu_1^2 m}{\Lambda_0^2}\right). \tag{19.173}$$

We must choose the functions F and E such that

$$F_{mn,kl}[\Lambda] = \sum_{s',t'} G_{mn,s't'}(\Lambda)\Delta_{t's',kl}(\Lambda_0). \tag{19.174}$$

$$E_{mn,kl}[\Lambda] = \sum_{s,t,s',t'} \Delta_{mn,st}(\Lambda_0)(G_{ts,t's'}(\Lambda) - G_{ts,t's'}(\Lambda_0))\Delta_{s't',kl}(\Lambda_0). \tag{19.175}$$

There remains to solve the Wilson–Polchinski equation (19.172).

19.3.2 Free propagator

We are dealing with the Laplacian

$$G_{mn,kl} = \left(\mu^2 + \mu_1^2(m + n - 1)\right)\delta_{n,k}\delta_{m,l} - \mu_1^2\sqrt{\omega(m - 1)(n - 1)}\ \delta_{n-1,k}\delta_{m-1,l}$$
$$- \mu_1^2\sqrt{\omega mn}\ \delta_{n+1,k}\delta_{m+1,l}. \tag{19.176}$$

The propagator is defined by

$$\sum_{l=1}^{\infty}\sum_{k=1}^{\infty}\Delta_{nm,lk} G_{kl,rs} = \sum_{l=1}^{\infty}\sum_{k=1}^{\infty}G_{mn,kl}\Delta_{lk,sr} = \delta_{ns}\delta_{mr}. \tag{19.177}$$

We consider the massless case $\mu = 0$. The indices m, n, k, l of each term contributing to the Laplacian $G_{mn,kl}$ are such that $m + k = n + l$. Thus the indices m, n, k, l of each term contributing to the propagator $\Delta_{lk,nm}$ are such that $m + k = n + l$. By using $m + k = n + l$ we get with $n = m + \alpha$ and $s = r + \beta$ the identity

$$\sum_{l=1}^{\infty} G_{mm+\alpha,l+\alpha l}\,\Delta_{ll+\alpha,r+\beta r} = \delta_{\alpha\beta}\delta_{mr}. \tag{19.178}$$

For $\alpha \neq \beta$ we get $\sum_l G_{mm+\alpha,l+\alpha l}\,\Delta_{ll+\alpha,r+\beta r} = 0$ from which we conclude that

$$\Delta_{ll+\alpha,r+\beta r} = 0, \quad \alpha \neq \beta. \tag{19.179}$$

For $\alpha = \beta$ we get the independent equations

$$\sum_{l=1}^{\infty} G_{mm+\alpha,l+\alpha l}\,\Delta_{ll+\alpha,r+\alpha r} = \delta_{mr}. \tag{19.180}$$

By defining $G_{mm+\alpha,l+\alpha l} = G^{(\alpha)}_{m,\,l}$, $\Delta_{ll+\alpha,r+\alpha r} = \Delta^{(\alpha)}_{l,\,r}$ we get for every α the ordinary matrix equation

$$\sum_{l=1}^{\infty} G^{(\alpha)}_{m,\,l}\Delta^{(\alpha)}_{l,\,r} = \delta_{mr}. \tag{19.181}$$

We introduce a cutoff $\mathcal{N}$. We have then to solve

$$\sum_{l=1}^{\mathcal{N}} G^{(\alpha)}_{m,\,l}\Delta^{(\alpha)}_{l,\,r} = \delta_{mr}. \tag{19.182}$$

As an example let us choose $\mathcal{N} = 4$. We compute

$$G^{(\alpha)}_{m,\,l} - \mu_1^2 v\delta_{m,l} = \mu_1^2(2m + \alpha - 1 - v)\delta_{m,l} - \mu_1^2\sqrt{\omega l(l + \alpha)}\;\delta_{m-1,l} - \mu_1^2\sqrt{\omega m(m + \alpha)}\;\delta_{m+1,l}$$

$$= \mu_1^2\begin{pmatrix} \alpha + 1 - v & -\sqrt{\omega(1 + \alpha)} & 0 & 0 \\ -\sqrt{\omega(1 + \alpha)} & \alpha + 3 - v & -\sqrt{2\omega(2 + \alpha)} & 0 \\ 0 & -\sqrt{2\omega(2 + \alpha)} & \alpha + 5 - v & -\sqrt{3\omega(3 + \alpha)} \\ 0 & 0 & -\sqrt{3\omega(3 + \alpha)} & \alpha + 7 - v \end{pmatrix}. \tag{19.183}$$

This can be rewritten as

$$G^{(\alpha)}_{m,\,l} - \mu_1^2 v\delta_{m,l} = \mu_1^2 TT^{+}. \tag{19.184}$$

$$T = \begin{pmatrix} \sqrt{\alpha + 1}\sqrt{A_1^{\alpha,\omega}(v)} & 0 & 0 & 0 \\ -\sqrt{\dfrac{\omega}{A_1^{\alpha,\omega}(v)}} & \sqrt{\alpha + 2}\sqrt{A_2^{\alpha,\omega}(v)} & 0 & 0 \\ 0 & -\sqrt{\dfrac{2\omega}{A_2^{\alpha,\omega}(v)}} & \sqrt{\alpha + 3}\sqrt{A_3^{\alpha,\omega}(v)} & 0 \\ 0 & 0 & -\sqrt{\dfrac{3\omega}{A_3^{\alpha,\omega}(v)}} & \sqrt{\alpha + 4}\sqrt{A_4^{\alpha,\omega}(v)} \end{pmatrix}. \tag{19.185}$$

$$A_n^{\alpha,\omega}(v) = \frac{1}{\alpha + n}(\alpha + 2n - 1 - v - \frac{(n-1)\omega}{A_{n-1}^{\alpha,\omega}(v)}), \quad n \geqslant 1. \tag{19.186}$$

This last equation can be put in the form

$$nL_n^{\alpha,\omega}(v) - (\alpha + 2n - 1 - v)L_{n-1}^{\alpha,\omega}(v) + \omega(\alpha + n - 1)L_{n-2}^{\alpha,\omega}(v) = 0, \tag{19.187}$$

where

$$A_n^{\alpha,\omega}(v) = \frac{n}{\alpha + n}\frac{L_n^{\alpha,\omega}(v)}{L_{n-1}^{\alpha,\omega}(v)}, \quad L_0^{\alpha,\omega} \equiv 1. \tag{19.188}$$

The functions $L_n^{\alpha,\omega}(v)$ for $\omega = 1$ are the Laguerre polynomials. Thus the functions $L_n^{\alpha,\omega}(v)$ are the deformed Laguerre polynomials.

The eigenvalues of the Laplacian are determined by the condition $\det(G_{m,l}^{(\alpha)} - \mu_1^2 v \delta_{m,l}) = 0$, which is equivalent to the condition $\det \ T = 0$. This leads to the condition $A_1 A_2 A_3 A_4 = 0$ or equivalently $L_4^{\alpha,\omega}(v) = 0$. In general we will get the condition $L_{\mathcal{N}}^{\alpha,\omega}(v) = 0$. In other words, the eigenvalues are the zeroes of the deformed Laguerre polynomial $L_{\mathcal{N}}^{\alpha,\omega}(v)$, viz

$$L_{\mathcal{N}}^{\alpha,\omega}(v_i^{(\alpha,\omega)}) = 0, \quad i = 1, \ldots, \mathcal{N}. \tag{19.189}$$

Let $U_i^{(\alpha,\omega)}$ be the eigenvector of $G_{m,l}^{(\alpha)}$ with eigenvalue $\mu_1^2 v_i^{(\alpha,\omega)}$. We can write

$$G_{m,l}^{(\alpha)} = \sum_{i=1}^{\mathcal{N}} \mu_1^2 v_i^{(\alpha,\omega)}(U_i^{(\alpha,\omega)})_{m+1}(U_i^{(\alpha,\omega)+})_{l+1}. \tag{19.190}$$

We must have the completeness relation

$$\delta_{m,l} = \sum_{i=1}^{\mathcal{N}}(U_i^{(\alpha,\omega)})_m(U_i^{(\alpha,\omega)+})_l. \tag{19.191}$$

We must also have the orthogonality relation

$$\delta_{i,j} = \sum_{l=1}^{\mathcal{N}}(U_j^{(\alpha,\omega)+})_l(U_i^{(\alpha,\omega)})_l. \tag{19.192}$$

The propagator is therefore given by

$$\Delta_{m,l}^{(\alpha)} = \sum_{i=1}^{\mathcal{N}} \frac{1}{\mu_1^2 v_i^{(\alpha,\omega)}}(U_i^{(\alpha,\omega)})_{m+1}(U_i^{(\alpha,\omega)+})_{l+1}. \tag{19.193}$$

By inserting the mass we obtain

$$\Delta_{m,l}^{(\alpha)} = \sum_{i=1}^{\mathcal{N}} \frac{1}{\mu^2 + \mu_1^2 v_i^{(\alpha,\omega)}}(U_i^{(\alpha,\omega)})_{m+1}(U_i^{(\alpha,\omega)+})_{l+1}. \tag{19.194}$$

It remains to find the eigenvectors $(U_i^{(\alpha,\omega)})_j$. Let us do this for $\mathcal{N}=4$. We find by using $A_4^{\alpha,\omega}(v_i^{(\alpha,\omega)})=0$ the result

$$(U_i^{(\alpha,\omega)})_3 = \frac{\alpha+7-v_i^{(\alpha,\omega)}}{\sqrt{3\omega(3+\alpha)}}(U_i^{(\alpha,\omega)})_4 = \sqrt{\frac{\omega(\alpha+3)}{3}}\frac{L_2^{\alpha,\omega}(v_i^{(\alpha,\omega)})}{L_3^{\alpha,\omega}(v_i^{(\alpha,\omega)})}(U_i^{(\alpha,\omega)})_4. \quad (19.195)$$

We also find

$$(U_i^{(\alpha,\omega)})_2 = \frac{(\alpha+7-v_i^{(\alpha,\omega)})(\alpha+5-v_i^{(\alpha,\omega)})-3\omega(3+\alpha)}{\sqrt{6\omega^2(2+\alpha)(3+\alpha)}}(U_i^{(\alpha,\omega)})_4$$
$$= \sqrt{\frac{\omega^2(\alpha+2)(\alpha+3)}{6}}\frac{L_1^{\alpha,\omega}(v_i^{(\alpha,\omega)})}{L_3^{\alpha,\omega}(v_i^{(\alpha,\omega)})}(U_i^{(\alpha,\omega)})_4. \quad (19.196)$$

$$(U_i^{(\alpha,\omega)})_1 = \frac{\sqrt{\omega(1+\alpha)}}{\alpha+1-v_i^{(\alpha,\omega)}}(U_i^{(\alpha,\omega)})_2$$
$$= \sqrt{\frac{\omega^3(\alpha+1)(\alpha+2)(\alpha+3)}{6}}\frac{1}{L_3^{\alpha,\omega}(v_i^{(\alpha,\omega)})}(U_i^{(\alpha,\omega)})_4. \quad (19.197)$$

Generalization of this result is given by

$$(U_i^{(\alpha,\omega)})_j = \sqrt{\frac{\omega^{\mathcal{N}}\Gamma(\alpha+\mathcal{N})\Gamma(j)}{\omega^j\Gamma(\alpha+j)\Gamma(\mathcal{N})}}\frac{L_{j-1}^{\alpha,\omega}(v_i^{(\alpha,\omega)})}{L_{\mathcal{N}-1}^{\alpha,\omega}(v_i^{(\alpha,\omega)})}(U_i^{(\alpha,\omega)})_{\mathcal{N}}. \quad (19.198)$$

Normalizing this vector to 1 we get

$$(U_i^{(\alpha,\omega)})_j = \sqrt{\frac{\dfrac{\Gamma(j)}{\omega^j\Gamma(\alpha+j)}\left(L_{j-1}^{\alpha,\omega}(v_i^{(\alpha,\omega)})\right)^2}{\displaystyle\sum_{h=1}^{\mathcal{N}}\frac{\Gamma(h)}{\omega^h\Gamma(\alpha+h)}\left(L_{h-1}^{\alpha,\omega}(v_i^{(\alpha,\omega)})\right)^2}}. \quad (19.199)$$

Clearly if $(U_i^{(\alpha,\omega)})_j$ are the components of the eigenvector associated with the eigenvalue i then $(U_j^{(\alpha,\omega)})_i$ are the components of the eigenvector associated with the eigenvalue j.

We must also have

$$(U_i^{(\alpha,\omega)+})_j = (U_i^{(\alpha,\omega)})_j. \quad (19.200)$$

19.3.3 Scaling exponents

The regularized propagator is

$$\Delta_{nm,lk}(\Lambda) = \Delta_{nm,lk}\prod_{i\in\{m,n,k,l\}}K_2\left(\frac{\mu_1^2 i}{\Lambda^2}\right). \quad (19.201)$$

We compute

$$Q_{nm,lk}(\Lambda) = \frac{1}{\nu_2}\Lambda\frac{\partial\Delta_{nm,lk}(\Lambda)}{\partial\Lambda}$$

$$= -\sum_{j\in\{m,n,k,l\}}\frac{2\mu_1^2 j}{\nu_2\Lambda^2}K_2'\left(\frac{\mu_1^2 j}{\Lambda^2}\right)\prod_{i\in\{m,n,k,l\}/\{j\}}K_2\left(\frac{\mu_1^2 i}{\Lambda^2}\right)\Delta_{nm,lk}.$$

(19.202)

The cutoff function in two dimensions is defined by

$$K(n,\,m;\,\Lambda) = K_2\left(\frac{\mu_1^2 n}{\Lambda^2}\right)K_2\left(\frac{\mu_1^2 m}{\Lambda^2}\right).$$

(19.203)

We will also use the notation $K_2(\frac{\mu_1^2 m}{\Lambda^2}) = K_2(m)$. We choose $K_2(x)$ to be a smooth function on $\mathbf{R}^+$, i.e. $K_2 \in \mathbf{R}^+$ such that

$$K_2(x) = 1,\quad 0 \leqslant x \leqslant 1;\quad K_2(x) = 0,\quad x \geqslant 2.$$

(19.204)

In other words, for finite Λ there is only a finite number of indices m, n, k, l for which $Q_{nm,lk} \neq 0$. Clearly the support of $K_2'(x)$ is $[1, 2]$. In other words the indices m, n, k, l are such that their maximum is in the range given by $1 \leqslant \frac{\mu_1^2 i}{\Lambda^2} \leqslant 2$. Hence $Q_{nm,lk}$ is non-zero only if

$$\frac{\Lambda^2}{\mu_1^2} \leqslant \max(m,\,n,\,k,\,l) \leqslant 2\frac{\Lambda^2}{\mu_1^2}.$$

(19.205)

The cutoff function $K_2(x)$ is also chosen such that the volume of the support of $Q_{nm,lk}$ scales as Λ^d, viz

$$\text{SCA} = \sum_m \text{sign}(\max_{n,l}|K(m,\,n;\,\Lambda)K(l+n-m,\,l;\,\Lambda)|) \leqslant C_d\left(\frac{\Lambda}{\mu_1}\right)^d.$$

(19.206)

Recall that the indices m, n, k, l of each term contributing to $Q_{nm,lk}$ are such that $n - m = k - l$. Note also that

$$Q_{nm,lk}(\Lambda) = \frac{1}{\nu_2}\Lambda\frac{\partial}{\partial\Lambda}\left(\ln\prod_{j\in\{m,n,k,l\}}K_2\left(\frac{\mu_1^2 j}{\Lambda^2}\right)\right)\Delta_{nm,lk}(\Lambda).$$

(19.207)

The support of $Q_{nm,lk}$ can be estimated by

$$\sum_{m,n,k,l}\text{sign}|Q_{nm,lk}(\Lambda)| \leqslant \text{const} \times \sum_m\text{sign}(\max_{n,l,k}|Q_{nm,lk}(\Lambda)|).$$

(19.208)

We compute

$$\sum_m \text{sign}(\max_{n,l,k}|Q_{nm,lk}(\Lambda)|) = \sum_m \text{sign}(\max_{n,l,k}|\Delta_{nm,lk}(\Lambda)|)$$
$$= \sum_m \text{sign}(\max_{n,l}|K(m, n; \Lambda)K(l + n - m, l; \Lambda)|) \qquad (19.209)$$
$$= \text{SCA}.$$

It is trivial to see that

$$\text{SCA} = \sum_m \text{sign}(\max_{n,l}|K_2(m)K_2(n)K_2(l)K_2(l + n - m)|)$$
$$= \sum_m \text{sign}(|K_2(m)K_2(n_0)K_2(l_0)K_2(n_0 + l_0 - m)|)$$
$$= \sum_{m=1}^{2\frac{\Lambda^2}{\mu_1^2}} 1 \qquad (19.210)$$
$$= 2\frac{\Lambda^2}{\mu_1^2}.$$

In other words, $d = 2$ as desired.

We define two exponents δ_0 and δ_1 by

$$\text{SCA}_0 = \max_{m,n,k,l}|Q_{nm;lk}(\Lambda)| \leqslant C_0\left(\frac{\mu_1}{\Lambda}\right)^{\delta_0} \delta_{n-m,k-l}. \qquad (19.211)$$

$$\text{SCA}_1 = \max_n\left(\sum_k \max_{m,l}|Q_{nm;lk}(\Lambda)| \leqslant C_1\left(\frac{\mu_1}{\Lambda}\right)^{\delta_1}\right). \qquad (19.212)$$

We compute δ_0 and δ_1. We have

$$|Q_{nm,lk}(\Lambda)| \leqslant \sum_{j\in\{m,n,k,l\}}\left|\frac{2\mu_1^2 j}{\nu_2\Lambda^2}K_2'\left(\frac{\mu_1^2 j}{\Lambda^2}\right)\prod_{i\in\{m,n,k,l\}/\{j\}} K_2\left(\frac{\mu_1^2 i}{\Lambda^2}\right)\Delta_{nm,lk}\right|$$
$$\leqslant \frac{\mu_1^2(1 + \sqrt{\omega})}{\pi}|\Delta_{nm,\,lk}^{\mathcal{C}}|\sum_{j\in\{m,n,k,l\}}\left|K_2'\left(\frac{\mu_1^2 j}{\Lambda^2}\right)\prod_{i\in\{m,n,k,l\}/\{j\}} K_2\left(\frac{\mu_1^2 i}{\Lambda^2}\right)\right|. \qquad (19.213)$$

In the above equation we have used the facts that $j \leqslant 2\frac{\Lambda^2}{\mu_1^2}$ and that $\frac{\Lambda^2}{\mu_1^2} \leqslant \max(m, n, k, l) \leqslant 2\frac{\Lambda^2}{\mu_1^2}$ and hence the propagator $\Delta_{nm,\,lk}^{\mathcal{C}}$ is given by

$$\Delta_{nm,\,lk}^{\mathcal{C}} = \Delta_{nm,lk}, \quad \text{if } \frac{\Lambda^2}{\mu_1^2} \leqslant \max(m, n, k, l) \leqslant 2\frac{\Lambda^2}{\mu_1^2}, \quad \text{otherwise } \Delta_{nm,\,lk}^{\mathcal{C}} = 0. \qquad (19.214)$$

Thus we get

$$|Q_{nm,lk}(\Lambda)| \leqslant \frac{4\mu_1^2(1 + \sqrt{\omega})}{\pi}|\Delta_{nm,\,lk}^{\mathcal{C}}|\max_x|K_2'(x)|(\max_x|K_2(x)|)^3. \qquad (19.215)$$

Assuming that $\max_x|K_2(x)| = 1$ we get

$$|Q_{nm,lk}(\Lambda)| \leqslant \frac{4\mu_1^2(1 + \sqrt{\omega})}{\pi}|\Delta^{\mathcal{C}}_{nm,\,lk}|\max_x|K_2'(x)|. \tag{19.216}$$

Thus

$$\max_{nm,lk}|Q_{nm,lk}(\Lambda)| \leqslant \frac{4\mu_1^2(1 + \sqrt{\omega})}{\pi}\max_{nm,lk}|\Delta^{\mathcal{C}}_{nm,\,lk}|\max_x|K_2'(x)|. \tag{19.217}$$

$$\max_n\left(\sum_k \max_{m,l}|Q_{nm;lk}(\Lambda)|\right) \leqslant \frac{4\mu_1^2(1 + \sqrt{\omega})}{\pi}\max_n\left(\sum_k \max_{m,l}|\Delta^{\mathcal{C}}_{nm,\,lk}|\right)\max_x|K_2'(x)|. \tag{19.218}$$

It remains to compute $\max_{nm,lk}|\Delta^{\mathcal{C}}_{nm,\,lk}|$ and $\max_n(\sum_k \max_{m,l}|\Delta^{\mathcal{C}}_{nm,\,lk}|)$. The starting point is the propagator

$$\Delta_{mm+\alpha,l+\alpha l} = \Delta^{(\alpha)}_{m,\,l} = \sum_{i=1}^{N}\frac{1}{\mu^2 + \mu_1^2 v_i^{(\alpha,\omega)}}\frac{\sqrt{\dfrac{m!\,l!}{\omega^{m+l}(\alpha+m)!(\alpha+l)!}}L_m^{\alpha,\omega}(v_i^{(\alpha,\omega)})L_l^{\alpha,\omega}(v_i^{(\alpha,\omega)})}{\displaystyle\sum_{h=0}^{N-1}\frac{h!}{\omega^h(\alpha+h)!}(L_h^{\alpha,\omega}(v_i^{(\alpha,\omega)}))^2}. \tag{19.219}$$

Let us just note that for $\omega = 1$ the eigenvalues $v_i^{(\alpha,\omega)}$ become the eigenvalues of the ordinary Laguerre polynomial $\mathcal{L}_N^\alpha(v)$. In other words, the eigenvalues $v_i^{(\alpha,\omega)}$ become continous variables v and the sum in the above equation becomes an integral.

We use the numerical estimations of Grosse and Wulkenhaar given by (with $\mathcal{C} = \Lambda^2/\mu_1^2$)

$$\begin{aligned}\max_{nm,lk}|\Delta^{\mathcal{C}}_{nm,\,lk}| &= \sqrt{\frac{3 - 2\omega}{\mu^4 + 4\mu^2\mu_1^2\mathcal{C} + 4\mu_1^2(1 - \omega)\mathcal{C}^2}}\,\delta_{m+k,n+l}\\[4pt] &= \sqrt{\frac{3 - 2\omega}{\mu^4 + 4\mu^2\Lambda^2 + 4(1 - \omega)\Lambda^4}}\,\delta_{m+k,n+l}.\end{aligned} \tag{19.220}$$

This was obtained numerically by (1) fixing $\mathcal{N}$ to a very large number (2) finding $\max_{nm,lk}|\Delta^{\mathcal{C}}_{nm,\,lk}|$ for indices m, n, k, l such that $\frac{\Lambda^2}{\mu_1^2} \leqslant \max(m, n, k, l) \leqslant 2\frac{\Lambda^2}{\mu_1^2}$ and for fixed parameters μ, μ_1 and ω using the exact formula (19.219), (3) varying Λ, (4) finding the best fit for the result and then (5) verify how the fit changes for other values of the parameters μ, μ_1 and ω.

We get then (with $m + k = n + l$, $0 \leqslant \omega \leqslant 1$ and Λ very large)

$$\begin{aligned}\max_{nm,lk}|Q_{nm,lk}(\Lambda)| &\leqslant \frac{4\mu_1^2(1 + \sqrt{\omega})}{\pi}\max_x|K_2'(x)|\sqrt{\frac{3 - 2\omega}{\mu^4 + 4\mu^2\Lambda^2 + 4(1 - \omega)\Lambda^4}}\\[6pt] &\leqslant \frac{\mu_1^2}{\Lambda^2\sqrt{1 - \omega}}\frac{4\sqrt{3}}{\pi}\max_x|K_2'(x)|,\quad \omega \neq 1 \tag{19.221}\\[6pt] &\leqslant \frac{\mu_1^2}{\mu\Lambda}\frac{4}{\pi}\max_x|K_2'(x)|,\quad \omega = 1.\end{aligned}$$

In other words, we have (with $C_0 = \frac{4}{\pi} \max_x |K_2'(x)|$)

$$\max_{nm,lk} |Q_{nm,lk}(\Lambda)| \leqslant C_0 \frac{\mu_1^2}{\Lambda^2 \sqrt{1 - \omega}}, \quad \omega \neq 1$$
$$\leqslant C_0 \frac{\mu_1^2}{\mu \Lambda}, \quad \omega = 1. \tag{19.222}$$

We use also their other numerical estimation

$$\max_n \left(\sum_k \max_{m,l} |\Delta_{nm,\,lk}^C| \right) = \frac{1}{\mu^2 + \mu_1^2 (1 - \omega)C}$$
$$= \frac{1}{\mu^2 + (1 - \omega)\Lambda^2}. \tag{19.223}$$

Thus (with $C_1 = \frac{8}{\pi} \max_x |K_2'(x)|$)

$$\max_n \left(\sum_k \max_{m,l} |Q_{nm;lk}(\Lambda)| \right) \leqslant \frac{4\mu_1^2 (1 + \sqrt{\omega})}{\pi} \frac{1}{\mu^2 + (1 - \omega)\Lambda^2} \max_x |K_2'(x)|$$
$$\leqslant C_1 \frac{\mu_1^2}{(1 - \omega)\Lambda^2}, \quad \omega \neq 1 \tag{19.224}$$
$$\leqslant C_1 \frac{\mu_1^2}{\mu^2}, \quad \omega = 1.$$

19.3.4 Projection to the irrelevant part

The vertices are given by

$$L_{m_1 n_1 \,\ldots\, m_N n_N}[\Lambda] = \frac{\partial^N L[M, \Lambda]}{\partial \tilde{M}_{m_1 n_1} \ldots \partial \tilde{M}_{m_N n_N}} \Big|_{M=0}. \tag{19.225}$$

In the solution of the theory given by Feynman or ribbon graphs the vertices are connected to each other by internal lines $\Delta_{mn,kl}(\Lambda)$ while they are connected to sources J_{mn} by external lines $\Delta_{mn,kl}(\Lambda_0)$. It is clear that, for finite Λ, loop summations are finite. Renormalizability is the requirement that the Polchinski equation admits a regular solution, which depends on a finite number of initial conditions.

We will deal with two scales. The initial scale Λ_0 and the renormalization scale Λ_R. In other words, $\Lambda_R \leqslant \Lambda \leqslant \Lambda_0$. So we can integrate from Λ_0 down to Λ or from Λ_R up to Λ.

The choice of initial conditions is at the same time determined by and required to prove the power-counting theorem. Thus, we have to make the correct ansatz for the initial interactions, which we then verify as the correct set of relevant and marginal interactions using the power-counting theorem.

The Grosse and Wulkenhaar ansatz for the initial interactions, i.e., the set of relevant and marginal interactions, in $d = 2$ dimensions is given by

$$L[M, \Lambda_0, \Lambda_0, \omega, \rho^0] = \nu_2 \sum_{m,n,k,l} \frac{1}{2} \rho^0_{[m]} \tilde{M}_{mn} \tilde{M}_{nm} + \nu_2 \frac{\lambda}{4!} \sum_{m,n,k,l} \tilde{M}_{mn} \tilde{M}_{nk} \tilde{M}_{kl} \tilde{M}_{lm}. \tag{19.226}$$

This choice of the initial interactions is dictated by the fact that only the planar two-point function at one-loop is divergent. Thus the interaction at $\Lambda = \Lambda_0$, i.e., the boundary condition must include a quadratic diagonal term. The coefficients $\rho^0_{[m]}$ are meant to renormalize the parameters μ^2 and μ_1^2, i.e.,

$$\rho^0_{[m]} = \rho^0_1 + 2\,m\rho^0_2. \tag{19.227}$$

The mass μ^2 should therefore be thought of as the renormalized mass and ρ^0_1 is the bare mass. Similarly ρ^0_2 should be thought of as the bare value of the coupling μ_1^2.

The off-diagonal quadratic term, which is proportional to ω, does not correspond to any divergences in two dimensions so it does not need to appear in the initial interaction. It is irrelevant.

The coupling constants $\rho^0_{[m]}$ and ω must depend on Λ_0 in such a way that the limit $\Lambda_0 \longrightarrow \infty$ exists. Let us say that the difference between the two cases is that ω appears only in the propagator whereas $\rho^0_{[m]}$ appears in the interaction.

The marginal part of the 4-point function is independent of the scale Λ_0, i.e., $\rho^0_4 = \lambda$. In other terms, the 4-point function is finite and thus it does not require a renormalization and hence λ is super-renormalizable.

The goal is to integrate the Polchinski equation (19.172) with the initial condition (19.226). The solution will depend on the current cutoff Λ, on the initial cutoff Λ_0 and on the parameters ω and $\rho^0_{[m]}$. As we have said $\omega = \omega(\Lambda_0)$ and $\rho^0_{[m]} = \rho^0_{[m]}(\Lambda_0)$. The solution will also depend on the renormalized couplings μ and λ but these are fixed numbers. The solution can always be put in the form

$$L[M, \Lambda, \Lambda_0, \omega, \rho^0] = \nu_2 \sum_{m,n,k,l} \frac{1}{2} \rho_{[m]}[\Lambda, \Lambda_0, \omega, \rho^0] \tilde{M}_{mn} \tilde{M}_{nm} + \nu_2 \frac{\lambda}{4!} \sum_{m,n,k,l} \tilde{M}_{mn} \tilde{M}_{nk} \tilde{M}_{kl} \tilde{M}_{lm} \tag{19.228}$$
$$+ \text{ other } M \text{ structures.}$$

We have $\rho_{[m]}[\Lambda_0, \Lambda_0, \omega, \rho^0] = \rho^0_{[m]}$.

The renormalization conditions are conditions that must be satisfied by the ρ-coefficients at the renormalization cutoff Λ_R, viz $\rho_{[m]}[\Lambda_R, \Lambda_0, \omega, \rho^0] = \text{constant}$. We will choose in particular

$$\rho_{[m]}[\Lambda_R, \Lambda_0, \omega, \rho^0] = 0. \tag{19.229}$$

Thus, in particular $\rho_{[0]}[\Lambda_R, \Lambda_0, \omega, \rho^0] = 0$ and hence μ^2 is the renormalized mass as pointed out above.

Although the initial interaction at $\Lambda = \Lambda_0$, which is given by (19.226), is very simple, the interaction $L[M, \Lambda, \Lambda_0, \omega, \rho^0]$ at lower scales $\Lambda < \Lambda_0$ can become very complicated. At the renormalization scale Λ_R, which is far below Λ_0, we expect that the interaction $L[M, \Lambda_R, \Lambda_0, \omega, \rho^0]$ becomes simple again. So no matter what the initial interaction $L[M, \Lambda_0, \Lambda_0, \omega, \rho^0]$ is the interaction $L[M, \Lambda_R, \Lambda_0, \omega, \rho^0]$ will be attracted to a p-dimensional submanifold in the infinite dimensional space of interactions.

The p renormalizable operators in two dimensions are given by $\rho_{[m]}[\Lambda, \Lambda_0, \omega, \rho^0]$. We should also include λ and ω to the set of renormalizable operators. A good system of coordinates of the space of interactions is given by the vertices $L_{m_1 n_1 \ldots m_N n_N}[\Lambda]$. We need to determine relevant and irrelevant parts of the action.

The basic starting objects are the vectors in the space of actions given by

$$\Lambda_0 \frac{\partial L[M, \Lambda, \Lambda_0, \omega, \rho^0]}{\partial \Lambda_0}. \tag{19.230}$$

$$\Lambda_0 \frac{\partial L[M, \Lambda, \Lambda_0, \omega, \rho^0]}{\partial \rho^0_{[n]}}. \tag{19.231}$$

$$\Lambda_0 \frac{\partial L[M, \Lambda, \Lambda_0, \omega, \rho^0]}{\partial \omega}. \tag{19.232}$$

We will fix Λ at the renormalization scale Λ_R, i.e., $\Lambda = \Lambda_R$ and vary the initial scale Λ_0. We need first to compute the following difference

$$L[M, \Lambda_R, \Lambda'_0, \omega' \rho^{0'}] - L[M, \Lambda_R, \Lambda''_0, \omega'' \rho^{0''}]. \tag{19.233}$$

We have the following

$$
\begin{aligned}
L[M, \Lambda_R, \Lambda'_0, \omega' \rho^{0'}] - L[M, \Lambda_R, \Lambda''_0, \omega'' \rho^{0''}] &= \int_{\Lambda''_0}^{\Lambda'_0} \frac{d\Lambda_0}{\Lambda_0} \left(\Lambda_0 \frac{dL[M, \Lambda_R, \Lambda_0, \omega, \rho^0]}{d\Lambda_0} \right) \\
&= \int_{\Lambda''_0}^{\Lambda'_0} \frac{d\Lambda_0}{\Lambda_0} \left(\Lambda_0 \frac{\partial L[M, \Lambda_R, \Lambda_0, \omega, \rho^0]}{\partial \Lambda_0} \right. \\
&\quad + \Lambda_0 \frac{d\omega}{d\Lambda_0} \frac{\partial L[M, \Lambda_R, \Lambda_0, \omega, \rho^0]}{\partial \omega} \\
&\quad \left. + \Lambda_0 \frac{d\rho^0_{[n]}}{d\Lambda_0} \frac{\partial L[M, \Lambda_R, \Lambda_0, \omega, \rho^0]}{\partial \rho^0_{[n]}} \right).
\end{aligned}
\tag{19.234}
$$

From the renormalization condition $\rho_{[m]}[\Lambda_R, \Lambda_0, \omega, \rho^0] = \text{constant}$ we have

$$
\begin{aligned}
d\rho_{[m]}[\Lambda_R, \Lambda_0, \omega, \rho^0] &= \frac{\partial \rho_{[m]}[\Lambda_R, \Lambda_0, \omega, \rho^0]}{\partial \Lambda_0} d\Lambda_0 + \frac{\partial \rho_{[m]}[\Lambda_R, \Lambda_0, \omega, \rho^0]}{\partial \omega} d\omega \\
&\quad + \frac{\partial \rho_{[m]}[\Lambda_R, \Lambda_0, \omega, \rho^0]}{\partial \rho^0_{[n]}} d\rho^0_{[n]}.
\end{aligned}
\tag{19.235}
$$

If we assume that $\frac{\partial \rho_{[m]}[\Lambda_R, \Lambda_0, \omega, \rho^0]}{\partial \rho^0_{[m]}}$ can be inverted then we can write

$$
\begin{aligned}
\frac{d\rho^0_{[n]}}{d\Lambda_0} &= - \frac{\partial \rho^0_{[n]}}{\partial \rho_{[m]}[\Lambda_R, \Lambda_0, \omega, \rho^0]} \frac{\partial \rho_{[m]}[\Lambda_R, \Lambda_0, \omega, \rho^0]}{\partial \Lambda_0} \\
&\quad - \frac{\partial \rho^0_{[n]}}{\partial \rho_{[m]}[\Lambda_R, \Lambda_0, \omega, \rho^0]} \frac{\partial \rho_{[m]}[\Lambda_R, \Lambda_0, \omega, \rho^0]}{\partial \omega} \frac{d\omega}{d\Lambda_0}.
\end{aligned}
\tag{19.236}
$$

We get then

$$L[M, \Lambda_R, \Lambda_0', \omega'\rho^{0'}] - L[M, \Lambda_R, \Lambda_0'', \omega''\rho^{0''}] = \int_{\Lambda_0''}^{\Lambda_0'} \frac{d\Lambda_0}{\Lambda_0} R[M, \Lambda_R, \Lambda_0, \omega, \rho^0] \qquad (19.237)$$

where R is defined by

$$\begin{aligned}
R[M, \Lambda, \Lambda_0, \omega, \rho^0] = {}&\Lambda_0 \frac{\partial L[M, \Lambda, \Lambda_0, \omega, \rho^0]}{\partial \Lambda_0} + \Lambda_0 \frac{d\omega}{d\Lambda_0} \frac{\partial L[M, \Lambda, \Lambda_0, \omega, \rho^0]}{\partial \omega} \\
&- \sum_{m,n} \Lambda_0 \frac{\partial \rho_{[n]}^0}{\partial \rho_{[m]}[\Lambda, \Lambda_0, \omega, \rho^0]} \frac{\partial \rho_{[m]}[\Lambda, \Lambda_0, \omega, \rho^0]}{\partial \Lambda_0} \frac{\partial L[M, \Lambda, \Lambda_0, \omega, \rho^0]}{\partial \rho_{[n]}^0} \\
&- \sum_{m,n} \Lambda_0 \frac{\partial \rho_{[n]}^0}{\partial \rho_{[m]}[\Lambda, \Lambda_0, \omega, \rho^0]} \frac{\partial \rho_{[m]}[\Lambda, \Lambda_0, \omega, \rho^0]}{\partial \omega} \frac{d\omega}{d\Lambda_0} \frac{\partial L[M, \Lambda, \Lambda_0, \omega, \rho^0]}{\partial \rho_{[n]}^0}.
\end{aligned} \qquad (19.238)$$

Since at small Λ the interaction $L[M, \Lambda, \Lambda_0, \omega, \rho^0]$ becomes simple of the form (19.226) we expect that R vanishes at small Λ near Λ_R, i.e., it becomes very small in the infrared. This is equivalent to the statement that the vectors (19.230), (19.231) and (19.232) become parallel to the relevant p-dimensional submanifold and that R is the linear combination which vanishes in the p relevant directions.

Using the fact that $\dfrac{\partial \rho_{[m]}[\Lambda_R, \Lambda_0, \omega, \rho^0]}{\partial \rho_{[m]}^0}$ is the inverse of $\dfrac{\partial \rho_{[m]}^0}{\partial \rho_{[m]}[\Lambda_R, \Lambda_0, \omega, \rho^0]}$ we can show that if we replace in R the interaction L by the quadratic effective term in (19.228) we get zero. It is trivial that the quartic term proportional to λ leads to zero since we are assuming that λ is just a number. In other words, since R is linear in L it is trivial to see that the splitting given in (19.228) leads for all Λ to a vanishing R on the ρ-operators. Thus R projects to the irrelevant part of the effective action. Let us also note from the above equation that ω appears more like the cutoff Λ_0 than like the coupling constants $\rho_{[n]}^0$.

19.3.5 Flow equations

We compute

$$\begin{aligned}
\Lambda \frac{\partial R}{\partial \Lambda} = {}&\Lambda_0 \frac{\partial(\Lambda \frac{\partial L}{\partial \Lambda})}{\partial \Lambda_0} + \Lambda_0 \frac{d\omega}{d\Lambda_0} \frac{\partial(\Lambda \frac{\partial L}{\partial \Lambda})}{\partial \omega} - \sum_{m,n} \Lambda_0 \frac{\partial \rho_{[n]}^0}{\partial \rho_{[m]}} \frac{\partial \rho_{[m]}}{\partial \Lambda_0} \frac{\partial(\Lambda \frac{\partial L}{\partial \Lambda})}{\partial \rho_{[n]}^0} \\
&- \sum_{m,n} \Lambda_0 \frac{\partial \rho_{[n]}^0}{\partial \rho_{[m]}} \frac{\partial \rho_{[m]}}{\partial \omega} \frac{d\omega}{d\Lambda_0} \frac{\partial(\Lambda \frac{\partial L}{\partial \Lambda})}{\partial \rho_{[n]}^0} - \sum_{m,n} \Lambda_0 \frac{\partial \rho_{[n]}^0}{\partial \rho_{[m]}} \frac{\partial(\Lambda \frac{\partial \rho_{[m]}}{\partial \Lambda})}{\partial \Lambda_0} \frac{\partial L}{\partial \rho_{[n]}^0} \\
&- \sum_{m,n} \Lambda_0 \frac{\partial \rho_{[n]}^0}{\partial \rho_{[m]}} \frac{\partial(\Lambda \frac{\partial \rho_{[m]}}{\partial \Lambda})}{\partial \omega} \frac{d\omega}{d\Lambda_0} \frac{\partial L}{\partial \rho_{[n]}^0} + \sum_{m,n,k,l} \Lambda_0 \frac{\partial \rho_{[n]}^0}{\partial \rho_{[k]}} \frac{\partial(\Lambda \frac{\partial \rho_{[k]}}{\partial \Lambda})}{\partial \rho_{[l]}^0} \frac{\partial \rho_{[l]}^0}{\partial \rho_{[m]}} \frac{\partial \rho_{[m]}}{\partial \Lambda_0} \frac{\partial L}{\partial \rho_{[n]}^0} \\
&+ \sum_{m,n,k,l} \Lambda_0 \frac{\partial \rho_{[n]}^0}{\partial \rho_{[k]}} \frac{\partial(\Lambda \frac{\partial \rho_{[k]}}{\partial \Lambda})}{\partial \rho_{[l]}^0} \frac{\partial \rho_{[l]}^0}{\partial \rho_{[m]}} \frac{\partial \rho_{[m]}}{\partial \omega} \frac{d\omega}{d\Lambda_0} \frac{\partial L}{\partial \rho_{[n]}^0}.
\end{aligned} \qquad (19.239)$$

In the last two terms we have used the identity

$$-\Lambda\frac{\partial}{\partial\Lambda}\left(\frac{\partial\rho_{[n]}^0}{\partial\rho_{[m]}}\right) = \frac{\partial\rho_{[n]}^0}{\partial\rho_{[k]}}\frac{\partial(\Lambda\frac{\partial\rho_{[k]}}{\partial\Lambda})}{\partial\rho_{[l]}^0}\frac{\partial\rho_{[l]}^0}{\partial\rho_{[m]}}. \tag{19.240}$$

We define

$$\frac{\partial L}{\partial\rho_{[m]}} = \sum_n \frac{\partial L}{\partial\rho_{[n]}^0}\frac{\partial\rho_{[n]}^0}{\partial\rho_{[m]}}. \tag{19.241}$$

Thus

$$\Lambda\frac{\partial R}{\partial\Lambda} = \Lambda_0\frac{\partial(\Lambda\frac{\partial L}{\Lambda\Lambda})}{\partial\Lambda_0} + \Lambda_0\frac{d\omega}{d\Lambda_0}\frac{\partial(\Lambda\frac{\partial L}{\partial\Lambda})}{\partial\omega} - \sum_m\Lambda_0\frac{\partial\rho_{[m]}}{\partial\Lambda_0}\frac{\partial(\Lambda\frac{\partial L}{\partial\Lambda})}{\partial\rho_{[m]}}$$
$$- \sum_m\Lambda_0\frac{\partial\rho_{[m]}}{\partial\omega}\frac{d\omega}{d\Lambda_0}\frac{\partial(\Lambda\frac{\partial L}{\partial\Lambda})}{\partial\rho_{[m]}} - \sum_m\Lambda_0\frac{\partial(\Lambda\frac{\partial\rho_{[m]}}{\partial\Lambda})}{\partial\Lambda_0}\frac{\partial L}{\partial\rho_{[m]}} - \sum_m\Lambda_0\frac{\partial(\Lambda\frac{\partial\rho_{[m]}}{\partial\Lambda})}{\partial\omega}\frac{d\omega}{d\Lambda_0}\frac{\partial L}{\partial\rho_{[m]}} \tag{19.242}$$
$$+ \sum_{m,k}\Lambda_0\frac{\partial(\Lambda\frac{\partial\rho_{[k]}}{\partial\Lambda})}{\partial\rho_{[m]}}\frac{\partial\rho_{[m]}}{\partial\Lambda_0}\frac{\partial L}{\partial\rho_{[k]}} + \sum_{m,k}\Lambda_0\frac{\partial(\Lambda\frac{\partial\rho_{[k]}}{\partial\Lambda})}{\partial\rho_{[m]}}\frac{\partial\rho_{[m]}}{\partial\omega}\frac{d\omega}{d\Lambda_0}\frac{\partial L}{\partial\rho_{[k]}}.$$

Using the Polchinski equation we define

$$M[L,\frac{\partial L}{\partial\Lambda_0}] \equiv \frac{\partial}{\partial\Lambda_0}(\Lambda\frac{\partial L}{\partial\Lambda}) = \frac{1}{2\nu_2}\int_p\sum_{m,n,k,l}\Lambda\frac{\partial\Delta_{nm,lk}(\Lambda)}{\partial\Lambda}\left[2\frac{\partial L}{\partial\tilde M_{kl}}\frac{\partial\left(\frac{\partial L}{\partial\Lambda_0}\right)}{\partial\tilde M_{mn}} - \frac{\partial^2\left(\frac{\partial L}{\partial\Lambda_0}\right)}{\partial\tilde M_{kl}\partial\tilde M_{mn}}\right]. \tag{19.243}$$

$$M[L,\frac{\partial L}{\partial\rho_{[q]}^0}] \equiv \frac{\partial}{\partial\rho_{[q]}^0}(\Lambda\frac{\partial L}{\partial\Lambda}) = \frac{1}{2\nu_2}\int_p\sum_{m,n,k,l}\Lambda\frac{\partial\Delta_{nm,lk}(\Lambda)}{\partial\Lambda}\left[2\frac{\partial L}{\partial\tilde M_{kl}}\frac{\partial\left(\frac{\partial L}{\partial\rho_{[q]}^0}\right)}{\partial\tilde M_{mn}} - \frac{\partial^2\left(\frac{\partial L}{\partial\rho_{[q]}^0}\right)}{\partial\tilde M_{kl}\partial\tilde M_{mn}}\right]. \tag{19.244}$$

$$M[L,\frac{\partial L}{\partial\omega}] \equiv \frac{\partial}{\partial\omega}(\Lambda\frac{\partial L}{\partial\Lambda}) = \frac{1}{2\nu_2}\int_p\sum_{m,n,k,l}\Lambda\frac{\partial\Delta_{nm,lk}(\Lambda)}{\partial\Lambda}\left[2\frac{\partial L}{\partial\tilde M_{kl}}\frac{\partial\left(\frac{\partial L}{\partial\omega}\right)}{\partial\tilde M_{mn}} - \frac{\partial^2\left(\frac{\partial L}{\partial\omega}\right)}{\partial\tilde M_{kl}\partial\tilde M_{mn}}\right]. \tag{19.245}$$

We can make the expansion

$$M[L,\ldots] = \frac{1}{2}\sum_{m,n}M_{[m]}[L,\ldots]\tilde M_{mn}\tilde M_{nm} + \text{other } M \text{ structures.} \tag{19.246}$$

Clearly we can make the identification

$$M_{[m]}[L,\frac{\partial L}{\partial\Lambda_0}] = \frac{\partial}{\partial\Lambda_0}(\Lambda\frac{\partial\rho_{[m]}}{\partial\Lambda}). \tag{19.247}$$

$$M_{[m]}[L, \frac{\partial L}{\partial \rho_{[n]}^0}] = \frac{\partial}{\partial \rho_{[n]}^0}(\Lambda \frac{\partial \rho_{[m]}}{\partial \Lambda}).$$
(19.248)

$$M_{[m]}[L, \frac{\partial L}{\partial \omega}] = \frac{\partial}{\partial \omega}(\Lambda \frac{\partial \rho_{[m]}}{\partial \Lambda}).$$
(19.249)

Recall that R is given by

$$\begin{aligned}
R &= \Lambda_0 \frac{\partial L}{\partial \Lambda_0} + \Lambda_0 \frac{d\omega}{d\Lambda_0} \frac{\partial L}{\partial \omega} - \sum_m \Lambda_0 \frac{\partial \rho_{[m]}}{\partial \Lambda_0} \frac{\partial L}{\partial \rho_{[m]}} - \sum_m \Lambda_0 \frac{\partial \rho_{[m]}}{\partial \omega} \frac{d\omega}{d\Lambda_0} \frac{\partial L}{\partial \rho_{[m]}} \\
&= \Lambda_0 \frac{\partial L}{\partial \Lambda_0} + \Lambda_0 \frac{d\omega}{d\Lambda_0} \frac{\partial L}{\partial \omega} - \sum_{m,n} \Lambda_0 \frac{\partial \rho_{[m]}}{\partial \Lambda_0} \frac{\partial \rho_{[n]}^0}{\partial \rho_{[m]}} \frac{\partial L}{\partial \rho_{[n]}^0} - \sum_{m,n} \Lambda_0 \frac{\partial \rho_{[m]}}{\partial \omega} \frac{d\omega}{d\Lambda_0} \frac{\partial \rho_{[n]}^0}{\partial \rho_{[m]}} \frac{\partial L}{\partial \rho_{[n]}^0} \\
&= \hat{R}(L).
\end{aligned}$$
(19.250)

The operator $\hat{R}$ is a linear combination of the operators $\partial/\partial \Lambda_0$, $\partial/\partial \rho_{[n]}^0$ and $\partial/\partial \omega$. M is linear in the second argument. Thus

$$M[L, R] = \Lambda_0 \frac{\partial(\Lambda \frac{\partial L}{\partial \Lambda})}{\partial \Lambda_0} + \Lambda_0 \frac{d\omega}{d\Lambda_0} \frac{\partial(\Lambda \frac{\partial L}{\partial \Lambda})}{\partial \omega} - \sum_m \Lambda_0 \frac{\partial \rho_{[m]}}{\partial \Lambda_0} \frac{\partial(\Lambda \frac{\partial L}{\partial \Lambda})}{\partial \rho_{[m]}} - \sum_m \Lambda_0 \frac{\partial \rho_{[m]}}{\partial \omega} \frac{d\omega}{d\Lambda_0} \frac{\partial(\Lambda \frac{\partial L}{\partial \Lambda})}{\partial \rho_{[m]}}.$$
(19.251)

Similarly

$$\begin{aligned}
M_{[m]}[L, R] &= \Lambda_0 \frac{\partial(\Lambda \frac{\partial \rho_{[m]}}{\partial \Lambda})}{\partial \Lambda_0} + \Lambda_0 \frac{d\omega}{d\Lambda_0} \frac{\partial(\Lambda \frac{\partial \rho_{[m]}}{\partial \Lambda})}{\partial \omega} - \sum_k \Lambda_0 \frac{\partial \rho_{[k]}}{\partial \Lambda_0} \frac{\partial(\Lambda \frac{\partial \rho_{[m]}}{\partial \Lambda})}{\partial \rho_{[k]}} \\
&\quad - \sum_k \Lambda_0 \frac{\partial \rho_{[k]}}{\partial \omega} \frac{d\omega}{d\Lambda_0} \frac{\partial(\Lambda \frac{\partial \rho_{[m]}}{\partial \Lambda})}{\partial \rho_{[k]}}.
\end{aligned}$$
(19.252)

Hence

$$\begin{aligned}
\Lambda \frac{\partial R}{\partial \Lambda} &= M[L, R] - \sum_m \Lambda_0 \frac{\partial(\Lambda \frac{\partial \rho_{[m]}}{\partial \Lambda})}{\partial \Lambda_0} \frac{\partial L}{\partial \rho_{[m]}} - \sum_m \Lambda_0 \frac{\partial(\Lambda \frac{\partial \rho_{[m]}}{\partial \Lambda})}{\partial \omega} \frac{d\omega}{d\Lambda_0} \frac{\partial L}{\partial \rho_{[m]}} \\
&\quad + \sum_{m,k} \Lambda_0 \frac{\partial(\Lambda \frac{\partial \rho_{[m]}}{\partial \Lambda})}{\partial \rho_{[k]}} \frac{\partial \rho_{[k]}}{\partial \Lambda_0} \frac{\partial L}{\partial \rho_{[m]}} + \sum_{m,k} \Lambda_0 \frac{\partial(\Lambda \frac{\partial \rho_{[m]}}{\partial \Lambda})}{\partial \rho_{[k]}} \frac{\partial \rho_{[k]}}{\partial \omega} \frac{d\omega}{d\Lambda_0} \frac{\partial L}{\partial \rho_{[m]}} \\
&= M[L, R] - \sum_m M_{[m]}[L, R] \frac{\partial L}{\partial \rho_{[m]}}.
\end{aligned}$$
(19.253)

We can also compute the Λ-scaling

$$\Lambda \frac{\partial}{\partial \Lambda}\left(\frac{\partial L}{\partial \rho_{[n]}}\right) = M\left[L, \frac{\partial L}{\partial \rho_{[n]}}\right] - \sum_m M_{[m]}\left[L, \frac{\partial L}{\partial \rho_{[n]}}\right] \frac{\partial L}{\partial \rho_{[m]}}.$$
(19.254)

19.3.6 Expansion in power series

With our normalization the effective action L is dimensionless. The dimensionless coupling is $\nu_2\lambda = \pi\theta\lambda$. The field $\tilde{M}$ is dimensionless in two dimensions. We expand L in the coupling constant $\nu_2\lambda$ and in the number of fields $\tilde{M}$ as

$$L = \sum_{V=1}^{\infty}(\nu_2\lambda)^{V-1}\sum_{N=2}^{\infty}\frac{1}{N!}\sum_{m_1n_1,\ldots,m_Nn_N} A^{(V)}_{m_1n_1,\,\ldots,\,m_Nn_N}\tilde{M}_{m_1n_1}\cdots\tilde{M}_{m_Nn_N}. \tag{19.255}$$

R is also dimensionless. We expand R as

$$R = \sum_{V=1}^{\infty}(\nu_2\lambda)^{V-1}\sum_{N=2}^{\infty}\frac{1}{N!}\sum_{m_1n_1,\ldots,m_Nn_N} R^{(V)}_{m_1n_1,\,\ldots,\,m_Nn_N}\tilde{M}_{m_1n_1}\cdots\tilde{M}_{m_Nn_N}. \tag{19.256}$$

$\partial L/\partial\rho_{[m]}$ is of dimension $1/M^2$ since $\rho_{[m]}$ is of dimension M^2. We expand it as

$$\frac{\partial L}{\partial\rho_{[m]}} = \frac{1}{\lambda}\sum_{V=0}^{\infty}(\nu_2\lambda)^{V}\sum_{N=2}^{\infty}\frac{1}{N!}\sum_{m_1n_1,\ldots,m_Nn_N} H^{(V)}_{m_1n_1,\,\ldots,\,m_Nn_N}\tilde{M}_{m_1n_1}\cdots\tilde{M}_{m_Nn_N}. \tag{19.257}$$

Define

$$Q_{nm,lk}(\Lambda) = \frac{1}{\nu_2}\Lambda\frac{\partial\Delta_{nm,lk}(\Lambda)}{\partial\Lambda}. \tag{19.258}$$

Compute

$$\frac{\partial L}{\partial\tilde{M}_{kl}} = \sum_{V=1}^{\infty}(\nu_2\lambda)^{V-1}\sum_{N=2}^{\infty}\frac{1}{(N-1)!}\sum_{m_1n_1,\ldots,m_{N-1}n_{N-1}} A^{(V)}_{m_1n_1,\,\ldots,\,m_{N-1}n_{N-1},\,kl}\tilde{M}_{m_1n_1}\cdots\tilde{M}_{m_{N-1}n_{N-1}}. \tag{19.259}$$

We need to compute

$$\frac{1}{2}\sum_{m,n,k,l}Q_{nm,lk}(\Lambda)\frac{\partial L}{\partial\tilde{M}_{kl}}\frac{\partial L}{\partial\tilde{M}_{mn}} \tag{19.260}$$

and

$$\frac{1}{2}\sum_{m,n,k,l}Q_{nm,lk}(\Lambda)\frac{\partial^2 L}{\partial\tilde{M}_{kl}\partial\tilde{M}_{mn}}. \tag{19.261}$$

We compute up to the third power in the fields the following

$$\begin{aligned}
\frac{1}{2}\sum_{m,n,k,l}Q_{nm,lk}(\Lambda)\frac{\partial L}{\partial\tilde{M}_{kl}}\frac{\partial L}{\partial\tilde{M}_{mn}} &= \sum_{V=2}^{\infty}(\nu_2\lambda)^{V-2}\frac{1}{2!}\sum_{m_1,n_1,m_2,n_2}\sum_{V_1=1}^{V-1}\sum_{m,n,k,l}\frac{1}{2}Q_{nm,lk}(\Lambda)\\
&\quad\times\left(A^{(V_1)}_{m_1n_1,\,mn}A^{(V-V_1)}_{m_2n_2,\,kl} + A^{(V_1)}_{m_2n_2,\,mn}A^{(V-V_1)}_{m_1n_1,\,kl}\right)\tilde{M}_{m_1n_1}\tilde{M}_{m_2n_2}\\
&\quad+ \sum_{V=2}^{\infty}(\nu_2\lambda)^{V-2}\frac{1}{3!}\sum_{m_1,n_1,m_2,n_2,m_3n_3}\sum_{V_1=1}^{V-1}\sum_{m,n,k,l}\frac{1}{2}Q_{nm,lk}(\Lambda)\\
&\quad\times\left(A^{(V_1)}_{m_1n_1,\,mn}A^{(V-V_1)}_{m_2n_2m_3n_3,\,kl} + A^{(V_1)}_{m_2n_2,\,mn}A^{(V-V_1)}_{m_1n_1m_3n_3,\,kl}\right.\\
&\quad\left.+ A^{(V_1)}_{m_3n_3,\,mn}A^{(V-V_1)}_{m_2n_2m_1n_1,\,kl}\right)\tilde{M}_{m_1n_1}\tilde{M}_{m_2n_2}\tilde{M}_{m_3n_3}\\
&\quad+ \sum_{V=2}^{\infty}(\nu_2\lambda)^{V-2}\frac{1}{3!}\sum_{m_1,n_1,m_2,n_2,m_3n_3}\sum_{V_1=1}^{V-1}\sum_{m,n,k,l}\frac{1}{2}Q_{nm,lk}(\Lambda)\\
&\quad\times\left(A^{(V_1)}_{m_1n_1m_2n_2,\,mn}A^{(V-V_1)}_{m_3n_3,\,kl} + A^{(V_1)}_{m_1n_1m_3n_3,\,mn}A^{(V-V_1)}_{m_2n_2,\,kl}\right.\\
&\quad\left.+ A^{(V_1)}_{m_3n_3m_2n_2,\,mn}A^{(V-V_1)}_{m_1n_1,\,kl}\right)\tilde{M}_{m_1n_1}\tilde{M}_{m_2n_2}\tilde{M}_{m_3n_3}.
\end{aligned} \tag{19.262}$$

In this above equation we have used the obvious identity

$$\sum_{V=2}^{\infty}\sum_{V_1=1}^{V-1} I^{V_1} J^{V-V_1} = \sum_{V=1}^{\infty}\sum_{V'=1}^{\infty} I^V J^{V'}. \tag{19.263}$$

This result needs to be compared with

$$\Lambda\frac{\partial L}{\partial \Lambda} = \sum_{V=1}^{\infty}(\nu_2\lambda)^{V-1}\frac{1}{2!}\sum_{m_1,n_1,m_2,n_2}\Lambda\frac{\partial A_{m_1 n_1,\, m_2 n_2}^{(V)}}{\partial \Lambda}\tilde{M}_{m_1 n_1}\tilde{M}_{m_2 n_2} + \cdots.$$

$$+ \sum_{V=1}^{\infty}(\nu_2\lambda)^{V-1}\frac{1}{3!}\sum_{m_1 n_1,m_2 n_2,m_3 n_3}\Lambda\frac{\partial A_{m_1 n_1,\, m_2 n_2,\, m_3 n_3}^{(V)}}{\partial \Lambda}\tilde{M}_{m_1 n_1}\tilde{M}_{m_2 n_2}\tilde{M}_{m_3 n_3} + \cdots.$$

$$= \sum_{V=2}^{\infty}(\nu_2\lambda)^{V-2}\frac{1}{2!}\sum_{m_1 n_1,m_2 n_2}\Lambda\frac{\partial A_{m_1 n_1,\, m_2 n_2}^{(V-1)}}{\partial \Lambda}\tilde{M}_{m_1 n_1}\tilde{M}_{m_2 n_2} + \cdots. \tag{19.264}$$

$$+ \sum_{V=2}^{\infty}(\nu_2\lambda)^{V-2}\frac{1}{3!}\sum_{m_1 n_1,m_2 n_2,m_3 n_3}\Lambda\frac{\partial A_{m_1 n_1,\, m_2 n_2,\, m_3 n_3}^{(V-1)}}{\partial \Lambda}\tilde{M}_{m_1 n_1}\tilde{M}_{m_2 n_2}\tilde{M}_{m_3 n_3} + \cdots.$$

We get

$$\Lambda\frac{\partial A_{m_1 n_1,\, m_2 n_2}^{(V-1)}}{\partial \Lambda} = \sum_{V_1=1}^{V-1}\sum_{m,n,k,l}\frac{1}{2}Q_{nm,lk}(\Lambda)\Big(A_{m_1 n_1,\, mn}^{(V_1)}A_{m_2 n_2,\, kl}^{(V-V_1)} + A_{m_2 n_2,\, mn}^{(V_1)}A_{m_1 n_1,\, kl}^{(V-V_1)}\Big). \tag{19.265}$$

$$\Lambda\frac{\partial A_{m_1 n_1,\, m_2 n_2,\, m_3 n_3}^{(V-1)}}{\partial \Lambda} = \sum_{V_1=1}^{V-1}\sum_{m,n,k,l}\frac{1}{2}Q_{nm,lk}(\Lambda)\Big(A_{m_1 n_1,\, mn}^{(V_1)}A_{m_2 n_2 m_3 n_3,\, kl}^{(V-V_1)} + \text{permutations}\Big)$$

$$+ \sum_{V_1=1}^{V-1}\sum_{m,n,k,l}\frac{1}{2}Q_{nm,lk}(\Lambda)\Big(A_{m_1 n_1 m_2 n_2,\, mn}^{(V_1)}A_{m_3 n_3,\, kl}^{(V-V_1)} + \text{permutations}\Big). \tag{19.266}$$

Thus we obtain the general result

$$\Lambda\frac{\partial A_{m_1 n_1,\, \ldots,\, m_N n_N}^{(V-1)}}{\partial \Lambda} = \sum_{N_1=2}^{N}\sum_{V_1=1}^{V-1}\sum_{m,n,k,l}\frac{1}{2}Q_{nm,lk}(\Lambda)\Big(A_{m_1 n_1,\, \ldots,\, m_{N_1}-1 n_{N_1}-1,\, mn}^{(V_1)}A_{m_{N_1} n_{N_1}\, \ldots,\, m_N n_N,\, kl}^{(V-V_1)}$$

$$+ \text{permutations}). \tag{19.267}$$

For every N_1 there are $N!/(N_1!(N-N_1)!)$ permutations of the indices $m_1 n_1,\ldots,$ $m_{N_1} n_{N_1},\ldots, m_N n_N$. However we should also include the effect of the second term in the Wilson–Polchinski equation. We compute

$$\sum_{m,n,k,l}Q_{nm,lk}(\Lambda)\frac{\partial^2 L}{\partial \tilde{M}_{kl}\partial \tilde{M}_{mn}} = \sum_{V=2}^{\infty}(\nu_2\lambda)^{V-2}\Bigg[\sum_{m,n,k,l}Q_{nm,lk}A_{mnkl}^{(V-1)} + \sum_{m,n,k,l}Q_{nm,lk}\sum_{m_1 n_1}A_{m_1 n_1 mnkl}^{(V-1)}\tilde{M}_{m_1 n_1}$$

$$+ \sum_{N=2}^{\infty}\frac{1}{N!}\sum_{m_1 n_1,\ldots,m_N n_N}\sum_{m,n,k,l}Q_{nm,lk}A_{m_1 n_1\,\ldots\,m_N n_N mnkl}^{(V-1)}\tilde{M}_{m_1 n_1}\cdots\tilde{M}_{m_N n_N}\Bigg]. \tag{19.268}$$

By neglecting the first two terms we get

$$\Lambda\frac{\partial A_{m_1 n_1,\, \ldots,\, m_N n_N}^{(V-1)}}{\partial \Lambda} = \sum_{N_1=2}^{N}\sum_{V_1=1}^{V-1}\sum_{m,n,k,l}\frac{1}{2}Q_{nm,lk}(\Lambda)\Big(A_{m_1 n_1,\, \ldots,\, m_{N_1}-1 n_{N_1}-1,\, mn}^{(V_1)}A_{m_{N_1} n_{N_1}\,\ldots,\, m_N n_N,\, kl}^{(V-V_1)}$$

$$+ \text{permutations}) - \frac{1}{2}\sum_{m,n,k,l}Q_{nm,lk}A_{m_1 n_1,\, \ldots,\, m_N n_N,\, mn,\, kl}^{(V-1)}. \tag{19.269}$$

We will set

$$H^{(V)}_{m_1 n_1, \dots, m_N n_N} = 0. \tag{19.270}$$

This is possible because as it turns out for $\omega < 1$ the flow equations of the functions $R^{(V)}_{m_1 n_1, \dots, m_N n_N}$ do not depend on the functions $H^{(V)}_{m_1 n_1, \dots, m_N n_N}$. Hence we do not need to evaluate the functions $H^{(V)}_{m_1 n_1, \dots, m_N n_N}$. For all practical purposes we can set them equal 0. The flow equations of $R^{(V)}_{m_1 n_1, \dots, m_N n_N}$ are derived from the Polchinski equation for R given by (19.253). We find

$$\Lambda \frac{\partial R^{(V-1)}_{m_1 n_1, \dots, m_N n_N}}{\partial \Lambda} = \sum_{N_1=2}^{N} \sum_{V_1=1}^{V-1} \sum_{m,n,k,l} Q_{nm,lk}(\Lambda) \Big(A^{(V_1)}_{m_1 n_1, \dots, m_{N_1} - 1, n_{N_1} - 1, mn} R^{(V-V_1)}_{m_{N_1} n_{N_1} \dots, m_N n_N, kl}$$

$$+ \text{permutations} \Big) - \frac{1}{2} \sum_{m,n,k,l} Q_{nm,lk} R^{(V-1)}_{m_1 n_1, \dots, m_N n_N, mn, kl}. \tag{19.271}$$

19.3.7 Ribbon graphs

The coefficient $L_{m_1 n_1 \dots m_N n_N}[\Lambda]$ is represented by a circle with N outgoing lines m_i and n incoming lines n_i (figure 19.1). The arrow are only added for bookkeeping so in fact there is no distinction between outgoing and incoming lines since we are dealing with a real scalar field. We will represent the differentiated propagator $Q_{nm,lk}(\Lambda)$ by a ribbon with two lines nk and ml (figure 19.2). We differentiate the Polchinski

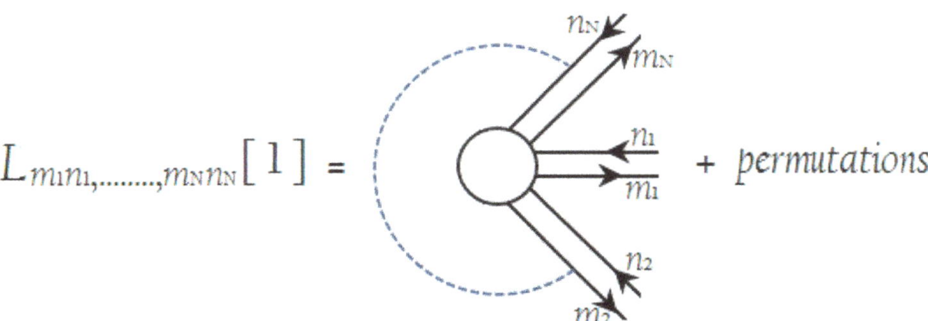

Figure 19.1. The coefficient $L_{m_1 n_1 \dots m_N n_N}[\Lambda]$.

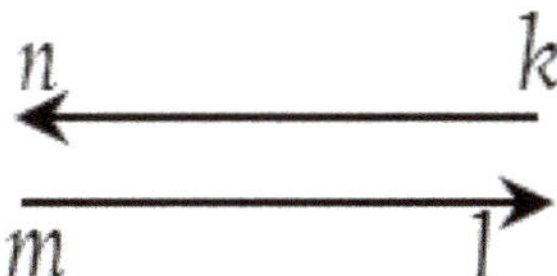

Figure 19.2. The differentiated propagator $Q_{nm,lk}(\Lambda)$.

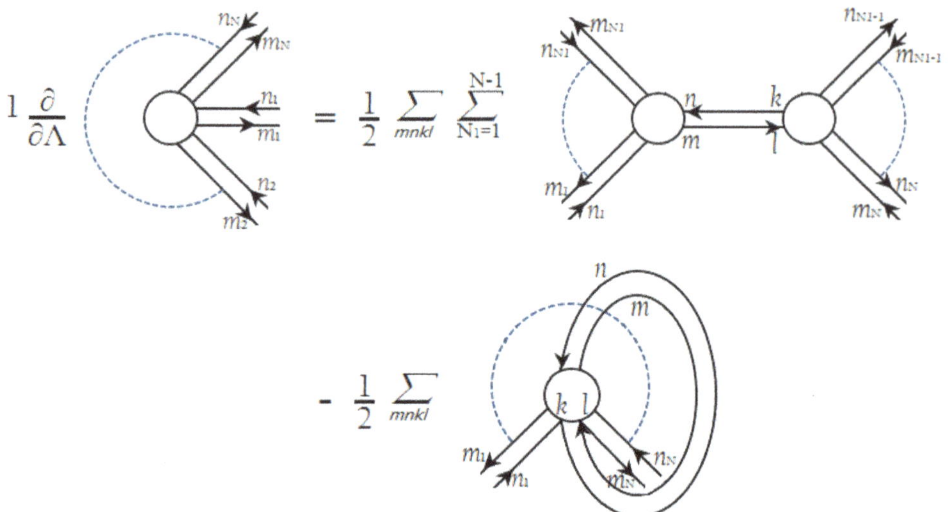

Figure 19.3. The Wilson–Polchinski equation (19.172) differentiated with respect to the fields $\phi_{m_1 n_1}, \ldots, \phi_{m_2 n_2}$.

equation (19.172) with respect to the fields $\phi_{m_1 n_1}, \ldots, \phi_{m_2 n_2}$[1]. The resulting equation is represented graphically on figure 19.3.

Generically the solution $L[\phi, \Lambda]$ of the Polchinski equation (19.172) can be given by an expansion in the coupling constant λ of the form

$$L[\phi, \Lambda] = \sum_{V=1}^{\infty} \lambda^V L^{(V)}[\phi, \Lambda].$$
(19.272)

For a ϕ^4-model we choose

$$L^{(1)}_{m_1 n_1, \, m_2 n_2}[\Lambda] = 0$$
(19.273)

and

$$L^{(1)}_{m_1 n_1, \, \ldots, \, m_N n_N}[\Lambda] = 0, \quad N > 4.$$
(19.274)

Inserting these into the graphical Polchinski equation (figure 19.3) we conclude that $L^{(1)}_{m_1 n_1, \, \ldots, \, m_4 n_4}[\Lambda]$ is independent of Λ. We can therefore identify $L^{(1)}_{m_1 n_1, \, \ldots, \, m_4 n_4}$ with the original ϕ^4 interaction, viz

$$L^{(1)}_{m_1 n_1, \, \ldots, \, m_4 n_4}[\Lambda_0] = \frac{1}{4!6}(\delta_{n_1 m_2} \delta_{n_2 m_3} \delta_{n_3 m_4} \delta_{n_4 m_1} + \text{permutations}).$$
(19.275)

This is represented graphically on figure (19.4).

The iterative solution of the graphical Polchinski equation in figure 19.3 with the initial conditions (19.274) and (19.275) is given in terms of ribbon graphs. Clearly

---

[1] In this section we will denote the field $\tilde{M}_{mn}$ by ϕ_{mn}.

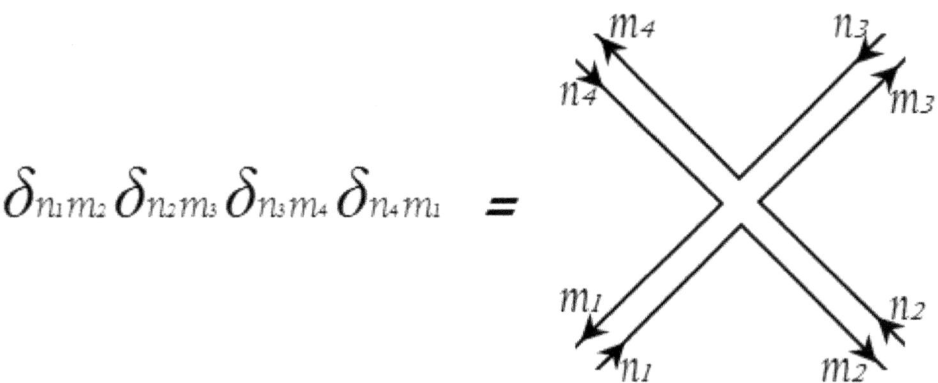

$$\delta_{n_1 m_2}\, \delta_{n_2 m_3}\, \delta_{n_3 m_4}\, \delta_{n_4 m_1} \;=\;$$

Figure 19.4. The coefficient $L^{(1)}_{m_1 n_1,\ \ldots,\ m_4 n_4}[\Lambda]$.

ribbon graphs can become very complicated with an arbitrary number of crossing of lines so they cannot be drawn on a plane.

A general ribbon graph can be drawn on a Riemannian surface of genus g. More precisely a given ribbon graph defines a topological Riemannian surface via Euler characteristic $\chi = 2 - 2g$. Euler characteristics can also be computed from the formula $\chi = V - E + F$ where V is the number of polyhedron vertices, E is the number of polyhedron edges and F is the number of faces. To apply this formula we consider closed ribbon graphs where the external lines are amputated, i.e., the external lines m_i are connected with n_i. The number of faces F is the number $\bar{L}$ of single-line loops of the closed graph, viz $F = \bar{L}$ whereas the number of edges E is equal to the number I of internal double lines of the graph, viz $E = I$. Thus we have

$$\chi = 2 - 2g = V - I + \bar{L}. \tag{19.276}$$

We can draw the ribbon graph and its Riemannian surface in a variety of equivalent ways provided V, I, $\bar{L}$ and g are kept the same. The Polchinski equation in figure 19.3 tells us which external legs of the vertices are connected. The ribbons are drawn between these legs in any convenient way. Thus, for example, there is no difference between overcrossing and undercrossing.

In a ribbon graph we have two kinds of loops. Some loops are such that they contain at least one external leg. These are the boundary components of the ribbon graph, which correspond to the holes of the Riemannian surface. Their number is B. Another type of loops do not contain any external leg. They are called inner loops. Their number is $\bar{L}_0 = \bar{L} - B$. Boundary components are composed of an ensemble of trajectories from an incoming index n_i to an outgoing index m_j. An example is shown on the first graph of figure 19.5. In this case $\bar{L} = B = 2$, i.e., there are two boundary components. The inner boundary component is composed of the trajectory $\overrightarrow{n_1 m_6} = m_1 n_1 m_6 n_6$. The outer component is composed of the two trajectories $\overrightarrow{n_5 m_2} = m_5 n_5 m_2 n_2$ and $\overrightarrow{n_3 m_4} = m_3 n_3 m_4 n_4$. We also have $V = 3$, $I = 3$. Other examples are given in the second and third graphs of figure 19.5.

We also need the concept of an external vertex, which is a vertex with at least one external line attached to it. Let V_e be the number of external vertices. There are four

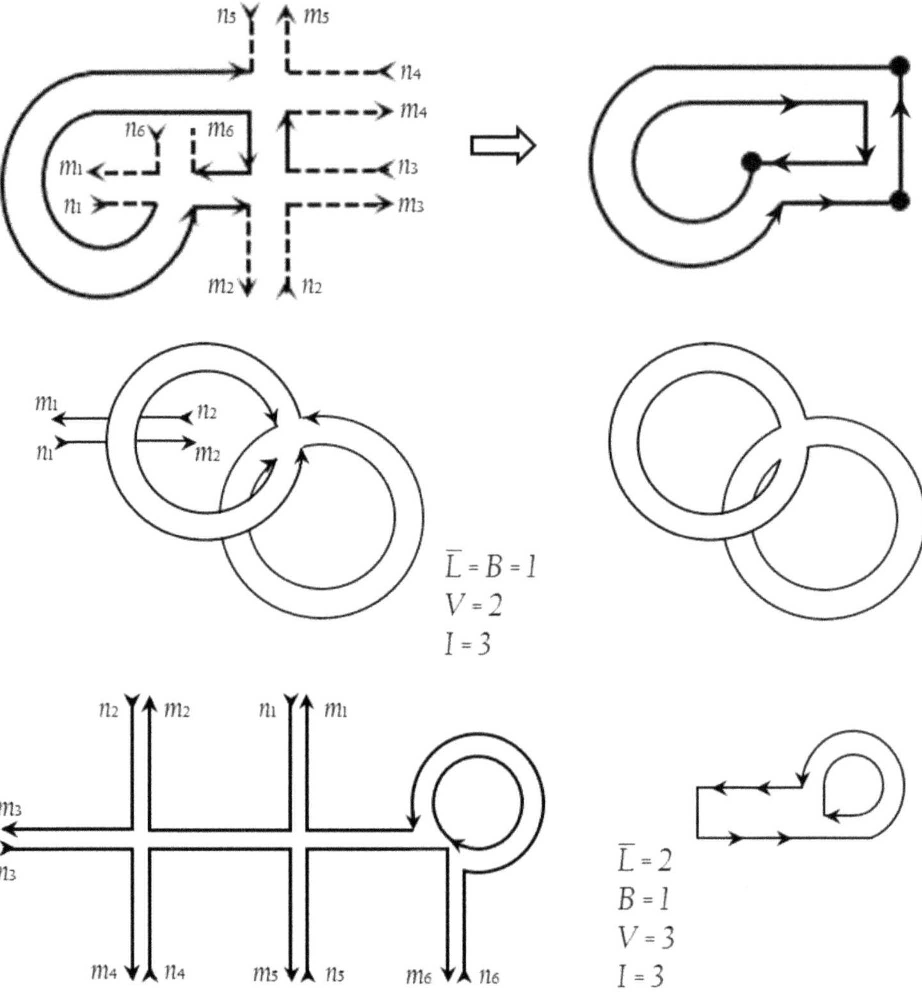

Figure 19.5. Examples of ribbon graphs.

possibilities for the arrangement of the external legs at an external vertex. Firstly, we can have only one external leg. Secondly, we can have three external legs. Thirdly, we can have two external legs with one starting point and one end point. Fourthly, we can have two external legs with two starting points and two end points. See the first graph of figure 19.6. Recall that only in external legs that the index m_i is connected with the index n_i. Thus in the first three possibilities we have only one starting point and one end point for trajectories inside a ribbon graph. These are called simple vertices. In the fourth possibility we have two starting points and two end points. This is called a composed vertex. This vertex can be decomposed as in the second graph of figure 19.6.

A given ribbon graph with V_c composed vertices can be decomposed into S segments, which is generically not equal to the total number of decompositions is

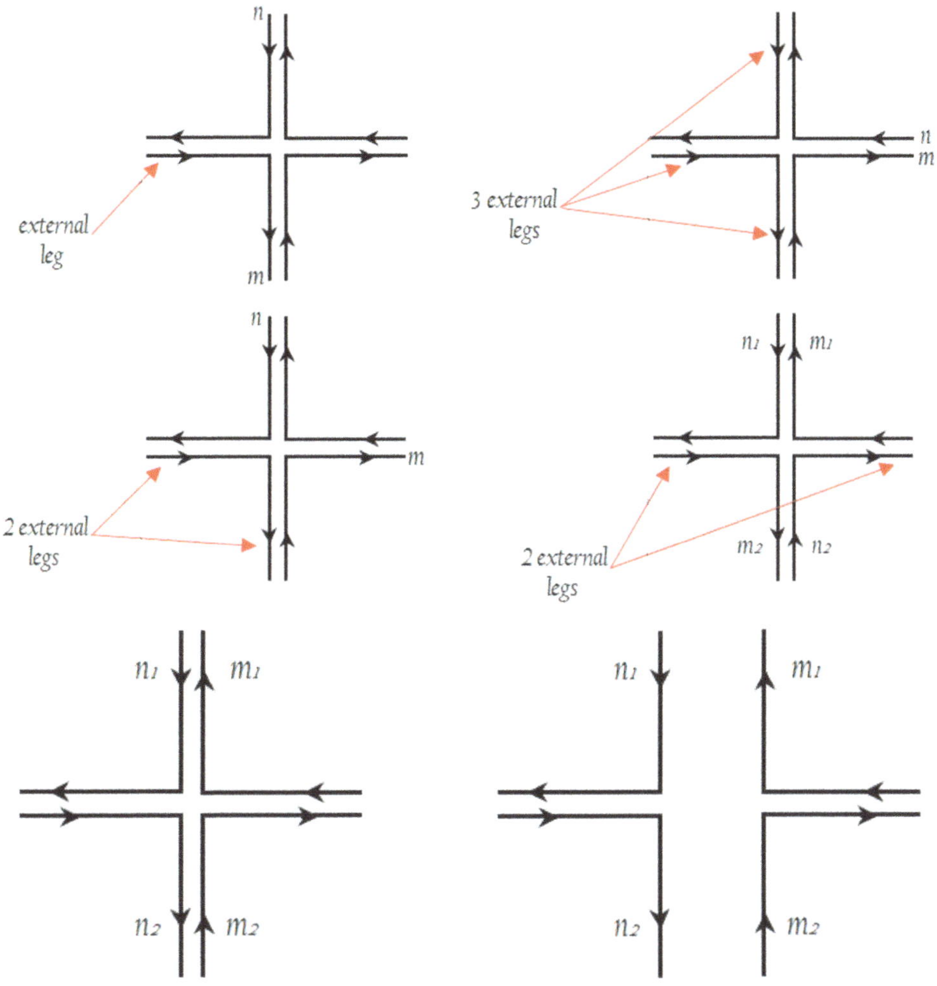

Figure 19.6. More examples of ribbon graphs.

$V_c + 1$. The segmentation index $\mathcal{I}$ of a given graph is the maximum number of decompositions of composed vertices which keep the graph connected. It is given by

$$\mathcal{I} = V_c - S + 1. \tag{19.277}$$

For the first graph of (19.5) we have $V_c = 0$, $S = 1$ and thus $\mathcal{I} = 0$. For the second graph of (19.5) we have $V_c = 1$, $S = 1$ and thus $\mathcal{I} = 1$. For the third graph of (19.5) we have $V_c = 1$, $S = 2$ and thus $\mathcal{I} = 0$.

We are going to fix all starting points of trajectories send sum over end points. Because of conservation of momentum not all summations are independent. In each segment the sum of outgoing indices is equal to the sum of incoming indices. The total number of end points in a ribbon graph is $V_e + V_c$ since the composed vertices contain two end points whereas the simple vertices contain one end point. Hence the number of independent index summations is

$$s \leqslant V_e + V_c - S = V_e + \mathcal{I} - 1. \tag{19.278}$$

19.3.8 Marginal, relevant and irrelevant operators

Renormalizability of a given model is the requirement that the Polchinski differential equation admits a regular solution, which depends only on a finite number of initial conditions. The Polchinski equation will be integrated between the two scales Λ_R and Λ or between the two scales Λ and Λ_0. The vertices $L_{m_1 n_1 \ldots m_N n_N}[\Lambda]$ are assumed to be decomposable into parts $L^{(i)}_{m_1 n_1 \ldots m_N n_N}[\Lambda]$, which are homogeneous. In other words, for any Λ in the interval $[\Lambda_R, \Lambda_0]$ these parts scale as

$$|\Lambda \frac{\partial L^{(i)}_{m_1 n_1 \ldots m_N n_N}[\Lambda]}{\partial \Lambda}| \leqslant \Lambda^{r_i} P^{q_i}[\ln \frac{\Lambda}{\Lambda_R}]. \tag{19.279}$$

In the above equation $P^q[x]$ is a polynomial of degree q in $x \geqslant 0$ such that $P^q[x] \geqslant 0$. Homogeneous parts of vertices $L^{(i)}_{m_1 n_1 \ldots m_N n_N}[\Lambda]$ with $r_i >$ are called relevant. If $r_i < 0$ the parts are called irrelevant whereas if $r_i = 0$ the parts are called marginal. Clearly we can write the vertices $L_{m_1 n_1 \ldots m_N n_N}[\Lambda]$ as

$$L^{(i)}_{m_1 n_1 \ldots m_N n_N}[\Lambda] = L^{(i)}_{m_1 n_1 \ldots m_N n_N}[\Lambda_0] - \int_\Lambda^{\Lambda_0} \frac{d\Lambda'}{\Lambda'} \left(\Lambda' \frac{\partial}{\partial \Lambda'} L^{(i)}_{m_1 n_1 \ldots m_N n_N}[\Lambda'] \right). \tag{19.280}$$

$$L^{(i)}_{m_1 n_1 \ldots m_N n_N}[\Lambda] = L^{(i)}_{m_1 n_1 \ldots m_N n_N}[\Lambda_R] + \int_{\Lambda_R}^{\Lambda} \frac{d\Lambda'}{\Lambda'} \left(\Lambda' \frac{\partial}{\partial \Lambda'} L^{(i)}_{m_1 n_1 \ldots m_N n_N}[\Lambda'] \right). \tag{19.281}$$

The largest contribution to these integrals comes from the term $\Lambda^{r-1}(\ln \Lambda / \Lambda_R)^q$. We use the integrals

$$\int dx\, x^{r-1} (\ln \frac{x}{x_R})^q = \frac{(-1)^q q!}{r^{q+1}} x^r \sum_{j=0}^{q} \frac{(-r \ln \frac{x}{x_R})^j}{j!} + \text{constant}, \quad r \neq 0. \tag{19.282}$$

$$\int dx\, x^{r-1} (\ln \frac{x}{x_R})^q = \frac{1}{q+1} (\ln \frac{x}{x_R})^{q+1} + \text{constant}, \quad r = 0. \tag{19.283}$$

First let us consider the case $r_i < 0$. We use for the homogeneous parts $L^{(i)}_{m_1 n_1 \ldots m_N n_N}[\Lambda]$ the definition (19.280). We get

$$|L^{(i)}_{m_1 n_1 \ldots m_N n_N}[\Lambda]| \leqslant |L^{(i)}_{m_1 n_1 \ldots m_N n_N}[\Lambda_0]| + \int_\Lambda^{\Lambda_0} \frac{d\Lambda'}{\Lambda'} |\left(\Lambda' \frac{\partial}{\partial \Lambda'} L^{(i)}_{m_1 n_1 \ldots m_N n_N}[\Lambda'] \right)|. \tag{19.284}$$

The contribution coming from the upper limit of the integral in the limit $\Lambda_0 \longrightarrow 0$ goes to zero. Thus we have

$$|L^{(i)}_{m_1 n_1 \ldots m_N n_N}[\Lambda]| \leqslant |L^{(i)}_{m_1 n_1 \ldots m_N n_N}[\Lambda_0]| + \int_\Lambda^\infty \frac{d\Lambda'}{\Lambda'} |\left(\Lambda' \frac{\partial}{\partial \Lambda'} L^{(i)}_{m_1 n_1 \ldots m_N n_N}[\Lambda'] \right)|$$

$$\leqslant |L^{(i)}_{m_1 n_1 \ldots m_N n_N}[\Lambda_0]| + \Lambda^{r_i} Q^{q_i}[\ln \frac{\Lambda}{\Lambda_R}]. \tag{19.285}$$

In the above equation $Q^q[x]$ is another polynomial of degree q in $x \geqslant 0$ such that $Q^q[x] \geqslant 0$. We require that for the boundary condition we have

$$|L^{(i)}_{m_1 n_1 \ldots m_N n_N}[\Lambda_0]| \leqslant \Lambda_0^{r_i} Q^{q_i}[\ln \frac{\Lambda_0}{\Lambda_R}], \quad r_i < 0. \tag{19.286}$$

We obtain

$$|L^{(i)}_{m_1 n_1 \ldots m_N n_N}[\Lambda]| \leqslant \Lambda^{r_i} Q^{q_i}[\ln \frac{\Lambda}{\Lambda_R}]. \tag{19.287}$$

Now we consider the case $r_i > 0$. In this case we use for the homogeneous parts $L^{(i)}_{m_1 n_1 \ldots m_N n_N}[\Lambda]$ the definition (19.281). We get

$$|L^{(i)}_{m_1 n_1 \ldots m_N n_N}[\Lambda]| \leqslant |L^{(i)}_{m_1 n_1 \ldots m_N n_N}[\Lambda_R]| + \int_{\Lambda_R}^\Lambda \frac{d\Lambda'}{\Lambda'} |\left(\Lambda' \frac{\partial}{\partial \Lambda'} L^{(i)}_{m_1 n_1 \ldots m_N n_N}[\Lambda'] \right)|. \tag{19.288}$$

The lower limit of the integral gives zero since $\ln \Lambda_R/\Lambda_R = 0$ for $j \neq 0$ whereas the term $j = 0$ will yield a contribution proportional to Λ_R^r which can be neglected compared to the contribution corresponding to the term $j = 0$ coming from the upper limit of the integral $\Lambda' = \Lambda$, which is proportional to Λ. We get then

$$|L^{(i)}_{m_1 n_1 \ldots m_N n_N}[\Lambda]| \leqslant |L^{(i)}_{m_1 n_1 \ldots m_N n_N}[\Lambda_R]| + \Lambda^{r_i} Q^{q_i}[\ln \frac{\Lambda}{\Lambda_R}]. \tag{19.289}$$

In summary for the irrelevant interactions we perform the integration from Λ_0 to Λ starting from an initial condition $|L^{(i)}_{m_1 n_1 \ldots m_N n_N}[\Lambda_0]| \leqslant \Lambda_0^{r_i} Q^{q_i}[\ln \frac{\Lambda_0}{\Lambda_R}]$. For relevant and marginal interactions we perform the integration from Λ_R to Λ, starting from an initial condition $|L^{(i)}_{m_1 n_1 \ldots m_N n_N}[\Lambda_R]| < \infty$. In all cases we have

$$|L^{(i)}_{m_1 n_1 \ldots m_N n_N}[\Lambda]| \leqslant \Lambda^{r_i} Q^{q_i}[\ln \frac{\Lambda}{\Lambda_R}]. \tag{19.290}$$

Note that the number of conditions imposed at Λ_R must be finite for the theory to be renormalizable. Conditions imposed at Λ_0 are not normalization conditions.

19.3.9 The power-counting theorem

This provides the power-counting behaviors of non-local matrix models with ϕ^4 interactions, which specify the class of all divergent functions in these theories. As an example all non-planar graphs in these models will turn out to be irrelevant.

The first ingredient is the differentiated propagator

$$Q_{nm,lk}(\Lambda) = \frac{1}{\nu_d^{\frac{d}{2}}} \Lambda \frac{\partial \Delta_{nm,lk}(\Lambda)}{\partial \Lambda}.$$

(19.291)

The cutoff function K is chosen so that only for a finite number of indices m,n,k and l we have $Q_{nm,lk}(\Lambda) \neq 0$. This function is normalized such that

$$\text{SCA} = \sum_m \text{sign}|K(m, \Lambda)| \leqslant C_d \left(\frac{\Lambda}{\mu}\right)^d, \quad \mu = \frac{1}{\nu_d^{\frac{1}{d}}}.$$

(19.292)

The two exponents δ_0 and δ_1 (related to the summed and differentiated cutoff propagator, respectively) are defined by

$$\text{SCA}_0 = \max_{m,n,k,l} |Q_{nm;lk}(\Lambda)| \leqslant C_0 \left(\frac{\mu}{\Lambda}\right)^{\delta_0} \delta_{n-m,k-l}.$$

(19.293)

$$\text{SCA}_1 = \max_n \left(\sum_k \max_{m,l} |Q_{nm;lk}(\Lambda)| \leqslant C_1 \left(\frac{\mu}{\Lambda}\right)^{\delta_1}.$$

(19.294)

We need also the product of SCA, SCA_0 given by

$$\text{SCA}_2 = \text{SCA}. \text{SCA}_0 \leqslant C_2 \left(\frac{\Lambda}{\mu}\right)^{\delta_2}.$$

(19.295)

Clearly $C_2 = C_d C_0$ and $\delta_2 = d - \delta_0$.

A non-local matrix model is called regular if we have $\delta_0 = \delta_1 = 2$ otherwise it is called anomalous.

We integrate the Wilson–Polchinski equation between the initial scale Λ_0 and the renormalization scale Λ_R. Recall that the Wilson–Polchinski equation is solved by Ribbon graphs, which are characterized by V (number of vertices), V_e (number of external vertices), B (number of boundary components), g (the genus) and $\mathcal{I}$ (segmentation index). It is necessary to sum over the $s < V_e + \mathcal{I} - 1$ indices of the external legs keeping the incoming indices fixed.

The homogeneous parts $A^{(i)}_{m_1 n_1 \,\ldots\, m_n n_N}[\Lambda]$ with $i = (V, V_e, B, g, \mathcal{I})$ of the interaction coefficients $A^{(V)}_{m_1 n_1 \,\ldots\, m_n n_N}[\Lambda]$ are defined by

$$A^{(V)}_{m_1 n_1 \,\ldots\, m_N n_N}[\Lambda] = \sum_{1 \leqslant V_e \leqslant V} \sum_{1 \leqslant B \leqslant N} \sum_{0 \leqslant g \leqslant 1 + \frac{V}{2} - \frac{N}{4} - \frac{B}{2}} \sum_{0 \leqslant \mathcal{I} \leqslant B - 1} A^{(i)}_{m_1 n_1 \,\ldots\, m_n n_N}[\Lambda].$$

(19.296)

The power-counting theorem for the interaction action $L[M, \Lambda]$ states that the homogeneous parts $A^{(i)}_{m_1 n_1 \,\ldots\, m_n n_N}[\Lambda]$ must scale as

$$\sum_{\varepsilon^s} A^{(i)}_{m_1 n_1 \,\ldots\, m_n n_N}[\Lambda] \leqslant \left(\frac{\Lambda}{\mu}\right)^{\delta_2(V - \frac{N}{2} + 2 - 2g - B)} \left(\frac{\Lambda}{\mu}\right)^{-\delta_1(V - V_e - \mathcal{I} + 2g + B - 1 + s)} \left(\frac{\Lambda}{\mu}\right)^{-\delta_0(V_e + \mathcal{I} - 1 - s)}$$
$$\times P^{2V - \frac{N}{2}}[\ln \frac{\Lambda}{\Lambda_R}].$$

(19.297)

We must have $2 \leqslant N \leqslant 2V + 2$ and $\sum_{i=1}^{N}(m_i - n_i) = 0$.

The set $\mathcal{E}^s$ is the set of s independent end points of trajectories, which are summed over keeping the starting points of these trajectories fixed.

For $V' < V$ and $2 \leqslant N' \leqslant 2V' + 2$ the initial conditions for relevant and marginal $A^{(i)}_{m_1 n_1 \, \ldots \, m_n n_N}[\Lambda]$ are imposed at Λ_R. For $V' = V$ and $N + 2 \leqslant N' \leqslant 2V + 2$ the initial conditions for relevant and marginal $A^{(i)}_{m_1 n_1 \, \ldots \, m_n n_N}[\Lambda]$ are imposed at Λ_0.

The choice of initial (boundary) conditions is at the same time determined by and required to prove this power-counting theorem. In other words, we have to make the correct ansatz for the initial interactions, which we then verify as the correct set of relevant and marginal interactions using the power-counting theorem.

This result is actually model independent (and thus it also applies in four dimensions) and as we can see it depends only on the exponents δ_0 and δ_1 and on the topological structure of the theory given by ribbon graphs.

19.3.10 Summary: Grosse–Wulkenhaar theorem

The theorem can be summarized as follows:
- In $d - 2$ the scaling exponents are given by (19.222) and (19.224), which needs to be substituted into the power-counting theorem (19.297).
- We can see from (19.222) and (19.224) that $\delta_0 = \delta_1 = 2$ for $\omega < 1$ (non-zero harmonic oscillator term) and $\delta_0 = 1$, $\delta_1 = 0$ for $\omega = 1$ (zero harmonic oscillator term). Hence, noncommutative ϕ^4 on $\mathbf{R}^2_\theta$ is expected to be regular (and thus renormalizable) for $\omega < 1$ and it is expected to be anomalous (and thus not renormalizable) for $\omega = 1$.
- We integrate the Wilson–Polchinski equation for the effective action $L[M, \Lambda]$ between the initial scale Λ_0 and the renormalization scale Λ_R. Recall that the Wilson–Polchinski equation is solved by ribbon graphs, which are characterized by V (number of vertices), V_e (number of external vertices), B (number of boundary components), g (the genus) and $\mathcal{I}$ (segmentation index). We need also to sum over external indices keeping fixed incoming indices.
- The choice of initial conditions is at the same time determined by and required to prove the power-counting theorem. The Grosse and Wulkenhaar ansatz for the initial interactions, i.e., the set of relevant and marginal interactions, in $d = 2$ dimensions, is dictated by the fact that only the planar two-point function at one-loop is divergent, is given by (19.226) with

$$\rho^0_{[m]} = \rho^0_1 + 2 \, m \rho^0_2. \tag{19.298}$$

The mass μ^2 should therefore be thought of as the renormalized mass. The coupling constant λ is super-renormalizable. The harmonic oscillator parameter ω does not correspond to any divergences in two dimensions so it does not need to appear in the initial interaction.

The coupling constants $\rho^0_{[m]}$ and ω must therefore depend on Λ_0 in such a way that the limit $\Lambda_0 \longrightarrow \infty$ exists.

- At the renormalization scale Λ_R, which is far below Λ_0, we expect that the interaction $L[M, \Lambda_R, \Lambda_0, \omega, \rho^0]$ becomes simple again. Indeed, after we integrate the Polchinski equation (19.172) from Λ_0 down to Λ_R with the initial condition (19.226) we obtain a solution of the form

$$L[M, \Lambda, \Lambda_0, \omega, \rho^0] = \nu 2 \sum_{m,n,k,l} \frac{1}{2} \rho^0_{[m]} \tilde{M}_{mn} \tilde{M}_{nm} + \nu 2 \frac{\lambda}{4!} \sum_{m,n,k,l} \tilde{M}_{mn} \tilde{M}_{nk} \tilde{M}_{kl} \tilde{M}_{lm} \tag{19.299}$$
$$+ \text{other } M \text{ structures.}$$

- The renormalization conditions are conditions which must be satisfied by the ρ-coefficients at the renormalization cutoff Λ_R, viz $\rho_{[m]}[\Lambda_R, \Lambda_0, \omega, \rho^0] = \text{constant}$. We will choose in particular

$$\rho_{[m]}[\Lambda_R, \Lambda_0, \omega, \rho^0] = 0. \tag{19.300}$$

Thus, in particular $\rho_{[0]}[\Lambda_R, \Lambda_0, \omega, \rho^0] = 0$ and hence μ^2 is the renormalized mass as pointed out above.

- The homogeneous parts $A^{(i)}_{m_1 n_1 \dots m_n n_N}[\Lambda]$ of the coefficients of the effective action $L[M, \Lambda]$ must scale as in (19.297) with $2 \leqslant N \leqslant 2V + 2$ and $\sum_{i=1}^{N}(m_i - n_i) = 0$. These homogeneous parts are therefore bounded and they describe a regularized noncommutative ϕ^4 on $\mathbf{R}^2_\theta$. The limit $\Lambda_0 \longrightarrow \infty$ exists.
- The homogeneous parts $R^{(i)}_{m_1 n_1 \dots m_n n_N}[\Lambda]$ of the coefficients of the varied effective action $R[M, \Lambda]$ given by (19.237) must also scale as in (19.297) times Λ^2/Λ_0^2 with $2 \leqslant N \leqslant 2V + 2$ and $\sum_{i=1}^{N}(m_i - n_i) = 0$. These homogeneous parts are therefore also bounded and they approach zero in the limit $\Lambda_0 \longrightarrow \infty$.
- Hence, noncommutative ϕ^4 on $\mathbf{R}^2_\theta$ is perturbatively renormalizable order by order in λ for $\omega < 1$, i.e., with a non-zero harmonic oscillator term.
- However, in the limit $\Lambda_0 \longrightarrow \infty$ we can send $\omega[\Lambda_0] \longrightarrow 1$ in such a way that a limit of the expansion coefficients $A^{(i)}_{m_1 n_1 \dots m_n n_N}$ also exists. A solution is given by

$$\omega[\Lambda_0] = 1 - \frac{1}{\left(1 + \ln \dfrac{\Lambda_0}{\Lambda_R}\right)^2}. \tag{19.301}$$

Thus noncommutative ϕ^4 on $\mathbf{R}^2_\theta$ is perturbatively renormalizable order by order in λ without a harmonic oscillator term.

19.4 Renormalization of scalar ϕ^4 in four dimensions with a harmonic oscillator term

19.4.1 The main result in 4 dimensions

Again, we will deal with two scales. The initial scale Λ_0, which is actually infinite for the noncommutative space $\mathbf{R}^4_\theta$ and the renormalization scale Λ_R at which renormalization conditions must be imposed in order to subtract divergences. In other words, $\Lambda_R \leqslant \Lambda \leqslant \Lambda_0$.

On $\mathbf{R}_\theta^4$ or equivalently $\mathbf{R}_\theta^2 \times \mathbf{R}_\theta^2$ we replace mn by $m_1 n_1 m_2 n_2$, kl by $k_1 l_1 k_2 l_2$, nk by $n_1 k_1 n_2 k_2$, kl by $k_1 l_1 k_2 l_2$ and lm by $l_1 m_1 l_2 m_2$ to obtain the Laplacian

$$
\begin{aligned}
G^{m_2 n_2, k_2 l_2}_{m_1 n_1,\, k_1 l_1} =\ & \left(\mu^2 + \mu_1^2(m_1 + n_1 + m_2 + n_2 - 2)\right)\delta_{n_1,k_1}\,\delta_{m_1,l_1}\,\delta_{n_2,k_2}\,\delta_{m_2,l_2} \\
& - \mu_1^2\sqrt{\omega}\left(\sqrt{(m_1 - 1)(n_1 - 1)}\ \delta_{n_1-1,k_1}\delta_{m_1-1,l_1} + \sqrt{m_1 n_1}\ \delta_{n_1+1,k_1}\delta_{m_1+1,l_1}\right) \\
& \times \delta_{n_2,k_2}\,\delta_{m_2,l_2} \\
& - \mu_1^2\sqrt{\omega}\left(\sqrt{(m_2 - 1)(n_2 - 1)}\ \delta_{n_2-1,k_2}\delta_{m_2-1,l_2} + \sqrt{m_2 n_2}\ \delta_{n_2+1,k_2}\delta_{m_2+1,l_2}\right) \\
& \times \delta_{n_1,k_1}\,\delta_{m_1,l_1}.
\end{aligned}
\tag{19.302}
$$

The propagator will now involve the orthogonal Meixner polynomials.

The power-counting theory (19.297) holds also in four dimensions since it is model independent. And we integrate the Wilson–Polchinski renormalization group equation (19.172) for the effective action $L[M, \Lambda]$ between the initial scale Λ_0 and the renormalization scale Λ_R with a solution given as before by ribbon graphs.

Again, a choice of initial conditions is at the same time determined by and required to prove the power-counting theorem. The Grosse–Wulkenhaar ansatz for the initial interactions, which we then verify as the correct set of relevant and marginal interactions using the power-counting theorem, in four dimensions is given by

$$
\begin{aligned}
L[M, \Lambda_0, \rho^0] =\ & \sqrt{\det(2\pi\tilde{\theta})} \sum_{m_1,n_1,m_2,n_2,k_1,l_1,k_2,l_2} \frac{1}{2} \tilde{M}^{m_2 n_2}_{m_1 n_1}\, \tilde{G}^{m_2 n_2, k_2 l_2}_{m_1 n_1,\, k_1 l_1}\, \tilde{M}^{k_2 l_2}_{k_1 l_1} \\
& + \sqrt{\det(2\pi\tilde{\theta})}\,\frac{\rho_4^0}{4!} \sum_{m_1,n_1,m_2,n_2,k_1,k_2,l_1,l_2} \tilde{M}^{m_2 n_2}_{m_1 n_1}\, \tilde{M}^{n_2 k_2}_{n_1 k_1}\, \tilde{M}^{k_2 l_2}_{k_1 l_1}\, \tilde{M}^{l_2 m_2}_{l_1 m_1}.
\end{aligned}
\tag{19.303}
$$

$$
\begin{aligned}
\tilde{G}_{m_1 n_1 m_2 n_2, k_1 l_1 k_2 l_2} =\ & \left(\rho_1^0 + \rho_2^0(m_1 + n_1 + m_2 + n_2)\right)\delta_{n_1,k_1}\,\delta_{n_2,k_2}\,\delta_{m_1,l_1}\,\delta_{m_2,l_2} \\
& - \rho_3^0\left(\sqrt{m_1 n_1}\ \delta_{n_1+1,k_1}\delta_{m_1+1,l_1}\delta_{n_2,k_2}\delta_{m_2,l_2} + \sqrt{m_2 n_2}\ \delta_{n_2+1,k_2}\delta_{m_2+1,l_2}\delta_{n_1,k_1}\delta_{m_1,l_1}\right).
\end{aligned}
\tag{19.304}
$$

This corresponds to the fact that in four dimensions we have four divergences: the mass parameter μ^2, the coupling constant λ, the field $\tilde{M}$ (wave function renormalization) and also the harmonic oscillator frequency Ω.

The main result here is that noncommutative ϕ^4 on $\mathbf{R}_\theta^4$ is perturbatively renormalizable order by order in λ for $\omega < 1$, i.e., with a non-zero harmonic oscillator term. In other words, a limit $\Lambda_0 \longrightarrow \infty$ exists for $\omega < 1$.

However, in contrast with the case of $\mathbf{R}_\theta^2$, we cannot in this case scale away the frequency $\omega[\Lambda_0]$ of the harmonic oscillator with the cutoff Λ_0. The theory on $\mathbf{R}_\theta^4$ is therefore very different from the usual commmutative ϕ^4 since the noncommutativity θ relevant at very short distances is reflected in the structure of space at large distances (the so-called UV/IR mixing).

Indeed, the harmonic oscillator frequency is seen from this renormalizability result to be a genuine physical parameter, which indicates that space at very large distances is not what it seems. This is illustrated by the discrete eigenvalues of the squared momentum, which are characterized by an equidistant spacing equal $4\Omega/\theta$ and thus the minimum momentum carried by the scalar field is $\sqrt{4\Omega/\theta}$. This signals that the size of the Universe in any direction is effectively finite of the order of $\sqrt{\theta/4\Omega}$.

19.4.2 The beta function

Another important result is the one-loop beta function of duality-covariant non-commutative ϕ^4 on $\mathbf{R}^4_\theta$, i.e. the theory with a non-zero harmonic oscillator term, which was calculated in [14]. The beta function of the coupling constant λ was found to be non-negative but vanishes for the self-dual theory with $\Omega = 1$. It is expected that the vanishing of the beta function at the self-dual point is exact to all orders in pertrubation theory.

   This result means in particular that the theory is not asymptotically free in the ultraviolet since the renormalization group flow of the coupling constant is bounded and thus the theory does not exhibit a Landau ghost, i.e. not trivial. In contrast the commutative ϕ^4 theory although also asymptotically free exhibits a Landau ghost.

19.4.3 Gaussian noncommutative fixed points

In this section we give an explicit derivation of a general class of Wilson–Polchinski renormalization group equations of duality-covariant noncommutative ϕ^4 on $\mathbf{R}^4_\theta$ following [15]. Then we introduce Gaussian noncommutative fixed points (also called floating fixed points in [16]) and their renormalized trajectories.

19.4.3.1 The Wilson–Polchinski renormalization group equation
We consider the case of four dimensions. The action is given explicitly by (with $\nu_4 = \sqrt{\det(2\pi\theta)}$, $\tilde{X}_i = 2(\theta^{-1})_{ij} X_j$ and $B^2\theta^2 = 4\Omega^2$)

$$
\begin{aligned}
S &= \int d^4x \left[\Phi\left(-\frac{1}{2}\partial_i^2 + \frac{1}{2}(B_{ij}x_j)^2 + \frac{\mu^2}{2}\right)\Phi + \frac{\lambda}{4!}\Phi*\Phi*\Phi*\Phi(x)\right] \\
&= \int d^4x \left[\Phi\left(-\frac{1}{2}\partial_i^2 + \frac{1}{2}\Omega^2\tilde{x}_i^2 + \frac{\mu^2}{2}\right)\Phi + \frac{\lambda}{4!}\Phi*\Phi*\Phi*\Phi\right] \\
&= \nu_4 Tr_{\mathcal{H}} \left[\Phi\left(-\frac{1}{2}\hat{\partial}_i^2 + \frac{1}{2}\Omega^2\hat{\tilde{X}}_i^2 + \frac{\mu^2}{2}\right)\Phi + \frac{\lambda}{4!}\Phi^4\right] \\
&= \nu_4 Tr_{\mathcal{H}} \left[\frac{2+\omega}{\theta^2}\Phi\hat{x}_i^2\Phi + \frac{\omega}{\theta^2}\Phi\hat{x}_i\Phi\hat{x}_i + \frac{\mu^2}{2}\Phi^2 + \frac{\lambda}{4!}\Phi^4\right].
\end{aligned}
\tag{19.305}
$$

In this section ω is defined by

$$\Omega^2 = 1 + \omega. \tag{19.306}$$

The field Φ can be expanded as $\Phi = \sum_{m_i,\,n_i=1}^{\infty} \tilde{M}_{m_1 n_1}^{m_2 n_2} \hat{\phi}_{m_1,n_1} \otimes \hat{\phi}^{m_2,\,n_2}$. The action in the matrix base reads

$$S[M] = \nu_4 \sum_{m_i,n_i,k_i,l_i} \left(\frac{1}{2} \tilde{M}^{m_2 n_2}_{m_1 n_1} \, \tilde{G}^{m_2 n_2, k_2 l_2}_{m_1 n_1,\, k_1 l_1} \, \tilde{M}^{k_2 l_2}_{k_1 l_1} + \frac{\lambda}{4!} \tilde{M}^{m_2 n_2}_{m_1 n_1} \, \tilde{M}^{n_2 k_2}_{n_1 k_1} \, \tilde{M}^{k_2 l_2}_{k_1 l_1} \, \tilde{M}^{l_2 m_2}_{l_1 m_1} \right). \tag{19.307}$$

The Laplacian is given by

$$\begin{aligned}
G^{m_2 n_2, k_2 l_2}_{m_1 n_1,\, k_1 l_1} &= \left(\mu^2 + 2\frac{2 + \omega}{\theta}(m_1 + n_1 + m_2 + n_2 - 2) \right) \delta_{n_1,k_1} \delta_{m_1,l_1} \delta_{n_2,k_2} \delta_{m_2,l_2} \\
&\quad + \frac{2\omega}{\theta}(\sqrt{(m_1 - 1)(n_1 - 1)} \, \delta_{n_1 - 1, k_1} \delta_{m_1 - 1, l_1} + \sqrt{m_1 n_1} \, \delta_{n_1 + 1, k_1} \delta_{m_1 + 1, l_1}) \\
&\quad \times \delta_{n_2, k_2} \delta_{m_2, l_2} \\
&\quad + \frac{2\omega}{\theta}(\sqrt{(m_2 - 1)(n_2 - 1)} \, \delta_{n_2 - 1, k_2} \delta_{m_2 - 1, l_2} + \sqrt{m_2 n_2} \, \delta_{n_2 + 1, k_2} \delta_{m_2 + 1, l_2}) \\
&\quad \times \delta_{n_1, k_1} \delta_{m_1, l_1}.
\end{aligned} \tag{19.308}$$

We split the action $S[M]$ into a kinetic term L_0 and an interaction L, i.e.,

$$S[M] = L_0 + L. \tag{19.309}$$

The kinetic part is

$$\begin{aligned}
L_0 &= \nu_4 Tr_{\mathcal{H}} \frac{2}{\theta^2} \Phi \hat{x}_i^2 \Phi \\
&= \nu_4 \sum_{m_i,n_i,k_i,l_i} \frac{1}{2} \tilde{M}^{m_2 n_2}_{m_1 n_1} \left(\frac{4}{\theta}(m_1 + n_1 + m_2 + n_2 - 2)\delta_{n_1,k_1} \delta_{m_1,l_1} \delta_{n_2,k_2} \delta_{m_2,l_2} \right) \tilde{M}^{k_2 l_2}_{k_1 l_1}.
\end{aligned} \tag{19.310}$$

The definition of L is obvious from the above equations. The bare propagator is given by

$$\Delta^{m_2 n_2, k_2 l_2}_{m_1 n_1,\, k_1 l_1} = \frac{\theta}{4(m_1 + n_1 + m_2 + n_2 - 2)} \delta_{n_1,k_1} \delta_{m_1,l_1} \delta_{n_2,k_2} \delta_{m_2,l_2}. \tag{19.311}$$

We introduce a cutoff Λ and a cutoff function $K_2(x)$ defined by $K_2(x) = 1$ for $x \leqslant 1$ and $K_2(x) \longrightarrow 0$ rapidly for $x > 1$. The regularized propagator is (with $\bar{\theta} = \theta \Lambda^2$)

$$\Delta^{m_2 n_2, k_2 l_2}_{m_1 n_1,\, k_1 l_1}(\Lambda) = \Delta^{m_2 n_2, k_2 l_2}_{m_1 n_1,\, k_1 l_1} \prod_{i \in \{m_i, n_i\}} K_2\!\left(\frac{4i}{\bar{\theta}} \right). \tag{19.312}$$

We consider now an action $S[M, \Lambda]$ given in terms of the inverse of the above regularized propagator and a general interaction $L[\Lambda]$ satisfying $L[\infty] = L$ by

$$S[M, \Lambda] = L_0[\Lambda] + L[\Lambda]. \tag{19.313}$$

The kinetic part is given by

$$\begin{aligned}
L_0[\Lambda] &= \nu_4 \sum_{m_i,n_i,k_i,l_i} \frac{1}{2} \tilde{M}^{m_2 n_2}_{m_1 n_1} \left(\frac{4}{\theta}(m_1 + n_1 + m_2 + n_2 - 2) \prod_{i \in \{m_i, n_i\}} K_2^{-1}\!\left(\frac{4i}{\bar{\theta}} \right) \delta_{n_1,k_1} \delta_{m_1,l_1} \delta_{n_2,k_2} \delta_{m_2,l_2} \right) \\
&\quad \times \tilde{M}^{k_2 l_2}_{k_1 l_1}.
\end{aligned} \tag{19.314}$$

The general interaction $L[\Lambda]$ must solve the Polchinski equation

$$\Lambda\frac{\partial L[\Lambda]}{\partial \Lambda} = \frac{1}{2\nu_4}\sum_{m_i,n_i,k_i,l_i}\Lambda\frac{\partial\Delta_{n_1m_1,\,l_1k_1}^{n_2m_2,k_2l_2}(\Lambda)}{\partial\Lambda}\left[\frac{\partial L[\Lambda]}{\partial\tilde{M}_{k_1l_1}^{k_2l_2}}\frac{\partial L[\Lambda]}{\partial\tilde{M}_{m_1n_1}^{m_2n_2}} - \frac{\partial^2 L[\Lambda]}{\partial\tilde{M}_{k_1l_1}^{k_2l_2}\partial\tilde{M}_{m_1n_1}^{m_2n_2}}\right] \quad (19.315)$$

For ease of notation we write this as follows:

$$\begin{aligned}\Lambda\frac{\partial L[\Lambda]}{\partial\Lambda} &= \frac{1}{2\nu_4}\sum_{m,n,k,l}\Lambda\frac{\partial\Delta_{nm,lk}(\Lambda)}{\partial\Lambda}\left[\frac{\partial L[\Lambda]}{\partial\tilde{M}_{kl}}\frac{\partial L[\Lambda]}{\partial\tilde{M}_{mn}} - \frac{\partial^2 L[\Lambda]}{\partial\tilde{M}_{kl}\partial\tilde{M}_{mn}}\right] \\ &= -\frac{1}{2\nu_4}\sum_{m,n,k,l}\dot{\Delta}_{nm,lk}(\Lambda)\left[\frac{\partial L[\Lambda]}{\partial\tilde{M}_{kl}}\frac{\partial L[\Lambda]}{\partial\tilde{M}_{mn}} - \frac{\partial^2 L[\Lambda]}{\partial\tilde{M}_{kl}\partial\tilde{M}_{mn}}\right].\end{aligned} \quad (19.316)$$

Let us define $\Sigma = S - 2L_0$. The above equation can be put into the form

$$\Lambda\frac{\partial S[M,\Lambda]}{\partial\Lambda} = -\frac{1}{2\nu_4}\sum_{m,n,k,l}\dot{\Delta}_{nm,lk}(\Lambda)\left[\frac{\partial S[M,\Lambda]}{\partial\tilde{M}_{kl}}\frac{\partial\Sigma[M,\Lambda]}{\partial\tilde{M}_{mn}} - \frac{\partial^2\Sigma[M,\Lambda]}{\partial\tilde{M}_{kl}\partial\tilde{M}_{mn}}\right]. \quad (19.317)$$

The starting point for the derivation of the Polchinski equation is

$$-\Lambda\frac{\partial}{\partial\Lambda}Z = \partial_t Z = 0. \quad (19.318)$$

$$Z = \int d\tilde{M}_{mn}\; e^{-S[M,\Lambda]}. \quad (19.319)$$

The partition function does not depend on the cutoff which means that the low-energy observables do not depend on the cutoff. The renormalization group time t is defined by $t = \ln\frac{\mu}{\Lambda}$ where μ is a fixed physical mass scale.

Let us now make a field redefinition $\tilde{M}_{mn}' = \tilde{M}_{mn} - \Theta_{mn}(\tilde{M}_{kl},t)$. We compute $d\tilde{M}_{mn} = (1 + \sum_{m,n}\frac{\partial\Theta_{mn}}{\partial\tilde{M}_{mn}'})d\tilde{M}_{mn}'$ and $e^{-S[M,\Lambda]} = e^{-S[M',\Lambda]}(1 - \sum_{m,n}\Theta_{mn}\frac{\partial S}{\partial\tilde{M}_{mn}'})$. Hence from the invariance of the partition function Z under field redifinition we obtain

$$0 = \int d\tilde{M}_{mn}\;\frac{\partial}{\partial\tilde{M}_{mn}}(\Theta_{mn}e^{-S[M,\Lambda]}). \quad (19.320)$$

If the field redefinition is coming from an infinitesimal change in the cutoff Λ then $\Theta = \Psi dt$. We get then

$$0 = \int d\tilde{M}_{mn}\;\frac{\partial}{\partial\tilde{M}_{mn}}(\Psi_{mn}e^{-S[M,\Lambda]}). \quad (19.321)$$

Putting (19.318) and (19.321) together we get the general exact renormalization group equation [15]

$$\partial_t e^{-S[M,\Lambda]} = \frac{\partial}{\partial\tilde{M}_{mn}}(\Psi_{mn}e^{-S[M,\Lambda]}). \quad (19.322)$$

Thus a change in the cutoff is actually compensated by a field redefinition. The functional Ψ_{mn} defines a Kadanoff blocking procedure. We can get the Polchinski equation (19.325) from (19.322) by choosing

$$\Psi_{mn} = \frac{1}{2\nu_4}\sum_{k,l}\dot{\Delta}_{nm,lk}(\Lambda)\frac{\partial\Sigma[M,\Lambda]}{\partial\tilde{M}_{kl}}. \tag{19.323}$$

Let us choose instead

$$\Psi_{mn} = \frac{Z}{2\nu_4}\sum_{k,l}\dot{\Delta}_{nm,lk}(\Lambda)\frac{\partial\Sigma[M,\Lambda]}{\partial\tilde{M}_{kl}}. \tag{19.324}$$

We get the Polchinski equation

$$\Lambda\frac{\partial S[M,\Lambda]}{\partial\Lambda} = -\frac{Z}{2\nu_4}\sum_{m,n,k,l}\dot{\Delta}_{nm,lk}(\Lambda)\left[\frac{\partial S[M,\Lambda]}{\partial\tilde{M}_{kl}}\frac{\partial\Sigma[M,\Lambda]}{\partial\tilde{M}_{mn}} - \frac{\partial^2\Sigma[M,\Lambda]}{\partial\tilde{M}_{kl}\partial\tilde{M}_{mn}}\right]. \tag{19.325}$$

We rescale to dimensionless variables by setting $\tilde{M}_{mn} = \tilde{m}_{mn}\sqrt{Z}\,\Lambda$ where Z is the field strength renormalization. We compute $(\tilde{m}\sqrt{Z}\,\Lambda)_{\Lambda+d\Lambda} = (\tilde{m}\sqrt{Z}\,\Lambda)_\Lambda + ((2+\eta)\tilde{m}\sqrt{Z}\,d\Lambda)/2$ where $\eta = \Lambda(\partial\ln Z/\partial\Lambda)$ is the anomalous dimension of the field. We compute

$$\begin{aligned}\Lambda\frac{\partial S[M,\Lambda]}{\partial\Lambda} &= \Lambda\,\mathrm{Lim}_{d\Lambda\longrightarrow 0}\frac{S[M,\Lambda+d\Lambda] - S[M,\Lambda]}{(\Lambda+d\Lambda) - \Lambda}\\[2mm] &= \Lambda\,\mathrm{Lim}_{d\Lambda\longrightarrow 0}\frac{S[(\tilde{m}\sqrt{Z}\,\Lambda)_{\Lambda+d\Lambda} - \frac{2+\eta}{2}\tilde{m}\sqrt{Z}\,d\Lambda,\,\Lambda+d\Lambda] - S[(\tilde{m}\sqrt{Z}\,\Lambda)_\Lambda,\,\Lambda]}{(\Lambda+d\Lambda) - \Lambda}\\[2mm] &= \Lambda\frac{\partial S[\tilde{m}\sqrt{Z}\,\Lambda,\,\Lambda]}{\partial\Lambda} - \frac{2+\eta}{2}\tilde{m}_{mn}\frac{\partial S[\tilde{m}\sqrt{Z}\,\Lambda,\,\Lambda]}{\partial\tilde{m}_{mn}}.\end{aligned} \tag{19.326}$$

The Polchinski equation becomes, given by (with $S[\tilde{m}\sqrt{Z}\,\Lambda,\Lambda] = S[\tilde{m},\Lambda]$ and $\Sigma[\tilde{m}\sqrt{Z}\,\Lambda,\Lambda] = \Sigma[\tilde{m},\Lambda]$)

$$\left(\partial_t + \frac{2+\eta}{2}\tilde{m}_{mn}\frac{\partial}{\partial\tilde{m}_{mn}}\right)S[\tilde{m},\Lambda] = \frac{1}{2\pi^2\bar{\theta}}\sum_{m,n,k,l}\dot{\Delta}_{nm,lk}(\Lambda)\left[\frac{\partial S[\tilde{m},\Lambda]}{\partial\tilde{m}_{kl}}\frac{\partial\Sigma[\tilde{m},\Lambda]}{\partial\tilde{m}_{mn}} - \frac{\partial^2\Sigma[\tilde{m},\Lambda]}{\partial\tilde{m}_{kl}\partial\tilde{m}_{mn}}\right]. \tag{19.327}$$

Let us recall that $\dot{\Delta}_{nm,lk}(\Lambda) = \partial_t\Delta_{nm,lk}(\Lambda)$. However, the propagator $\Delta_{mn,kl}(\Lambda) = \Delta_{nm,lk}(\Lambda)$ is defined now without the θ in front, viz

$$\begin{aligned}\Delta^{m_2 n_2,k_2 l_2}_{m_1 n_1,\,k_1 l_1}(\Lambda) &= \frac{1}{4(m_1 + n_1 + m_2 + n_2 - 2)}\delta_{n_1,k_1}\delta_{m_1,l_1}\delta_{n_2,k_2}\delta_{m_2,l_2}\prod_{i\in\{m_i,n_i\}}K_2\left(\frac{4i}{\bar{\theta}}\right)\\[2mm] &= \Delta^{m_2 n_2,k_2 l_2}_{m_1 n_1,\,k_1 l_1}(\bar{\theta}).\end{aligned} \tag{19.328}$$

In the following it is understood that all other rescalings are performed using Λ and not θ in analogy with $\bar{\theta}$ and $\tilde{m}$.

19.4.3.2 Gaussian noncommutative fixed points

In commutative field theory critical fixed points are defined by the condition

$$\partial_t|_\phi\, S_*[\phi] = 0. \tag{19.329}$$

These fixed points correspond to conformal field theories, which are non-perturbatively renormalizable. Since conformal field theory is scale independent, one can send in a trivial way the cutoff Λ_0 (the bare scale) to infinity. Let us recall here that there is also another scale, which is the renormalization scale Λ_R. The non-perturbative renormalizability of a theory is equivalent to the statement that having integrated out momenta between the bare scale Λ_0 and the effective scale Λ the limit $\Lambda_0 \longrightarrow \infty$ can be shown to exist, which means that all divergences can be absorbed in a finite number of parameters.

In noncommutative field theory critical floating points or noncommutative fixed points are defined by the condition [16]

$$\partial_t|_{\phi,\bar{\theta}}\, S_*[\phi] = 0. \tag{19.330}$$

In other words, floating points or noncommutative fixed points are points that are independent of the scale up to dependence on $\bar{\theta}$, i.e., we allow dependence of $S_*[\phi]$ on $\bar{\theta}$, which is taken to be fixed when differentiating with respect to Λ. Thus we do not allow dependence of $S_*[\phi]$ on $\theta\Lambda_0^2 = \bar{\theta}\Lambda_0^2/\Lambda^2$.

Let us write the kinetic term L_0 as

$$L_0[\Lambda] = \frac{\overline{V_4}}{\bar{\theta}} Z \sum_{m_i,n_i,k_i,l_i} \frac{1}{2}\tilde{m}_{m_1n_1}^{m_2n_2}(\Delta^{-1})_{m_1n1,\,k_1l_1}^{m_2n_2,k_2l_2}\,\tilde{m}_{k_1l_1}^{k_2l_2}. \tag{19.331}$$

It is trivial to observe that (with $Z=1$)

$$\partial_t|_{\phi,\bar{\theta}}\, L_0[\Lambda] = 0. \tag{19.332}$$

Let us next consider an interaction of the form

$$L[\Lambda] = \frac{\overline{V_4}}{\bar{\theta}} Z \sum_{m_i,n_i,k_i,l_i} \frac{1}{2}\tilde{m}_{m_1n_1}^{m_2n_2}\,\Omega_{m_1n1,\,k_1l_1}^{m_2n_2,k_2l_2}\,\tilde{m}_{k_1l_1}^{k_2l_2}. \tag{19.333}$$

We set $F = \Delta^{-1} + \Omega$. By using the fact that $\bar{\theta} = \theta\Lambda^2$, i.e. $\partial_t\bar{\theta} = -2\bar{\theta}$ we compute

$$\left(\partial_t + \frac{2+\eta}{2}\tilde{m}_{mn}\frac{\partial}{\partial\tilde{m}_{mn}}\right)S[\tilde{m},\Lambda] = \frac{\overline{V_4}}{\bar{\theta}} Z \sum_{m_i,n_i,k_i,l_i} \frac{1}{2}\tilde{m}_{m_1n_1}^{m_2n_2}\,(\partial_t F)_{m_1n1,\,k_1l_1}^{m_2n_2,k_2l_2}\,\tilde{m}_{k_1l_1}^{k_2l_2}, \tag{19.334}$$

and

$$\frac{1}{2\pi^2\bar{\theta}} \sum_{m,n,k,l} \dot{\Delta}_{nm,lk}(\Lambda)\left[\frac{\partial S[\tilde{m},\Lambda]}{\partial\tilde{m}_{kl}}\frac{\partial\Sigma[\tilde{m},\Lambda]}{\partial\tilde{m}_{mn}} - \frac{\partial^2\Sigma[\tilde{m},\Lambda]}{\partial\tilde{m}_{kl}\partial\tilde{m}_{mn}}\right] =$$
$$\frac{\overline{V_4}}{\bar{\theta}} Z \sum_{m_i,n_i,k_i,l_i} \frac{1}{2}\tilde{m}_{m_1n_1}^{m_2n_2}\,(ZF\dot{\Delta}(-2\Delta^{-1}+F))_{m_1n1,\,k_1l_1}^{m_2n_2,k_2l_2}\,\tilde{m}_{k_1l_1}^{k_2l_2}. \tag{19.335}$$

In the last equation we have neglected as usual a constant term, i.e. the vacuum energy term. We compute $\dot{\Delta}_{mn,kl}(\Lambda) = \Delta_{mn,kl}(\Lambda)\,\partial_t\ln\prod_{i\in\{m_i,n_i\}} K_2(\frac{4i}{\theta\Lambda^2})$. We write this fact as $\dot{\Delta} = \Delta K$ where K is a diagonal matrix given by $K_{m_1n_1,\,k_1l_1}^{m_2n_2,k_2l_2} = \partial_t\ln\prod_{i\in\{m_i,n_i\}} K_2(\frac{4i}{\theta\Lambda^2})\delta_{n_1,k_1}\,\delta_{m_1,l_1}\,\delta_{n_2,k_2}\,\delta_{m_2,l_2}$. By comparing the above two equations we get the condition

$$\partial_t F = ZF\dot{\Delta}(-2\Delta^{-1} + F). \tag{19.336}$$

A solution is given by (with $Z = 1$, i.e. $\eta = 0$)

$$F = \frac{1}{1 - W^{-1}\Delta}\Delta^{-1}. \tag{19.337}$$

The constant W is independent of $\bar{\theta}$. We have therefore a line of equivalent Gaussian floating/fixed points linked by field reparametrizations.

19.4.3.3 Renormalized trajectories

We again write the action as

$$S = \nu_4 Tr_{\mathcal{H}}\left[(2 + \omega)\Phi\tilde{x}_i^2\Phi + \omega\Phi\tilde{x}_i\Phi\tilde{x}_i + \frac{\mu_0^2}{2}\Phi^2 + \frac{\lambda}{4!}\Phi^4\right]. \tag{19.338}$$

We have defined $\tilde{x}_i = \hat{x}_i/\theta$, which is of order $1/\sqrt{\theta}$.

Let $S_*[\phi]$ be the action at the noncommutative fixed point. It may depend on $\bar{\theta} = \theta\Lambda^2$ so we write it as $S_*[\phi][\bar{\theta}]$. The renormalization group time is $t = \ln\mu/\Lambda$ where Λ is the cutoff and μ is a physical mass scale. In the vicinity of the noncommutative fixed point the action is

$$\begin{aligned}
S[\phi](t, \bar{\theta}) &= S_*[\phi](\bar{\theta}) + T_t[\phi](\bar{\theta}) \\
&= S_*[\phi](\bar{\theta}) + \bar{\nu}_4\sum_i \alpha_i e^{\zeta_i t} \mathcal{Q}_i[\phi](\bar{\theta}).
\end{aligned} \tag{19.339}$$

The Polchinski equation is

$$\left(\partial_t + \frac{2 + \eta}{2}\tilde{m}_{mn}\frac{\partial}{\partial\tilde{m}_{mn}}\right)S[\tilde{m}, \Lambda] = \frac{1}{2\pi^2\bar{\theta}}\sum_{m,n,k,l}\dot{\Delta}_{nm,lk}(\Lambda)\left[\frac{\partial S[\tilde{m}, \Lambda]}{\partial\tilde{m}_{kl}}\frac{\partial\Sigma[\tilde{m}, \Lambda]}{\partial\tilde{m}_{mn}} - \frac{\partial^2\Sigma[\tilde{m}, \Lambda]}{\partial\tilde{m}_{kl}\partial\tilde{m}_{mn}}\right]. \tag{19.340}$$

We have $S = S_* + T = L_0 + T$ and $\Sigma = S - 2L_0 = -L_0 + T$. We know that

$$\partial_t S_*[\tilde{m}, \Lambda] = 0 \tag{19.341}$$

and

$$\frac{2 + \eta}{2}\tilde{m}_{mn}\frac{\partial}{\partial\tilde{m}_{mn}}L_0[\tilde{m}, \Lambda] = \frac{1}{2\pi^2\bar{\theta}}\sum_{m,n,k,l}\dot{\Delta}_{nm,lk}(\Lambda)\left[-\frac{\partial L_0[\tilde{m}, \Lambda]}{\partial\tilde{m}_{kl}}\frac{\partial L_0[\tilde{m}, \Lambda]}{\partial\tilde{m}_{mn}} + \frac{\partial^2 L_0[\tilde{m}, \Lambda]}{\partial\tilde{m}_{kl}\partial\tilde{m}_{mn}}\right]. \tag{19.342}$$

Since $\dot{\Delta}_{nm,lk} = \Delta_{nm,lk}K$ and $\Delta_{nm,lk} = \Delta_{lk,nm}$ we obtain

$$\left(\partial_t + \frac{2 + \eta}{2}\tilde{m}_{mn}\frac{\partial}{\partial\tilde{m}_{mn}}\right)T[\tilde{m}, \Lambda] = \frac{1}{2\pi^2\bar{\theta}}\sum_{m,n,k,l}\dot{\Delta}_{nm,lk}(\Lambda)\left[\frac{\partial T[\tilde{m}, \Lambda]}{\partial\tilde{m}_{kl}}\frac{\partial T[\tilde{m}, \Lambda]}{\partial\tilde{m}_{mn}} - \frac{\partial^2 T[\tilde{m}, \Lambda]}{\partial\tilde{m}_{kl}\partial\tilde{m}_{mn}}\right]. \tag{19.343}$$

Also $\eta = \Lambda\partial\ln Z/\partial\Lambda = 0$ since $Z = 1$. Furthermore, by neglecting quadratic terms in T we get

$$\left(\partial_t + \tilde{m}_{mn}\frac{\partial}{\partial\tilde{m}_{mn}}\right)T[\tilde{m}, \Lambda] = -\frac{1}{2\pi^2\bar{\theta}}\sum_{m,n,k,l}\dot{\Delta}_{nm,lk}(\Lambda)\frac{\partial^2 T[\tilde{m}, \Lambda]}{\partial\tilde{m}_{kl}\partial\tilde{m}_{mn}}. \tag{19.344}$$

We have $\partial_t = -\Lambda\partial_\Lambda$, $\partial_t\bar{\theta} = -2\bar{\theta}$ and $\partial_t = -2\bar{\theta}\partial_{\bar{\theta}}$. Thus

$$\partial_t T[\tilde{m}, \Lambda] = \bar{\nu}_4\sum_i \alpha_i e^{\zeta_i t}(\zeta_i - 4 - 2\bar{\theta}\partial_{\bar{\theta}})Q_i[\phi](\bar{\theta}). \tag{19.345}$$

We obtain therefore

$$(\zeta_i - 4 - 2\bar{\theta}\partial_{\bar{\theta}} + \tilde{m}_{mn}\frac{\partial}{\partial\tilde{m}_{mn}})Q_i[\phi](\bar{\theta}) = -\frac{1}{2\pi^2\bar{\theta}}\sum_{m,n,k,l}\dot{\Delta}_{nm,lk}(\Lambda)\frac{\partial^2 Q_i[\phi](\bar{\theta})}{\partial\tilde{m}_{kl}\partial\tilde{m}_{mn}} = -\dot{V}Q_i. \tag{19.346}$$

Define

$$\begin{aligned}
Q_i' &= \exp\left(\frac{1}{2\pi^2\bar{\theta}}\sum_{m,n,k,l}\Delta_{nm,lk}(\Lambda)\frac{\partial^2}{\partial\tilde{m}_{kl}\partial\tilde{m}_{mn}}\right)Q_i \\
&= \exp\left(\frac{1}{2\pi^2\bar{\theta}}\sum_{m,n,k,l}\Delta_{nm,lk}(\Lambda)\frac{\partial^2}{\partial\tilde{M}_{kl}\partial\tilde{M}_{mn}}\right)Q_i = \exp(V)Q_i.
\end{aligned} \tag{19.347}$$

From the second expression it is obvious that

$$(\zeta_i - 4 - 2\bar{\theta}\partial_{\bar{\theta}} + \tilde{m}_{mn}\frac{\partial}{\partial\tilde{m}_{mn}})Q_i' = e^V(\zeta_i - 4 - 2\bar{\theta}\partial_{\bar{\theta}} + \tilde{m}_{mn}\frac{\partial}{\partial\tilde{m}_{mn}} + \dot{V})Q_i \tag{19.348}$$
$$= 0.$$

The operators Q_i' are single-trace terms built out of $2J$'s ϕ and 2ξ's $\tilde{x}_i$ of the generic form

$$Q_i' = \bar{\theta}^s Q'^{\rho}_{\mu_1 a_1 b_1, \dots, \mu_{2\xi} a_{2\xi} b_{2\xi}; c_1 d_1, \dots, c_{2J} d_{2J}} \tilde{x}^{\mu_1}_{a_1 b_1} \cdots \tilde{x}^{\mu_{2\xi}}_{a_{2\xi} b_{2\xi}} \tilde{m}_{c_1 d_1} \cdots \tilde{m}_{c_{2J} d_{2J}}. \tag{19.349}$$

The index ρ denotes the independent operators which we can form from $2J$'s ϕ and 2ξ's $\tilde{x}_i$. Thus the index i stands for $(s, \rho, 2\xi, 2J)$. We compute $-2\bar{\theta}\partial_{\bar{\theta}}Q_i' = -2(s - \xi)Q_i'$. Hence we get immediately the constraint

$$\zeta_i - 2(s - \xi) = 4 - 2J. \tag{19.350}$$

Clearly the different operators with different values of s but with the same number of ϕ and $\tilde{x}_i$ are actually the same operator. We write then

$$Q_i' = \bar{\theta}^s Q_j', \quad j = (\rho, 2\xi, 2J). \tag{19.351}$$

We have then

$$\begin{aligned}
S[\phi](t, \bar{\theta}) &= S_*[\phi](\bar{\theta}) + \bar{\nu}_4\sum_i \alpha_i e^{\zeta_i t}Q_i[\phi](\bar{\theta}) \\
&= S_*[\phi](\bar{\theta}) + \bar{\nu}_4\sum_j \int ds\alpha_j(s)e^{\zeta_j t}\bar{\theta}^s Q_j[\phi](\bar{\theta})
\end{aligned} \tag{19.352}$$

where

$$Q_j = \exp(-V)Q_j' = \exp\left(-\frac{1}{2\pi^2\bar{\theta}}\sum_{m,n,k,l}\Delta_{nm,lk}(\Lambda)\frac{\partial^2}{\partial\tilde{m}_{kl}\partial\tilde{m}_{mn}}\right)Q_j'. \tag{19.353}$$

Introducing an appropriate norm we write the above result as

$$S[\phi](t, \bar{\theta}) = S_*[\phi](\bar{\theta}) + \bar{\nu}_4 \sum_j \int ds \alpha_j(s) e^{\zeta_j t \bar{\theta}^s} ||\mathcal{Q}_j|| \hat{\mathcal{Q}}_j[\phi](\bar{\theta})$$
$$= S_*[\phi](\bar{\theta}) + \bar{\nu}_4 \sum_j \hat{g}_j[\alpha](t, \bar{\theta}) \hat{\mathcal{Q}}_j[\phi](\bar{\theta}). \tag{19.354}$$

$$\hat{g}_j[\alpha](t, \bar{\theta}) = \int ds \alpha_j(s) e^{\zeta_j t \bar{\theta}^s} ||\mathcal{Q}_j||. \tag{19.355}$$

The norm will be given by the Gaussian functional integration

$$(a, b) = \frac{\int \mathcal{D}\tilde{m} \, \exp\left(-\frac{1}{2}\pi^2\bar{\theta} \sum_{mnkl} \tilde{m}_{mn} W_{mn,lk} \tilde{m}_{lk}\right) ab}{\int \mathcal{D}\tilde{m} \, \exp\left(-\frac{1}{2}\pi^2\bar{\theta} \sum_{mnkl} \tilde{m}_{mn} W_{mn,lk} \tilde{m}_{lk}\right)}. \tag{19.356}$$

In the above equation W is defined by

$$W^{-1} = \Delta. \tag{19.357}$$

Examples are now in order.
- **Example 1: The self-dual kinetic term** Let us consider operators with $2J = 2$, i.e., with two fields $\tilde{m}_{mn}$ and $\tilde{m}_{kl}$. These read explicitly

$$\mathcal{Q}_2^{(\rho, 2\zeta)} = \exp(-V) \sum_{mnkl} \mathcal{Q}'^{(\rho, 2\zeta)}_{mnkl} \tilde{m}_{mn} \tilde{m}_{kl}$$
$$= \left[1 - \frac{1}{2\pi^2\bar{\theta}} \sum_{m,n,k,l} \Delta_{nm,lk}(\Lambda) \frac{\partial^2}{\partial \tilde{m}_{kl} \partial \tilde{m}_{mn}}\right] \sum_{mnkl} \mathcal{Q}'^{(\rho, 2\zeta)}_{mnkl} \tilde{m}_{mn} \tilde{m}_{kl} \tag{19.358}$$
$$= \sum_{mnkl} \mathcal{Q}'^{(\rho, 2\zeta)}_{mnkl} \left[\tilde{m}_{mn} \tilde{m}_{kl} - \frac{1}{\pi^2\bar{\theta}} \Delta_{nm,lk}\right].$$

We compute the inner product

$$(\mathcal{Q}_2^{(\rho, 2\zeta)}, \mathcal{Q}_2^{(\rho', 2\zeta')}) = \sum_{mnkl} \sum_{m'n'k'l'} \mathcal{Q}'^{(\rho, 2\zeta)}_{mnkl} \mathcal{Q}'^{(\rho', 2\zeta')}_{m'n'k'l'} (\tilde{m}_{mn} \tilde{m}_{kl} - \frac{1}{\pi^2\bar{\theta}} \Delta_{nm,lk}, \tilde{m}_{m'n'} \tilde{m}_{k'l'}$$
$$- \frac{1}{\pi^2\bar{\theta}} \Delta_{n'm',l'k'})$$
$$= \sum_{mnkl} \sum_{m'n'k'l'} \mathcal{Q}'^{(\rho, 2\zeta)}_{mnkl} \mathcal{Q}'^{(\rho', 2\zeta')}_{m'n'k'l'} \left(-\frac{1}{\pi^2\bar{\theta}} \Delta_{n'm',l'k'} \langle \tilde{m}_{mn} \tilde{m}_{kl} \rangle \right. \tag{19.359}$$
$$- \frac{1}{\pi^2\bar{\theta}} \Delta_{nm,lk} \langle \tilde{m}_{m'n'} \tilde{m}_{k'l'} \rangle + \langle \tilde{m}_{mn} \tilde{m}_{kl} \tilde{m}_{m'n'} \tilde{m}_{k'l'} \rangle$$
$$\left. + \frac{1}{(\pi^2\bar{\theta})^2} \Delta_{nm,lk} \Delta_{n'm',l'k'}\right).$$

The expectation values are defined with respect to the partition function

$$Z[J] = \int \mathcal{D}\tilde{m}\, \exp\left(-\frac{1}{2}\pi^2\bar{\theta}\sum_{mnkl}\tilde{m}_{mn}W_{mn,lk}\tilde{m}_{lk} - \sum_{mn}J_{mn}\tilde{m}_{nm}\right)$$
$$= \exp\left(\frac{1}{2\pi^2\bar{\theta}}\sum_{mn}J_{nm}\,W^{-1}_{nm,\,kl}J_{kl}\right)\int \mathcal{D}\tilde{m}\,\exp\left(-\frac{1}{2}\pi^2\bar{\theta}\sum_{mnkl}\tilde{m}_{mn}W_{mn,lk}\tilde{m}_{lk}\right). \tag{19.360}$$

Thus

$$\langle\tilde{m}_{mn}\tilde{m}_{kl}\rangle = \frac{1}{Z[J]}\frac{\partial^2 Z[J]}{\partial J_{nm}\,\partial J_{lk}}\Big|_{J=0} = \frac{1}{\pi^2\bar{\theta}}\Delta_{nm,lk}. \tag{19.361}$$

$$\langle\tilde{m}_{mn}\tilde{m}_{kl}\tilde{m}_{m'n'}\tilde{m}_{k'l'}\rangle = \frac{1}{Z[J]}\frac{\partial^4 Z[J]}{\partial J_{nm}\,\partial J_{lk}\,\partial J_{n'm'}\,\partial J_{l'k'}}\Big|_{J=0}$$
$$= \frac{1}{(\pi^2\bar{\theta})^2}[\Delta_{nm,lk}\Delta_{n'm',l'k'} + \Delta_{nm,n'm'}\Delta_{lk,l'k'} + \Delta_{nm,l'k'}\Delta_{n'm',lk}].$$

Hence we obtain

$$(\mathcal{Q}_2^{(\rho,\,2\varsigma)},\,\mathcal{Q}_2^{(\rho',\,2\varsigma')}) = \frac{1}{(\pi^2\bar{\theta})^2}\sum_{mnklm'n'k'l'}\mathcal{Q}_{mnkl}^{'(\rho,\,2\varsigma)}\mathcal{Q}_{m'n'k'l'}^{'(\rho',\,2\varsigma')}[\Delta_{nm,n'm'}\Delta_{lk,l'k'} + \Delta_{nm,l'k'}\Delta_{n'm',lk}].$$

The self-dual kinetic term is an example. We have

$$L_0 = \nu_4 Tr_{\mathcal{H}}\frac{2}{\theta^2}\Phi\hat{x}_i^2\Phi$$
$$= \nu_4\sum_{m_i,n_i,k_i,l_i}\frac{1}{2}\tilde{M}_{m_1n_1}^{m_2n_2}\left(\frac{4}{\theta}(m_1 + n_1 + m_2 + n_2 - 2)\delta_{n_1,k_1}\delta_{m_1,l_1}\delta_{n_2,k_2}\delta_{m_2,l_2}\right)\tilde{M}_{k_1l_1}^{k_2l_2} \tag{19.362}$$
$$= \bar{\nu}_4\sum_{mnkl}\tilde{m}_{mn}\left(\frac{2}{\theta}(m + n - 2)\delta_{n,k}\delta_{m,l}\right)\tilde{m}_{kl}.$$

In other words, we have

$$\mathcal{Q}_2^{'(\rho,\,2\varsigma)} = \frac{2}{\bar{\theta}}(m + n - 2)\delta_{n,k}\delta_{m,l}. \tag{19.363}$$

We compute

$$(\mathcal{Q}_2^{(\rho,\,2\varsigma)},\,\mathcal{Q}_2^{(\rho,\,2\varsigma)}) = \frac{1}{(\pi^2\bar{\theta})^2}\cdot 2\cdot\left(\frac{2}{\bar{\theta}}\right)^2\sum_{mnkl}(m + n - 2)(k + l - 2)\Delta_{nm,kl}\Delta_{mn,lk}$$
$$= \frac{1}{2\pi^4\bar{\theta}^4}\sum_{mn}c^2(m,\,n,\,\bar{\theta}). \tag{19.364}$$

We have used

$$\Delta_{mn,kl}(\Lambda) = \frac{1}{4(m + n - 2)}\delta_{nk}\delta_{ml}\,c(m,\,n,\,\bar{\theta}), \quad c(m,\,n,\,\bar{\theta}) = \prod_{i\in\{m_i,n_i\}}K_2\left(\frac{4i}{\bar{\theta}}\right). \tag{19.365}$$

In the large Λ limit the above sum can be evaluated by setting the cutoff function equal 1, replacing the sums by integrals and taking $\bar{\theta}$ as the upper limit of the integrals, i.e.,

$$
\begin{aligned}
(\mathcal{Q}_2^{(\rho,\,2\zeta)},\ \mathcal{Q}_2^{(\rho,\,2\zeta)}) &= \frac{1}{2\pi^4\bar{\theta}^4} \int_0^{\bar{\theta}} dm_1 \int_0^{\bar{\theta}} dm_2 \int_0^{\bar{\theta}} dn_1 \int_0^{\bar{\theta}} dn_2 \\
&= \frac{1}{2\pi^4}.
\end{aligned}
\tag{19.366}
$$

The self-dual kinetic operator has a constant norm in the large Λ limit. The corresponding coupling is

$$
\hat{g}_j[\alpha](t,\,\bar{\theta}) = \int ds\,\alpha_j(s)e^{\zeta_j t\bar{\theta}^s} = \int ds\,\alpha_j(s)e^{-(\zeta_j - 2s)\ln\Lambda}e^{s\ln\theta + \zeta_j\ln\mu\bar{\theta}^s}.
\tag{19.367}
$$

For this operator $2\xi = 2$ since it is constructed out of two $\tilde{x}_i$. Thus $\xi = 1$. The identity $\zeta_i - 2(s - \xi) = 4 - 2J$ becomes $\zeta_i - 2s = 0$. In other words, the overall dependence on Λ vanishes, which means that this is a marginal coupling.

In general marginal and relevant couplings are given by the constraint

$$
\lambda_i = \zeta_i - 2s \geqslant 0 \quad \Leftrightarrow \quad 2J + 2\xi \leqslant 4.
\tag{19.368}
$$

For marginal and relevant couplings taking the limit $\Lambda \longrightarrow 0$ brings us back into the Gaussian noncommutative fixed point since the couplings vanish in this limit. If the noncommutative renormalization group eigenvalue λ_i is negative we obtain an irrelevant interaction.

- **Example 2: The mass term** This is given by

$$
\begin{aligned}
L_0 &= \nu_4 Tr_{\mathcal{H}} \frac{\mu_0^2}{2}\Phi\Phi \\
&= \nu_4 \sum_{m_i,n_i,k_i,l_i} \frac{\mu_0^2}{2}\tilde{M}_{m_1 n_1}^{m_2 n_2}(\delta_{n_1,k_1}\delta_{m_1,l_1}\delta_{n_2,k_2}\delta_{m_2,l_2})\tilde{M}_{k_1 l_1}^{k_2 l_2} \\
&= \bar{\nu}_4 \sum_{mnkl} \tilde{m}_{mn}\left(\frac{\bar{\mu}_0^2}{2}\delta_{nk}\delta_{m,l}\right)\tilde{m}_{kl}.
\end{aligned}
\tag{19.369}
$$

In this case

$$
\mathcal{Q}_2^{\prime(\rho,\,2\zeta)} = \frac{\bar{\mu}_0^2}{2}\delta_{n,k}\delta_{m,l}.
\tag{19.370}
$$

We compute now

$$
\begin{aligned}
(\mathcal{Q}_2^{(\rho,\,2\zeta)},\ \mathcal{Q}_2^{(\rho,\,2\zeta)}) &= \frac{2}{(\pi^2\bar{\theta})^2}\left(\frac{\bar{\mu}_0^2}{2}\right)^2 \sum_{mnkl} \Delta_{nm,kl}\Delta_{mn,lk} \\
&= \frac{2}{(\pi^2\bar{\theta})^2}\left(\frac{\bar{\mu}_0^2}{2}\right)^2 \frac{1}{16}\sum_{mn} \frac{c^2(m,\,n,\,\bar{\theta})}{(m + n - 2)^2}.
\end{aligned}
\tag{19.371}
$$

Again we verify that in the large Λ limit we obtain

$$(\mathcal{Q}_2^{(\rho,\,2\zeta)},\,\mathcal{Q}_2^{(\rho,\,2\zeta)}) = \frac{2}{(\pi^2\bar{\theta})^2}\left(\frac{\bar{\mu}_0^2}{2}\right)^2\frac{1}{16}\int dm_1\int dn_1\int dm_2\int dn_2 \frac{c^2(m_1,\,n_1,\,m_2,\,n_2,\,\bar{\theta})}{(m_1+n_1+m_2+n_2-2)^2}$$

$$= \frac{2}{(\pi^2\bar{\theta})^2}\left(\frac{\bar{\mu}_0^2}{2}\right)^2\frac{1}{16}\int_0^{\bar{\theta}} dm_1\int_0^{\bar{\theta}} dn_1\int_0^{\bar{\theta}} dm_2\int_0^{\bar{\theta}} dn_2\frac{1}{(m_1+n_1+m_2+n_2-2)^2}$$

$$= \text{constant}.$$

(19.372)

The mass term has also a constant norm in the large Λ limit. Thus the corresponding coupling is

$$\hat{g}_j[\alpha](t,\,\bar{\theta}) = \int ds\alpha_j(s)e^{\zeta_j t\bar{\theta}^s} = \int ds\alpha_j(s)e^{-(\zeta_j\,-\,2s)\ln\Lambda}e^{s\,\ln\theta+\zeta_j\,\ln\mu\bar{\theta}^s}. \tag{19.373}$$

Since for the mass operator $\xi = 0$ we will have $\zeta_i - 2s = 2$ and hence

$$\hat{g}_j[\alpha](t,\,\bar{\theta}) = O\left(\frac{1}{\Lambda^2}\right). \tag{19.374}$$

The mass is relevant with a renormalization group eigenvalue equal $+2$.

In summary, we obtain at the Gaussian floating or noncommutative fixed point the following behavior:
- The mass operator ($2J = 2$, $2\xi = 0$) is relevant.
- The self-dual kinetic term ($2J = 2$, $2\xi = 2$) is marginal.
- The harmonic osillator term, i.e. the off-diagonal kinetic term ($2J = 2$, $2\xi = 2$) is marginal.
- The ϕ^4 coupling ($2J = 4$, $2\xi = 0$) is marginal.
- All other couplings are irrelevant.

The ϕ^4 coupling at the self-dual point is exactly marginal as shown by the vanishing of the beta function to all orders in perturbation theory.

References

[1] Ydri B 2017 *Lectures on Matrix Field Theory* Lecture Notes in Physics **929** (Berlin: Springer) 1

[2] Grosse H and Wulkenhaar R 2005 Power counting theorem for nonlocal matrix models and renormalization *Commun. Math. Phys.* **254** 91

[3] Grosse H and Wulkenhaar R 2005 Renormalization of phi**4 theory on noncommutative R**4 in the matrix base *Commun. Math. Phys.* **256** 305

[4] Grosse H and Wulkenhaar R 2003 Renormalization of phi**4 theory on noncommutative R**2 in the matrix base *JHEP* **0312** 019

[5] Connes A 1994 *Noncommutative Geometry* (London: Academic)

[6] Weyl H 1931 *The Theory of Groups and Quantum Mechanics* (New York: Dover)

[7] Groenewold H J 1946 On the principles of elementary quantum mechanics *Physica* **12** 405

[8] Moyal J E 1949 Quantum mechanics as a statistical theory *Proc. Cambridge Phil. Soc.* **45** 99

[9] Langmann E and Szabo R J 2002 Duality in scalar field theory on noncommutative phase spaces *Phys. Lett.* B **533** 168

[10] Langmann E 2003 Interacting fermions on noncommutative spaces: exactly solvable quantum field theories in 2n+1 dimensions *Nucl. Phys.* B **654** 404

[11] Gracia-Bondia J M and Varilly J C 1988 Algebras of distributions suitable for phase space quantum mechanics. 1 *J. Math. Phys.* **29** 869

[12] Langmann E, Szabo R J and Zarembo K 2004 Exact solution of quantum field theory on noncommutative phase spaces *JHEP* **0401** 017

[13] Langmann E, Szabo R J and Zarembo K 2003 Exact solution of noncommutative field theory in background magnetic fields *Phys. Lett.* B **569** 95

[14] Grosse H and Wulkenhaar R 2004 The beta function in duality covariant noncommutative phi**4 theory *Eur. Phys. J.* C **35** 277

[15] Latorre J I and Morris T R 2000 Exact scheme independence *JHEP* **0011** 004

[16] Gurau R and Rosten O J 2009 Wilsonian renormalization of noncommutative scalar field theory *JHEP* **0902** 064

IOP Publishing

A Modern Course in Quantum Field Theory, Volume 2 (Second Edition)
Advanced topics
Badis Ydri

Chapter 20

The neural network method for quantum field theory

In this chapter we propose neural networks as a potentially very powerful alternative to the Monte Carlo method for the non-perturbative study of quantum field theory. In the first section we give a brief overview of the Monte Carlo method following mostly the presentation of [1]. In section two we summarize the main four quantum scalar field theories on commutative and noncommutative spaces, which we have already discussed in great detail in this book. In section three we introduce systematically neural networks and in particular the restricted Boltzmann machine [5, 7–9]. In section four we apply the restricted Boltzmann machine to the noncommutative phi-four theory in the matrix basis and draw an interesting universality conjecture.

20.1 The Monte Carlo method

In this first section we briefly review how the Monte Carlo method is applied to quantum scalar field theory.

20.1.1 The Feynman path integral

- We start with a point particle given by the coordinate $q(t) = x(t)$. The solution of the Schrodinger equation is given by

$$\begin{aligned}
|\psi_s(t)\rangle &= \int dx \langle x|\psi_s(t)\rangle |x\rangle \\
&= \int dx \int dx_0\, G(x, t; x_0, t_0)\langle x_0|\psi_s(t_0)\rangle |x\rangle.
\end{aligned} \tag{20.1}$$

doi:10.1088/978-0-7503-5834-7ch20

We have used

$$\int dx\,|x\rangle\langle x| = 1$$

$$|\psi_s(t)\rangle = U(t, t_0)|\psi_s(t_0)\rangle, \quad U(t, t_0) = \exp\left(-\frac{i}{\hbar}H(t - t_0)\right).$$

(20.2)

The Green function or propagator $G(x, t; x_0, t_0)$ is the transition amplitude from the point x_0 at time t_0 to the point x at time t. It is given explicitly by

$$G(x, t; x_0, t_0) = \langle x, t|x_0, t_0\rangle = \langle x|e^{-\frac{i}{\hbar}H(t-t_0)}|x_0\rangle.$$

(20.3)

- Next, we discretize the time interval as $t_j = t_0 + j\varepsilon$, $\varepsilon = (t - t_0)/N$, $j = 0, 1, \ldots, N$. The corresponding coordinates are $x_0, x_1, \ldots, x_N \equiv x$ while the corresponding momenta are $p_0, p_1, \ldots, p_{N-1}$ (where each p_j corresponds to the interval $[x_j, x_{j+1}]$). The propagator is then given by

$$G(x, t; x_0, t_0) = \int dx_1 dx_2 \cdots dx_{N-1} \prod_{j=0}^{N-1} \langle x_{j+1}, t_{j+1}|x_j, t_j\rangle.$$

(20.4)

We compute (with $\langle p|x\rangle = \exp(-ipx/\hbar)/\sqrt{2\pi\hbar}$)

$$\langle x_{j+1}, t_{j+1}|x_j, t_j\rangle = \left\langle x_{j+1}\left|\left(1 - \frac{i}{\hbar}H\varepsilon\right)\right|x_j\right\rangle.$$

(20.5)

In the limit $N \longrightarrow \infty$, $\varepsilon \longrightarrow 0$ keeping $t - t_0 = $ fixed we obtain the propagator

$$G = \int \frac{dp_0}{2\pi\hbar}\frac{dp_1 dx_1}{2\pi\hbar} \cdots \frac{dp_{N-1}dx_{N-1}}{2\pi\hbar}\, e^{\frac{i}{\hbar}\sum_{j=0}^{N-1}(p_j\dot{x}_j - H(p_j, x_j))\varepsilon}$$

$$= \int \mathcal{D}p\mathcal{D}x\, e^{\frac{i}{\hbar}\int_{t_0}^{t} ds(p\dot{x}-H(p, x))}.$$

(20.6)

- We are interested in Hamiltonians of the form

$$H = \frac{p^2}{2m} + V(x).$$

(20.7)

We obtain for this class of Hamiltonians a representation of the propagator defined by the celebrated Feynman path integral given explicitly by

$$G(x, t; x_0, t_0) = N\int \mathcal{D}x\, e^{\frac{i}{\hbar}S[x]}, \quad S[x] = \int dt\, L(x, \dot{x})$$

$$= \int dt\left(\frac{m}{2}\dot{x}^2 - V(x)\right).$$

(20.8)

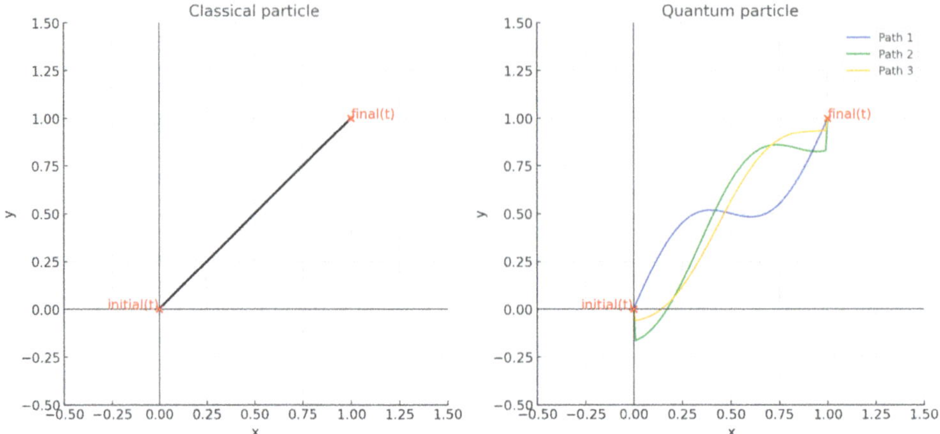

Figure 20.1. The classical versus quantum paths.

In summary, the propagator is given, in terms of an action $S[x]$ and a measure $\mathcal{D}x$, by the relation

$$G(x, t; x_0, t_0) = \mathcal{N} \int \mathcal{D}x \; e^{\frac{i}{\hbar}S[x]}. \tag{20.9}$$

The action is given explicitly by

$$S[x] = \int dt \; L(x, \dot{x}) = \int dt \left(\frac{m}{2}\dot{x}^2 - V(x) \right). \tag{20.10}$$

The measure $\mathcal{D}x$ is defined, in terms of a lattice of points, by a sequence of path integration given by

$$\mathcal{D}x = dx_1 dx_2 \cdots dx_{N-1}, \quad N \longrightarrow \infty, \, \varepsilon \longrightarrow 0, \, t - t_0 = \text{fixed}. \tag{20.11}$$

Each path $x = x(t)$ is thus associated a probability amplitude $\exp(iS[x]/\hbar)$ (figure 20.1).

- We have immediately the following generalization (with T being the time-ordering operator). The n-correlators are given by

$$\langle x, t | T(X(t_1) \ldots X(t_n)) | x_0, t_0 \rangle = \mathcal{N} \int \mathcal{D}x \; x(t_1) \ldots x(t_n) \; e^{\frac{i}{\hbar}S[x]}. \tag{20.12}$$

We work in the basis $|n\rangle$ defined by $H|n\rangle = E_n|n\rangle$. The above matrix elements are then given by

$$\sum_{n,m} e^{-itE_n + it_0 E_m} \langle x|n\rangle \langle m|x_0\rangle \langle n|T(X(t_1) \ldots X(t_n))|m\rangle$$

$$= \mathcal{N} \int \mathcal{D}x \; x(t_1) \ldots x(t_n) \; e^{\frac{i}{\hbar}S[x]}. \tag{20.13}$$

In the limit $t_0 \longrightarrow -\infty$ and $t \longrightarrow \infty$ only the ground state contributes to the correlators and vacuum-to-vacuum amplitude, viz

$$\langle 0| T(X(t_1) \ \ldots \ X(t_n))|0\rangle = \mathcal{N}' \int \mathcal{D}x \ x(t_1) \ \ldots \ x(t_n) \ e^{\frac{i}{\hbar}S[x]}$$

$$\langle 0|0\rangle = \mathcal{N}' \int \mathcal{D}x \ e^{\frac{i}{\hbar}S[x]}. \tag{20.14}$$

Alternatively, we write this result as

$$\langle 0| T(X(t_1) \ \ldots \ X(t_n))|0\rangle = \frac{\int \mathcal{D}x \ x(t_1) \ \ldots \ x(t_n) \ e^{\frac{i}{\hbar}S[x]}}{\int \mathcal{D}x \ e^{\frac{i}{\hbar}S[x]}}. \tag{20.15}$$

The path integral $Z[J]$ in the presence of a source $J(t)$ is given by the vacuum-to-vacuum amplitude

$$Z[J] = \int \mathcal{D}x \ e^{\frac{i}{\hbar}S[x]+\frac{i}{\hbar}\int \, dt J(t)x(t)}. \tag{20.16}$$

This is a generating functional of correlators, i.e.,

$$\langle 0| T(X(t_1) \ \ldots \ X(t_n))|0\rangle = \frac{1}{Z[0]}\left(\frac{\hbar}{i}\right)^n \frac{\delta^n Z[J]}{\delta J(t_1) \ \ldots \ \delta J(t_n)} \Big|_{J=0}. \tag{20.17}$$

20.1.2 Scalar phi-four theory and its lattice regularization

- A generalization to a scalar field theory in a d-dimensional Minkowski spacetime is given by the replacement $q(t) \longrightarrow \phi(t, \vec{x})$ where ϕ is the scalar field. The corresponding path integral (vacuum-to-vacuum amplitude) is given by

$$Z[J] = \int \mathcal{D}\phi \ e^{\frac{i}{\hbar}S[\phi]+\frac{i}{\hbar}\int \, d^d x J(x)\phi(x)}. \tag{20.18}$$

The scalar phi-four action is given explicitly by

$$S[\phi] = \int d^d x \left[\frac{1}{2}\partial_\mu\phi\partial^\mu\phi - \frac{1}{2}m^2\phi^2 - \frac{\lambda}{4}\phi^4\right]. \tag{20.19}$$

The measure $\mathcal{D}\phi$ must be be defined using a regulator. The n-point correlators are given immediately by

$$\langle 0| T(\Phi(x_1) \ \ldots \ \Phi(x_n))|0\rangle = \frac{1}{Z[0]}\left(\frac{\hbar}{i}\right)^n \frac{\delta^n Z[J]}{\delta J(x_1) \ \ldots \ \delta J(x_n)} \Big|_{J=0}$$

$$= \frac{\int \mathcal{D}\phi \ \phi(x_1) \ \ldots \ \phi(x_n) \ e^{\frac{i}{\hbar}S[\phi]}}{\int \mathcal{D}\phi \ e^{\frac{i}{\hbar}S[\phi]}}. \tag{20.20}$$

- We need to go from a quantum field theory to a statistical field theory (also called an Euclidean field theory) in order to define the theory more rigorously. This will also make the theory accessible to the Monte Carlo method.

1. The above theory needs to be defined in an Euclidean space via the so-called Wick rotation. The imaginary time formulation is obtained by the substitution $t \longrightarrow -i\tau$.
2. Hence $\partial_\mu \phi \partial^\mu \phi \longrightarrow -(\partial_\mu \phi)^2$ and $iS \longrightarrow -S_E$ where the Euclidean action is given by

$$S_E[\phi] = \int d^d x \left[\frac{1}{2}(\partial_\mu \phi)^2 + \frac{1}{2}m^2\phi^2 + \frac{\lambda}{4}\phi^4 \right]. \tag{20.21}$$

The Euclidean path integral and n-point functions are given by

$$Z_E[J] = \int \mathcal{D}\phi \; e^{-\frac{1}{\hbar}S_E[\phi] + \frac{1}{\hbar}\int d^d x J(x)\phi(x)}. \tag{20.22}$$

$$\langle 0 | T(\Phi(x_1) \ldots \Phi(x_n)) | 0 \rangle_E = \frac{\hbar^n}{Z[0]} \frac{\delta^n Z_E[J]}{\delta J(x_1) \ldots \delta J(x_n)} \Big|_{J=0}. \tag{20.23}$$

3. Each field configuration $\phi(x)$ is associated a Boltzmann probability given by

$$P[\Phi] = \frac{\exp(-S_E[\phi]/\hbar)}{Z_E[J]}. \tag{20.24}$$

- The vector phi-four theory is a generalization of the scalar phi-four theory to an $O(N)$-symmetric theory.
    1. The field ϕ is an $O(N)$-vector, i.e., with components ϕ^i, $i = 1, 2,..., N$.
    2. The Euclidean action is

$$S[\phi] = -\int d^d x \left(\frac{1}{2}(\partial_\mu \phi^i)^2 + \frac{1}{2}m^2\phi^i\phi^i + \frac{\lambda}{4}(\phi^i\phi^i)^2 \right). \tag{20.25}$$

The Euclidean path integral becomes given by

$$Z_E[J] = \int \mathcal{D}\phi \; e^{-\frac{1}{\hbar}S_E[\phi] + \frac{1}{\hbar}\int d^d x J^i(x)\phi^i(x)}.$$

- We define now a lattice regularization (20.2).
    1. We introduce now a lattice regularization with a lattice spacing a and points $n = (n_1, n_2,..., n_d)$. We have:

$$x = an, \quad \int d^d x = a^d \sum_n, \quad \phi^i(x) = \phi^i_n, \quad \partial_\mu \phi^i = \frac{\phi^i_{n+\hat{\mu}} - \phi^i_n}{a}. \tag{20.26}$$

    2. The lattice action is given by

$$S[\phi] = \sum_n \left(2\kappa \sum_\mu \Phi^i_n \Phi^i_{n+\hat{\mu}} - \Phi^i_n \Phi^i_n - g(\Phi^i_n \Phi^i_n - 1)^2 \right). \tag{20.27}$$

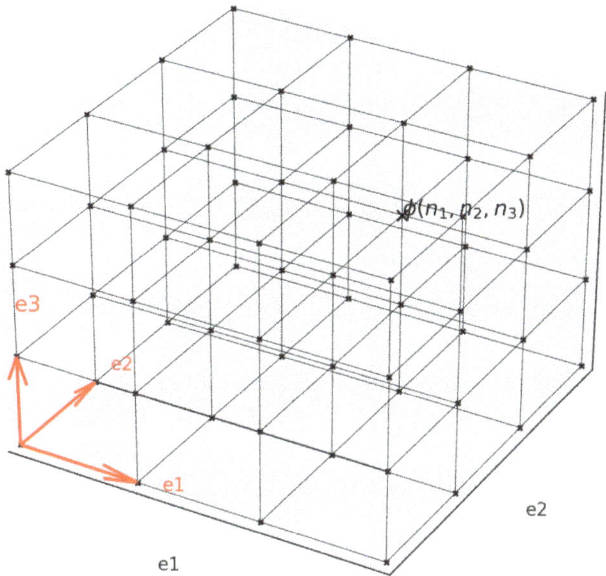

Figure 20.2. A cubic lattice.

3. The mass parameter m^2 is replaced by the so-called hopping parameter κ and the coupling constant λ is replaced by the coupling constant g where

$$m^2 a^2 = \frac{1 - 2g}{\kappa} - 2d, \quad \frac{\lambda}{a^{d-4}} = \frac{g}{\kappa^2}. \tag{20.28}$$

4. The fields ϕ_n^i and Φ_n^i are related by

$$\phi_n^i = \sqrt{\frac{2\kappa}{a^{d-2}}}\, \Phi_n^i. \tag{20.29}$$

5. The partition function is given by

$$Z = \int \prod_{n,i} d\Phi_n^i \; e^{S[\phi]} = \int d\mu(\Phi) \; e^{2\kappa \sum_n \sum_\mu \Phi_n^i \Phi_{n+\hat\mu}^i}. \tag{20.30}$$

6. The measure $d\mu(\phi)$ is given explicitly by

$$\begin{aligned}
d\mu(\Phi) &= \prod_{n,i} d\Phi_n^i \; e^{-\sum_n \left(\Phi_n^i \Phi_n^i + g(\Phi_n^i \Phi_n^i - 1)^2 \right)} \\
&= \prod_n \left(d^N \vec{\Phi}_n \; e^{-\vec{\Phi}_n^2 - g(\vec{\Phi}_n^2 - 1)^2} \right) \\
&\equiv \prod_n d\mu(\Phi_n).
\end{aligned} \tag{20.31}$$

20.1.3 Thermal field theory

Euclidean time, with periodic boundary condition, corresponds to a temperature. To show this, we consider the case of a point particle for which we have already derived the formula

$$\langle x|e^{-iH(t-t_0)}|x_0\rangle = \mathcal{N}\int \mathcal{D}x \; e^{iS[x]}. \tag{20.32}$$

After Euclidean rotation we obtain

$$\langle x|e^{-H(\tau-\tau_0)}|x_0\rangle = \mathcal{N}\int \mathcal{D}x \; e^{-S_E[x]}. \tag{20.33}$$

The thermodynamical partition function is then given by

$$\begin{aligned}
Z &= Tr e^{-\beta H}, \quad \beta = \tau - \tau_0 \\
&= \int dx \langle x|e^{-\beta H}|x\rangle \\
&= \mathcal{N}\int_{\text{periodic}} \mathcal{D}x \; e^{-S_E[x]}, \quad x \equiv x_0 \iff x(\tau_0 + \beta) = x(\tau_0).
\end{aligned} \tag{20.34}$$

Here, the temperature (Euclidean time) is restricted to a finite interval $[0, \beta]$ and we have periodic boundary conditions imposed on the fields (on the coordinate $x(\tau)$). Indeed, the integration measure is over all closed paths with a time period equals the inverse temperature while the Euclidean action S_E is given explicitly by

$$S_E = \int_0^\beta d\tau L_E(x, \dot{x}). \tag{20.35}$$

20.1.4 The canonical ensemble

- The energy of an isolated system is fixed and thus in this case we can use the microcanonical ensemble.

  However, for a system in equilibrium with a heat bath, the temperature is fixed and we can use the canonical ensemble, which is much more practical than the microcanonical ensemble. Fortunately, most systems are in equilibrium with a heat bath.

  The probability of finding a system in equilibrium with a heat bath at temperature T in a microstate or configuration s with energy E_s is defined by the Boltzmann distribution given by

$$P_s = \frac{1}{Z}e^{-\beta E_s}, \quad Z = \sum_s e^{-\beta E_s}, \quad \beta = \frac{1}{k_B T}. \tag{20.36}$$

  The free energy F of the system is given, in terms of the partition function Z, by

$$F = -k_B T \ln Z. \tag{20.37}$$

In equilibrium the free energy is minimum. All other thermodynamical quantities can be given by various derivatives of Z.

- For quantum dynamical field systems in Euclidean spacetime the partition function is given by the so-called path integral. In this case, the energy E_s of the microstate s is replaced with the action $S[\Phi]$ of the configuration Φ, while the sum over all microstates $\sum_s$ is replaced with the integral over all field configurations $\int \mathcal{D}\Phi$ where $\mathcal{D}\Phi$ is an appropriate measure on the space of field configurations.

The partition function of a quantum dynamical field system is then given by the path integral

$$Z = \int \mathcal{D}\Phi \; e^{-\beta S[\Phi]}. \tag{20.38}$$

All observables of a quantum field theory are derived from this path integral.

The goal is to calculate the partition function or the path integral of a given physical system, and since analytic solutions are difficult and rare, there remains only the numerical approach, which involves the evaluation of multidimensional integrals and hence the need of the Monte Carlo method.

20.1.5 Importance sampling and Metropolis algorithm

Importance sampling: The numerical error in any Monte Carlo integration is proportional to the standard deviation (the variance) and is inversely proportional to the number of samples. Thus, in order to reduce the error we should either reduce the variance or increase the number of samples. The first option is preferable since it does not require any extra computer time.

Importance sampling allows us to reduce the standard deviation of the integrand and hence the error by sampling more often the important regions of the integral where the integrand is largest.

As a concrete and simplified example, let us consider the one-dimensional integral

$$F = \int_a^b dx \; f(x). \tag{20.39}$$

We introduce the probability distribution $p(x)$ such that $1 = \int_a^b dx \; p(x)$ and write the integral as

$$F = \int_a^b dx \; p(x) \; \frac{f(x)}{p(x)}. \tag{20.40}$$

We evaluate this integral by sampling according to $p(x)$, i.e., we find a set of N random numbers x_i distributed according to $p(x)$ then approximate the integral by the sum

$$F_N = \frac{1}{N} \sum_{i=1}^N \frac{f(x_i)}{p(x_i)}. \tag{20.41}$$

The probability distribution $p(x)$ is chosen such that the function $f(x)/p(x)$ is slowly varying, which reduces the corresponding standard deviation.

The expectation value $\langle A \rangle$ of any physical quantity A is computed in the canonical ensemble by the formula

$$\langle A \rangle = \frac{\sum\limits_{s} A_s e^{-\beta E_s}}{\sum\limits_{s} e^{-\beta E_s}}. \tag{20.42}$$

In general the number of microstates N is very large. But in any Monte Carlo simulation we can only generate a very small number n of the total number N of the microstates. In other words, $\langle A \rangle$ will be approximated with

$$\langle A \rangle \simeq \langle A \rangle_n = \frac{\sum\limits_{s=1}^{n} A_s e^{-\beta E_s}}{\sum\limits_{s=1}^{n} e^{-\beta E_s}}. \tag{20.43}$$

The calculation of $\langle A \rangle_n$ proceeds therefore to choosing at random a microstate s, then compute E_s, A_s and $e^{-\beta E_s}$, and finally evaluating the contribution of this microstate to the expectation value $\langle A \rangle_n$.

This is clearly highly inefficient since the microstate s is very improbable and therefore its contribution is negligible.

Metropolis algorithm: The celebrated Metropolis algorithm is an importance sampling algorithm, which circumvents this problem as follows:

- We introduce a probability distribution p_s and rewrite $\langle A \rangle$ as

$$\langle A \rangle = \frac{\sum\limits_{s} \frac{A_s}{p_s} e^{-\beta E_s} p_s}{\sum\limits_{s} \frac{1}{p_s} e^{-\beta E_s} p_s}. \tag{20.44}$$

- We generate s with probabilities p_s and approximate $\langle A \rangle$ as

$$\langle A \rangle \simeq \langle A \rangle_n = \frac{\sum\limits_{s=1}^{n} \frac{A_s}{p_s} e^{-\beta E_s}}{\sum\limits_{s=1}^{n} \frac{1}{p_s} e^{-\beta E_s}}. \tag{20.45}$$

- The Metropolis algorithm is importance sampling where p_s is given by the Boltzmann distribution, i.e.,

$$p_s = \frac{e^{-\beta E_s}}{\sum\limits_{s=1}^{n} e^{-\beta E_s}}, \quad \text{Boltzmann distribution.} \tag{20.46}$$

- We get then the arithmetic average

$$\langle A \rangle_n = \frac{1}{n}\sum_{s=1}^{n} A_s.$$

(20.47)

In practice, the Metropolis algorithm, in the case of spin systems such as the Ising model, proceeds as follows:

(1) Choose an initial microstate.
(2) Choose a spin at random and flip it.
(3) Compute $\Delta E = E_{\text{trial}} - E_{\text{old}}$. This is the change in the energy of the system due to the trial flip.
(4) Check if $\Delta E \leqslant 0$.
   - If $\Delta E \leqslant 0$ then the trial microstate is accepted.
   - If $\Delta E > 0$ we compute the ratio of probabilities $w = e^{-\beta \Delta E}$.
   - Choose a uniform random number r in the interval $[0, 1]$.
   - Verify if $r \leqslant w$. In this case the trial microstate is accepted, otherwise it is rejected.
(5) Repeat steps (2) through (4) until all spins of the system are tested. This sweep counts as one unit of Monte Carlo time.
(6) Repeat steps (2) through (5) a sufficient number of times until thermalization, i.e. equilibrium is reached.
(7) Compute the physical quantities of interest in n thermalized microstates. This can be done periodically in order to reduce correlation between the data points.
(8) Compute averages.

The Metropolis algorithm is guaranteed to lead to a sequence of states which are distributed according to the Boltzmann distribution. Indeed, it is clear that steps (2) through (4) in the Metropolis algorithm correspond to a transition probability between the microstates $\{s_i\}$ and $\{s_j\}$ given by

$$W(i \longrightarrow j) = \min(1, e^{-\beta \Delta E}), \quad \Delta E = E_j - E_i.$$

(20.48)

This probability function satisfies the so-called detailed balance (reversibility) condition given by

$$W(i \longrightarrow j)\, e^{-\beta E_i} = W(j \longrightarrow i)\, e^{-\beta E_j}.$$

(20.49)

Any other probability function W, which satisfies this condition will also generate a sequence of states distributed according to the Boltzmann distribution. By summing now over the index j in the above equation and using $\sum_j W(i \longrightarrow j) = 1$ we get

$$e^{-\beta E_i} = \sum_j W(j \longrightarrow i)\, e^{-\beta E_j}.$$

(20.50)

The Boltzmann distribution is therefore an eigenvector of W. In other words, W leaves the equilibrium ensemble in equilibrium. Thus, detailed balance guarantees that the Metropolis algorithm will sample the Boltzmann distribution.

20.1.6 Statistical errors

Jackknife method:

Error estimation can be done via the Jackknife method, which works as follows:
1. We are given a set of $T = 2^P$ data points $f(i)$.
2. We proceed by removing z elements from the set. We end up with what we call a bin.
3. We can have $n = T/z$ sets (or bins).
4. The minimum number of data points we can remove is $z = 1$ and the maximum number is $z = T - 1$.
5. For a fixed partition given by z the average of the elements of the ith bin is

$$\langle y(j) \rangle_i = \frac{1}{T-z}\left(\sum_{j=1}^{T} f(j) - \sum_{j=1}^{z} f((i-1)z + j) \right), \quad i = 1, n. \tag{20.51}$$

6. The error of a fixed partition z is computed as follows

$$e(z) = \sqrt{\frac{n-1}{n}\sum_{i=1}^{n}(\langle y(j) \rangle_i - \langle f \rangle)^2}, \quad \langle f \rangle = \frac{1}{T}\sum_{j=1}^{T} f(j). \tag{20.52}$$

7. The true error is the largest value $e(z)$.

On auto-correlation: A measurement set is typically characterized by auto-correlation, i.e. the different measurements are not really statistically independent. The auto-correlation time can be estimated as follows:
1. Let f be some observable with values $f_i \equiv f(\phi_i)$. The average value $\langle f \rangle$ and the statistical error δf of the observable f are given by the equations

$$\langle f \rangle = \frac{1}{T}\sum_{i=1}^{T} f_i, \quad \delta f = \frac{\sigma}{\sqrt{T}}, \quad \sigma^2 = \langle f^2 \rangle - \langle f \rangle^2. \tag{20.53}$$

2. This estimate is valid provided the thermalized configurations $\phi_1, \phi_2, \ldots, \phi_T$ are statistically uncorrelated.
3. In general, two consecutive configurations will be dependent and the average number of configurations which separate two configurations is called the auto-correlation time.
4. The auto-correlation function Γ_j and the normalized auto-correlation function ρ_j for the observable f are defined by

$$\Gamma_j = \frac{1}{T-j}\sum_{i=1}^{T-j}(f_i - \langle f\rangle)(f_{i+j} - \langle f\rangle), \quad \rho_j = \frac{\Gamma_j}{\Gamma_0}.$$ (20.54)

5. The statistical error in the average $\langle f\rangle$ will be given by

$$\delta f = \frac{\sigma}{\sqrt{T}}\sqrt{2\tau_{\mathrm{int}}}.$$ (20.55)

6. The so-called integrated auto-correlation time τ_{int} is given in terms of the normalized auto-correlation function ρ_j by the formula

$$\tau_{\mathrm{int}} = \frac{1}{2} + \sum_{j=1}^{\infty}\rho_j \simeq \frac{1}{2} + \sum_{j=1}^{M}\rho_j, \quad \delta\tau_{\mathrm{int}} = \sqrt{\frac{4M+2}{T}}\,\tau_{\mathrm{int}}.$$ (20.56)

7. The auto-correlation function Γ_j, for large j, can not be precisely determined, and hence, one must truncate the sum over j in τ_{int} at some cutoff M, in order to not increase the error $\delta\tau_{\mathrm{int}}$ in τ_{int} by simply summing up noise. The integrated auto-correlation time τ_{int} should then be defined by

$$\tau_{\mathrm{int}} = \frac{1}{2} + \sum_{j=1}^{M}\rho_j.$$ (20.57)

8. The value M is chosen as the first integer between 1 and T such that $M \geqslant 4\tau_{\mathrm{int}} + 1$.

20.2 The four theories of quantum field theory

20.2.1 The phi-four scalar theory

For a pedagogical and detailed discussion of the phi-four theory see chapters 4, 7 and in particular chapter 9 of this book. In this subsection, we summarize the salient features that are important for our discussion here. Then, we give a new discussion of the renormalization group equation and continuum limit of the phi-four theory in two dimensions.

We consider Euclidean phi-four theory in d dimensions with $O(N)$ symmetry with parameters $\mu_0^2 \equiv m^2$ and λ. The lattice action is given by

$$S = \sum_n\left(-2\kappa\sum_\mu \Phi_n^i\Phi_{n+\hat\mu}^i + \Phi_n^i\Phi_n^i + g(\Phi_n^i\Phi_n^i - 1)^2\right).$$ (20.58)

The lattice parameters κ and g are given in terms of the dimensionless continuum parameters $\mu_{0L}^2 \equiv \mu_0^2 a^2$ and $\lambda_L \equiv \lambda/a^{d-4}$ by the relations

$$\kappa = \frac{\sqrt{8\lambda_L + (\mu_{0L}^2 + 4)^2} - (\mu_{0L}^2 + 4)}{4\lambda_L}, \quad g = \kappa^2 \lambda_L. \tag{20.59}$$

The phase diagram will be drawn in the $\mu_{0L}^2 - \lambda_L$ plane. The limit $g \longrightarrow \infty$ of the $O(1)$ model is precisely the Ising model in d dimensions.

The mean-field approximation: First, we discuss the mean-field approximation and phase structure of the theory.

1. There are two phases in this model. A disordered (paramagnetic) phase characterized by $\langle \Phi_n^i \rangle = 0$ and an ordered (ferromagnetic) phase characterized by $\langle \Phi_n^i \rangle = v_i \neq 0$.

2. In the mean-field approximation the spins Φ_n^i at the $2d$ nearest neighbors of each spin Φ_n^i are replaced by the average $v^i = \langle \Phi_n^i \rangle$, viz

$$\frac{\sum_{\mu}(\Phi_{n+\hat{\mu}}^i + \Phi_{n-\hat{\mu}}^i)}{2d} = v^i. \tag{20.60}$$

3. By using the path integral definition of $\langle \Phi_n^i \rangle$ we obtain a condition on v^i given explicitly by

$$v^i = \frac{\int d\mu(\Phi) \ \Phi_n^i e^{4\kappa d \sum_n \Phi_n^i v^i}}{\int d\mu(\Phi) \ e^{4\kappa d \sum_n \Phi_n^i v^i}}. \tag{20.61}$$

4. The solution in the limit $g \longrightarrow 0$ is

$$v^i = 2\kappa d v^i + \cdots \Rightarrow \kappa_c = \frac{1}{2d}. \tag{20.62}$$

5. The solution in the limit $g \longrightarrow \infty$ is

$$v^i = \frac{4\kappa d v^i}{N} + \cdots \Rightarrow \kappa_c = \frac{N}{4d}. \tag{20.63}$$

6. The critical line $\kappa_c = \kappa_c(g)$ interpolates in the $\kappa - g$ plane between the above two lines (figure 20.3).

7. For example, for the $O(1)$ model in the limit $g \longrightarrow \infty$ the solution must satisfy the condition

$$v = \tanh 4\kappa d v \Rightarrow \frac{1}{3}(4d)^2 \kappa^3 v^2 = \kappa - \kappa_c. \tag{20.64}$$

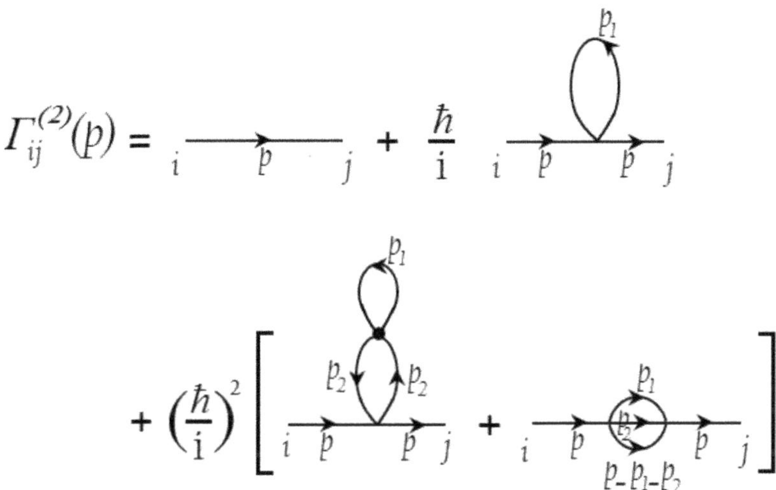

Figure 20.3. The 2-point proper vertex.

For $\kappa < \kappa_c$ (where $\kappa_c = 1/4d$) there is only one solution at $v = 0$ whereas for $\kappa > \kappa_c$ there are two solutions $v \neq 0$. Clearly, for $\kappa \longrightarrow \kappa_c$ the solution behaves as $v \longrightarrow 0$.

8. In summary we have the two phases:

$$\kappa > \kappa_c: \quad \text{broken, ordered, ferromagnetic.} \tag{20.65}$$

$$\kappa < \kappa_c: \quad \text{symmetric, disordered, paramagnetic.} \tag{20.66}$$

The one-loop approximation in four dimensions: Next, we discuss the one-loop result in four dimensions.

1. The mean-field approximation is exact in dimension 4, i.e. the critical behavior is fully controlled by the Gaussian fixed point.
2. Indeed, the critical value for $d = 4$ at $g = 0$, which is $\kappa_c = 1/8$ for all N, can be derived from the one-loop order of the continuum phi-four theory with $O(N)$ symmetry.
3. The renormalized mass at two-loop order is given by the Feynman diagrams contributing to the 2-point proper vertex $\Gamma^{(2)}(0)$.
4. At one-loop order only the tadpole diagram contributes. We have explicitly the renormalized mass

$$\begin{aligned}
a^2 m_R^2 &= a^2 m^2 + \frac{(N+2)\lambda}{16\pi^2} + \frac{(N+2)\lambda}{16\pi^2} a^2 m^2 \ln a^2 m^2 \\
&\quad + \frac{(N+2)\lambda}{16\pi^2} a^2 m^2 \mathbf{C} + a^2 \times \text{finite terms.}
\end{aligned} \tag{20.67}$$

Here, $a = 1/\Lambda$.

5. In the continuum limit $a \longrightarrow 0$ (and since the renormalized mass is finite) we have the result

$$a^2 m^2 \longrightarrow -\frac{(N+2)\lambda}{16\pi^2} - \frac{(N+2)\lambda}{16\pi^2} a^2 m^2 \ln a^2 m^2$$
$$- \frac{(N+2)\lambda}{16\pi^2} a^2 m^2 \mathbf{C} - a^2 \times \text{finite terms.} \tag{20.68}$$

6. We obtain the critical line

$$a^2 m^2 \longrightarrow a^2 m_c^2 = -\frac{r_0}{2}\lambda + O(\lambda^2). \tag{20.69}$$

Here, $r_0 = (N+2)/8\pi^2$.

7. The critical line for small values of the coupling constant when expressed in terms of the lattice parameters κ and g is given by

$$\kappa \longrightarrow \kappa_c = \frac{1}{8} + \left(\frac{r_0}{2} - \frac{1}{4}\right)g + O(g^2). \tag{20.70}$$

8. The continuum limit $a \longrightarrow 0$ corresponds precisely to the limit in which the mass approaches its critical value. This happens for every value of the coupling constant and hence the continuum limit $a \longrightarrow 0$ is the limit in which we approach the critical line. The continuum limit is therefore a second-order phase transition.

RG equation and continuum limit of phi-four in two dimensions: Now, we discuss the continuum limit in two dimensions using the renormalization group (RG) equation.

1. In two dimensions the Φ^4 theory requires only mass renormalization (mass diverges logarithmically) while the quartic coupling constant is finite (recall that in four dimensions, three renormalizations are required: the mass, the quartic coupling and the wave function renormalizations).

2. Thus, the lattice parameters $\mu_{0L}^2 \equiv \mu_0^2 a^2$ and $\lambda_L \equiv \lambda a^2$ go to zero in the continuum limit $a \longrightarrow 0$.

3. We know that mass renormalization is due to the tadpole diagram, which is the only divergent Feynman diagram in the theory and takes the form of a simple reparametrization given by

$$\mu_0^2 = \mu^2 - \delta\mu^2. \tag{20.71}$$

μ^2 is the renormalized mass parameter and $\delta\mu^2$ is the counter term, which is fixed via an appropriate renormalization condition.

4. The ultraviolet divergence $\ln\Lambda$ of μ_0^2 is contained in $\delta\mu^2$ while the renormalization condition will split the finite part of μ_0^2 between μ^2 and $\delta\mu^2$.

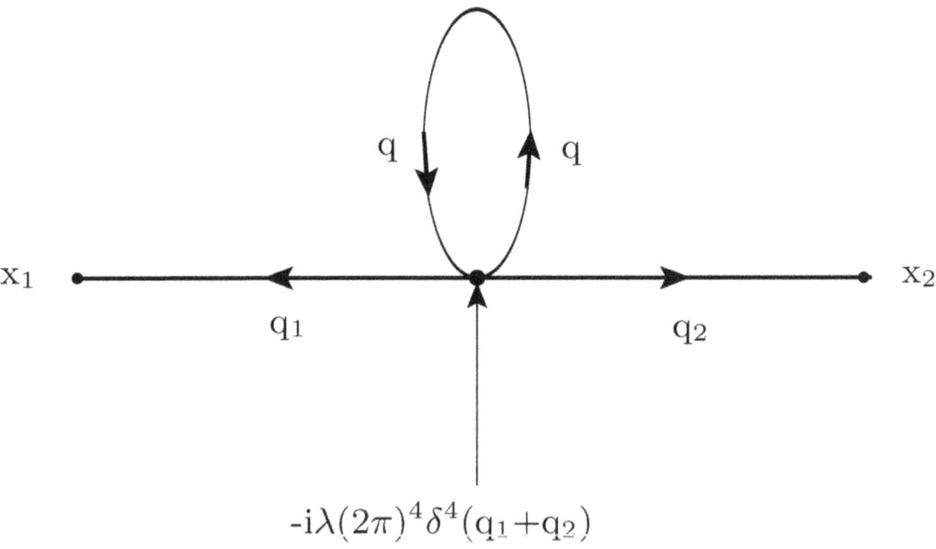

Figure 20.4. The tadpole diagram.

5. The choice of the renormalization condition can be quite arbitrary. A convenient choice of the renormalization condition suitable for Monte Carlo measurements and which distinguishes between the two phases of the theory is given by the usual normal ordering prescription (figure 20.4).

6. Quantization at one-loop can be done via Hartree–Fock resummation.

7. Quantization at one-loop (together with a self-consistent Hartree treatment) gives then in the symmetric phase where $\mu^2 > 0$ the 2-point function

$$
\begin{aligned}
\Gamma^{(2)}(p) &= p^2 + \mu_0^2 + 3\lambda \int \frac{d^2k}{(2\pi)^2}\frac{1}{k^2 + \mu_0^2}, \quad \text{one\,--\,loop result} \\
&= p^2 + \mu_0^2 + 3\lambda \int \frac{d^2k}{(2\pi)^2}\frac{1}{\Gamma^{(2)}(k)}, \quad \text{Hartree–Fock} \\
&= p^2 + \mu^2 + 3\lambda \int \frac{d^2k}{(2\pi)^2}\frac{1}{\Gamma^{(2)}(k)} - \delta\mu^2, \quad \text{renormalization} \\
&= p^2 + \mu^2 + 3\lambda \int \frac{d^2k}{(2\pi)^2}\frac{1}{k^2 + \mu^2} - \delta\mu^2 + \text{higher loop effects.}
\end{aligned}
\tag{20.72}
$$

8. The 2-point proper vertex is then given by (symmetric phase)

$$
\Gamma^{(2)}(p) = p^2 + \mu^2 + \Sigma(p), \quad \Sigma(p) = 3\lambda A_{\mu^2} - \delta\mu^2 + \cdots.
\tag{20.73}
$$

A_{μ^2} is precisely the value of the tadpole diagram given by

$$A_{\mu^2} = \int \frac{d^2k}{(2\pi)^2} \frac{1}{k^2 + \mu^2}. \tag{20.74}$$

9. The renormalization condition, which is equivalent to normal ordering, is equivalent to the choice

$$\delta\mu^2 = 3\lambda A_{\mu^2}. \tag{20.75}$$

10. A dimensionless coupling constant can be defined by

$$f = \frac{\lambda}{\mu^2}. \tag{20.76}$$

11. The action becomes

$$S[\phi] = \int d^2x \left(\frac{1}{2}(\partial_\mu \phi)^2 + \frac{1}{2}\mu^2(1 - 3fA_{\mu^2})\phi^2 + \frac{f\mu^2}{4}\phi^4 \right). \tag{20.77}$$

12. For sufficiently small f the exact effective potential is well approximated by the classical potential with a single minimum at $\phi_{\mathrm{cl}} = 0$.
13. For larger f, the coefficient of the mass term in the above action can become negative and as a consequence a transition to the broken symmetry phase is possible, although in this regime the effective potential is no longer well approximated by the classical potential.
14. Indeed, a transition to the broken symmetry phase was shown to be present in [35], where a duality between the strong coupling regime of the above action and a weakly coupled theory normal-ordered with respect to the broken phase was explicitly constructed.
15. In summary, we have the renormalized mass $\mu_0^2 = \mu^2 - \delta\mu^2$ with the normal ordering prescription $\delta\mu^2 = 3\lambda A_{\mu^2}$.
16. On a finite volume lattice with periodic boundary conditions the renormalization condition $\mu^2 - \mu_0^2 = 3\lambda A_{\mu^2}$ becomes

$$\mu_L^2 - \mu_{0L}^2 = \frac{3\lambda_L}{L^2} \sum_{n_1,n_2=1}^{L} \frac{1}{4\sin^2 \pi n_1/L + 4\sin^2 \pi n_2/L + \mu_L^2}. \tag{20.78}$$

17. Thus, given the critical value of μ_{0L}^2 for every value of λ_L we need to determine the corresponding critical value of μ_L^2.
18. The continuum limit $a \longrightarrow 0$ is then given by extrapolating the results into the origin, i.e., taking $\lambda_L = a^2\lambda \longrightarrow 0$, $\mu_L^2 = a^2\mu^2 \longrightarrow 0$ in order to determine the critical value

$$f_* = \lim_{\lambda_L, \mu_L^2 \to 0} \frac{\lambda_L}{\mu_{L*}^2}. \tag{20.79}$$

This can be done using the Newton-Raphson algorithm.

Monte Carol simulation of phi-four theory in two dimensions: For completeness, we also discuss here the continuum limit in two dimensions using the Monte Carlo method.

1. The Ising model ($N=1$) in two dimensions shares with the more realistic case in three dimensions (which describes all second-order phase transitions in nature) the same critical behavior.

2. However, in two dimensions the model admits a quasi-exact solution in terms of quantum field theory (super-renormalizability) and the renormalization group equation (the Wilson–Fisher fixed point), which in the limit $g \longrightarrow \infty$ reduces to the remarkable Onsager's exact solution [36].

3. In two dimensions the lattice points are labeled by $n = (n_1, n_2)$ where $n_1 = 1, \ldots, L_1$ and $n_2 = 1, \ldots, L_2$.

4. We will impose periodic boundary conditions, i.e., $\Phi_{n_1 + L_1, n_2} \equiv \Phi_{n_1, n_2}$ and $\Phi_{n_1, n_2 + L_2} \equiv \Phi_{n_1, n_2}$.

5. The lattice action for Φ^4 theory in two dimensions with $O(N)$ symmetry reads explicitly

$$S[\phi] = $$

$$\sum_{n_1=1}^{L_1} \sum_{n_2=1}^{L_2} \left(-2\kappa \Phi_{n_1,n_2}^i (\Phi_{n_1+1,n_2}^i + \Phi_{n_1,n_2+1}^i) + \Phi_{n_1,n_2}^i \Phi_{n_1,n_2}^i + g(\Phi_{n_1,n_2}^i \Phi_{n_1,n_2}^i - 1)^2 \right). \tag{20.80}$$

6. The Boltzmann weight and the partition function are given by

$$\mathcal{P}[\Phi] = \frac{\exp(-S[\Phi])}{Z}, \quad Z = \int \mathcal{D}\Phi \, \exp(-S[\Phi]). \tag{20.81}$$

7. Metropolis algorithm: It generates a collection of thermalized configurations Φ_n^i according to the Boltzmann probability distribution as follows:

   • We propose small changes in the field configurations:

$$\Phi_{p_1, p_2}^j \longrightarrow \Phi_{p_1, p_2}^j + h. \tag{20.82}$$

   • We compute the variation of the action under these changes:

$$\Delta S_{p_1, p_2}^j(h) = -2\kappa h \left(\Phi_{p_1+1, p_2}^j + \Phi_{p_1-1, p_2}^j + \Phi_{p_1, p_2+1}^j + \Phi_{p_1, p_2-1}^j \right)$$

$$+ (1 - 2g)\left(h^2 + 2h\Phi_{p_1, p_2}^j \right) \tag{20.83}$$

$$+ g\left(h^2 + 2h\Phi_{p_1, p_2}^j \right)\left(h^2 + 2h\Phi_{p_1, p_2}^j + 2\Phi_{p_1, p_2}^i \Phi_{p_1, p_2}^i \right).$$

- We accept/reject these changes using precisely the Boltzmann probability distribution

$$W(\Phi^j_{p_1, p_2} \longrightarrow \Phi^j_{p_1, p_2} + h) = \min(1, \exp(-\Delta S^j_{p_1, p_2}(h)). \qquad (20.84)$$

8. In the actual code (written in Fortran, which is our preferred programming language) we will go through the following steps:
   - We write a subroutine in which we compute the energy (action) and other primary observables such as the magnetization given by

$$M = \frac{1}{NL_1 L_2} \langle m \rangle, \quad m = |\sum_{i,n_1,n_2} \Phi^i_{n_1, n_2}|. \qquad (20.85)$$

   - We write a subroutine in which we compute the variation of the action, which depends on j, p_1, p_2 and h.
   - Include a random number generator.
   - Implement the Metropolis algorithm. So we change all the fields at all lattice points successively and accept/reject according to the Boltzmann probability. The value of the variation h is drawn randomly and its range is tuned so that the acceptance rate is fixed around 1/3.
   - We write a subroutine which computes the average, the error and the auto-correlation time for arbitrary observables.
9. The main program consists roughly of the following parts:
   - The random number generator, the physical parameters of the model and the field configuration are initialized.
   - A thermalization process which changes the system until thermalization (or equilibrium) is reached.
   - A measurement process in which a set of thermalized configurations Φ^i_n is collected and primary observables are measured.
   - A part which measures secondary observables such as second moments like the specific heat (associated with the energy) and susceptibility (associated with the magnetization). These are given by

$$C_v = \langle S^2 \rangle - \langle S \rangle^2, \quad \chi = \langle m^2 \rangle - \langle m \rangle^2. \qquad (20.86)$$

10. The ultimate goal is to construct the phase diagram of the theory. Hence, we are required to perform the following tasks:
   - Construct the lattice phase diagram in the $\mu^2_{0L} - \lambda_L$ plane (Monte Carlo).
   - Take the infinite volume limit defined by $l_1 = L_1 a \longrightarrow \infty$ and $l_2 = L_2 a \longrightarrow \infty$ with a fixed a (Monte Carlo and extrapolation).

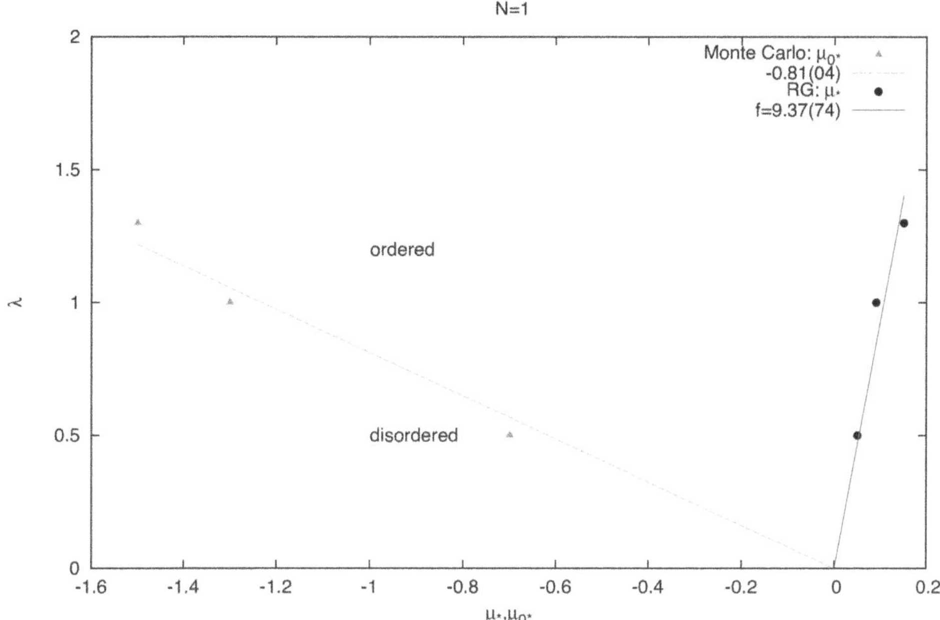

Figure 20.5. The phase diagram.

- Construct the continuum phase diagram in terms of the renormalized parameters obtained in the continuum limit $a \longrightarrow 0$ (renormalization group equation and extrapolation).

The result is shown in figure 20.5.

20.2.2 The Ising model

For a pedagogical and detailed discussion of the Ising model see chapter 11, and references therein, of this book.

We consider a d-dimensional periodic lattice with l points in every direction. In every lattice site we put a spin variable s_i, which can take either the value $+1$ or -1. A configuration of this system of $N = l^d$ spins is therefore specified by a set of numbers $\{s_i\}$. The energy of the configuration $\{s_i\}$ is given by

$$E_I(s) = -J\sum_{\langle i,j \rangle} s_i s_j - \sum_{i=1}^{N} J_i s_i. \tag{20.87}$$

A constant external magnetic field is introduced by $H \equiv J_i$. The symbol $\langle i, j \rangle$ stands for nearest-neighbor spins. The sum over $\langle i, j \rangle$ extends over $\gamma N / 2$ terms where γ is the number of nearest neighbors. In 2, 3, 4 dimensions $\gamma = 4, 6, 8$. The parameter J is the exchange interaction energy between the spins i and j.

The partition function is given by (with $\beta = 1/k_B T$)

$$Z(J) = \sum_{s_1}\sum_{s_2} \cdots \sum_{s_N} e^{-\beta E_I(s)}. \tag{20.88}$$

There are 2^N terms in the sum each corresponding to a distinct microstate of the system.

The Ising model can be rewritten as a scalar field theory on a d-dimensional Euclidean lattice with a partition function

$$Z(J) = \int \prod_n d\mu(\Phi_n)\ e^{\sum_{n,m}\Phi_n V_{nm}\Phi_m + \sum_n J_n\Phi_n}. \tag{20.89}$$

The positive matrix V_{nm} is defined by

$$V_{nm} = \kappa\sum_{\hat{\mu}}(\delta_{m,n+\hat{\mu}} + \delta_{m,n-\hat{\mu}}), \quad \kappa = J\beta. \tag{20.90}$$

The measure is defined by

$$\int d\mu(\Phi_n)f(\Phi_n) = \frac{1}{2}(f(+1) + f(-1))\int d\mu(\Phi_n). \tag{20.91}$$

The mean-field approximation: Now, we discuss the mean-field approximation of the Ising model.

1. In the mean-field approximation we replace V_{ij} by $W_{ij} = V_{ij}/L$ and we replace every spin Φ_n by $\hat{\Phi}_n = \sum_{l=1}^{L}\Phi_n^l$, i.e., by the sum of L spins Φ_n^l which are assumed to be distributed with the same probability $d\mu(\Phi_n^l)$.
2. Next, we perform a Hubbard transformation and then apply the saddle point method.
3. The QFT vacuum energy (the thermodynamical free energy) of the Ising model in the mean-field approximation is given by

$$\begin{aligned} W(J) &= \frac{1}{L}\ln Z[J] \\ &= -\frac{1}{4}\sum_{n,m}(\phi_n - J_n)V_{nm}^{-1}(\phi_m - J_m) - \sum_n A(\phi_n). \end{aligned} \tag{20.92}$$

The field ϕ_n satisfies the saddle point equation (best magnetic field approximation)

$$\phi_n + 2\sum_m V_{nm}\frac{dA}{d\phi_m} = J_n. \tag{20.93}$$

The potential $A(\phi_n)$ is given explicitly for the Ising model by

$$A(\phi_n) = -\ln\cosh\phi_n. \tag{20.94}$$

4. The magnetization (effective field, order parameter) is conjugate to the magnetic field J_n, viz

$$M_n = \frac{\partial W}{\partial J_n}$$

$$= \tanh \phi_n \Rightarrow \phi_n = \frac{1}{2}\ln(1 + M_n) - \frac{1}{2}\ln(1 - M_n). \tag{20.95}$$

5. The QFT effective action (the thermodynamical Gibbs energy) is the Legendre transform of $W(J)$ defined by

$$\Gamma(M) = \sum_n M_n J_n - W(J)$$

$$= -\sum_{n,m} M_n V_{nm} M_m + \sum_n B(M_n). \tag{20.96}$$

The function $B(M_n)$ is the Legendre transform of $A(\phi_n)$ given by

$$B(M_n) = M_n \phi_n + A(\phi_n). \tag{20.97}$$

6. In systems where translation is a symmetry we can assume that the magnetization is uniform, i.e., $M_n = M = \text{constant}$ and as a consequence the effective potential per degree of freedom is given by

$$\frac{\Gamma(M)}{\mathcal{N}} = -vM^2 + B(M). \tag{20.98}$$

The effective potential $\Gamma(M)$ is a convex function of M, i.e. for M, M_1 and M_2 such that $M = xM_1 + (1 - x)M_2$ with $0 < x < 1$ we must have

$$\Gamma(M) \leqslant x\Gamma(M_1) + (1 - x)\Gamma(M_2). \tag{20.99}$$

Thus, a linear interpolation is always greater than the potential, i.e., $\Gamma(M)$ is an increasing function of M.

7. Hence, A is a convex function in the variable ϕ whereas B is a convex function in the variable M.

8. Here, we are faced with two distinct cases:
   - The **first-order phase transitions**, which are characterized as follows:
     (a) For high T (small v) the effective potential is dominated by $B(M)$, which is a convex function. The minimum is $M = 0$.
     (b) We start decreasing T by increasing v. At some $T = T_c$ new minima of $\Gamma(M)$ appear, which are degenerate with $M = 0$.
     (c) For $T < T_c$ the new minima become absolute and as a consequence M jumps discontinuously from 0 to a finite value.
     (d) In this case Γ'' at the minimum is always strictly positive and as a consequence the correlation length is always finite.
   - The **second-order phase transitions** are even more important and they characterized as follows:

(a) The minimum at $M = 0$ becomes at some $T = T_c$ a maximum and simultaneously new minima appear, which start moving away from the origin as we decrease T.

(b) The critical temperature T_c is defined by $\Gamma''(0) = 0$.

(c) Above T_c we have only the solution $M = 0$ whereas below T_c we have two minima moving continuously away from $M = 0$.

(d) In this case the magnetization remains continuous at $T = T_c$ (a continuous transition) (figure 20.6).

(e) The correlation length diverges at $T = T_c$.

20.2.3 The real quartic matrix model

For a good introduction to the subject of random matrix theory see [39, 40]. In [41] an interesting discussion of the connection between random matrix theory and statistical mechanics is given. The main results discussed here can be found in [37] and [38].

The real quartic matrix model depends on a single Hermitian matrix M and is given by the action

$$V = BTrM^2 + CTrM^4 = \frac{N}{g}\left(-TrM^2 + \frac{1}{4}TrM^4\right).$$
(20.100)

The model depends actually on a single coupling g such that (by scaling the field M as $M \longrightarrow \alpha M$)

$$B\alpha^2 = -\frac{N}{g}, \quad C\alpha^4 = \frac{N}{4g} \Rightarrow B^2 = \frac{4NC}{g}.$$
(20.101)

The partition function is given by

$$Z = \int dM \ e^{-V}.$$
(20.102)

We can now diagonalize the scalar matrix M as

$$M = U\Lambda U^{-1} \Rightarrow dM = d\Lambda dU \Delta^2(\Lambda).$$
(20.103)

Clearly, dU is the Haar measure over the group $SU(N)$ whereas $\Delta^2(\Lambda)$ is the Vandermonde determinant given explicitly by

$$\Delta^2(\Lambda) = \prod_{i>j} (\lambda_i - \lambda_j)^2.$$
(20.104)

The partition function becomes

$$Z = \int d\Lambda \ \exp(-N^2 V_{\text{eff}}).$$
(20.105)

The effective potential is given explicitly by

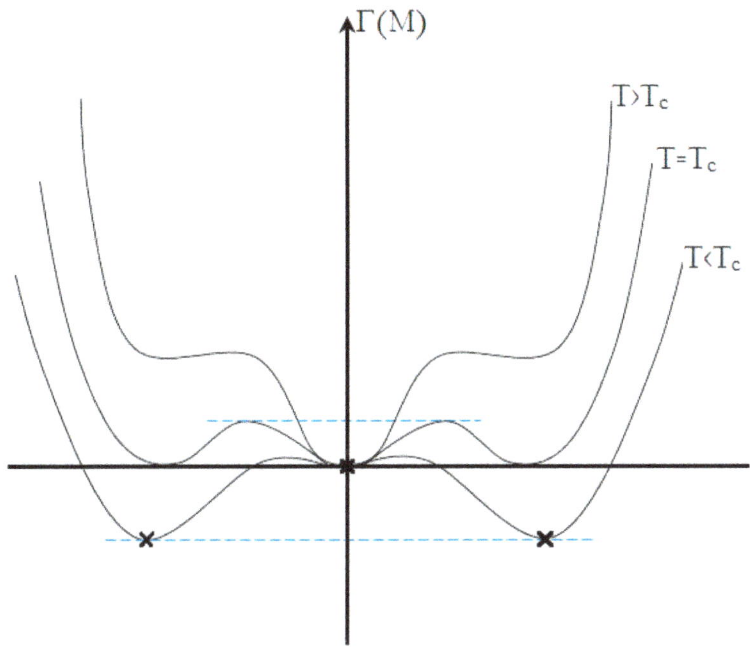

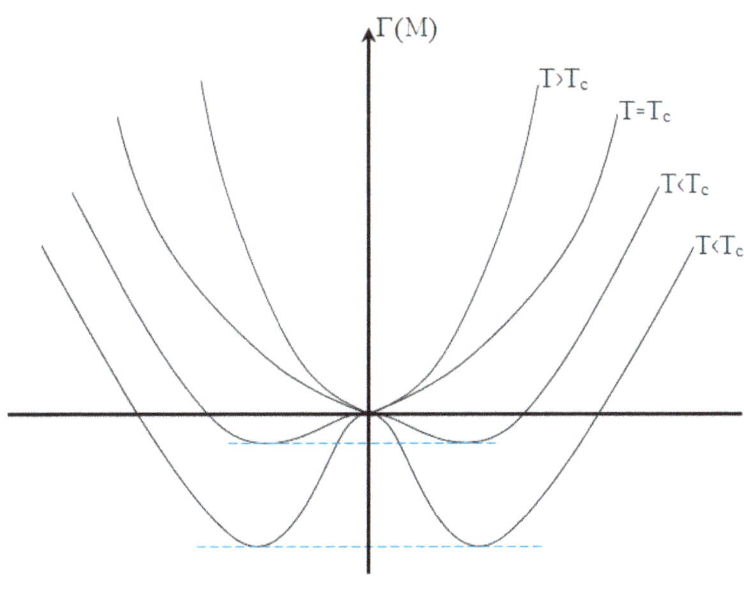

Figure 20.6. The first-order and second-order phase transitions.

$$V_{\text{eff}} = \frac{1}{N}\sum_{i=1}^{N}V(\lambda_i) - \frac{1}{2N^2}\sum_{i\neq j}\ln(\lambda_i - \lambda_j)^2, \quad V = \tilde{B}\lambda_i^2 + \tilde{C}\lambda_i^4. \tag{20.106}$$

The scaled parameters are determined in the large N limit by

$$\tilde{B} = \frac{B}{N^{3/2}}, \quad \tilde{C} = \frac{C}{N^2}. \tag{20.107}$$

This is a gaz of eigenvalues with Coulomb repulsive interaction.

In the large N limit we use the saddle point method. The dominant configuration is a solution of the equation

$$\frac{dV_{\text{eff}}}{d\lambda_i} = 0. \tag{20.108}$$

We introduce a resolvent and a density of eigenvalues by

$$W(z) = \frac{1}{N}Tr\frac{1}{z-M} = \int d\lambda\rho(\lambda)\frac{1}{z-\lambda} \sim \frac{1}{z}, \quad z \longrightarrow \infty. \tag{20.109}$$

This satisfies the loop equation

$$W^2(z) = V'(z)W(z) - P(z), \quad P(z) = \frac{1}{N}\sum_i\frac{V'(z) - V'(\lambda_i)}{z-\lambda_i}. \tag{20.110}$$

The solution is immediately given by

$$\begin{aligned}
W(z) &= \frac{1}{2}(V'(z) - \sqrt{V'^2(z) - 4P(z)}), \quad \rho(\lambda) \\
&= -\frac{1}{2\pi i}(W(\lambda + i0) - W(\lambda - i0)).
\end{aligned} \tag{20.111}$$

Wigner semi-circle law: First, we need to discuss the case of free theories and the Wigner semi-circle law.

1. A large class of distributions are one-cut solutions corresponding to potentials with at least one deep well. In this case the support of the eigenvalue distribution has only one connected component (cut).
2. Let us consider as an example the quadratic potential (representing free fields) $NS = NTrV$ given by

$$NS = NTrV = \frac{Ng_2}{2}TrM^2, \quad B = \frac{Ng_2}{2}. \tag{20.112}$$

The resolvent in this case is given by

$$W(z) = \frac{1}{2}(g_2 z - i\sqrt{4g_2 - g_2^2 z^2}). \tag{20.113}$$

The eigenvalue density is then given by

$$\rho(\lambda) = -\frac{1}{\pi}\,\text{Im}\,W(\lambda) = \frac{B}{N\pi}\sqrt{a^2 - z^2}, \quad a^2 = \frac{4}{g_2} = \frac{2N}{B}. \tag{20.114}$$

3. Free noncommutative field theory is also dominated by the Wigner's semi-circle law. Indeed, we can verify on fuzzy $\mathbf{CP}_l^n$ the correspondence

$$\int_{\Lambda,V} d^d x \left(\frac{1}{2}(\partial_i\Phi)^2 + \frac{m^2}{2}\Phi^2\right) = \frac{V}{\mathcal{N}}Tr\left(-\frac{1}{2}[L_i,\,\Phi]^2 + \frac{m^2}{2}\Phi^2\right) \longrightarrow \frac{2\mathcal{N}}{\alpha_0^2}Tr\Phi^2. \tag{20.115}$$

For example, on fuzzy $\mathbf{CP}_l^1$, we have $\mathcal{N} = N = l + 1$, $V = 4\pi R^2$, $\Lambda = N/R$ and the square of the largest eigenvalue is replaced as

$$a^2 = \frac{2N}{B} = \frac{\Lambda^2}{\pi m^2} \longrightarrow \alpha_0^2 = \frac{1}{\pi}\ln\left(1 + \frac{\Lambda^2}{m^2}\right). \tag{20.116}$$

We can also show that the kinetic term can be replaced by a multitrace potential to any degree of accuracy desired.

Phase structure: We return to the real quartic matrix model and discuss now its rich phase structure.

1. There are two stable phases and one metastable phase in this model, which are given in terms of an eigenvalue distribution $\rho(\lambda)$ as follows.
2. Disordered phase characterized by the eigenvalue distribution

$$\rho(\lambda) = \frac{1}{N\pi}(2C\lambda^2 + B + Cr^2)\sqrt{r^2 - \lambda^2}. \tag{20.117}$$

This is a single cut solution with the cut defined by

$$-r \leqslant \lambda \leqslant r, \quad r^2 = \frac{1}{3C}(-B + \sqrt{B^2 + 12NC}). \tag{20.118}$$

3. Non-uniform ordered phase characterized by the eigenvalues distribution

$$\rho(\lambda) = \frac{2C|\lambda|}{N\pi}\sqrt{(\lambda^2 - r_-^2)(r_+^2 - \lambda^2)}. \tag{20.119}$$

Here there are two cuts defined by

$$r_- \leqslant |\lambda| \leqslant r_+, \quad r_{\mp}^2 = \frac{1}{2C}(-B \mp 2\sqrt{NC}). \tag{20.120}$$

4. A third-order transition between the above two phases occurs at the critical point

$$B_c^2 = 4NC \longleftrightarrow B_c = -2\sqrt{NC} \longleftrightarrow -\tilde{B}_c = 2\sqrt{\tilde{C}}. \tag{20.121}$$

5. This critical line separates the disordered phase with $\langle M \rangle = 0$ for $-\tilde{B} < -\tilde{B}_c$ from the non-uniform ordered phase with $\langle M \rangle \neq 0$ for $-\tilde{B} > -\tilde{B}_c$.

6. For $C = 0$ (perturbative regime) the eigenvalues distribution is given by the Wigner semi-circle law, viz

$$\rho(\lambda) = \frac{\tilde{B}}{\pi}\sqrt{\delta^2 - \lambda^2}, \quad \delta^2 = \frac{2}{\tilde{B}}. \tag{20.122}$$

7. At $\tilde{C} = -\tilde{B}^2/(12C)$ the square root behavior of $\rho(\lambda)$ becomes $(\delta^2 - \lambda^2)^{3/2}$. This is the pure gravity critical point.

8. There is another non-perturbative one-cut solution (but metastable), which is the Ising or uniform order. It occurs for $\mu^2 < 0$ and $g < \mu^4/60$ where $\mu^2 = 2B$ and $g = NC$. It is given by

$$\rho(\lambda) = \frac{1}{\pi}(2g\lambda^2 + 2gr_2\lambda + 2gr_2^2 + gr_1^2 + \frac{\mu^2}{2})\sqrt{(r_1 + r_2 - \lambda)(\lambda + r_1 - r_2)}. \tag{20.123}$$

Here, r_1 and r_2 are given by

$$r_1^2 = \frac{-\mu^2 - \sqrt{\mu^4 - 60g}}{15g}, \quad r_2^2 = \frac{-3\mu^2 + 2\sqrt{\mu^4 - 60g}}{20g}. \tag{20.124}$$

Monte Carlo simulation: The Monte Carlo method can also be applied to this model.

1. This theory can be simulated using the Metropolis algorithm.

2. Under the change $\lambda_i \longrightarrow \lambda_i + h$ the effective potential changes as $V_{\text{eff}} \longrightarrow V_{\text{eff}} + \Delta V_{i,h}$ where

$$\Delta V_{i,h} = B\Delta S_2 + C\Delta S_4 + \Delta S_{\text{Vand}}. \tag{20.125}$$

The variations ΔS_2, ΔS_4 and ΔS_{Vand} are given explicitly by

$$\Delta S_2 = h^2 + 2h\lambda_i, \quad \Delta S_4 = 6h^2\lambda_i^2 + 4h\lambda_i^3 + 4h^3\lambda_i + h^4, \quad \Delta S_{\text{Vand}}$$
$$= -2\sum_{j \neq i} \ln|1 + \frac{h}{\lambda_i - \lambda_j}|. \tag{20.126}$$

3. The Metropolis accept/reject step:

$$W(\lambda_i \longrightarrow \lambda_i + h) = \min(1, \exp(-\Delta V_{i,h})). \tag{20.127}$$

4. Measurement: we can measure the Schwinger–Dyson identity $\langle 2BTrM^2 + 4CTrM^4 \rangle = N^2$ (as calibration) and the observables $\langle V \rangle$, $C_v = \langle (V - \langle V \rangle)^2 \rangle$, $m = |TrM|$ and $\chi = \langle (m - \langle m \rangle)^2 \rangle$.

20.2.4 The noncommutative phi-four theory

Next, we are interested in a noncommutative scalar Φ^4 theory. As a non-perturbatively well-defined example, we will take the underlying noncommutative space to be the fuzzy projective space $\mathbf{CP}^n_l$, which is obtained by Berezin quantization [42] of the commutative projective space $\mathbf{CP}^n = SU(n+1)/U(n)$.

Fuzzy projective spaces: Fuzzy $\mathbf{CP}^n_l$ is given by the spectral triple [43, 44] (see also [45])

$$\mathbf{CP}^n_l = (\mathbf{H}_{n,l}, \ \mathrm{Mat}_{N_{n,l}}, \ \Delta_{n,l}). \tag{20.128}$$

$\mathbf{H}_{n,l}$ is the Hilbert space associated with the irreducible representation of $su(n+1)$, which is given by the totally symmetrized tensor product of l fundamental representations, viz $(l, 0, 0, \ldots, 0)$. The dimension of this representation is given by

$$N_{n,l} = \frac{(n+l)!}{n!\,l!}. \tag{20.129}$$

The Hilbert space $\mathbf{H}_{n,l}$ is thus acted on by the complete matrix algebra $\mathrm{Mat}_{N_{n,l}}$ of finite dimension $N_{n,l}$ with inner product defined by

$$(f, g) = \frac{1}{N_{n,l}} Trf^\dagger g. \tag{20.130}$$

The $\Delta_{n,l}$ is the Laplace operator on fuzzy $\mathbf{CP}^n_l$ defined, in terms of the generators L_i of $su(n+1)$ in the representation $(l, 0, 0, \ldots, 0)$, by the quadratic Casimir

$$\Delta_{n,l}(f) = [L_i, [L_i, f]]. \tag{20.131}$$

Derivation on fuzzy $\mathbf{CP}^n_l$ are given precisely by the adjoint action of the generators L_i, i.e.,

$$\mathrm{ad}L_i(f) = [L_i, f] = (L_i^L - L_i^R)(f). \tag{20.132}$$

This shows explicitly that the space of functions on fuzzy $\mathbf{CP}^n_l$, which is the matrix algebra $\mathrm{Mat}_{d_{n,l}}$, decomposes under the action of $SU(n+1)$ as the direct sum of the irreducible representations $(m, \ldots, m)$ of $SU(n+1)$,viz

$$(l, \ldots, 0) \otimes \overline{(l, \ldots, 0)} = \oplus_{m=0}^{l}(m, 0, \ldots, 0, m). \tag{20.133}$$

Hence, the matrix algebra $\mathrm{Mat}_{N_{n,l}}$ is the endomorphism $\mathbf{Hom}(V_{n,l}) = V_{n,l} \otimes V_{n,l}^*$ where $V_{n,l}$ is the vector space associated with the representation $(l, 0, \ldots, 0)$.

The polarization tensors $T_{m\sigma}$, which transform in the irreducible representation $(m, \ldots, m)$, are the eigenmatrices of the Laplace $\Delta_{n,l}$ with eigenvalue λ_m and degeneracy $d_{n,m}$ given by

$$\lambda_m = 2m(m + n) \tag{20.134}$$

and

$$d_{n,m} = \frac{n(2m + n)((m + n - 1)!)^2}{(n!)^2(m!)^2}.$$ (20.135)

Indeed, the index σ in $T_{m\sigma}$ denotes the other quantum numbers required to specify the representation $(m,..., m)$.

The above spectrum is precisely the spectrum of the Laplace operator on commutative $\mathbf{CP}^n$ only cutoff at $m = l$. The commutative limit is therefore $l \longrightarrow \infty$ in which the eigenmatrices $T_{m\sigma}$ go over to the correct eigenfunctions $Y_{m\sigma}$ (spherical harmonics) on $\mathbf{CP}^n$.

The coordinate functions X_i on fuzzy $\mathbf{CP}^n_l$ are obtained from suitably rescaling the generators L_i of $su(n + 1)$ in the representation $(l, 0,..., 0)$, namely $X_i = a_l L_i$. We have then the commutation relations

$$[X_i, X_j] = ia_l f_{ijk} X_k.$$ (20.136)

The data contained in the spectral triple $(\mathbf{H}_{n,l}, \mathrm{Mat}_{N_{n,l}}, \Delta_{n,l})$ is sufficient to write down an Euclidean action for a scalar field theory on fuzzy $\mathbf{CP}^n_l$. A scalar field Φ is a Hermitian $N \times N$ matrix in $\mathrm{Mat}_{N_{n,l}}$, i.e. $N \equiv N_{n,l}$. An action on fuzzy $\mathbf{CP}^n_l$ is then given by

$$S = Tr(\Phi C_2 \Phi + r\Phi^2 + g\Phi^4).$$ (20.137)

Here, $C_2 \equiv \Delta_{n,l}$. The partition function of the model is given by

$$Z = \int d\mu_D(\Phi)\exp(-\beta S[\Phi]).$$ (20.138)

$d\mu_D(\Phi)$ is Dyson measure on Mat_N.

The multitrace approach: We can diagonalize the matrix Φ as $\Phi = U\Lambda U^\dagger$. We also write $\Phi = \Phi_\mu T_\mu$ or $\Phi_\mu = Tr\Phi T_\mu$ where T_μ, $\mu = 1,..., N^2$ are the generators of $u(N)$. The partition function then becomes

$$Z = \int \prod_{i=1}^{N} d\lambda_i \Delta^2(\Lambda)\exp\left(-\beta Tr(r\Lambda^2 + g\Lambda^4)\right)$$
$$\times \int d\mu_H(U)\exp\left(-\beta K_{\mu\nu} Tr U\Lambda U^\dagger T_\mu Tr U\Lambda U^\dagger T_\nu\right).$$ (20.139)

The Vandermonde $\Delta^2(\Lambda)$ is given by

$$\Delta^2(\Lambda) = \prod_{i=1}^{N} (\lambda_i - \lambda_j)^2.$$ (20.140)

$d\mu_H(U)$ is the Haar measure on $U(N)$. The kinetic matrix $K_{\mu\nu}$ is given, on the other hand, by

$$K_{\mu\nu} = -Tr[L_i, T_\mu][L_i, T_\nu].$$ (20.141)

We can now perform a hopping parameter expansion to perform the integral over U (by using the properties $(A \otimes B)(C \otimes D) = (AC \otimes BD)$ and $Tr(A \otimes B) = TrATrB$ and the orthogonality relation of the Haar measure) as follows [46, 47]

$$
\begin{aligned}
Z &= \int \prod_{i=1}^{N} d\lambda_i \Delta^2(\Lambda) \exp\left(-\beta Tr(r\Lambda^2 + g\Lambda^4)\right) \sum_k \frac{(-\beta)^k}{k!} \int d\mu_H(U) \prod_{i=1}^{k} K_{\mu_i \nu_i} Tr U\Lambda U^\dagger T_{\mu_i} Tr U\Lambda U^\dagger T_{\nu_i} \\
&= \int \prod_{i=1}^{N} d\lambda_i \Delta^2(\Lambda) \exp\left(-\beta Tr(r\Lambda^2 + g\Lambda^4)\right) \sum_k \frac{(-\beta)^k}{k!} K_{\mu_1 \nu_1} \cdots K_{\mu_k \nu_k} \\
&\quad \times \int d\mu_H(U) Tr((U \otimes \cdots \otimes U)(\Lambda \otimes \cdots \otimes \Lambda)(U^\dagger \otimes \cdots \otimes U^\dagger)(T_{\mu_1} \otimes T_{\nu_1} \otimes \cdots \otimes T_{\mu_k} \otimes T_{\nu_k})) \\
&= \int \prod_{i=1}^{N} d\lambda_i \Delta^2(\Lambda) \exp\left(-\beta Tr(r\Lambda^2 + g\Lambda^4)\right) \sum_k \frac{(-\beta)^k}{k!} K_{\mu_1 \nu_1} \cdots K_{\mu_k \nu_k} \\
&\quad \times \sum_\rho \frac{1}{\dim(\rho)} Tr_\rho(\Lambda \otimes \cdots \otimes \Lambda) Tr(T_{\mu_1} \otimes T_{\nu_1} \otimes \cdots \otimes T_{\mu_k} \otimes T_{\nu_k}).
\end{aligned}
\tag{20.142}
$$

This leads directly to the multitrace approach to noncommutative phi-four theory.

1. After a very long and involved calculation we find upto the order $O(\beta^2)$ the multitrace matrix model [46, 47]

$$
Z = \int \prod_{i=1}^{N} d\lambda_i \Delta^2(\Lambda) \exp\left(-\beta Tr(r\Lambda^2 + g\Lambda^4)\right) \exp\left(-\beta(S_1 + S_2)\right).
\tag{20.143}
$$

The actions S_1 and S_2 are given explicitly by

$$
S_1 = \mathcal{J}_1 = a_{2,0} Tr\Lambda^2 + a_{1,1}(Tr\Lambda)^2.
\tag{20.144}
$$

and

$$
\begin{aligned}
S_2 &= \frac{\beta}{2}(\mathcal{J}_1^2 - \mathcal{J}_2) \\
\mathcal{J}_2 &= a_{4,0} Tr\Lambda^4 + a_{3,1} Tr\Lambda^3 Tr\Lambda + a_{2,2}(Tr\Lambda^2)^2 \\
&\quad + a_{2,1,1} Tr\Lambda^2(Tr\Lambda)^2 + a_{1,1,1,1}(Tr\Lambda)^4.
\end{aligned}
\tag{20.145}
$$

2. It was argued in [49] that the non-vanishing of the multitrace term proportional to $a_{3,1}$ is a necessary and a sufficient condition for a stable Ising phase and a stable emergent background geometry.
3. By neglecting all multitrace terms we find that the quartic matrix model $NTr(\tilde{r}\Phi^2 + \tilde{g}\Phi^4)$ (where $\tilde{r} = r/N^{3/2}$ and $\tilde{g} = g/N^2$) in the limit of very weak coupling $\tilde{g} \longrightarrow 0$ is dominated by the Wigner semi-circle law given by

$$
\rho(\lambda) = \frac{\tilde{r}}{\pi} \sqrt{\frac{2}{\tilde{r}} - \lambda^2}.
\tag{20.146}
$$

4. However, by including all multitrace terms we get a noncommutative Φ^4 on fuzzy $\mathbf{CP}_l^n$, which is also dominated in the limit of weak coupling $\tilde{g} \longrightarrow 0$ by a Wigner semi-circle corresponding to the following simple model [48]

$$S = \frac{2N}{\alpha_0^2(m)} Tr\Phi^2. \tag{20.147}$$

In other words, we must make in the above Wigner semi-circle law the following replacement

$$\tilde{r} \longrightarrow \frac{2}{\alpha_0^2(m)}. \tag{20.148}$$

Here, α_0 is the maximum eigenvalue and it is given in terms of the cutoff $\Lambda \sim l$ on fuzzy $\mathbf{CP}_l^n$ and the mass $m^2 \sim r$ by the relation [48]

$$\alpha_0^2(m) = 4c(m, \Lambda)\Lambda^{d-2}. \tag{20.149}$$

We have explicitly

$$c(m, \Lambda) = \frac{1}{16\pi^2}\left(1 - \frac{m^2}{\Lambda^2}\ln\left(1 + \frac{\Lambda^2}{m^2}\right)\right), \quad d = 4, n = 2. \tag{20.150}$$

$$c(m, \Lambda) = \frac{1}{4\pi}\ln\left(1 + \frac{\Lambda^2}{m^2}\right), \quad d = 2, n = 1. \tag{20.151}$$

Thus, by studying the multitrace approximation at weak coupling we can determine when the Wigner behavior goes from the behavior of the pure matrix model $d = 0$ to the behavior of noncommutative matrix models at $d = 2$ and $d = 4$ as we vary for example the parameter $a_{3,1}$. The emergence of the correct behavior is an indication of a stable Ising phase and as a consequence a stable emergent background geometry.

5. A very promising two-body interaction approximating the multitrace effective action of noncommutative phi-four theory is also discussed in [50, 51]. A recent review of fuzzy field theory, their related matrix models and their multitrace approximations is also given in [52].

Phase diagram of noncommutative phi-four: Finally, we would like to discuss here how to construct the phase diagram of noncommutative scalar phi-four via a minimal coupling to a noncommutative gauge field.

1. Indeed, another approach to the effective action on fuzzy $\mathbf{CP}_l^n$ is based on the algorithm proposed in [53].
2. Instead of the scalar action (20.137) we consider a scalar field coupled to a gauge field X_a on fuzzy $\mathbf{CP}_l^n$ given by the action

$$S_m = -Tr[X_a, \Phi]^2 + Tr(r\Phi^2 + g\Phi^4) + S_G[X]. \tag{20.152}$$

For the fuzzy sphere case $\mathbf{CP}_l^1$, the gauge action S_G is given explicitly by

$$S_G = NTr\left(-\frac{1}{4}[X_a, X_b]^2 + \frac{2i\alpha}{3}\varepsilon_{abc}X_a X_b X_c\right) + NTr(MTr(X_a^2)^2 + \beta X_a^2). \tag{20.153}$$

Similar actions exist for higher fuzzy $\mathbf{CP}^n_l$.

We can now diagonalize the scalar field by means of a $U(N)$ gauge transformation, viz $\Phi = U\Lambda U^+$, where the unitary matrix U can then be integrated out from the path integral.

3. In this algorithm, we thus trade off the Monte Carlo sampling of the unitary matrix U, in the original model, with the Monte Carlo sampling of a gauge field on the fuzzy $\mathbf{CP}^n_l$, which we know is much more efficient using ordinary Metropolis.

4. Indeed, the phase structure of noncommutative phi-four is very complicated as it involves transitions between vacuum states, with very low probability distributions, and as a consequence, they are extremely difficult to sample correctly with the Metropolis algorithm. In particular, the non-uniform-to-uniform transition is virtually unobservable in ordinary Metropolis, due to the absence of tunneling between the identity matrix, corresponding to the uniform phase, and the other idempotent matrices, corresponding to the non-uniform phase. These are the three known phases in this model: the usual Ising transition between disorder and uniform order; A matrix transition between disorder and a non-uniform ordered phase; and a (very hard to observe) transition between uniform order and non-uniform order. The three phases meet at a triple point [54, 55]. The non-uniform phase, in which rotational invariance is spontaneously broken, is simply absent in the commutative theory. The non-uniform phase is the analogue of the stripe phase observed on the Moyal–Weyl spaces [56], whereas the disorder-to-non-uniform-order transition is the generalization to the fuzzy sphere of the one-cut-to-two-cut transition, which is observed in the real quartic matrix model [37, 38].

In summary, we have

$$\langle \Phi \rangle = 0 \quad \text{disordered phase.} \tag{20.154}$$

$$\langle \Phi \rangle = \pm\sqrt{-\frac{r}{2g}}\,1_N \quad \text{Ising (uniform) phase.} \tag{20.155}$$

$$\langle \Phi \rangle = \pm\sqrt{-\frac{r}{2g}}\,\gamma \quad \text{matrix (nonuniform or stripe) phase.} \tag{20.156}$$

5. In here we will supplement this algorithm by a further assumption as follows. We have the path integral (we set $\beta = 1$ here)

$$Z = \int d\Lambda \int \prod_a dX_a \exp(-S_g)\exp\left(Tr[X_a, \Lambda]^2\right)\exp\left(-Tr(r\Lambda^2 + g\Lambda^4) + \ln \Delta_N(\Lambda)\right)$$

$$= Z_G \int d\Lambda \,\langle \exp\left(Tr[X_a, \Lambda]^2\right)\rangle_G \exp\left[-Tr(r\Lambda^2 + g\Lambda^4) + \ln \Delta_N(\Lambda)\right] \tag{20.157}$$

$$\simeq Z_G \int d\Lambda \,\exp\left(Tr[X_a^{\text{eff}}, \Lambda]^2\right)\exp\left[-Tr(r\Lambda^2 + g\Lambda^4) + \ln \Delta_N(\Lambda)\right].$$

The scalar action, from the first line, is then given explicitly by

$$S[\Lambda] = - Tr[X_a, \Lambda]^2 + Tr(r\Lambda^2 + g\Lambda^4) - \ln \Delta_N(\Lambda)$$
$$= - 2\sum_{ij}(X_a)_{ij}(X_a)_{ji}\lambda_i\lambda_j + Nc_2\sum_i \lambda_i^2 + \sum_i(r\lambda_i^2 + g\lambda_i^4) - \sum_{i\neq j}\ln|\lambda_i - \lambda_j|. \quad (20.158)$$

The scalar action, from the last line, is given by

$$S[\Lambda] = - Tr[X_a^{\text{eff}}, \Lambda]^2 + Tr(r\Lambda^2 + g\Lambda^4) - \ln \Delta_N(\Lambda)$$
$$= - 2\sum_{ij}(X_a^{\text{eff}})_{ij}(X_a^{\text{eff}})_{ji}\lambda_i\lambda_j + Nc_2\sum_i \lambda_i^2 + \sum_i(r\lambda_i^2 + g\lambda_i^4) - \sum_{i\neq j}\ln|\lambda_i - \lambda_j|. \quad (20.159)$$

Here, X_a^{eff} is a thermalized configuration around the global minimum of the gauge action in the fuzzy $\mathbf{CP}_l^n$ phase. The difference between the first and last lines of (20.157) is the fact that the last line is just an approximation. In fact, sufficiently deep in the fuzzy $\mathbf{CP}_l^n$ phase, we can simply replace X_a^{eff} with the background configuration L_a itself, i.e. we choose

$$X_a^{\text{eff}} = L_a. \quad (20.160)$$

20.3 The restricted Boltzmann machine (RBM)

20.3.1 Artificial neural networks

An artificial neural network is a computational model of the brain wherein neurons interact by exchanging signals allowing the net to learn.

Artificial neural networks consist the backbone of machine learning (ML), which is a subfield of artificial intelligence (AI), which in turn is a subfield of computer science.

The McCulloch–Pitts model (1943) was the first to propose a simplified computational model of how neurons work in the brain to perform complex tasks. Warren McCulloch (a neuroscientist) and Walter Pitts (a logician) demonstrated that neural networks could essentially act as a universal Turing machine.

In the human brain, which is composed of billions of interconnected neurons communicating with each other by exchanging electrical pulses, the accumulated incoming signal at each neuron must exceed an activation threshold in order for the neuron to activate (in this case the neuron fires yielding a non-zero output) otherwise the neuron will remain inactive (that is, the neuron does not fire and we have zero output).

Similarly, an artifical neural network is in general a graph with vertices (where the neurons are located), which are connected by directed or undirected edges (links between the neurons). This is very reminiscent of spacetime lattices of gauge theories.

An artificial neuron i in the neural network will fire, i.e., output the value y_i, according to some activation function f_i, which takes as input the sum of the received signals $x_{i1}, x_{i2},...,x_{in}$ from n neurons weighted by some numbers $w_{i1}, w_{i2},...,w_{in}$. The weight w_{ij} measures the strength of the link between the neurons i and j. Thus, the fired output y_i is equal to the value $f_i(\sum_j w_{ij} x_j + b_i)$ of the activation function f_i. We have then the formula

$$y_i \equiv f_i(z_i) = z_i \equiv \sum_j w_{ij} x_j + b_i. \tag{20.161}$$

We have also included a bias b_i for the artificial neuron i, which is needed in case of zero weights or inputs.

Two neural networks, which are of particular interest to us, due to their intimate connection to theoretical physics, are the Hopfield networks [2] and the restricted Boltzmann machines [7]. These two systems are essentially Ising models at zero and finite temperatures, respectively, which is a fact that makes their sampling straight-forward using the methods of computational physics. The training of these models admits also a very clear physical interpretation using the principles of statistical physics.

For a pedagogical introduction to neural networks from the perspective of computational physics, see the two modern courses [19, 20].

20.3.2 Definition of the RBM (energy and activation function)

As a more detailed example of neural networks, we will consider here Boltzmann machines, which are artificial neural networks used to acquire (learn) probability distributions of data sets. They are also generative models, i.e., after learning the underlying probability distribution of the input data they can generate new samples.

Boltzmann machines are stochastic (not deterministic) neural networks.

Boltzmann machine turns into a Hopfield network if we replace its stochastic updating rule (see below) by a deterministic one (which is given by the step function).

The Boltzmann machines were first studied in the context of statistical mechanics as solvable spin-glass models by Sherrington and Kirkpatrick in 1975 [3]. These machines were then popularized starting in 1985 by Hinton and collaborators [4, 5].

Boltzmann machines are essentially Ising models and they belong to the family of energy-based neural networks, which learn the probability distribution of the data by looking at the energy of the system.

A Boltzmann machine consists typically of two layers. The input visible layer and a hidden layer. Each neuron in the visible layer is linked to every neuron in the hidden layer. All of the layers' individual neurons are also linked. These neurons can be taken to be binary units, i.e., they can take only the values $+1$ and -1. The visible neurons are designed to interact with the environment while the hidden neurons serve to encode dependencies internally.

There are three types of Boltzmann machines are: 1) Deep belief networks (DBN), 2) Restricted Boltzmann machines (RBM), 3) Deep Boltzmann machines (DBM).

We will concentrate on the restricted Boltzmann machine in which there is no intra-layer connectivity. See figure 20.7.

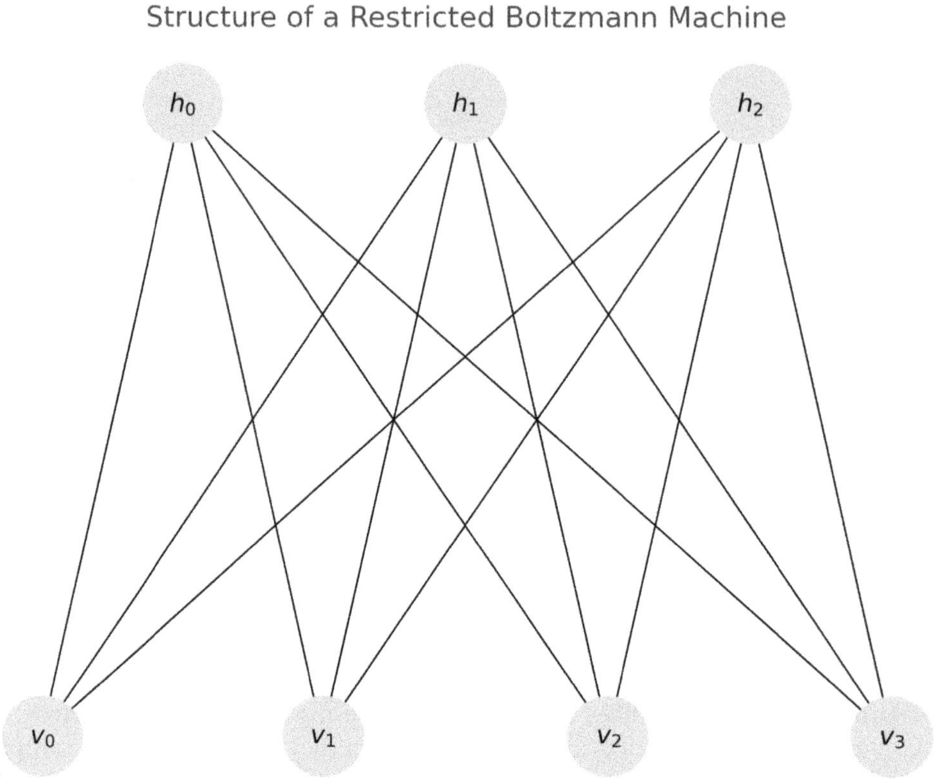

Structure of a Restricted Boltzmann Machine

Figure 20.7. The restricted Boltzmann machine.

The energy function of the restricted Boltzmann machine with visible neurons denoted v_i (with bias denoted by a_i) and hidden neurons denoted h_i (with bias denoted by b_i) is given by the Ising-like energy

$$E(v, h) = -\sum_{i,j} v_i w_{ij} h_j - \sum_i a_i v_i - \sum_i b_i h_i. \tag{20.162}$$

The coupling w_{ij} represents the strength of the connection between neurons i and j, i.e., it represents the weight of the link ij. The states v_i and h_i can only take the binary values 0, 1.

The probability of the configuration (v_i, h_i) is given by the Boltzmann distribution (and hence the name Boltzmann in Boltzmann machine) with temperature $T = 1$, viz

$$p(v, h) = \frac{\exp(-E(v, h))}{Z}, \quad Z = \sum_{v,h} \exp(-E(v, h)). \tag{20.163}$$

This is the joint probability distribution.

The restricted Boltzmann machine is really nothing but an Ising model. The weights w_{ij} describe the interactions between the visible spins v_i and the hidden spins

h_j whereas the visible/hidden bias a_i/b_i represents a magnetic field that encourages the visible/hidden spins to flip.

The parameters W_{ij}, a_i and b_i define the restricted Boltzmann machine and they are clearly tunable.

In addition to the joint probability distribution $p(v, h)$ we have also the marginal probability distribution over the visible neurons given by summing over all hidden neurons, viz

$$p(v) = \sum_h p(v, h) = \sum_h \frac{\exp(-E(v, h))}{Z}. \tag{20.164}$$

This represents the probability distribution of the data as modeled by the restricted Boltzmann machine.

We need also the conditional probability for activation of the neurons. The conditional probability that the hidden neuron h_j activates, given the visible neurons, is given by the logistic function

$$p(h_j = 1|v) = \sigma\left(b_j + \sum_i w_{ij} v_i\right). \tag{20.165}$$

Similarly, the conditional probability that the visible neuron v_i activates, given the hidden neurons, is given by the logistic function

$$p(v_i = 1|h) = \sigma\left(a_i + \sum_j w_{ij} h_j\right). \tag{20.166}$$

Here, σ is the sigmoid function: $\sigma(x) = (1 + \exp(-x))^{-1}$. This is the activation function f considered earlier.

This means that the activation process of the hidden neurons in the restricted Boltzmann machine is a stochastic process, i.e., it is a probabilistic process dependent on the input from the visible neurons and the configuration of the weights and biases of the restricted Boltzmann machine. This is the reason why the restricted Boltzmann machine is a stochastic neural network.

In practice, whether a hidden neuron activates or not involves generating a random number r and comparing it to the activation probability $p(h_j = 1|v)$. If the random number r is less than the calculated probability, the hidden neuron is set to 1 otherwise it is set to 0. This is the stochastic nature of activation inherent in Boltzmann machines.

20.3.3 Probabilities of the RBM

The joint probability distribution of the RBM is given by

$$p(v, h) = \frac{\exp(-E(v, h))}{Z}, \quad Z = \sum_{v,h} \exp(-E(v, h)). \tag{20.167}$$

In the following, the neurons v_i and h_i are thought of as spins, which can take the two values of spin up $+1$ and spin down -1.

The conditional probability that the state of the visible neurons is v, given that the state h of the hidden neurons is fixed, is given by Bayes' theorem, viz

$$p(v|h) = \frac{p(v, h)}{p(h)}. \tag{20.168}$$

This is a probability distribution over the variable v since by construction we have $\sum_v p(v|h) = 1$.

Similarly, the conditional probability that the state of the hidden neurons is h, given that the state v of the visible neurons is fixed, is given by Bayes' theorem, viz

$$p(h|v) = \frac{p(h, v)}{p(v)}, \quad p(h, v) = p(v, h). \tag{20.169}$$

This is a probability distribution over the variable h.

The two complete conditional probabilities $p(v|h)$ and $p(h|v)$ are related to the individual conditional probabilities $p(v_i|h)$ and $p(h_i|v)$ by the relations

$$p(v|h) = \prod_i p(v_i|h) = \prod_i \frac{e^{m_i v_i}}{e^{m_i v_i} + e^{-m_i v_i}}, \quad m_i = \sum_j w_{ij} h_j + a_i. \tag{20.170}$$

$$p(h|v) = \prod_i p(h_i|v) = \prod_i \frac{e^{m_i h_i}}{e^{m_i h_i} + e^{-m_i h_i}}, \quad m_i = \sum_j v_j w_{ji} + b_i. \tag{20.171}$$

This factorization property means in particular that the visible/hidden variables, for a fixed configuration of the hidden/visible variables, are independent and thus they can be sampled independently and efficiently using the Markov Chain Monte Carlo method known as Gibbs sampling.

20.3.4 Markov chain Gibbs sampling of the RBM

A restricted Boltzmann machine (RBM) consists of N_v visible units v_i (visible layer of neurons/spins) connected to N_h hidden units h_i (hidden layer of neurons/spins) with an energy function given by

$$E(v, h) = -\sum_{i=1}^{N_v}\sum_{j=1}^{N_h} v_i w_{ij} h_j - \sum_{i=1}^{N_v} a_i v_i - \sum_{i=1}^{N_h} b_i h_i. \tag{20.172}$$

Strictly speaking, this is a binary RBM since the units can only take the two values ± 1. We will also be interested in the Gaussian restricted Boltzmann machine (GRBM) in which the visible units are real numbers drawn from a Gaussian distribution.

The state of this machine is characterized by the weights w_{ij}, between the visible unit i and the hidden unit j, and the visible and hidden biases a_i and b_i. The weights w_{ij} are chosen randomly close to zero, while the visible biases a_i can be chosen according to the probability distribution of the visible units, and the hidden biases b_i are typically initialized to zero.

The visible dimension N_v is fixed by the dimensionality of the input data whereas the hidden dimension N_h is chosen based on the expected complexity hidden in the data.

Building an RBM neural network consists of sampling the underlying probability distribution but also it consists of 'training'. Training an RBM simply means that we must tune the parameters (w, a, b) appearing in the energy function in such a way that the RBM's probability distribution match the distribution of a training sample, which is the data set used to initialize the values of the visible units.

In our code, we need then to save the N_v-dimensional and N_h-dimensional vectors of visible and hidden states v_i and h_i, the dimensions N_v and N_h, the $N_v - N_h$ matrix of weights w_{ij}, the N_v-dimensional visible bias vector a_i and the N_h-dimensional hidden bias vector b_i.

The goal is to sample a sequence of configurations (v, h) of visible and hidden units with the joint probability distribution $p(v, h)$ starting from an initial visible configuration. We can use the Metropolis algorithm or any other Markov Chain Monte Carlo algorithm. However, an efficient method for sampling the restricted Boltzmann machine, due is clearly given by the Gibbs sampling algorithm. This is due to the bipartite structure of the RBM, i.e., it is due to the factorization property of the conditional probabilities $p(v|h)$ and $p(h|v)$ shown in equations (20.170) and (20.171).

Gibbs sampling goes as follows.

- Step 1: First, given the initial visible configuration v, we go through each hidden spin i and sets equal to $h_i = +1$ with probability $p(h_i = +1|v) = \exp(m_i)/(\exp(m_i) + \exp(-m_i))$ or sets it equal to $h_i = -1$ with probability $1 - p(h_i = +1|v)$ where m_i is defined in equation (20.171). Thus, we get a new hidden configuration h.
- Step 2: Second, given the newly obtained hidden configuration h, we go through each visible spin i and sets it equal to $v_i = +1$ with probability $p(v_i = +1|h) = \exp(m_i)/(\exp(m_i) + \exp(-m_i))$ or sets it equal to $v_i = -1$ with probability $1 - p(v_i = +1|h)$ where m_i is defined in equation (20.170). Thus, we get a new visible configuration v.
- Step 3: We repeat steps 1 and 2 many times k.
- These three steps, for large enough k, generate a sequence of configurations (v, h) distributed with the joint probability distribution $p(v, h)$ and v, h will be distributed according to the marginal distributions $p(v)$, $p(h)$, respectively.

20.3.5 The training and unsupervised learning of RBM (CD) algorithm

The Markov chain Gibbs sampling is only a part of the algorithm underpinning the restricted Boltzmann machine. The other part, in fact the most important part, is a training procedure in which the parameters of the RBM are themselves treated as fields and modified. In contrast to Gibbs sampling, in which we are only attempting to thermalize the system by changing the fields (given here by the spins), the goal of training is to approximate an external probability distribution by changing both the spins (through the Gibbs sampling) as well as the parameters of the RBM (through the method of gradient descent). The combined algorithm is called the contrastive divergence algorithm **CD** [8, 9].

The parameters of the RBM should then be thought of as fields on equal footing as the spin degrees of freedom. This makes the restricted Boltzmann machine quite distinct from the Ising model in which the parameters of the model (like exchange interaction, temperature and magnetic field) are treated as fixed physical parameters.

Thus, we need to sample the RBM while tuning or optimizing its parameters $\theta = (w_{ij}, a_i, b_i)$ at the same time. This is what we call training, which is also called unsupervised learning since this training/learning is done dynamically starting from an initial input consisting of unlabeled data.

The RBM is characterized by the marginal probability distribution $p(v)$ of the visible units. This is called the 'model distribution'.

Let $q(v)$ be the probability distribution underlying an input data set. This data set is the initial set of configurations of the visible units and is called the 'training data'. Naturally, $q(v)$ is called the 'training data'.

The goal of training is to adjust the parameters θ of the RBM in such a way that the model distribution $p(v)$ approximates the data distribution $q(v)$ as best as possible.

In here the training data distribution $q(v)$ should be thought of as the unknown and we are trying to find an approximation of it which is the closest to the model distribution $p(v)$. This is really what is meant by training the restricted Boltzmann machine. In fact, the goal is to let the RBM learn on its own, i.e. dynamically the training data probability distribution. This is the reason why this training is termed unsupervised learning.

Furthermore, the restricted Boltzmann machine, after it had learned the probability distribution underlying the dataset, it can generate new data samples, which are distributed according to this learned probability distribution. The restricted Boltzmann machine is therefore termed a generative model of unsupervised learning.

The two distributions $p(v)$ and $q(v)$ are as close to each other as possible if their relative or Shannon entropy becomes as small as possible, i.e., it reaches a local minimum. This relative entropy (also called the Kullback–Leibler divergence) is defined by the objective function

$$O(\theta) = \sum_v q(v) \log \frac{q(v)}{p(v)} = \sum_v q(v) \log q(v) - \sum_v q(v) \log p(v). \qquad (20.173)$$

This is also related to the so-called log-likelihood function (the θ-dependent second term in the above equation). Both the Kullback–Leibler divergence and the negative log-likelihood can play the role of loss/cost function.

Thus, we will start at some point θ^0 in the parameter space, compute the derivative of the entropy $O(\theta)$ with respect to the parameter θ, and then evaluate this derivative at the point θ^0. This derivative is the gradient of the objective function $O(\theta)$ at θ^0 and points in the direction of the steepest ascent of the function in its parameter space. To minimize the objective function $O(\theta)$ we need therefore to move in the opposite direction, which is the direction of steepest descent. In other words, we should consider the change

$$\theta^0 \longrightarrow \theta^1 = \theta^0 - \eta\frac{\partial O}{\partial\theta^0}. \tag{20.174}$$

The step size η in the above equation is called the learning rate.

Clearly, the derivative or gradient of the objective function decreases in magnitude, i.e., $O(\theta^1) < O(\theta^0)$ since we are moving in the direction of steepest descent. This can be seen explicitly as follows

$$\begin{aligned}
O(\theta^1) &= O(\theta^0 - \eta\frac{\partial O}{\partial\theta^0}) \\
&= O(\theta^0) - \eta\frac{\partial O}{\partial\theta^0} \cdot \frac{\partial O}{\partial\theta^0} \\
&= O(\theta^0) - \eta|\frac{\partial O}{\partial\theta^0}|^2 \\
&< O(\theta^0).
\end{aligned} \tag{20.175}$$

By iterating the above equation, i.e., $\theta^0 \longrightarrow \theta^1 \longrightarrow \theta^2\ldots$ we are guaranteed to reach the local minimum of the objective function $O(\theta)$.

Thus, the minimum of the objective function $O(\theta)$, where the model and data distributions $p(v)$ and $q(v)$ are as closest as possible, can be reached by changing the parameters w_{ij}, a_i and b_i iteratively as follows:

$$\begin{aligned}
w_{ij} &\longrightarrow w_{ij}' = w_{ij} - \eta\frac{\partial O}{\partial w_{ij}} \\
a_i &\longrightarrow a_i' = a_i - \eta\frac{\partial O}{\partial a_i} \\
b_i &\longrightarrow b_i' = b_i - \eta\frac{\partial O}{\partial b_i}.
\end{aligned} \tag{20.176}$$

In the above equations, the derivative can be computed in a straightforward way to obtain

$$\begin{aligned}
\frac{\partial O}{\partial w_{ij}} &= \langle v_i h_j\rangle_{\text{model}} - \langle v_i h_j\rangle_{\text{data}} \\
\frac{\partial O}{\partial a_i} &= \langle v_i\rangle_{\text{model}} - \langle v_i\rangle_{\text{data}} \\
\frac{\partial O}{\partial b_i} &= \langle h_i\rangle_{\text{model}} - \langle h_i\rangle_{\text{data}}.
\end{aligned} \tag{20.177}$$

The data-dependent term is known as the positive phase of the gradient while the model-dependent term is known as the negative phase.

The expectation value $\langle\rangle_{\text{model}}$ is computed with respect to the RBM states, i.e., with respect to the joint distribution $p(h, v) = p(h|v)p(v)$ whereas the expectation value $\langle\rangle_{\text{data}}$ is computed with respect to the RBM states given that the visible states are distributed according to the training distribution, i.e., with respect to $p(h|v)q(v)$. More precisely, we have

$$
\begin{aligned}
\langle f\rangle_{\text{model}} &= \sum_{v,h} f(v, h)p(h|v)p(v) = \langle\langle f(v, h)\rangle_{h\sim p(h|v)}\rangle_{v\sim p(v)} \\
\langle f\rangle_{\text{data}} &= \sum_{v,h} f(v, h)p(h|v)q(v) = \langle\langle f(v, h)\rangle_{h\sim p(h|v)}\rangle_{v\sim q(v)}.
\end{aligned}
\tag{20.178}
$$

Th notation $x \sim p(x)$ means that x is randomly drawn from $p(x)$.

Hence, we can see from the above equations, that the model expectation value $\langle\rangle_{\text{model}}$ is computed by means of Gibbs sampling whereas the data expectation $\langle\rangle_{\text{data}}$ is computed by taking the average over all samples in the dataset.

The restricted Boltzmann machine is therefore an algorithm which combines the Markov chain Gibbs sampling with gradient descent in order to update the weights and biases as follows.

1. **First:** We initialize the parameters w_{ij}, a_i and b_i. The initial gradients are set to zero, viz

$$
\delta w_{ij} = 0, \quad \delta a_i = 0, \quad \delta b_i = 0.
\tag{20.179}
$$

2. **Second:** We sample M visible configurations $\{v^1, v^2,..., v^M\}$ from the training dataset. This is called a mini-batch (since it is only a small subset of the dataset).

3. **Third:** For each initial visible configuration v^n we sample a corresponding initial hidden configuration h^n according to the conditional probability $p(h|v)$. The gradients are changed as follows

$$
\delta w_{ij} = \delta w_{ij} - v_i^n h_j^n, \quad \delta a_i = \delta a_i - v_i^n, \quad \delta b_i = \delta b_i - h_i^n.
\tag{20.180}
$$

This is the positive (observation) phase.

4. **Fourth:** Starting from $v^n \equiv v^{n0}$ we perform k iterations of Gibbs sampling back and forth: $v^{n0} \longrightarrow h^{n1} \longrightarrow v^{n1} \longrightarrow h^{n2} \longrightarrow v^{n2} \cdots \longrightarrow h^{nk} \longrightarrow v^{nk}$. We use the last configuration to change the gradients as follows

$$
\delta w_{ij} = \delta w_{ij} + v_i^{nk} h_j^{nk}, \quad \delta a_i = \delta a_i + v_i^{nk}, \quad \delta b_i = \delta b_i + h_i^{nk}.
\tag{20.181}
$$

This is the negative (reconstruction) phase.

5. **Fifth:** We go through all the configurations in the mini-batch $\{v^1, v^2,..., v^M\}$ and repeat steps 3-4.

6. **Six:** The parameters of the RBM are then changed as

$$
w_{ij} = w_{ij} - \frac{\eta\delta w_{ij}}{M}, \quad a_i = a_i - \frac{\eta\delta a_i}{M}, \quad b_i = b_i - \frac{\eta\delta b_i}{M}.
\tag{20.182}
$$

7. **Seven:** We repeat steps 1-6 for another batch until all batches of the input/ initial dataset are exhausted.
8. The Gibbs sampling typically only equilibrate for sufficiently large k and thus the computation of the negative phase can be very extensive. We can run the Gibbs sampling only for a small number of steps k which seems to work quite well, even with $k = 1$, which is amazing considering the fact that Monte Carlo equilibration is very hard in general (this seems to be due to some cancellation of errors).

The above algorithm is called the k steps contrastive divergence algorithm (denoted by **CD–k**). This is essentially the algorithm underlying the restricted Boltzmann machine and it is the analogue of the backpropagation algorithm used in feedforward neural networks.

20.3.6 The RBM learning of the Ising model

The restricted Boltzmann machine is an energy-based, two-layer, stochastic neural network which is also generative [6, 7]. This means that after learning, if learning is acquired, the restricted Boltzmann machine can generate the input data probability distribution without any further simulation.

The restricted Boltzmann machine can be thought of as a computational system, i.e., an algorithm or as a physical system. Here, we will think of it as a computational system, which we can use to simulate physical systems such as the Ising model. In fact, RBM as an algorithm for simulating physical systems may potentially surpass the Monte Carlo method. See for example [15–18].

The Ising model in two dimensions on a periodic square lattice of linear dimension L, i.e., with $L^2 \equiv N$ spins s_i, is given by the energy function and the probability distribution

$$E(s) = -J\sum_{\langle i,j \rangle} s_i s_j, \quad p_{\text{Ising}}(s) = \frac{1}{Z_{\text{Ising}}} \exp(-\beta E(s)). \tag{20.183}$$

Here, $\langle i, j \rangle$ indicates sum over nearest-neighbor pairs only. The spins are binary variables, i.e., $s_i = \pm 1$ and the exchanged interaction J is taken to be positive for ferromagnetic behavior (we can simply set $J = 1$). The only physical parameter of the system is the temperature T defined by $\beta = 1/k_B T$ (we can also set the Boltzmann constant equal one, viz $k_B = 1$).

There are two phases in this model. A disordered (paramagnetic) phase at high temperature and an ordered (ferromagnetic) phase at low temperature.

The restricted Boltzmann machine can be used to learn the phase structure of the Ising model.

In the following we review the interesting study [11]. This goes as follows:

1. First, we should perform Monte Carlo simulation of the Ising model by using the Markov chain Metropolis algorithm (with a single flip) in order to generate a sample of configurations $\vec{s} \equiv (s_1, s_2,..., s_N)^T$ distributed according to the Boltzmann distribution $p_{\text{Ising}}(s)$. For a sufficiently large lattice L,

we generate thermalized Ising configurations for n_T different values of the temperature T. For each T, we collect M thermalized configurations in a dataset D_T. In [11] they take $L = 64$, $n_T = 16$ and $M = 50000$.

2. We will associate a data probability distribution $p_D(s)$ with the dataset D_T. Clearly, $p_D \equiv p_{\text{Ising}}$ if thermalization is achieved.

3. Second, we take the visible layer of the RBM to be constituted of $N_v \equiv N$ visible units v_j, which are identified with the spins s_i of the Ising model, i.e., $v_i \equiv s_i$. The number of the hidden units N_h should be taken sufficiently large but not too large. An optimal value observed in [11] is $N_h = 400$. The energy function and the probability distribution are given by

$$E(v, h) = -\sum_{i=1}^{N_v}\sum_{j=1}^{N_h} h_i W_{ij} v_j - \sum_{i=1}^{N_v} a_i v_i - \sum_{i=1}^{N_h} b_i h_i, \quad p_\theta(v, h)$$

$$= \frac{1}{Z_\theta}\exp(-E(v, h)). \tag{20.184}$$

Here, the interaction is restricted between the hidden units from the one hand and the visible units from the other hand. There is no inter-layer interaction. The weight matrix is denoted here by W_{ij} (which is different from the notation of the previous section).

4. The parameters of the restricted Boltzmann machine are given by $\theta = (W_{ij}, a_i, b_i)$. The weight is initialized using Glorot normal initialization [12] while the biases are initialized at 0.

5. The goal of the restricted Boltzmann machine learning is to approximate the data probability distribution $p_D(v \equiv s)$, which contains M thermalized samples of the Ising model, with some RBM distribution p_θ, i.e. with an RBM with some tuned parameters θ.

6. This is done via minimization with respect to θ of the Kullbach-Leibler (KL) divergence given by

$$O(\theta) = \langle \ln p_D(v)\rangle_{v \sim p_D} - \langle \ln p_\theta(v)\rangle_{v \sim p_D}. \tag{20.185}$$

Here, $p_\theta(v)$ is the marginal probability distribution. Equivalently, we can minimize the negative log-likelihood defined by [13]

$$L(\theta) = -\langle \ln p_\theta(v)\rangle_{v \sim p_D}. \tag{20.186}$$

The minimum of this negative log-likelihood is the maximum of the likelihood estimate for the dataset $D_{M_0} = \{\vec{v}^{(1)}, \vec{v}^{(2)},..., \vec{v}^{(M_0)}\}$ under the model marginal distribution $p_\theta(v)$, viz $P(D_{M_0}) = \prod_{i=1}^{M_0} p_\theta(v^{(i)})$.

7. The training is done for each dataset D_T, i.e., we have in fact n_T restricted Boltzmann machines.

8. As we have said, the RBM is a generative stochastic neural network. This means that after learning, if learning is acquired, the restricted Boltzmann machine can generate thermalized Ising configurations without any further simulation.

9. The contrastive divergence algorithm with k steps, i.e., **CD–k** is used to train the restricted Boltzmann machine with learning rate η and batch size M_0. This algorithm uses Gibbs sampling inside a gradient descent procedure to update the RBM's parameters θ. In [11] they use $k = 5$, $\eta = 10^{-4}$ and $M_0 = 128$.

10. In conclusion, the restricted Boltzmann machine is therefore an algorithm which combines the Markov chain Gibbs sampling with gradient descent in order to update the weights and biases by following the k steps contrastive divergence algorithm [7–10].

The main results reported in [11] are as follows:

1. **First**:

   - It is observed that the elements W_{ij} of the weight matrix follow a Gaussian distribution of zero mean. The behavior of the biases seems to be trivial and thus it will not be discussed here any further.
   - The $N_h - N_v$ weight matrix W_{ij} of the **RBM** is viewed as either N_h row vectors $(w_i)_j = W_{ij}$ which are N_v-dimensional or N_v column vectors $(\bar{w}_i)_j = W_{ji}$ which are N_h-dimensional. We have then the vectors

   $$\vec{w}_i = (W_{i1}, W_{i2}, \ldots, W_{iN_v}), \quad \vec{\bar{w}}_i = (W_{1i}, W_{2i}, \ldots, W_{N_h i})^T. \tag{20.187}$$

   - The filters reconstruct the hidden units (filters map visible to hidden units) and the inverse filters reconstruct the visible units (inverse filters map hidden to visible units).
   - Indeed, the filter sums $m_i = \sum_{j=1}^{N_v} W_{ij}/N_v$ play the role of magnetization order parameters for the filters $\vec{w}_i$. We have then N_h magnetization order parameters given by

   $$m_i = \frac{1}{N_v}\sum_{j=1}^{N_v} W_{ij}, \quad i = 1, \ldots, N_h. \tag{20.188}$$

   - It is observed that the probability distribution function of the N_h filter sums m_i behaves exactly as one expects from the magnetization of the Ising model. Indeed, at low temperature in the ferromagnetic phase, this probability distribution function presents two peaks symmetric around zero (corresponding to the uniform order). Whereas at high temperature in the paramagnetic phase this probability distribution function presents a single peak around 0 (corresponding to the disorder). At the critical temperature this probability distribution function shows long-range critical fluctuations. The critical temperature is precisely captured by the peak in the susceptibility, which is defined by

   $$\chi = \sum_i (\langle |m_i|^2\rangle - \langle |m_i|\rangle^2). \tag{20.189}$$

- In contrast, the distribution of the N_v inverse filter sums $\bar{m}_i = \sum_{j=1}^{N_h} W_{ji}/N_h$ is Gaussian centered about 0. However, the correlation between the Ising spins is captured by the correlation between inverse filters.

2. **Second:**
   - The distribution of the hidden units h_i consists of equal numbers of $+1$ and -1 (since the distribution of the filter sums m_i is symmetric about 0). Indeed, the distribution of the hidden magnetization $m_h = \sum_{i=1}^{N_h} h_i/N_h$ is observed to be symmetric about zero.
   - The filter sums $m_i \neq 0$ (bimodal distribution symmetric around 0) indicate therefore a ferromagnetic phase while the filter sums $m_i = 0$ (unimodal distribution centered around 0) indicate a paramagnetic phase. See figure 20.8.
   - Thus, in the ferromagnetic phase, the components $(w_i)_j = W_{ij}$ of the filters $\vec{w}_i$ are either positive/negative matching the spin up/down of the visible layer $\vec{v} \equiv \vec{s}$. In other words, we must have $\vec{w}_i \cdot \vec{v} > 0$ (filters match Ising spins).
   - In the paramagnetic phase, the filters $\vec{w}_i$ present a random mixture consisting of equal numbers of positive and negative components,

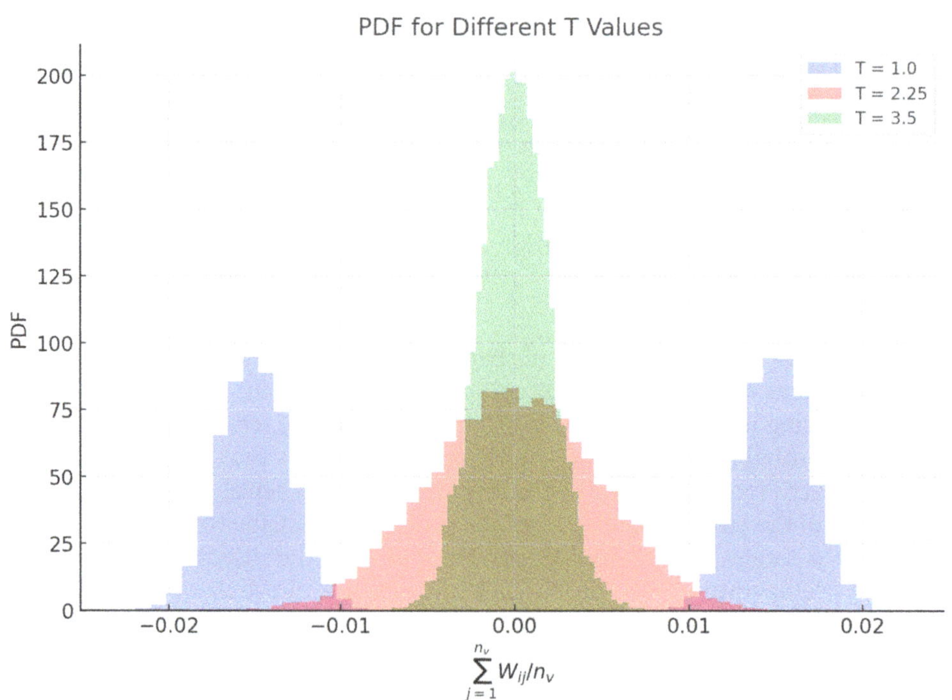

Figure 20.8. Sketch of PDF of filter magnetization for different temperatures.

Table 20.1. In the case $\sum(\mathbf{w}_i^T) > 0$, it is more probable that a visible layer pattern $\mathbf{v}$ with magnetization $m > 0$ (or $m < 0$) is encoded by a hidden unit $h_i = +1$ (or $h_i = -1$). In the case $\sum(\mathbf{w}_i^T) < 0$, the encoding is opposite.

	$\sum(\mathbf{w}_i^T) > 0$	$\sum(\mathbf{w}_i^T) < 0$
$m > 0$	$h_i = +1$	$h_i = -1$
$m < 0$	$h_i = -1$	$h_i = +1$

which also match the components of the visible layer, i.e., we must also have $\vec{w}_i \cdot \vec{v} > 0$.

- Furthermore, the term in the **RBM** energy $\sum_{i,j} h_i W_{ij} v_j = \sum_i h_i(\vec{w}_i \cdot \vec{v})$ is more positive if the signs of the hidden units h_i match the signs of the scalar products $\vec{w}_i \cdot \vec{v}$, i.e., $h_i = +1$.
- But we can also have $\vec{w}_i \cdot \vec{v} < 0$ (filters anti-match the Ising spins). Yet, the term in the **RBM** energy $\sum_{i,j} h_i W_{ij} v_j = \sum_i h_i(\vec{w}_i \cdot \vec{v})$ is still more positive if the signs of the hidden units h_i still match the signs of the scalar products $\vec{w}_i \cdot \vec{v}$, i.e., $h_i = -1$.
- In summary, for a given visible layer $\vec{v} \equiv \vec{s}$ of magnetization m, the sign of the scalar product $\vec{w}_i \cdot \vec{v}$ and the sign of the hidden unit h_i are essentially determined by the sign of the corresponding filter magnetization $m_i = \sum_{j=1}^{N_v} W_{ij}/N_v$.
- Thus, a hidden unit $h_i = +1$ can encode a visible layer $\vec{v} \equiv \vec{s}$ with magnetization $m > 0$ which matches the filter $\vec{w}_i$ with magnetization $m_i > 0$, but it can also encode a visible layer $\vec{v} \equiv \vec{s}$ with magnetization $m < 0$ which matches the filter $\vec{w}_i$ with magnetization $m_i < 0$.
- Similarly, a hidden unit $h_i = -1$ can encode a visible layer $\vec{v} \equiv \vec{s}$ with magnetization $m > 0$, which anti-matches the filter $\vec{w}_i$ with magnetization $m_i < 0$, but it can also encode a visible layer $\vec{v} \equiv \vec{s}$ with magnetization $m < 0$, which anti-matches the filter $\vec{w}_i$ with magnetization $m_i > 0$ (table 20.1).

3. **Third:** The restricted Boltzmann machines, after training and unsupervised learning of the Ising model probability distribution, can be used to generate new thermalized Ising configurations starting, for various temperatures, from some well chosen hidden layers $\vec{h}^{(0)}$. The internal energy, specific heat and magnetization of the Ising model can all be reconstructed by one-step RBM's.

4. **Fourth:** The marginal probability distribution $p_\theta(v)$ can be associated with a visible energy $E_\theta(v)$ by the formula

$$p_\theta(v) = \frac{1}{Z_\theta} \exp(-E_\theta(v)). \tag{20.190}$$

This quantity differs from the physical energy βE of the Ising model by a temperature-dependent coefficient. This energy can be approximated by the function

$$E_\theta(v) \simeq -\sum_{i=1}^{N_h} |\sum_{j=1}^{N_v} W_{ij}| = -N_v \sum_{i=1}^{N_h} |m_i|. \tag{20.191}$$

The expectation value of the visible energy $E_\theta(v)$ increases with temperature, which reflects the fact that $|m_i|$ decreases with temperature as it should be. The variance $\langle E_\theta^2 \rangle - \langle E_\theta \rangle^2$ presents a peak at the phase transition between ferromagnetic and paramagnetic orders.

5. **Fifth:** Similarly, the physical entropy $S = -k_B \langle \ln p_{\text{Ising}}(s) \rangle$ is also related to the negative-log-likelihood $L(\theta) = -\langle \ln p_\theta(v) \rangle_{v \sim p_D}$ after training. The negative-log-likelihood $L(\theta)$ is approximated by the pseudo-negative-log-likelihood $\tilde{L}(\theta)$ defined in terms of the conditional probability $p_\theta(v_i|v_{j \neq i})$ by [14]

$$L(\theta) \simeq \tilde{L}(\theta) \simeq -\langle \sum_{i=1}^{N_v} \ln p_\theta(v_i|v_{j \neq i}) \rangle_{v \sim p_D} \simeq -\langle N_v \ln p_\theta(v_{i_0}|v_{j \neq i_0}) \rangle_{v \sim p_D}. \tag{20.192}$$

The pseudo-negative-log-likelihood $\tilde{L}(\theta)$ of the different restricted Boltzmann machines, which are associated with different temperatures, provide a good approximation of the actual entropy of the Ising model.

Indeed, pseudo-negative-log-likelihood $\tilde{L}(\theta)$ is observed to be increasing with temperature and with perfect agreement with the entropy of the Ising model.

The specific heat can then be evaluated from the pseudo-negative-log-likelihood $\tilde{L}(\theta)$ by taking the numerical derivative according to the formula $C_v = Td\tilde{L}/dT$. The peak of C_v determines the critical temperature T_c and the result obtained from the restricted Boltzmann machines agrees very well with the known Ising model result.

20.4 The Wigner restricted Boltzmann machine (W-RBM)

20.4.1 The restricted Boltzmann machine (RBM): a summary

The Boltzmann machines were first studied in the context of statictical mechanics as solvable spin-glass models by Sherrington and Kirkpatrick in 1975 [3]. These machines were then popularized as neural networks starting in 1985 by Hinton and collaborators [4, 5]. Boltzmann machines are intimately connected to Hopfield networks [2].

A Boltzmann machine consists typically of a visible layer and a hidden layer. Each neuron in the visible layer is linked to every neuron in the hidden layer. All of the layers' individual neurons are also linked. These neurons can be taken to be binary units, i.e., they can take only the values $+1$ and -1.

The restricted Boltzmann machine (RBM) is a special case of Boltzmann machines in which there is no intra-layer connectivity [6].

The restricted Boltzmann machine is an energy-based, two-layer, stochastic and generative neural network.

The energy function of the restricted Boltzmann machine with visible neurons denoted v_i (with bias denoted by a_i) and hidden neurons denoted h_i (with bias denoted by b_i) is given by the Ising-like energy

$$E(v, h) = -\sum_{i,j} v_i w_{ij} h_j - \sum_i a_i v_i - \sum_i b_i h_i. \tag{20.193}$$

The coupling w_{ij} is the weight of the connection between neurons i and j.

The joint probability of the configuration (v_i, h_i) is given by the Boltzmann distribution (and hence the name 'Boltzmann' in Boltzmann machine) with temperature $T = 1$, viz

$$p(v, h) = \frac{\exp(-E(v, h))}{Z}, \quad Z = \sum_{v,h} \exp(-E(v, h)). \tag{20.194}$$

The activation process of the neurons in the restricted Boltzmann machine is a stochastic process, which is the reason why the restricted Boltzmann machine is a stochastic neural network.

Indeed, the conditional probability that the hidden/visible neuron h_j/v_i activates, given the visible/hidden neurons state v/h, is given by the logistic function

$$p(h_j = 1|v) = \sigma\left(b_j + \sum_i w_{ij} v_i\right), \quad p(v_i = 1|h) = \sigma\left(a_i + \sum_j w_{ij} h_j\right). \tag{20.195}$$

The RBM is characterized by the marginal probability distribution $p(v)$ of the visible units. This is called the 'model distribution'.

The input dataset is characterized by an unknown probability distribution denoted by $q(v)$, which is called the 'training data'. This dataset is the initial set of configurations of the visible units.

The visible dimension N_v is fixed by the dimensionality of the input data whereas the hidden dimension N_h is chosen based on the expected complexity hidden in the data.

The goal of the so-called 'learning' of the RBM is to sample a sequence of configurations (v, h) of visible and hidden units with the joint probability distribution $p(v, h)$, starting from an initial visible configuration drawn from the input dataset, while at the same time optimizing the parameters $\theta = (w_{ij}, a_i, b_i)$ of the RBM in such a way that the model distribution $p(v)$ approximates the data distribution $q(v)$ as close as possible. This optimization is called 'training' of the RBM.

After 'training' is completed, we say that the RBM has 'learned' the probability distribution $q(v)$. This means that the restricted Boltzmann machine can generate the input data probability distribution $q(v)$ without any further simulation/calculation. This is the reason why the restricted Boltzmann machine is a generative neural network.

Thus, we need to sample the **RBM** while tuning or optimizing its parameters $\theta = (w_{ij}, a_i, b_i)$ at the same time. This is the 'training' of the **RBM**, which is also called 'unsupervised learning' since this training/learning is done dynamically starting from an initial input consisting of 'unlabeled' data.

The sampling of the **RBM** is typically done by means of the Markov chain Gibbs sampling.

However, we are simultaneously required to approximate the data probability distribution $q(v)$ with some **RBM** distribution $p_\theta(v)$, i.e., with a tuned parameters θ. This is done via minimization with respect to θ of the so-called Kullbach-Leibler (KL) divergence given by

$$O(\theta) = \langle \ln q(v) \rangle_{v \sim q(v)} - \langle \ln p_\theta(v) \rangle_{v \sim q(v)}. \tag{20.196}$$

Indeed, the two distributions $p(v)$ and $q(v)$ are as close to each other as possible if their relative or Shannon entropy becomes as small as possible, i.e., it reaches a local minimum. This relative entropy is precisely the above Kullback–Leibler divergence.

Thus, the parameters θ are optimized via a gradient descent procedure $\theta^0 \longrightarrow \theta^1 \longrightarrow \theta^2 \ldots$ where each iteration is of the form (where η is the learning rate)

$$\theta^i \longrightarrow \theta^{i+1} = \theta^i - \eta \frac{\partial O}{\partial \theta^i}. \tag{20.197}$$

Explicitly, the derivative is given as

$$\frac{\partial O}{\partial w_{ij}} = \langle v_i h_j \rangle_{\text{model}} - \langle v_i h_j \rangle_{\text{data}}$$

$$\frac{\partial O}{\partial a_i} = \langle v_i \rangle_{\text{model}} - \langle v_i \rangle_{\text{data}} \tag{20.198}$$

$$\frac{\partial O}{\partial b_i} = \langle h_i \rangle_{\text{model}} - \langle h_i \rangle_{\text{data}}.$$

The data-dependent term is known as the positive phase of the gradient while the model-dependent term is known as the negative phase. The expectation value $\langle \rangle_{\text{model}}$ is computed with respect to the **RBM** states, i.e., with respect to the joint distribution $p(h, v) = p(h|v)p(v)$ whereas the expectation value $\langle \rangle_{\text{data}}$ is computed with respect to the **RBM** states given that the visible states are distributed according to the training distribution, i.e., with respect to $p(h|v)q(v)$. Hence, the model expectation value $\langle \rangle_{\text{model}}$ is computed by means of Gibbs sampling whereas the data expectation $\langle \rangle_{\text{data}}$ is computed by taking the average over all samples in the dataset.

The restricted Boltzmann machine is therefore an algorithm, which combines the Markov chain Gibbs sampling with gradient descent in order to update the weights and biases.

The Gibbs sampling typically only equilibrate for sufficiently large number of steps k and thus the computation of the negative phase can be very extensive. We can run the Gibbs sampling only for a small number of steps k, which seems to work quite well, even with $k = 1$, which is amazing considering the fact that Monte Carlo equilibration is very hard in general (this seems to be due to some cancellation of errors).

In conclusion, the restricted Boltzmann machine is an algorithm which combines the Markov chain Gibbs sampling with gradient descent in order to update the weights and biases. The combined algorithm is called the contrastive divergence algorithm **CD–k** [7–10].

In fact, the restricted Boltzmann machine as an algorithm for simulating physical systems may potentially surpass the Monte Carlo method. See for example [15–18].

20.4.2 Gaussian restricted Boltzmann machine (G-RBM) and scalar phi-four theory

The Euclidean ϕ^4 theory in two dimensions is given by the action

$$S[\phi] = \int d^2x \left(\frac{1}{2}(\partial_\mu \phi)^2 + \frac{1}{2} m^2 \phi^2 + \frac{\lambda}{4} \phi^4 \right). \tag{20.199}$$

We will employ a periodic lattice of linear dimension $L = a.\,N$ where a is the lattice spacing and N^2 is the number of lattice sites. We set $x = an$, $\int d^2x = a^2 \sum_n$, $\phi(x) = \phi_n$ and $\partial_\mu \phi = (\phi_{n+\hat\mu} - \phi_n)/a$. The lattice action reads

$$S[\Phi] = \sum_n \left(-2\kappa \sum_\mu \Phi_n \Phi_{n+\hat\mu} + \Phi_n^2 + g(\Phi_n^2 - 1)^2 \right). \tag{20.200}$$

The mass parameter m^2 is replaced by the so-called hopping parameter κ while the coupling constant λ is replaced by the dimensionless coupling constant g where

$$m^2 a^2 = \frac{1 - 2g}{\kappa} - 4, \quad a^2 \lambda = \frac{g}{\kappa^2}. \tag{20.201}$$

The fields ϕ_n and Φ_n are related by

$$\phi_n = \sqrt{2\kappa}\, \Phi_n. \tag{20.202}$$

This is a generalized Ising model. Indeed, we can rewrite the above action as

$$S[\Phi] = \sum_n \left(-2\kappa \sum_\mu \Phi_n \Phi_{n+\hat\mu} + \Phi_n^2 + g(\Phi_n^2 - 1)^2 \right)$$
$$= \beta H(s) \tag{20.203}$$
$$H(s) = -J \sum_{\langle i,j \rangle} s_i s_j + \frac{1}{\beta} \sum_i \left(s_i^2 + g(s_i^2 - 1)^2 \right).$$

In this equation we have set

$$\Phi_n \equiv s_i, \quad 2\kappa \equiv \beta J \tag{20.204}$$

and the map between the lattice point and the spin label is given by

$$n = (n_1, n_2) \longrightarrow i = (n_1 - 1)L + n_2 \tag{20.205}$$

Here, $\langle i, j \rangle$ denotes summation over nearest-neighbor pairs only.

In summary, we have the scalar phi-four theory given by the action

$$S_{\text{Scalar}}[s] = -2\kappa \sum_{\langle i,j \rangle} s_i s_j + \sum_i \left(s_i^2 + g(s_i^2 - 1)^2 \right). \tag{20.206}$$

In the limit $g \longrightarrow \infty$ the dominant configurations are such that $s_i^2 = 1$, i.e. $s_i = \pm 1$ and thus we end up with the Ising model action

$$\beta E_{\text{Ising}}(s) = -J \sum_{\langle i,j \rangle} s_i s_j. \tag{20.207}$$

The phase diagrams of the Ising model and the phi-four theory contain the same phases. In other words, the ferromagnetic phase in the Ising model corresponds to the uniform ordered phase in the phi-four theory. The paramagnetic phase in the Ising model corresponds to the disordered phase in the phi-four theory. The only difference lies in the fact that the coexistence curve is a point given by the critical temperature in the Ising model while in the phi-four theory the coexistence curve is a critical line separating the two-dimensional phase diagram into two regions.

The free theory: Let us discuss first the free theory $g = 0$.

The only real difference between the Ising model and the free scalar field theory lies in the fact that the Ising model employs discrete variables while the free scalar field theory employs continuous variable, which is Gaussian.

We have seen that the Ising model can be captured by the ordinary restricted Boltzmann machine, which is a binary-binary machine, i.e., both visible and hidden units are binary taking values of either 0 (spin down -1) or 1 (spin up $+1$). The energy function of the RBM is given by

$$E_{\text{RBM}}(v, h) = -\sum_i a_i v_i - \sum_i b_i h_i - \sum_{i,j} v_i w_{ij} h_j. \tag{20.208}$$

Similarly, it is only natural to assume that the free scalar field theory should be captured by the Gaussian restricted Boltzmann machine (G-RBM) which is a real-binary machine, i.e., the visible units take real values while the hidden units remain binary. In fact, the visible units are distributed according to a Gaussian distribution, which is something that characterizes the free scalar field. The energy function reads in this case

$$E_{\text{G-RBM}}(v, h) = \sum_i \frac{1}{2\sigma_i^2}(v_i - a_i)^2 - \sum_i b_i h_i - \sum_{i,j} \frac{v_i}{\sigma_i} w_{ij} h_j. \tag{20.209}$$

Here, σ_i is the standard deviation of the Gaussian noise for the visible unit i. The rest are as in the case of the binary-binary RBM.

This modified energy function leads to a modification of the activation function of the visible units, which becomes Gaussian while that of the hidden units remains given by the logistic function. Explicitly, we have the activation functions (conditional probabilities)

$$p(h_j = 1|v) = \sigma(m_j), \quad m_j = \sum_i \frac{v_i}{\sigma_i} w_{ij} + b_j$$

$$p(v_i = v_i^0|h) = \mathcal{N}(v_i|m_i, \sigma_i^2), \quad m_i = \sum_j w_{ij} h_j + a_i. \tag{20.210}$$

Despite all this, both the RBM and the G-RBM are generative stochastic neural networks used for unsupervised learning.

Interactions: Next, we consider the interacting theory $g \neq 0$.

We add interaction by modifying the energy function of the G-RBM in an obvious way, i.e., by adding the quartic potential to the energy. Thus, we obtain the phi-four restricted Boltzmann machine (ϕ^4-RBM) given by the energy function

$$E_{\text{G-RBM}}(v, h) = \sum_i \frac{1}{2\sigma_i^2}(v_i - a_i)^2 + \sum_i g(v_i^2 - 1)^2 - \sum_i b_i h_i - \sum_{i,j} \frac{v_i}{\sigma_i} w_{ij} h_j. \tag{20.211}$$

We conjecture that the phase structure of the phi-four theory, similarly to the case of the Ising model, is largely encoded in the weight matrix $W = w^T$.

The phase structure of the Ising model as probed by the RBM was considered in [11]:

1. It is observed that the elements W_{ij} of the weight matrix follow a Gaussian distribution of zero mean.

2. The $N_h - N_v$ weight matrix W_{ij} of the RBM can be viewed as N_h row vectors $(w_i)_j = W_{ij}$, which are N_v-dimensional.

3. It is observed that the so-called filter sums $m_i = \sum_{j=1}^{N_v} W_{ij}/N_v$ play the role of magnetization order parameters. Indeed, the probability distribution function of the filter sums m_i behaves exactly as one expects from the magnetization of the Ising model. Indeed, at low temperature in the ferromagnetic phase, this probability distribution function presents two peaks symmetric around zero (corresponding to the uniform order). Whereas at high temperature in the paramagnetic phase this probability distribution function presents a single peak around 0 (corresponding to the disorder). At the critical temperature this probability distribution function shows long-range critical fluctuations.

4. The filter sums $m_i \neq 0$ (bimodal distribution symmetric around 0) indicate therefore a ferromagnetic phase while the filter sums $m_i = 0$ (unimodal distribution centered around 0) indicate a paramagnetic phase.

5. Thus, in the ferromagnetic phase, the components $(w_i)_j = W_{ij}$ of the filters $\vec{w}_i$ are either positive/negative matching the spin up/down of the visible layer $\vec{v} \equiv \vec{s}$. In other words, we must have $\vec{w}_i \cdot \vec{v} > 0$ (filters match Ising spins).

6. In the paramagnetic phase, the filters $\vec{w}_i$ present a random mixture consisting of equal numbers of positive and negative components, which also match the components of the visible layer, i.e. we must also have $\vec{w}_i \cdot \vec{v} > 0$.

7. Furthermore, the term in the RBM energy $\sum_{i,j} h_i W_{ij} v_j = \sum_i h_i(\vec{w}_i \cdot \vec{v})$ is more positive if the signs of the hidden units h_i match the signs of the scalar products $\vec{w}_i \cdot \vec{v}$, i.e., $h_i = +1$.

8. But we can also have $\vec{w}_i \cdot \vec{v} < 0$ (filters anti-match the Ising spins). Yet, the term in the RBM energy $\sum_{i,j} h_i W_{ij} v_j = \sum_i h_i(\vec{w}_i \cdot \vec{v})$ is still more positive if the signs of the hidden units h_i still match the signs of the scalar products $\vec{w}_i \cdot \vec{v}$, i.e., $h_i = -1$.

9. In summary, for a given visible layer $\vec{v} \equiv \vec{s}$ of magnetization m, the sign of the scalar product $\vec{w}_i \cdot \vec{v}$ and the sign of the hidden unit h_i are essentially determined by the sign of the corresponding filter magnetization $m_i = \sum_{j=1}^{N_v} W_{ij}/N_v$.

Finally, we should note that the Gaussian restricted Boltzmann machine should capture the phi-four theory in all dimensions. The only difference between the various dimensions seems to lie in the fact that the input datasets, used for training and learning, should consist of thermalized configurations of the phi-four theory in those various dimensions.

20.4.3 Wigner restricted Boltzmann machine (W-RBM) and noncommutative phi-four

The noncommutative phi-four scalar field theories on fuzzy spaces are non-pure matrix models, which take the generic form

$$S = Tr(\Phi[L_a, [L_a, \Phi]] + B\Phi^2 + C\Phi^4). \tag{20.212}$$

By diagonalizing the matrix Φ, the potential is diagonalized completely, and thus we obtain an effective action given by

$$Z = \int \prod_{i=1}^{N} d\lambda_i \, \Delta^2(\Lambda) \exp\left(-Tr(B\Lambda^2 + C\Lambda^4)\right)\exp(-\Delta S_{\text{eff}}). \tag{20.213}$$

Here, Λ is the diagonal matrix of eigenvalues λ_i. The Vandermonde determinant $\Delta^2(\Lambda)$ arises from the diagonalization of the measure and gives rise to a logarithmic contribution to the effective potential. The effective action ΔS_{eff} corresponds to the diagonalization of the kinetic action and is a complicated function of the eigenvalues which involves all possible multitrace terms [33, 34]. The multitrace approach is reviewed in [21].

Here, another approach to the effective action of noncommutative phi-four scalar field theories on fuzzy spaces is considered. This approach is based on the algorithm proposed in [22] in which we consider a scalar field coupled to a gauge field X_a on fuzzy spaces. The action is given explicitly by

$$S_m = -Tr[X_a, \Phi]^2 + Tr(B\Phi^2 + C\Phi^4) + S_G[X]. \tag{20.214}$$

Here, X_a is the gauge field with corresponding gauge actions $S_G[X]$ on various fuzzy spaces, which are well known. For example, noncommutative gauge actions on fuzzy $\mathbf{CP}^n$ are constructed in [27] (and references therein).

We can now diagonalize the scalar field by means of a $U(N)$ gauge transformation, viz $\Phi = U\Lambda U^+$, where the unitary matrix U can then be integrated out from the path integral.

In this algorithm, we thus trade off the Monte Carlo sampling of the unitary matrix U, in the original model, with the Monte Carlo sampling of a gauge field on fuzzy spaces, which we know is much more efficient using ordinary Metropolis. We obtain the path integral

$$Z = Z_G \int d\Lambda \; \langle \exp(Tr[X_a, \Lambda]^2) \rangle_G \exp\left[-Tr(B\Lambda^2 + C\Lambda^4) + \ln \Delta_N(\Lambda)\right]. \quad (20.215)$$

The expectation value $\langle \exp(Tr[X_a, \Lambda]^2) \rangle_G$ is computed with respect to the gauge action S_G and Z_G is the corresponding partition function.

In here we will supplement this algorithm by a further approximation as follows. We simply use the path integral

$$Z = Z_G \int d\Lambda \; \exp\left(Tr[X_a^{\text{eff}}, \Lambda]^2\right) \exp\left[-Tr(B\Lambda^2 + C\Lambda^4) + \ln \Delta_N(\Lambda)\right]. \quad (20.216)$$

Here, X_a^{eff} is a thermalized configuration around the global minimum of the gauge action in the geometric phase of the theory where the fuzzy space is stable. In fact, sufficiently deep in the geometric phase, we can simply replace X_a^{eff} with the background configuration L_a itself, i.e., we choose

$$X_a^{\text{eff}} = L_a. \quad (20.217)$$

The scalar action is then given by

$$\begin{aligned}
S[\Lambda] = & -Tr[X_a^{\text{eff}}, \Lambda]^2 + Tr(B\Lambda^2 + C\Lambda^4) - \ln \Delta_N(\Lambda) \\
= & -2\sum_{ij}(X_a^{\text{eff}})_{ij}(X_a^{\text{eff}})_{ji}\lambda_i\lambda_j \\
& + Nc_2\sum_i \lambda_i^2 + \sum_i(B\lambda_i^2 + C\lambda_i^4) - \sum_{i\neq j}\ln|\lambda_i - \lambda_j|.
\end{aligned} \quad (20.218)$$

The phase structure of noncommutative phi-four is very complicated as it involves transitions between three phases, which meet a triple point [23, 24]. These phase transitions consist of the usual Ising transition between disorder and uniform order, a matrix transition between disorder and a non-uniform ordered phase, and a (very hard to observe) transition between uniform order and non-uniform order. The disorder-to-non-uniform-order transition is the generalization of the one-cut-to-two-cut transition observed in the real quartic matrix model [25, 26]. Clearly, the real quartic matrix model truncation/reduction of this model is obtained by dropping the kinetic term.

In each phase we write the order parameter in the noncommutative phi-four theory and the corresponding eigenvalues distribution in the real quartic matrix model as follows

'disordered phase'

$$\langle \Phi \rangle_{\text{NC}} = 0 \longrightarrow \rho_{\text{M}}(\lambda) = \frac{1}{N\pi}(2C\lambda^2 + B + C\delta^2)\sqrt{r^2 - \lambda^2}: \quad \text{stable.} \quad (20.219)$$

'matrix (nonuniform/stripe) phase'

$$\langle\Phi\rangle_{\mathrm{NC}} = \pm\sqrt{-\frac{B}{2C}}\,\gamma \longrightarrow \rho_{\mathrm{M}}(\lambda) = \frac{2C|\lambda|}{N\pi}\sqrt{(\lambda^2 - r_-^2)(r_+^2 - \lambda^2)}: \quad \text{stable.} \tag{20.220}$$

'Ising (uniform) phase'

$$\langle\Phi\rangle_{\mathrm{NC}} = \pm\sqrt{-\frac{B}{2C}}\,1_N \longrightarrow$$

$$\rho_{\mathrm{M}}(\lambda) = \frac{1}{\pi}(2NC\lambda^2 + 2NCr_2\lambda + 2NCr_2^2 + NCr_1^2 + 2B^2)$$

$$\sqrt{(r_1 + r_2 - \lambda)(\lambda + r_1 - r_2)}: \quad \text{metastable.} \tag{20.221}$$

remark that the uniform order is only metastable in the pure matrix model context but in the noncommutative theory context it is quite stable. The cuts r, $r_\pm$ and r_i are functions of the parameters B and C, which we do not write here. See [26].

We are therefore interested in the noncommutative scalar phi-four theory given by the following action

$$S_{\mathrm{NC}}[\lambda] = -\sum_{i,j} w_{ij}\lambda_i\lambda_j - \frac{1}{2}\sum_{i\neq j}\ln(\lambda_i - \lambda_j)^2$$
$$+ \sum_i \left(B'\lambda_i^2 + C\lambda_i^2\right), \quad w_{ij} = 2(L_a)_{ij}(L_a)_{ji}. \tag{20.222}$$

The pure matrix model truncation/reduction of this model is obtained by dropping the kinetic term. We obtain the matrix scalar phi-four theory given by the real quartic matrix model with action

$$S_{\mathrm{M}}[\lambda] = -\frac{1}{2}\sum_{i\neq j}\ln(\lambda_i - \lambda_j)^2 + \sum_i\left(B\lambda_i^2 + C\lambda_i^2\right). \tag{20.223}$$

Three crucial remarks are in order:

1. First, the logarithmic potential, which arises from the path integral's measure over Hermitian $N \times N$ matrices, should really be thought of as part of the potential term. This logarithmic potential, in the large N saddle point approximation of the model, will give rise to the Wigner's semi-circle law [30, 31]. This law captures the perturbative regime and even the geometry of the theory [32]. For $w = 0$ and $C = 0$, we have then

$$\rho(\lambda) = \frac{B}{N\pi}\sqrt{r^2 - z^2}, \quad r^2 = \frac{2N}{B}. \tag{20.224}$$

2. Second, the hopping term seems to be absent from the pure matrix model. This is in contrast with the commutative case where the exact hopping term is present in both the Ising model and the phi-four theory. This indicates that the real quartic matrix model exists in a separate universality class and that

the noncommutative phi-four theory interpolates between the Ising universality class (commutative phi-four) and the matrix universality class (real quartic matrix model) [28, 29]. This is also the reason why the Ising uniform phase is metastable in the real quartic matrix model.

3. Third, the real quartic matrix model can still be formally associated with a weight matrix, which is completely diagonal in the matrix indices, viz $w_{ij} = \delta_{ij}$. We have then

$$
\begin{aligned}
S_{\mathrm{M}}[\lambda] = & -\sum_{i,j} w_{ij} \lambda_i \lambda_j - \frac{1}{2} \sum_{i \neq j} \ln(\lambda_i - \lambda_j)^2 \\
& + \sum_i \left((B + 1)\lambda_i^2 + C\lambda_i^2 \right), \quad w_{ij} = \delta_{ij}.
\end{aligned}
\tag{20.225}
$$

In analogy with the Ising model and the phi-four theory, we will approximate the real quartic matrix model with a restricted Boltzmann machine, which will involve, as parts of the energy of the visible units, a Gaussian noise (corresponding to the quadratic mass term), a quartic interaction term (corresponding to the quartic coupling constant C), but also it will involve a logarithmic potential (corresponding to the measure over Hermitian $N \times N$ matrices in the path integral). This machine is termed the Wigner restricted Boltzmann machine (W-RBM) and it is given by the energy function

$$
\begin{aligned}
E_{\mathrm{W-RBM}}(v, h) = & -\frac{1}{2} \sum_{i \neq j} \ln(v_i - v_j)^2 + \sum_i \frac{1}{2\sigma_i^2}(v_i - a_i)^2 \\
& + \sum_i C v_i^4 - \sum_i b_i h_i - \sum_{i,j} \frac{v_i}{\sigma_i} W_{ij} h_j.
\end{aligned}
\tag{20.226}
$$

We generate the input datasets for the W-RBM either by Monte Carlo simulation of the real quartic matrix model or by directly sampling the stable eigenvalues distributions (20.219) and (20.220). The second choice is more natural and in fact much more efficient (since we have this exact solution at our disposal we may as well use it). This option is not available to us in the commutative case where we need to generate thermalized configurations of the Ising model and the phi-four scalar field theory by means of the Monte Carlo method.

The above W-RBM will capture the real quartic matrix model in the same way that the RBM and G-RBM capture the Ising model and the phi-four theory, respectively.

In fact, the above Wigner restricted Boltzmann machine should also describe noncommutative phi-four theories on fuzzy spaces, which are the ones that have genuine hopping terms in their actions similarly to the commutative cases. We claim that by generating input datasets from the eigenvalues distributions (20.219), (20.220) and (20.221), i.e., including configurations from the metastable phase, the W-RBM will approach a universal noncommutative phi-four theory to which all phi-four theories converge, under the renormalization group equations (RGE), regardless of their dimensions.

Indeed, the metastable uniform distribution (20.221) does not know about dimension but it can be made stable by the weight matrix term of the W-RBM. This is a fundamental difference with the commutative case where dimension plays a crucial role.

We conjecture therefore that there is a single universal noncommutative phi-four field theory corresponding to a noncommutative fixed point of the RGE, i.e., all noncommutative phi-four theories in various dimensions are really equivalent, due to their underlying matrix structure, to this fixed point or universal noncommutative phi-four theory, which is defined precisely by the above Wigner restricted Boltzmann machine.

The noncommutative phi-four field theory lies in its own universality class, which is distinct from both the Ising universality class (commutative phi-four) and the matrix universality class (real quartic matrix model). This universality class is best described by the Wigner restricted Boltzmann machine.

References

[1] Ydri B 2017 *Computational Physics: An Introduction to Monte Carlo Simulations of Matrix Field Theory* (Singapore: World Scientific) [arXiv:1506.02567 [hep-lat]]

[2] Hopfield J J Neural networks and physical systems with emergent collective computational abilities *Proc. Natl Acad. Sci.* **79** 2554–8

[3] Sherrington D and Kirkpatrick S 1975 Solvable model of a spin-glass *Phys. Rev. Lett.* **35** 1792–6

[4] Ackley D H, Hinton G E and Sejnowski T J 1985 A learning algorithm for Boltzmann machines *Cogn. Sci.* **9** 147–69

[5] Hinton E 1668 Boltzmann machine *Scholarpedia* **2**

[6] Smolensky P 1986 *Information Processing in Dynamical Systems: Foundations of Harmony Theory in Parallel Distributed Processing: Explorations in the Microstructure of Cognition* (Cambridge, MA: MIT Press) pp 194–281

[7] Hinton G E and Salakhutdinov R R 2006 Reducing the dimensionality of data with neural networks *Science* **313** 504

[8] Hinton G E 2002 Training products of experts by minimizing contrastive divergence *Neural Comput.* **14** 1771–800

[9] Hinton G E 2012 *A Practical Guide to Training Restricted Boltzmann Machines in Neural Networks: Tricks of the Trade* (Berlin: Springer) pp 599–619

[10] Hinton G E 2002 Training products of experts by minimizing contrastive divergence *Neural Comput.* **14** 1771–800

[11] Gu J and Zhang K 2022 Thermodynamics of the Ising model encoded in restricted Boltzmann machines *Entropy* **24** 1201

[12] Glorot X and Bengio Y 2010 Understanding the difficulty of training deep feedforward neural networks *Proceedings of the 13th Int. Conf. on Artificial Intelligence and Statistics (JMLR Workshop and Conference Proceedings)* pp 249–56

[13] Murphy K P 2012 *Machine Learning: A Probabilistic Perspective* (Boston, MA: MIT Press)

[14] Besag J 1975 Statistical analysis of non-lattice data *J. R. Stat. Soc.: Ser. D (The Statistician)* **24** 179

[15] Stepanov O A and Amosov O S 2007 The comparison of the Monte-Carlo method and neural networks algorithms in nonlinear estimation problems *IFAC Proc.* **40** 392–7

[16] Kappen H J 1994 Using Boltzmann Machines for probability estimation: a general framework for neural network learning *Machine Intelligence and Pattern Recognition* **vol 16** ed E S Gelsema and L S Kanal (Amsterdam: Elsevier) 299–312

[17] Stosic D, Stosic D and Stosic B 2022 Ising models of deep neural networks arXiv:2209.08678 [cond-mat.stat-mech].

[18] Barra A, Genovese G, Sollich P and Tantari D 2017 Phase transitions in Restricted Boltzmann Machines with generic priors *Phys. Rev. E* **96** 042156

[19] Clark B 2025 Computing in physics https://courses.physics.illinois.edu/PHYS446/.

[20] Hjorth-Jensen M 2021 Advanced topics in computational physics https://github.com/CompPhysics/ComputationalPhysics2

[21] Šubjaková M and Tekel J 2020 Fuzzy field theories and related matrix models *PoS CORFU2019* **189** [arXiv:2006.12605 [hep-th]]

[22] Ydri B 2014 New algorithm and phase diagram of noncommutative ϕ^4 on the fuzzy sphere *JHEP* **03** 065

[23] Garcia Flores F, Martin X and O'Connor D 2009 Simulation of a scalar field on a fuzzy sphere *Int. J. Mod. Phys.* A **24** 3917. See also [24]

[24] Garcia Flores F, O'Connor D and Martin X 2006 Simulating the scalar field on the fuzzy sphere *PoS LAT* **2005** 262 arXiv:hep-lat/0601012

[25] Brezin E, Itzykson C, Parisi G and Zuber J B 1978 Planar diagrams *Commun. Math. Phys.* **59** 35

[26] Shimamune Y 1982 On the phase structure of large N matrix models and gauge models *Phys. Lett.* B **108** 407

[27] Ydri B 2018 *Matrix Models of String Theory* (Bristol: IOP Publishing)

[28] Ydri B, Ahmim R and Bouchareb A 2015 Wilson RG of noncommutative Φ_4^4 *Int. J. Mod. Phys.* A **30** 1550195

[29] Ydri B and Bouchareb A 2012 The fate of the Wilson-Fisher fixed point in non-commutative ϕ^4 *J. Math. Phys.* **53** 102301 [arXiv:1206.5653 [hep-th]]

[30] Wigner E 1955 Characteristic vectors of bordered matrices with infinite dimensions *Ann. Math.* **62** 548–56

[31] Wigner E 1958 On the distribution of the roots of certain symmetric matrices *Ann. Math.* **67** 325–8

[32] Ydri B, Khaled R and Soudani C 2022 Quantized noncommutative geometry from multitrace matrix models *Int. J. Mod. Phys.* A **37** 2250052

[33] Saemann C 2010 The multitrace matrix model of scalar field theory on fuzzy CP^n *SIGMA* **6** 050

[34] O'Connor D and Saemann C 2007 Fuzzy scalar field theory as a multitrace matrix model *JHEP* **0708** 066

[35] Chang S J 1976 The existence of a second order phase transition in the two-dimensional phi**4 field theory *Phys. Rev.* D **13** 2778 [erratum: *Phys. Rev. D* **16** 1979 (1977)]

[36] Onsager L 1944 Crystal statistics. 1. A two-dimensional model with an order disorder transition *Phys. Rev.* **65** 117–49

[37] Brezin E, Itzykson C, Parisi G and Zuber J B 1978 Planar diagrams *Commun. Math. Phys.* **59** 35

[38] Shimamune Y 1982 On the phase structure of large N matrix models and gauge models *Phys. Lett.* B **108** 407

[39] Eynard B 2015 *Random matrices Cours de Physique Theorique de Saclay* https://www.ipht.fr/pisp/bertrand-eynard/

[40] Eynard B, Kimura T and Ribault S 2015 Random matrices arXiv:1510.04430 [math-ph]

[41] Ch Angles d'Auriac J and Maillard J M 2003 Random matrix theory in lattice statistical mechanics *Physica A: Stat. Mech. Appl.* **321** 325–33

[42] Berezin F A 1975 General concept of quantization *Commun. Math. Phys.* **40** 153

[43] Frohlich J and Gawedzki K 1993 Conformal field theory and geometry of strings *Vancouver 1993, Proc., Mathematical Quantum Theory* **vol 1** 57–97 and Preprint—Gawedzki, K. (rec. Nov.93) 44 [hep-th/9310187]

[44] Connes A 1994 *Noncommutative Geometry* (London: Academic)

[45] Dolan B P, O'Connor D and Presnajder P 2002 Matrix phi**4 models on the fuzzy sphere and their continuum limits *JHEP* **0203** 013

[46] Saemann C 2010 The multitrace matrix model of scalar field theory on fuzzy CP^n *SIGMA* **6** 050

[47] O'Connor D and Saemann C 2007 Fuzzy scalar field theory as a multitrace matrix model *JHEP* **0708** 066

[48] Steinacker H 2005 A non-perturbative approach to non-commutative scalar field theory *JHEP* **0503** 075

[49] Ydri B, Soudani C and Rouag A 2017 Quantum gravity as a multitrace matrix model *Int. J. Mod. Phys.* A **32** 1750180

[50] Šubjaková M and Tekel J 2018 Matrix models of fuzzy field theories *PoS CORFU2017* **144** [arXiv:1802.05188 [hep-th]]

[51] Šubjaková M and Tekel J 2020 Multitrace matrix models of fuzzy field theories *PoS CORFU2019* **234** 13577

[52] Šubjaková M and Tekel J 2020 Fuzzy field theories and related matrix models *PoS CORFU2019* **189** [arXiv:2006.12605 [hep-th]]

[53] Ydri B 2014 New algorithm and phase diagram of noncommutative ϕ^4 on the fuzzy sphere *JHEP* **03** 065

[54] Garcia Flores F, Martin X and O'Connor D 2009 Simulation of a scalar field on a fuzzy sphere *Int. J. Mod. Phys.* A **24** 3917 See also [55].

[55] Garcia Flores F, O'Connor D and Martin X 2006 Simulating the scalar field on the fuzzy sphere *PoS LAT* **2005** 262 [hep-lat/0601012]

[56] Gubser S S and Sondhi S L 2001 Phase structure of noncommutative scalar field theories *Nucl. Phys.* B **605** 395

IOP Publishing

A Modern Course in Quantum Field Theory, Volume 2 (Second Edition)
Advanced topics
Badis Ydri

Appendix A

Lie algebra representation theory: a primer

In this primer I follow the excellent pedagogical presentation of Zuber [1].

A.1 The Cartan subalgebra

We consider a semi-simple[1] Lie algebra $\mathbf{g}$ of finite dimension over the complex numbers $\mathbf{C}$. The Cartan subalgebra is the maximal Abelian subalgebra $\mathbf{h}$ of $\mathbf{g}$ such that all its elements are diagonalizable in the adjoint representation. The elements of $\mathbf{g}$ can be chosen to be Hermitian. The Cartan subalgebra is not unique and different choices are related by an automorphism of the algebra, i.e., as $h \longrightarrow ghg^{-1}$. The dimension l of $\mathbf{h}$ is called the rank of $\mathbf{g}$. For example, for $su(n)$ the Cartan subalgebra can be generated by the following $l = n - 1$ diagonal traceless matrices

$$H_1 = (1, -1,..., 0), \quad H_2 = (0, 1, -1,..., 0),..., \quad H_{n-1} = (0,..., 1, -1). \quad \text{(A.1)}$$

Let $H_i, i = 1,..., l$ be the basis elements of $\mathbf{h}$, which are chosen to be Hermitian. By definition we have

$$[H_i, H_j] = 0 \Leftrightarrow [\mathrm{ad}H_i, \mathrm{ad}H_j] = 0. \quad \text{(A.2)}$$

Thus from $\mathrm{ad}H_i H_j = [H_i, H_j] = 0$ we see that H_j are eigenvectors of $\mathrm{ad}H_i$ with eigenvalues 0. The other linearly independent eigenvectors with eigenvalues α_i, not all vanishing, will be denoted by E_α, viz

$$\mathrm{ad}H_i E_\alpha = [H_i, E_\alpha] = \alpha_i E_\alpha. \quad \text{(A.3)}$$

These eigenvalues are real since $\mathrm{ad}H_i$ are Hermitian after multiplication by i. We compute for $H = \sum_i h^i H_i$ that

---

[1] A semi-simple algebra has no *Abelian ideal* but $\{0\}$, whereas a simple algebra has no *ideal* but $\{0\}$. In other words, the semi-simple is more general since it can have ideals. Thus any semi-simple algebra can be decomposed into direct sum of simple algebras.

doi:10.1088/978-0-7503-5834-7ch21　　　　　　

$$\mathrm{ad}H E_\alpha = \alpha(H) E_\alpha, \quad \alpha(H) = \sum_i h^i \alpha_i. \tag{A.4}$$

Obviously, $\alpha(H) = \sum_i h^i \alpha_i$ is a linear form on $\mathbf{h}$. Linear forms on a vector space E form the dual vector space E^*. Thus the set of all $\alpha(H)$ form the dual space $\mathbf{h}^*$ of $\mathbf{h}$. This dual space is called the root space and $\alpha(H)$ is called a root. We remark that

$$\alpha(H_i) = \alpha_i. \tag{A.5}$$

The total number of the eigenvectors H_j and E_α of H_i is equal to the dimension d of the adjoint representation of the Lie algebra $\mathbf{g}$. Since the number of the H_j is equal to the rank l and the roots are not degenerate we conclude that the number of the roots is $d - l$. This number is even since if α is a root $-\alpha$ is also a root.

The Killing form in the basis (H_i, E_α) of $\mathbf{g}$ is given by

$$(H_i, E_\alpha) = 0, \quad (E_\alpha, E_\beta) = 0 \ \text{ unless } \ \alpha + \beta = 0. \tag{A.6}$$

The Killing form is defined by the trace in the adjoint representation in an obvious way, viz

$$(H_i, E_\alpha) = \mathrm{tr} \ \mathrm{ad}H_i \mathrm{ad}E_\alpha. \tag{A.7}$$

The Killing form on $\mathbf{h}$ is non-degenerate, i.e., there exists an isomorphism between $\mathbf{h}$ and $\mathbf{h}^*$, which allows us to associate to every element $\alpha \in \mathbf{h}^*$ a unique element $H_\alpha = \sum_i h_\alpha^i H_i \in \mathbf{H}$ as follows

$$(H_\alpha, H) = \sum_i h_\alpha^i \sum_j h^j (H_i, H_j) = \sum_i h_\alpha^i \sum_j h^j g_{ij} = \sum_j h^j \alpha_j = \alpha(H) \equiv \alpha. \tag{A.8}$$

In the above equation, we have used the fact that the metric $g_{ij} = (H_i, H_j)$ is invertible since the Killing form was assumed to be non-degenerate, and $\alpha_j = \sum_i h_\alpha^i g_{ij}$. The bilinear form on the space of roots $\mathbf{H}^*$ can then be defined by

$$\langle \alpha, \beta \rangle = (H_\alpha, H_\beta). \tag{A.9}$$

In order to compute the commutation relations of the E_α, we compute using Jacobi identity the following

$$\begin{aligned}
\mathrm{ad}H_i[E_\alpha, E_\beta] &= [H_i, [E_\alpha, E_\beta]] \\
&= -[E_\beta, [H_i, E_\alpha]] - [E_\alpha, [E_\beta, H_i]] \\
&= (\alpha + \beta)_i [E_\alpha, E_\beta].
\end{aligned} \tag{A.10}$$

We have three possibilities. If $\alpha + \beta$ is a root then $[E_\alpha, E_\beta] = N_{\alpha\beta} E_{\alpha+\beta}$. If $\alpha + \beta \neq 0$ is not a root then $[E_\alpha, E_\beta] = 0$. If $\alpha + \beta = 0$ then $[E_\alpha, E_{-\alpha}] = H \in \mathbf{h}$. We compute

$$\begin{aligned}
(H_i, [E_\alpha, E_{-\alpha}]) &= \mathrm{tr} \ \mathrm{ad}H_i[\mathrm{ad}E_\alpha, \mathrm{ad}E_{-\alpha}] \\
&= \alpha_i \mathrm{tr} \ \mathrm{ad}E_\alpha \mathrm{ad}E_{-\alpha} \\
&= \alpha_i (E_\alpha, E_{-\alpha}) \\
&= (H_i, H_\alpha)(E_\alpha, E_{-\alpha}) \Rightarrow [E_\alpha, E_{-\alpha}] = (E_\alpha, E_{-\alpha})H_\alpha.
\end{aligned} \tag{A.11}$$

In summary, we have obtained the commutation relations

$$[H_i, H_j] = 0, \quad [H_i, E_\alpha] = \alpha_i E_\alpha \tag{A.12}$$

$$
\begin{aligned}
&[E_\alpha, E_\beta] = N_{\alpha\beta} E_{\alpha+\beta}, \quad \alpha + \beta = \text{root} \\
&[E_\alpha, E_{-\alpha}] = (E_\alpha, E_{-\alpha}) H_\alpha, \quad \alpha + \beta = 0 \\
&[E_\alpha, E_{-\alpha}] = 0, \quad \text{otherwise.}
\end{aligned}
\tag{A.13}
$$

Since the restriction of the Killing form to $\mathbf{h}$ is positive definite we can choose the normalization

$$(H_i, H_j) = \delta_{ij}, \quad (E_\alpha, E_\beta) = \delta_{\alpha+\beta,0} \tag{A.14}$$

Thus $g_{ij} = \delta_{ij}$, $\alpha_i = h_\alpha^i$ and hence

$$H_\alpha = \sum_i \alpha_i H_i, \quad \langle \alpha, \alpha \rangle = (H_\alpha, H_\alpha) = \sum_i \alpha_i^2. \tag{A.15}$$

We get finally the commutation relations

$$[H_\alpha, E_{\pm\alpha}] = \pm\langle \alpha, \alpha \rangle E_{\pm\alpha}, \quad [E_\alpha, E_{-\alpha}] = H_\alpha. \tag{A.16}$$

Thus we get an $su(2)$ Lie algebra for every root.

A.2 Roots, Cartan matrix and Dynkin diagrams

We have established that we have $d - l$ roots. Their inner product is inherited from the Killing form, viz $\langle \alpha, \beta \rangle = (H_\alpha, H_\beta) = \sum_i \alpha_i \beta_i$. It is obvious that $H_\alpha/\langle \alpha, \alpha \rangle$ plays the role of J_z in $su(2)$, whereas the E_α and $E_{-\alpha}$ play the role of the raising and lowering operators J_+ and J_-, respectively, in $su(2)$. We also compute

$$\mathrm{ad} H_\alpha E_\beta = [H_\alpha, E_\beta] = \langle \alpha, \beta \rangle E_\beta. \tag{A.17}$$

Let $p \leqslant 0$ be the smallest integer such that $(\mathrm{ad} E_{-\alpha})^{|p|} E_\beta$ is non-zero, i.e., $(\mathrm{ad} E_{-\alpha})^{|p|+1} E_\beta = 0$. This means that $E_{-k\alpha+\beta}$, $k = 1,\ldots, |p|$, correspond to non-zero roots $-k\alpha + \beta$. Let $q \geqslant 0$ be the largest integer such that $(\mathrm{ad} E_\alpha)^q E_\beta$ is non-zero, i.e., $(\mathrm{ad} E_\alpha)^{q+1} E_\beta = 0$. This means that $E_{k\alpha+\beta}$, $k = 1,\ldots, q$, correspond to non-zero roots $k\alpha + \beta$. The set $\{E_{\beta'}; \ \beta' = \beta + q\alpha,\ldots, \beta,\ldots, \beta + p\alpha\}$ is called the α-chain through β. Similarly to $su(2)$, the $E_{\beta'}$ form a representation of the $su(2)$ algebra generated by $E_{\pm\alpha}$ and H_α. Furthermore, the eigenvalues corresponding to the highest state $E_{\beta+q\alpha}$ and the lowest state $E_{\beta+p\alpha}$ given, respectively,

$$[H_\alpha, E_{\beta+q\alpha}] = \langle \alpha, \beta + q\alpha \rangle E_{\beta+q\alpha}, \quad [H_\alpha, E_{\beta+p\alpha}] = \langle \alpha, \beta + p\alpha \rangle E_{\beta+p\alpha}, \tag{A.18}$$

are opposite to each other, i.e.,

$$\langle \alpha, \beta + q\alpha \rangle = -\langle \alpha, \beta + p\alpha \rangle \Rightarrow 2\frac{\langle \alpha, \beta \rangle}{\langle \alpha, \alpha \rangle} = -q - p = m \in \mathbf{Z}. \tag{A.19}$$

The set of all roots, which are not linearly independent in $\mathbf{h}^*$, is denoted by Δ and it is decomposed into positive and negative roots. Obviously, the opposite of a positive root is a negative root and vice versa. A basis in Δ can be given by the so-called simple roots denoted by α_i, $i = 1,\ldots, l$, such that the positive (negative) roots are linear combinations of simple roots with positive (negative) integer coefficients. A simple root can not be rewritten as the sum of two positive roots. The choice of simple roots is not unique and different choices are related by the Weyl group, which leaves the set of roots globally invariant.

Let α and β be two simple roots. Then obviously $\alpha - \beta$ can not be a root and as a consequence $p = 0$ and $m = -q \leqslant 0$. In other words, $\langle \alpha, \beta \rangle \leqslant 0$. The Cartan matrix is defined in terms of the simple roots by

$$C_{ij} = 2\frac{\langle \alpha_i, \alpha_j \rangle}{\langle \alpha_j, \alpha_j \rangle}. \tag{A.20}$$

This is not a symmetric matrix. We compute for $i \neq j$, using $\langle \alpha, \beta \rangle = |\alpha||\beta|\cos\theta$, that

$$C_{ij} = 2\frac{\langle \alpha_i, \alpha_j \rangle}{\langle \alpha_j, \alpha_j \rangle} = 2\frac{|\alpha_i|}{|\alpha_j|}\cos\theta_{ij} = m_i \leqslant 0. \tag{A.21}$$

$$C_{ji} = 2\frac{\langle \alpha_j, \alpha_i \rangle}{\langle \alpha_i, \alpha_i \rangle} = 2\frac{|\alpha_j|}{|\alpha_i|}\cos\theta_{ij} = m_j \leqslant 0. \tag{A.22}$$

By multiplying the above last two equations and using the Schwarz inequality $\langle \alpha, \beta \rangle^2 < \langle \alpha, \alpha \rangle\langle \beta, \beta \rangle$ we obtain

$$m_i m_j < 4. \tag{A.23}$$

The value $m_i m_j = 4$ is forbidden because $\alpha_i \neq -\alpha_j$. We also compute by dividing the above two equations the following

$$\frac{|\alpha_i|}{|\alpha_j|} = \sqrt{\frac{m_i}{m_j}}, \quad \cos\theta_{ij} = -\frac{1}{2}\sqrt{m_i m_j}. \tag{A.24}$$

The only allowed angles are therefore $\pi/2$, $2\pi/3$, $3\pi/4$ and $5\pi/6$ corresponding to the cosine equals 0, $-1/2$, $-\sqrt{2}/2$, $-\sqrt{3}/2$. The allowed ratios of lengths of roots are therefore 0, 1, $\sqrt{2}$ and $\sqrt{3}$.

If the set of roots decomposes into two mutually orthogonal subsets then the corresponding semi-simple algebra $\mathbf{g}$ decomposes into the direct sum of two simple algebras corresponding to the orthogonal sets. In the following we will only consider simple algebras following [1].

For rank 1 there is one complex simple Lie algebra, which is $A_1 = sl(2, \mathbf{C})$. For rank 2 there are two complex simple Lie algebras, which are $A_2 = sl(3, \mathbf{C})$ and $B_2 = so(5, \mathbf{C})$ and one complex semi-simple Lie algebra which is $D_2 = so(4, \mathbf{C})$. There is also another rank 2 complex simple algebra G_2, which is an exceptional algebra of dimension 14. For higher ranks, the Cartan classification of complex simple algebra is given by

$$A_l = sl(l+1, \mathbf{C}), \quad B_l = so(2l+1, \mathbf{C}), \quad C_l = sp(2l, \mathbf{C}), \quad D_l = so(2l, \mathbf{C}). \quad (A.25)$$

These four infinite families are the so-called classical Lie algebra. The unique real compact forms of these algebras are given, respectively, by $A_l = su(l+1)$, $B_l = so(2l+1)$, $C_l = usp(2l)$, $D_l = so(2l)$.

Furthermore, the full Cartan classification of the exceptional complex simple algebra is given by the following five cases

$$E_6(78), \quad E_7(133), \quad E_8(248), \quad F_4(52), \quad G_2(14). \quad (A.26)$$

The number in brackets is the dimension.

For higher ranks we use Dynkin diagram, which encodes the Cartan matrix to visualize the root system. The Dynkin diagram is constructed as follows
- Each simple root is represented by a vertex.
- Any two roots such that $\langle \alpha_i, \alpha_j \rangle \neq 0$ are linked by a line.
- The line is simple if

$$C_{ij} = C_{ji} = -1, \quad \left(\theta_{ij} = \frac{2\pi}{3}, \quad \frac{|\alpha_i|}{|\alpha_j|} = 1\right). \quad (A.27)$$

- The line is double if

$$C_{ij} = -2, =C_{ji} = -1, \quad \left(\theta_{ij} = \frac{3\pi}{4}, \quad \frac{|\alpha_i|}{|\alpha_j|} = \sqrt{2}\right). \quad (A.28)$$

- The line is triple if

$$C_{ij} = -3, =C_{ji} = -1, \quad \left(\theta_{ij} = \frac{5\pi}{6}, \quad \frac{|\alpha_i|}{|\alpha_j|} = \sqrt{3}\right). \quad (A.29)$$

- The line carries an arrow from i to j if $|\alpha_i| > |\alpha_j|$.

A.3 Weights, Dynkin labels and representations

The roots studied so far provide one particular finite dimensional irreducible unitary representation of the algebra $\mathbf{g}$ known as the adjoint representation. A general finite dimensional irreducible unitary representation of the algebra $\mathbf{g}$ is given by the so-called weights. We start with the Cartan subalgebra generated by the commuting elements H_i. These can be obviously diagonalized simultaneously. Let $|\lambda_a\rangle$ be the eigenvectors of the Cartan elements H_i with eigenvalues λ_i, viz

$$H_i|\lambda_a\rangle = \lambda_i|\lambda_a\rangle. \quad (A.30)$$

The vector of eigenvalues $\lambda = (\lambda_1, \ldots, \lambda_l)$ is called a weight and is associated with the vectors $|\lambda_a\rangle$ where the index a denotes possible degeneracy of the eigenvalue λ. Obviously, for $H = \sum_i h^i H_i$ we have

$$H|\lambda_a\rangle = \lambda(H)|\lambda_a\rangle, \quad \lambda(H) = \sum_i h^i \lambda_i. \tag{A.31}$$

Thus, λ is a linear form on $\mathbf{h}$ and thus an element in $\mathbf{h}^*$ which is the space of roots. Since H is Hermitian the weights are real. The set of weights in a given representation forms what we call a weight diagram.

Let us call the representation space E. The dimension of the representation space E is equal to the total number of λ_a including their multiplicities. Since there is an $su(2)$ Lie algebra for every root α, the space E contains subspaces corresponding to the $d - l\,su(2)$ Lie algebras $\{H_\alpha, E_\alpha, E_{-\alpha}\}$ satisfying $[H_\alpha, E_{\pm\alpha}] = \pm\langle \alpha, \alpha\rangle E_{\pm\alpha}$, $[E_\alpha, E_{-\alpha}] = H_\alpha$.

We compute immediately

$$H_i E_{\pm\alpha}|\lambda_a\rangle = (\lambda \pm \alpha)_i E_{\pm\alpha}|\lambda_a\rangle. \tag{A.32}$$

Thus the vector $E_\alpha|\lambda_a\rangle$ is an eigenvector associated with the eigenvalue $\lambda + \alpha$. In other words, all vectors in the representation space E are obtained from each other by the action of the $E_{\pm\alpha}$. We also conclude that any two weights of the same representation can only differ by a linear combination of roots with integer coefficients.

The weight λ will be associated with the operator E_λ. Let $p' \leqslant 0$ be the smallest integer such that $(E_{-\alpha})^{|p'|}|\lambda_a\rangle$ is non-zero, i.e., $(E_{-\alpha})^{|p'|+1}|\lambda_a\rangle = 0$. This means that $(E_{-\alpha})^k|\lambda_a\rangle$, $k = 1, \ldots, |p'|$, correspond to non-zero weights $-k\alpha + \lambda$. Let $q' \geqslant 0$ be the largest integer such that $(E_\alpha)^{q'}|\lambda_a\rangle$ is non-zero, i.e., $(E_\alpha)^{q'+1}|\lambda_a\rangle = 0$. This means that $E_{k\alpha+\lambda}|\lambda_a\rangle$, $k = 1, \ldots, q'$, correspond to non-zero weights $k\alpha + \lambda$. The weights $\{\beta' = \lambda - |p'|\alpha, \ldots, \lambda, \ldots, \lambda + q'\alpha\}$ will be associated with the operator $E_{\beta'}$. Similarly to before, the operators $\{E_{\beta'}\}$ form a representation of the $su(2)$ algebra generated by $E_{\pm\alpha}$ and H_α. Furthermore, the eigenvalues corresponding to the highest state $E_{\lambda+q'\alpha}$ and the lowest state $E_{\lambda+p'\alpha}$ given respectively

$$[H_\alpha, E_{\lambda+q'\alpha}] = \langle \alpha, \lambda + q'\alpha\rangle E_{\lambda+q'\alpha}, \quad [H_\alpha, E_{\lambda+p'\alpha}] = \langle \alpha, \lambda + p'\alpha\rangle E_{\lambda+p'\alpha}, \tag{A.33}$$

are opposite to each other, i.e.,

$$\langle \alpha, \lambda + q'\alpha\rangle = -\langle \alpha, \lambda + p'\alpha\rangle \Rightarrow 2\frac{\langle \alpha, \lambda\rangle}{\langle \alpha, \alpha\rangle} = -q' - p' = m' \in \mathbf{Z}. \tag{A.34}$$

Since we are dealing with an $su(2)$ algebra, the number of the states $\{E_{\beta'}\}$ is exactly given by $q' - p' + 1 = 2j + 1$, i.e. $q' - p' = 2j$.

In the above space we can clearly introduce an ordering given by $\lambda' > \lambda$ if $\lambda' - \lambda = \sum_i n_i \alpha_i$ where n_i are positive integers. There exists therefore a highest weight state denoted by $|\Lambda\rangle$ such that if α is any positive root we have

$$E_\alpha |\Lambda\rangle = 0. \tag{A.35}$$

In this case $q' = 0$ and $p' = -2j$ and hence

$$2\langle \Lambda, \alpha \rangle = j\langle \alpha, \alpha \rangle > 0. \tag{A.36}$$

We define the Dynkin labels of a weight λ by

$$\lambda_i = 2\frac{\langle \lambda, \alpha_i \rangle}{\langle \alpha_i, \alpha_i \rangle} \in \mathbf{Z}. \tag{A.37}$$

The α are simple roots and therefore there are l Dynkin labels. If we choose the weight λ to be the highest weight Λ we obtain

$$\lambda_i = 2\frac{\langle \Lambda, \alpha_i \rangle}{\langle \alpha_i, \alpha_i \rangle} = j \in \mathbf{N}. \tag{A.38}$$

We define the fundamental weights Λ_i by the formula

$$\delta_{ij} = 2\frac{\langle \Lambda_i, \alpha_i \rangle}{\langle \alpha_i, \alpha_i \rangle}. \tag{A.39}$$

Clearly, there are l of them and they provide a basis in $\mathbf{h}^*$. Also, it is obvious that any highest weight state can be rewritten as a linear combination of the fundamental weights with coefficients given by the Dynkin labels, viz

$$\Lambda = \sum_{i=1}^{l} \lambda_i \Lambda_i, \quad \lambda_i \in \mathbf{N}. \tag{A.40}$$

Any irreducible unitary representation is completely characterized by its highest weight state. The fundamental weights define the so-called fundamental representations. Hence, we have l fundamental representations characterized by Λ_i.

The dimension and the Casimir operators of a given irreducible representation characterized by a highest weight state Λ are given in terms of the Weyl vector defined by the half sum of the positive roots or by the sum of the fundamental roots as follows

$$\rho = \frac{1}{2}\sum_{\alpha>0}\alpha = \sum_{i}\Lambda_i. \tag{A.41}$$

The dimension and the quadratic Casimir of the irreducible representation Λ are given, respectively, by the Weyl formulas

$$\dim(\Lambda) = \prod_{\alpha>0} \frac{\langle \Lambda + \rho, \alpha \rangle}{\langle \rho, \alpha \rangle}. \tag{A.42}$$

$$C_2(\Lambda) = \frac{1}{2}\langle \Lambda, \Lambda + 2\rho \rangle. \tag{A.43}$$

A.4 Explicit construction of Lie algebra representations

A.4.1 $A_l = su(l+1)$

In this case the dimension of $\mathbf{h}^*$ is $l = n - 1$, i.e. $\mathbf{h}^* = \mathbf{R}^{n-1}$. Let $\hat{e}_i$, $i = 1,\ldots, n$, be an orthonormal basis in $\mathbf{R}^n$ satisfying

$$\langle \hat{e}_i, \hat{e}_j \rangle = \delta_{ij}. \tag{A.44}$$

We define any hyperplane in $\mathbf{R}^n$ by the vector normal to it. We consider the hyperplane given by the normal vector

$$\hat{\rho} = \sum_{i=1}^{n} \hat{e}_i. \tag{A.45}$$

We project the $\hat{e}_i$ on this hyperplane to obtain vectors e_i given by

$$e_i = \hat{e}_i - \frac{1}{n}\hat{\rho}. \tag{A.46}$$

Indeed, we check that $\langle \hat{\rho}, e_i \rangle = 0$. Furthermore, we check that

$$\langle e_i, e_j \rangle = \delta_{ij} - \frac{1}{n}. \tag{A.47}$$

We also check that $\sum_i e_i = 0$. In other words, the e_i are not linearly independent and they form a basis in $\mathbf{R}^{n-1}$. We define the $l = n - 1$ simple roots and the $(d - l)/2 = n(n-1)/2$ positive roots, respectively, by

$$\alpha_i = \alpha_{ii+1} = e_i - e_{i+1}, \quad i = 1,\ldots, l. \tag{A.48}$$

$$\alpha_{ij} = e_i - e_j, \quad i < j = 1,\ldots, n. \tag{A.49}$$

The simple roots satisfy

$$\langle \alpha_i, \alpha_j \rangle = 2\delta_{ij} - \delta_{ij+1} - \delta_{i+1j}. \tag{A.50}$$

From this equation, we can infer directly the value of the Cartan matrix. We have

$$C_{ij} = \langle \alpha_i, \alpha_j \rangle = 2\delta_{ij} - \delta_{ij+1} - \delta_{i+1j}. \tag{A.51}$$

In order to compute the Weyl vector we divide the positive roots as $\{\alpha_{1i} = e_1 - e_i, \; i = 2,\ldots, n\}$ ($n - 1$ roots), $\{\alpha_{2i} = e_2 - e_i, i = 3,\ldots, n\}$ ($n - 2$ roots), $\{\alpha_{3i} = e_3 - e_i, i = 4,\ldots, n\}$ ($n - 3$ roots),..., $\{\alpha_{n-2i} = e_{n-2} - e_i, i = n - 1, n\}$ (2 roots), $\{\alpha_{n-1i} = e_{n-1} - e_i, i = n\}$ (1 root). We have then

$$2\rho = \sum_{\alpha>0} \alpha = \sum_{i<j} \alpha_{ij}$$
$$= (n - 1)e_1 + (n - 3)e_2 + (n - 5)e_3 + \cdots + (n - 2i + 1)e_i + \cdots -(n - 1)e_n. \tag{A.52}$$

The index i in the above formula goes from $i = 1$ to n. Now, if we re-express the e_i in terms of the simple roots α_i we find that the coefficient of each α_i is the sum of the coefficients of e_j with $j \leqslant i$, viz

$$\sum_{j=1}^{i}(n - 2j + 1) = i(n - i), \quad i = 1,\ldots, n - 1 \tag{A.53}$$

We obtain then

$$2\rho = (n - 1)\alpha_1 + 2(n - 2)\alpha_2 + 3(n - 3)\alpha_3 + \cdots + i(n - i)\alpha_i + \cdots + (n - 1)\alpha_{n-1}. \tag{A.54}$$

By using $\sum_i e_i = 0$ we can also rewrite the Weyl vector as

$$\rho = \sum_{i=1}^{l}(n - i)e_i. \tag{A.55}$$

From equation (A.50) we have $\langle \alpha_i, \alpha_i \rangle = 2$. Thus the fundamental weights must be defined by the condition $\langle \Lambda_j, \alpha_i \rangle = \delta_{ij}$. A solution is given by

$$\Lambda_i = \sum_{j=1}^{i}e_j. \tag{A.56}$$

We verify immediately that $\langle \Lambda_j, \alpha_i \rangle = \sum_{k=1}^{j}\delta_{ki} - \sum_{k=1}^{j-1}\delta_{ki}$. For $i=j$ we get $1 - 0$, for $i > j$ we get $0 - 0$ whereas for $i < j$ we get $1 - 1$. Hence

$$2\frac{\langle \Lambda_j, \alpha_i \rangle}{\langle \alpha_i, \alpha_i \rangle} = \delta_{ij}. \tag{A.57}$$

Explicitly, we have

$$\begin{aligned}
e_1 &= \Lambda_1 \\
e_i &= \Lambda_i - \Lambda_{i-1}, \quad i = 2,\ldots, n - 1 \\
e_n &= -\Lambda_{n-1}.
\end{aligned} \tag{A.58}$$

Furthermore, we compute

$$\langle \Lambda_j, \alpha_i \rangle = \frac{i(n - j)}{n}, \quad i \leqslant j. \tag{A.59}$$

We need now to compute the dimension of the irreducible representation with highest weight state $\Lambda = \sum_{i=1}^{l}\lambda_i\Lambda_i$. We have

$$\begin{aligned}
\langle \Lambda, \alpha_{ij} \rangle &= \langle \Lambda, e_i \rangle - \langle \Lambda, e_j \rangle \\
&= \sum_{k=1}^{l}\lambda_k \sum_{k'=1}^{k}(\delta_{k'i} - \frac{1}{n}) - (i \longrightarrow j) \\
&= \sum_{p=1}^{l}f_p(\delta_{pi} - \frac{1}{n}) - (i \longrightarrow j) \\
&= f_i - f_j.
\end{aligned} \tag{A.60}$$

The f_i is defined by the equation

$$f_i = \sum_{q=i}^{l}\lambda_q, \quad i = 1,\ldots, l; \; f_n = 0. \tag{A.61}$$

Further, we compute in the same way

$$\langle \rho, \alpha_{ij} \rangle = \langle \rho, e_i \rangle - \langle \rho, e_j \rangle$$
$$= \frac{1}{2} \sum_{k=1}^{l} (n - 2k + 1) \langle e_k, e_i \rangle - (i \longrightarrow j) \tag{A.62}$$
$$= j - i.$$

Hence, we find the dimension

$$\dim(\Lambda) = \prod_{i<j} \frac{f_i - f_j + j - i}{j - i}. \tag{A.63}$$

The Casimir operator is given by

$$C_2(\Lambda) = \frac{1}{2} \sum_{i=1}^{l} \sum_{j=1}^{l} \lambda_i (\lambda_j + 2) \langle \Lambda_i, \Lambda_j \rangle. \tag{A.64}$$

We compute

$$\langle \Lambda_i, \Lambda_j \rangle = \frac{i(n - j)}{n}, \quad i \leqslant j. \tag{A.65}$$

$$\langle \Lambda_i, \Lambda_j \rangle = \frac{j(n - i)}{n}, \quad i > j. \tag{A.66}$$

Example: $su(4)$

In this case $n = 4$, $d = n^2 - 1 = 15$, $l = 3$ and thus $\mathbf{h}^* = \mathbf{R}^3$. We have $(d - l)/2 = 6$ positive roots. The simple roots are

$$\alpha_1 = \alpha_{12} = e_1 - e_2, \quad \alpha_2 = \alpha_{23} = e_2 - e_3, \quad \alpha_3 = \alpha_{34} = e_3 - e_4. \tag{A.67}$$

The other positive roots are

$$\alpha_{13} = e_1 - e_3, \quad \alpha_{14} = e_1 - e_4, \quad \alpha_{24} = e_2 - e_4. \tag{A.68}$$

The Weyl vector is (using also $\sum_i e_i = 0$)

$$2\rho = 3e_1 + e_2 - e_3 - 3e_4 = 6e_1 + 4e_2 + 2e_3$$
$$= 3\alpha_1 + 4\alpha_2 + 3\alpha_3. \tag{A.69}$$

The weights are

$$\Lambda_1 = e_1, \quad \Lambda_2 = e_1 + e_2, \quad \Lambda_3 = e_1 + e_2 + e_3. \tag{A.70}$$

The Weyl vector can also be rewritten as

$$\rho = \Lambda_1 + \Lambda_2 + \Lambda_3. \tag{A.71}$$

The Casimir operator and the dimension are given by

$$C_2(\Lambda) = \frac{1}{8} \lambda_1 (3\lambda_1 + 2\lambda_2 + \lambda_3 + 12) + \frac{1}{4} \lambda_2 (\lambda_1 + 2\lambda_2 + \lambda_3 + 8)$$
$$+ \frac{1}{8} \lambda_3 (\lambda_1 + 2\lambda_2 + 3\lambda_3 + 12). \tag{A.72}$$

$$\dim(\Lambda) = \frac{1}{12}(\lambda_1 + 1)(\lambda_2 + 1)(\lambda_3 + 1)(\lambda_1 + \lambda_2 + 2)(\lambda_2 + \lambda_3 + 2)(\lambda_1 + \lambda_2 + \lambda_3 + 3). \tag{A.73}$$

For example $C_2(n, 0, n) = n(n + 3)$ and $\dim(n, 0, n) = (n + 1)^2(n + 2)^2(2n + 3)/12$.

A.4.2 $B_l = so(2l + 1)$

In this case the dimension is $d = l(2l + 1)$ and the rank is l, i.e., $\mathbf{h}^* = \mathbf{R}^l$. We choose a basis in $\mathbf{R}^l$ given by e_i, $i = 1,\dots, l$, such that $\langle e_i, e_j \rangle = \delta_{ij}$. The l simple roots are

$$\alpha_i = e_i - e_{i+1}, \quad i = 1,\dots, l - 1; \quad \alpha_l = e_l. \tag{A.74}$$

There are $(d - l)/2 = l^2$ positive roots. Here they are

$$e_i = \sum_{k=i}^{l} \alpha_k, \quad i = 1,\dots, l. \tag{A.75}$$

$$e_i - e_j = \sum_{k=i}^{j-1} \alpha_k, \quad 1 \leqslant i < j \leqslant l. \tag{A.76}$$

$$e_i + e_j = \sum_{k=i}^{j-1} \alpha_k + 2\sum_{k=j}^{l} \alpha_k, \quad 1 \leqslant i < j \leqslant l. \tag{A.77}$$

The Weyl vector is the half sum of the positive roots. We get

$$\begin{aligned}
2\rho &= \sum_{i=1}^{l} e_i + \sum_{i<j}(e_i - e_j) + \sum_{i<j}(e_i + e_j) \\
&= \sum_{i=1}^{l} e_i(2l - 2i + 1) \\
&= \sum_{i=1}^{l}\sum_{k=i}^{l} \alpha_k(2l - 2i + 1) \\
&= \sum_{i=1}^{l}\sum_{k=1}^{i} \alpha_i(2l - 2k + 1) \\
&= \sum_{i=1}^{l} \alpha_i \cdot i(2l - i).
\end{aligned} \tag{A.78}$$

The non-zero elements of the Cartan matrix are given by

$$\langle \alpha_i, \alpha_i \rangle = 2, \quad i = 1,\dots, l - 1. \tag{A.79}$$

$$\langle \alpha_l, \alpha_l \rangle = 1. \tag{A.80}$$

$$\langle \alpha_i, \alpha_j \rangle = -1, \quad 1 \leqslant i = j + 1, \ j = i + 1 \leqslant l. \tag{A.81}$$

The fundamental weights are defined by

$$\Lambda_i = \sum_{j=1}^{i} e_j, \quad i = 1,\dots, l-1; \quad \Lambda_l = \frac{1}{2}\sum_{j=1}^{l} e_j. \tag{A.82}$$

We can re-express the basis vectors in terms of the fundamental weights as

$$e_1 = \Lambda_1; \quad e_i = \Lambda_i - \Lambda_{i-1}, \quad 2 \leqslant i \leqslant l-1; \quad e_l = 2\Lambda_l - \Lambda_{l-1}. \tag{A.83}$$

The Dynkin labels of the roots are given by

$$\alpha_1 = 2\Lambda_1 - \Lambda_2 = (2, -1,\dots). \tag{A.84}$$

$$\alpha_i = -\Lambda_{i-1} + 2\Lambda_i - \Lambda_{i+1} = (\dots, -1, 2, -1,\dots), \quad i = 2,\dots, l-2. \tag{A.85}$$

$$\alpha_{l-1} = -\Lambda_{l-2} + 2\Lambda_{l-1} - 2\Lambda_l = (\dots, -1, 2, -2). \tag{A.86}$$

$$\alpha_l = -\Lambda_{l-1} + 2\Lambda_l = (\dots, -1, 2). \tag{A.87}$$

In general, irreducible representations of $so(2l+1)$ are characterized by the highest weight vectors Λ, which can be rewritten as $\Lambda = \sum_i n_i e_i = (n_1,\dots, n_l)$ with $n_1 \geqslant \cdots \geqslant n_{l-1} \geqslant n_l \geqslant 0$ with dimensions (see [2] page 407)

$$\dim(n_1,\dots, n_l) = \prod_{i<j} \frac{l_i^2 - l_j^2}{m_i^2 - m_j^2} \prod_i \frac{l_i}{m_i} \tag{A.88}$$

$$l_i = n_i + l - i + \frac{1}{2}, \quad m_i = l - i + \frac{1}{2}.$$

Example:
$so(5)$ In this case $l = 2$ and thus $\mathbf{h}^* = \mathbf{R}^2$. The simple roots are

$$\alpha_1 = e_1 - e_2, \quad \alpha_2 = e_2. \tag{A.89}$$

There are four positive roots. These are the two simple roots plus

$$e_1 = \alpha_1 + \alpha_2, \quad e_1 + e_2 = \alpha_1 + 2\alpha_2. \tag{A.90}$$

The Weyl vector is

$$\rho = \frac{3}{2}e_1 + \frac{1}{2}e_2 = \frac{3}{2}\alpha_1 + 2\alpha_2. \tag{A.91}$$

The fundamental weights are

$$\Lambda_1 = e_1 = \alpha_1 + \alpha_2$$
$$\Lambda_2 = \frac{1}{2}(e_1 + e_2) = \frac{1}{2}\alpha_1 + \alpha_2. \tag{A.92}$$

A straightforward calculation gives the dimension and the Casimir

$$\dim(\Lambda) = \frac{1}{6}(\lambda_1 + 1)(\lambda_2 + 1)(\lambda_1 + \lambda_2 + 2)(2\lambda_1 + \lambda_2 + 3). \tag{A.93}$$

$$C_2(\Lambda) = \frac{1}{4}\lambda_1(2\lambda_1 + \lambda_2 + 6) + \frac{1}{4}\lambda_2(\lambda_1 + \lambda_2 + 4). \tag{A.94}$$

The highest weight state $\Lambda = \sum_i \lambda_i \Lambda_i$ is usually expressed as $\Lambda = \sum_i n_i e_i$ where the spin quantum numbers n_i are defined in terms of the Dynkin labels λ_i by $\lambda_1 = n_1 - n_2$ and $\lambda_2 = 2n_2$. The dimension and the Casimir operator become in terms of n_i given by

$$\dim(\Lambda) = \frac{1}{6}(2n_1 + 3)(2n_2 + 1)(n_1 - n_2 + 1)(n_1 + n_2 + 2). \tag{A.95}$$

$$C_2(\Lambda) = \frac{1}{2}n_1(n_1 + 3) + \frac{1}{2}n_2(n_2 + 1). \tag{A.96}$$

A.4.3 $D_l = so(2l)$

In this case the dimension is $d = l(2l - 1)$ and the rank is l, i.e., $\mathbf{h}^* = \mathbf{R}^l$. We choose a basis in $\mathbf{R}^l$ given by e_i, $i = 1,\ldots, l$, such that $\langle e_i, e_j \rangle = \delta_{ij}$. The l simple roots are

$$\alpha_i = e_i - e_{i+1}, \quad i = 1,\ldots, l - 1; \quad \alpha_l = e_{l-1} + e_l. \tag{A.97}$$

There are $(d - l)/2 = l(l - 1)$ positive roots. Here they are

$$e_i - e_j = \sum_{k=i}^{j-1} \alpha_k, \quad 1 \leqslant i < j \leqslant l. \tag{A.98}$$

$$e_i + e_j = \sum_{k=i}^{j-1} \alpha_k + 2\sum_{k=j}^{l-2} \alpha_k + \alpha_{l-1} + \alpha_l, \quad 1 \leqslant i < j \leqslant l - 1. \tag{A.99}$$

$$e_i + e_l = \sum_{k=i}^{l-2} \alpha_k + \alpha_l, \quad 1 \leqslant i \leqslant l - 1. \tag{A.100}$$

The Weyl vector is the half sum of the positive roots. We get

$$
\begin{aligned}
2\rho &= \sum_{i=1}^{l-1} 2(l - i)e_i \\
&= \sum_{i=1}^{l-2} \alpha_i \cdot i(2l - i - 1) + \frac{l(l - 1)}{2}(\alpha_{l-1} + \alpha_l).
\end{aligned}
\tag{A.101}
$$

The non-zero elements of the Cartan matrix are given by

$$\langle \alpha_i, \alpha_i \rangle = 2, \quad i = 1,\ldots, l. \tag{A.102}$$

$$\langle \alpha_i, \alpha_j \rangle = -1, \quad i = l, \ j = l - 2; \quad i = l - 2, \ j = l. \tag{A.103}$$

$$\langle \alpha_i, \alpha_j \rangle = -1, \quad 1 \leqslant i = j+1, \quad j = i+1 \leqslant l-2. \tag{A.104}$$

The fundamental weights are defined by

$$\Lambda_i = \sum_{j=1}^{i} e_j$$
$$= \alpha_1 + 2\alpha_2 + \cdots + (i-1)\alpha_{i-1} + i(\alpha_i + \cdots + \alpha_{l-2}) + \frac{i}{2}(\alpha_{l-1} + \alpha_l), \quad i = 1,\ldots, l-2. \tag{A.105}$$

$$\Lambda_{l-1} = \frac{1}{2}(e_1 + \cdots + e_{l-1} - e_l)$$
$$= \frac{1}{2}(\alpha_1 + 2\alpha_2 + \cdots + (l-2)\alpha_{l-2}) + \frac{l}{4}\alpha_{l-1} + \frac{l-2}{4}\alpha_l. \tag{A.106}$$

$$\Lambda_l = \frac{1}{2}(e_1 + \cdots + e_{l-1} + e_l)$$
$$= \frac{1}{2}(\alpha_1 + 2\alpha_2 + \cdots + (l-2)\alpha_{l-2}) + \frac{l-2}{4}\alpha_{l-1} + \frac{l}{4}\alpha_l. \tag{A.107}$$

In general, irreducible representations of $so(2l)$ are characterized by the highest weight vectors Λ, which can be rewritten as $\Lambda = \sum_i n_i e_i = (n_1,\ldots, n_l)$ with $n_1 \geqslant \cdots \geqslant n_{l-1} \geqslant |n_l| \geqslant 0$ with dimensions (see [2] page 409)

$$\dim(n_1,\ldots, n_l) = \prod_{i<j} \frac{l_i^2 - l_j^2}{m_i^2 - m_j^2} \tag{A.108}$$
$$l_i = n_i + l - i, \, m_i = l - i.$$

Example: $so(6)$
In this case $l = 3$ and thus $\mathbf{h}^* = \mathbf{R}^3$. The simple roots are

$$\alpha_1 = e_1 - e_2, \quad \alpha_2 = e_2 - e_3, \quad \alpha_3 = e_2 + e_3. \tag{A.109}$$

There are six positive roots. These are the three simple roots plus

$$e_1 - e_3 = \alpha_1 + \alpha_2, \quad e_1 + e_3 = \alpha_1 + \alpha_3, \quad e_1 + e_2 = \alpha_1 + \alpha_2 + \alpha_3. \tag{A.110}$$

The Weyl vector is

$$\rho = 2e_1 + e_2 = 2\alpha_1 + \frac{3}{2}\alpha_2 + \frac{3}{2}\alpha_3. \tag{A.111}$$

The Cartan matrix is

$$C = \begin{pmatrix} 2 & -1 & -1 \\ -1 & 2 & 0 \\ -1 & 0 & 2 \end{pmatrix} \tag{A.112}$$

The fundamental weights are

$$\Lambda_1 = e_1 = \alpha_1 + \frac{1}{2}\alpha_2 + \frac{1}{2}\alpha_3$$

$$\Lambda_2 = \frac{1}{2}(e_1 + e_2 - e_3) = \frac{1}{2}\alpha_1 + \frac{3}{4}\alpha_2 + \frac{1}{4}\alpha_3 \tag{A.113}$$

$$\Lambda_3 = \frac{1}{2}(e_1 + e_2 + e_3) = \frac{1}{2}\alpha_1 + \frac{1}{4}\alpha_2 + \frac{3}{4}\alpha_3.$$

A straightforward calculation gives the dimension and the Casimir

$$\dim(\Lambda) = \frac{1}{12}(\lambda_1 + 1)(\lambda_2 + 1)(\lambda_3 + 1)(\lambda_1 + \lambda_2 + 2)(\lambda_1 + \lambda_3 + 2)(\lambda_1 + \lambda_2 + \lambda_3 + 3). \tag{A.114}$$

$$C_2(\Lambda) = \frac{1}{4}\lambda_1(2\lambda_1 + \lambda_2 + \lambda_3 + 8) + \frac{1}{8}\lambda_2(2\lambda_1 + 3\lambda_2 + \lambda_3 + 12) + \frac{1}{8}\lambda_3(2\lambda_1 + \lambda_2 + 3\lambda_3 + 12). \tag{A.115}$$

In terms of the spin quantum numbers n_i defined by $\lambda_1 = n_1 - n_2$, $\lambda_2 = n_2 - n_3$ and $\lambda_3 = n_2 + n_3$ we have

$$\dim(\Lambda) = \frac{1}{12}((n_1 + 2)^2 - n_3^2)((n_1 + 2)^2 - (n_2 + 1)^2)((n_2 + 1)^2 - n_3^2). \tag{A.116}$$

$$C_2(\Lambda) = \frac{1}{2}n_1(n_1 + 4) + \frac{1}{2}n_2(n_2 + 2) + \frac{1}{2}n_3^2. \tag{A.117}$$

References

[1] Zuber J B 2011   Invariances in physics and group theory Parcours de Physique Théorique, Fall https://www.lpthe.jussieu.fr/~zuber/Cours/Cours-M2-2012_e.pdf
[2] Fulton W and Harris J 1991 *Representation Theory: A First Course Graduate Texts in Mathematics* **vol 129** (New York: Springer)

IOP Publishing

A Modern Course in Quantum Field Theory, Volume 2 (Second Edition)
Advanced topics

Badis Ydri

Appendix B

On homotopy theory

- **Compactification:** In one dimension we can compactify $\mathbf{R}$ by adding one point (at) ∞ to obtain the one-sphere S^1. Conversely, by removing one point from the circle we get essentially $\mathbf{R}$. This is called Alexandroff one-point compactification. Generalization to higher dimensions is obvious.

- **Topological spaces and manifolds:** A topological space is a collection of open subsets (they do not contain any of their boundary points) with certain properties, which allow the introduction of the concept of continuity (smoothness). Manifolds have the added property of differentiability.

- **Homeomorphic:** Two spaces are said to be homeomorphic if they can be mapped continuously and bijectively onto each other. Two homeomorphic spaces are topologically identical and possess the same connectedness properties (i.e., they are homotopically equivalent).

- **Homotopic equivalence:** Two spaces X and Y are homotopically equivalent if there exists continuous mappings $f: X \longrightarrow Y$ and $g: Y \longrightarrow X$ such that $g \circ f = 1_X$, $f \circ g = 1_Y$. A very important example is the sphere S^n and the punctured $\mathbf{R}^{n+1}$ (i.e. $\mathbf{R}^{n+1}/\{0\})^1$.

- **Homotopy:** A homotopy (or a deformation) of a smooth map f between two smooth manifolds X and Y is a smooth map $F: X \times I \longrightarrow Y$, $I = [0, 1]$ with the property $F(x, 0) = f(x)$. The maps $f_t(x) = F(x, t)$ are said to be homotopic.

---

[1] This can be showed using the stereographic projection.

- **Homotopy classes:** The relation homotopy divides the set of smooth maps between two smooth manifolds X and Y into equivalence classes called homotopy classes.

- **Connected spaces:** An arcwise connected space is one in which every two points are connected by some path.

- **Loops:** A loop is a closed path. A loop through a point $x_0 \in M$ is a map $\alpha: [0, 1] \longrightarrow M$ such that $\alpha(0) = \alpha(1) = x_0$.

  A product of two loops α and β is a loop $\gamma = \alpha * \beta$, which corresponds to traversing the original loops consecutively, viz $\gamma(t) = \alpha(2t), 0 \leqslant t \leqslant 1/2$ and $\gamma(t) = \beta(2t - 1), 1/2 \leqslant t \leqslant 1$. The inverse loop α^{-1} corresponds to traversing the loop α in the opposite direction. The constant loop is obviously given by $c(t) = x_0$ for all t.

  Two loops are said to be homotopic, and we write $\alpha \sim \beta$, if they can be continuously deformed into each other. Thus, there must exist a mapping $H: [0, 1] \times [0, 1] \longrightarrow M$, which satisfy $H(s, 0) = \alpha(s)$, $H(s, 1) = \beta(s)$ and $H(0, t) = H(1, t) = x_0$.

- **The fundamental group:** The fundamental group of M, denoted $\pi_1(M, x_0)$, consists of all the equivalence (homotopy) classes of loops through $x_0 \in M$. The product of homotopy classes is given by the product of their representatives and thus $\pi_1(M, x_0)$ is a group where the neutral element is given by the constant loop. For arcwise connected space the fundamental group is independent of the base point x_0.

- **Homotopic equivalence:** Homotopically equivalent spaces have the same fundamental group.

- **Simply connected:** In a simply connected space every loop can be contracted to a point. The fundamental group in this case is trivial, viz $\pi_1 = 0$.

- **The circle:** Let us consider $M = S^1$. The maps $\theta: S^1 \longrightarrow S^1$ are phases. We can divide the first S^1 by 2π to get the interval $[0, 1]$. An arbitrary phase on S^1 always satisfy $\theta(0) = 0$, $\theta(2\pi) = 2\pi m$. This can be continuously deformed to the linear function $m\phi$. To see this consider the map

$$H(\phi, t) = (1 - t)\theta(\phi) + t\phi \frac{\theta(2\pi)}{2\pi}. \tag{B.1}$$

This satisfies $H(0, t) = \theta(0) = 0$, $H(2\pi, t) = \theta(2\pi)$. In other words, $H(\phi, t)$ is a homotopy and as a consequence $\theta(\phi)$ is in the same equivalence class as $m\phi$. The set of homotopy (equivalence) classes is therefore Z. The fundamental group of the circle is

$$\pi_1(S^1) = Z. \tag{B.2}$$

- **Higher spheres:** All higher spheres are simply connected and as a consequence

$$\pi_1(S^n) = 0, \quad n \geqslant 2. \tag{B.3}$$

The reason is very simple. Any loop in $\mathbf{R}^{n+1}/\{0\}$ can always avoid the point defect at the origin and be shrunk to a point.

- **Torus:** The fundamental group of a product of spaces X and Y is $\pi_1(X \otimes Y) = \pi_1(X) \otimes \pi_1(Y)$. Thus for a two-dimensional torus we have

$$\pi_1(T) = Z \otimes Z. \tag{B.4}$$

- **Higher homotopy groups:** The fundamental group uses loops and their behavior under deformations to characterize topological properties. Furthermore the fundamental group (using loops) cannot detect point defects in dimensions higher than two. The higher homotopy groups uses generalizations of the one-dimensional loop to detect point defects in higher dimensions.

- **The n-cubes and n-loops:** An n-cube is defined by

$$I^n = \{(s_1,\ldots, s_n)|0 \leqslant s_i \leqslant 1 \text{ all } s_i\}. \tag{B.5}$$

The boundary is defined by

$$\partial I^n = \{(s_1,\ldots, s_n) \in I^n|s_i = 0 \text{ or } s_i = 1\}. \tag{B.6}$$

An n-loop is a continuous map from the n-cube to the topological space X, viz

$$\alpha \colon I^n \longrightarrow X, \tag{B.7}$$

which satisfies

$$\alpha(s) = x_0, \quad s \in \partial I^n. \tag{B.8}$$

In other words, all the points on the boundary are mapped to a single point $x_0 \in X$. Thus, n-loops are topologically equivalent to n-spheres.

As before a homotopy is a continuous deformation of the above n-loop. We define

$$F \colon I^n \times I \longrightarrow X. \tag{B.9}$$

We demand

$$F(s_1, s_2,\ldots, 0) = \alpha(s_1, s_2,\ldots, s_n), \quad F(s_1, s_2,\ldots, 0) = \beta(s_1, s_2,\ldots, s_n). \tag{B.10}$$

$$F(s_1, s_2,\ldots, t) = x_0, \quad (s_1, s_2,\ldots, s_n) \in \partial I^n. \tag{B.11}$$

The two n-loops α, β are therefore homotopic, viz $\alpha \sim \beta$.

- **Higher homotopy groups:** Again the homotopy relation defines an equivalence relation and as a consequence the space of n-loops is turned into a set of equivalence classes. For arcwise connected spaces (i.e. the base point x_0 is irrelevant) the set of equivalence classes is denoted by $\pi_n(X)$ and it is a group. The higher homotopy groups for $n > 1$ are all Abelian as opposed to the fundamental group which can be non-Abelian. Some of the most important examples

$$\pi_n(S^n) = Z. \tag{B.12}$$

$$\pi_m(S^n) = 0, \quad m < n. \tag{B.13}$$